Polymer Chemistry

POLYMER CHEMISTRY
An Introduction

SECOND EDITION

Malcolm P. Stevens
University of Hartford

New York   Oxford
OXFORD UNIVERSITY PRESS
1990

Oxford University Press

Oxford   New York   Toronto
Delhi   Bombay   Calcutta   Madras   Karachi
Petaling Jaya   Singapore   Hong Kong   Tokyo
Nairobi   Dar es Salaam   Cape Town
Melbourne   Auckland

and associated companies in
Berlin   Ibadan

Copyright © 1990 by Oxford University Press, Inc.

Copyright © 1975 by Malcolm P. Stevens

Published by Oxford University Press, Inc.,
200 Madison Avenue, New York, New York 10016

Oxford is a registered trademark of Oxford University Press

Library of Congress Cataloging-in-Publication Data
Stevens, Malcolm P., 1934–
Polymer chemistry: an introduction/Malcolm P. Stevens. —2nd ed.
p.   cm.   Bibliography: p.
Includes index.
ISBN 0-19-505759-7
1. Polymers and polymerization.   I. Title.
QD381.S73   1990   547.7–dc20   89-31080

9 8 7 6 5 4 3 2 1

Printed in the United States of America
on acid-free paper

To Marcia, Jeff, and Phil

Preface

Polymer chemistry is a marvelous subject. It draws upon all the traditional subdivisions of chemistry: organic, inorganic, physical, analytical, and biochemistry. And it includes an occasional side trip into physics, engineering, and even business economics. What makes the subject particularly appealing to the student is that it deals with materials that not only are commercially important, but are a part of the everyday experience. Polymers also represent the area of chemistry in which the student is most likely to find employment on graduation. Yet paradoxically, polymers have traditionally been among the most neglected topics in both the undergraduate and graduate curriculum.

Some years ago the renowned polymer chemist Carl S. Marvel remarked: "Polymer chemistry has become such an important part of chemical technology, and polymers have come to play such a role in everyday living, that no chemist can consider himself adequately trained in his science without some introduction to this field" (*Journal of Chemical Education*, January 1965). When the late Professor Marvel made that comment—indeed, when the first edition of this book was published in 1975—few chemistry departments offered an elective in polymer chemistry. Much has changed in the intervening years. The subject has gained increased respectability in academic circles. The American Chemical Society's Committee on Professional Training now includes polymer chemistry among the electives that students should be "strongly encouraged" to take as part of their undergraduate training; and guidelines have been published (*Journal of Chemical Education*, **59**, 652, 1982) outlining topics that should be included in such courses. The ACS also offers a standardized examination in polymer chemistry. Paralleling this recognition of the importance of polymer education, an increasing number of polymer research centers are being established on college and university campuses.

This book, like its predecessor, is intended as an introductory text for a course in polymer chemistry at the advanced undergraduate or beginning graduate level. It should also serve as an introduction to the field for the industrial chemist with no prior training in polymer chemistry, or as a reference source for the practicing polymer chemist. I assume that students using this book will have completed undergraduate courses in organic and physical chemistry and will be familiar with the more commonly used spectroscopic and chromatographic methods of analysis and characterization.

I have taken ACS guidelines and reviewers' suggestions into account in preparing this edition. Those familiar with the first edition will notice some significant changes:

1. There is more balance between organic and physical subject matter, in keeping with the recognition that polymer studies play a central role in *materials* science. In Part I, morphology and properties have been separated into two chapters. Topics that have become increasingly important in recent years (liquid crystal polymers, conducting polymers, thermoplastic elastomers, degradable polymers, polymer blends) have been added. (The significance of polymer blends was reflected in a 1985 *Business Week* article entitled "Why Polymers Have Become Red Hot").

2. A new chapter covering step-reaction and ring-opening polymerization has been added as an introduction to Part III. Both topics are first introduced in Chapter 1, but a more detailed treatment is reserved for the later chapter. Similarly, chain-reaction polymerization is treated more thoroughly in Part II.

3. Greater emphasis is placed on commercially important polymers. This emphasis begins in Chapter 1 and is reinforced in later chapters. Tables of commercial polymers are included where appropriate to ensure that students, while grappling with the fundamentals of polymer chemistry, do not lose sight of the significant role polymers play in the chemical industry. Although I have mentioned trade names on occasion, I have resisted the temptation to include separate listings. Products come and go from the industrial marketplace with such regularity that trade name compilations achieve almost instantaneous obsolescence. I recommend *Polymer Yearbook* for comprehensive up-to-date listings (*see* Appendix B).

4. The chapter dealing with polymer characterization has been substantially revised to reflect new developments, particularly in the areas of surface analysis and Fourier transform infrared and nuclear magnetic resonance, including spectroscopic characterization of solids.

5. The new technique of group transfer polymerization of vinyl monomers has been included in Part II.

6. End-of-chapter review exercises have been added. Some of these constitute a straightforward review of material covered in the chapter; others are drawn from the polymer literature to illustrate genuine research problems. References are provided for many of them. A number of the exercises are designed to reinforce concepts covered in previous chapters, or to give students practice in writing reactions and deducing repeating units.

7. Two new appendices have been added, one listing abbreviations commonly used for polymers and the other listing the available periodical and encyclopedic polymer literature. The first-edition appendix covering laboratory sources has been updated, and experiments from *Journal of Chemical Education* have been classified according to type (synthesis, characterization, etc.). The laboratory-sources appendix affords the instructor a wide choice of experiments for use in a special polymer chemistry laboratory or for inte-

grating polymer chemistry into the standard laboratory courses in the undergraduate curriculum. A number of the *Journal of Chemical Education* experiments make excellent lecture demonstrations.

Despite these changes, the format and scope of the first edition remain essentially intact, including the treatment of commercially important nonvinyl polymers according to "family." Thus, polyethers are found collectively in Chapter 11, regardless of structure or method of synthesis; polyesters appear in Chapter 12; and so on. Vinyl polymers are treated separately in Part 2. Inorganic and partially inorganic polymers and a variety of "miscellaneous" organic polymers are discussed in Chapters 15 and 16, respectively. Chapter 17 deals with natural polymers, with the emphasis on commercial rather than biochemical aspects.

A word about end-of-chapter references: These are provided as a mechanism for the interested student to delve deeper into the literature on any given topic. As such they comprise in large measure review articles or monographs. To have listed others would have made the reference sections unwieldy. Original sources may, of course, be obtained from the reviews. Original references are used (1) where there is historical interest, (2) where the work in question might not have appeared in a review, and (3) where more useful information may be available, or where the reference relates to a review exercise.

With a work of this size, almost certainly some errors have escaped my scrutiny, but I hope they are few in number. I would be grateful to anyone who alerts me to errors or inconsistencies, or who might offer suggestions for improving the text.

Many of the figures and data in this book are reproduced with the permission of publishers, industrial firms, and individuals, and I wish to acknowledge their courtesy in allowing me use of these materials. I am particularly grateful to the following individuals: David Burge of Wescan Instruments for Figure 2.3; Frank Bovey of Bell Laboratories for Figures 5.3 and 5.4; James Martin and Alvin Tanner of Instron Corporation for Figure 5.22; the late Professor Jerome Vinograd of California Institute of Technology for Figure 17.12; and Dr. Leon Barstow of the University of Arizona and Protein Technology for Figure 17.14. I am also indebted to Professor Ed Gray of the University of Hartford for his comments and constructive criticism of Chapter 5; to Professor Harry Workman of the University of Hartford for the computer-generated Figure 6.2; and to Professors Peter Kovacic of the University of Wisconsin, Milwaukee, and William Johns of Washington State University, Pullman, for providing me with updated information on poly-(p-phenylene) and urea–formaldehyde polymers, respectively. My thanks, too, to colleagues at various institutions who provided critical reviews of the manuscript.

Parts of this edition were written during a sabbatical leave at Colorado State University, and I am indebted to the late Professor John Stille and his

x *Preface*

colleagues for their hospitality during that period. For transforming rough
notes into orderly word processing files I will be forever grateful to my wife,
Marcia Reed Stevens, and to Betty Lord-Wood. To Associate Editor Irene
Pavitt and her colleagues at Oxford University Press, my thanks for their very
fine work and cooperation. It's been a pleasure working with them. A special
debt of gratitude is due my wife and family for their patience and understanding during this seemingly endless project. I dedicate the book to them.

West Hartford, Conn.   M.P.S.
February 1989

Contents

3. Chemical structure and polymer morphology, 70

4. Chemical structure and polymer properties, 110

5. Evaluation, characterization, and analysis of polymers, 146

PART II   VINYL POLYMERS

6. Free radical vinyl polymerization, 189

Part I
POLYMER STRUCTURE
AND PROPERTIES

1
Basic principles

1.1 Introduction and historical development

We live in a polymer age. Plastics, fibers, elastomers, coatings, adhesives, rubber, protein, cellulose—these are all common terms in our modern vocabulary, and all a part of the fascinating world of polymer chemistry. Innumerable examples of synthetic polymers may be cited, some everyday ones, others esoteric: polyester and nylon textile fibers; high-strength polyamide fibers for lightweight bulletproof vests; polyethylene plastic for milk bottles; polyurethane plastic for an artificial heart; rubber for automobile tires; fluorinated phosphazene elastomers that remain flexible in arctic environments. Whatever example or application one might select for purposes of illustration, an underlying consideration is that the particular polymer, for reasons of its unique properties or its economy, or both, is used because it does the job better than other available materials.

The purpose of this book is to provide an understanding of the chemistry of polymeric materials—how these materials differ from nonpolymers, how they are synthesized, and how they may be modified to assume a range of chemical and physical properties. While the experimental techniques for handling polymers may be somewhat different from those used with low-molecular-weight compounds, the chemistry of polymers will, in most cases, be familiar to the student who has completed introductory courses in organic and physical chemistry. The major adjustment the student has to make in beginning a study of polymers is coming to grips with the reality that these materials exhibit certain properties, especially macroscopic ones, that differ markedly from those of the low-molecular-weight compounds usually encountered in undergraduate courses.

Polymers are large molecules made up of simple repeating units. The name is derived from the Greek *poly*, meaning "many," and *mer*, meaning "part." *Macromolecule* is a term synonymous with polymer. Polymers are synthesized from simple molecules called *monomers* ("single part"). A few common monomers, together with the structure of a representative unit of the

3

corresponding polymer, are given in the following reactions:

Ethylene	$CH_2{=}CH_2$	$\longrightarrow$	$-\!\!\left[CH_2CH_2\right]\!\!-$ (1.1)
Tetrafluoro-ethylene	$CF_2{=}CF_2$	$\longrightarrow$	$-\!\!\left[CF_2CF_2\right]\!\!-$ (1.2)

$$CH_2{=}CH \qquad\qquad \longrightarrow \qquad -\!\!\left[CH_2CH\right]\!\!- \qquad (1.3)$$

Styrene

Methyl methacrylate
$$CH_2{=}\overset{\displaystyle CH_3}{\underset{}{C}}{-}CO_2CH_3 \qquad \longrightarrow \qquad -\!\!\left[CH_2\overset{\displaystyle CH_3}{\underset{\displaystyle CO_2CH_3}{C}}\right]\!\!-$$

$$(1.4)$$

1,3-Butadiene $CH_2{=}CH{-}CH{=}CH_2 \qquad \longrightarrow$

$$-\!\!\left[CH_2CH{=}CHCH_2\right]\!\!- \quad (1.5)$$

Ethylene oxide $\overset{\displaystyle O}{CH_2{-}CH_2} \qquad\qquad \longrightarrow \quad -\!\!\left[OCH_2CH_2\right]\!\!-$

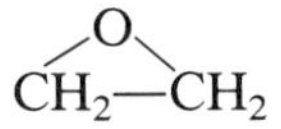

$$(1.6)$$

Ethylene glycol $HOCH_2CH_2OH \qquad \overset{(-H_2O)}{\longrightarrow} \quad -\!\!\left[OCH_2CH_2\right]\!\!-$

$$(1.7)$$

4-Hydroxymethyl-benzoic acid $HOCH_2{-}\!\!\bigcirc\!\!{-}CO_2H \quad \overset{(-H_2O)}{\longrightarrow}$

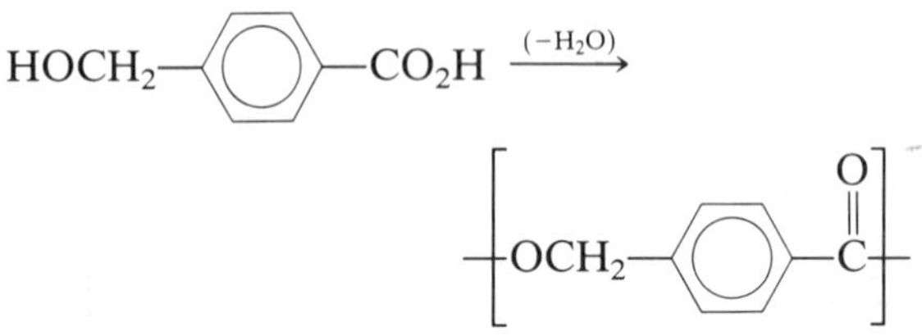

$$-\!\!\left[OCH_2{-}\!\!\bigcirc\!\!{-}\overset{\displaystyle O}{\underset{}{C}}\right]\!\!- \quad (1.8)$$

It should be noted that in the first six of these eight examples the representative polymer units contain the same atoms as the corresponding monomers, whereas in the last two they contain fewer atoms because a by-product (water) is formed in the polymerization process. Of the former type, the first five (prepared by polymerization involving double bonds) are commonly referred to as *addition polymers*, arising from *addition polymerization*. The last two are called *condensation polymers*, arising from *condensation polymerization*. This terminology is deeply ingrained in the language of polymer chemistry, but it can lead to confusion. Consider, for example, the polymerization of ethylene oxide (1.6). This is an example of what is appropriately called *ring-opening polymerization*. No by-product is formed, yet the reaction yields a polymer having the same structure as that in reaction (1.7).

Is $-\text{[OCH}_2\text{CH}_2]-$ an addition polymer or a condensation polymer? This ambiguity in defining polymer types will be considered later.

Each of the reactions shown should be familiar. Mechanistically, the first four involve addition to a double bond. The fifth involves 1,4-addition to a conjugated diene. Ring-opening of ethylene oxide is characteristic of highly strained three-membered rings. Ethylene glycol yields a polyether by dehydration*; 4-hydroxymethylbenzoic acid undergoes ordinary esterification to form polyester. It should be noted that the fundamental difference between polymerization reactions such as (1.7) and (1.8) and simple organic reactions is that difunctional or polyfunctional compounds are necessary for polymer formation. Thus, ethyl alcohol and benzoic acid react to form ethyl benzoate (1.9), whereas ethylene glycol and terephthalic acid give polyester (1.10).

$$\text{(ring)}-\text{CO}_2\text{H} + \text{HOCH}_2\text{CH}_3 \xrightarrow{(-\text{H}_2\text{O})} \text{(ring)}-\text{CO}_2\text{CH}_2\text{CH}_3 \tag{1.9}$$

$$\text{HO}_2\text{C}-\text{(ring)}-\text{CO}_2\text{H} + \text{HOCH}_2\text{CH}_2\text{OH} \xrightarrow{(-\text{H}_2\text{O})}$$

$$\left[\begin{array}{c} \overset{O}{\underset{\|}{C}}-\text{(ring)}-\overset{O}{\underset{\|}{C}}\text{OCH}_2\text{CH}_2\text{O} \end{array} \right] \tag{1.10}$$

Polymers of the above type are well characterized, even to the point of understanding their stereochemistry and in some cases the conformations of the polymer chains. But this accumulation of knowledge is the result of a considerable research effort spanning more than a century.[1-3] The word *polymer* was first used by the Swedish chemist Berzelius in 1833.[4] Throughout the 19th century chemists worked with macromolecules without having any clear understanding of their structure. Some modified natural polymers were, in fact, commercialized. Nitrated cellulose (known by the misnomer of nitrocellulose), for example, was marketed under such names as Celluloid and guncotton. As long ago as 1839 the polymerization of styrene was reported,[5] and during the 1860s the synthesis of poly(ethylene glycol) and poly(ethylene succinate) was published even with the correct structures.[6] (Nomenclature of polymers will be discussed later.)

$$-\text{[C}_2\text{H}_4\text{O]}- \qquad \text{Poly(ethylene glycol)}$$

$$\left[\text{OCH}_2\text{CH}_2\overset{O}{\underset{\|}{O C}}\text{CH}_2\text{CH}_2\overset{O}{\underset{\|}{C}} \right] \qquad \text{Poly(ethylene succinate)}$$

*Dehydration of ethylene glycol gives only low-molecular-weight polymer because of the preponderance of side reactions, and the product has little practical utility. This is discussed more fully in Chapter 11.

About the same time, isoprene was isolated as a degradation product of rubber,[7] although how the isoprene was incorporated into the polymer was not understood at the time. Numerous other examples of macromolecular chemistry may be found in the 19th-century chemical literature.[8]

The first truly synthetic polymer to be used on a commercial scale was a phenol–formaldehyde resin. Developed in the early 1900s by the Belgian-born chemist Leo Baekeland[9] (who had already earned considerable success with his invention of light-sensitive photographic paper), it was known commercially as Bakelite. By the decade of the 1920s, Bakelite had found its way into a wide spectrum of consumer products, and its inventor had achieved the ultimate in visibility—he was featured on a cover of *Time*. Other polymers, notably alkyd (polyester) paints and polybutadiene rubber, were introduced about the same time. Yet despite such commercial successes, most scientists had no clear concept of polymer structure. The prevailing theory was that polymers were aggregates of small molecules, much like colloids, but were held together by some mysterious secondary force.

This aggregation or association theory eventually gave way, with no small amount of resistance, to the theories of the German chemist Hermann Staudinger, who attributed the remarkable properties of polymers to ordinary intermolecular forces between molecules of very high molecular weight. Staudinger suggested the linear chain structures of paraformaldehyde and polystyrene[10]:

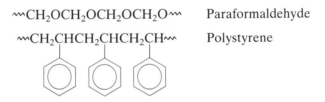

$\sim\!\!\sim\!\!CH_2OCH_2OCH_2OCH_2O\!\!\sim\!\!\sim$ Paraformaldehyde

$\sim\!\!\sim\!\!CH_2CHCH_2CHCH_2CH\!\!\sim\!\!\sim$ Polystyrene

Staudinger also introduced the term *makromolekül*.[11] In recognition of his contributions, Staudinger was awarded the Nobel Prize in Chemistry in 1953. In the 1930s, the brilliant work of the American chemist Wallace Hume Carothers placed the theories of Staudinger on a firm experimental basis and led to the commercial development of neoprene rubber and polyamide (nylon) fibers.[12]

World War II led to significant advances in polymer chemistry, particularly with the development of synthetic rubber when the natural rubber-growing regions of the Far East became inaccessible to the Allies. Among the more significant developments of the postwar years was the discovery by Karl Ziegler[13] in Germany of new coordination catalysts for initiating polymerization reactions and the application by Giulio Natta in Italy of these new systems to development of polymers having controlled stereochemistry.[14] Their work has revolutionized the polymer industry, for these so-called stereoregular polymers have mechanical properties superior in most instances to those of nonstereoregular polymers. The importance of their discoveries was recognized by the award of the Nobel Prize in Chemistry jointly to

Ziegler and Natta in 1963. Equally significant was the work of Paul Flory[15] (Nobel Prize 1974), who established a quantitative basis for polymer behavior, whether it be the physical properties of macromolecules in solution or in bulk or such chemical phenomena as crosslinking and chain transfer (concepts to be encountered later in this text).

More recent years have seen a number of important advances in polymer science, which will be elaborated on in this and later chapters. Examples include:

Polymers having excellent thermal and oxidative stability, for use in high-performance aerospace applications

Engineering plastics—polymers designed to replace metals

High-strength aromatic fibers, some based on liquid crystal technology, for use in a variety of applications from tire cord to cables for anchoring oceanic oil-drilling platforms

Nonflammable polymers, including some that emit a minimum of smoke or toxic fumes

Degradable polymers, which not only help reduce the volume of unsightly plastics waste but also allow controlled release of drugs or agricultural chemicals

Polymers for a broad spectrum of medical applications, from degradable sutures to artificial organs

Conducting polymers—polymers that exhibit electrical conductivities comparable to those of metals

Polymers that serve as insoluble supports for catalysts or for automated protein or nucleic acid synthesis (Bruce Merrifield, who originated solid-phase protein synthesis, was awarded the Nobel Prize in Chemistry in 1984)

This list, by no means exhaustive, clearly illustrates that polymer chemistry is an exciting field with almost limitless possibilities. The possibilities are exemplified in the tremendous impact polymers have had in the field of communications,[16,17] from the metalized plastic tapes for information storage to the printed circuits of modern computers. In the coming decades the influence of polymers will undoubtedly be greater in this diverse industry than in any other; we can foresee the day when scientists may be able to mimic the kind of transmission of electrical impulses that occurs with natural polymers in living systems.

Current polymer literature (Appendix B) is replete with reports of new polymers, improvements in old ones, better methods of characterizing or processing polymers, and new applications. That serious students of polymer chemistry and practicing polymer scientists should stay abreast of current developments is obvious.

1.2 Definitions

As already mentioned, the term *polymer* refers to a large molecule—a macromolecule—whose structure depends on the monomer or monomers used in its preparation. If only a few monomer units are joined together, the resulting low-molecular-weight polymer is called an *oligomer* (Greek *oligos* "few"). Since all synthetic polymers are prepared by linking monomers together, it follows that a certain chemical unit will repeat itself over and over again. Such a unit is set within brackets or parentheses and is referred to as the *repeating unit* (or *repeat unit*). One could reasonably argue that the repeating units shown in reactions (1.1) and (1.2) are $+CH_2+$ and $+CF_2+$, respectively; however, it is more conventional to define repeating units in terms of monomer structure, whereas the *smallest* possible repeating units are referred to as *base units*. (Repeating units are sometimes called *monomeric units*.) *End groups* are the structural units that terminate polymer chains. When end groups are specified, they are shown outside the brackets; for example,

$$CH_3CH_2\!-\!\!\left[CH_2CH_2\right]\!-\!CH\!=\!CH_2$$

In the case of monomers polymerized through an ethylene or vinyl group (i.e., vinyl monomers), as in reactions (1.1) through (1.4) and arguably (1.5), or of monomers that polymerize by ring-opening reactions (e.g., 1.6), the repeating units contain the same atoms as the monomer. Thus the molecular weights of monomer and repeating units are equal. For polymers prepared from difunctional or polyfunctional monomers where a by-product is formed, as in reactions (1.7), (1.8), and (1.10), the repeating units contain fewer atoms than the monomers. It may be noted that the polymer chain (or *backbone*) of vinyl polymers consists only of carbon atoms with other atoms or groups attached; such is referred to as a *a homochain* polymer. *Heterochain* polymers contain more than one atom type in the backbone.

The *degree of polymerization* (DP) refers to the total number of structural units, including end groups, and hence is related to both chain length and molecular weight. Consider, for example, the polymerization of vinyl acetate (an important industrial monomer) in reaction (1.11):

$$\begin{array}{ccc} nCH_2\!=\!CH & & \left[CH_2\!-\!CH\right] \\ \quad\quad | & \longrightarrow & \quad\quad\; | \\ CH_3CO & & \left[\;CH_3CO\;\right]_{n-2} \\ \quad\; \| & & \quad\;\; \| \\ \quad\; O & & \quad\;\; O \end{array} \qquad (1.11)$$

DP in this case is equivalent to n (note that two monomer units are at the chain ends), and the molecular weight of the macromolecule is the product of DP and the molecular weight of the structural unit. For a DP of 500, for example,

$$\text{Molecular weight} = 500 \times 86 = 43,000$$

Because polymer chains within a given polymer sample are almost always of varying lengths (except for certain natural polymers like proteins), we normally refer to the *average degree of polymerization* ($\overline{DP}$).

If the polymer is prepared from a single monomer, A, the product is referred to as a *homopolymer*. If more than one monomer is employed, the

$$-A—A—A—A—A—A—A— \qquad \text{Homopolymer}$$

product is a *copolymer*. If monomers A and B are polymerized together, four arrangements are possible in the polymer structure. If the two structural units alternate in a linear fashion, the product is called an *alternating copolymer*, whereas if the distrubution is random it is called a *random copolymer*. A third arrangement is where blocks of A and B appear together. Such an arrange-

$$-A—B—A—B—A—B—A—B— \qquad \text{Alternating copolymer}$$

$$-A—A—B—A—B—B—A—B— \qquad \text{Random copolymer}$$

ment is referred to as a *block copolymer*. There are any number of possibe

$$-A—A—A—A—B—B—B—B— \qquad \text{Block copolymer}$$

block copolymers. Where blocks of A and B alternate in the backbone, the polymer is designated an $+$AB$+$ block copolymer. If the backbone consists only of one strand of each monomer, it is an AB type. Other possibilities include ABA (a central B block with terminal A blocks) and ABC (three different blocks). Finally, a nonlinear block arrangement is possible, consisting essentially of one polymer with another polymer branching from it. This is called a *graft copolymer*. As will be seen later, certain monomer combinations

$$
\begin{array}{l}
-A—A—A—A—A—A—A—A— \\
\quad\;\; | \\
\quad\;\; B \qquad\qquad\qquad\qquad\qquad \text{Graft copolymer}\\
\quad\;\; | \\
\quad\;\; B—B—B—B—B—
\end{array}
$$

display a tendency toward alternation during copolymerization, whereas formation of block and graft copolymers requires special techniques. Polyesters of the type shown in equation (1.10) may also be considered copolymers, since two monomers, dibasic acid and glycol, are employed in the synthesis. More commonly, however, the term copolymer is reserved, in cases of this type, for those polymers having more than one kind of repeating unit—for example, a *copolyester* prepared from two different dibasic acids and a glycol.

One can also describe polymers as *linear*, *branched*, and *network* (Figure 1.1). A linear polymer has no branching other than the pendant groups associated with the monomer (for example, the phenyl group of polystyrene). Graft copolymers, on the other hand, are examples of branched polymers. It should be stressed, however, that a branched polymer is not necessarily a graft copolymer. Low-density polyethylene (discussed later) is a common example of a branched homopolymer whereby chain branching arises as a result of side reactions during the polymerization process.

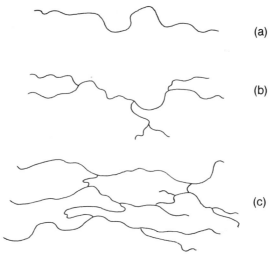

Figure 1.1. Representation of polymer types: (a) linear, (b) branched, and (c) network.

Some more unusual types of branched polymers are represented in Figure 1.2. *Star* polymers contain three or more polymer chains emanating from a central structural unit. *Comb* polymers contain pendant chains (which may or may not be of equal length) and are related structurally to graft copolymers. Such a polymer might be formed, for example, by polymerizing a long-chain vinyl monomer. *Ladder* polymers are made up of regular recurring fused-ring structures or, in the case of *semiladder* or *stepladder* polymers, of cyclic moieties linked with open-chain units.

Network polymers arise when polymer chains are linked together or when polyfunctional instead of difunctional monomers are used. As an example of the former, one can cite the vulcanization of rubber in which linear rubber

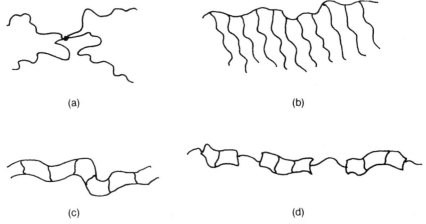

Figure 1.2. Representation of (a) star, (b) comb, (c) ladder, and (d) semiladder (or stepladder) polymers.

molecules are linked together through sulfur atoms. If the polyfunctional monomer glycerol is used instead of ethylene glycol in equation (1.10), a three-dimensional network polyester results (1.12). Reactions of the above types will be discussed in detail in later chapters.

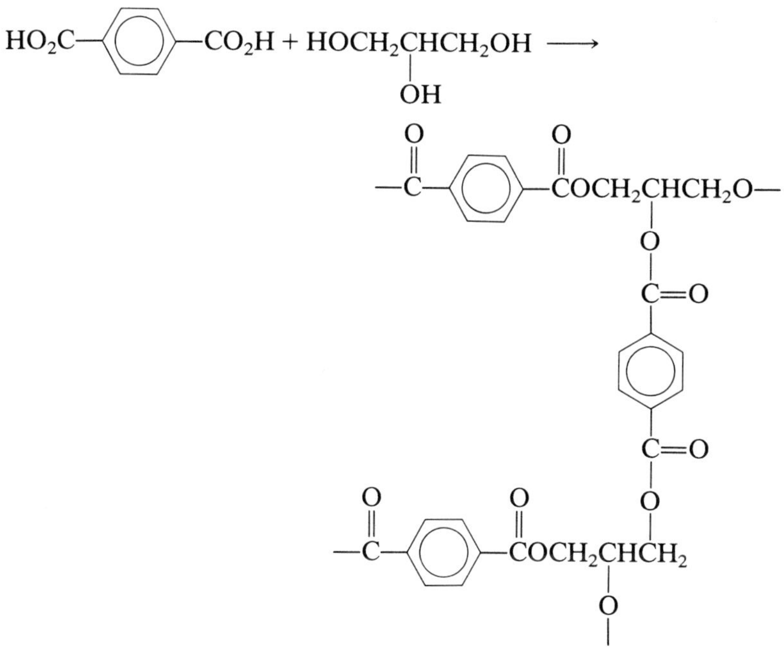

$$HO_2C-\langle\bigcirc\rangle-CO_2H + HOCH_2CHCH_2OH \longrightarrow$$
$$\underset{OH}{|}$$

(1.12)

Network polymers are also commonly referred to as *crosslinked polymers.* Because of crosslinking, the polymer chains lose their ability to flow past one another and the material exhibits a considerable degree of dimensional stability. The polymer will not melt or flow and cannot, therefore, be molded. Such polymers are said to be *thermosetting* or *thermoset*. To manufacture useful articles out of thermosetting polymers, one must accomplish the crosslinking reaction in place or temporarily disrupt the crosslinking to allow the polymer to flow. Thermosetting polymers are also insoluble because the crosslinking causes a tremendous increase in molecular weight. At most, thermosetting polymers only swell in the presence of solvent, as solvent molecules penetrate the network. It is interesting to reflect that an article made with a thermosetting polymer may be considered one gigantic molecule if one can believe that all polymer chains present are linked together!

Polymers that are not crosslinked (linear or branched) can usually be dissolved in some solvent, and in most instances they will melt and flow. Such materials are said to be *thermoplastic.*

With the exception of the terms thermosetting and thermoplastic, which are based on physical properties, the above definitions arise from differences in chemical structure. Other classifications based on physical and/or mechanical properties or on end use divide polymers into *rubbers* (or *elastomers*), *plastics, fibers, coatings,* and *adhesives.* (Many polymer scientists prefer to reserve the term rubber for the natural material and elastomer for the synthetic. The latter term is also generally applied to any lightly crosslinked polymer that exhibits "elastic memory.") This type of terminology is more for the convenience of the manufacturer or customer because individual polymers can fit into more than one category. Styrene–butadiene copolymer, for example, serves both as a rubber and, in the form of latex (i.e., an aqueous emulsion), as a water-based paint. The polyester poly(ethylene terephthalate) is useful as a plastic film and as a fiber. One can also vary the ingredients in polyesters to give the polymers rubberlike properties or properties desirable for coatings (toughness, flexibility, weather resistance). In general, we can identify the properties of a rubber with segmental mobility of the polymer backbone, low resistance to stress, and resilience. Plastics have more glasslike properties with high resistance to stress and a lack of segmental mobility. Plastics may be brittle or tough, depending on the molecular structure and the cohesive forces holding the molecules together (Chapter 3). To be of any commercial value, fibers must exhibit the greatest strength of these various polymer types. Coatings and adhesives are usually complex mixtures of polymers, solvents, and various additives. These classifications are based solely on end use.

1.3   Polymerization processes

Traditionally, polymers have been classified into two main groups, *addition polymers* and *condensation polymers.* This classification, first proposed by Carothers,[18] is based on whether the repeating unit of the polymer contains the same atoms as the monomer. An addition polymer has the same atoms as the monomer in its repeating unit, whereas condensation polymers contain fewer because of the formation of by-products during the polymerization process. The corresponding polymerization processes would then be called addition polymerization and condensation polymerization. As was mentioned earlier, this classification can lead to confusion, for it was recognized in later years that many important types of polymers can be prepared by both addition and condensation processes. Equations (1.6) and (1.7), the formation of a polyether from ethylene oxide and ethylene glycol, respectively, illustrate one example of this behavior. Others, which involve equally important classes of polymers, are given in the following examples:

1. Polyester from lactone (1.13) and ω-hydroxycarboxylic acid (1.14):

$$\xrightarrow{\text{H+}} \left[\overset{O}{\overset{\|}{C}}(CH_2)_5O \right] \tag{1.13}$$

$$HO(CH_2)_5CO_2H \xrightarrow{\text{heat}} \left[\overset{O}{\overset{\|}{C}}(CH_2)_5O \right] + H_2O \tag{1.14}$$

2. Polyamide from lactam (1.15) and ω-amino acid (1.16):

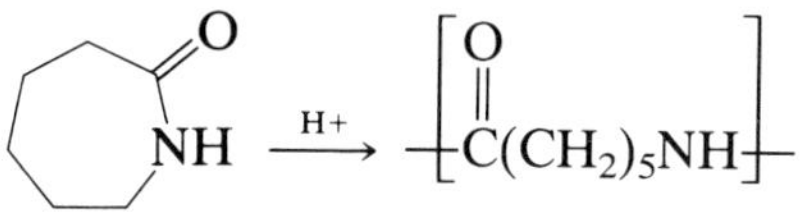

$$\text{(lactam ring with NH and C=O)} \xrightarrow{\text{H+}} \left[\begin{matrix} O \\ \parallel \\ C(CH_2)_5NH \end{matrix} \right] \qquad (1.15)$$

$$H_2N(CH_2)_5CO_2H \xrightarrow{\text{heat}} \left[\begin{matrix} O \\ \parallel \\ C(CH_2)_5NH \end{matrix} \right] + H_2O \qquad (1.16)$$

3. Polyurethane from diisocyanate and diol (1.17) and diamine and bis-chloroformate (1.18):

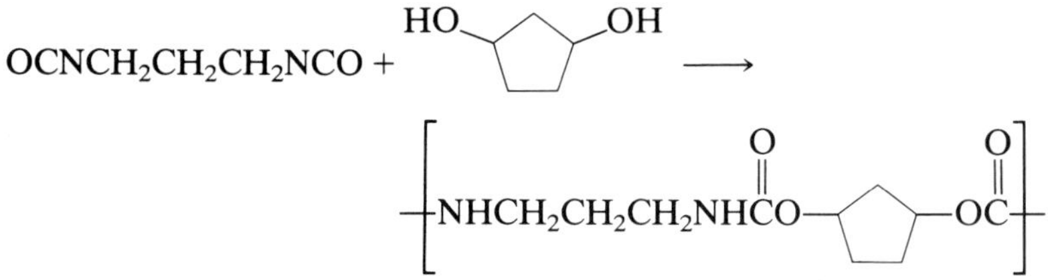

$$OCNCH_2CH_2CH_2NCO + \quad \text{HO} \diagup \diagdown \text{OH} \quad \longrightarrow$$

$$\left[NHCH_2CH_2CH_2NH\overset{O}{\overset{\parallel}{C}}O-\diagup \diagdown -O\overset{O}{\overset{\parallel}{C}} \right] \qquad (1.17)$$

$$H_2NCH_2CH_2CH_2NH_2 + \quad Cl\overset{O}{\overset{\parallel}{C}}O \diagup \diagdown O\overset{O}{\overset{\parallel}{C}}Cl \longrightarrow$$

$$\left[NHCH_2CH_2CH_2NH\overset{O}{\overset{\parallel}{C}}O-\diagup \diagdown -O\overset{O}{\overset{\parallel}{C}} \right] + 2HCl \qquad (1.18)$$

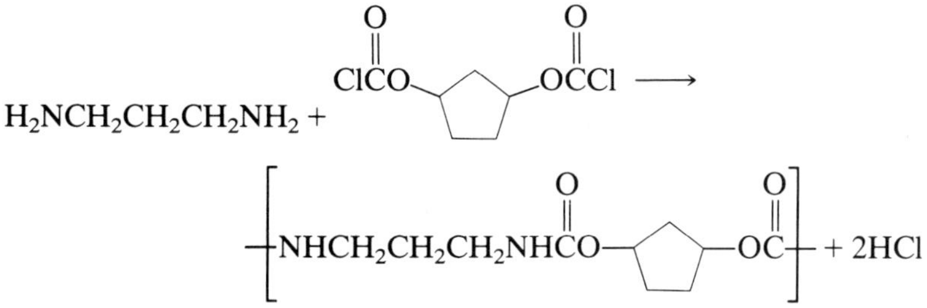

4. Hydrocarbon polymer from ethylene (1.19) and α,ω-dibromide by the Wurtz reaction (1.20):

$$CH_2{=}CH_2 \xrightarrow{\text{initiator}} [CH_2CH_2] \qquad (1.19)$$

$$BrCH_2(CH_2)_8CH_2Br \xrightarrow{\text{2Na}} [CH_2CH_2]_5 + 2NaBr \qquad (1.20)$$

Polymers having identical repeating units but formed by entirely different reactions do not necessarily have identical properties. On the contrary, physical and mechanical properties may differ markedly because different polymerization processes may give rise to differences in molecular weight, end groups, stereochemistry, or possibly chain branching. (How structural variations influence properties is considered in later chapters.) Spectroscopic properties and elemental analyses will, however, be the same, apart from small differences arising from different end groups, such differences becoming less significant as the molecular weight increases.

In more recent years the emphasis has changed to classifying polymers according to whether the polymerization occurs in a stepwise fashion (*step reaction* or *step growth*) or by propagating from a growing chain (*chain reaction* or *chain growth*). Let us now examine the basic differences between the two.

1.4   Step-reaction polymerization

There are two approaches to preparing linear step-reaction polymers, one having both reactive functional groups in one molecule:

$$A—B \rightarrow —[A—B]—$$

and the other having two difunctional monomers:

$$A—A + B—B \rightarrow —[A—A—B—B]—$$

Reactions (1.14) and (1.16) illustrate the former, whereas (1.17) and (1.18) are of the latter type.

Polyesterification, whether between diol and dibasic acid or intermolecularly between hydroxy acid molecules, is an example of a step-reaction process. The esterification reaction occurs anywhere in the monomer matrix where two monomer molecules possessing the proper orientation and energy collide. Once the ester has formed (with loss of a water molecule), it, too, can react further by virtue of its still-reactive hydroxyl and carboxyl groups. The net effect is that monomer molecules are consumed rapidly without any concomitant large increase in molecular weight. Figure 1.3 schematically illustrates this phenomenon for a system of 12 monomer molecules containing functional groups A and B (e.g., a hydroxy acid). When half the monomer molecules are consumed in the manner shown (Figure 1.3b), $\overline{DP}$ of the polymer is 2 (1.3 if all molecules are counted). Figure 1.3c shows 75% of the monomer molecules consumed with $\overline{DP}$ of the polymer molecules still only

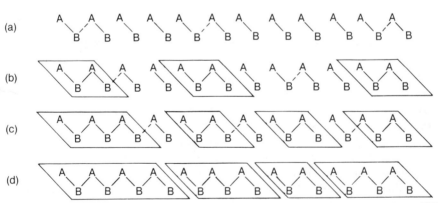

Figure 1.3. Step-reaction polymerization: (a) unreacted monomer; (b) 50% reacted, $\overline{DP}$ (polymer) = 2; (c) 75% reacted, $\overline{DP}$ (polymer) = 2.25; and (d) 100% reacted, $\overline{DP}$ (polymer) = 3. (Dotted lines represent reacting species.)

2.25 (1.7 based on all molecules). With all the monomer molecules reacted (Figure 1.3d), $\overline{DP}$ is still low. But each polymer (or oligomer) molecule still has reactive end groups, hence the polymerization reaction will continue in stepwise fashion with each step being mechanistically the same. The polymerization rate will decrease as the functional groups are consumed. Thus molecular weight of the polymer increases slowly even at high levels of monomer conversion, and it will continue to increase until the concentration of reactive end groups becomes too low for the reaction to proceed at a significant rate. Other factors, such as increasing viscosity of the reaction medium, which reduces the mobility of the reactive chain ends and increases the difficulty of removing by-products, may also contribute to a rate reduction.

Carothers developed a simple equation for relating $\overline{DP}$ to percent conversion of monomer. If one assumes that there are N_0 molecules initially and N molecules (total) after a given reaction period, then the amount reacted is $N_0 - N$. The reaction conversion, p, is then given by the expression

$$p = \frac{N_0 - N}{N_0}$$

or

$$N = N_0(1 - p)$$

The average number of repeating units of all molecules present, $\overline{DP}$, is equal to N_0/N. Substituting this in the above expression reduces the latter to

$$\overline{DP} = \frac{1}{1 - p}$$

This simple equation demonstrates one fundamental aspect of step-reaction polymerization—that very high conversions are necessary to achieve practical molecular weights. At 98% conversion, for example,

$$\overline{DP} = \frac{1}{1 - 0.98}$$

and $\overline{DP}$ is still only 50. Thus very high $\overline{DP}$'s are generally not obtainable for linear polymers by step polymerization.

It can also be shown, by using similar types of relationships, that in the A—A + B—B type of polymerization, an exact stoichiometric balance and very pure monomers are necessary to achieve high molecular weights. If some monofunctional impurity is present, its reaction will limit the molecular weight by rendering a chain end inactive. Similarly, high-purity monomers are necessary in the A—B type of polycondensation, and it follows that high-yield reactions are the only practical ones for polymer formation, because side reactions will upset the stoichiometric balance. These topics, together with step-reaction kinetics, will be considered in more detail in Chapter 10.

1.5   Chain-reaction polymerization

Chain-reaction polymerization involves two distinct kinetic steps, *initiation* and *propagation*. The first requires an *initiator* to start the reaction. The initiator might, for example, be a free radical ($R\cdot$) to initiate polymerization of a monomer such as ethylene:

Initiation

$$R\cdot + CH_2 = CH_2 \rightarrow RCH_2CH_2\cdot$$

Propagation

$$RCH_2CH_2\cdot + CH_2 = CH_2 \rightarrow RCH_2CH_2CH_2CH_2\cdot$$

Alternatively, it might be an anion ($B:^-$) to initiate ring-opening polymerization of a cyclic monomer such as ethylene oxide:

Initiation

$$B:^- + CH_2\overset{\displaystyle O}{-}CH_2 \longrightarrow BCH_2CH_2O^-$$

Propagation

$$BCH_2CH_2O^- + CH_2\overset{\displaystyle O}{-}CH_2 \longrightarrow BCH_2CH_2OCH_2CH_2O^-$$

Other types of initiators (that are discussed in later chapters) include cationic and complex coordination compounds. (Note that it is incorrect to refer to initiating species as catalysts if they are consumed in the reaction.) In both addition and ring-opening polymerization, the reaction propagates at a reactive chain end and continues until a *termination* reaction renders the chain end inactive (e. g., combination of radicals), or until monomer is completely consumed. The probability of termination reactions occurring increases as monomer concentration decreases. Because polymerization occurs at the chain end, molecular weight increases rapidly even though relatively large amounts of monomer remain unreacted. This is shown schematically for a vinyl monomer in Figure 1.4, where $\overline{DP}$ of the polymer molecule is already 6 when only half the monomer molecules have reacted, and is 9 when 75% have reacted. Kinetics and mechanisms of chain-reaction polymerization are discussed in Chapter 6.

   Figures 1.3 and 1.4 illustrate one of the fundamental differences between step-reaction and chain-reaction polymerization. In the former, molecular weight increases slowly while monomer is consumed rapidly; in the latter, high molecular weights are reached rapidly at low monomer conversions. Furthermore, in chain-reaction processes, initiation, propagation, and termination are all different reactions occurring at different rates and, once termination has occurred, the polymerization cannot proceed further. If, on

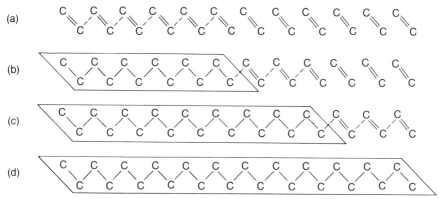

Figure 1.4. Chain-reaction polymerization: (a) unreacted monomer; (b) 50% reacted, $\overline{DP}$ (polymer) = 6; (c) 75% reacted, $\overline{DP}$ (polymer) = 9; and (d) 100% reacted. $\overline{DP}$ (polymer) = 12. (Dotted lines represent reacting species.)

the other hand, monomer is completely consumed but *no* termination has taken place, polymerization will continue if more monomer is added. By contrast, in step-reaction polymerization, initiation and propagation reactions are for all practical purposes the same, and once the polymerization reaction ends, the polymer still has reactive end groups. Major differences between the two polymerization processes are summarized in Table 1.1.

1.6 Step-reaction addition and chain-reaction condensation

Most step-reaction polymerizations are condensation processes and most chain-reaction polymerizations are addition processes, but there are exceptions. Reaction (1.17), for example, illustrates a step-growth process that is

Table 1.1. Comparison of step-reaction and chain-reaction polymerization

Step reaction	Chain reaction
Growth occurs throughout matrix by reaction between monomers, oligomers, and polymers	Growth occurs by successive addition of monomer units to limited number of growing chains
$\overline{DP}^a$ low to moderate	$\overline{DP}$ can be very high
Monomer consumed rapidly while molecular weight increases slowly	Monomer consumed relatively slowly, but molecular weight increases rapidly
No initiator needed; same reaction mechanism throughout	Initiation and propagation mechanisms different
No termination step; end groups still reactive	Usually chain-terminating step involved
Polymerization rate decreases steadily as functional groups consumed	Polymerization rate increases initially as initiator units generated; remains relatively constant until monomer depleted

$^a \overline{DP}$, average degree of polymerization.

also an addition reaction. Another example, a polymer formed by the Diels–
Alder reaction between a bisdiene, 1,6-bis(cyclopentadienyl)hexane, and
benzoquinone, is shown in reaction (1.21). The boron trifluoride-catalyzed
polymerization of diazomethane,

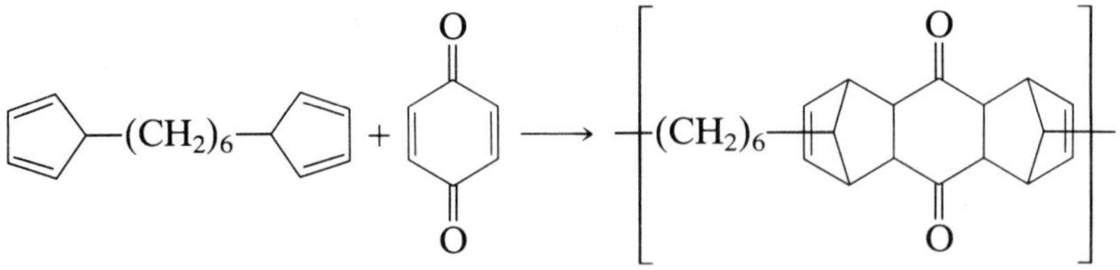

$$(1.21)$$

reaction (1.22), however, is a case of a chain-growth condensation reaction.

$$CH_2N_2 \xrightarrow{BF_3} \text{—}[CH_2]\text{—} + N_2 \qquad (1.22)$$

In later discussions of different types of polymers, the more logical classification according to step reaction and chain reaction has been followed,
although occasional references to condensation and addition reactions may
appear.

1.7 Nomenclature

Polymer nomenclature is complicated for a variety of reasons. First, many
polymer names are based on names of the corresponding monomers, and
although this *source-based* system is widely accepted, one frequently encoun-
ters variations in format. Second, although the Macromolecular Nomenclature Commission of the International Union of Pure and Applied Chemistry
has grappled valiantly with the complexities of polymer structure and has
proposed perfectly logical rules,[19] the IUPAC system is not widely used
except in reference works. Third, some polymer structures are so complex,
particularly where branching or crosslinking is involved, that naming them
defies all but the most persistent of nomenclature purists. And finally,
polymer science has some fairly sharply defined boundaries; rubber chemists,
for example, employ terminology that might be unintelligible to a plastics or
textiles chemist. All of this takes on added significance in this age of com-
puter-based information storage and retrieval, where a standardized set of
abbreviations is a basic necessity.[20] Given the magnitude of the problem,
therefore, we will attempt in this section to clarify the most generally
accepted terminology.

As should be apparent from the previous sections, polymer types, or
families, are named according to the functional group present in the repeating
unit, with the prefix *poly*: polyesters, polyamides, polyethers, and so on.
Polyamides are unusual in that they are also called *nylons*, a term that

originated as a trade name but then evolved over the years into general use. If more than one functional group is present, the polymer is named accordingly; for example, polyetherimide.

Where polymer structure is complex and not easily definable, the family is usually named for the monomers employed, as with phenol–formaldehyde polymers. Part III of this book discusses a variety of polymers according to the family classification. Polymers derived from vinyl monomers are referred to collectively as vinyl polymers; their chemistry is the subject of Part II of this book. We will here consider vinyl nomenclature first.

1.7.1 Vinyl polymers

The most widely accepted method of naming vinyl polymers is to place the prefix *poly* before the name of the corresponding monomer; for example,

$-[CH_2CH_2]-$ Polyethylene

$-[CF_2CF_2]-$ Polytetrafluoroethylene

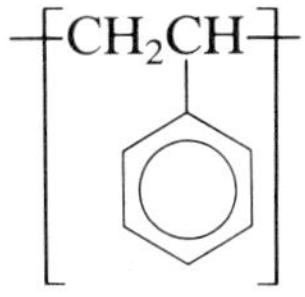

$-[CH_2CH]-$ Polystyrene

(Note that *common* names of the monomers are used, not ethene or phenyl-ethene.) By this system, the polymer $+CH_2+$ prepared according to reaction (1.22) is called polymethylene to distinguish it from the commercial polymer prepared from ethylene. (In the British literature, polyethylene is frequently called *polythene*.) Occasionally a trade name may be employed in casual usage, for example Teflon (du Pont) for polytetrafluoroethylene, but this occurs only where such names are widely used.

If the monomer name consists of more than one word, or if the name is preceded by a letter or number, that name is enclosed in parentheses and prefixed with *poly*:

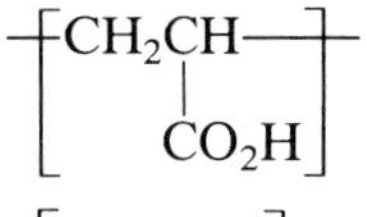

$-[CH_2CH]-$
$\quad\quad CO_2H$ Poly(acrylic acid)

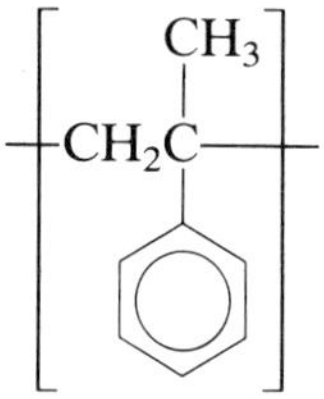

$-[CH_2C]-$
$\quad\quad CH_3$ Poly(α-methylstyrene)

$-[CH_2CH]-$
$\quad\quad CH_2CH_2CH_3$ Poly(1-pentene)

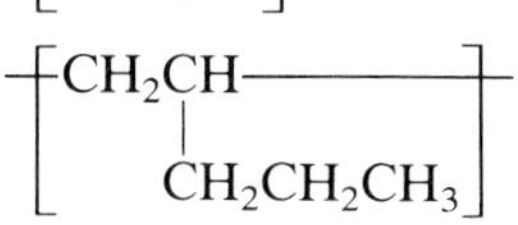

Unfortunately, one still sees parentheses omitted—polyacrylic acid, for example—especially in trade literature. This nomenclature is no longer considered acceptable and should be avoided.

The IUPAC recommends that names be derived from the structure of the base unit, or *constitutional repeating unit* (CRU) as this organization prefers to call it, according to the following steps:

1. The smallest structural unit (CRU) is identified.
2. Subunits of the CRU are assigned seniority on the basis of point of attachment, and are written with seniority decreasing from left to right; thus polystyrene is written

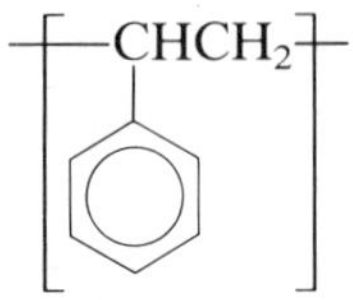

$$\left[\!\!\!\begin{array}{c} -CHCH_2- \\ | \\ \bigcirc \end{array}\!\!\!\right]$$

3. Substituents are numbered from left to right.
4. The name is placed in parentheses (or brackets and parentheses, where necessary), and prefixed with *poly*.

The preceding polymers would thus be named as follows:

Source name	*IUPAC name*
Polyethylene	Poly(methylene)
Polytetrafluoroethylene	Poly(difluoromethylene)
Polystyrene	Poly(1-phenylethylene)
Poly(acrylic acid)	Poly(1-carboxylatoethylene)
Poly(α-methylstyrene)	Poly(1-methyl-1-phenylethylene)
Poly(1-pentene)	Poly[1-(1-propyl)ethylene]

In the IUPAC system, parentheses are *always* used following the prefix; also ethylene (rather than ethene) is the acceptable structural name. Furthermore, the IUPAC system makes no distinction between polyethylene and polymethylene.

For polymers having substituents on each carbon of the backbone, the CRU is named according to stated IUPAC organic rules, and the corresponding IUPAC polymer name is written accordingly. For example,

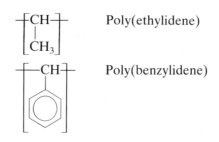

Poly(ethylidene)

Poly(benzylidene)

Such polymers are relatively rare, hence IUPAC names have more readily been adopted.

Diene monomers create special difficulties because they can undergo both 1,2- and 1,4-addition. Thus poly(1,3-butadiene) may have two possible repeating units:

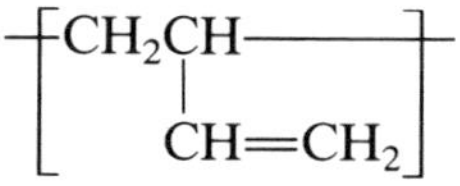

$$\left[\begin{array}{c} CH_2CH \\ | \\ CH=CH_2 \end{array}\right] \qquad -[CH_2CH=CHCH_2]-$$

1,2-addition   1,4-addition

with the latter capable of exhibiting *cis-trans* isomerism. According to the IUPAC rules, the CRU of the 1,4-addition product would be written in the less familiar format:

$$-[CH=CHCH_2CH_2]-$$

Assuming each butadiene polymer were to consist of a single repeating unit, the names would be as follows:

	Source name	*IUPAC name*
1,2-addition	1,2-Poly(1,3-butadiene) or 1,2-Polybutadiene	Poly(1-vinylethylene)
1,4-addition	1,4-Poly(1,3-butadiene) or 1,4-polybutadiene	Poly(1-butenylene)

The prefixes *cis* and *trans* are used as appropriate for the 1,4 polymer. In polydiene synthesis, both repeating units and stereochemistries are invariably present in the final product, with one predominating depending on the reaction conditions. (This is explored more fully in Part II. Other aspects of polymer stereochemistry are discussed in Chapter 4.) Some additional examples of vinyl polymer nomenclature are given in Table 1.2.

Having two such nomenclature systems can be confusing at first. The source-based system is the one of choice among most polymer chemists, notwithstanding the fact that such important reference works as *Chemical Abstracts* and *Polymer Handbook* have adopted the IUPAC system. The IUPAC is not dogmatic on the point; it recognizes the use of source-based names where no ambiguity arises, but it *advocates* a more systematic approach.

1.7.2   Vinyl copolymers

By definition, copolymers are polymers derived from more than one species of monomer. (The term *terpolymer* is sometimes used to denote a copolymer prepared from three different monomers.) For example, a copolymer prepared from the monomers styrene and methyl methacrylate quite logically is

Table 1.2. Representative nomenclature of vinyl polymers

Monomer structure	Monomer name	Polymer repeating unit	Source name	IUPAC[a] name
$CH_3CH=CH_2$	Propylene	$+CH_2CH+$ $\mid$ CH_3	Polypropylene	Poly(propylene)
$CH_2=CHCl$	Vinyl chloride	$+CH_2CH+$ $\mid$ Cl	Poly(vinyl chloride)	Poly(1-chloroethylene)
$CH_2=C(CH_3)_2$	Isobutylene	CH_3 $\mid$ $+CH_2C+$ $\mid$ CH_3	Polyisobutylene	Poly(1,1-dimethylethylene)
$CH_2=CHCN$	Acrylonitrile	$+CH_2CH+$ $\mid$ CN	Polyacrylonitrile	Poly(1-cyanoethylene)
CH_3 $\mid$ $CH_2=CCH_3$	Methyl methacrylate	CH_3 $\mid$ $+CH_2C+$ $\mid$ CO_2CH_3	Poly(methyl methacrylate)	Poly[1-(methoxycarbonyl)-1-methylethylene]

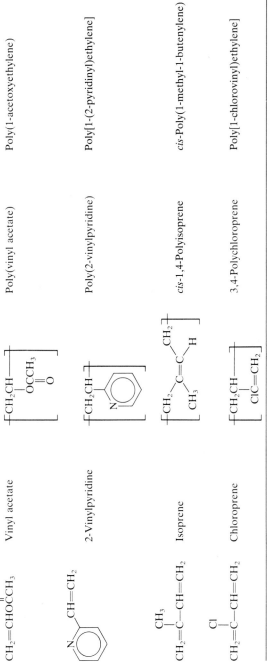

Vinyl acetate Poly(vinyl acetate) Poly(1-acetoxyethylene)

2-Vinylpyridine Poly(2-vinylpyridine) Poly[1-(2-pyridinyl)ethylene]

Isoprene cis-1,4-Polyisoprene cis-Poly(1-methyl-1-butenylene)

Chloroprene 3,4-Polychloroprene Poly[1-chlorovinyl)ethylene]

[a] IUPAC, International Union of Pure and Applied Chemistry.

called styrene–methyl methacrylate copolymer, using source-based terminology. The IUPAC recognizes that defining a particular constitutional repeating unit for copolymers is frequently impossible; even when it is possible, as with alternating, block, and graft copolymers, structure-based nomenclature can be very unwieldy. The IUPAC, therefore, recommends source-based names for copolymers.[21] In the IUPAC system the polymer example just given is called by either of the two names

Poly[styrene-*co*-(methyl methacrylate)]

Copoly(styrene/methyl methacrylate)

The first is more systematic; the second is more concise. For very complex copolymers, the more systematic approach is recommended. In both names the prefix *co* implies that no distribution of the monomeric repeating units is specified.*

Alternating, block, and graft copolymers of the same monomers are named as follows:

Systematic
Poly[styrene-*alt*-(methyl methacrylate)]
Polystyrene-*block*-poly(methyl methacrylate)
Polystyrene-*graft*-poly(methyl methacrylate)

Concise
Alt-copoly(styrene/methyl methacrylate)
Block-copoly(styrene/methyl methacrylate)
Graft-copoly(styrene/methyl methacrylate)

In the systematic name for block copolymers, the term *block* may be replaced with a long dash, thus,

Polystyrene——poly(methyl methacrylate)

With graft copolymers, the order of monomer names is important. As written above, the name specifies that poly(methyl methacrylate) is grafted onto a polystyrene backbone. If the reverse were the case, the monomer names would, of course, be reversed. Some additional examples of copolymer names, systematic followed by concise, are the following:

Poly[styrene-*co*-butadiene-*co*-(vinyl acetate)]

Copoly(styrene/butadiene/vinyl acetate]

(a copolymer of the three named monomers with no specified distribution of

---

*If the monomer sequence is known to obey some specified statistical distribution, for example, Bernoulli, the term *stat* is recommended instead of *co*.

repeating units)

> Polystyrene-*block*-polyisoprene-*block*-polystyrene
>
> *Block*-copoly(styrene/isoprene/styrene)

(an ABA block copolymer with polystyrene terminal blocks)

> Poly[styrene-*alt*-(maleic anhydride)-*block*-poly(vinyl chloride)]
>
> *Block*-copoly[*alt*-co(styrene/maleic anhydride/vinyl chloride)]

[a block copolymer, one block made up of an alternating copolymer of styrene and maleic anhydride, and the other of poly(vinyl chloride)]

> Polybutadiene-*block*-(polystyrene-*graft*-polyacrylonitrile)
>
> *Block*-copoly[butadiene/*graft*-co(styrene/acrylonitrile)]

(a block copolymer of styrene and butadiene with polyacrylonitrile grafted to the polystyrene block)

The last two copolymers could be correctly named with the blocks in reverse order.

1.7.3 Nonvinyl polymers

Nomenclature of nonvinyl polymers is much more complicated because of the great variety of possible repeating units. Three of the more important commercial polymer types—polyethers, polyesters, and polyamides—are included in Table 1.3 under the appropriate subheadings.

Polyethers are named according to monomer used (source) or structure (IUPAC). In the latter case, seniority is given to the oxygen atom, and each subunit is named in sequence.

Like polyethers, polyesters can be formed by ring opening as well as by step-reaction polymerization, although the former is less common. Ring opening of cyclic esters (lactones) yields products that are named according to the source or the ester repeating unit in the common nomenclature, as shown for β-propiolactone. The IUPAC name is based on two subunits of the CRU, with the carbonyl oxygen named as an *oxo* substituent:

$$-\text{O}-\overset{\overset{\displaystyle \text{O}}{\|}}{\text{C}}\text{CH}_2\text{CH}_2-$$

oxy   1-oxotrimethylene

The presence of the 1-oxo substituent requires that the subunit be enclosed with parentheses; also, with an alkyl chain larger than ethylene, the number of methylene groups is indicated. In the case of the polyester formed by step-reaction polymerization of 10-hydroxydecanoic acid, the product is

Table 1.3. Representative nomenclature of nonvinyl polymers

Monomer structure	Monomer name	Polymer repeating unit	Source or common name	IUPAC name
Polyethers				
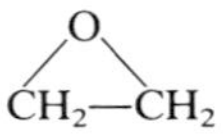 (epoxide) CH₂—CH₂ with O	Ethylene oxide	$-[CH_2CH_2O]-$	Poly(ethylene oxide)	Poly(oxyethylene)
$HOCH_2CH_2OH$	Ethylene glycol	$-[CH_2CH_2O]-$	Poly(ethylene glycol)	Poly(oxyethylene)
CH_2O	Formaldehyde	$-[CH_2O]-$	Polyformaldehyde	Poly(oxymethylene)
CH_3CHO	Acetaldehyde	$\left[\begin{array}{c}CHO \\ | \\ CH_3\end{array}\right]$	Polyacetaldehyde	Poly(oxyethylidene)
Polyesters				
(β-propiolactone ring)	β-Propiolactone	$\left[OCH_2CH_2\overset{\displaystyle O}{\overset{\|}{C}}\right]$	Poly(β-propiolactone) or poly(3-propionate)	Poly[oxy(1-oxotrimethylene)]
$HO(CH_2)_9CO_2H$	10-Hydroxydecanoic acid	$\left[O(CH_2)_9\overset{\displaystyle O}{\overset{\|}{C}}\right]$	Poly(10-decanoate)	Poly[oxy(1-oxodecamethylene)]
$HOCH_2CH_2OH$ 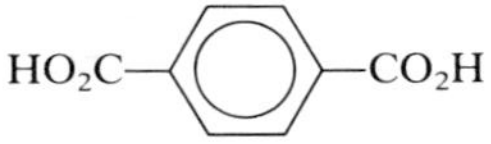 + $HO_2C\!-\!C_6H_4\!-\!CO_2H$	Ethylene glycol + Terephthalic acid	$\left[OCH_2CH_2O\overset{\displaystyle O}{\overset{\|}{C}}\!-\!C_6H_4\!-\!\overset{\displaystyle O}{\overset{\|}{C}}\right]$	Poly(ethylene terephthalate)	Poly(oxyethyleneoxyterephthaloyl)

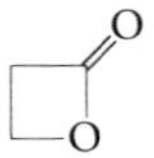

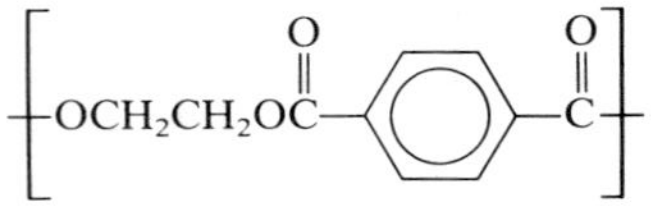

Polyamides

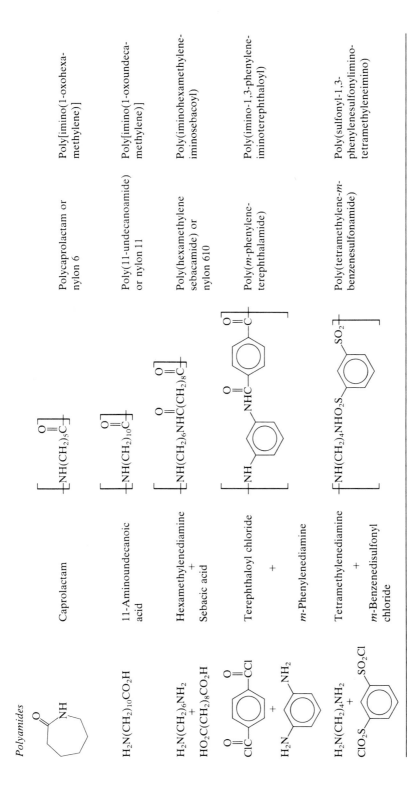

Monomer(s)	Polymer	Polymer name
Caprolactam	$\left[\text{NH(CH}_2\text{)}_5\overset{\text{O}}{\overset{\|}{\text{C}}}\right]$	Polycaprolactam or nylon 6 — Poly[imino(1-oxohexamethylene)]
11-Aminoundecanoic acid — $\text{H}_2\text{N(CH}_2\text{)}_{10}\text{CO}_2\text{H}$	$\left[\text{NH(CH}_2\text{)}_{10}\overset{\text{O}}{\overset{\|}{\text{C}}}\right]$	Poly(11-undecanoamide) or nylon 11 — Poly[imino(1-oxoundecamethylene)]
Hexamethylenediamine + Sebacic acid — $\text{H}_2\text{N(CH}_2\text{)}_6\text{NH}_2$ + $\text{HO}_2\text{C(CH}_2\text{)}_8\text{CO}_2\text{H}$	$\left[\text{NH(CH}_2\text{)}_6\text{NHC(CH}_2\text{)}_8\overset{\text{O}}{\text{C}}\right]$	Poly(hexamethylene sebacamide) or nylon 610 — Poly(iminohexamethylene iminosebacoyl)
Terephthaloyl chloride + m-Phenylenediamine	(aromatic diamide unit)	Poly(m-phenyleneterephthalamide) — Poly(imino-1,3-phenylene iminoterephthaloyl)
Tetramethylenediamine + m-Benzenedisulfonyl chloride	$\left[\text{NH(CH}_2\text{)}_4\text{NHO}_2\text{S (ring) SO}_2\right]$	Poly(tetramethylene-m-benzenesulfonamide) — Poly(sulfonyl-1,3-phenylenesulfonyliminotetramethyleneimino)

named in analogous fashion. Occasionally one may see the name poly(10-hydroxydecanoic acid), but this is less common.

For the polyester prepared from ethylene glycol and terephthalic acid, the IUPAC recognizes the larger terephthaloyl structural unit in accordance with IUPAC organic nomenclature rules:

$$\underbrace{-O-}_{\text{oxy}}\ \underbrace{CH_2CH_2-}_{\text{ethylene}}\ \underbrace{O-}_{\text{oxy}}\ \underbrace{C-\!\!\!\langle\bigcirc\rangle\!\!\!-C-}_{\text{terephthaloyl}}$$

Where no recognized larger unit is present, the carbonyl group is named separately, as with the commercially important polycarbonate

$$\left[-OCO-\!\!\!\langle\bigcirc\rangle\!\!\!-\overset{CH_3}{\underset{CH_3}{C}}\!\!\!-\!\!\!\langle\bigcirc\rangle\!\!\!-\right]$$

which bears the IUPAC name poly[oxycarbonyloxy(1,4-phenylene)isopropylidene(1,4-phenylene)]. More commonly this polymer is called bisphenol A polycarbonate after the trivial name for the bisphenol monomer used in its synthesis.

Polyamides are named like polyesters with two important distinctions. In the IUPAC system, the term *imino* denotes the -NH- group; and in the common nomenclature, the term *nylon* is applied to those polymers that contain only methylene groups in the backbone in addition to the amide functions (although the generic term nylon is often used to denote all polyamides). The first three polyamides in Table 1.3 are illustrative. For polyamides prepared by ring-opening of cyclic amides (lactams) or by step polymerization of amino acids, the number following the nylon name indicates the number of carbon atoms in the repeating unit. For those prepared from diamines and dicarboxylic acids, the number indicates the number of carbons in the diamine and diacid moieties, respectively. Thus nylon 610 (or 6-10) is prepared from 1,6-diaminohexane and 1,10-decanedioic acid (sebacic acid).

The fourth polyamide illustrates how a benzene-containing polymer is named; the fifth is an example of a polyamide of a sulfonic acid. In this case, the functional group is named separately (sulfonyl) because no appropriate larger subunit is recognized. Where more than one heteroatom appears in the backbone, the CRU is named with the heteroatoms in descending order of seniority: O, S, N, P; thus sulfonyl is named before imino.

1.7.4   Nonvinyl copolymers

The IUPAC also recommends source-based nomenclature for nonvinyl copolymers. A copolyester formed, for example, from a 2:1:1-molar ratio of the monomers ethylene glycol, terephthalic acid, and isophthalic acid could be named most simply,

Poly(ethylene terephthalate-*co*-ethylene isophthalate)

A copolyamide prepared from a mixture of 6-aminohexanoic acid and 11-aminoundecanoic acid would be called by either of the following names:

Poly[(6-aminohexanoic acid)-*co*-(11-aminoundecanoic acid)]

Poly[(6-aminohexanoamide)-*co*(11-aminoundecanoamide)]

1.7.5 Abbreviations

Not surprisingly, given the complexities of polymer structure, polymer scientists are very fond of using abbreviations [PVC for poly(vinyl chloride), HDPE and LDPE for high- and low-density polyethylene, PET for poly(ethylene terephthalate), etc.]. Not only are abbreviations used in journal articles as well as trade literature, but efforts are under way to standardize them. Commonly used abbreviations will be noted where appropriate and are listed in Appendix A.

1.8 Industrial polymers

One of the greatest challenges confronting students of chemistry is to relate the abstracts of chemical formulas and physical principles to the applied reality of the chemical industry and consumer products. No other branch of chemistry allows this chasm to be bridged more tangibly than that of polymer chemistry. In no other area of chemistry can one relate chemical structure to physical or mechanical properties so readily. This will be a continuing theme throughout the book and will be addressed in some detail in Chapters 3 and 4. In preparation for this most intellectually satisfying aspect of polymer chemistry, let us take a brief look at the polymer industry—which of the myriad of known polymers have achieved commmercial success, how they are used, and in which direction the industry is moving. This overview of what is the single biggest division of the chemical industry is intended to serve as an introduction to later chapters, where some of the polymers described here are discussed more fully. An important point for the student to keep in mind is that chemical structures encountered in this book frequently bear direct relation to articles of commerce and are by no means just laboratory curiosities.

There are three major classifications of the polymer industry: plastics, fibers, and rubber (or elastomers). The differentiation (and resultant end use) of these three types of polymers is based to a large degree on a particular mechanical property of polymers called *modulus*, which in more common terms means stiffness. Fibers have the highest modulus, rubbers the lowest. (Chapter 4 elaborates on this point.) Included in all three divisions are synthetic and natural products. The latter include cellulose-based plastics, cellulose (cotton, sisal, etc.) and protein (wool, silk) fibers, and natural rubber. Two other polymer-based industries are also delineated by end use: coatings and adhesives.

A graphic depiction of the growth of the polymer industry is given in Figure 1.5. Although the figure shows only one division of polymers, it illustrates clearly that plastics are destined to replace metals more and more as we approach the end of the century. There are sound economic reasons for the trend. Plastics weigh less and are generally more corrosion resistant. Like metals, they can be blended (alloyed) to improve their physical properties. And of particular significance with increasing costs of energy, they can be

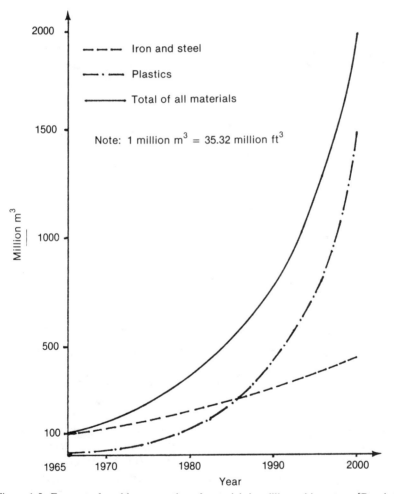

Figure 1.5. Forecast of world consumption of materials in million cubic meters. [Reprinted with permission from V. Hopp and I. Hennig, *Handbook of Applied Chemistry*, copyright 1983, courtesy of McGraw-Hill.]

manufactured and processed with lower energy input than either metals or glass.

World consumption of synthetic polymers is now in the neighborhood of 70 million metric tons per year, approximately 56% of which are plastics, 18% fibers, and 11% synthetic rubber. The balance is made up of coatings and adhesives.[22] Organizations such as the Society of the Plastics Industry (SPI), the Textile Economics Bureau, and the Rubber Manufacturers Association regularly provide detailed statistics on polymer production. Summaries are published frequently in industry-oriented chemistry and engineering journals.

1.8.1   Plastics

Plastics are divided into two major classifications on the basis of economic considerations and end use: *commodity* and *engineering*.[23,24] Commodity plastics are characterized by high volume and low cost; they may be compared with steel and aluminum in the metals industry. They are used frequently in the form of disposable items such as packaging film, but also find application in durable goods. Engineering plastics are higher in cost and much lower in volume, but have superior mechanical properties and greater durability. They compete with metals, ceramics, and glass in a variety of applications.

Commodity plastics (Table 1.4) consist principally of four major thermoplastic polymers: polyethylene, polypropylene, poly(vinyl chloride), and polystyrene. Polyethylene is divided into low-density (< 0.94 g/cm^3) and high-density (> 0.94 g/cm^3) product. (This difference in densities arises from structure: High-density polyethylene is essentially a linear polymer and low-density polyethylene is branched. How the two are formed is discussed in Part 2.) Commodity plastics represent about 90% of all thermoplastics production, with the remainder divided among styrene-butadiene and acrylonitrile–butadiene–styrene (ABS) copolymers, polyamides, and polyesters.

Table 1.4. Commodity plastics

Type	Abbreviation	Major uses
Low-density polyethylene	LDPE	Packaging film, wire and cable insulation, toys, flexible bottles, housewares, coatings
High-density polyethylene	HDPE	Bottles, drums, pipe, conduit, sheet, film, wire and cable insulation
Polypropylene	PP	Automobile and appliance parts, rope, cordage, webbing, carpeting, film
Poly(vinyl chloride)	PVC	Construction, rigid pipe, flooring, wire and cable insulation, film and sheet
Polystyrene	PS	Packaging (foam and film), foam insulation, appliances, housewares, toys

Table 1.5. Principal engineering plastics

Type	Abbreviation	Chapter where discussed
Acetal[a]	POM	11
Polyamide[b]	—	13
Poly(amideimide)	PAI	13
Polyarylate	—	12
Polycarbonate	PC	12
Polyester[c]	—	12
Polyetheretherketone	PEEK	11
Polyetherimide	PEI	11
Polyimide	PI	13
Poly(phenylene oxide)	PPO	11
Poly(phenylene sulfide)	PPS	11
Polysulfone[d]	—	11

[a] Common name for polyformaldehyde. Abbreviation refers to poly(oxymethylene).
[b] Principally nylons 6 and 66.
[c] Principally poly(ethylene terephthalate) (PET) and poly(butylene terephthalate) (PBT).
[d] Several types marketed.

Table 1.6. Principal thermosetting plastics

Type	Abbreviation	Typical uses	Chapter where discussed
Phenol–formaldehyde	PF	Electrical and electronic equipment, automobile parts, utensil handles, plywood adhesives, particle board binder	14
Urea–formaldehyde	UF	Similar to PF polymers; also treatment of textiles, coatings	14
Unsaturated polyester	—	Construction, automobile parts, boat hulls, marine accessories, corrosion-resistant ducting, pipe, tanks, etc., business equipment	12
Epoxy	—	Protective coatings, adhesives, electrical and electronics applications, industrial flooring, highway paving materials, composites	11
Melamine–formaldehyde	MF	Similar to UF polymers; decorative panels, counter and table tops, dinnerware	14

Principle engineering plastics[25] are listed alphabetically in Table 1.5 along with the chapters in which their chemistry is discussed. World-wide consumption of engineering plastics as of the late 1980s is about 1.5×10^9 kg/year, of which polyamide, polycarbonate, acetal, poly(phenylene oxide), and polyester represent about 99% of the market. Not listed are engineering grades of acrylonitrile–butadiene–styrene copolymers, a variety of fluorinated polymers, and an increasing number of copolymers and polymer blends.

There is considerable overlap in markets for engineering plastics, but primarily these plastics find use in transportation (automobiles, trucks, aircraft), construction (housing, plumbing, hardware), electrical and electronic goods (business machines, computers), industrial machinery, and consumer goods. Besides the ones listed, an increasing number of copolymers and polymer blends are especially tailored to improve properties. Engineering plastics are a rapidly growing market with consumption projected to increase by up to 10% per year in the foreseeable future.

Almost all the plastics described thus far are thermoplastic. Table 1.6 lists principal thermosetting plastics in decreasing order of consumption. Of these, only certain grades of epoxy qualify as engineering plastics. Phenol–and urea–formaldehyde polymers and unsaturated polyesters represent about 90% of production. The ratio of thermoplastic to thermosetting plastics production is roughly 6 : 1.

1.8.2   *Fibers*[26]

Fibers are characterized by having high strength and modulus, good elongation (stretchability), good thermal stability (enough to withstand ironing, for example), spinnability (the ability to be converted to filaments), and a host of other properties depending on whether they are to be used in textiles, tire cord, rope and cable, and so on. A partial list of properties might include dyeability, chemical resistance, insect and mildew resistance, crease resistance, and luster.

There are two principal natural fibers: cotton and wool, the former the polysaccharide cellulose, and the latter a protein. (Silk, another protein fiber, is produced in much smaller quantities.) World production of cotton is roughly five times that of wool on a weight basis. Natural fibers are covered in greater detail in Chapter 17. Following the pioneering work of Carothers at the du Pont Corporation, and with the increasing availability of low-cost raw materials, the synthetic fibers industry developed rapidly as demand for textiles outstripped production of cotton and wool. At the present time, world production of all fibers exceeds 30 million metric tons a year, with approximately 50% being synthetic. Production of synthetic fibers overtook that of cotton in the late 1970s.

Synthetic fibers are classified as *cellulosic* and *noncellulosic*. The principal types are described in Table 1.7. Besides those listed, some high-performance specialty fibers are produced, but these represent a very small percentage of

Table 1.7. Principal synthetic fibers

Type	Description
Cellulosic	
Acetate rayon	Cellulose acetate
Viscose rayon	Regenerated cellulose
Noncellulosic	
Polyester	Principally poly(ethylene terephthalate)
Nylon	Includes nylon 66, nylon 6, and variety of other aliphatic and aromatic polyamides
Olefin	Includes polypropylene and copolymers of vinyl chloride, with lesser amounts of acrylonitrile, vinyl acetate, or vinylidene chloride ($CH_2 = CCl_2$) (copolymers consisting of more than 85% vinyl chloride are called *vinyon* fibers)
Acrylic	Contain at least 80% acrylonitrile; included are *modacrylic* fibers comprising acrylonitrile and about 20% vinyl chloride or vinylidene chloride

total fibers production. Polyester and nylon account for about 70% of the total.

1.8.3   Rubber (elastomers)[27]

Rubbers, or elastomers, are polymers that exhibit resilience, or the ability to stretch and retract rapidly. Most have a network structure. In recent years some important varieties of nonnetwork elastomers, referred to as *thermoplastic elastomers*, have been developed. These materials, which owe their elastomeric properties to ionic or secondary bonding forces, are decribed in subsequent chapters.

Natural rubber exists in different forms, but by far the most important is composed almost entirely of *cis*-1,4-polyisoprene (Table 1.2). Synthetic rubber, the production of which surpassed that of natural rubber in the 1950s, comes in a variety of forms, the most important of which are described in Table 1.8. Thermoplastic elastomers are already making inroads into markets currently dominated by these materials. Styrene–butadiene, polybutadiene, and ethylene–propylene account for about 70% of total production.

1.8.4   Coatings and adhesives

Coatings and adhesives are as old as civilization. Since the first prehistoric artist daubed colored clay on the walls of a cave, the science of coatings[28–30] developed very slowly. Early painters discovered the value of oils for portraits and landscapes, and some experimented with eggs for more durable murals on the exterior walls of churches and monasteries. Later, natural varnishes were employed as decorative coatings, then lacquers such as cellulose nitrate. With the development of synthetic polymers in the 20th century, the coatings

Table 1.8. Principal types of synthetic rubber

Type	Description
Styrene–butadiene	Copolymer of the two monomers in various proportions depending on properties desired; called SBR for styrene-butadiene rubber
Polybutadiene	Consists almost entirely of the *cis*-1,4 polymer
Ethylene–propylene	Often abbreviated EPDM for ethylene-propylene-diene monomer; made up principally of ethylene and propylene units with small amounts of a diene to provide unsaturation
Polychloroprene	Principally the *trans*-1,4 polymer, but also some *cis*-1,4 and-1,2 polymer; also known as *neoprene* rubber
Polyisoprene	Mainly the *cis*-1,4 polymer; sometimes called "synthetic natural rubber"
Nitrile	Copolymer of acrylonitrile and butadiene, mainly the latter
Butyl	Copolymer of isobutylene and isoprene, with only small amounts of latter
Silicone	Contains inorganic backbone of alternating oxygen and methylated silicon atoms; also called polysiloxane (Chap. 15)
Urethane	Elastomers prepared by linking polyethers through urethane groups (Chap. 13)

industry evolved from the earliest polyester (alkyd) synthetic varnishes and paints introduced in the 1920s to the more recent latex paints consisting of polymers emulsified in water. First came the latex wall paints based on styrene–butadiene copolymers, then the exterior latex paints employing poly-(vinyl acetate) and a variety of poly(acrylate ester)s.

A similar history can be traced with adhesives,[31,32] from the earliest uses of bitumens and natural gums and resins, even blood. Starch and cellulose nitrate adhesives came much later. Like coatings, adhesives came into their own with the development of synthetic polymers, particularly the phenol–formaldehyde and urea–formaldehyde polymers that are still used extensively, especially in the wood industries (plywood, particleboard, etc.). Later came the epoxies, cyanoacrylates, and so on. In more recent years the adhesives industry has encompassed most of the major polymer types, including some high-temperature-resistant varieties for use in aerospace applications.

Both the coatings and adhesives industries are far more complex than the polymers used. Formulations include solvents, fillers, stabilizers, and a variety of other additives depending on application.

References

1. H. Staudinger, *From Organic Chemistry to Macromolecules*, Wiley-Interscience, New York, 1970.

2. H. Morawetz, *Polymers—The Origins and Growth of a Science*, Wiley-Interscience, New York, 1985.
3. R. B. Seymour, *History of Polymer Science and Technology*, Dekker, New York, 1982.
4. J. J. Berzelius, *Jahresberichte*, **12**, 63 (1833).
5. E. Simon, *Liebigs Ann. Chem.*, **31**, 265 (1839).
6. A-V. Lourenço, *Compt. Rend.*, **51**, 365 (1860); *Ann. Chim. Phys.*, **3**, 67, 257 (1863).
7. J. C. F. Williams, *J. Chem. Soc.*, **15**, Pt. 10 (1862).
8. G. A. Stahl, in *Polymer Science Overview* (G. A. Stahl, ed.), ACS Symp. Ser. 175, American Chemical Society, Washington, D.C., 1981, Chap. 3.
9. L. H. Baekeland, *Ind. Eng. Chem.*, **5**, 506 (1913).
10. H. Staudinger, *Chem. Ber.*, **53**, 1073 (1920).
11. H. Staudinger, *Chem. Ber.*, **57**, 1203 (1924).
12. H. Mark and G. S. Whitby (eds.), *Collected Papers of Wallace Hume Carothers on High Polymeric Substances*, Wiley-Interscience, New York, 1940.
13. K. Ziegler, E. Holzkamp, H. Breil, and H. Martin, *Angew. Chem.*, **67**, 541 (1955).
14. G. Natta, P. Pino, P. Corradini, F. Danusso, E. Mantica, G. Mazzanti, and G. Moraglio, *J. Am. Chem. Soc.*, **77**, 1708 (1955).
15. P. J. Flory, *Principles of Polymer Chemistry*, 2nd ed., Cornell University Press, Ithaca, N.Y., 1953.
16. W. O. Baker, in *Polymer Science Overview* (G. A. Stahl, ed.) ACS Symp. Ser. 175, American Chemical Society, Washington, DC., 1981, Chap. 13.
17. T. Davidson (ed.), *Polymers in Electronics*, ACS Symp. Ser. 242, American Chemical Society, Washington, D.C., 1984.
18. W. H. Carothers, *J. Am. Chem. Soc.*, **51**, 2548 (1928).
19. International Union of Pure and Applied Chemistry, Macromolecular Nomenclature Commission, *Macromolecules*, **6**, 149 (1973); *Pure Appl. Chem.*, **40**, 479 (1974).
20. H. Feuerberg, *J. Polym. Sci., Polym. Lett. Ed.*, **22**, 413 (1984).
21. W. Ring, I. Mita, A. D. Jenkins, and N. M. Bikales, *Pure Appl. Chem.*, **57**, 1427 (1985).
22. K. Weissermel and H. Cherdron, *Angew. Chem. Int. Ed. Engl.*, **22**, 764 (1983).
23. N. Platzer, *J. Appl. Polym. Sci. Symp.*, **36**, v (1981).
24. L. H. Gillespie, in *High Performance Polymers: Their Origin and Development* (R. B. Seymour and G. S. Kirshenbaum, eds.), Elsevier, New York, 1986, pp. 9–15.
25. J. M. Margolis (ed.), *Engineering Thermoplastics: Properties and Applications*, Dekker, New York, 1985.
26. R. W. Mancrieff, *Man-Made Fibres*, 6th ed., Wiley, New York, 1975.
27. M. Morton (ed.), *Rubber Technology*, 2nd ed., Van Nostrand Reinhold, New York, 1973.
28. S. Paul, *Surface Coatings*, Wiley-Interscience, New York, 1985.
29. G. P. A. Turner, *Paint Chemistry*, Chapman and Hall, London, 1980.
30. R. Lambourne (ed.), *Paints and Surface Coatings—Theory and Practice*, Halstead Press, New York, 1987.
31. E. J. Bruno (ed.), *Adhesives in Modern Manufacturing*, Society of Manufacturing Engineers, Dearborn, Mich., 1970.
32. L. -H. Lee (ed.), *Adhesive Chemistry*, Plenum Press, New York, 1985.

Review exercises

1. Following is an alphabetical listing of important terms encountered in this chapter. Their meaning must be clear to you before you proceed further into the book. Write a concise definition of each, using examples as appropriate.

(a)	addition polymerization	(p)	homochain polymer
(b)	alternating copolymer	(q)	homopolymer
(c)	block copolymer	(r)	ladder polymer
(d)	branched polymer	(s)	linear polymer
(e)	comb polymer	(t)	macromolecule
(f)	commodity plastic	(u)	monomer
(g)	condensation polymerization	(v)	nonvinyl monomer
(h)	copolymerization	(w)	oligomer
(i)	crosslinking	(x)	repeating unit
(j)	degree of polymerization	(y)	ring-opening polymerization
(k)	elastomer	(z)	star polymer
(l)	end group	(aa)	thermoplastic
(m)	engineering plastic	(bb)	thermosetting
(n)	graft copolymer	(cc)	vinyl monomer
(o)	heterochain polymer		

2. In your own words, explain the difference between step-reaction polymerization and chain-reaction polymerization. What is the advantage of this terminology over the more traditional addition and condensation polymerization?

3. Comment on the correctness of the statement: The number of end groups in a polymer molecule is two.

4. What is the $\overline{DP}$ of (a) a sample of poly(methyl methacrylate) (Table 1.2) of average molecular weight 50,000, and (b) a sample of poly(tetramethylene-*m*-benzenesulfonamide) (Table 1.3) of average molecular weight 26,000?

5. What percent conversion is represented by the polymer in question 4(b)?

6. What would be the molecular weight of the polyamide formed from 11-aminoundecanoamide (Table 1.3) if the reaction were carried to 96% conversion? To 99.9% conversion?

7. Write the name and structure of the monomers needed to synthesize the following vinyl polymers:

(a) $-\!\left[CH_2CHF\right]\!-$ (b) $-\!\left[CH_2CH\right]\!-$ with $CH_2CH_2CH_3$ (c) $-\!\left[CH_2C\right]\!-$ with CH_3 above and CH_3 below

(d) $-\!\left[CH_2C\right]\!-$ with CN above and CN below

(e) $-\!\left[CH_2CH\right]\!-$ with pyridine ring (N)

(f) $-\!\left[CH_2CH\right]\!-$ with chlorobenzene ring (Cl)

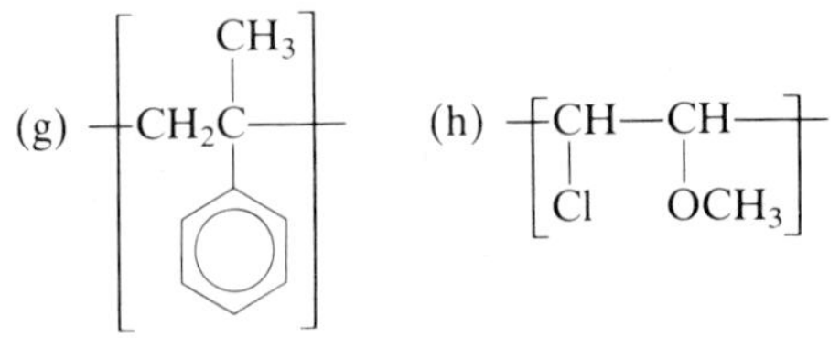

(g) $\left[CH_2C \underset{\displaystyle \bigcirc}{\overset{\displaystyle CH_3}{\underset{|}{\overset{|}{|}}}} \right]$

(h) $\left[CH\underset{|}{\overset{}{-}}CH\underset{|}{\overset{}{-}} \\ \quad Cl \quad\; OCH_3 \right]$

8. Write the name and structure of the monomers needed to synthesize the following nonvinyl polymers. In some instances, more than one answer may be correct. (You might wish to review organic nomenclature before attempting to name some of these monomers.)

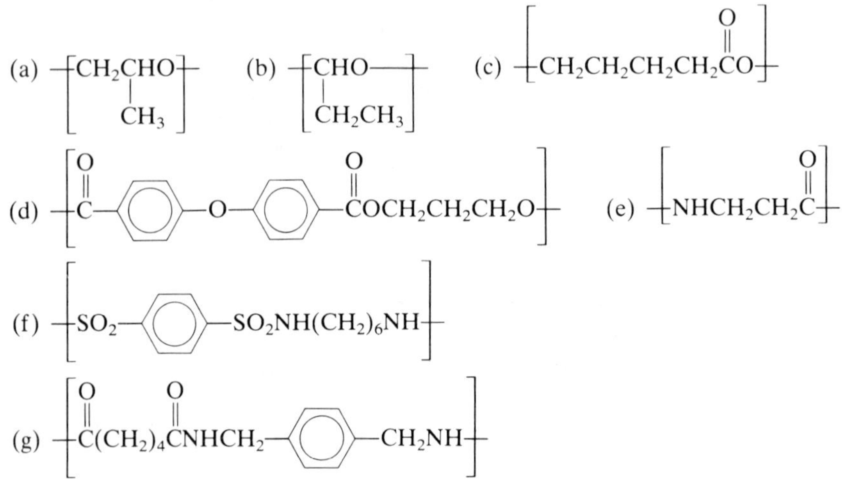

(a) $\left[CH_2CHO \atop \quad\;\; CH_3 \right]$

(b) $\left[CHO \atop CH_2CH_3 \right]$

(c) $\left[CH_2CH_2CH_2CH_2\overset{\displaystyle O}{\overset{\displaystyle \|}{C}}O \right]$

(d) $\left[\overset{O}{\overset{\|}{C}}-\bigcirc-O-\bigcirc-\overset{O}{\overset{\|}{C}}OCH_2CH_2CH_2O \right]$

(e) $\left[NHCH_2CH_2\overset{\displaystyle O}{\overset{\displaystyle \|}{C}} \right]$

(f) $\left[SO_2-\bigcirc-SO_2NH(CH_2)_6NH \right]$

(g) $\left[\overset{O}{\overset{\|}{C}}(CH_2)_4\overset{O}{\overset{\|}{C}}NHCH_2-\bigcirc-CH_2NH \right]$

9. Write common or source-based names and IUPAC-recommended names for each of the polymers in exercises 7 and 8.

10. Write all possible head-to-tail repeating units for polyisoprene and give each one an acceptable source-based and IUPAC name.

11. An inconsistency arises in using source-based nomenclature with the commercially important poly(vinyl alcohol), $\left[CH_2CH \atop \quad\; OH \right]$. What is the inconsistency? (You can check your answer in Chapter 9.)

12. The first commercially available polyamide fiber was poly(iminohexamethyleneiminoadipoyl). Write the structure, and suggest two alternate names.

13. A polyester fiber used for surgical sutures is poly[oxy(1-oxoethylene)]. Write the structure and suggest an alternate name.

14. In the following schematic copolymer diagrams, A = acrylonitrile, B = butadiene, M = methyl methacrylate, S = styrene, and V = vinyl acetate. Write an acceptable name for each.

(a) MMMMMMMMSVSVSVSVSVSV

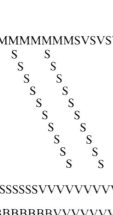

(b) SSSSSSSSVVVVVVVVVSSSSSSSSS

(c) BBBBBBBBVVVVVVVVVV

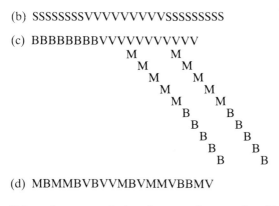

(d) MBMMBVBVVMBVMMVBBMV

15. Using the same designations as in exercise 14, write schematics for an —⫟AB—block copolymer of styrene and methyl methacrylate, and an ABC block copolymer of vinyl acetate, butadiene, and acrylonitrile.

16. Using appropriate letters or symbols, draw schematic structures for the following copolymers:
    (a) Poly[propylene-*co*-(methyl methacrylate)]
    (b) Copoly(isoprene/styrene/acrylonitrile)
    (c) Polystyrene-*block*-[butadiene-*alt*-(vinyl acetate)]
    (d) *Block*-copoly[*graft-co*(ethylene/styrene)/α-methylstyrene]
    (e) *Block*-copoly(acrylic acid/1-butene/vinyl chloride)

2
Molecular weight and polymer solutions

2.1 Number average and weight average molecular weight

Molecular weight is an extremely important variable because it relates directly to a polymer's physical properties. In general, the higher the molecular weight, the tougher the polymer; however (as will be shown in the next chapter), too high a molecular weight can lead to processing difficulties. What defines an optimum molecular weight depends in large measure on the chemical structure of the polymer and the application for which it is intended. In this chapter we will be concerned with molecular weight definitions, methods of determining molecular weight, and methods of determining the distribution of molecular weights in a polymer sample. We will reserve our discussion of molecular structure for the following chapter. We will also be concerned with solution properties of polymers, because solubility is prerequisite to molecular weight determination.

Given that polymers are high-molecular-weight compounds, we must begin with the question, what do we *mean* by high molecular weight? Where does low molecular weight end and high begin? There is no simple answer, because what constitutes a "low" molecular weight for a sample of polyethylene, for example, might be ideal for a sample of polyamide. Furthermore, some polymers are deliberately prepared with low molecular weight (even as oligomers) to facilitate initial processing, molecular weight being increased at a subsequent processing stage. In general, however, we think of polymers as having molecular weights that run from the low thousands up to the millions, with optimum molecular weight depending on chemical structure and application. Vinyl polymers of any commercial import normally have molecular weights in the range 10^5 to 10^6. Polymers having very polar functional groups, such as polyamides, may have molecular weights as low as 15,000 to 20,000.

To determine molecular weights of simple (nonpolymeric) compounds, we employ the familiar techniques of mass spectrometry, freezing-point depression (cryoscopy), boiling-point elevation (ebulliometry), and, where suitable functional groups are present, titration (for example, neutralization or saponification equivalents). Determining molecular weights of polymers, however, is considerably more complex for two basic reasons. First, in any polymerization process, it is virtually impossible for all growing polymer chains to terminate at the same size; hence one must necessarily deal with *average* molecular weights. (Certain natural polymers having discrete molecular weights provide exceptions to this rule.) Second, the techniques of cryoscopy, ebulliometry, and titration are effective only with relatively low-molecular-weight polymers; more sophisticated methods must be used for polymers with molecular weights higher than about 40,000. The conventional techniques of mass spectrometry have not been used extensively in the polymer field beyond the characterization of polymer degradation products because of the requirements for volatilizable samples. Recent years have seen some exciting new developments in *field desorption* that have extended mass spectrometry into the macromolecular range. Such developments are, however, in their infancy and do not enjoy the routine applicability of the more traditional methods of molecular weight determination.

Techniques more commonly used for determining molecular weights of polymers include osmometry, light scattering, and ultracentrifugation, although titration (end-group analysis), cryoscopy, and ebulliometry are also used in some applications.[1,2] The most convenient method for routinely determining molecular weights involves measuring solution viscosities, but this is not an absolute method and can be used only in conjunction with one of the techniques of measuring absolute molecular weights.

Molecular weight values obtained depend in large measure on the method of measurement. Methods that depend on end-group analysis or colligative properties (freezing-point depression, boiling-point elevation, osmotic pressure) give rise to what is known as the *number average molecular weight* because the number of molecules of each weight in the sample are counted. The total weight of a polymer sample, w, is the sum of the weights of each molecular species present:

$$w = \sum_{i=1}^{\infty} w_i = \sum_{i=1}^{\infty} N_i M_i$$

where N and M are the number of moles and molecular weight, respectively, of each species i. Number average molecular weight, $\bar{M}_n$, is the weight of sample per mole:

$$\bar{M}_n = \frac{w}{\sum_{i=1}^{\infty} N_i} = \frac{\sum_{i=1}^{\infty} M_i N_i}{\sum_{i=1}^{\infty} N_i}$$

Suppose, for example, we have a polymer sample consisting of 9 mol of molecular weight 30,000 and 5 mol of molecular weight 50,000:

$$\bar{M}_n = \frac{(9 \times 30,000) + (5 \times 50,000)}{(9 + 5)} = 37,000$$

Suppose, instead, our sample consists of 9 g of molecular weight 30,000 and 5 g of molecular weight 50,000:

$$\bar{M}_n = \frac{(9+5)}{(9/30,000)+(5/50,000)} = 35,000$$

Light scattering and ultracentrifugation, on the other hand, are methods of determining molecular weight based on mass or polarizability of the species present. The greater the mass, the greater is the contribution to the measurement. In contrast to number average molecular weight (which is the summation of the *mole* fraction of each species times its molecular weight), these methods sum the *weight* fraction of each species times its molecular weight. The value thus obtained is called the *weight average molecular weight*, $\bar{M}_w$, and is expressed mathematically as

$$\bar{M}_w = \frac{\sum\limits_{i=1}^{\infty} w_i M_i}{\sum\limits_{i=1}^{\infty} w_i} = \frac{\sum\limits_{i=1}^{\infty} N_i M_i^2}{\sum\limits_{i=1}^{\infty} N_i M_i}$$

Consider the same two samples described previously. Nine mol of 30,000 molecular weight and 5 mol of 50,000 molecular weight:

$$\bar{M}_w = \frac{9(30,000)^2 + 5(50,000)^2}{9(30,000) + 5(50,000)} = 40,000$$

Substituting grams for moles:

$$\bar{M}_w = \frac{9(30,000) + 5(50,000)}{(9+5)} = 37,000$$

In each instance, we see that $\bar{M}_w$ is greater than $\bar{M}_n$.

In measurements of colligative properties, each molecule contributes equally regardless of weight, whereas with light scattering, the larger molecules contribute more because they scatter light more effectively. It is for this reason that weight average molecular weights are always greater than number average molecular weights except, of course, when all molecules are of the same weight; then $\bar{M}_w = \bar{M}_n$. The narrower the molecular weight range, the closer are the values of $\bar{M}_w$ and $\bar{M}_n$, and the ratio $\bar{M}_w/\bar{M}_n$ may thus be used as an indication of the breadth of the molecular weight range in a polymer sample. This ratio is called the *polydispersity index*, and any system having a range of molecular weights is said to be *polydisperse*.

In our discussions of molecular weight measurement (Section 2.3), detailed descriptions of apparatus and derivations of working equations are not given. Students who wish to explore these areas in more detail are encouraged to consult the references provided. First, however, we will look briefly at *solution* properties of polymers because the various methods of determining molecular weight or molecular weight distribution depend on solubility and polymer-solvent interactions.

2.2 Polymer solutions[3]

Dissolving a polymer is unlike dissolving low-molecular-weight compounds because of the vastly different dimensions of solvent and polymer molecules. Solution occurs in two stages. Initially the solvent molecules diffuse through the polymer matrix to form a swollen, solvated mass called a *gel*. (Any student who has stoppered with a rubber stopper a flask containing an organic solvent has undoubtedly experienced the swelling effects associated with gel formation.) In the second stage, the gel breaks up and the molecules are dispersed into a true solution. Dissolution is often a slow process. While some polymers dissolve readily in certain solvents, others may require prolonged periods of heating at temperatures near the melting point of the polymer. Network polymers do not dissolve, but usually swell in the presence of solvent. Even some linear polymers defy all attempts to dissolve them, hence lack of solubility does not necessarily mean that the polymer is of the network type.

How does one choose a solvent? The simplest way is to consult a polymer handbook, where extensive lists of solvents and nonsolvents for numerous polymers are compiled.[4a] Detailed studies of polymer solubilities using thermodynamic principles have led to semiempirical relationships for predicting solubility. Any solution process is governed by the free-energy relationship.

$$\Delta G = \Delta H - T\Delta S$$

When a polymer dissolves spontaneously, the free energy of solution, ΔG, is negative. The entropy of solution, ΔS, invariably has a positive value arising from increased conformational mobility of the polymer chains. Therefore the magnitude of the enthalpy of solution, ΔH, determines the sign of ΔG. It has been proposed that the heat of mixing, ΔH_{mix}, for a binary system is related to concentration and energy parameters by the expression

$$\Delta H_{mix} = V_{mix}\left[\left(\frac{\Delta E_1}{V_1}\right)^{1/2} - \left(\frac{\Delta E_2}{V_2}\right)^{1/2}\right]^2 \phi_1\phi_2$$

where V_{mix} is the total volume of the mixture, V_1 and V_2 are molar volumes (molecular weight/density) of the two components, ϕ_1 and ϕ_2 are their volume fractions, and ΔE_1 and ΔE_2 are the energies of vaporization. The terms $\Delta E_1/V_1$ and $\Delta E_2/V_2$ are called the *cohesive energy densities*. If $(\Delta E/V)^{1/2}$ is replaced by the symbol δ, the equation is written more simply:

$$\Delta H_{mix} = V_{mix}(\delta_1 - \delta_2)^2 \phi_1\phi_2$$

The symbol δ is called the *solubility parameter*. Clearly, for the polymer to dissolve (negative ΔG), ΔH_{mix} must be small; therefore, $(\delta_1 - \delta_2)^2$ must also be small. In other words, δ_1 and δ_2 should be of about equal magnitude. Where $\delta_1 = \delta_2$, solubility is governed solely by entropy effects. Predictions of solubility are therefore based on finding solvents and polymers with similar solubility parameters, which requires a means of determining cohesive energy densities.

Cohesive energy density is the energy needed to remove a molecule from its nearest neighbors, thus is analogous to heat of vaporization per volume for a volatile compound. For the solvent, δ_1 can be calculated directly from the latent heat of vaporization (ΔH_{vap}) using the relationship

$$\Delta E = \Delta H_{vap} - RT$$

where R is the gas constant, and T is the temperature in kelvins. Thus

$$\delta_1 = \left(\frac{H_{vap} - RT}{v} \right)^{1/2}$$

Since polymers have negligible vapor pressure, the most convenient method of determining δ_2 is to use *group molar attraction constants*. These are constants derived from studies of low-molecular-weight compounds that lead to numerical values for various molecular groupings on the basis of considerations of intermolecular forces. Two sets of values (designated G) have been suggested, one by Small, [5] derived from heats of vaporization, and the other by Hoy,[6] based on vapor pressure measurements. Typical G values are given in Table 2.1. Clearly there are significant differences between the Small and Hoy values. Which set one uses is normally determined by the method used to determine δ_1 for the solvent.

G values are additive for a given structure, and are related to δ by

$$\delta = \frac{d\Sigma G}{M}$$

Table 2.1. Representative group molar attraction
constants

Group	G [(cal cm^3)$^{1/2}$ mol^{-1}]	
	Small[a]	Hoy[b]
CH$_3$—	214	147.3
—CH$_2$—	133	131.5
$>$CH—	28	85.99
$>$C$<$	−93	32.03
=CH$_2$	190	126.5
=CH—	19	84.51
—C$_6$H$_5$ (phenyl)	735	—
—CH= (aromatic)	—	117.1
$>$C=O (ketone)	275	262.7
—CO$_2$— (ester)	310	326.6

[a] According to Small.[5]
[b] According to Hoy.[6]

where d is density and M is molecular weight. For polystyrene, for example, which has a density (obtained from a handbook[4b]) of 1.05, a repeating unit mass of 104, and structure

$$\left[\text{CH}_2\text{—CH} \right]$$

δ is calculated, using Small's G values, as

$$\delta = \frac{1.05(133 + 28 + 735)}{104} = 9.0$$

or using Hoy's data,

$$\delta = \frac{1.05[131.5 + 85.99 + 6(117.1)]}{104} = 9.3$$

We note that solvent δ's may be calculated similarly. A major problem with solubility parameters as discussed above is that they do not take into account strong dipolar forces like hydrogen bonding. Proposals have been made,[7] however, to separate solubility parameters into components reflecting dispersion forces, dipole–dipole attractions, and hydrogen bonding. Compilations of δ for solvents and polymers are available in the literature.[4c]

Once a polymer-solvent system has been selected, another consideration is how the polymer molecules behave in that solvent. Particularly important from the standpoint of molecular weight determinations is the resultant size, or *hydrodynamic volume*, of the polymer molecules in solution. Assuming polymer molecules of a given molecular weight are fully separated from one another by solvent, the hydrodynamic volume will depend on a variety of factors, including interactions between solvent and polymer molecules, chain branching, conformational effects arising from the polarity and steric bulk of the substituent groups, and restricted rotation caused by resonance, for example, of the type common to polyamides,

$$\overset{\text{O}}{\underset{}{\overset{\|}{\text{—C—NH—}}}} \longleftrightarrow \overset{\text{O}^-}{\underset{}{\overset{|}{\text{—C}=\overset{+}{\text{N}}\text{H—}}}}$$

Because of Brownian motion, molecules are changing shape continuously. Hence any method of trying to predict molecular size must necessarily be based on statistical considerations and average dimensions. If a molecule were fully extended, its size could easily be computed from a knowledge of bond lengths and bond angles. Such is not the case, however, with most common polymers; therefore size is generally expressed in terms of the mean-square average distance between chain ends, $\bar{r}^2$, for a linear polymer, or the square average radius of gyration about the center of gravity, $\bar{s}^2$, for a branched polymer. Figure 2.1 illustrates the meaning of r and s from the perspective of a coiled structure of an individual polymer molecule having its center of gravity at the origin. The average shape of the coiled molecule is spherical. The greater the affinity of solvent for polymer, the larger will be the sphere, that is, the hydrodynamic volume. As the solvent–polymer interaction decreases, intramolecular interactions become more important, leading to a contraction of the hydrodynamic volume.

It is convenient to express r and s in terms of two factors: an *unperturbed dimension* (r_0 or s_0) and an *expansion factor* (α). Thus,

$$\bar{r}^2 = r_0^2 \alpha^2$$

$$\bar{s}^2 = s_0^2 \alpha^2$$

The unperturbed dimension refers to the size of the macromolecule exclusive of solvent effects. It arises from a combination of free rotation and intramolecular steric and polar interactions. The expansion factor, on the other hand, arises from interactions between solvent and polymer. For a linear polymer, $\bar{r}^2 = 6\bar{s}^2$. Since

$$\alpha = \frac{(\bar{r}^2)^{1/2}}{(\bar{r}_0^2)^{1/2}}$$

it follows that α will be greater than unity in a "good" solvent, and the actual (perturbed) dimensions will exceed the unperturbed dimensions. The greater the value of α, the "better" the solvent. For the special case where $\alpha = 1$, the polymer assumes its unperturbed dimensions and behaves as an "ideal" statistical coil.

Because solubility properties vary with temperature in a given solvent, α is temperature dependent. For a given polymer in a given solvent, the lowest temperature at which $\alpha = 1$ is called the *theta* (θ) *temperature* (or *Flory temperature*), and the solvent is then called a *theta solvent*. Additionally, the

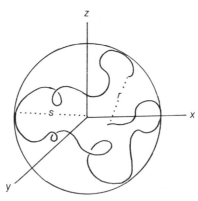

Figure 2.1. Schematic representation of a molecular coil. r = end-to-end distance; s = radius of gyration.

polymer is said to be in a *theta state*. It is convenient to think of the theta state as that in which the polymer is on the brink of becoming insoluble; in other words, the solvent is having a minimal solvation effect on the dissolved molecules. Any further diminution of this effect causes the attractive forces between polymer molecules to predominate, and the polymer precipitates. Extensive compilations of theta solvents and temperatures are available in the literature.[4d]

From the standpoint of molecular weight determinations, the significance of these parameters is that they can be related to dilute solution viscosity according to the Flory–Fox equation,

$$[\eta] = \frac{\phi(\bar{r}^2)^{3/2}}{\bar{M}}$$

where $[\eta]$ is the intrinsic viscosity (to be defined later), $\bar{M}$ is the average molecular weight, and ϕ is a proportionality constant (called the Flory constant) equal to approximately 3×10^{24} mol^{-1}. Substituting $\bar{r}_0^2\alpha^2$ for $\bar{r}^2$, we obtain

$$[\eta] = \frac{\phi(r_0^2\alpha^2)^{3/2}}{\bar{M}}$$

which can be rearranged to

$$[\eta] = \phi(r_0^2 M^{-1})^{3/2}\bar{M}^{1/2}\alpha^3$$

Since $\bar{r}_0$ and $\bar{M}$ are constants, we can set $K = \phi(\bar{r}_0^2\bar{M}^{-1})^{3/2}$, then

$$[\eta] = K\bar{M}^{1/2}\alpha^3$$

At the theta temperature, $\alpha = 1$ and

$$[\eta] = K\bar{M}^{1/2}$$

For conditions other than the theta temperature, the equation has the form

$$[\eta] = K\bar{M}^a$$

where a is a constant that varies with polymer, solvent, and temperature. Known as the *Mark–Houwink–Sakurada equation*, this is an important relationship which, as we shall see in Section 2.5, is used to relate dilute solution viscosity to molecular weight.

Apart from molecular weight determinations, important practical considerations arise from solubility effects. As one moves in the direction of "good" solvent to "poor," and intramolecular forces become more important, the polymer molecules shrink in volume. This increasing compactness leads to reduced "drag" and hence a lower viscosity. Paint chemists are particularly concerned with solvent effects because viscosity is of fundamental importance in spray or brush applications of solvent-based paints or lacquers. Fiber manufacturers can also reduce energy costs involved in spinning fibers from solution (see Chapter 4) by reducing solution viscosity.

2.3   Measurement of number average molecular weight

2.3.1   End-group analysis[8a,9a,10a]

The number average molecular weight of any linear polymer having end groups that can be measured by chemical or physical means can theoretically be determined if the method of measurement is sensitive enough. It must be remembered, however, that end groups are present in very low concentrations. Presently available techniques allow a practical upper limit of molecular weight measurement of about 50,000. Some of the current methods of end-group determination include (1) titration, using either indicators or potentiometric techniques; (2) elemental analysis of element-specific end groups; (3) measurement of activity of a radioactive-tagged end group; and (4) ultraviolet spectroscopic determination of an end group with a characterizable chromophore. Infrared and nuclear magnetic resonance (NMR) spectroscopic techniques are of more limited use.

Points to be kept in mind in applying end-group analysis are the following:

1. The method cannot be applied to branched polymers unless the number of branches is known with certainty; thus it is practically limited to linear polymers.
2. In a linear polymer there are twice as many end groups as polymer molecules.
3. If the polymer contains different groups at each end of the chain and only one characterizable end group is being measured, the number of this type is equal to the number of polymer molecules.
4. Measurement of molecular weight by end group analysis is only meaningful when the mechanisms of initiation and termination are well understood.

As a typical example, let us consider the unsaturated polyesters (Chapter 12). These are linear polymers of relatively low molecular weight prepared from anhydrides or dicarboxylic acids and diols, and they are normally crosslinked in a subsequent fabrication step. (Saturated polyesters of the type used in film and fiber applications are usually of too high molecular weight for end-group analysis.)

To determine the number average molecular weight of the linear polyester before crosslinking, one can titrate the carboxyl and hydroxyl end groups by standard methods.[9a] In the case of carboxyl, a weighed sample of polymer is dissolved in an appropriate solvent such as acetone and titrated with standard base to a phenolphthalein end point. For hydroxyl, a sample is acetylated with excess acetic anhydride, and liberated acetic acid, together with carboxyl end groups, is similarly titrated. From the two titrations, one obtains the number of milliequivalents of carboxyl and hydroxyl in the sample. Moles of polymer per gram is then given by the expression

$$\text{Moles of polymer per gram} = \frac{\text{meq COOH} + \text{meq OH}}{2 \times 1000 \times \text{sample weight}}$$

(The 2 in the denominator is to take into account that two end groups are being counted per molecule.)

$$\text{Molecular weight} = \frac{1}{\text{moles of polymer per gram}}$$

In polyester processing, the *acid number*, defined as the number of milligrams of base required to neutralize 1 g of polyester, is frequently used to monitor reaction progress.

Other titratable end groups include amino groups in polyamides, acetyl groups in acetyl-terminated polyamides, isocyanate in polyurethanes, and epoxide in epoxy polymers. A number of factors can complicate end-group analysis: lack of solubility, high solution viscosity, severe steric hindrance. But where these difficulties can be surmounted, end-group analysis is preferred for polymers in the 5,000 to 10,000 molecular weight range.

2.3.2 *Membrane osmometry*[8b,9b,10b]

Of the various methods of number average molecular weight determination based on colligative properties, membrane osmometry is most useful. When pure solvent is separated from a solution by a barrier that allows solvent but not solute molecules to pass through, then solvent will pass through the barrier until a hydrostatic pressure develops that prevents further passage, or in other words, until equilibrium is established. This pressure is the osmotic pressure. Barriers of this type are called *semi-permeable membranes*.

A schematic representation of an osmometer is shown in Figure 2.2. Semi-permeable membranes are usually constructed from polymeric materials such as rubber, nitrocellulose, cellulose acetate, and poly(vinyl alcohol). Osmotic pressure can be determined by allowing the system to reach equilibrium and measuring the hydrostatic head that develops. This is referred to as the *static equilibrium method*. Alternatively, one can apply a counterpressure to the measuring tube connected to the solution compartment, which prevents flow of solvent and maintains equal liquid levels in the two measuring tubes. This is called the *dynamic equilibrium method*. Because the former method usually requires long periods of time to attain equilibrium, the dynamic method is preferred.

A variety of dynamic membrane osmometers are manufactured, which usually encompass a horizontal membrane separating solution and solvent cells. The type shown schematically in Figure 2.3 measures osmotic pressure directly via a strain gauge transducer attached to a flexible diaphragm in the solvent cell.

Osmotic pressure is related to molecular weight by the van't Hoff equation extrapolated to zero concentration:

$$\left(\frac{\pi}{C}\right)_{C=0} = \frac{RT}{\overline{M}_n} + A_2 C$$

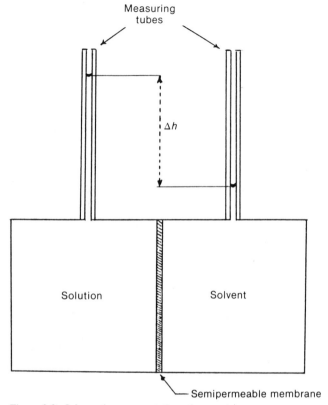

Measuring
tubes

Δh

Solution

Solvent

Semipermeable membrane

Figure 2.2. Schematic representation of a membrane osmometer.

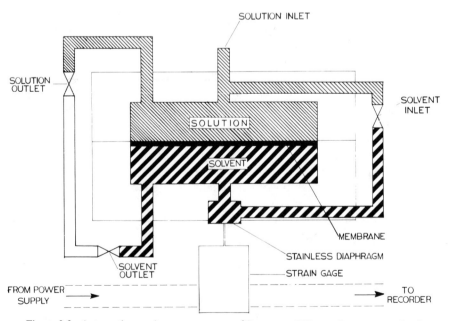

SOLUTION INLET

SOLUTION
OUTLET

SOLVENT
INLET

SOLUTION

SOLVENT

MEMBRANE

SOLVENT
OUTLET

STAINLESS DIAPHRAGM

STRAIN GAGE

FROM POWER
SUPPLY

TO
RECORDER

Figure 2.3. Automatic membrane osmometer. [Courtesy of Wescan Instruments, Inc.]

where π, the osmotic pressure, is given by

$$\pi = \rho g \Delta H$$

where R is the gas constant, 0.082 L atm mol^{-1} K^{-1} (CGS) or 8.314 J mol^{-1} K^{-1} (SI); T is the temperature, in kelvins; C is the concentration, in grams per liter; ρ is the solvent density, in grams per cubic centimeter; g is the acceleration due to gravity, 0.981 m/s^2; Δh is the difference in heights of solvent and solution, in centimeters; and A_2 is the second virial coefficient (a measure of the interaction between solvent and polymer).

A plot of *reduced osmotic pressure*, π/C, versus concentration (Figure 2.4) is linear with the intercept equal to $RT/\bar{M}_n$ and slope equal to A_2. [Units for π/C are dyne L g^{-1} cm^{-1} (CGS) or J kg^{-1} (SI).] Because A_2 is a measure of solvent–polymer interaction, the slope is zero at the theta temperature. Thus osmotic pressure measurements may be used to determine theta conditions.

The major source of error in membrane osmometry arises from low-molecular-weight species diffusing through the membrane. It explains why molecular weight values obtained are generally lower than those obtained by other colligative property measurements where *all* species present are counted. Membrane osmometry is generally useful over a molecular weight range of about 50,000 to 2 million, the limiting factors being the permeability of the membrane at the lower end and the smallest measurable osmotic pressure at the upper. Because most polymers of commercial significance fall in this range, membrane osmometry is the most widely used method of determining absolute number average molecular weights.

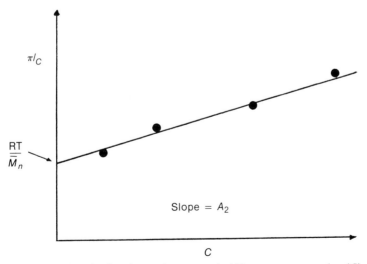

Figure 2.4. Plot of reduced osmotic pressure (π/C) versus concentration (C).

2.3.3   Cryoscopy and ebulliometry[8c,9c,10c,11a]

The techniques employed in freezing-point depression (cryoscopy) and boiling-point elevation (ebulliometry) are analogous to those for low-molecular-weight compounds, and the thermodynamic relationships, derived for infinitely dilute solutions, are similar:

$$\left(\frac{\Delta T_f}{C}\right)_{C=0} = \frac{RT^2}{\rho \Delta H_f \bar{M}_n} + A_2 C \qquad \text{for freezing-point depression}$$

$$\left(\frac{\Delta T_b}{C}\right)_{C=0} = \frac{RT^2}{\rho \Delta H_v \bar{M}_n} + A_2 C \qquad \text{for boiling-point elevation}$$

In these expressions C is the concentration in grams per cubic centimeter; T is the freezing point or boiling point of the solvent in kelvins; R is the gas constant; ρ is the solvent density; ΔH_f and ΔH_v are the latent heats of fusion and vaporization, respectively, per gram of solvent; $\bar{M}_n$ is the number average molecular weight and A_2 is the second virial coefficient. Molecular weights are determined from plots of $\Delta T_f/C$ (or $\Delta T_b/C$) versus concentration, analogous to those in osmometry.

The major limitation of cryoscopy and ebulliometry lies in the sensitivity of the methods of measuring the freezing-point depression, ΔT_f, or the boiling-point elevation, ΔT_b. As molecular weight increases, ΔT_f or ΔT_b becomes progressively smaller so that, even with the most sensitive thermistors (some are sensitive to $1 \times 10^{-4}°C$), the upper limit of measurable molecular weight is about 40,000. These methods, together with that of vapor pressure osmometry (discussed below), are, however, preferred when molecular weights are below 20,000.

2.3.4.   Vapor pressure osmometry[5c,8c,9c]

This method is generally useful for determining molecular weights below about 25,000. While no membrane is involved in vapor pressure osmometry, the thermodynamic principle involved is analogous to that of membrane osmometry. A drop of solvent and a drop of solution are placed by syringe on matched thermistors in an insulated chamber saturated with solvent vapor. Condensation heats the solution thermistor until the vapor pressure is increased to that of the pure solvent. The temperature change, measured by the change in resistance of the thermistor, is related to solution molality, and hence molecular weight, by the expression

$$\Delta T = \left(\frac{RT^2}{\lambda 100}\right) m$$

where λ is the heat of vaporization per gram of solvent and m is molality.

2.3.5 *Refractive index measurements[12]*

It has been demonstrated that, for certain polymers, a linear relationship exists between index of refraction and inverse number average molecular weight. While the method appears to be most suitable for low-molecular-weight polymers, it does have the advantage of simplicity.

2.4 Measurement of weight average molecular weight

2.4.1 *Light scattering[8d,9d]*

Apart from osmometry, light scattering is the most widely used method of obtaining absolute molecular weights. It relies on the fact that light, when passing through a solvent or solution, loses energy by absorption, conversion to heat, and scattering. The scattering for a pure liquid arises from finite nonhomogeneities in the distribution of molecules within adjacent areas that give rise to differences in density. For solutions, additional scattering is introduced by the solvent molecules. The amplitude or intensity of the scattered light depends on several factors, the most important being concentration, size, and polarizability of the scattering molecules. Refractive index also depends on concentration and amplitude of vibration. To evaluate the turbidity arising from scattering, one combines equations derived from scattering and index of refraction measurements. Turbidity, τ, is related to concentration, c, by the expression

$$\tau = Hc\bar{M}_w$$

where

$$H = \frac{32\pi^3}{3} \frac{n_0^2(dn/dc)^2}{\lambda^4 N_0}$$

and n_0 is the refractive index of solvent, λ is the wavelength of the incident light, and N_0 is Avogadro's number. The expression dn/dc, referred to as the *specific refractive increment*, is obtained by measuring the slope of the refractive index as a function of concentration, and it is constant for a given polymer, solvent, and temperature. As molecular size approaches in magnitude the wavelength of the light, corrections must be made for interference between scattered light coming from different parts of the molecules.

To determine molecular weight, the expression for turbidity is rewritten as

$$\frac{Hc}{\tau} = \frac{1}{\bar{M}P(\theta)} + 2A_2C$$

where $P(\theta)$ is a function of the angle, θ, at which τ is measured, a function that depends on the shape of the molecules in solution. A_2 is the second virial coefficient. Turbidity is then measured at different concentrations as well as at different angles, the latter to compensate for variations in molecular shape.

The experimental data are then extrapolated to both zero concentration and zero angle, where $P(\theta)$ is equal to 1. Such double extrapolations, shown in Figure 2.5, are called *Zimm plots* (after Bruno Zimm, who originated the method). The factor k on the abscissa is an arbitrary constant. The intercept corresponds to $1/\bar{M}_w$.

A major problem in light scattering is to obtain perfectly clear, dust-free solutions. This is usually accomplished by ultracentrifugation or careful filtration. Despite such difficulties, the light-scattering method is widely used for obtaining weight average molecular weights between 10,000 and 10,000,000. Older light-scattering photometers employed high-pressure mercury lamps and filters to obtain a monochromatic beam. These have been supplanted with laser light sources. A schematic of a laser light-scattering photometer is given in Figure 2.6.

2.4.2   *Ultracentrifugation*[8e,13,14a]

The ultracentrifuge is by far the most intricate and expensive instrument for determining molecular weight. It is not as widely used as light scattering or osmometry in determining molecular weights of synthetic polymers, but it has been used extensively with natural polymers, particularly proteins. The technique is based on the principle that molecules, under the influence of a strong centrifugal field, distribute themselves according to size perpendicularly to the axis of rotation, a process called *sedimentation*, the rate of which is proportional to molecular mass. Centrifugation is accomplished in an evacuated chamber in a cell set in a rotor, both of which are provided with windows so that optical methods, such as index of refraction measurements or interfer-

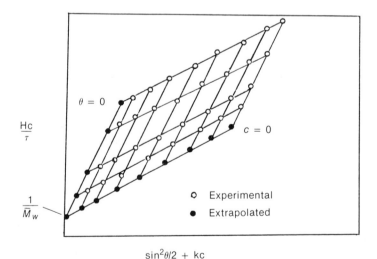

$$\sin^2\theta/2 + kc$$

Figure 2.5. Zimm plot of light-scattering data.

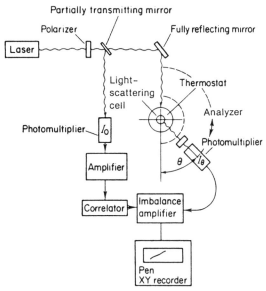

Figure 2.6. Schematic of a laser light-scattering photometer. [From Rabek,[8d] copyright 1980. Reprinted by permission of John Wiley & Sons, Ltd.]

ometry, can be used to observe concentration gradients within the polymer solution. The basic components of an ultracentrifuge are shown in Figure 2.7.

There are two approaches to obtaining molecular weights with the ultracentrifuge. One, the *sedimentation equilibrium* method, involves rotating the polymer solution at low velocities for long periods until equilibrium is established between sedimentation and diffusion. Change in refractive index between two points in the cell (the meniscus and the cell bottom for a heterogeneous sample, to include all species) is related to concentration change, and weight average molecular weight is given by the expression

$$\bar{M}_w \doteq \frac{2RT\ln(C_2/C_1)}{(1 - \bar{v}\rho)\omega^2(r_2^2 - r_1^2)}$$

where C_1 and C_2 are concentrations at distances r_1 and r_2, respectively, from the center of rotation to the point of observation in the cell; $\bar{v}$ is the specific volume of the polymer; ρ is the density of the solution; and ω is the angular velocity of rotation.

The other approach is to operate the ultracentrifuge at high speeds (up to 70,000 rpm) and relate the sedimentation rate to average molecular weight. This is called the *sedimentation velocity* method. The rate of sedimentation is defined by the sedimentation constant, s, which is related to the particle mass by the expression

$$s = \frac{1}{\omega^2 r}\frac{dr}{dt} = \frac{m(1 - \bar{v}\rho)}{f}$$

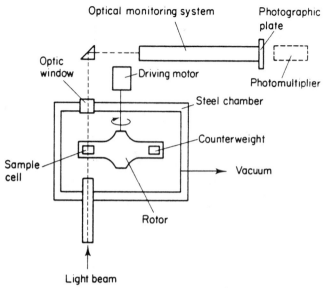

Figure 2.7. Schematic of an ultracentrifuge. [From Rabek,[8e] copyright 1980. Reprinted by permission of John Wiley & Sons, Ltd.]

where dr/dt is the sedimentation velocity, m is the mass, and f is the frictional coefficient. For random coil polymers, f is related to the diffusion coefficient, D, at infinite dilution by the equation

$$D = \frac{kT}{f}$$

Average molecular weight can then be obtained from the expression

$$\frac{D}{s} = \frac{RT}{\bar{M}_w(1 - \bar{v}\rho)}$$

This type of measurement is most effective with monodisperse systems such as proteins, and only approximate values are obtainable with polydisperse synthetic polymers.

2.4.3   *Field desorption mass spectrometry (FDMS)*

Field desorption is a method of detaching molecular ions directly from the solid to the gaseous state without decomposition.[15] Typically, the polymer sample is coated on carbon filaments that are attached to the end of a sharp-edged anode at the inlet of a mass spectrometer. The sample is then subjected to extremely large electric fields (of the order of 10^5 volts per centimeter). As the anode is heated, molecular ions desorb from the sample and are analyzed by the spectrometer. The mass spectrum thus provides a count of the various molecular masses present in a polydisperse sample. The great potential of FDMS lies in its ability to measure both $\bar{M}_n$ and $\bar{M}_w$ as well as the molecular weight distribution.[16] FDMS will almost certainly assume greater prominence as the instrumentation is developed for more routine use.

2.5 Viscometry

Measurements of dilute solution viscosity provide the simplest and most widely used technique for routinely determining molecular weights.[8f,9e,14b] It is not an absolute method; each polymer system must first be calibrated with absolute molecular weight determinations (usually by light scattering) run on fractionated polymer samples. Viscosities are measured at concentrations of about 0.5 g/100 mL of solvent by determining the flow time of a certain volume of solution through a capillary of fixed length. Flow time in seconds is recorded as the time for the meniscus to pass between two designated marks on the viscometer. Viscosities are run at constant temperature, usually $30.0 \pm 0.01°C$.

Two typical viscometers are shown in Figure 2.8. Of the two, the Ubbelohde type is more convenient to use in that it is not necessary to have exact volumes of solution to obtain reproducible results. Furthermore, additional solvent can be added (assuming the reservoir is large enough); thus concentration can be reduced without having to empty and refill the viscometer. Whichever type is used, it is necessary to filter the polymer solutions into the viscometer because dust particles affect flow time. Filtration is conveniently accomplished by micro filter units that replace the needle on hypodermic syringes.

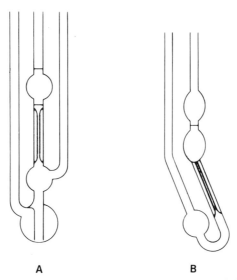

A   B

Figure 2.8. Capillary viscometers: (A) Ubbelohde, and (B) Cannon-Fenske.

Computer-driven viscometer modules are now commercially available. They measure capillary flow time photoelectrically and provide automatic dilution and mixing as well as calculation and read-out of viscosity.

Viscosity can be expressed in several ways (Table 2.2). Common names are, at the present time, more widely used than the IUPAC-recommended[17] names. (The latter were adopted by the IUPAC to avoid inconsistencies arising from "viscosity" designations where there were no units of viscosity.)

Relative viscosity (viscosity ratio) (η_{rel}) is the ratio of solution viscosity to solvent viscosity, which is proportional to a first approximation for dilute solutions to the ratio of the corresponding flow times. Viscosity units (commonly expressed as poises) or flow times cancel out in the various viscosity expressions. *Specific viscosity* (η_{sp}) is the fractional increase in viscosity. Both η_{rel} and η_{sp} are dimensionless. As concentration increases, so does viscosity. Hence to eliminate concentration effects, the specific viscosity is divided by concentration and extrapolated to zero concentration to give the *intrinsic viscosity*, [η]. Not uncommonly viscosities are determined at a single concentration and the *inherent viscosity* (η_{inh}) is used as an approximate indication of molecular weight. Inherent viscosity extrapolates to the same [η]. Concentration, C, in the above expressions is in grams per 100 mL of solvent or in grams per cubic centimeter, more commonly the former. Thus reduced, inherent and intrinsic viscosities have units of deciliters per gram or, less commonly, cubic centimeters per gram. Obviously, units of concentration must be specified when viscosity data are reported.

Intrinsic viscosity is the most useful of the various viscosity designations because it can be related to molecular weight by the Mark–Houwink–Sakurada equation:

$$[\eta] = K\bar{M}_v{}^a$$

Table 2.2. Dilute solution viscosity designations[a]

Common name	IUPAC name	Definition
Relative viscosity	Viscosity ratio	$\eta_{rel} = \dfrac{\eta}{\eta_0} = \dfrac{t}{t_0}$
Specific viscosity	—	$\eta_{sp} = \dfrac{\eta - \eta_0}{\eta_0} = \dfrac{t - t_0}{t_0} = \eta_{rel} - 1$
Reduced viscosity	Viscosity number	$\eta_{red} = \dfrac{\eta_{sp}}{C} = \dfrac{\eta_{rel} - 1}{C}$
Inherent viscosity	Logarithmic viscosity number	$\eta_{inh} = \dfrac{\ln \eta_{rel}}{C}$
Intrinsic viscosity	Limiting viscosity number	$[\eta] = \left(\dfrac{\eta_{sp}}{C}\right)_{C=0} = (\eta_{inh})_{C=0}$

[a] Concentrations (most commonly expressed in grams per/100 mL of solvent) of about 0.5 g/dL.

where $\bar{M}_v$ is the *viscosity average molecular weight*, defined as

$$\bar{M}_v = \frac{\sum\limits_{i=1}^{\infty} N_i M_i^{1+a}}{\sum\limits_{i=1}^{\infty} N_i M_i}$$

The constants K and a are the intercept and slope, respectively, of a plot of log $[\eta]$ versus log $\bar{M}_w$ or log $\bar{M}_n$ of a series of fractionated polymer samples. Such plots are linear (except at low molecular weights) for linear polymers, thus

$$\log[\eta] = \log K + a \log \bar{M}$$

Viscosity average molecular weights lie between those of the corresponding $\bar{M}_w$ and $\bar{M}_n$, but closer to the former. Hence better results are obtained if K and a are determined with fractionated samples of measured $\bar{M}_w$. To evaluate K and a requires considerable manipulation, but fortunately a wide range of values representing a broad spectrum of polymers, solvents, and temperatures has been published.[4e] For most common polymers, values of a vary between 0.5 (for a randomly coiled polymer in a theta solvent) and 0.8; for more rodlike extended-chain polymers where the hydrodynamic volume is relatively large, a may be as high as 1.0, in which case $\bar{M}_v = \bar{M}_w$. K values generally vary between 10^{-3} and 0.5. A few representative values of a and K are given in Table 2.3.

Factors that may complicate the application of the Mark–Houwink–Sakurada relationship are chain branching, too broad a molecular weight distribution in the samples used to determine K and a, solvation of polymer molecules, and the presence of alternating or block sequences in the polymer backbone. Chain entanglement is not usually a problem at such high dilution unless molecular weights are extremely large.

Other types of viscosity measurements based on the principle of mechanical shearing[14b] are also employed, most commonly with concentrated polymer solutions or undiluted polymer; these methods, however, are more applicable to flow properties of polymers (Chapter 3), not molecular weight determinations. Relative methods of determining molecular weight on the basis of polymer fractionation are discussed in the next section.

2.6 Molecular weight distribution

Molecular weight distribution is an important characteristic of polymers because, like molecular weight, it can significantly affect polymer properties. Just as low-molecular-weight polystyrene behaves differently from the high-molecular-weight material, a sample of polystyrene having a narrow molecular weight range will exhibit different properties from one having a broad range, even if the average molecular weights of the two samples are the same.

Table 2.3. Representative viscosity-molecular weight constants[a]

Polymer	Solvent	Temperature, °C	Molecular weight range × 10^{-4}	$K^b × 10^3$	a^b
Polystyrene (atactic)[c]	Cyclohexane	35[d]	8–42[e]	80	0.50
	Cyclohexane	50	4–137[e]	26.9	0.599
	Benzene	25	3–61[f]	9.52	0.74
Polyethylene (low pressure)	Decalin	135	3–100[e]	67.7	0.67
Poly(vinyl chloride)	Benzyl alcohol	155.4[d]	4–35[e]	156	0.50
	Cyclohexanone	20[d]	7–13[f]	13.7	1.0
Polybutadiene					
98% cis-1,4, 2% -1,2	Toluene	30[d]	5–50[f]	30.5	0.725
97% trans-1,4, 3% -1,2	Toluene	30	5–16[f]	29.4	0.753
Polyacrylonitrile	DMF[g]	25	5–27[e]	16.6	0.81
	DMF	25	3–100[f]	39.2	0.75
Poly(methyl methacrylate-co-styrene)					
30–70 mol %	1-Chlorobutane	30	5–55[e]	17.6	0.67
71–29 mol %	1-Chlorobutane	30	4.8–81[e]	24.9	0.63
Poly(ethylene terephthalate)	m-Cresol	25	0.04–1.2[f]	0.77	0.95
Nylon 66	m-Cresol	25	1.4–5[f]	240	0.61

[a] Values taken from reference 4e.
[b] See text for explanation of these constants.
[c] Atactic defined in Chapter 3.
[d] θ temperature.
[e] Weight average.
[f] Number average.
[g] N,N-dimethylformamide.

Broadly speaking, techniques for determining molecular weight distribution involve fractionation of the polymer sample and comparison of the fractions thus obtained with samples of known absolute molecular weight by means of some calibration procedure. A variety of methods have been developed.

2.6.1   Gel permeation chromatography (GPC)[8g,14c,18,19a]

This is by far the most widely used method of determining molecular weight distribution. A column chromatography technique, GPC may be used preparatively to obtain narrow molecular weight fractions. Separation is accomplished on a column packed with a highly porous material that separates the polymer molecules according to size, a phenomenon often referred to as *molecular sieving*. (The general technique is sometimes called *size exclusion chromatography*.) Current thinking is that separation is based on the hydrodynamic volume of the molecules rather than on the molecular weight per se.[20] Small molecules are able to diffuse into the pores of the column packing more efficiently, and hence they travel through the column more slowly. Higher molecular weight fractions are thus eluted first.

The essential features of a gel permeation chromatograph are shown in Figure 2.9. Column packing materials come in a variety of forms, but most commonly they consist of fine, semirigid beads of polystyrene crosslinked with divinylbenzene and swollen with solvent, or of rigid porous beads of glass or silica. Detection of polymer fractions in the eluant is most commonly accomplished with refractive index or spectroscopic (ultraviolet and infrared) detectors. Commercially available instruments provide automated sample injection and fraction collection, rapid flow-through by high-pressure pumping, and computer-assisted treatment of data. Closely related to GPC is *gel filtration chromatography* (GFC), a technique used principally to characterize natural polymers in aqueous solution. GFC columns are packed with hydrophilic gels, usually crosslinked dextrans or crosslinked polyacrylamides. As in the case of GPC, separation in GFC columns occurs by molecular sieving, but adsorption, ion exchange, and ion exclusion also play a role.

A typical gel permeation chromatogram (Figure 2.10) plots detector response against the volume of dilute polymer solution that passes through the column (the elution volume, V_r). To obtain molecular weights at a given retention volume, the chromatogram may be compared with a reference chromatogram obtained with fractions of known average molecular weight in the same solvent and at the same temperature. For purposes of comparison, the elution band is divided into "counts" of a specific volume, usually 2.5 or 5.0 mL, as shown, with the height above the baseline being proportional to the amount of polymer (N_1M_1) eluted.

The major problem with calibrating a particular GPC column for a particular polymer is that few standard samples of narrow molecular weight distribution are available commercially. Polystyrene standards having polydispersity indexes close to unity are available over a wide range of molecular weights

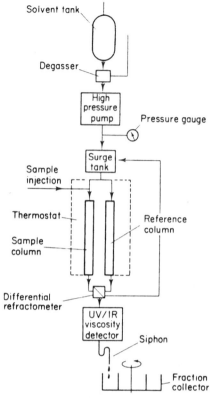

Figure 2.9. Schematic representation of a gel permeation chromatograph. [From Rabek,[8g] copyright 1980. Reprinted by permission of John Wiley & Sons, Ltd.]

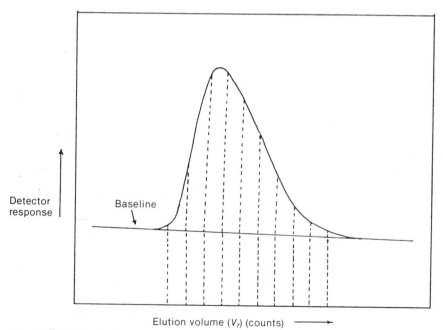

Figure 2.10. Typical gel permeation chromatogram. Dotted lines represent volume "counts."

(600–2.5 million) and these are often used; if one is dealing with a polymer other than polystyrene, however, the molecular weights thus obtained are at best approximate and may in some instances be seriously in error. To circumvent this difficulty, the *universal calibration* method is employed.

The universal calibration is based on the observation[21] that the product of intrinsic viscosity (limiting viscosity number) and molecular weight is independent of polymer type. This product, $[\eta]M$, is called the *universal calibration parameter*. As shown in Figure 2.11, a plot of log ($[\eta]M$) versus elution volume in tetrahydrofuran (THF) solvent yields a single curve, approximately linear, for a widely disparate group of polymers. Thus log ($[\eta]M$) may be considered a constant for all polymers *for a given column, temperature, and elution volume*. If we assume that the reference polymer (e.g., polystyrene) is polymer 1 and the polymer to be fractionated is polymer 2, it follows that

$$[\eta]_1 M_1 = [\eta]_2 M_2$$

From the Mark–Houwink–Sakurada relationship,

$$[\eta]_1 = K_1 M_1{}^a$$

$$[\eta]_2 = K_2 M_2{}^a$$

Combining these equations and solving for log M_2, we obtain

$$\log M_2 = \left(\frac{1}{1 + a_2}\right)\log\left(\frac{K_1}{K_2}\right) + \left(\frac{1 + a_1}{1 + a_2}\right)\log M_1$$

To determine the molecular weight (M_2) at a given retention volume, the column must first be calibrated with the standard polystyrene fractions (same solvent, same temperature). This gives a relationship of the type shown in Figure 2.12. Such semilogarithmic calibration plots are generally linear over a broad range of molecular weights, with deviations from linearity occurring at high and low molecular weights, particularly the former. Constants K and a are normally obtainable from a polymer handbook.[4e] Substituting the value of M_1 for a particular retention volume from the calibration plot and the values of K and a in the above equation, M_2 can be readily calculated. GPC thus provides a rapid and convenient method of obtaining molecular weight distribution once the appropriate calibrations have been worked out.

2.6.2 *Fractional solution*[8h,22]

In its simplest form, fractional solution might involve extracting a polymer in a Soxhlet-type apparatus to dissolve fractions of increasing molecular weight with time of extraction. More practically it has been adapted as a chromatographic procedure. The chromatographic column is packed with an inert support such as fine sand or glass beads coated with polymer. Nonsolvent is run through the column followed by increasing amounts of solvent mixed with the nonsolvent. Low-molecular-weight polymer dissolves first, followed by increasingly higher molecular weight fractions. This is the reverse of GPC,

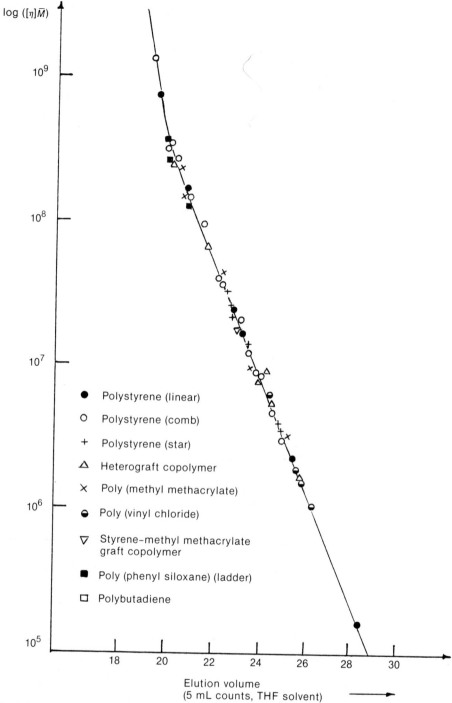

Figure 2.11. Universal calibration for gel permeation chromatography. THF, tetrahydrofuran. [From Grubisic, Rempp, and Benoit,[19] copyright 1976. Reprinted by permission of John Wiley & Sons, Inc.]

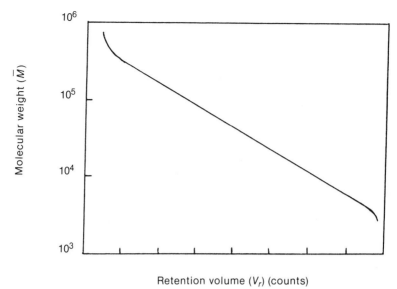

Figure 2.12. Typical semilogarithmic calibration plot of molecular weight versus retention volume.

therefore, in that low-molecular-weight fractions are eluted first. Typical combinations of solvent and nonsolvent are 2-butanone and methanol, respectively, for polystyrene; and xylene and 1-propanol, respectively, for polyethylene. Application of thermal gradients along the column—a method called *solvent-gradient chromatography*—is a variation of fractional solution.

2.6.3 *Fractional precipitation*[8h,22]

This is essentially the reverse of fractional solution. The usual procedure is to add small portions of nonsolvent to a dilute homogeneous solution of polydisperse polymer until high-molecular-weight material precipitates. The precipitate is removed by filtration or decantation and more nonsolvent is added to the solution. The procedure is repeated as many times as necessary to obtain the desired fractionation and all polymer has been recovered.

A variation of fractional precipitation is *turbidimetry*.[4h,20b] As nonsolvent is added to polymer solution, the resultant turbidity, recorded as scattered light in a photocell, is correlated with molecular weight. If temperature changes are used to effect precipitation, the technique is called *thermal gradient turbidimetry*.

2.6.4 *Thin-layer chromatography (TLC)*[17c,23–25]

TLC on alumina- or silica- gel-coated plates has been investigated, not only as a method of determining molecular weight distribution, but also for separating and characterizing polymer types, such as block and graft copolymers.[26]

There are two principal mechanisms by which polymers are separated on TLC plates: adsorption and precipitation. (Molecular sieving may also occur if a macroporous coating is used.) Adsorption relies on competition between solvent and solute for the adsorbent; it is a function primarily of chemical structure. Precipitation involves fractionation by molecular weight; it is the principal mechanism in determining molecular weight distribution.

There are three ways to effect fractionation once the TLC plate has been spotted with polymer: (1) A uniform mix of solvent and nonsolvent is used as the eluting medium. As the solvent boundary moves up the plate, its composition changes due to a combination of differential adsorption and evaporation. As the boundary becomes enriched in nonsolvent, high-molecular-weight fractions begin to precipitate. (2) A nonsolvent is used initially, and solvent is added in increasing amounts during elution. Low-molecular-weight fractions begin moving out first, followed by increasingly higher molecular weight fractions. (3) Solvent is used initially to move out the entire spot, then nonsolvent is added in increasing amounts, causing the high-molecular-weight fractions to precipitate first.

The obvious advantage of TLC is its low cost and simplicity. As such, it holds promise as a routine method for preliminary screening of polymer samples or monitoring of polymerization processes. *Paper chromatography* has also been investigated as a method of determining molecular weight distribution,[27] but much less extensively.

2.6.5   *Ultracentrifugation*[13]

This method involves observations of the sedimentation boundary during sedimentation velocity (high-speed centrifugation) experiments. The sedimentation boundary is the moving boundary caused by movement of polymer through solvent during centrifugation. The boundary broadens, partly because of diffusion and partly because of polydispersity, and this broadening can be related to molecular weight distribution. This technique offers the advantage of being able to resolve fractions of very narrow range, but has the disadvantages of high equipment cost and complexity.

References

1. S. G. Weissberg, S. Rothman, and M. Wales, in *Analytical Chemistry of Polymers*, Part 2 (G. M. Kline, ed.), Wiley-Interscience, New York, 1962, Chap. 1.
2. N. C. Billingham, *Molar Mass Measurements in Polymer Science*, Halsted Press, New York, 1977.
3. H. Morawetz, *Polymers in Solution*, 2nd ed., Wiley-Interscience, New York, 1975.
4. J. Brandrup and E. H. Immergut (eds.), *Polymer Handbook*, 2nd ed., Wiley-Interscience, New York, 1975: (a) O. Fuchs and H.-H. Suhr, pp. IV-241ff; (b) J. F. Rudd, p. V-59; (c) H. Burrell, pp. IV-337ff; (d) H.-G. Elias and

H. G. Bührer, pp. IV-157ff; (e) M. Kurata, Y. Tsunashima, M. Iwama, and K. Kamada, pp. IV-1ff.

5. P. M. Small, *J. Appl. Chem.*, **3**, 71 (1953).
6. K. L. Hoy, *J. Paint Tech.*, **42**, 76 (1970).
7. C. M. Hansen, *J. Paint. Tech.*, **39**, 104, 505, 511 (1967).
8. J. F. Rabek, *Experimental Methods in Polymer Chemistry*, Wiley-Interscience, New York, 1980: (a) Chap. 7; (b) Chap. 5; (c) Chap. 6; (d) Chap. 13; (e) Chap. 8; (f) Chap 9; (g) Chap. 25; (h) Chap. 4.
9. P. W. Allen (ed.), *Techniques of Polymer Characterization*, Butterworth, London, 1959: (a) G. F. Price, Chap. 7; (b) H. T. Hookway, Chap. 3; (c) D. F. Rushman, Chap. 4; (d) F. W. Peaker, Chap. 5; (e) P. F. Onyan, Chap. 6.
10. P. E. Slade Jr. (ed.), *Polymer Molecular Weights*, Part 1, Dekker, New York, 1975: (a) R. C. Garmon, Chap. 5; (b) R. D. Ulrich, Chap. 2 (c) C. A. Glover, Chap. 4; (d) E. P. Cassassa and G. C. Berry, Chap. 5.
11. L. S. Bark and N. S. Allen (eds.), *Analysis of Polymer Systems*, Applied Science, London, 1982: (a) G. Davison, Chap. 7; (b) A. R. Cooper, Chap. 8.
12. R. A. Rhein and D. D. Lawson, *CHEMTECH*, 122 (Feb. 1971).
13. J. W. Williams, *Ultracentrifugation of Macromolecules*, Academic Press, New York, 1972.
14. P. E. Slade Jr. (ed.), *Polymer Molecular Weights*, Part 2, Dekker, New York, 1975: (a) Th. G. Scholte, Chap. 8; (b) D. K. Carpenter and L. Westerman, Chap. 7; (c) A. C. Ouano, E. M. Barrall II, and J. F. Johnson, Chap. 6.
15. W. D. Reynolds, *Anal. Chem.*, **51**, 283A (1979).
16. R. P. Lattimer, D. J. Harmon, and G. E. Hansen, *Anal. Chem.* **52**, 1808 (1980).
17. International Union of Pure and Applied Chemistry, *J. Polym. Sci.*, **8**, 257 (1952).
18. K. H. Altgelt and L. E. Segal (eds.), *Gel Permeation Chromatography*, Dekker, New York, 1971.
19. L. H. Tung (ed.), *Fractionation of Synthetic Polymers*, Dekker, New York, 1977: (a) L. H. Tung and J. C. Moore, Chap. 6; (b) H. -G. Elias, Chap. 4; (c) H. Inagaki, Chap. 7
20. J. C. Moore, in *Liquid Chromatography of Polymers and Related Materials*, Part 3 (J. Cazes, ed.), Dekker, New York, 1981, p. 1ff.
21. Z. Grubisic, P. Rempp, and H. Benoit, *J. Polym. Sci.*, **B5**, 753 (1967).
22. R. M. Screaton, in *Newer Methods of Polymer Characterization* (B. Ke, ed.), Wiley-Interscience, New York, 1964, Chap. 11.
23. E. P. Otocka, in *Polymer Molecular Weight Methods*, Adv. Chem. Ser. 125, American Chemical Society, Washington, D.C., 1973, p. 55.
24. B. G. Benenkii and E. S. Gankina, *J. Chromatog. Rev.*, **141**, 13 (1977).
25. D. W. Armstrong and K. H. Bul, *Anal. Chem.*, **54**, 706 (1981).
26. H. Inagaki, T. Kotaka, and T.-I. Min, *Pure Appl. Chem.*, **46**, 61 (1976).
27. M. I. Siling, V. Ya. Kovner, Yu. Pvyrsky, and O. F. Alkayeva, *J. Chromatog.*, **101**, 83 (1974).

Review exercises

1. Suppose you wished to make a "model" of a linear polyethylene having a molecular weight of about 170,000 (a reasonable number for a commercial product) using paper clips to represent the repeating unit. How many paper clips would you have to string together?

2. Increasing temperature in general reduces the viscosity of polymer solutions. How might the magnitude of this effect compare in a "poor" solvent and a "good" solvent? (The principle invoked by this exercise is the basis for multiviscosity motor oils.)

3. An experimental method of determining solubility parameters of polymers involves taking lightly crosslinked (network) samples of the polymers in question and immersing them in different solvents whose solubility parameters are known. Explain the rationale behind this method.

4. Determine solubility parameters for amorphous polypropylene and poly(methyl methacrylate). (Densities of the two polymers are 0.905 and 1.18 g/mL, respectively.) Consult a polymer handbook and suggest suitable solvents for the two polymers on the basis of solubility parameters.

5. From the practical standpoint, is it better to use a "good" solvent or a "poor" solvent when measuring polymer molecular weight? Explain.

6. What would be the number average and weight average molecular weight of a sample of propylene oligomer that consists of 5 mol of pentamer and 10 mol of hexamer?

7. Calculate $\bar{M}_n$, $\bar{M}_w$ and the polydispersity index for a hypothetical polymer sample that contains equimolar amounts of polymer having molecular weights of 30,000, 60,000, and 90,000.

8. In exercise 7, change the words "equimolar amounts" to "equal weights" and recalculate.

9. A 0.5000-g sample of an unsaturated polyester resin was reacted with excess acetic anhydride. Titration of the reaction mixture with 0.0102 M KOH required 8.17 mL to reach the end point. What is the number average molecular weight of the polyester? Would this method be suitable for determining any polyester? Explain.

10. What is the $\overline{\text{DP}}$ of a sample of polyester prepared from 4-hydroxybenzoic acid if the acid number, determined with standard KOH solution, is 11.2?

11. Calculate the freezing point depression, boiling point elevation, and osmotic pressure (in millimeters of water at 25°C) for 1.00% weight per volume aqueous solutions of three samples of a polymer having number average molecular weights of 5000, 50,000, and 100,000. On the basis of your answer, what conclusions might you draw about the relative effectiveness of each method for determining molecular weight?

12. Explain how one might experimentally determine the Mark–Houwink–Sakurada constants K and a for a given polymer.

13. What would be the viscosity average molecular weight of a hypothetical polymer if the intrinsic viscosity were 2.3 dL/g under theta conditions? Assume a K value of 3.6×10^{-3}.

14. What is the approximate $\overline{DP}$ of a polyimide of structure

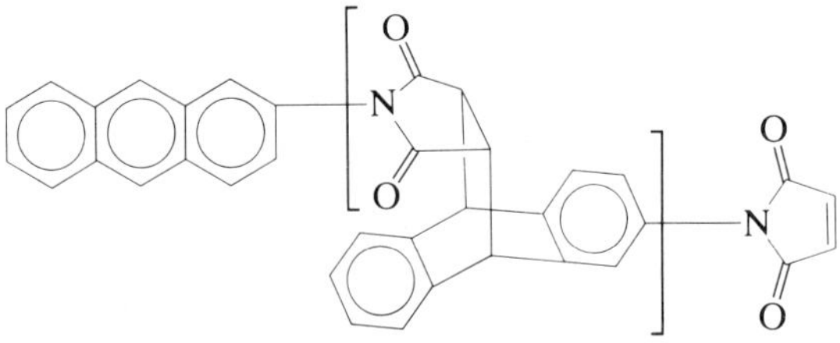

if a 0.50-g/L solution in a 1-cm cell exhibits an anthracene absorbance of 0.28 at 360 nm? Assume a molar absorptivity of 7800 for the anthracene unit. (M. P. Stevens, *J. Polym. Sci., Polym. Lett. Ed.*, **22**, 467, 1984)

3
Chemical structure and polymer morphology

3.1 Introduction

If asked to list desirable properties of a plastic, one would probably include toughness and durability and possibly, depending on the application, transparency, weather resistance, and heat and flame resistance. For a fiber one would expect high tensile strength, spinnability, and dyeability; for a rubber, resilience; and so on. Because it is properties such as these that make polymers commercially useful, the effect of chemical structure on polymer properties is without doubt the most important aspect of polymer chemistry. It is a subject that recurs throughout the book, and this chapter and the next serve as an introduction to the various factors that contribute to mechanical, physical, or chemical properties in polymeric materials.

We will begin in this chapter with polymer *morphology*—the structure, arrangement, and physical form of polymer molecules—and how morphology relates to chemical structure. In the next chapter we will be concerned more with properties as they relate to commercial application: mechanical strength, thermal and chemical stability, flammability, and so on. The two areas are related. Mechanical and thermal properties of polymers hinge to a large degree on molecular orientation (i.e., morphology) and on chemical constitution.

Generally speaking, two morphologies are characteristic of polymers— *amorphous* and *crystalline*. The former is a physical state characterized by almost complete lack of order among the molecules. The latter refers to the situation where polymer molecules are oriented, or aligned, in a regular array analogous—to a degree—to crystal lattice packing in nonpolymeric solids. Because polymers for all practical purposes never achieve 100% crystallinity, it is more practical to categorize them as amorphous and *semicrystalline*. We will look at characteristics of both amorphous and crystalline (or semicrystalline) states and at the various structural features that give rise to them, including intermolecular and intramolecular forces, stereochemistry, and chemical composition. We will also be concerned with the phenomenon of

liquid crystallinity, an interesting morphology that walks the tightrope between the purely crystalline solid and the purely amorphous liquid. Finally,
we will look at the effect of crosslinking and the increasingly significant
science of *polymer blends*, where morphology is of the utmost importance.

3.2 Molecular weight and intermolecular forces

The ultimate properties of any polymer, whether plastic, fiber, or rubber,
result from a combination of molecular weight and chemical structure.[1,2a]
Polymers have to be of a certain molecular weight before they have useful
mechanical properties, and the particular molecular weight necessary depends in large measure on chemical structure. These mechanical properties
result from attractive forces between molecules: dipole–dipole interaction
(including hydrogen bonding), induction forces, dispersion, or London forces
between nonpolar molecules. Or they may result from ionic bonding and
ion-dipole interactions with polymers containing ionic groups. Some of these
intermolecular forces are depicted in Figure 3.1.

Induction (induced dipole) interactions and London forces, particularly the
latter, are much weaker than those depicted; nevertheless they become
increasingly significant as molecular weight increases. Indeed, with nonpolar
polymers such as polyethylene, it is the London forces that give rise to the
molecular cohesion and resultant mechanical strength.

As a demonstration of the importance of secondary bonding forces and
their relation to molecular weight, polyesters can be compared with
polyamides (Figure 3.1). For a polyester and a polyamide having comparable
structure—that is, with R being the same in both cases—a higher molecular

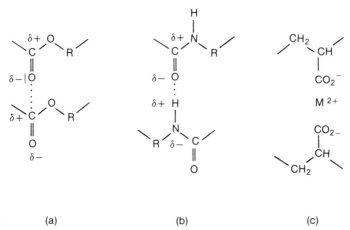

(a)   (b)   (c)

Figure 3.1. Intermolecular forces in polar polymers: (a) dipole-dipole in a polyester; (b) hydrogen bonding in a polyamide; and (c) ionic in a carboxyl-containing polymer.

weight is necessary to achieve good fiber properties in the polyester. The reason lies in the ability of polyamide molecules to associate through hydrogen bonding. Notwithstanding the polarity of the ester group, dipole-dipole interaction is clearly much stronger with polyamides. Similarly, both polyesters and polyamides exhibit strength characteristics at lower molecular weights than does polyethylene. At molecular weights up to about 10,000, polyethylene behaves more like a waxy solid, whereas polyamides are hard, brittle solids at molecular weights as low as 1000 to 2000.

Intermolecular forces drop off very rapidly with distance of separation. It is essential, therefore, that polymer molecules be able to pack together closely to achieve maximum cohesive strength. This phenomenon may be demonstrated very simply by stretching a rubber band. In the unstretched state, the rubber molecules are randomly distributed and the rubber has a low modulus; it exhibits little resistance to a force applied to it. It is also susceptible to attack by organic solvents. On stretching, however, the molecules become aligned along a common axis, which allows them to pack together more efficiently. At 600% elongation, the rubber band has a modulus about 2000 times greater than in the unstretched state and is considerably more resistant to solvents. Disregarding the structural features that give rise to elastic properties, the unstretched and stretched forms of rubber are, to a rough approximation, analogous to the amorphous and crystalline forms of polymers.

3.3   The amorphous state—rheology

The amorphous state[3a] is characteristic of those polymers in the solid state that, for reasons of structure, exhibit no tendency toward crystallinity. It is characteristic of *all* polymers at temperatures above their melting points (except under the special circumstances where liquid crystals may form). If a molten polymer retains its amorphous nature upon cooling to the solid state, the process is called *vitrification* (as opposed to crystallization). In the vitrified amorphous state, the polymer resembles a glass. The amorphous state has been aptly described by one authority[4] as resembling, on a molecular scale, a bowl of cooked spaghetti. The major difference between the solid and liquid amorphous states is that with the former, molecular motion is restricted to very short-range vibrations and rotations, whereas in the molten state there is considerable segmental motion or conformational freedom arising from rotation about chemical bonds. The molten state has been likened[5] on a molecular scale to a can of worms, all intertwined and wriggling about, except that the average "worm" would be extremely long relative to its cross-sectional area. To fully describe the amorphous state, taking into account bond lengths and bond angles, chain conformations, and chain entanglements, requires a statistical treatment[6] that is beyond the scope of this book.

When an amorphous polymer achieves a certain degree of rotational free-

dom, it can be deformed. If there is sufficient freedom, the polymer flows—
the molecules begin to move past one another. The science of deformation
and flow is called *rheology*[7,8] and is of fundamental importance in industrial
applications. In the traditional sense, rheology is of more interest to the
engineer or physicist. It is, after all, the engineer who designs the equipment
that converts polymers, usually in the molten state, into useful objects. It is
the chemist, however, who designs the molecules the engineer needs; therefore the two should be on speaking terms. The polymer chemist needs at least
a rudimentary understanding of this most important aspect of the amorphous
state. A more complete understanding than is provided here requires rigorous
mathematical modeling.

To cause a polymer to deform or flow requires the application of a force. If
a force is applied, then withdrawn quickly, the polymer molecules tend to
revert to their previous, undisturbed configuration, a process called *relaxation*. In other words, the amorphous liquid exhibits a certain *elastic* quality.
This elasticity comes about because the molecules were disrupted from what
was a thermodynamically favorable arrangement. If the force is applied
gradually and consistently, the molecules begin to flow irreversibly. Because
of chain entanglement and frictional effects, the flowing liquid will be very
viscous. This combination of properties—elasticity and viscous flow—is why
polymers are referred to as *viscoelastic* materials.[9] A consequence of viscoelasticity is illustrated in Figure 3.2, which depicts molten polymer being
forced through a narrow orifice, or die. As the molecules flow rapidly through
the die opening, they are compressed. When they emerge, the resultant
reduction in pressure causes the molecules to "rebound" to a degree, hence
the thickness of the emergent polymer melt will be greater than the width of
the die opening. This *die swell*, as it is called, must be taken into account by
engineers who design polymer processing machinery. Silly Putty, a siloxane
polymer (Chapter 15), is ideal for demonstrating the time dependence of
viscoelasticity. If dropped, it bounces; but it can be shaped by the slow
application of pressure.

A variety of forces are applicable to polymer deformation, but the one the
rheologist is most concerned with is *shear* (or *tangential stress*). Shear is a
force applied to one side of a surface in a direction parallel to the surface. If a
rectangle is subjected to shear, for example, it becomes a parallelogram.

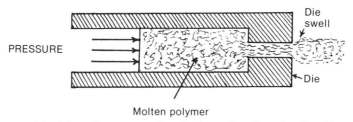

Figure 3.2. Schematic representation of polymer flow through a die orifice.

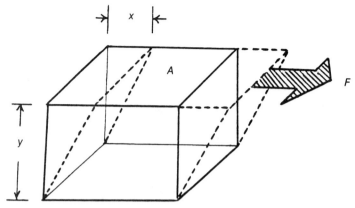

Figure 3.3. Representation of shear (tangential stress).

Shear applied to a rectangular body is illustrated in Figure 3.3. Shear stress (τ) is defined as the force (F) in dynes (or newtons) per unit of surface area (A) in square centimeters (or square meters); that is,

$$\tau = \frac{F}{A}$$

If the solid object is a piece of graphite, the graphite layers, which are held together by secondary bonding forces, slide over each other much as a deck of playing cards moves when shear is applied. It is this property that makes graphite a good lubricant.

When a polymer is in the molten state, shear causes the molecules to flow past one another. Such is the case when a polymer melt is stirred or subjected to pressure, as in Figure 3.2. How readily the molecules flow is a function of temperature (how much kinetic energy they possess); molecular weight (how much the molecules are entangled); and molecular structure (how strong the intermolecular forces are). To shape a plastic or spin a fiber, the polymer molecules have to flow readily—the more readily, the faster the desired object or fiber can be fabricated, and the less costly the process. Viscosity, whether in solution (as is discussed in Chapter 2) or in the melt, is a measure of resistance to flow.

Shear strain, γ, is the amount of deformation of one plane with respect to another. Referring to Figure 3.3,

$$\gamma = \frac{X}{Y}$$

Resistance to shear is the *shear modulus*, G, which is the ratio of shear stress to shear strain,

$$G = \frac{\tau}{\gamma}$$

Shear rate, ($\dot{\gamma}$), also called the *velocity gradient*, is the rate at which the planes (in Figure 3.3) or molecules (in an amorphous liquid) flow relative to one another; that is,

$$\dot{\gamma} = \frac{d\gamma}{dt}$$

If the shear stress increases in proportion to the shear rate, the liquid is said to be *Newtonian* (or ideal) because it follows Newton's law of viscosity, namely

$$\tau = \eta\dot{\gamma}$$

where the proportionality constant, η, is the viscosity. Units of viscosity are poises (dyne-s/cm^2) or, in the SI system, Pascal-seconds (Pa-s = newton-s/m^2). For comparison, a typical gas such as air has a viscosity of the order of 10^{-5}, water about 10^{-3}, glycerine about 1, and molten polymer about 10^2 to 10^6 Pa-s. Viscosity is related to temperature by an Arrhenius-type equation:

$$\eta = Ae^{-E_a/RT}$$

where A is a constant related to molecular motion and E_a is the activation energy for viscous flow. E_a is determined mainly by localized segmental motion of the polymer chains, as shown by its relative insensitivity to molecular weight but high dependence on chain structure and branching. The more bulky the chain branch or substituent, the higher the E_a; or the bulkier the group, the more sensitive the polymer viscosity to changes in temperature.

Deviations from ordinary Newtonian behavior are common in polymer flow. One type results when an initial resistance to flow must be overcome, but once flow begins it follows Newtonian behavior. Such a liquid is called a *Bingham Newtonian fluid* (named for Eugene Cook Bingham who developed much of the modern theory of rheology), and is defined by

$$\tau = \tau_c + \eta\dot{\gamma}$$

where τ_c is the critical shear stress, or threshold stress, needed to initiate flow. What causes Bingham-type behavior? Most probably it arises from some type of structured arrangement of the molecules arising from conformational and secondary bonding forces that must be disrupted by the application of an initial stress (τ_c) before the molecules begin to flow.

Other deviations, which are non-Newtonian, occur where shear stress does not increase in direct proportion to shear rate. The deviation may be in the direction of *thinning* (also called pseudoplastic) or *thickening* (dilatant). These various possibilities are illustrated in Figure 3.4. Most commonly, polymers exhibit shear thinning. As might be expected, a Bingham fluid might also exhibit non-Newtonian thinning or thickening. In cases of non-Newtonian flow, it is conventional to define viscosity as the slope of a secant drawn from the origin to any point on the stress-rate curve (shown by the dotted lines in Figure 3.4), with the resultant shear rate at that point specified.

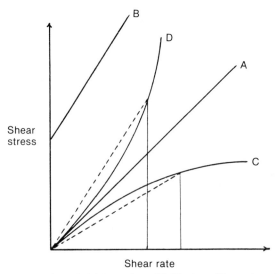

Figure 3.4. Types of shear flow: (A) Newtonian; (B) Bingham Newtonian; (C) shear thinning (pseudoplastic); and (D) shear thickening (dilatant).

Related to non-Newtonian flow with shear thinning is the behavior of *thixotropic* liquids, which have gel-like properties or a high viscosity at low stress but thin out and hence become more workable on being stirred. (Commercial paints are usually thixotropic.) The basic difference between the two is that shear thinning is dependent on shear rate, whereas thixotropic behavior is independent of shear rate but dependent on time at a fixed shear rate.

Flow behavior may also be expressed conveniently in terms of what is referred to as a power law equation:

$$\tau = A\dot{\gamma}^B$$

where A is a constant and B an index of flow. For a Newtonian fluid, $B = 1$ and $A = \eta$. A plot of log τ versus log $\dot{\gamma}$ (Figure 3.5) is linear with slope equal to B and intercept equal to log A.

An obvious question is, *why* do polymers characteristically exhibit shear thinning at high shear rates? On the molecular level this may be explained in terms of molecular entanglement. In the amorphous state there is considerable entanglement of the chains, and while low shear rates may disrupt this to a degree, the mass retains its entangled character. As shear rate increases, disruption may occur faster than the chains can reentangle, and the resultant decreasing entanglement allows the molecules to flow with less resistance, hence the decrease in viscosity. This is shown schematically in Figure 3.6. (Figure 3.2 illustrates the molecules disentangling as they flow rapidly

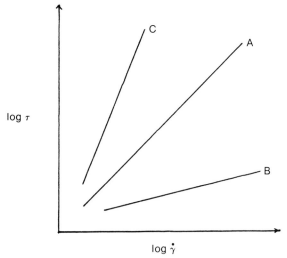

Figure 3.5. Power law plots: (A) Newtonian (slope = 1); (B) shear thinning (slope <1); and (C) shear thickening (slope >1).

through the die orifice, with subsequent reentanglement in the die swell region.)

If shear rate is increased sufficiently, chain rupture may occur with a resultant drop in molecular weight. (As will be shown later, this may actually be used to advantage to form block copolymers or polymer blends.) The much less common shear thickening may be envisioned as arising from an *increase* in molecular entanglement resulting from some peculiarity of molecular structure as shear rate increases; for example, a reasonably ordered arrangement of molecules might become more disordered, hence more entangled, as the shear rate increases.

Much has been said to this point about molecular entanglement, which is clearly going to increase as the molecular weight increases. Not unexpectedly, therefore, molecular weight is a critical variable in rheology. Interestingly, studies have shown that with flexible chain polymers, there exists a critical

Increasing shear rate

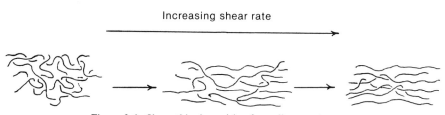

Figure 3.6. Shear thinning arising from disentanglement.

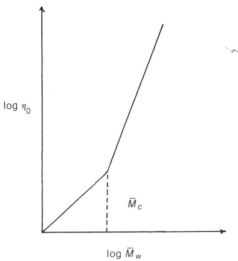

Figure 3.7. Effect of molecular weight on Newtonian viscosity.

molecular weight ($\bar{M}_c$) for entanglement to begin. This is illustrated schematically in Figure 3.7, which plots the log of Newtonian (or zero) viscosity (η_0) against the log of weight average molecular weight. Below $\bar{M}_c$ the slope is unity, followed by a sharp increase in η_0. For most common polymers, $\bar{M}_c$ falls in the range 4,000 to 15,000. Although $\bar{M}_c$ varies from one polymer to another, it has been shown, by eliminating mass effects arising from substituents on the chains, that $\bar{M}_c$ is a function of chain length, and chain length is remarkably constant from one polymer to another, corresponding to a $\overline{DP}$ of about 600. In other words, a critical chain length, rather than a critical molecular weight per se, is necessary for entanglement.

Considering the fact that η_0 is increasing *logarithmically* with molecular weight in Figure 3.7, it should be clear why molecular weight control is important in polymer processing. One needs a molecular weight high enough to attain good mechanical properties but not so high that the molten polymer is too viscous to be processed economically. Molecular weight *distribution* also influences flow behavior. In general, the broader the range, the lower the shear rate at which shear thinning develops. However—and this again is important in processing—the narrower the molecular weight distribution, the more pronounced will be the shear thinning once it is manifested, because of the absence of the more highly tangled high-molecular-weight constituents. This is illustrated in Figure 3.8.

Chain branching is another factor that influences flow. The more highly branched a given polymer, the lower will be its hydrodynamic volume and the lower its degree of entanglement at a given molecular weight. We can make the general observation, therefore, that viscosity is higher with linear than with branched polymers at a given shear rate and molecular weight. This does

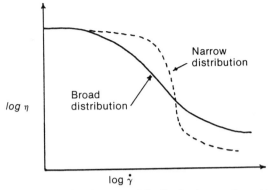

Figure 3.8. Effect of molecular weight distribution on shear thinning.

not mean that chain branching is necessarily desirable. As will be seen later, branching results in weaker secondary bonding forces and possibly poorer mechanical properties.

One other factor important in flow is polymer conformation (i.e., shape). In our earlier discussions of the Mark–Houwink–Sakurada equation,

$$[\eta] = KM_v{}^a$$

we observed that a varies from 0.5 for a random coil to about 1 for a more rodlike extended shape. While this applies to infinitely dilute solutions, one might expect the same kind of frictional effects in molten polymer, regardless of any reduced entanglement. This is, in effect, realized. More rigid polymers are significantly more viscous, except in the unusual situation where chain conformation results in liquid crystal behavior. (This will be considered later in the chapter.)

One final point needs to be considered: How does one determine viscosity of a polymer melt?[10a] One type of viscometer, the cone-plate rotational viscometer, is shown in Figure 3.9. The molten polymer is contained between the bottom plate and the cone, which is rotated at a constant velocity (Ω). Shear stress (τ) is defined as

$$\tau = \frac{3M}{2\pi R^3}$$

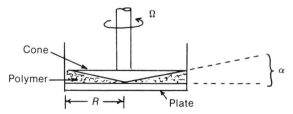

Figure 3.9. Cone-plate rotational viscometer.

where M is the torque in dynes per centimeter (CGS) or in newtons per meter (SI), and R is the cone radius in centimeters or meters. Shear rate ($\dot{\gamma}$) is given by

$$\dot{\gamma} = \frac{\Omega}{\alpha}$$

where Ω is the angular velocity in degrees per second (CGS) or in radians per second (SI) and α is the cone angle in degrees or radians. Viscosity is then

$$\eta = \frac{\tau}{\dot{\gamma}} = \frac{3\alpha M}{2\pi R^3 \Omega} = \frac{kM}{\Omega}$$

where

$$k = \frac{3\alpha}{2\pi R^3}$$

a constant defined by viscometer design. Other types of viscometers consist of rotating cylinders immersed in the viscous fluid, or steel capillaries through which the molten polymer is forced at constant pressure or constant flow.

3.4   Glass transition temperature[11,12a]

One of the most important characteristics of the amorphous state is the behavior of a polymer during its transition from solid to liquid. If an amorphous glass is heated, the kinetic energy of the molecules increases. Motion is still restricted, however, to short-range vibrations and rotations so long as the polymer retains its glasslike structure. As temperature is increased further, there comes a point where a decided change takes place; the polymer loses its glasslike properties and assumes those more commonly identified with a rubber. The temperature at which this takes place is called the *glass transition temperature* (T_g). (Less commonly it is called the *second-order transition temperature*, to differentiate it from the thermodynamic first-order transition of melting.) If heating is continued, the polymer will eventually lose its elastomeric properties and melt to a flowable liquid.

The importance of the glass transition temperature cannot be overemphasized. It is one of the fundamental characteristics as it relates to polymer properties and processing. What takes place at the molecular level at the glass transition temperature? As one might expect, the transition is accompanied by more long-range molecular motion—greater rotational freedom and consequently more segmental motion of the chains. It is estimated that between 20 and 50 chain atoms are involved in this segmental movement at the T_g. Clearly for this increased motion to take place, the space between the atoms (the *free volume*) must increase, which gives rise to an increase in the specific volume. The temperature at which this change in specific volume takes place, usually observed by dilatometry (volume measurement), may be used as a measure of T_g.

Other changes of a macroscopic nature occur at the glass transition. There is an enthalpy change, which may be measured by calorimetry. The modulus, or stiffness, decreases appreciably, the decrease readily detected by mechanical measurements. Refractive index and thermal conductivity change. We will not be concerned here with details of how T_g is measured (this is covered in Chapter 5), but rather with the nature of the phenomenon and the various factors that influence the temperature at which it occurs.

What determines the glass transition temperature? Clearly chemical structure is most important, and we will examine that in some detail. Other variables, however, make T_g an approximate value at best. Method of measurement is one: A range of several degrees may be observed between a thermal and a mechanical measurement. Difficulties in obtaining reproducible samples is another. The T_g of a polymer may even change with time because of such "aging" factors as oxidation or exposure to ultraviolet light. Finally there is molecular weight. The higher the molecular weight, the fewer the chain ends, which leads to a lower free volume. Polystyrene's T_g, for example, increases from about 40°C at an $\bar{M}_n$ of 3,000 to about 100°C at 300,000.

These various factors combine to make T_g an approximation, and any literature compilation must, therefore, be treated with some caution.[13] It is not uncommon to see discrepancies of as much as 30°C for a given polymer. With this in mind, let us now turn our attention to the effect of chemical structure. While structural effects are not always predictable, one can begin with some generalizations. Given that T_g is a function of rotational freedom, whatever restricts rotation should increase T_g. Any compound having carbon–carbon single bonds is capable of existing in an infinite number of conformations unless freedom of rotation is impaired. In most instances, energy barriers to rotation are low, but even with ethane there is a torsional barrier of about 3 kcal/mol, which results in a preference for the staggered conformation (**1**). As the hydrogens are replaced with more bulky groups, the energy barrier increases. Butane exists mainly in the *anti* conformation (**2**) which is some 5 kcal/mol more stable than the most energetic conformation in which the methyls are eclipsed. This may be carried to extreme cases where rotation is blocked, as with the *ortho-ortho'* substituted biphenyls (**3**), which are capable of existing as configurational isomers if the substituent groups are large enough.

$$R \neq R'$$

| 1 | 2 | 3 |

In the case of polymers, therefore, it follows that the bulkier the substituents on the polymer backbone, the less will be the rotational freedom and the higher the T_g. This is true to a degree; however, as shown with polymers **4** through **23** in Table 3.1, other factors come into play. The first grouping in the table (**4–11**) contains polymers with increasing length of the pendant group. As expected, polypropylene has a higher T_g than polyethylene; but then there is a decrease as the number of methylene groups in R increases, followed by an increase at even longer substituent chain lengths. This behavior is no doubt due to a plasticizing effect of the flexible side chain that is offset at higher

Table 3.1. Glass transition temperature (T_g) of representative vinyl polymers[a]

Number	Polymer $-\!\!\left[CH_2CH\right]\!\!-$ R	R	T_g (°C)
4	Polyethylene	H	−20
5	Polypropylene	CH_3	5
6	Poly(1-butene)	CH_3CH_2	−24
7	Poly(1-pentene)	$n\text{-}C_3H_7$	−40
8	Poly(1-hexene)	$n\text{-}C_4H_9$	−50
9	Poly(1-heptene)	$n\text{-}C_5H_{11}$	−31
10	Poly(1-decene)	$n\text{-}C_8H_{17}$	−41
11	Poly(1-dodecene)	$n\text{-}C_{10}H_{21}$	−6
12	Poly(3-methyl-l-butene)	$CH(CH_3)_2$	50
13	Poly(4-methyl-l-pentene)	$CH_2CH(CH_3)_2$	29
14	Poly(3,3-dimethyl-l-butene)	$C(CH_3)_3$	64
15	Poly(5-methyl-1-hexene)	$CH_2CH_2CH(CH_3)_2$	−14
16	Poly(4,4-dimethyl-l-pentene)	$CH_2C(CH_3)_3$	59
17	Poly(vinyl *n*-butyl ether)	$O\text{-}C_4H_9\text{-}n$	−55
18	Poly(vinyl *t*-butyl ether)	$O\text{-}C_4H_9\text{-}t$	88
19	Poly(vinyl chloride)	Cl	81
20	Poly(vinyl alcohol)	OH	85
21	Polystyrene	C_6H_5	100
22	Poly(2-vinylnaphthalene)	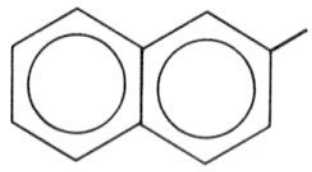	151
23	Poly(4-vinylpyridine)	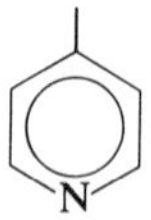	142

[a] Data from Lee and Rutherford.[13a]

chain lengths by entanglement or by side chain crystallization. (In the next chapter we shall see that plasticizers are very often added to polymers to lower the T_g.) Effect of size is also apparent in comparing polypropylene (**5**) with polystyrene (**21**) and poly(2-vinylnaphthalene) (**22**).

The second grouping in Table 3.1 (**12–16**) contains branched substituent groups. A comparison with polymers in the first grouping, which are structural isomers, shows clearly that branching increases T_g significantly. (Compare **12** with **7**, **13** and **14** with **8**, **15** and **16** with **9**.) A further comparison of **13** with **14** and **15** with **16**, as well as the two vinyl ether polymers (**17** and **18**) in the third grouping, confirms this trend.

In addition to size, polarity influences T_g, as shown by poly(vinyl chloride) (**19**) and poly(vinyl alcohol) (**20**). Chloro and hydroxy groups are comparable in size to methyl, but the more polar groups lead to higher T_g's because of increased dipole–dipole interaction. In the case of **20**, one might also envision decreased rotational freedom arising from intramolecular hydrogen bonding. Polarity effects are also evident in comparing polystyrene (**21**) and poly(4-vinylpyridine) (**23**).

Table 3.2 lists T_g's for some representative nonvinyl polymers. A comparison of polyesters and polyamides of similar structure (**25** and **29**) illustrates the importance of hydrogen bonding in increasing T_g. Also significant is the chain-stiffening effect of the benzene ring. (Compare polyesters **24** with **26** and **25** with **27**.)

The high T_g of **28**, a special type of polyester called a polycarbonate, reflects a further reduction in the flexibility of the polymer backbone. The monomer used to prepare **28**, 2,2-bis(4-hydroxyphenyl)propane, **34** (commonly called bisphenol A), finds use in a variety of polymer formulations where a high T_g is desirable. An extreme case of restricted rotation is shown with the semiladder

34

polybenzimidazole, **33**. These various polymer types are discussed in more detail in later chapters.

The effect of stereochemistry on T_g is shown in Table 3.3. From the limited data provided it is impossible to make generalizations as to which geometric configuration results in higher glass transition temperatures. Indeed, in the case of 1,4-polyisoprene, it has been demonstrated that there is very little difference between *cis* and *trans* isomers even when there are variations in morphology.[14]

For a polymer to serve as a useful plastic, its glass transition temperature must be appropriately higher than the temperature of its intended work

Table 3.2. Glass transition temperature (T_g) of representative nonvinyl polymers[a]

Number	Polymer	$T_g(°C)$
24	$-\overset{O}{\overset{\|}{C}}(CH_2)_4\overset{O}{\overset{\|}{C}}O(CH_2)_2O-$	−63
25	$-\overset{O}{\overset{\|}{C}}(CH_2)_4\overset{O}{\overset{\|}{C}}O(CH_2)_4O-$	−118
26	$-\overset{O}{\overset{\|}{C}}-\langle\bigcirc\rangle-\overset{O}{\overset{\|}{C}}O(CH_2)_2O-$	69
27	$-\overset{O}{\overset{\|}{C}}-\langle\bigcirc\rangle-\overset{O}{\overset{\|}{C}}O(CH_2)_4O-$	80
28	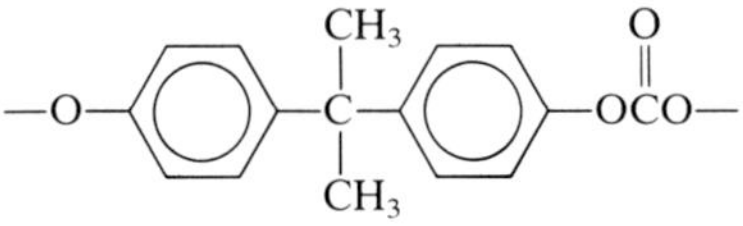 $-O-\langle\bigcirc\rangle-\overset{CH_3}{\underset{CH_3}{\overset{\|}{\underset{\|}{C}}}}-\langle\bigcirc\rangle-O\overset{O}{\overset{\|}{C}}O-$	149
29	$-\overset{O}{\overset{\|}{C}}(CH_2)_4\overset{O}{\overset{\|}{C}}NH(CH_2)_6NH-$	57
30	$-\overset{O}{\overset{\|}{C}}(CH_2)_4\overset{O}{\overset{\|}{C}}NH(CH_2)_{10}NH-$	50
31	$-NH(CH_2)_5\overset{O}{\overset{\|}{C}}-$	77
32	$-NH(CH_2)_{10}\overset{O}{\overset{\|}{C}}-$	46
33	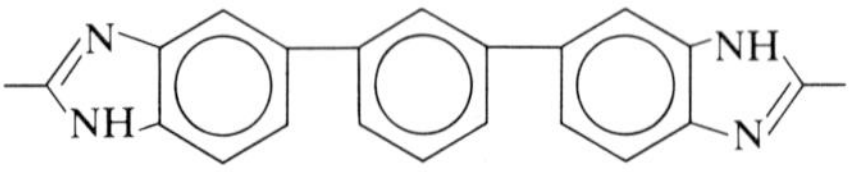 (benzimidazole–phenylene–phenylene–benzimidazole structure)	429

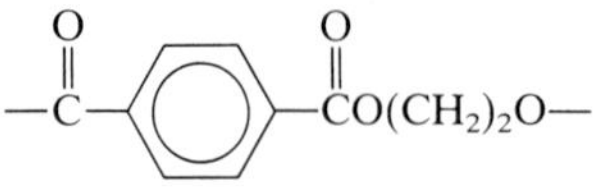

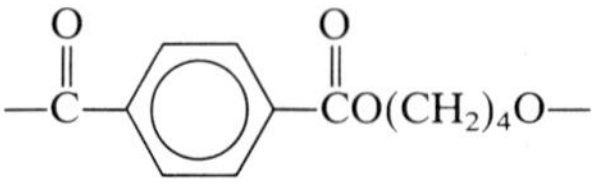

[a] Data from Lee and Rutherford.[13a]

Table 3.3. Glass transition temperature (T_g) of diene polymers

	T_g (°C)	
Polymer	cis	trans
1,4-Polybutadiene[a]	−102	−58
1,4-Polyisoprene[b]	−67	−70
1,4-Polychloroprene[a]	−20	−40

[a] Data from Lee and Rutherford.[13a]
[b] Data from Burfield and Lim.[14]

environment. A plastic used in manufacturing coffee cups obviously needs a T_g that is well above the temperature of hot coffee. Glass transition temperatures of elastomers, on the other hand, are *below* room temperature.

3.5 Stereochemistry

Before we examine the crystalline morphology, it behooves us to consider one of the most important structural parameters that leads to crystallinity, namely stereoregularity. That Ziegler and Natta shared the 1963 Nobel Prize in Chemistry for their work in this area attests to its significance.

For reasons discussed later, most commercially important vinyl monomers are monosubstituted ($CH_2 = CHR$) or 1,1-disubstituted ($CH_2 = CR_2$, where the R groups may be the same or different). Furthermore, such monomers undergo primarily *head-to-tail polymerization* (reactions 3.1 and 3.2) to yield polymers having the substituent groups on alternate carbon atoms. This is not

$$CH_2{=}CHR \longrightarrow -CH_2-\overset{\overset{R}{|}}{CH}-CH_2-\overset{\overset{R}{|}}{CH}-CH_2-\overset{\overset{R}{|}}{CH}- \quad (3.1)$$

$$CH_2{=}CR_2 \longrightarrow -CH_2-\overset{\overset{R}{|}}{\underset{\underset{R}{|}}{C}}-CH_2-\overset{\overset{R}{|}}{\underset{\underset{R}{|}}{C}}-CH_2-\overset{\overset{R}{|}}{\underset{\underset{R}{|}}{C}}- \quad (3.2)$$

to say that *head-to-head* or *tail-to-tail* units do not form; they do, but in most instances they contribute very little to the overall polymer structure. Given the fact that each substituent-bearing carbon on the polymer backbone may be a chiral (asymmetric) center, a number of stereochemical variations are possible.[15-17]

For head-to-tail polymers prepared from monomers of type $CH_2 = CHR$, there are two possible stereoregular arrangements—one in which each chiral center has the same configuration, and one in which alternate chiral centers have the same configuration.[15] Natta proposed the terms *isotactic* for the former and *syndiotactic* for the latter.* A completely random distribution is

---

* To quote Natta:

Comme nous attribuons au phénomène observé une importance fondamentale pour la connaissance d'une vaste catégorie de macromolécules, dont on a commencé une production destinée à de grands développements, nous avons voulu, pour en rendre plus facile la description, assigner aux atomes de carbone asymétriques ayant une configuration égale, le terme "isotaxiques," du grec *i'σος* = égal et *ταττω* = disposer en ordre." (G. Natta, *J. Polymer Sci.*, **16**, 143, 1955)

We propose to call all vinyl polymers with alternating D- and L-configurations of their substituents ... "syndyotactic" polymers, using the Greek words *tatto* (put in order) and *syndyo* (every two). (G. Natta and P. Corradini, *J. Polymer Sci.*, **20**, 251, 1956)

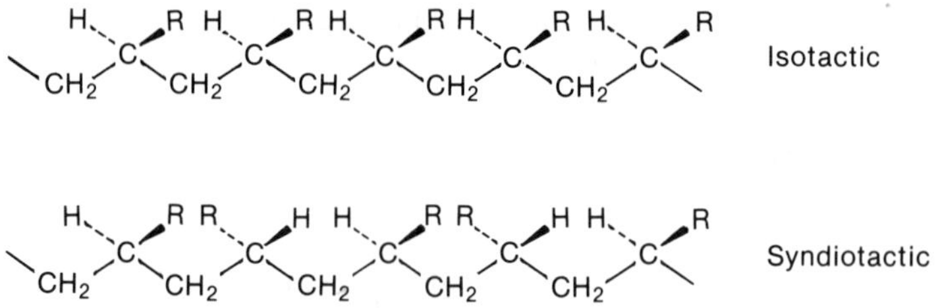

Isotactic

Syndiotactic

Figure 3.10. Stereoregular polymers derived from monomer CH_2=CHR.

called *atactic* or *heterotactic*. The two stereoregular arrangements are represented as idealized extended zig-zag chains in Figure 3.10. As will be seen later, all three types can be synthesized. Where blocks of stereoregular units are formed along the backbone, the term *stereoblock* describes the polymer.

Analogous structures may be drawn for polymers prepared from 1,1-disubstituted monomers having different subtituent groups (e. g., methyl methacrylate, **35**), but not from those having identical groups (e. g., vinylidene chloride, **36**), which yield polymers having no chiral centers.

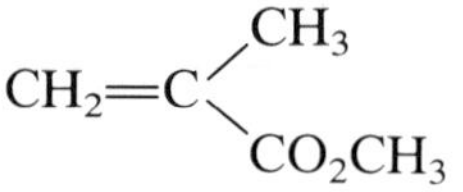

35

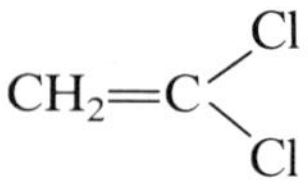

36

It is convenient to think of isotactic and syndiotactic in terms of the *meso* and *racemic* structures familiarly used with low-molecular-weight compounds.[18] Let us consider a segment of the polymer chain consisting of two monomeric units. Such is referred to as a *tactic dyad* or *tactic placement*. An isotactic dyad (**37**) has the *meso* configuration (designated *m*), whereas a syndiotactic dyad (**38**) is *racemic* (*r*). We compare, for example, *meso* (**39**)

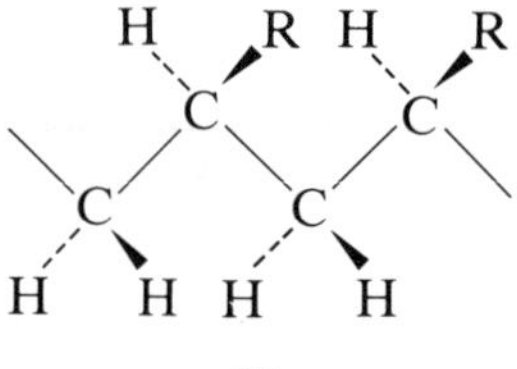

37

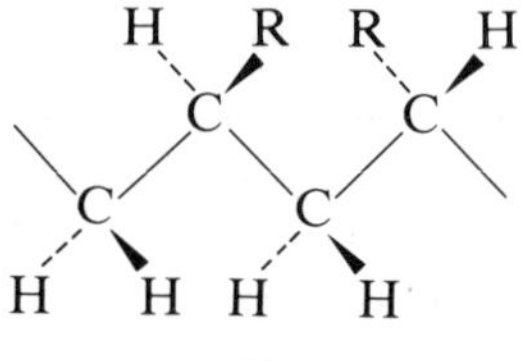

38

and *racemic* (**40** + **41**) 2,4-dichloropentane. The *r* and *m* designations are particularly useful in classifying heterotactic polymers. Tactic *triads*, for ex-

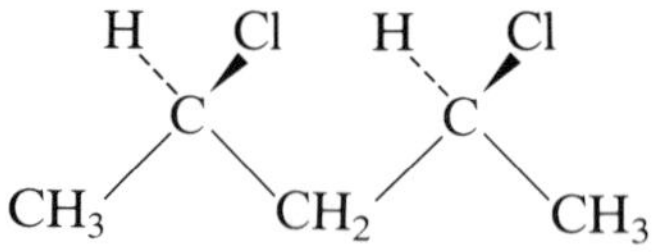

meso

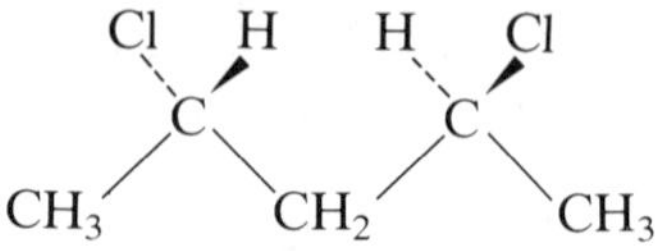

enantiomer

39

40

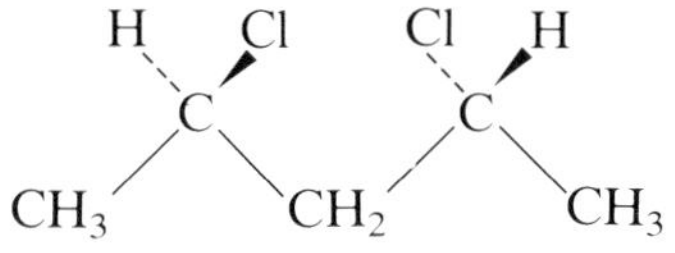

enantiomer

41

ample, have three possible sequences, *mm* (**42**), *rr* (**43**), and *mr* (or *rm*) (**44**), corresponding to isotactic, syndiotactic, and heterotactic, respectively. Tetrads have six: *rrr*, *rrm* (or *mrr*), *rmr*, *mrm*, *mmr* (or *rmm*), and *mmm*. Later we shall see how NMR spectroscopy may be used to identify such sequences

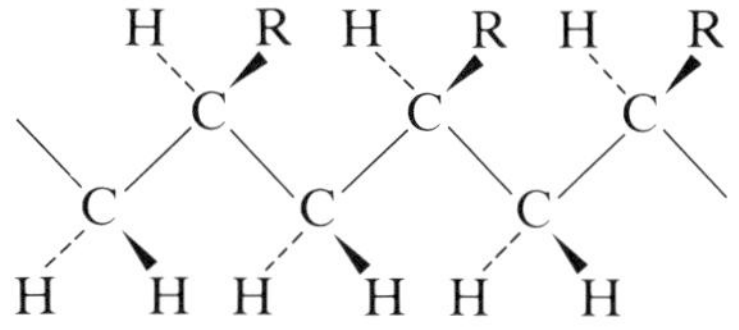

mm

42

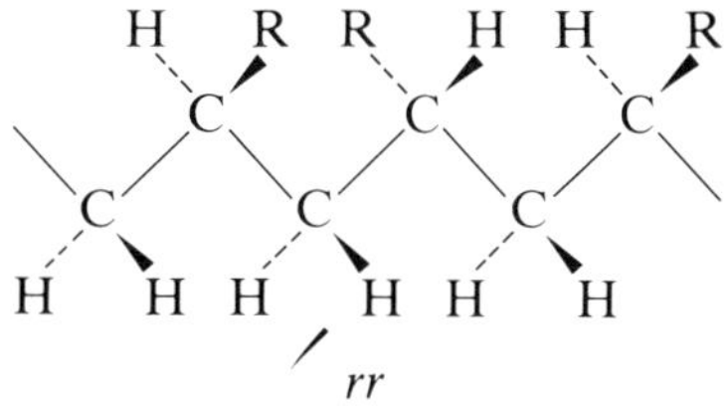

rr

43

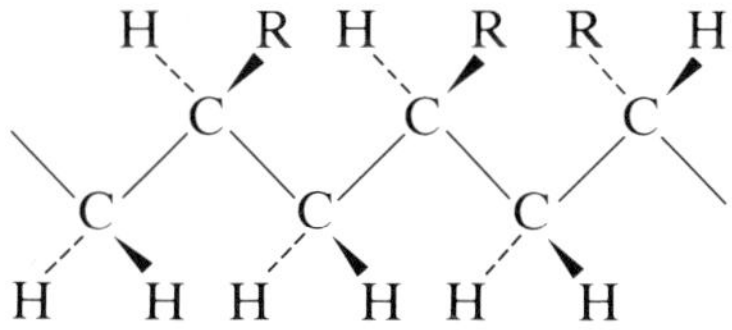

mr

44

in heterotactic polymers, which, in turn, sheds light on the mechanism by which tactic units are formed.

A variety of other stereochemistries are possible. Monomers of type RCH=CHR (both R groups the same) may also yield isotactic or syndiotactic polymer (Figure 3.11) having adjacent chiral centers. If the R groups are different, however, two types of isotactic and one syndiotactic are possible (Figure 3.12). As is the case with low-molecular-weight compounds, the prefixes *threo* and *erythro* refer to similar groups being on the opposite or the same side when the structures are depicted in the conformationally eclipsed Fisher projection:

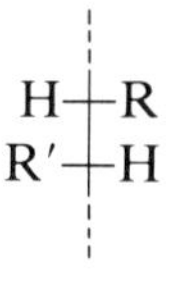

threo

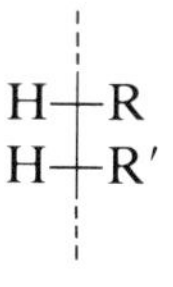

erythro

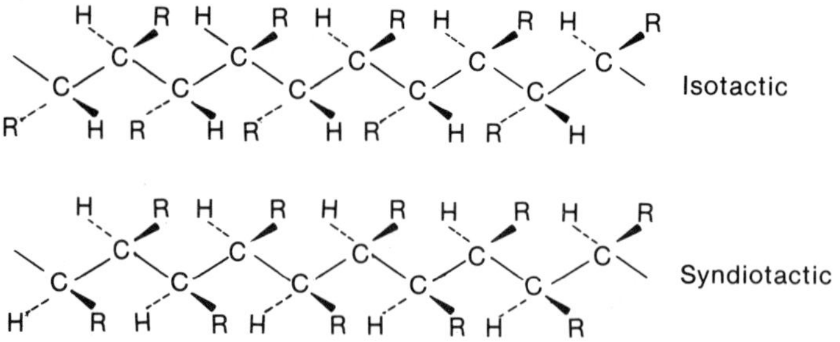

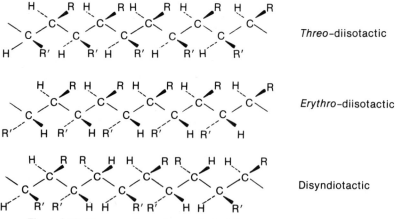

Figure 3.11. Stereoregular polymers derived from monomer RCH=CHR.

Only one syndiotactic form exists because the *threo* and *erythro* structures are identical except for the end groups.

If ethylene and RCH=CHR undergo alternating copolymerization, the resultant polymer (**45**) (which is, in effect, a head-to-head polymer of CH_2=CHR) may exist in four stereoregular forms (Figure 3.13). Similarly,

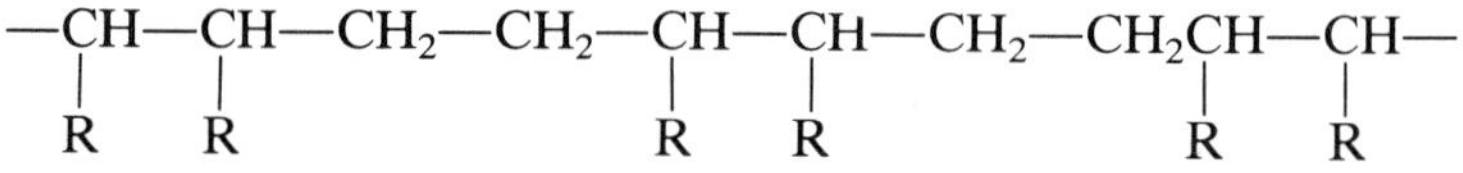

$$-CH-CH-CH_2-CH_2-CH-CH-CH_2-CH_2CH-CH-$$

$$\quad\; R \quad\; R \qquad\qquad R \quad\; R \qquad\qquad R \quad\; R$$

45

cycloalkenes such as cyclobutene can, in principle, form four stereoregular structures (Figure 3.14). Substituted dienes of type RCH=CH—CH=CHR form 1,4-polymer (**46**), which are *tritactic*—the double bond and two chiral carbons representing three sites of tacticity. Figure 3.15 illustrates two of the many possible stereoregular structures. Such polymers contain true chiral

Threo-diisotactic

Erythro-diisotactic

Disyndiotactic

Figure 3.12. Stereoregular polymers derived from monomer RCH=CHR'.

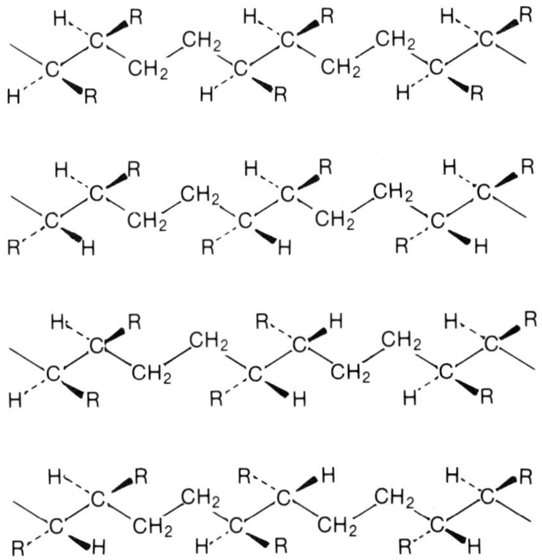

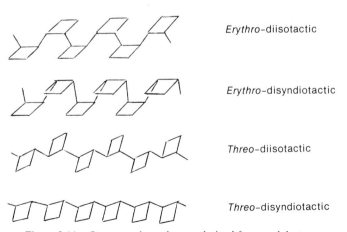

Figure 3.13. Stereoregular alternating copolymers of CH_2=CH_2 and RCH=CHR.

centers and may exhibit optical activity if chiral initiators are used in the polymerization.

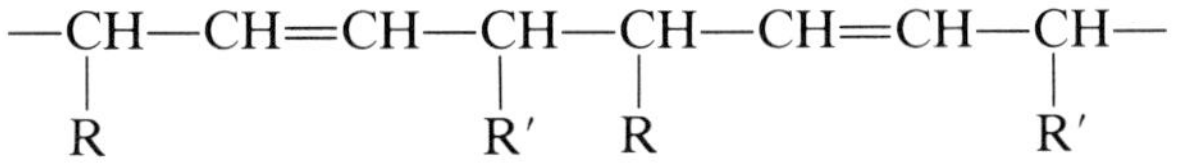

$$-CH-CH=CH-CH-CH-CH=CH-CH-$$

46

None of the polymers just described exist in the extended-chain form; rather, steric effects most often lead to a helical conformation. A case in point

Figure 3.14. Stereoregular polymers derived from cyclobutene.

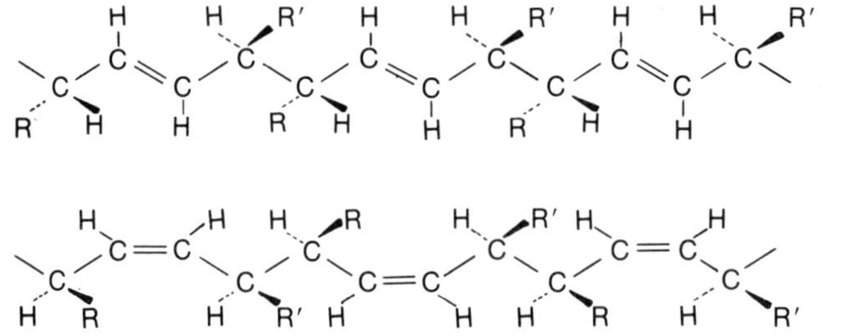

Trans-erythro-diisotactic

Cis-threo-disyndiotactic

Figure 3.15. Two stereoregular 1,4-polymers derived from monomer RCH=CH—CH=CHR'.

is the commercially important isotactic polypropylene in which each methyl is displaced 120° from the next.

Which of the various stereoregular structures predominates, and the extent of stereoregularity, depend on a variety of factors: whether the monomer double bond is *cis* or *trans*, whether the double bond opens in a *cis* or *trans* manner during polymerization, type of initiator, polymerization temperature, and solvent polarity. All of these are explored later in the book. Some of the structures given are hypothetical; only the *erythro* forms of polycyclobutene, for example, have been identified. An added complication with cyclic monomers is that ring-opening may compete with polymerization through the double bond. Furthermore, with polymers containing two chiral centers, it is not inconceivable to have different tacticities for each. Where only one of the two chiral centers is stereoregular, the term *hemitactic* (i. e., hemiisotactic or hemisyndiotactic) has been proposed.[19]

Most important from the standpoint of morphology is that in most instances, regardless of the type of stereochemistry, the tendency for crystalline polymer to form increases with increasing stereoregularity.

Before we leave the subject of stereochemistry, we should reflect a moment on the glass transition temperature. Given that stereoregularity leads to

Table 3.4. Glass transition temperature (T_g) of polymers of varying tacticity[a]

Polymer	T_g (°C)		
	Syndiotactic	Atactic	Isotactic
Poly(methyl methacrylate)	105	105	38
Poly(ethyl methacrylate)	65	65	12
Poly(t-butyl methacrylate)	114	118	7
Polypropylene[b]	−4	−6	−18
Polystyrene	100[c]		99[c]

[a] Data from Lee and Rutherford[13a] unless otherwise noted.

[b] Data from Burfield and Doi.[20]

[c] Data from Ishihara, Kuramoto, and Uoi. *Macromolecules*, **21**, 3356 (1988).

crystallinity, does it also lead to higher T_g's? Prevailing evidence[20] suggests that for a given polymer, syndiotactic usually exhibits a higher T_g than isotactic, as shown by the data in Table 3.4. The atactic T_g is generally equal to or close to that of the syndiotactic. Differences are not necessarily pronounced, however, and are subject to variations in the test sample's morphology. It must be remembered that the glass transition is a characteristic of the *amorphous* state, which is still present to varying degrees in semicrystalline polymers. Glass transitions are not always easy to detect in highly crystalline polymers because of the relatively small amount of amorphous material present; nevertheless, it is these amorphous regions that are undergoing the transition.

3.6   Crystallinity[2b,12b,21-23]

When a polymer has a highly stereoregular structure with little or no chain branching, or when it contains highly polar groups that give rise to very strong dipole–dipole interactions, it may exist in crystalline form. Such crystallinity is unlike that of low-molecular-weight compounds, but exists instead in regions of the polymer matrix where polymer molecules order themselves in a thermodynamically favorable alignment. The *fringed micelle* model defines crystallinity in terms of such ordered regions, called *crystallites*, in which any particular polymer chain may extend through a number of crystallites, as shown in Figure 3.16.

Crystallinity is induced in any number of ways—cooling of molten polymer, evaporation of polymer solutions, or heating of a polymer under vacuum

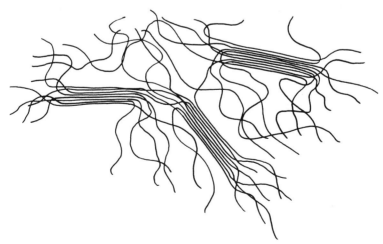

Figure 3.16. Schematic representation of a polymer matrix showing crystalline regions (fringed micelle model).

or in an inert atmosphere (to prevent oxidation) at a specified temperature, a process called *annealing*. In some instances crystallization may be brought about by stretching a polymer sample at a temperature above its glass transition temperature, a process called *drawing*. One might envision drawing as removing the "kinks" from the polymer chains and extending them along a common axis to promote closer packing. Stretching certain varieties of clear plastic film occasionally allows visual observation of the onset of crystallinity as opaque regions in the film. Drawing is a common commercial process for improving mechanical properties of films and fibers. Each of these methods of inducing crystallization allows the polymer molecules the vibrational and rotational freedom to reorient themselves into a crystalline morphology.

The individual crystalline regions thus formed are interconnected by amorphous regions. Crystalline regions may contain occasional kinks or folds, called *defects*, in the polymer chains. Studies of polymer single crystals using x-ray diffraction techniques indicate a plateletlike structure of about 100 Å thickness composed of polymer chains folded back on themselves to give parallel blocks perpendicular to the face of the crystal. This is called the *folded-chain lamella* model. As depicted in Figure 3.17, the parallel chains may be connected with adjacent reentry folds of uniform or nonuniform thickness or with a more amorphous nonadjacent folding (referred to as the *switchback* model). Which of three lamella models depicted in Figure 3.17 best represents reality, and what the true nature of the amorphous regions between lamellae is, are not known with certainty. Less common are *extended-chain crystals*,[24] characterized by an absence of folding and a crystal size dependent on the extended-chain length of the polymer molecules. Extended-chain crystals, which often take the form of well-developed needles, are usually formed with polymers of relatively low molecular weight by slow crystallization from the melt or under high pressure. Stereoregularity does not automatically lead to crystallinity; a molten stereoregular polymer may, on rapid cooling, solidify in a metastable amorphous state.

Onset of crystallinity is called *nucleation*. Nucleation may occur randomly throughout the matrix as polymer molecules begin to align (homogeneous

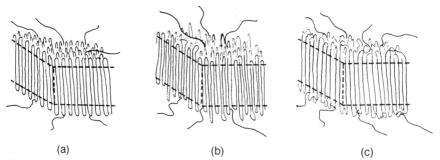

| (a) | (b) | (c) |

Figure 3.17. Folded chain lamella model: (a) regular adjacent folds; (b) irregular adjacent folds; and (c) nonadjacent switchback.

nucleation) or may occur at the interface of a foreign impurity, perhaps a purposely added nucleating agent such as finely divided silica (heterogeneous nucleation). A considerable variety of crystalline morphologies have been identified by x-ray analysis, three of which are shown in Figure 3.18. *Spherulites* are aggregates of small hairlike strands, called *fibrils*, arranged as clusters in an essentially radial pattern. The spherulitic morphology occurs commonly when nucleation originates in molten polymer or in concentrated polymer solutions in the absence of any applied stress. Drawing forces the spherulitic fibrils into the *drawn fibrillar morphology*. *Epitaxial* crystallinity, which usually arises during crystallization in stirred solutions or melts, is characterized by one crystalline growth on another, as in the so-called shish-kebab morphology that contains lamella growth on long fibrils.

Crystalline polymers are generally tougher, stiffer, more opaque, more resistant to solvents, and of higher density than their amorphous counterparts. (Recall from Chapter 1 the discussion of high- and low-density polyethylene; the linear, high-density form has a high degree of order and is semicrystalline). The higher the degree of crystallinity, the more pronounced the properties. The superior mechanical properties are a reflection of the greater cohesive strength arising from more effective intermolecular secondary forces among the closely packed molecules. Opaqueness arises from scattering of light by the crystallites.

Because crystalline polymers are more resistant to solvents, the fraction of a sample that fails to dissolve may be used as a crude measure of the degree of crystallinity or tacticity. Other methods are based on density measurements, x-ray diffraction, and spectroscopic and thermal measurements.[10b]

Unlike amorphous polymers, which liquify gradually over a broad temperature range above the glass transition temperature, crystalline (or semicrystalline) polymers melt over a relatively narrow range which may be observed most conveniently by thermal analysis (Chapter 5) or by observing the disappearance of birefringence (double refraction) under a polarizing microscope.[10b] The value thus obtained is called the *crystalline melting point*, T_m.

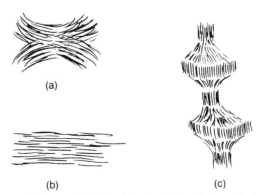

(a)

(b)   (c)

Figure 3.18. Some crystalline morphologies: (a) spherulitic; (b) drawn fibrillar; and (c) epitaxial (shish kebab).

Sometimes T_m is referred to as the *first-order transition temperature*. Like T_g, the crystalline melting point is not an exact value, but may vary a few degrees from one sample to another because of method of measurement or rate of heating.[13b] Because the same intermolecular forces are responsible for the magnitude of T_g and T_m, it follows that the values should parallel one another to an extent. As a rough approximation, T_g's are generally about one half to two thirds the value of the corresponding T_m, in kelvins. Table 3.5, which lists T_g and T_m values for a few representative polymers, illustrates this relationship.

The importance of stereoregularity in promoting crystallinity may be seen in the free radical polymerization of vinylidene chloride (1,1-dichloroethene) and vinylidene cyanide (1,1-dicyanoethene) to give highly crystalline polymers **47** and **48**, respectively. Under similar conditions, the monosubstituted

$$\left[CH_2-\overset{\displaystyle Cl}{\underset{\displaystyle Cl}{C}}\right] \qquad \left[CH_2-\overset{\displaystyle CN}{\underset{\displaystyle CN}{C}}\right]$$

$$\textbf{47} \qquad\qquad \textbf{48}$$

analogs, vinyl chloride and acrylonitrile, yield atactic polymer with a very low degree of crystallinity. Similarly, polyesters **49** and **51** are crystalline, whereas polyester **50**, which contains a chiral (asymmetric) carbon is not.[25] Stereoregularity is not always a prerequisite to crystallinity. Atactic poly(vinyl alcohol) is semicrystalline because the small size of the OH group allows it to fit into the crystalline lattice with the hydrogen atoms.

Table 3.5. Approximate glass transition temperatures (T_g) and crystalline melting temperatures (T_m) of some common polymers[a]

Polymer	T_g		T_m	
	°C	K	°C	K
cis-1,4-Polyisoprene	−73	200	26	299
trans-1,4-Polyisoprene	−58	215	74	347
cis-1,4-Polychloroprene	−20	253	70	343
trans-1,4-Polyisoprene	−40	233	101	374
Poly(methyl vinyl ether)	−31	242	144	417
Poly(vinyl chloride)	81	354	265	538
Polystyrene (isotactic)	100	373	242	515
Poly(ethylene adipate)	−63	210	54	327
Poly(hexamethylene adipamide)	57	330	268	541
Poly(ethylene terephthalate)	69	342	268	541

[a] Data from Lee and Rutherford[13a] and Miller[13b] (average of reported values).

$$\left[OCH_2CH_2CH_2O\overset{\overset{\displaystyle O}{\|}}{C}-\!\!\bigcirc\!\!-\overset{\overset{\displaystyle O}{\|}}{C} \right]$$ Crystalline (melting point 220°C)

49

$$\left[OCH_2\overset{\overset{\displaystyle CH_3}{|}}{CH}CH_2O\overset{\overset{\displaystyle O}{\|}}{C}-\!\!\bigcirc\!\!-\overset{\overset{\displaystyle O}{\|}}{C} \right]$$ Noncrystalline

50

$$\left[OCH_2\overset{\overset{\displaystyle CH_3}{|}}{\underset{\underset{\displaystyle CH_3}{|}}{C}}CH_2O\overset{\overset{\displaystyle O}{\|}}{C}-\!\!\bigcirc\!\!-\overset{\overset{\displaystyle O}{\|}}{C} \right]$$ Crystalline (melting point 140°C)

51

From the standpoint of polymer properties, crystallinity is extremely important because of its influence on mechanical properties. This aspect of structure-property relationships is explored in the following chapter.

3.7 Liquid crystallinity

Liquid crystals[26,27] are neither true liquids nor true solids, hence they have been referred to as a fourth state of matter. Pure liquids are *isotropic*—their molecules lack order. Crystalline materials are *anisotropic*. Liquid crystallinity occurs when molecules become aligned in a crystalline array while still in the liquid state; in other words, the liquid exhibits anisotropic behavior. The ordered regions in the liquid are called *mesophases*. As might be expected intuitively, the types of molecules that exhibit a propensity for forming mesophases are those with a relatively rigid, elongated, or disklike structure. With low-molecular-weight compounds, liquid crystallinity has been recognized for more than a century, although detailed studies and commercial application did not begin until the late 1950s and early 1960s. Liquid crystalline behavior was not recognized in polymers until the 1970s.

Liquid crystals exhibit certain properties not found in liquid or crystalline solid states. For example, their morphology may be influenced by external magnetic or electrical fields, sometimes they change color with temperature, and some exhibit extremely high optical rotation. At the same time they exhibit the fluidity of liquids and the opaqueness of crystalline solids. With low-molecular-weight compounds, a wide variety of ordered liquid crystalline arrangements have been identified; they vary according to temperature, chemical structure, and whether they exist in the melt or in solution. The interested student may pursue this topic further in the references provided. Our concern here is specifically with polymeric liquid crystals.[28-33]

There are two major classifications of liquid crystals: *lyotropic* and *thermotropic*. Lyotropic liquid crystals form under the influence of solvent. Thermotropic liquid crystals form in the melt. These classifications are broken down further (nematic, smectic, cholesteric, etc.) according to how the molecules are oriented in the mesophase. From the standpoint of polymer applications, two properties of liquid crystals are of major interest: the effect of order on viscosity, and the ability of the polymer to retain its ordered configuration in the solid state. Among the first polymers observed to exhibit liquid crystalline behavior were copolyesters prepared from terephthalic acid (**52**), ethylene glycol (**53**), and *p*-hydroxybenzoic acid (**54**).[34] It was found that as the amount

$$HO_2C \text{—}\bigcirc\text{—} CO_2H \qquad HOCH_2CH_2OH \qquad HO\text{—}\bigcirc\text{—}CO_2H$$

52 **53** **54**

of **54** was increased in the polymer, the melt viscosity initially increased, which was expected because of the decreased flexibility caused by incorporation of the "rigid" *p*-hydroxybenzoate unit. At levels of about 30 mol %, however, the melt viscosity began to *decrease*, reaching a minimum at about 60 to 70 mol %. This is shown in Figure 3.19 at four different shear rates. Furthermore, as the melt viscosity began to decrease, the melt's appearance changed from clear to opaque. The decrease in viscosity and opaqueness arise

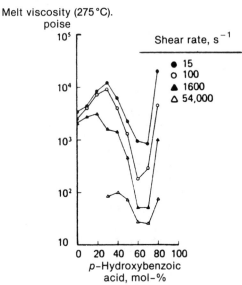

Figure 3.19. Effect of *p*-hydroxybenzoic acid concentration on melt viscosity of a terephthalic acid/*p*-hydroxybenzoic acid/ethylene glycol copolyester. [From Jackson and Kuhfuss,[34] copyright 1976. Reprinted by permission of John Wiley & Sons, Inc.]

from the onset of a thermotropic liquid crystalline morphology due to the increased backbone rigidity. The viscosity effect results from the rigid polymeric mesophases becoming aligned in the direction of flow, thus minimizing frictional drag. The implication from the standpoint of polymer processing is obvious: the lower the viscosity, the more readily the polymer can be fabricated into a useful plastic or fiber.

Equally important was the observation that the ordered arrangement of polymer molecules in the melt was retained upon cooling, which was manifested in greatly improved mechanical properties. Thus liquid crystalline behavior is advantageous from the standpoint of both processing and properties. Thermotropic liquid crystalline copolyesters of similar structure are now available commercially.

A commercially available lyotropic liquid crystalline polymer is the aromatic polyamide, **55** (du Pont trade name Kevlar). A sulfuric acid solution of

55

the polymer, which exhibits the liquid crystal phase, is extruded to form a fiber (see Chapter 4), which results in further alignment of the molecules. The product, once the sulfuric acid is removed, is a fiber with a more uniform alignment than could be obtained simply by drawing; as a consequence it has much better mechanical properties. Tensile strength of Kevlar, for example, is considerably higher than that of steel, whereas its density is much lower. Although most Kevlar produced is used in tire cord, the polymer also finds use in specialty clothing. Modern lightweight bullet-proof vests contain up to 18 layers of woven Kevlar cloth.

Major drawbacks to the type of rigid polymers that exhibit liquid crystalline behavior are that they have a very high melting point and are difficult, if not impossible, to dissolve in the usual organic solvents. One approach to circumventing these difficulties is to separate the rigid backbone groups (called *mesogens*), which are responsible for the mesophases, with *flexible spacers* such as **56, 57**, and **58**, or to attach mesogens with flexible spacers to the polymer

56   57   58

backbone. Both types are illustrated schematically in Figure 3.20. Although the use of flexible spacers does render the polymer more tractable, mechanical properties usually suffer.

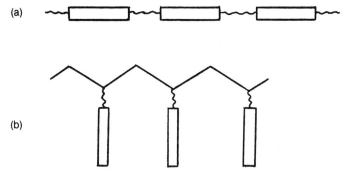

Figure 3.20. Representations of mesogenic groups (▭) and flexible spacers (⌇) in (a) the backbone and (b) the side chain of liquid crystalline polymers.

3.8 Chemical crosslinking

The most severe mechanism for decreasing molecular freedom is *chemical crosslinking*—linking the polymer chains together through covalent or ionic bonds to form a network.[35] Occasionally the term *curing* is used to denote crosslinking. There are a number of ways crosslinking can be brought about, but basically they fall into two categories: (1) crosslinking during polymerization by use of polyfunctional instead of difunctional monomers, and (2) crosslinking in a separate processing step after the linear (or branched) polymer is formed. The crosslinks may contain the same structural features as the main chains, which is usually the case with the former, or they may have an entirely different structure, which is more characteristic of the latter. In later chapters we will explore the chemistry of both covalent and ionic crosslinking.

A number of extreme changes accompany crosslinking. If previously soluble, the polymer will no longer dissolve (except in the case of some ionically crosslinked polymers). In the presence of solvent, a crosslinked polymer swells as solvent molecules penetrate the network. The degree of swelling depends on the affinity of solvent and polymer for one another, as well as on the level of crosslinking. It may be recalled (Chapter 2) that a solvent-swollen crosslinked polymer is called a gel. If the gel particles are very small (300–1000 μm), they are called *microgels*. Microgels behave as tightly packed spheres that can be suspended in solvents. Microgels have attracted considerable interest in recent years with the development of solid-phase synthesis and techniques for immobilizing catalysts. Covalently crosslinked polymers also lose their flow properties. They may still undergo deformation, but the deformation will be reversible; that is, the polymer will exhibit *elastic* properties. Ionically crosslinked polymers *will* flow at elevated temperatures, however.

With network polymers it is common to speak of the *crosslink density*, that is, the number of crosslinked monomer units per main chain. Crosslink density (Γ) is defined mathematically as

$$\Gamma = \frac{(\bar{M}_n)_0}{(\bar{M}_n)_c}$$

where $(\bar{M}_n)_0$ is the number average molecular weight of uncrosslinked poly-mer and $(\bar{M}_n)_c$ is the number average molecular weight between crosslinks. The higher the crosslink density, the more rigid the polymer. Very high crosslink densities lead to embrittlement. Because crosslinking reduces seg-mental motion, it is frequently employed to increase the glass transition temperature.

Elastomers are characterized by having a very *low* crosslink density—about one crosslink per 100 monomer units—together with highly flexible main chains to allow large deformations. Elastomeric behavior[12c,36] depends, of course, on polymer structure and crosslink density; but it also depends on morphology and molecular weight. Some of the morphological features that contribute nothing to the ideal elastic behavior of the network (and are thus termed *defects*) are chain ends, loops, and entanglements. One approach to improving elastic properties has been to design elastomers with reactive chain ends (called *telechelic polymers*)* so that the chain ends become incorporated into the network on crosslinking.[37]

3.9 Physical crosslinking

When polymer chemists use the term *crosslinking*, they invariably mean covalent chemical crosslinking. Covalent crosslinking has certain disadvan-tages, however. Once crosslinked, a polymer cannot be dissolved or molded. Scrap crosslinked polymer cannot be recycled. Polymer chemists have ex-plored ways to circumvent this problem. One approach has been to investi-gate thermally labile crosslinks, that is, chemical crosslinks that break apart on heating and reform on cooling. Ionic crosslinks fall into this category. Attempts to develop reversible covalent crosslinking are described in Chap-ter 9. The other approach has been to introduce strong secondary bonding attraction between polymer chains such that the polymer exhibits properties of a thermosetting material while remaining thermoplastic. Crystalline po-lymers fit into this category. Because of the very strong secondary forces arising from close chain packing, many of the mechanical and solution prop-erties of crystalline polymers resemble those of crosslinked amorphous po-lymers. Certain materials intermolecularly associated through hydrogen bonds also behave like crosslinked polymers. Gelatin, an animal-derived protein that exhibits elastomeric properties, is an example.

In recent years the technology of block copolymers has been applied to the area of physical crosslinking.[38,39] The method involves synthesis of block

---

* From the Greek *tele* (far) and *chele* (claw).

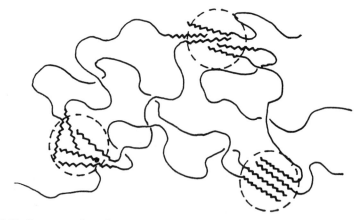

Figure 3.21. Representation of aggregation in an ABA block thermoplastic elastomer ($\sim\sim\sim$ represents end blocks, $\bigcirc$ represents microdomains).

copolymers of the ABA type in which the A and B blocks differ substantially in structure. Consider, for example, a long-chain "flexible" polymer such as polybutadiene, capped at each chain end with short blocks of a "rigid" polymer such as polystyrene. Because polybutadiene and polystyrene are inherently immiscible (incompatible), the polystyrene blocks tend to aggregate and form separate phases (microdomains) within the polymer matrix, as shown in Figure 3.21. If the structure of the end blocks is stereoregular, the aggregations may form crystalline microdomains. The aggregations impart a significant degree of elastic behavior, yet the copolymers still exhibit the flow properties of thermoplastics. Appropriately such materials are referred to as *thermoplastic elastomers*.[40] A number of thermoplastic elastomers have been commercialized.

3.10   Polymer blends

By definition, any physical mixture of two or more different polymers or copolymers that are not linked by covalent bonds is a *polymer blend*, or *polyblend*.[41-47] Some scientists like to make an analogy with metals and call such mixtures *polymer alloys*.[48] The concept of blending polymers is not new; the rubber industry has used it for decades. In recent years, however, there has been a resurgence of interest arising primarily from the demand for engineering plastics and specialty elastomers and fibers. There are sound economic reasons for this interest. Development of a new polymer to meet a specific need is a costly enterprise. If the desired properties can be realized simply by mixing two or more existing polymers, there is an obvious pecuniary advantage.

A number of technologies have been devised to prepare polyblends; these are summarized in Table 3.6. It so happens that most polymers are not compatible. Rather, they separate into discrete phases on being mixed, although an increasing number of completely miscible blends are being developed. Differences between the two types are manifested in appearance—miscible blends are usually clear, immiscible blends are opaque—and in such properties as glass transition temperature—miscible blends exhibit a single T_g intermediate between those of the individual components, whereas immiscible blends exhibit separate T_g's characteristic of each component. Miscibility is by no means prerequisite to commercial utility.

Homogeneous polymer blends are more convenient from the standpoint of being able to predict properties or processing characteristics. If additives are used, for example, there are no problems of migration from one phase to another. Physical or mechanical properties usually reflect, to a degree, the weighted average of the properties of each component. For a binary homogenous blend, this can be quantified for a particular property (P) using the semiempirical relationship

$$P = P_1\Phi_1 + P_2\Phi_2 + I\Phi_1\Phi_2$$

Table 3.6. Types of polyblends

Type	Description
Mechanical blends	Polymers are mixed at temperatures above T_g or T_m for amorphous and semicrystalline polymers, respectively
Mechanochemical blends	Polymers are mixed at shear rates high enough to cause degradation. Resultant free radicals combine to form complex mixtures including block and graft components
Solution-cast blends	Polymers are dissolved in common solvent and solvent is removed
Latex blends	Fine dispersions of polymers in water (latexes) are mixed, and the mixed polymers are coagulated
Chemical blends	
Interpenetrating polymer networks (IPN)	Crosslinked polymer is swollen with different monomer, then monomer is polymerized and crosslinked
Semiinterpenetrating polymer networks (semi-IPN)	Polyfunctional monomer is mixed with thermoplastic polymer, then monomer is polymerized to network polymer (also called pseudo-IPN)
Simultaneous interpenetrating polymer networks (SIN)	Different monomers are mixed, then homopolymerized and crosslinked simultaneously, but by noninteracting mechanisms
Interpenetrating elastomeric networks (IEN)	Latex polyblend is crosslinked after coagulation

where Φ is the volume fraction in the mix and I is an interaction term that may be negative, zero, or positive. If the properties are strictly additive, $I = 0$. If I is positive, the property in question is better than the weighted average, and the blend is said to be *synergistic* for that property. If I is negative, the property is worse than the weighted average (*nonsynergistic*). This is illustrated in Figure 3.22 by a plot of property versus composition. What gives rise to synergism or nonsynergism? The former might result from a very favorable dipole-dipole attraction between the polymer components. The latter might arise when a favorable intermolecular interaction is disrupted, for example, the prevention or reduction of crystallinity.

One example of a commercially important miscible polyblend is the engineering plastic Noryl (General Electric), which is composed of polystyrene **(21)**, an inexpensive polymer, and poly(oxy-2,6-dimethyl-1,4-phenylene) **(59)**, a relatively expensive polyether known commonly as poly(phenylene oxide) or PPO (also a General Electric trade name). For the most part, the properties of Noryl are additive. For example, Noryl has poorer thermal stability than the polyether alone, but is easier to process. Its single glass transition temperature increases with increasing polyether content. In terms of tensile strength (Chapter 4), however, the polyblend is synergistic.[49]

$$\left[\begin{array}{c} CH_3 \\ \langle\bigcirc\rangle - O \\ CH_3 \end{array} \right]$$

59

Morphology can have a dramatic effect on properties of polymer blends. A blend of low-density polyethylene (LDPE) with the terpolymer ethylene-

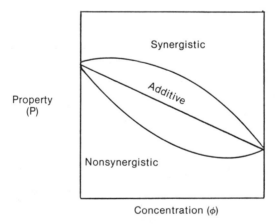

Figure 3.22. Hypothetical property-concentration relationships for a binary miscible polyblend.

propylene-diene monomer rubber (EPDM) exhibits a synergistic effect on tensile strength if the EPDM is partially crystalline, but a nonsynergistic effect if the EPDM is amorphous.[50] The synergism apparently arises from a tendency for crystallites in the LDPE to nucleate crystallization of ethylene segments in the EPDM. Stereochemistry is also important: syndiotactic poly(methyl methacrylate) is miscible with poly(vinyl chloride) at certain concentrations, whereas the isotactic form is immiscible over the entire composition range.[51]

A major problem with developing miscible polyblends is trying to predict miscibility. One approach is to use dipole moment differences[52]: the smaller the difference, the more compatible the polymers. Attempts to predict miscibility using solubility parameters of the type described in Chapter 2 have been largely unsuccessful because strong dipolar interactions are not taken into account. The importance of these interactions has been demonstrated, for example with miscible mixtures of poly(vinylidene fluoride) (**60**) and poly(acrylic esters) (**61**), where enthalpies of mixing have been correlated with the concentration of carbonyl oxygen in the mixture.[53] Compatibility may thus arise from dipolar interactions of type **62**.

$$-\!\!\left[CH_2CF_2\right]\!\!- \qquad -\!\!\left[CH_2CH\right]\!\!- \qquad$$

$$\begin{array}{c} OR \\ | \\ \overset{\displaystyle\diagdown}{CH}\!-\!\overset{\delta^+}{C}\!=\!\overset{\delta^-}{O}\cdots\cdots\overset{\delta^+}{C}\overset{F}{\underset{F}{\diagup}}\,\delta^- \end{array}$$

$$\overset{|}{CO_2R} \qquad \qquad \overset{\displaystyle\diagup}{CH_2}\diagdown \qquad \overset{\displaystyle\diagdown}{CH_2}\diagdown$$

60   **61**   **62**

Predictions of properties for immiscible polyblends is much more complicated. Attempts to develop additivity relationships are complicated by the effects of varying morphologies that might arise as a result of processing variables. Frequently, one polymer will constitute a continuous phase while the second will be dispersed as a noncontinuous phase in the form of fibrils, spheres, lamellae, and so on. Which polymer is the continuous phase largely determines properties. For example, a 50:50 blend of polystyrene (a hard, glassy polymer) and polybutadiene (an elastomer) will be hard if polystyrene is the continuous phase, but soft if polystyrene is the dispersed phase. Alternatively, an immiscible polyblend may have both components dispersed as continuous phases.

The major problem with immiscible blends is the often poor physical attraction at phase boundaries, which can lead to phase separation under stress with resultant poor mechanical properties. To improve compatibility between immiscible phases, a number of ingenious approaches have been adopted. One is through formation of interpenetrating networks,[45,54,55] as described in Table 3.6. In such cases the polymers are phsyically "locked" together by the interdispersed three-dimensional network, a phenomenon referred to as *topological bonding*.[54] Such mixtures still undergo phase

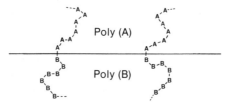

Figure 3.23. Representation of the use of an AB block copolymer to improve interfacial adhesion in an immiscible polyblend.

separation into microdomains that vary in size according to the degree of immiscibility.

Another approach is to incorporate *compatibilizers* into the blend to improve adhesion between phases.[56a] Let's consider, for example, a blend of poly(A) and poly(B), the homopolymers being made from monomers A and B. If a block copolymer of A and B is added to the blend, the natural affinities of the blocks for their respective homopolymers will serve to localize the copolymer at the phase boundary and help "glue" the phases together, as depicted in Figure 3.23. One drawback is the relatively high cost of block copolymers. Not all immiscible blends need an interfacial agent. Commercial varieties of *high-impact* polypropylene consist of immiscible blends of polypropylene and ethylene-propylene copolymer, which have sufficient interphase bonding arising from the natural affinity of the homopolymer for propylene segments in the copolymer.

The most successful approach to compatibilizing polymer blends from the commercial standpoint is to generate graft copolymers in situ that help bind the immiscible phases together. ABS (acrylonitrile–butadiene–styrene) plastics[57] are an example of this methodology. Typically, an amorphous styrene–butadiene copolymer is dissolved in styrene and acrylonitrile and the monomers are subsequently copolymerized. In the course of polymerization, chain-transfer reactions occur (the mechanisms of which will be described later) to produce grafts of one copolymer to the other. The number of grafts is small, but is sufficient to provide the necessary interfacial adhesion. ABS polymers are widely used as engineering plastics.[58]

Proper control of phase morphology is of much greater importance with immiscible blends. Despite this drawback, they have found much wider acceptance than their less ubiquitous miscible cousins, and include both engineering plastics [59,60] and fibers.[56b,61]

References

1. R. D. Deanin, *Polymer Structure, Properties and Applications*, Cahners Books, Boston, 1972.

2. M. L. Miller, *The Structure of Polymers*, Rheinhold, New York, 1966: (a) Chap. 4; (b) Chap. 10.
3. A. Ledwith and A. M. North (eds), *Molecular Behavior and the Development of Polymeric Materials*, Wiley-Interscience, New York, 1975: (a) R. N. Hayward, Chap. 12; (b) C. W. Bunn, Chap. 10.
4. A. V. Tobolsky, *Properties and Structure of Polymers*, Wiley-Interscience, New York, 1960.
5. G. L. Wilkes, *J. Chem. Educ.*, **58**, 880 (1981).
6. D. J. Williams, *Polymer Science and Engineering*, Prentice-Hall, Englewood Cliffs, N.J., 1971, Chap. 8.
7. L. E. Nielsen, *Polymer Rheology*, Dekker, New York, 1977.
8. F. N. Cogswell, *Polymer Melt Rheology*, George Godwin, London, 1981.
9. J. J. Aklonis and W. MacKnight, *Introduction to Polymer Viscoelasticity*, 2nd ed., Wiley-Interscience, New York, 1983.
10. J. F. Rabek, *Experimental Methods in Polymer Chemistry*, Wiley-Interscience, New York, 1980: (a) Chap. 9; (b) Chap. 26.
11. R. F. Boyer, in *Encyclopedia of Polymer Science and Technology* (N. R. Bikales, ed.), Suppl. Vol., Wiley-Interscience, New York 1979.
12. A. D. Jenkins (ed.), *Polymer Science*, Vol. 1, North-Holland, Amsterdam, 1972: (a) J. A. Brydson, Chap. 3; (b) A. Sharples, Chap. 4; (c) K. J. Smith, Jr., Chap. 5.
13. J. Brandrup and E. H. Immergut (eds.), *Polymer Handbook*, 2nd ed., Wiley-Interscience, New York, 1975; (a) W. A. Lee and R. A. Rutherford, pp. III-139ff; (b) R. L. Miller, pp. III-51ff.
14. D. R. Burfield and K. Lim, *Macromolecules*, **16**, 1170 (1983).
15. G. Natta and F. Danusso (eds.), *Stereoregular Polymers: Stereoregular Polymers and Stereospecific Polymerization*, Pergamon Press, New York, 1967.
16. P. Corradini, in *The Stereochemistry of Macromolecules*, Vol. 3 (A. D. Ketley, ed.), Dekker, New York, 1986, Chap. 1.
17. M. L. Huggins, G. Natta, V. Desreux, and H. Mark, *J. Polym. Sci.*, **56**, 152 (1962).
18. F. A. Bovey and T. K. Kwei, in *Macromolecules: An Introduction to Polymer Science* (F. A. Bovey and F. H. Winslow, eds.), Academic Press, New York, 1979, Chap. 3.
19. M. Farina, G. DiSilvestro, P. Sozzani, and B. Savare, *Macromolecules*, **18**, 923 (1985).
20. B. R. Burfield and Y. Doi, *Macromolecules*, **16**, 702 (1983).
21. P. H. Geil, E. Baer, and Y. Wada, *The Solid State of Polymers*, Dekker, New York, 1974.
22. L. Mandelkern, *Crystallization of Polymers*, McGraw-Hill, New York, 1964.
23. A. Kellor, in *Macromolecules* (H. Benoit and P. Rempp, eds.), Pergamon Press, New York, 1982, pp. 171ff.
24. B. Wunderlich and L. Melillo, *Makromol. Chem.*, **118**, 250 (1986).
25. R. W. Lenz, *Organic Chemistry of Synthetic High Polymers*, Wiley-Interscience, New York, 1967, Chap. 4.
26. G. H. Brown and P. C. Crooker, *Chem. Eng. News*, Jan. 31, 1983, p. 24.
27. S. Chandrasekhar, *Liquid Crystals*, Cambridge University Press, London, 1977.
28. M. Gordon (ed.), *Liquid Crystal Polymers, Adv. Polym. Sci.*, **59, 60, 61**, Springer-Verlag, New York, 1984.

29. A. Blumstein (ed.), *Liquid Crystalline Order in Polymers*, Academic Press, New York, 1978.
30. A. Ciferri, W. R. Krigbaum, and R. B. Meyer (eds.), *Polymer Liquid Crystals*, Academic Press, New York, 1982.
31. S. K. Varshney, *J. Macromol. Sci., Rev. Macromol. Chem. Phys.*, **C26**, 551 (1986).
32. R. W. Lenz, *Polym. J.*, **17**, 105 (1985).
33. H. Finkelmann, *Angew. Chem. Int. Ed. English*, **26**, 816 (1987).
34. W. J. Jackson Jr. and H. F. Kuhfuss, *J. Polym. Sci., Polym. Chem. Ed.*, **14**, 2043 (1976).
35. A. J. Chompff and S. Newman, *Polymer Networks: Structural and Mechanical Properties*, Plenum Press, New York, 1971.
36. L. R. G. Treolar, *The Physics of Rubber Elasticity*, 3rd ed., Claredon, Oxford, 1975.
37. C. A. Uranek, L. H. Hsieh, and O. G. Buck, *J. Polym. Sci.*, **46**, 535 (1960).
38. N. R. Legge, *CHEMTECH*, **13**, 630 (1983).
39. G. Holden, E. T. Bishop, and N. R. Legge, *J. Polym. Sci.*, **C26**, 37 (1969).
40. N .R. Legge (ed.), Thermoplastic Elastomers: A Comprehensive Review, Hanser, Munich, 1987.
41. O. Olbasi, L. M. Robeson, and M. T. Shaw, *Polymer-Polymer Miscibility*, Academic Press, New York, 1979.
42. J. A. Manson and L.H. Sperling, *Polymer Blends and Composites*, Plenum Press, New York, 1976.
43. S. L. Cooper and G. M. Estes (eds.), *Multiphase Polymers*, Adv. Chem. Ser. 176, American Chemical Society, Washington, D. C., 1979, pp. 313ff.
44. D. R. Paul and S. Newman (eds.), *Polymer Blends*, Vols. 1 and 2, Academic Press, New York, 1978.
45. L. H. Sperling, *Macromol. Rev.*, **12**, 141 (1977).
46. E. N. Kresge, *J. Appl. Polym. Sci., Appl. Polym. Symp.*, **39**, 37 (1984).
47. H. Rudolph, *Makromol. Chem., Makromol. Symp.*, **16**, 57 (1988).
48. B. J. Schmitt, *Angew. Chem. Int. Ed. English*, **18**, 273 (1979).
49. A. F. Yee, *Polym. Eng. Sci.*, **17**, 213 (1977).
50. G. H. Lindsay, C. J. Singleton, C. J. Carman, and R. W. Smithe, in *Multiphase Polymers* (S. L. Cooper and G. M. Estes, eds.), Adv. Chem. Ser. 176, American Chemical Society, Washington, D.C., 1979, p. 367.
51. J. W. Schurer, A. de Boer, and G. Challa, *Polymer,* **16**, 201 (1975).
52. A. Dondos and E. Pierri, *Polym. Bull.*, **16**, 567 (1986).
53. R. E. Bernstein, D. R. Paul, and J. W. Barlow, *Polym. Eng. Sci.,* **18**, 683 (1978).
54. H. I. Frisch, *Br. Polym. J.*, **17**, 149 (1985).
55. L. H. Sperling, *Interpenetrating Polymer Networks and Related Materials*, Plenum Press, New York, 1981.
56. D. R. Paul, in Polymer Blends, Vol. 2 (D. R. Paul and S. Newman, eds.), Academic Press, New York, 1978: (a) Chap. 12; (b) Chap. 16.
57. C. H. Basdekis, *ABS Plastics*, Reinhold, New York, 1964.
58. N. Platzer, *CHEMTECH*, **7**, 634 (1977).
59. D. R. Paul and J. W. Barlow, in *Multiphase Polymers* (S. L. Cooper and G. M. Estes, eds.), Adv. Chem. Ser. 176, American Chemical Society, Washington, D.C., 1979, p. 315.

60. B. D. Hanrahan, S. R. Angeli, and J. Runt, *Polym. Bull.*, **14**, 399 (1985).
61. S. P. Hersh, in *High Technology Fibers*, Part A (M. Lewin and J. Preston, eds.), Dekker, New York, 1985, Chap. 1.

Review exercises

1. Write concise definitions of the following terms:

   (a) annealing
   (b) atactic
   (c) crosslink density
   (d) erythro
   (e) glass transition temperature
   (f) head-to-tail polymer
   (g) hemitactic
   (h) interpenetrating polymer network
   (i) isotactic
   (j) lyotropic liquid crystal
   (k) mesogen
   (l) mesophase
   (m) nucleation
   (n) polyblend
   (o) rheology
   (p) shear stress
   (q) shear thinning
   (r) spherulite
   (s) stereoblock polymer
   (t) syndiotactic
   (u) telechelic polymer
   (v) thermoplastic elastomer
   (w) thermotropic liquid crystal
   (x) threo
   (y) viscoelasticity
   (z) vitrification

2. Explain the difference between shear thinning and thixotropy.

3. Draw the *trans-threo*-diisotactic and *cis-erythro*-disyndiotactic forms of the polymer depicted in Figure 3.15.

4. Draw a section of the chain (four or five repeating units) of each of the following head-to-tail polymers:

   (a) syndiotactic poly(vinyl alcohol)
   (b) isotactic poly(α-methylstyrene)
   (c) atactic poly(1,2-butadiene)
   (d) syndiotactic poly(1,2-difluoroethylene)
   (e) *threo*-diisotactic poly(ethyl β-chlorovinyl ether)
   (f) *cis*-1,4-polychloroprene
   (g) *threo*-diisotactic polycyclohexene
   (h) *cis-erythro*-diisotactic-1,4-poly(1-bromo-1,3-pentadiene)
   (i) *trans*-isotactic-1,4-poly(2-methyl-1,3-pentadiene)

5. What constitutional (i.e., not source-based) name could you assign to the polymer in question 4(i) if the double bond were to be hydrogenated? (M. Farina et al., *Macromolecules*, **15**, 1451, 1982)

6. (a) Draw structures analogous to structures **37** and **38** for the six possible tetrads of poly(methyl methacrylate). (b) How many pentads are possible? Write the *mr* designations (not structures) of each.

7. Which polymer in each of the following pairs would you expect to exhibit the higher glass transition temperature? Explain your choice in each case.

(a) 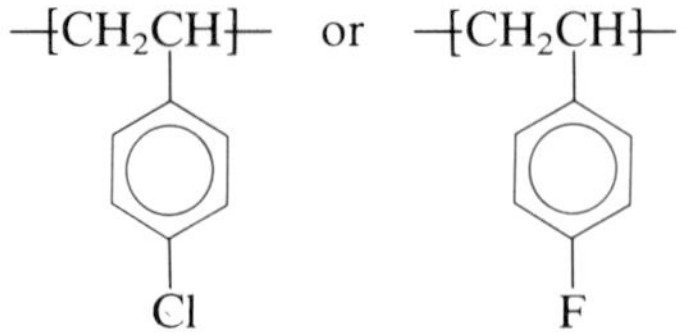 $-\!\!\{CH_2CH\}\!\!-$ or $-\!\!\{CH_2CH\}\!\!-$

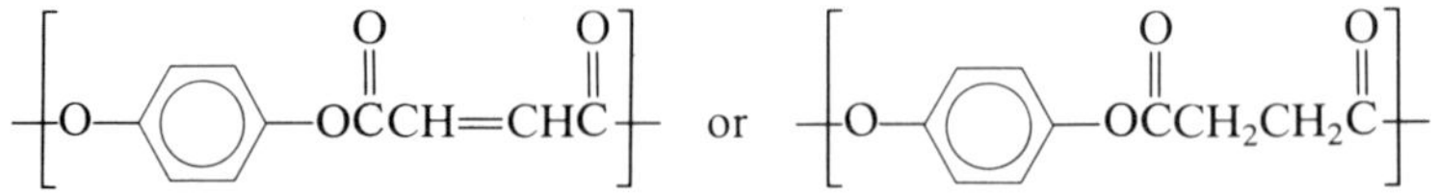

(b) $\{O\!-\!\!\bigcirc\!\!-\!OCCH\!=\!CHC\}$ or $\{O\!-\!\!\bigcirc\!\!-\!OCCH_2CH_2C\}$

(c) $-\!\!\{CH_2CH_2NH\}\!\!-$ or $-\!\!\{CH_2CH_2O\}\!\!-$

(d) 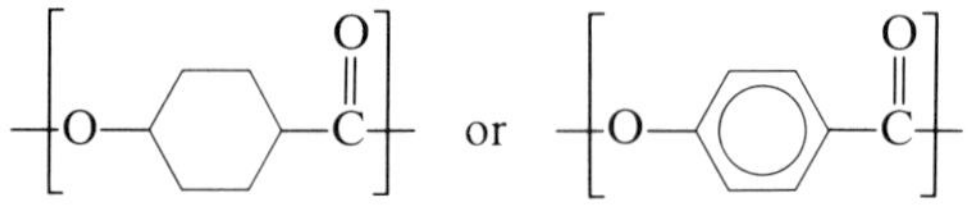   or

(e) 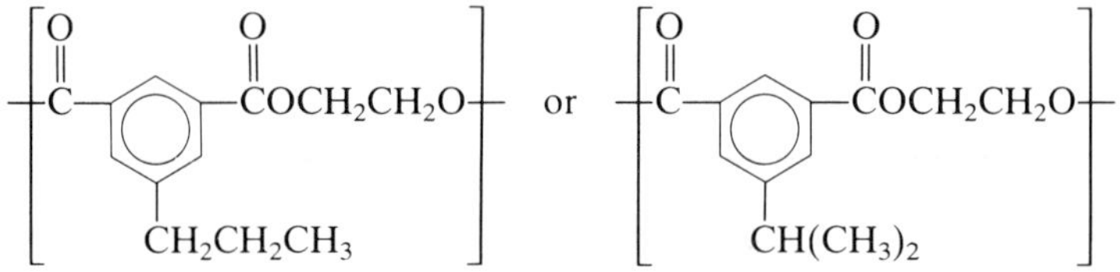   or

8. Which of the following "isomeric" polymers would you expect to exhibit the greater crystallinity? Explain.

(a) 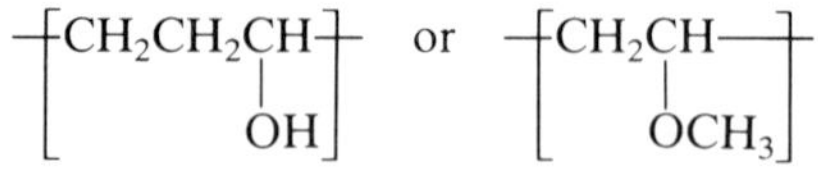   or

(b) 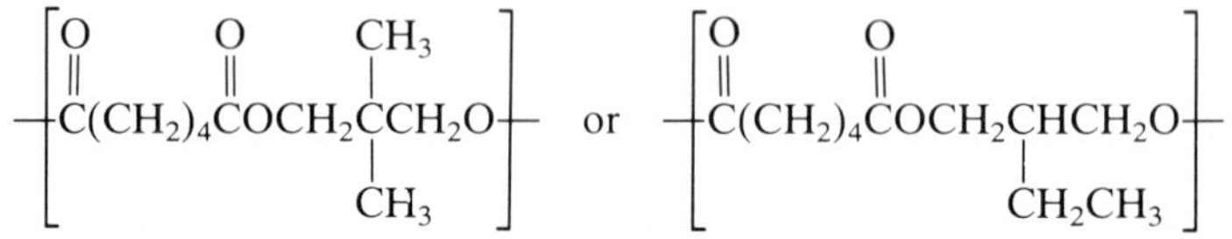   or

(c) $-\!\!\{CH_2CF_2\}\!\!-$ or $-\!\!\{CHFCHF\}\!\!-$

9. When poly(tetramethylene adipate) (a polymer of interest for "disappearing" surgical sutures) is subjected to hydrolytic or enzymatic degradation, the polymer's percent crystallinity initially increases. What does this suggest about the degradation process? (A.-C. Albertson and O. Ljunquist, *J. Macromol. Sci.-Chem.*, **A23**, 393, 1986)

10. One method of determining percent crystallinity in a polymer involves inverse gas chromatography, that is, the polymer in question is used as the stationary liquid phase coated on the column packing. Speculate on the rationale underlying this method.

11. When a metal spring, suspended with a weight attached at the bottom, is heated, the spring elongates further because more volume is needed to accommodate the increased atomic vibrations. When a strip of elastomer is similarly suspended, it *contracts* on heating. On the basis of what you have learned in this chapter, suggest a plausible explanation for this difference in behavior.

12. Speculate on why polyester clothing is more wrinkle resistant than cotton. Why is a *hot* iron used to press out the wrinkles.

4

Chemical structure and polymer properties

4.1 Introduction

In the last chapter we saw how chemical structure influences polymer morphology. Now we will examine the relationship between chemical structure
and the properties of major interest in commercial applications. With mechanical properties, in particular, we shall see that the interrelationship between
chemical structure and morphology is of overriding importance. Other properties—thermal, chemical, and electrical—also depend to an extent on morphology, but more so on chemical makeup. In many cases the properties
under discussion are significant in specialty products that represent only a
small fraction of the total synthetic polymer production. Nevertheless, such
markets are expanding and, importantly, they illustrate how synthetic polymers can often be tailored for a specific application. We will also look briefly
at *additives*—compounds added to polymers to *modify* their properties.
Given the fact that virtually all commercial polymer products contain one or
more additives, this is clearly an important facet of structure-property relationships.

We will begin with a brief digression from chemistry to see how polymers
are converted into the commercial products that benefit from the various
properties of interest. While a detailed study of fabrication methods is outside
the scope of this book and more relevant to engineering texts, it is instructive
to progress from our earlier discussion of polymer flow to recognize how
products of the laboratory are transformed into useful articles.

4.2 Fabrication methods

Polymers are most commonly shaped by one of three basic techniques:
molding, extrusion, or casting.[1] All three are accomplished at much lower
temperatures than those needed for the shaping of steel, aluminium, or glass,
and hence energy efficiency is one of the most attractive features of polymer

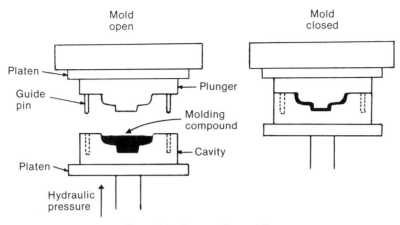

Figure 4.1. Compression molding.

processing. Polymers have one inherent disadvantage, however: they have poor thermal conductivity and melt slowly. To minimize this problem, rotating screw devices are used to transport the polymer, thus allowing granular polymer pellets or powder to be heated uniformly by a combination of external heaters and frictional heat generated during transport.

There are two fundamental molding processes: *compression molding* and *injection molding*. Compression molding uses heat and pressure to force molten polymer, introduced between the mating surfaces of a movable mold, into the shape of the mold. In injection molding, molten polymer is compressed into a closed mold cavity. The two methods are shown diagrammatically in Figures 4.1 and 4.2. The former is more commonly used with thermosetting polymers, which cannot be melted once crosslinking has taken

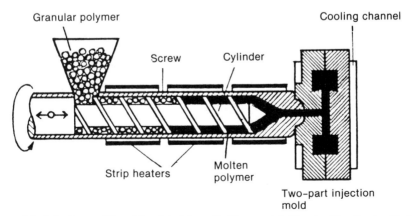

Figure 4.2. Injection molding. [Reprinted from V. Hopp and I. Hennig, *Handbook of Industrial Chemistry*, copyright 1983, courtesy of McGraw-Hill.]

place. Crosslinking is therefore accomplished in situ in the mold cavity. In some cases the polymer is partially polymerized so that it will still flow under pressure (this is called *B-staging*); then heat is applied to effect the final crosslinking. Other processes use *preforms* (also called *prepregs*) made of polymeric binder and a reinforcing material such as glass fiber, which are preshaped to fit the mold; then liquid polymer is poured over the preform. Molding pressure distributes the polymer through the preform, and heat activates the crosslinking. The end product is a *fiber-reinforced plastic*.

Injection molding is more commonly applied to thermoplastics. A hydraulic ram or, as illustrated in Figure 4.2, a screw is used to feed the polymer to the mold. Injection mold is generally faster than compression molding.

In recent years, a new type of injection molding called *reaction injection molding* (RIM) has become important for fabricating thermosetting polymers.[2-4] RIM differs from its predecessors in that the finished product is made directly from monomers or low-molecular-weight polymeric precursors, which are rapidly mixed and injected into the mold even as the polymerization reaction is taking place. A schematic of a RIM process is shown in Figure 4.3. For RIM to be successful, the monomers must be fast reacting, and reaction rates must be carefully synchronized with the molding process. There are, however, obvious advantages. Synthesis of polymer prior to molding is eliminated, and the energy requirements for handling of monomers are much lower than those for viscous polymers. Much of the early development of RIM was applied to polyurethanes (Chapter 13), but other polymer types and polymer blends also benefit from this technology. Reinforcing fillers are sometimes injected along with the reactants, a process called *reinforced reaction injection molding* (RRIM).

Another type of molding, particularly useful for manufacturing bottles, is *blow molding*. In this process, polymer tubing (called *parison*) is blown by compressed air or drawn by vacuum into the shape of the mold (Figure 4.4). *Casting*, a much simpler process, involves pouring molten polymer into a mold and allowing the product to cool.

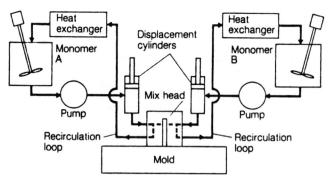

Figure 4.3. Basic components of a reaction injection molding (RIM) processing system. [From *Modern Plastics Encyclopedia, 1984–85.* Reprinted with permission of Modern Plastics.]

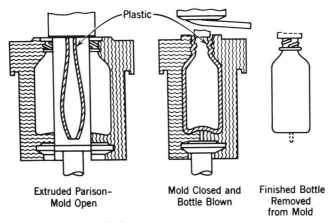

Extruded Parison–   Mold Closed and   Finished Bottle
Mold Open   Bottle Blown   Removed
from Mold

Figure 4.4. Blow molding. [Courtesy of the Society of the Plastics Industry, Inc.]

Extrusion involves forcing compacted, molten polymer through a die shaped to give the desired object. An extruder resembles the injection molding apparatus shown in Figure 4.2 except that the mold cavity is replaced with a die, as shown earlier in Figure 3.2. Extrusion is especially useful for making elongated objects such as rods or pipe.

Polymer film is made by casting from solution or by passing polymer under high pressure between hot rollers, a process known as *calendering*. Calendering is also used to make thick sheets —for floor tile, for example. Other methods of making film include extrusion through a slit die or through a ring-shaped die. In the latter process, pressure is maintained within the resultant plastic tube to expand it to a cylindrical film that can be cut to flat sheets or processed further as parison in blow molding.

Fibers are made by *spinning*, a process resembling extrusion. Polymer is forced under pressure through a perforated plate called a *spinneret*, as depicted in Figure 4.5. When several threads are combined, the material is called *tow*. Spinning is done either with molten polymer (*melt spinning*) or with polymer solution. In the case of melt spinning, the filaments are cooled rapidly by cold air in the process tank. If polymer solution is used, the solvent is removed in the process tank by evaporation with heat (*dry spinning*) or by leaching with another liquid (*wet spinning*). With some polymers, particularly cellulose (Chapter 17) which does not dissolve in the more common solvents, it may be more advantageous to convert the polymer chemically to a more soluble form and then to regenerate it in the process tank by chemical treatment of the filaments.

There are a variety of more specialized fabrication methods, such as *filament winding* to make fiber-reinforced composites, or *blowing* to make foams. For a more comprehensive discussion, the reader is referred to texts[5,6] that deal with fabrication processes in more detail.

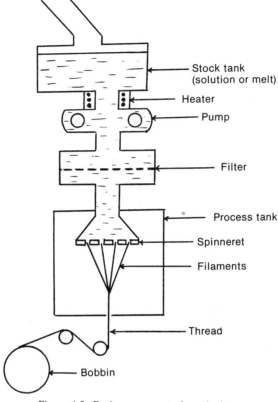

Figure 4.5. Basic components for spinning.

4.3   Mechanical properties

Let's consider a typical linear polymer having a uniform repeating unit and discrete end groups. At low molecular weight, the end groups contribute significantly to the overall structure, and this is manifested in such properties as density, refractive index, and spectroscopic absorption, which vary with molecular weight. When a certain molecular weight is attained, say 15,000, the concentration of end groups becomes negligible and these properties become constant, independent of any further increase in molecular weight. This is not the case with mechanical properties, which depend on intermolecular forces. Mechanical properties are much more dependent on molecular weight over a very broad range, although they too level off at the higher end of the molecular weight spectrum. Where they level off depends on structure. For a polyolefin such as polyethylene, where dispersion forces are responsible for mechanical properties, the leveling off occurs at relatively high molecular weight (above 10^5), whereas with very polar polymers such as polyamides, it may occur at a molecular weight as low as about 20,000 to 50,000. This dependence of properties on molecular weight is shown for a hypothetical polymer in Figure 4.6. (The assumption is made here that the "nonmechanical" property increases with molecular weight.) The working range of commercial polymers shown in the figure is at a molecular weight that affords good mechanical properties at a workable viscosity.

For the great bulk of commercial polymers, mechanical properties are of fundamental interest.[7-9] Although other properties, such as flame resistance, thermal stability, and chemical resistance, are of concern in more specialized applications, all polymers, regardless of use, must exhibit a specified range of mechanical properties suitable for that application. Among the dozens of interest to polymer manufacturers, *tensile, compressive*, and *flexural strength* (and their corresponding moduli) and *impact resistance* are most important. Related properties include hardness, abrasion resistance, and tear resistance.

To what exactly do these properties refer? In all cases they are a measure of how much stress a sample will withstand before the sample "fails." Tensile strength refers to resistance to stretching. Compressive strength is, in a sense, the opposite of tensile strength; it is the extent to which a sample can be compressed before it fails. Flexural strength is a measure of resistance to breaking, or snapping, when a sample is bent (flexed). Impact strength is a measure of "toughness"—how well a sample will withstand the sudden onset of stress, like a hammer blow. Another mechanical property of interest is *fatigue*—how well a sample will withstand repeated applications of tensile,

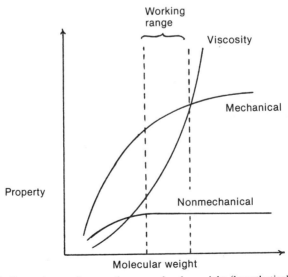

Figure 4.6. Dependence of properties on molecular weight (hypothetical polymer).

flexural, or compressive stress. How these various properties are measured is described in the next chapter. Here we are more concerned with the fundamentals of mechanical properties as manifested in stress-strain behavior. Although each mechanical property has its own idiosyncrasies, all such properties result from the same characteristics of structure and morphology and the manner in which the polymeric matrix undergoes molecular reorientation in response to stress. For purposes of our discussion, we will focus on tensile properties.

Tensile strength is determined by stretching a strip of polymer of uniform dimensions. The tensile stress, σ, is the force applied, F, divided by the cross-sectional area, A; that is,

$$\sigma = \frac{F}{A}$$

in units of dynes per square centimeter (CGS) or newtons per square meter (SI) (or pounds per square inch, psi, in the English system). Tensile strain, ϵ, is the change in sample length divided by the original length:

$$\epsilon = \frac{\Delta l}{l}$$

The ratio of stress to strain is the tensile modulus, E,

$$E = \frac{\sigma}{\epsilon}$$

which is a measure of the resistance to tensile stress. (Note that these definitions parallel those of shear stress described in Section 3.3.) Because strain is dimensionless, modulus carries the same units as stress.

A distinction among fibers, plastics, and elastomers is often expressed in terms of stress-strain curves, as shown in Figure 4.7. Both plastics and fibers exhibit a steep slope (high modulus), but the fiber can sustain greater stress before breaking (end of curve). The elastomer has a low modulus initially, but once in the stretched state its modulus increases sharply. A general stress-strain curve for a plastic such as polyethylene demonstrating various elements of tensile behavior is shown in Figure 4.8. Initially the modulus is high, until a point is reached where the plastic "yields," or deforms. Prior to the yield point, the *elongation* is reversible. At the yield point, enough stress has been applied to cause the molecules to untangle and flow over one another, and further elongation is irreversible. Eventually the sample breaks. The behavior of the polymer—or shape of the curve—beyond the yield point depends to a large degree on the polymer's initial morphology. If the polymer is amorphous or of low crystallinity, application of stress may increase the crystallinity, which in turn increases the modulus. This is the principle behind drawing. Highly crystalline polymers exhibit little change in morphology on drawing and break soon after the yield point, which is the case with the fiber in Figure 4.7.

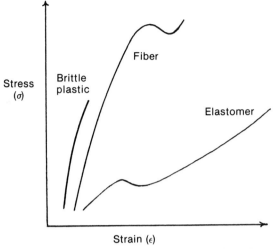

Figure 4.7. Characteristic tensile stress-strain behavior.

The stress-strain behaviors depicted in Figures 4.7 and 4.8 are at constant temperature. Mechanical properties are, however, very temperature dependent. Let's assume, for example, that we take a strip of plastic similar to the one used to generate the curve shown in Figure 4.8, cool it to a low temperature, and then apply a constant stress. From the elongation ($\Delta l/l$) we can determine the modulus. As the temperature is increased, the modulus will remain high until the glass transition temperature is reached, at which point the modulus drops sharply. This is shown in Figure 4.9. The modulus scale is

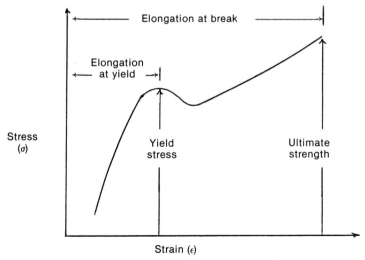

Figure 4.8. General tensile-strain curve for a typical thermoplastic.

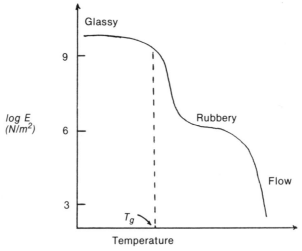

Figure 4.9. Effect of temperature on tensile modulus of an amorphous thermoplastic. Log E, modulus scale; T_g glass transition temperature.

logarithmic. Very high moduli (of the order of 3×10^9 N/m^2) are typical in the glassy state, but these decline precipitously once the molecules gain more freedom of movement. The sample remains rubbery until the temperature is sufficiently high to cause flow. As discussed earlier, room temperature would lie somewhere along the glassy plateau for a plastic and along the rubbery plateau for an elastomer.

Figure 4.10 illustrates the modulus-temperature behavior for a variety of polymer types. A thermoplastic amorphous polymer shows an expected molecular weight dependence in the flow region. The higher the molecular weight, the higher the temperature necessary to overcome the increased molecular entanglements. A crosslinked polymer, on the other hand, does not flow. The higher the crosslink density, the greater will be the modulus (the less the elongation) in the rubbery state. Semicrystalline thermoplastic polymers behave much like crosslinked polymers below the melting tempera- ture, T_m, because of the very strong intermolecular forces arising from close chain packing. How sharp a break occurs at the glass transition temperature depends on the degree of crystallinity. Above T_m, of course, the crystalline polymer will flow.

In addition to temperature dependence, there are two important *time-* dependent properties: *creep* (or *cold flow*) and *stress relaxation*. Creep is a measure of the change in strain when a polymer sample is subjected to a constant stress (gravity, for example). Stress relaxation refers to the decrease in stress when a sample is elongated rapidly to constant strain. Both phe- nomena are characteristic of viscoelastic materials and arise from the slippage of polymer molecules. Plots of stress change versus time bear a distinct resemblance to modulus-temperature plots of the type shown in Figure 4.9. This is not unexpected; increasing temperature, in effect, speeds up the

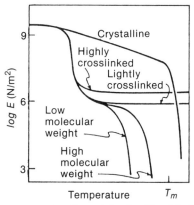

Figure 4.10. Effect of temperature on tensile modulus (log E scale) of various types of polymers. T_m, crystalline melting temperature. [Reprinted with permission from J. J. Aklonis, *J. Chem. Educ.*, **58**, 11 (1981).]

molecular disentanglement. Creep and stress relaxation are of critical concern to plastics manufacturers for the long-term mechanical properties of their materials.

The various characteristics of tensile behavior are readily understandable in terms of the structure-morphology relationships described in Chapter 3, and can be extrapolated to other kinds of mechanical properties. It is difficult to make generalizations when relating structure to properties because so many variables are involved—molecular weight, molecular weight distribution, morphology, method of sample preparation, number and type of additives. One might reasonably assume, however, that the more "flexible" the polymer backbone, the lower will be the tensile, flexural, or compressive strength. Anything that contributes to chain stiffening—bulky side groups, cyclic units, crosslinking, crystallinity—ought to increase these properties at the expense of tensile elongation. Trends in impact resistance are more difficult to predict. Chain stiffening in most instances leads to increased embrittlement and lower impact strength. On the other hand, conformational factors that allow rapid molecular reorientation at the onset of stress lead to improved impact strength. Measures taken to improve impact strength more often than not lead to a diminution of other mechanical properties.

Table 4.1 gives some representative mechanical properties for a few important commercial homopolymers.[10a] The wide range in properties reflects the many variables involved in physical testing. Worth noting is the remarkably high impact strength of polycarbonate. This, together with other strength characteristics, is a key factor in this relatively high-cost material being used widely as an engineering plastic. The significant increase in tensile properties of the mostly linear, crystalline high-density polyethylene, compared with those of the branched low-density polyethylene, arises from close chain packing in the former. Techniques have been developed to refine polyethylene's crystalline morphology further for use in high-strength fiber

Table 4.1. Mechanical properties of common homopolymers[a]

Property	Tensile properties at break				Compressive strength[b] (MPa)	Flexural strength[b] (MPa)	Impact strength[c] (N/cm)
Polymer	Strength[b] (MPa)	Modulus[b] (MPa)	Elongation (%)				
Polyethylene, low density	8.3–31	172–283	100–650		—	—	No break
Polyethylene, high density	22–31	1070–1090	10–1200		20–25	—	0.23–2.3
Polypropylene	31–41	1170–1720	100–600		38–55	41–55	0.23–0.57
Poly(vinyl chloride)	41–52	2410–4140	40–80		55–90	69–110	0.23–1.3
Polystyrene	36–52	2280–3280	1.2–2.5		83–90	69–101	0.20–0.26
Poly(methyl methacrylate)	48–76	2240–3240	2–10		72–124	72–131	0.17–0.34
Polytetrafluoroethylene	14–34	400–552	200–400		12	—	1.7
Nylon 66	76–83	—	60–300		103	42–117	0.46–1.2
Poly(ethylene terephthalate)	48–72	2760–4140	50–300		76–103	96–124	0.14–0.37
Polycarbonate	66	2380	110		86	93	9.1

[a] Values taken from Agranoff,[10a] converted to SI units, and rounded off.
[b] To convert megapascals to pounds per square inch multiply by 145.
[c] Izod notched impact test (see Chap. 5). To convert newtons per centimeter to foot-pounds per inch, multiply by 1.75.

Table 4.2. Fiber properties[a]

Fiber type	Tenacity[b] (N/tex)	Specific gravity
Natural		
Cotton	0.26–0.44	1.50
Wool	0.09–0.15	1.30
Synthetic		
Polyester	0.35–0.53	1.38
Nylon	0.40–0.71	1.14
Aromatic polyamide (aramid)[c]	1.80–2.0	1.44
Polybenzimidazole	0.27	1.43
Polypropylene	0.44–0.79	0.90
Polyethylene (high strength)	2.65[d]	0.95
Inorganic[e]		
Glass	0.53–0.66	2.56
Steel	0.31	7.7

[a] Unless otherwise noted, data taken from L. Rebenfeld, in *Encyclopedia of Polymer Science and Engineering* (H. F. Mark, N. M. Bikales, C. G. Overberger, G. Menges, and J. I. Kroschwitz, eds.), Vol. 6, Wiley-Interscience, New York, 1986, pp. 647–733.

[b] To convert newtons per tex to grams per denier, multiply by 11.3.

[c] Kevlar (see Chap. 3, structure **55**.)

[d] From *Chem. Eng. News*, **63**(8), 7 (1985).

[e] From V. L. Erlich, in *Encyclopedia of Polymer Science and Technology* (H. F. Mark, N. G. Gaylord, and N. M. Bikales, eds.). Vol. 9, Wiley-Interscience, New York, 1968, p. 422.

applications. Called *gel-spinning*, the method involves converting very-high-molecular-weight ($>10^6$) linear polymer to a gel with solvent, followed by spinning and drawing.[11] Gel formation provides the necessary molecular freedom for achieving a high degree of alignment. Similar results have been obtained with specially prepared polyethylene films without resorting to gel formation.[12] The drawable films are deposited directly in a disentangled crystalline state during polymerization onto initiator-treated surfaces. Properties of high-strength polyethylene are compared with properties of other fibers in Table 4.2. Tensile strength is given as *tenacity*, the common tensile parameter used by fiber scientists. Units of tenacity are newtons per tex, where tex is defined as the weight in grams of 1000 meters of the fiber.* The

*A more traditional unit of tenacity is grams per denier, where denier is defined as the weight in grams of 9000 m of the fiber. The denier evolved from units adopted for numbering raw silk at the Paris International Congress in 1900. The tex is used by the American Society for Testing and Materials (ASTM) as its standard linear density unit, and is also acceptable (along with kilograms per meter) in the SI system.

only fiber listed with strength comparable to that of high-strength poly-ethylene is the aromatic polyamide prepared by the liquid crystal method described in Chapter 3.

4.4   Thermal stability

When organic substances are heated to high temperatures they have a ten-dency to form aromatic compounds. It follows that aromatic polymers should be resistant to high temperatures. A wide variety of polymers having aromatic repeating units have been developed in recent years,[13–16] much of the impetus arising from the need of the aerospace industries for high-performance ma-terials that will withstand extremes of temperature. For a polymer to be considered "thermally stable" or "heat-resistant," it should not decompose below 400°C and should retain its useful service properties at temperatures near the decomposition temperature. As might be expected, such polymers must have high glass transition or crystalline melting temperatures. Some representative thermally stable polymers, along with their initial decomposi-tion temperatures, are given in Table 4.3. (Chemistry of these polymers is covered in Part III.)

Thermal stability is primarily a function of bond energy. When the temper-ature increases to the point where vibrational energy causes bond rupture, the polymer degrades. In the case of cyclic repeating units, breaking of one bond in a ring does not lead to a decrease in molecular weight, and the probability of two bonds breaking within one ring is low. Thus, ladder or semiladder polymers are expected to have higher thermal stabilities than open-chain polymers.[17] The decomposition temperatures given in Table 4.3 were deter-mined in inert atmospheres. Decomposition in air—a measure of a polymer's *thermooxidative stability*—generally follows a different mechanism. The pres-ence of oxygen, however, in most cases has little effect on the initial decom-position temperature, hence bond rupture is primarily a thermal rather than an oxidative process.

One individual, the late Carl S. Marvel, stands out as the leader in the field of thermally stable polymers; with justification he has been called "the father of high-temperature polymers."[18] Marvel's pioneering work led to commer-cial development of polybenzimidazole (**1**)[19] (Celanese trade name PBI). In fiber form, **1** has been used as a fabric for astronauts' clothing, one of a number of applications in aerospace programs. A wide variety of aromatic

1

Table 4.3. Representative thermally stable polymers[a]

Type	Structure	Decomposition temperature(°C)[b]
Poly(p-phenylene)	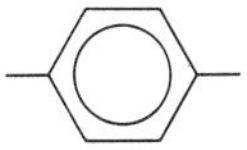	660
Polybenzimidazole	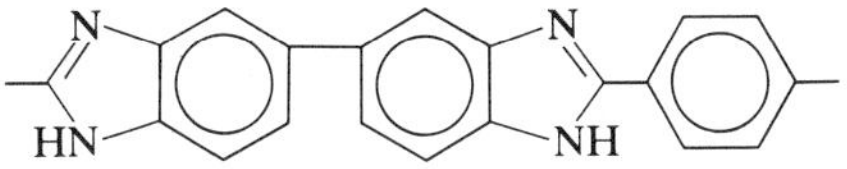	650
Polyquinoxaline	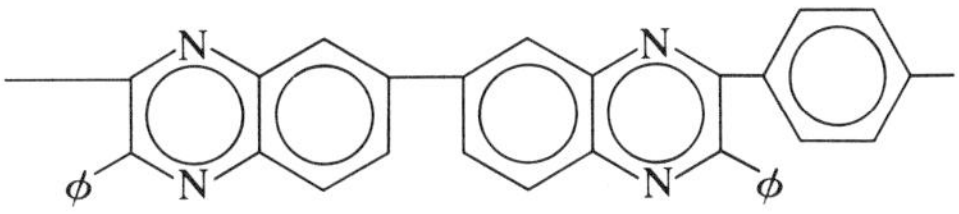	640
Polyoxazole	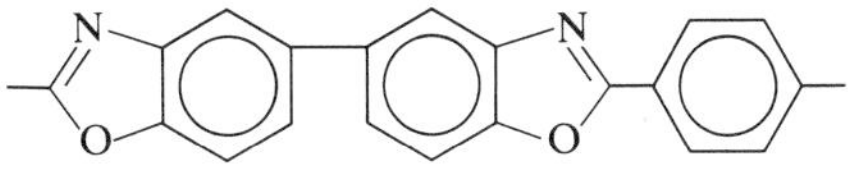	620
Polyimide	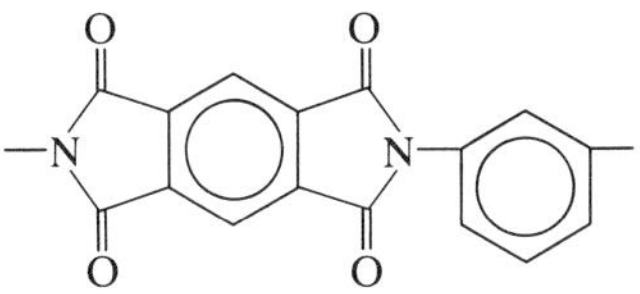	585[c]
Poly(phenylene oxide)	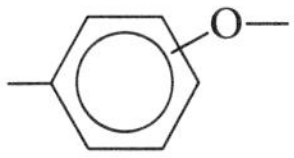	570
Polythiadiazole	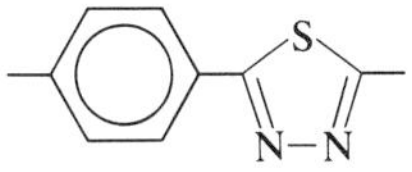	490
Poly(phenylene sulfide)	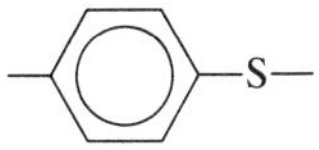	490

[a] Data from Korshak.[15]
[b] Nitrogen atmosphere unless otherwise indicated.
[c] Helium atmosphere.

and thermally stable organometallic polymers have been developed, many of which will be encountered later in the book. Few have found commercial acceptance because of a combination of high cost and poor processability. Because of their rigid backbone structure, aromatic polymers characteristically exhibit very high glass transition temperatures, high melt viscosities, and low solubility and are therefore more intractable than most other kinds of polymers. It is generally recognized that the upper limits of thermal stability of organic polymers have already been achieved, hence the emphasis in recent years has been to introduce structural variations that allow improved

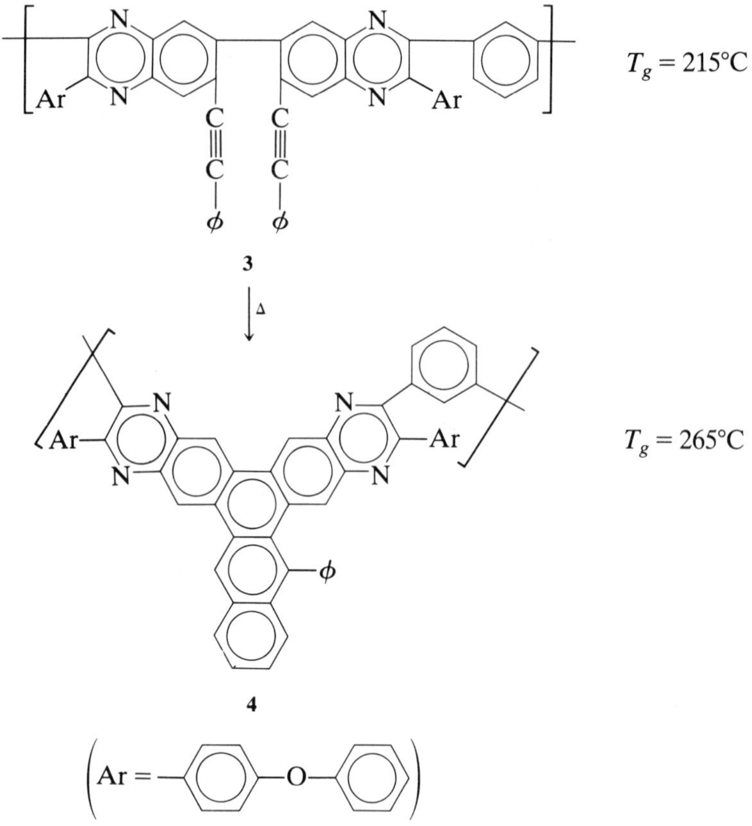

$T_g = 215°C$

3

$\downarrow \Delta$

$T_g = 265°C$

4

$\left(Ar = \underset{}{\bigcirc}\!\!-\!\!O\!\!-\!\!\bigcirc \right)$

Scheme 4.1. Increasing T_g of a polyquinoxaline by intramolecular cycloaddition.

processability. Incorporation of "flexibilizing" groups such as ether or sulfone into the backbone is one strategy. Although these measures often lead to greater solubility and lower viscosity, thermal stability usually suffers.

Another approach has been to introduce cyclic aromatic groups that lie *perpendicular* to the planar aromatic backbone, as in the polybenzimidazole **2**.[20] Such structures, referred to as *cardo polymers* (from the Latin *cardo*, loop),[21,22] usually exhibit improved solubility with no sacrifice of thermal properties.

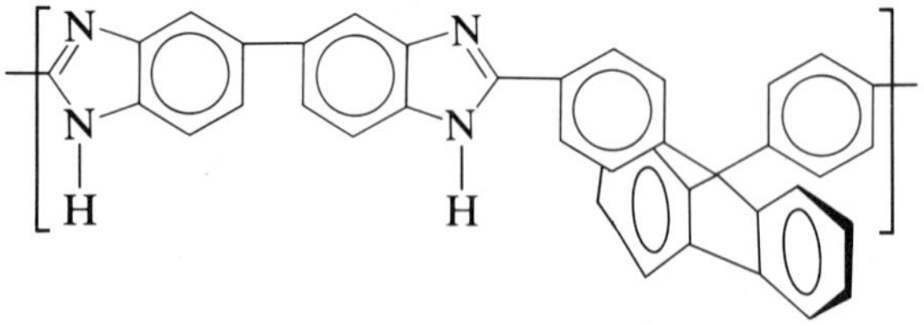

2

Introduction on the backbone of reactive groups that undergo intramolecular cycloaddition on heating is another way to improve processability. The rationale is that longer flow times are possible because little or no crosslinking takes place; rather, the glass transition temperature increases due to chain stiffening. An example, shown in Scheme 4.1, is the conversion of the aromatic polyquinoxaline **3** into **4** via complex cycloaddition reactions of the phenylethynyl substituents, resulting in a 50° increase in the glass transition temperature.[23] (Polyquinoxaline synthesis is given in Chapter 16.)

The most productive approach from the standpoint of commercial development has been the synthesis of aromatic oligomers or prepolymers capped with reactive end groups.[24] The end-capped oligomers melt at relatively low

Table 4.4. Reactive end groups for converting aromatic oligomers to network polymers[a]

Type	Structure
Ethynyl	$-C{\equiv}CH$
Phenylethynyl	$-C{\equiv}C\phi$
Phenylbutadiynyl	$-C{\equiv}C-C{\equiv}C\phi$
Phenylbutenynyl	$-C{\equiv}C-CH{=}CH\phi$
Biphenylene	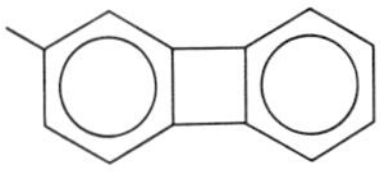
Styryl	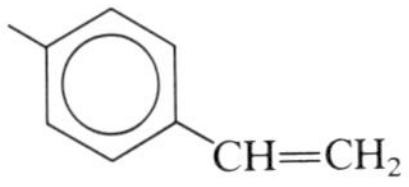
Maleimide	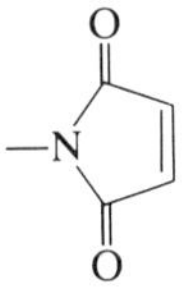
Nadimide (5-norbornene-2,3-dicarboximide)	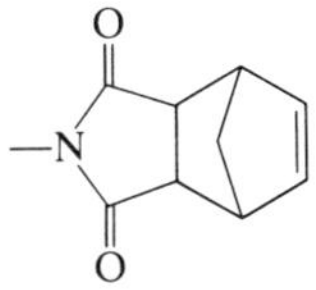
Cyanate	$-OCN$
N-Cyanourea	$-NHCNHCN$

[a] Data from Harris and Spinelli,[24] Hergenrother,[25] Harris et al.,[26] Droske and Stille,[27] Stille and Droske,[28]. Keszler et al.,[29] Delos et al.,[30] Stenzenberger et al.,[31] and Varma et al.[32]

temperature and are soluble in a variety of solvents. On heating they are converted to thermally stable network polymers. Representative end groups are given in Table 4.4.[24–32] The chemistry of network polymer formation usually involves cycloaddition or addition polymerization reactions of end groups. Examples will be encountered in later chapters.

4.5 Flammability and flame resistance

Because synthetic polymers are used increasingly in construction and transportation, considerable effort has been expended to develop nonflammable polymers, as well as to understand the mechanism of flame propagation and flame retardation.[33–35] Other concerns include the suppression of smoke and toxic gases formed during combustion, and the development of nonflammable textile fibers.

Some polymers are inherently nonflammable, for example, poly(vinyl chloride) and other polymers having a high halogen content. Others, such as polycarbonates and polyurethanes, will burn as long as a source of flame is present, but stop burning when the flame is removed. Such polymers are said to be *self-extinguishing*. Most polymers burn readily. Burning occurs in a series of steps. First, an external heat source increases the polymer temperature to a point where it begins to decompose and release combustible gases. Once the gases ignite, the temperature increases until the release of combustibles is rapid enough for combustion to be self-sustaining so long as sufficient oxygen is available to support the combustion process. With some polymers, such as polystyrene and poly(methyl methacrylate), the combustible gases may be high in monomer because of thermally induced depolymerization of the polymer chains. Monomer breaks down further to lower molecular weight combustible products, including hydrogen, as it diffuses toward the flame. Where depolymerization does not occur, surface oxidation plays a role in generation of combustible gases. A simplified picture of polymer burning, similar to that of a candle,[36] is depicted in Figure 4.11 as a closed cycle in

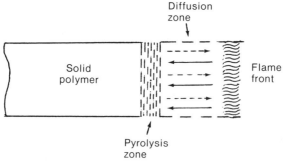

Figure 4.11. Representation of polymer combustion. --→, gas diffusion; ←—, heat flux. [Adapted from Factor.[36]]

which combustion of the diffusing gases generates heat that radiates back to sustain the pyrolysis.

Approaches to promoting flame resistance in polymers focus on three strategies: (1) retarding the combustion process in the vapor phase, (2) causing "char" formation in the pyrolysis zone, and (3) adding materials that decompose either to give nonflammable gases or endothermically to cool the pyrolysis zone.

In the vapor phase, combustion occurs by a complex series of free radical propagation and radical transfer reactions. Retardation of combustion may, therefore, be brought about by incorporating into the polymer *radical traps*— compounds that reduce the concentration of radicals in the vapor. Halogenated compounds are particularly effective because the hydrogen halide that is released reacts with free radicals to form less reactive halogen atoms. The commercially available anhydride, **5**, is one example.[37] Polyesters prepared

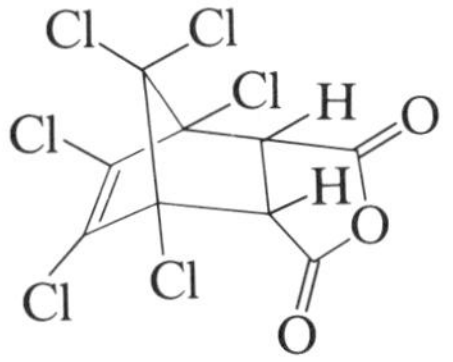

5

with **5** decompose on heating by the retrograde Diels–Alder reaction,

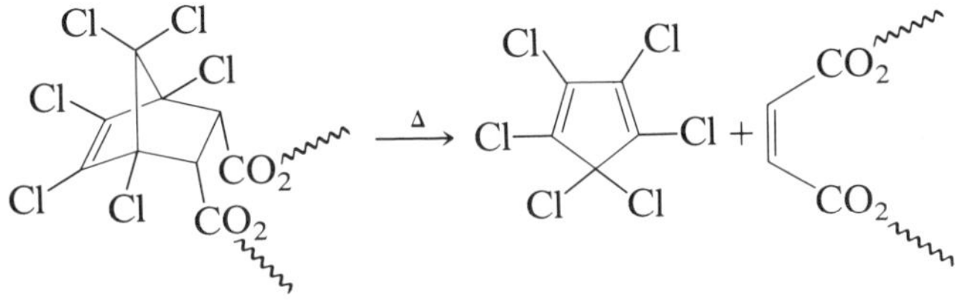

6

and the resultant hexachlorocyclopentadiene, **6**, suppresses radical formation in the vapor phase and inhibits free radical depolymerization in the pyrolysis zone. Antimony oxides are often used in combination with halogen compounds because of a synergistic effect believed to arise from formation of antimony halides. A serious disadvantage of halogen compounds is the toxicity of the hydrogen halide formed during burning.

Char formation at the polymer surface reduces flammability by acting as a barrier (principally carbonaceous) to inhibit gaseous products from diffusing to the flame, and to shield the polymer surface from the heat flux. Aromatic polymers have a natural tendency toward char formation, which accounts for their generally low flammability. Crosslinking may also increase char formation; introduction of chloromethyl groups onto polystyrene, for example, is believed to reduce the burning rate of that polymer primarily by promoting crosslinking during pyrolysis.[38] Certain additives, notably phosphorus-containing compounds, also increase char formation.[39,40] Phosphorus compounds are effective, for example, in reducing the flammability of cellulose by promoting dehydration to yield unsaturated compounds that subsequently polymerize to a crosslinked char. Thus cellulosic fibers, especially cotton, are made flame resistant by grafting with phosphorus-containing monomers. Flame retardants that promote formation of a carbonaceous char are referred to as *intumescent* flame retardants.[41]

The third strategy for reducing flammability involves using compounds such as hydrated alumina, $Al_2O_3 \cdot 3H_2O$, that evolve water endothermically to cool the pyrolysis zone, or sodium bicarbonate, which decomposes to form carbon dioxide which in turn dilutes the combustible gases.

Equally important in considerations of flammability are the generation of toxic decomposition products and smoke. Among the former are hydrogen cyanide evolved from burning nitrogen-containing polymers such as polyamides and polyurethanes (widely used in the manufacture of carpeting, bedding, and furniture upholstery), and hydrogen chloride from poly(vinyl chloride) (used in flooring, wall panels, and siding). Almost all polymers evolve copious amounts of carbon monoxide. Smoke arises from formation of acetylene and benzene, which condense to form soot particles. Ironically, flame retardants that operate in the vapor phase probably promote smoke formation by removing radicals that might otherwise oxidize the soot-forming particles. Although materials such as hydrated alumina are often effective in reducing smoke, toxic by-products of combustion continue to be a serious impediment to more widespread use of polymers in construction.

4.6   Chemical resistance

One of the problems that oil companies face with their huge petroleum storage tanks is rusting away of the metal bottom from underneath. Some companies have remedied this by spray-coating the inside floor of the tanks with glass fiber-reinforced unsaturated polyester (Chapter 12). The hard plastic coating significantly lengthens the tank's lifetime and avoids the expense of having to replace the tank bottom with steel, despite the fact that the polymer backbone contains hydrolyzable ester groups.

Not all polyesters have good hydrolytic stability, and people who coat tank bottoms are obviously going to choose a polymer that has proven to be highly resistant to water. With polyesters, two general approaches have been used to increase chemical resistance—one to increase the steric hindrance about the ester groups, and the other to reduce the number of ester groups per unit chain length. Both increase the hydrophobic nature of the polyesters. Two glycols that have found commercial application in chemically resistant polyester formulations are 2,2-4-trimethylpentane-1,3-diol (**7**) and the bisphenol A-propylene oxide derivative **8**.

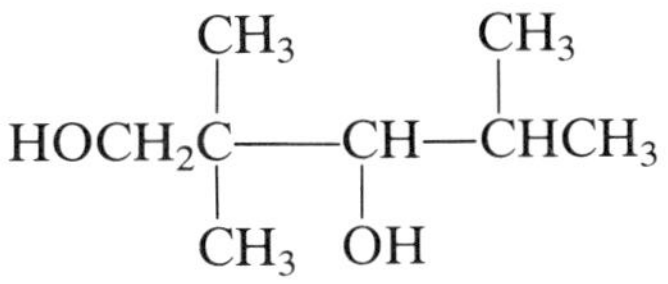

$$
\underset{\underset{\textstyle CH_3 \quad OH}{|\qquad\quad|}}{HOCH_2C}\text{—}\overset{\overset{\textstyle CH_3}{|}}{CH}\text{—}\overset{\overset{\textstyle CH_3}{|}}{CHCH_3}
$$

7

$$
HO\overset{\overset{\textstyle CH_3}{|}}{CH}CH_2O\text{—}\bigcirc\text{—}\overset{\overset{\textstyle CH_3}{|}}{\underset{\underset{\textstyle CH_3}{|}}{C}}\text{—}\bigcirc\text{—}OCH_2\overset{\overset{\textstyle CH_3}{|}}{CH}OH
$$

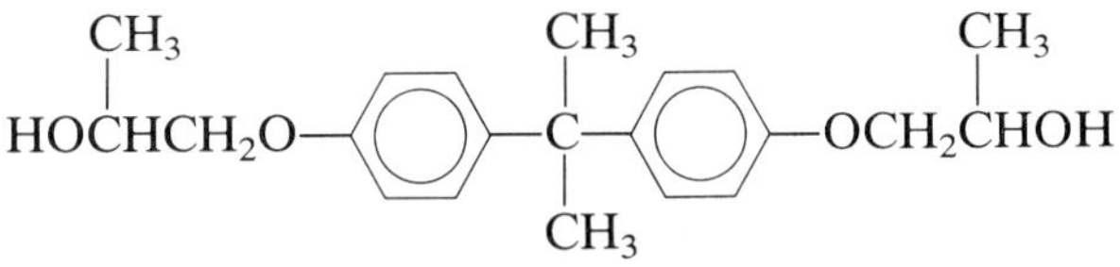

8

Another approach has been to reduce the hydrophilic nature of the end groups.[42] Phenyl isocyanate, for example, converts hydroxyl groups of polyesters to urethanes,

$$
\text{\Large\leftsquigarrow}OH + \phi NCO \longrightarrow \text{\Large\leftsquigarrow}O\overset{\overset{\textstyle O}{\|}}{C}NH\phi
$$

and carboxyl to amide,

$$
\text{\Large\leftsquigarrow}CO_2H + \phi NCO \longrightarrow \text{\Large\leftsquigarrow}\overset{\overset{\textstyle O}{\|}}{C}NH\phi + CO_2
$$

These examples illustrates how one type of polymer can be made more resistant to corrosive environments by the application of relatively straight-forward chemical principles.[43,44] Fluorine has proved to be an element that imparts both water and solvent resistance to a variety of polymers.[45] The inorganic polyphosphazene **9** (Chapter 15), which is very unstable in the presence of moisture, is rendered highly moisture resistant by conversion to **10**.[46] The fluorine, in effect, provides a water-resistant sheath to protect the phosphorus-nitrogen backbone. An ethylene-chlorotrifluoroethylene copolymer is marketed as a chemically resistant coating for underground cables. Because of their chemical inertness, a variety of fluorinated polymers including polytetrafluoroethylene **11**, poly(vinylidene fluoride) **12**, and copolymers such as poly[hexafluoropropylene-*co*-(vinylidene fluoride)] **13** have been developed commercially for use as gaskets, sealants, valves, and so on, where resistance to lubricating fluids is necessary.[47]

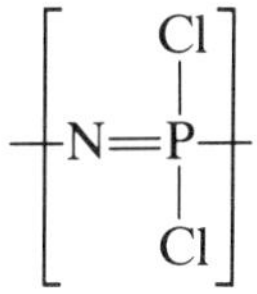

$$
\left[\begin{array}{c} Cl \\ | \\ \text{—}N{=}P\text{—} \\ | \\ Cl \end{array}\right] \qquad\qquad \left[\begin{array}{c} OCH_2CF_3 \\ | \\ \text{—}N{=}P\text{———} \\ | \\ OCH_2CF_3 \end{array}\right]
$$

9 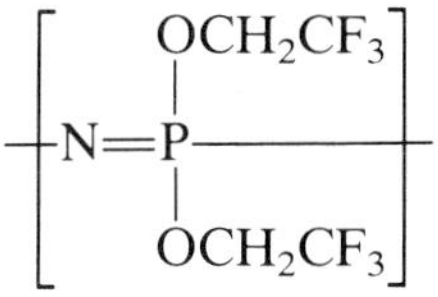   **10**

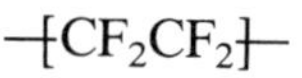

$$-\!\!+\!CF_2CF_2\!+\!\!- \qquad -\!\!+\!CH_2CF_2\!+\!\!- \qquad \text{\textasciitilde\textasciitilde}CF_2CF\text{\textasciitilde\textasciitilde\textasciitilde}CH_2CF_2\text{\textasciitilde\textasciitilde}$$

$$\underset{CF_3}{|}$$

11	**12**	**13**

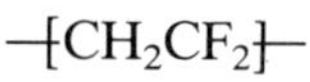

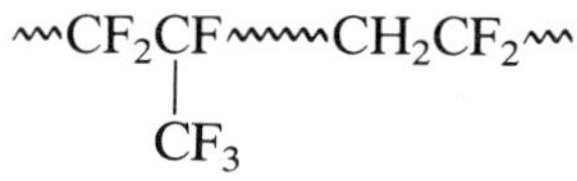

Ozone, formed by the action of ultraviolet light or electrical discharge on oxygen, degrades polymers containing double bonds in the backbone by ozonolysis followed by hydrolysis:

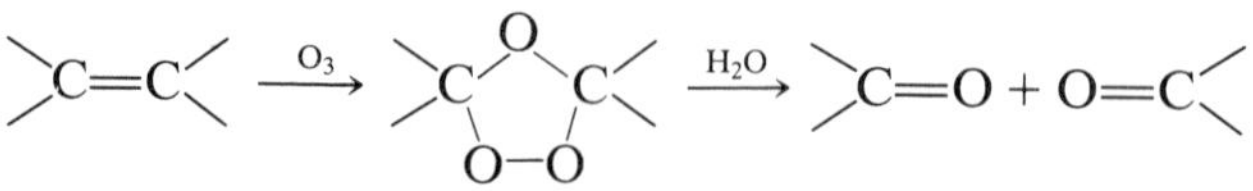

This can be a serious problem with many elastomers where double bonds are necessary to effect crosslinking. One way to circumvent this is to replace the more commonly used 1,3-butadiene or isoprene in the elastomer with a cyclic diene such as cyclopentadiene. This results in a cyclopentene unit (**14**) in the backbone by 1,4-addition. Ozonolysis may still occur, but the backbone remains intact.

14

Morphology is also an important variable in chemical resistance. Crystalline polymers are invariably more resistant than their amorphous counterparts because close chain packing reduces permeability. Similarly, crosslinking increases solvent resistance. One industry in which crosslinking is particularly important from the standpoint of chemical resistance is microelectronics.[48,49] One step in the manufacture of printed circuits[50,51] involves coating a substrate with a polymer that crosslinks under the influence of light or ionizing radiation. A *mask* carrying the pattern to be transferred to the substrate is placed over the coated surface, then the surface is irradiated. The pattern allows radiation through, and those portions of the polymer thus exposed undergo crosslinking. When the mask is removed, the unexposed parts of the polymer are dissolved with solvent, leaving behind the desired pattern. Such a process, illustrated schematically in Figure 4.12(a), is referred to as a *negative resist*—the exposed portion of the surface becomes *resist*ant to solvent.

Sunlight is another agent that can bring about polymer degradation. Monomers containing ultraviolet-absorbing chromophores such as 2,4-dihydroxy-4'-vinylbenzophenone (**15**) have been incorporated into vinyl poly-

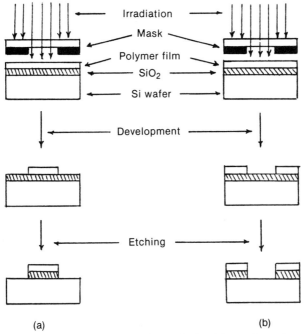

Figure 4.12. Schematic of a typical procedure for producing (a) negative resists and (b) positive resists in the manufacture of integrated circuits.

mers to improve light-stability.[52] Photolytic and chemical degradation of polymers is discussed more fully in Chapter 9.

15

4.7 Degradability

Most polymers are very durable; indeed, it is this property that allows them to compete with such other durable materials as glass and metals. As discussed in the previous section, research has focused on increasing the lifetime of certain polymers. Longevity can lead to problems, however. In recent years conservationists have become increasingly concerned with polymeric waste littering the landscape. Although synthetic polymers represent, at present, a relatively small percentage of the total solid waste, they are nevertheless highly visible, particularly in view of their widespread use in packaging. As a result, attention has shifted to the opposite end of the durability spectrum— to the synthesis of polymers that are degradable by the environmental effects of sunlight and soil microorganisms.[53,54]

Polymers can be made to degrade photochemically by incorporation of carbonyl groups that absorb ultraviolet (UV) radiation to form excited states energetic enough to undergo bond cleavage. Such processes (referred to as Norrish type II reactions) occur as follows[55]:

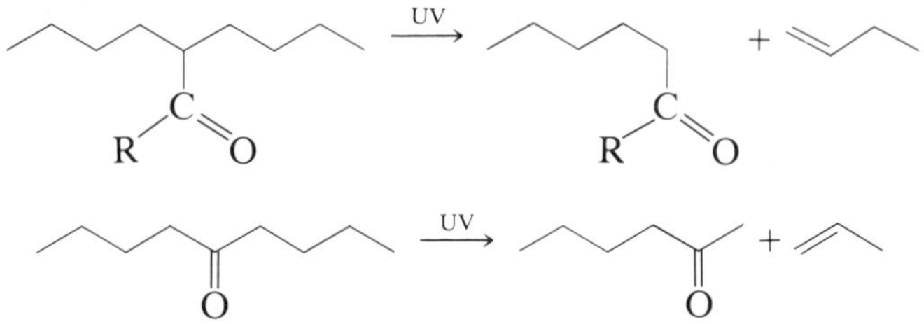

Commercially available photodegradable packing materials employ this technology. Similar degradation reactions occur with polyketoesters and polyketoamines.[56] Microorganisms degrade polymers by catalyzing hydrolysis and oxidation. The lower the molecular weight, the more rapidly the polymer degrades. A combination of light-sensitive and hydrolyzable functional groups are thus more effective in degrading high-molecular-weight polymers in natural environments.[57]

Although the initial impetus for degradable polymers arose from ecological concerns, considerable research is now directed toward resist technology and *controlled release* applications. In the former, degradable polymers are used for *positive resists*, which work in a sense opposite to that of the resists described in the previous section; that is, radiation promotes degradation of the resist exposed by the mask, leaving the unexposed coating intact.[50,51] This is illustrated in Figure 4.12(b).

Controlled release refers to the use of polymers containing agents of agricultural, medicinal, or pharmaceutical activity, which are released into the environment of interest at relatively constant rates over prolonged periods.[58–63] In the agricultural field, degradable mulches to promote crop growth are composed of combinations of natural polymers (which degrade readily in the presence of soil microorganisms) and synthetic polymers. Examples are starch-*graft*-poly(methyl acrylate)[64] and block copolymers of amylose or cellulose with polyesters.[65] At the completion of the growing season, such mulches may be plowed directly into the soil along with crop residues.

Another application involves binding of agriculture chemicals in polymer formulations for slow release at a rate effective for their intended purpose without the risk of the reagents being washed away by rain or irrigation. The herbicide 2,4-dichlorophenoxyacetic acid (2,4-D), for example, has been incorporated into polymers either as a chelate with iron (**16**)[66] or as a hydrolyzable pendant ester group on a vinyl polymer (**17**).[58a] A different strategy for controlled release based on polymer permeability rather than degradation

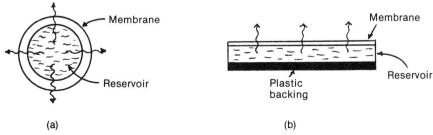

16   **17**

is to encapsulate the active reagent within a polymeric membrane or in a *strip*, as shown in Figure 4.13. Ideally the reagent is contained in the reservoir as a saturated solution with excess in suspension. This allows diffusion through the membrane at a constant rate without loss of activity. Alternatively, the reagent may be dispersed in a polymer matrix and released to the environment by diffusion or extraction. A variety of membrane and matrix devices are commercially available.[58b] Pheromone release strips for insect control[58c] and household fly and cockroach strips for release of insecticide[58d] are also in commercial use.

Encapsulated pharmaceuticals have been available for many years. Reservoir strips called *transdermal patches* are used to release drugs through the skin, for example, nitroglycerin to treat angina or scopolamine to combat motion sickness. Implanted polymeric matrix devices, including some made with degradable polymers to release the drug through surface erosion, are under investigation. Degradable polyesters are already used as disappearing

Figure 4.13. Membrane-controlled release devices: (a) microencapsulation, and (b) strip.

surgical sutures. The inorganic polyphosphazene (**9**), with amino acids, esters, and steroids in place of the chlorine atoms,[67] are of interest for slow release of the steroid, the other degradation products (amino acid, ester, phosphate, and ammonia) being nontoxic. (By contrast, the fluorinated polyphosphazene, **10**, is of interest for artificial organs and other prosthetic devices because of its inertness and compatibility with living tissue.)

4.8   Electrical conductivity

Most polymers are good insulators. Some, for example, poly(N-vinyl-carbazole)[68] (**18**), are *photoconducting*, that is, they conduct electricity to a

$$\left[\!\!-CH_2CH-\!\!\right]$$

18

small degree under the influence of iight, and are used in the electrophotography (photocopying) industry.[69] Other polymers undergo pyrolysis to yield chars that exhibit moderate conductivity.[70] A major discovery of the 1970s was that certain polymers, notably poly(sulfur nitride) (**19**) and polyacetylene (**20**) can be made highly conducting in the presence of certain additives, called *dopants*.[71-77] This major breakthrough initiated a major research effort to

$$-\!\!\left[S\!=\!N\right]\!\!-\qquad-\!\!\left[CH\!=\!CH\right]\!\!-$$

   19   **20**

elucidate the mechanism of conduction and to apply it to practical light-weight battery technology. Although a complete understanding of the conduction mechanism remains elusive, certain structural features are known to influence the level of conductivity. These include:

1. Delocalization. An extended conjugated system is usually necessary for backbone conductivity; however, charge may be transferred in some cases (e.g., **18**) through pendant groups.
2. Doping. Dopants may be electron acceptors such as arsenic pentafluoride or halogen, or electron donors such as alkali metals. Conductivity varies with dopant concentration. Doping may also effect rearrangement of the double bonds of nonconjugated polymers (e.g., polyisoprene) into a conjugated conducting mode.[78]
3. Morphology. Conduction is influenced by configurational and conformational factors, as well as crystallinity. Conductivity of polyacetylene film in the direction of molecular alignment is increased significantly by stretching.

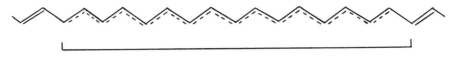

Soliton

Figure 4.14. Proposed conducting unit of polyacetylene. Soliton may be neutral (radical), positive (carbocation), or negative (carbanion).

Conductivity (σ) is usually expressed in units of reciprocal ohms per centimeter. Materials are classified as insulators ($\sigma < 10^{-8}$ ohm^{-1}cm^{-1}), semiconductors ($\sigma = 10^{-7}$–10^{-1} ohm^{-1}cm^{-1}), and conductors ($\sigma > [10^2$ ohm^{-1}cm^{-1}). Ordinary poly(sulfur nitride), **19**, has a conductivity of about 100 ohm^{-1}cm^{-1}, although at very low temperatures (<0.3 K) it exhibits superconductivity. Polyacetylene's intrinsic conductivity is 1.7×10^{-9} ohm^{-1}cm^{-1} for the *cis* isomer (**21**) and 4.4×10^{-5} ohm^{-1}cm^{-1} (weakly semiconducting) for the *trans* isomer (**22**). Doping increases poly-

21   **22**

acetylene's conductivity dramatically into the range of metal conductors. Doping apparently converts the red metallic-colored *cis* isomer to blue metallic *trans*. The dopant forms a charge transfer complex with the polymer that is believed to give rise to highly delocalized cation radicals or anion radicals depending on whether the dopant is electron accepting or donating, respectively; for example,

$$2\text{-}\!\!\left[\text{CH}=\text{CH}\right]\!\! + 3\ I_2 \longrightarrow 2\ \left[\text{CH}\text{=}\!\!\!\text{=}\text{CH}\right]^{+\cdot} + 2\ I_3^-$$

$$\left[\text{CH}=\text{CH}\right]\! + \text{Na} \longrightarrow \left[\text{CH}\text{=}\!\!\!\text{=}\text{CH}\right]^{-\cdot} + \text{Na}^+$$

According to one theory of polyacetylene conductivity, the delocalized regions, called solitons* (Figure 4.14), extend about 15 bond lengths, at which point the energy gain arising from lengthening of double bonds and shortening of single bonds apparently outweighs stabilization arising from delocalization. Conduction is believed to involve movement of electrons intramolecularly and intermolecularly via the positive or negative solitons. Intermolecular conduction is highly dependent on crystallinity, which accounts for the increase in polyacetylene's conductivity (to as much as 1.5×10^5 ohm^{-1}cm^{-1}) when films of the polymer are properly oriented.

* From *solitary wave solutions*, a mathematical treatment of conductivity.

Table 4.5. Conductivities of doped polymers and metals[a]

Material	Dopant	Density (d) $(g\ cm^{-3})$	Conductivity (σ) $(ohm^{-1}\ cm^{-1})$	Specific conductivity (σ/d) $(cm^2\ ohm^{-1}\ g^{-1})$
Copper	—	8.92	5.8×10^5	6.5×10^4
Polyacetylene	AsF$_5$	0.8	1.5×10^5	1.9×10^5
Gold	—	19.3	4.1×10^5	2.1×10^4
Mercury	—	13.5	1×10^4	7.4×10^2
Poly(sulfur nitride)	Br$_2$	2.67	3.8×10^4	1.4×10^4
Poly(p-phenylene)	AsF$_5$	1.39	500	360
Polypyrrole	I$_2$	1.48	100	68
Polyphthalocyaninesiloxane	I$_2$	1.4	1.4	1
Poly(phenylene sulfide)	AsF$_5$	1.35	1	0.74

[a] Data from Mort,[73] Seymour,[74] and Wegner,[75] and from *Chem. Eng. News*, June 22, 1987, p. 20.

Other polymers that exhibit conduction on doping include poly(p-phenylene) (**23**), poly(phenylene sulfide) (**24**), polypyrrole (**25**), and metal phthalocyanine polymers (**26**), which lack a conjugated backbone but undergo electron transport through the cofacially "stacked" heterocyclic rings.[76] Conductivities of these various polymers compared with those of metal conductors are given in Table 4.5. *Specific conductivities* (conductivity per unit density) are of interest when weight is an important consideration, as in batteries. Polyacetylene, being readily available and exhibiting high conductivity, has received the most attention.[79] Polyacetylene is, however, less stable than the other organic polymers and is difficult to process, although strategies for improving the processability are being developed (see Chapter 9).

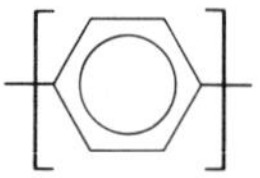

23

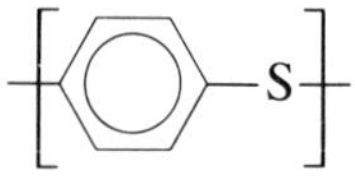

24

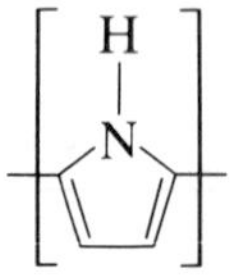

25

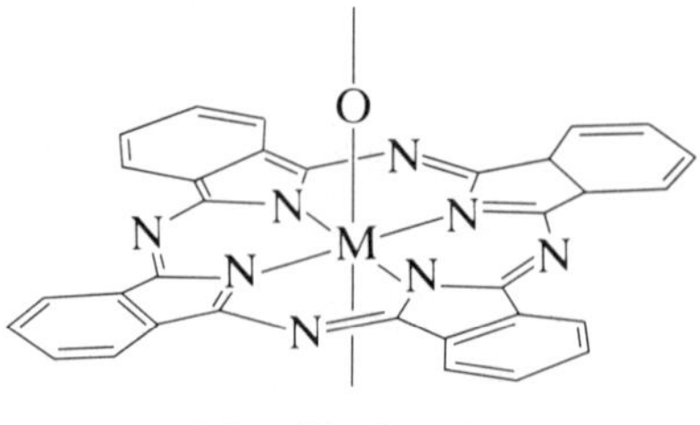

M = Si, Ge, Sn

26

Another area of conductivity where polymers are of interest is that of solid electrolytes.[80] *Polyelectrolytes* for solid battery applications generally consist of polymers such as poly(ethylene oxide) (**27**) or the polyphosphazene (**28**) in combination with metal salts. A 4:1 ratio of **27** to $NaBF_4$ is typical. To function as an effective polyelectrolyte, the polymer should be highly amorphous and have a low glass transition temperature to allow the freedom of molecular movement necessary for ion transport.

$$-\!\!\left[CH_2CH_2O\right]\!\!- \qquad -\!\!\left[\begin{array}{c}(OCH_2CH_2)_2OCH_3\\ |\\ P\!\!=\!\!N \\ |\\ (OCH_2CH_2)_2OCH_3\end{array}\right]\!\!-$$

27   28

4.9   Additives[81–83]

Virtually every polymer in commercial use contains additives, often a combination. Their purpose is twofold: (1) to alter the properties of the polymer and (2) to enhance processability. Those with the former purpose range from pigments and odorants used for aesthetic reasons to plasticizers for modifying mechanical properties. Those with the latter purpose vary from lubricants to prevent sticking to fabrication machinery to compounds that alter the chemical structure, such as crosslinking agents. Additives may be completely miscible or, as is the case with reinforcing agents, completely immiscible. Sometimes additives reduce cost. Wood flour, a by-product of saw mills, is added to phenol-formaldehyde polymers as a cost-reducing *extender*, as well as to reduce brittleness. Additives are a major industry; more than 10^9 kg are consumed in the United States alone every year.

Table 4.6 lists commercially important additives according to function. Some appear in more than one category. Plasticizers, for example, increase flexibility, but also reduce the melt viscosity to facilitate molding or extruding. Slip agents minimize adhesion to processing equipment, and at the same time prevent film or sheet from sticking together. Some of the additives listed, such as nucleating agents and flame retardants, have already been mentioned. Others, including crosslinking agents, promoters, emulsifiers, antioxidants, and blowing agents, are covered in later chapters. A number of the additives, particularly surface modifiers such as waxes and surfactants, have limited solubility in the polymers, which causes them to migrate to the surface where their activity is needed.

Plasticizers[84,85] are the most widely used, with di-2-ethylhexyl phthalate (**29**) being the cheapest "general purpose" plasticizer. Other commonly used plasticizers are listed in Table 4.7. About 80% of all plasticizer consumption goes into a single polymer—poly(vinyl chloride). It is believed that the

Table 4.6. Polymer additives

Type	Function
Mechanical property modifiers	
Plasticizers	Increase flexibility
Impact modifiers	Improve impact strength
Reinforcing fillers	Increase strength properties
Nucleating agents	Modify crystalline morphology
Surface property modifiers	
Slip and antiblocking agents	Prevent film and sheet sticking
Lubricants	Prevent sticking to machinery
Antistatic agents	Prevent static charge on surfaces
Coupling agents	Improve bonding between polymer and filler
Wetting agents	Stabilize dispersions of filler
Antifogging agents	Disperse moisture droplets on films
Chemical property modifiers	
Flame retardants	Reduce flammability
Ultraviolet stabilizers	improve light stability
Antioxidants	Prevent oxidative degradation
Biocides	Prevent mildew
Aesthetic property modifiers	
Dyes and pigments	Impart color
Odorants	Add fragrance
Deodorants	Prevent development of odor
Nucleating agents	Improve light transmission
Processing modifiers	
Plasticizers	Reduce melt viscosity
Slip agents and lubricants	Prevent sticking to processing machinery
Low-profile additives	Prevent shrinkage and warpage
Thickening agents	Increase viscosity of polymer solutions or dispersions
Heat stabilizers	Prevent degradation during processing
Defoaming agents	Reduce foaming
Blowing agents	Manufacture stable foams
Emulsifiers	Stabilize polymer emulsions
Crosslinking (curing) agents	Crosslink polymer
Promoters	Speed up crosslinking (curing)

Table 4.7. Commonly used plasticizers

Aromatic
   Di-2-ethylhexyl phthalate
   Di-*n*-octyl phthalate
   Di-*i*-octyl phthalate
   Di-*i*-decyl phthalate
   Di-*n*-undecyl phthalate
   Di-*n*-tridecyl phthalate
   Tri-2-ethylhexyl trimellitate
Aliphatic
   Di-2-ethylhexyl adipate
   Di-2-ethylhexl sebacate
   Di-2-ethylhexyl azelate
Epoxy
   Epoxidized linseed oil
   Epoxidized soya oil
Polymeric
   Poly(alkylene adipates, sebacates, or azelates)
Fire retardant
   Chlorinated paraffins
   Phosphate esters

thermal motion of the low-molecular-weight plasticizers increases the polymer's free volume, allowing more "elbow room" for increased long-range segmental motion of the polymer molecules. For some applications, plasticizer comprises almost 50% of the formulation and, in the case of poly(vinyl chloride), the glass transition temperature may be reduced from about 80°C for unplasticized polymer to below 0°C. For reasons of compatibility, plasticizers are not used with highly crystalline polymers.

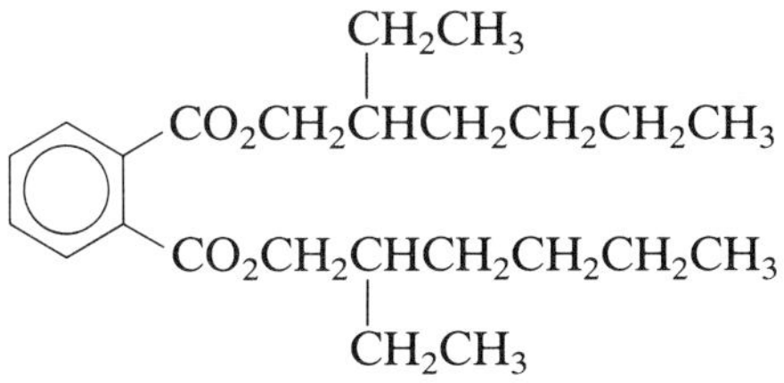

$$CH_2CH_3$$
$$|$$
$$CO_2CH_2CHCH_2CH_2CH_2CH_3$$

$$CO_2CH_2CHCH_2CH_2CH_2CH_3$$
$$|$$
$$CH_2CH_3$$

29

A number of criteria are considered in choosing a plasticizer, including cost, compatibility, stability, processability, and *permanence*. Permanence refers to how well the plasticizer remains in the polymer: Will it vaporize on heating or be extracted if the plastic comes into contact with solvents or lubricants? An example is the use of phthalate esters in automobile interiors. The "new car smell" enjoyed by many car owners results mainly from plasticizer vaporized in the closed car interior, and actually advertises the deterioration of the vinyl upholstery. If the car is left in the sun and becomes very hot, enough plasticizer may be released to fog the windows with a greasy film. In time the upholstery becomes brittle and cracks. For this reason, higher molecular weight phthalates are commonly used for modern car interiors.

Permanence is a phenomenon related to compatibility. Solubility factors, discussed in Chapter 2 for polymer solutions, apply equally as well to plasticizers, except that now the polymer is the major constituent of the solution. Thus a plasticizer having "poor" solubility will reduce the viscosity of a polymer more than one having "good" solubility at equal levels of dilution; however, this reduced compatibility also results in less permanence. Obviously a balance must be struck between the two. In some instances even water functions as a plasticizer, particularly where highly hydrophilic functional groups are present in the polymer. Cotton (cellulose) fibers (Chapter 17) and polyamides (Chapter 13) represent two common polymers in which the effect of water must be taken into account by polymer manufacturers.

Aliphatic esters are usually used for low temperature flexibility, and epoxy plasticizers are used where good heat and light stability are desirable. Polymeric plasticizers are obviously less volatile, but their relatively high cost limits their use. Polymers plasticized with polymeric plasticizers are, in effect, polyblends[85] (Section 3.10).

The plasticizers just described are *external* plasticizers. *Internal* plasticization is brought about by chemically binding the flexibilizing moiety to the polymer.

Impact modifiers are elastomeric polymers such as polybutadiene. *Impact polystyrene*, for example, is an immiscible polyblend of polystyrene and 5 to 10% polybutadiene. Impact polystyrene is much tougher than polystyrene alone, and accounts for more than half the polystyrene produced.

Reinforcing fillers come in two forms—fibrous and granular (or powder). Carbon black, used to reinforce natural and synthetic rubber, is an example of the latter. Glass fiber is widely used to increase strength properties, particularly in combination with unsaturated polyesters. Such materials are termed *composites*.[87–89] In recent years the development of very-high-strength composites[90] using aromatic polyamide and graphite fibers in combination with engineering plastics has resulted in a wide range of products from tennis rackets and fishing poles to high-performance aerospace components.

The remaining additives listed in Table 4.6 comprise a much smaller percentage of the overall polymer formulation. Details of their composition and function may be obtained from the references provided.

References

1. D. H. Morton-Jones, *Polymer Processing*, Chapman and Hall, New York, 1989.
2. W. E. Becker (ed.), *Reaction Injection Molding*, Van Nostrand Reinhold, New York, 1979.
3. J. E. Kresta (eds.), *Reaction Injection Molding*, ACS Symp. Ser. 270, American Chemical Society, Washington, D. C., 1985.
4. C. Macosko, *Fundamentals of Reaction Injection Molding*, Hanser, Munich, 1988.
5. F. W. Billmeyer, *Textbook of Polymer Science*, 3rd ed., Wiley-Interscience, New York, 1984, Part 6.
6. F. Rodriguez, *Principles of Polymer Systems*, McGraw-Hill, New York, 1970, Chap. 12.
7. I. M. Ward, *Mechanical Properties of Polymers*, Wiley-Interscience, New York, 1971.
8. A. Rudin, *The Elements of Polymer Science and Engineering*, Academic Press, New York, 1982.
9. R. P. Kampour and R. E. Roberton, in *Polymer Science* (A. D. Jenkins, ed.), Vol. 1, North-Holland, Amsterdam, 1972, Chap. 11.
10. J. Agranoff (ed.), *Modern Plastics Encyclopedia, 1984–1985*, Vol. 61, McGraw-Hill, New York, 1984: (a) pp. 449–704; (b) pp. 106ff.
11. V. A. Marichin, L. P. Mjasnikova, D. Zenke, R. Hirte, and P. Weigel, *Polym. Bull.*, **12**, 287 (1984).
12. P. Smith, H. D. Chanzy, and B. P. Rotzinger, *Polym. Commun.*, **26**, 258 (1985).
13. P. E. Cassidy, *Thermally Stable Polymers*, Dekker, New York, 1980.
14. J. P. Critchley, G. J. Knight, and W. W. Wright, *Heat-Resistant Polymers*, Plenum Press, New York, 1983.
15. V. V. Korshak, *Heat-Resistant Polymers*, Halstead Press, New York, 1972.
16. C. Arnold Jr., *J. Polym. Sci., Macromol. Rev.*, **14**, 265 (1979).
17. C. G. Overberger and J. A. Moore, *Adv. Polym. Sci.*, **7**, 113 (1970).
18. B. C. Anderson and R. D. Lipscomb, *Macromolecules*, **17**, 1641 (1984).
19. G. M. Moelter, R. F. Tetrault, and N. Heffland, *Polym. News*, **9**, 134 (1983).
20. P. R. Srinivasan, V. Mahadevan, and M. Srinivasan, *J. Polym. Sci., Polym. Chem. Ed.*, **20**, 3095 (1982).
21. J. K. Stille, R. M. Harris, and S. M. Padaki, *Macromolecules*, **14**, 486 (1981).
22. V. V. Korshak, S. V. Vinogradova, and Y. S. Vygodskii, *Rev. Macromol. Chem.*, **12**, 45 (1974–1975).
23. F. L. Hedberg and F. E. Arnold, *J. Polym. Sci., Polym. Chem. Ed.*, **14**, 2607 (1976).
24. F. W. Harris and H. J. Spinelli (eds.), *Reactive Oligomers*, ACS Symp. Ser. 282, American Chemical Society, Washington, D.C., 1985.
25. P. N. Hergenrother, *J. Macromol. Sci., Rev. Macromol. Chem.*, **C19**, 1 (1980).
26. F. W. Harris, A. Pamidimukkala, R. Gupta, S. Das, T. Wu, and G. Mock, *J. Macromol. Sci.-Chem.*, **A21**, 1117 (1984).
27. J. P. Droske and J. K. Stille, *Macromolecules*, **17**, 1 (1984).
28. J. K. Stille and J. P. Droske, *J. Macromol. Sci.-Chem.*, **A21**, 913 (1984).
29. B. Keszler, V. S.C. Chang, and J. P. Kennedy, *J. Macromol. Sci. Chem.*, **A21**, 307 (1984).

30. S. E. Delos, R. K. Shellenberg, J. E. Smedley, and D. E. Kranbuehl, *J. Appl. Polym. Sci.*, **27**, 4295 (1982).
31. H. D. Stenzenberger, M. Hertzog, W. Römer, R. Scheiblich, and N. J. Reeves, *Br. Polym. J.*, **15**, 1 (1983).
32. I. K. Varma, Sangita, and D. S. Varma, *J. Polym. Sci., Polym. Chem. Ed.*, **22**, 1419 (1984).
33. J. W. Lyons, *The Chemistry and Uses of Fire Retardants*, Wiley-Interscience, New York, 1970.
34. M. Lewin, S. M. Atlas, and E. M. Pearce, *Flame Retardant Polymeric Materials*, Plenum Press, New York, 1975.
35. E. M. Pearce, Y. P. Khanna, and D. Raucher, in *Thermal Characterization of Polymeric Materials* (E. A. Turi, ed.), Academic Press, New York, 1981, Chap. 8.
36. A. Factor, *J. Chem. Educ.*, **51**, 453 (1974).
37. C. T. Vijayakumar and J. K. Fink, *J. Appl. Polym. Sci.*, **27**, 1629 (1982).
38. Y. P. Khanna and E. M. Pearce, *Flame Retard. Polym. Mat.*, **2**, 43 (1978).
39. R. Liepins, J. R. Surles, N. Morosoff, V. Stannett, J. J. Duffy, and F. H. Day, *J. Appl. Polym. Sci.*, **22**, 2403 (1978).
40. J. A. Albright and C. K. Kmiec, *J. Appl. Polym. Sci.*, **22**, 2451 (1978).
41. G. Montaudo, E. Scamporrino, and D. Vitalini, *J. Polym. Sci., Polym. Chem. Ed.*, **21**, 3361 (1983).
42. W. K. Seifert, M. P. Stevens, J. D. Gardner, and D. F. Percival, *J. Appl. Polym. Sci.*, **9**, 1681 (1965).
43. R. B. Seymour and R. H. Steiner, *Plastics for Corrosion Resistant Applications*, Reinhold, New York, 1954.
44. Yu. V. Moiseev and G. E. Zaikov, *Chemical Resistance of Polymers in Aggressive Media*, Consultants Bureau, Plenum Press, New York, 1987.
45. R. E. Uschold, *Polym. J.*, **17**, 253 (1985).
46. H. R. Allcock, *Chem. Rev.*, **72**, 315 (1972).
47. C. A. Sperata, in *Modern Plastics Encyclopedia*, 1984–85, Vol. 61 (J. Agranoff, ed.), McGraw-Hill, New York, 1984, p. 24.
48. T. Davidson (ed.), *Polymers in Electronics*, ACS Symp. Ser. 242, American Chemical Society, Washington, D. C., 1984.
49. M. J. Bowden and S. R. Turner (eds.), *Polymers for High Technology: Electronics and Photonics*, ACS Symp. Ser. 346, American Chemical Society, Washington, D.C., 1987.
50. H. Steppen, G. Buhr, and H. Vollmann, *Angew. Chem. Int. Ed. English*, **21**, 455 (1982).
51. W. Schnabel and H. Sotobayashi, *Prog. Polym. Sci.*, **9**, 297 (1983).
52. D. Bailey, D. Tirrell, C. Pinazzi, and O. Vogl, *Macromolecules*, **11**, 312 (1978).
53. J. E. Guillet (ed.), *Polymers and Ecological Problems*, Plenum Press, New York, 1973.
54. G. S. Kumar, V. Kalpagam, and U.S. Nandi, *J. Macromol. Sci. Rev. Macromol. Chem. Phys.*, **C22**, 225 (1982–1983).
55. B. Rånby and J. F. Rabek, *Photodegradation, Photooxidation and Photostabilization of Polymers*, Wiley-Interscience, New York, 1975, pp. 48–49.
56. S. J. Huang and C. A. Byrne, *J. Appl. Polym. Sci.*, **27**, 2467 (1982).
57. S. J. Huang, C. A. Byrne, and J. A. Pavlisko, in *Modification of Polymers* (C. E. Carraher Jr. and M. Tsuda, eds.), ACS Symp. Ser. 121, American Chemical Society, Washington, D.C. 1980, p. 299.

58. D. R. Paul and F. W. Harris (eds.), *Controlled Release Polymeric Formulations*, ACS Symp. Ser. 33, American Chemical Society, Washington, D.C., 1976: (a) F. W. Harris, A. E. Aulabaugh, R. D. Case, M. K. Dykes, and W. A. Feld, pp. 222ff; (b) D. R. Paul, pp. 1ff; (c) B. A. Bierl, E. D. DeVilbis, and J. R. Plimmer, pp. 265ff; (d) A. F. Kydonieus, A. R. Quisumbing, and S. Hyman, pp. 295ff.
59. J. Kopeček and K. Ulbrich, *Prog. Polym. Sci.*, **9**, 1 (1983).
60. R. Duncan and J. Kopeček, *Adv. Polym. Sci.*, **57**, 51 (1984).
61. C. L. McCormick, K. W. Anderson, and B. H. Hutchinson, *J. Macromol. Sci.-Rev. Macromol. Chem. Phys.*, **C22**, 57 (1982–1983).
62. C. G. Gebelein and F. F. Koblitz (eds.), *Biomedical and Dental Applications of Polymers*, Plenum Press, New York, 1981.
63. C. G. Gebelein and C. E. Carraher Jr. (eds.), *Polymeric Materials in Medication*, Plenum Press, New York, 1985.
64. R. J. Dennenberg, R. J. Bothast, and T. P. Abbott, *J. Appl. Polym. Sci.*, **22**, 459 (1978).
65. K. S. Lee, V. T. Stannett, and R. D. Gilbert, *J. Polym. Sci., Polym. Chem. Ed.*, **20**, 997 (1982).
66. M. L. Beasley and R. L. Collins, *Science*, **169**, 769 (1970).
67. H. R. Allock, *J. Polym. Sci., Polym. Symp.*, **70**, 71 (1983).
68. R. C. Penwell, B. N. Ganguly, and T. W. Smith, *Macromol. Rev.*, **13**, 63 (1978).
69. J. Weigl, *Angew. Chem. Int. Ed. English*, **16**, 374 (1977).
70. T. M. Keller, *J. Polym. Sci., Polym. Chem.*, **25**, 2569 (1987).
71. T. A. Skotheim (ed.), *Handbook of Conducting Polymers*, Vols. 1 and 2, Dekker, New York, 1986.
72. C. K. Chiang, M. A. Druy, S. C. Gau, A. J. Heeger, E. J. Louis, A. G. Mac-Diarmid, Y. W. Park, and H. Shirakawa, *J. Am. Chem. Soc.*, **100**, 1013 (1978).
73. J. Mort, *Science*, **208**, 819 (1980).
74. R. B. Seymour (ed.), *Conductive Polymers*, Plenum Press, New York, 1981.
75. G. Wegner, *Makromol. Chem., Macromol. Symp.*, **1**, 151 (1986).
76. C. W. Dirk, T. Inabe, J. W. Lyding, K. F. Schoch Jr., C. W. Kannewurf, and T. J. Marks, *J. Polym. Sci., Polym. Symp.*, **70**, 1 (1983).
77. R. H. Baughman, J. L. Bredes, R. R. Chance, R. L. Elsenbaumer, and L. W. Shacklette, *Chem. Rev.*, **82**, 209 (1982).
78. M. Thakur, *Macromolecules*, **21**, 661 (1988).
79. J. C. W. Chien, *Polyacetylene*, Academic Press, New York, 1984.
80. H. Charadame, in *Macromolecules* (H. Benoit and P. Rempp, eds.), Pergamon Press, New York, 1982, p. 226.
81. L. Mascia, *The Role of Additives in Plastics*, Halstead Press, New York, 1975.
82. J. Štěpek and H. Daoust, *Additives for Plastics*, Springer-Verlag, New York, 1983.
83. J. T. Lutz (ed.), *Thermoplastic Polymer Additives*, Dekker, New York, 1988.
84. I. Mellan, *Industrial Plasticizers*, Macmillan, New York, 1963.
85. J. K. Sears and J. R. Darby, *The Technology of Plasticizers*, SPE Monograph Series, Wiley-Interscience, New York, 1982.
86. C. F. Hammer, in *Polymer Blends*, Vol. 2 (D. R. Paul and S. Newman, eds.), Academic Press, New York, 1978, Chap. 17.
87. R. P. Sheldon, *Composite Polymeric Materials*, Applied Science, New York, 1982.

88. D. Hull, *Introduction to Composite Materials*, Cambridge University Press, Cambridge, 1981.
89. R. B. Seymour and R. D. Deanin (eds.), *History of Polymeric Composites*, VNU Science Press, Utrecht, 1987.
90. P. Beardmore, J. J. Harwood, K. R. Kinsman, and R. E. Robertson, *Science*, **208**, 833 (1980).

Review exercises

1. Explain the difference between
   (a) tensile strength and flexural strength.
   (b) tensile stress and tensile modulus.
   (c) extrusion and spinning.
   (d) injection molding and compression molding.
   (e) creep and stress relaxation.
   (f) thermal stability and thermooxidative stability.
   (g) positive resist and negative resist.

2. Explain, in terms of structure and morphology, what is happening in the various regions of the stress-strain curve depicted in Figure 4.8.

3. What effect would you expect each of the following to have on creep:
   (a) temperature
   (b) molecular weight
   (c) crosslinking

4. What effect would you expect oxidation to have on stress relaxation over prolonged periods of time?

5. Explain, in terms of structure and morphology, the rationale behind the three strategies described for improving processability of thermally stable polymers: (a) cardo polymers, (b) intramolecular cyclization, (c) reactive oligomers.

6. Poly(vinyl alcohol) is significantly less flammable than the "isomeric" polyoxyethylene. Suggest a reason.

7. The cyanate group (Table 4.4) trimerizes on heating to triazole:

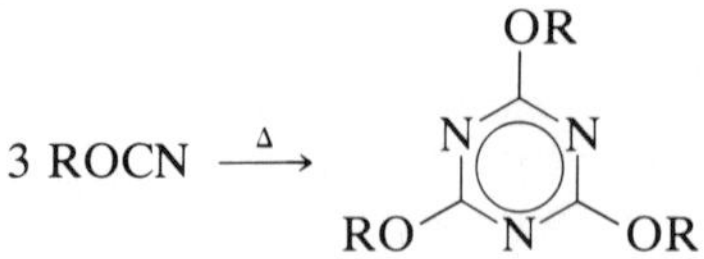

$$3 \text{ ROCN} \xrightarrow{\Delta}$$

Describe how one might make a semi-IPN (see Chap. 3) employing a cyanate-terminated monomer or oligomer. (Allied Chemical has developed such a product: see *Plast. Technol.*, March, 1985, p. 26; or H. A. Witcoff, *CHEMTECH*, **17**, 156, 1987.)

8. In elastomer synthesis, 1,4-hexadiene may be used to impart the unsaturation needed for crosslinking. (J. Lal and P. H. Sandstrom, *J. Polym. Sci., Polym. Lett. Ed.*, **13**, 83, 1975). What advantage might this compound offer over butadiene or isoprene?

9. Tygon plastic tubing is commonly used in the laboratory for transporting water or gases. If organic solvents are run through the tubing, it soon becomes stiff. What is the most likely reason?

10. An exercise at the end of Chapter 3 dealt with ironing polyester and cotton fabrics. Why is steam commonly used in ironing? In the same vein, why does paper (almost 100% cellulose) turn "limp" in very humid weather?

5

Evaluation, characterization, and analysis of polymers

5.1 Introduction

Characterizing polymers is much more complex than characterizing low-molecular-weight compounds.[1,2] Apart from their basic chemical make-up, a variety of configurational and conformational variables relate to their usefulness. If the polymer contains pendant groups, are the groups stereoregular or randomly oriented? Is the polymer linear or branched? If more than one monomeric unit is identified, are the units distributed randomly or in blocks or grafts? What is the average molecular weight, and does the average represent a narrow or a wide (polydisperse) distribution? Is the polymer crosslinked? What is the nature of the end groups? Which additives are present, and at what concentration? If a polymer is being considered for a particular application, how well will it stand up to its use environment? Is it stable enough to withstand the mechanical and thermal stresses of extrusion or molding? Is the polymer tough or brittle? How tough? Will it retain its properties on being heated? To answer this multitude of questions, a broad range of evaluative and analytical procedures must be brought into play.[3]

This chapter presents a necessarily brief introduction to many of the analytical and evaluative tools available to the polymer chemist (excluding those covered in Chapter 2 for determining molecular weight). Major focus is placed on spectroscopic and thermal methods because these are used most frequently by the polymer chemist. We shall also explore techniques for surface analysis, a discipline that has assumed increasing importance in the polymer field in recent years. Mechanical and electrical testing are discussed briefly. Because test methods may vary substantially from one laboratory to another, organizations such as the American Society for Testing and Materials (ASTM) and the British Standards Institute (BSI) provide detailed standardized tests that are used throughout the polymer industry.

It is assumed here that readers, in the course of their studies, have received adequate exposure to the fundamentals of chromatography and spectroscopy [infrared, ultraviolet, and proton and carbon nuclear magnetic resonance (NMR)], hence the discussion in this chapter is concerned only with the application of these techniques to polymer characterization. Fundamentals of the less commonly encountered instrumental methods are provided. To keep things in perspective, it should be remembered that "wet" chemical analysis lay at the heart of early structural studies of polymers and is still a valuable complement to the instrumental techniques of today. We begin, therefore, with a short discussion of chemical analysis.

5.2 Chemical methods of analysis[2,4]

Among the earliest questions of concern to polymer chemists were those of backbone structure, for example the spacing of substituent groups in vinyl polymers and the location of double bonds in polydienes. That the head-to-tail structure predominates in poly(vinyl alcohol) was demonstrated convincingly by reaction with periodic acid or lead tetraacetate, which causes oxidative cleavage of 1,2-diols[5]:

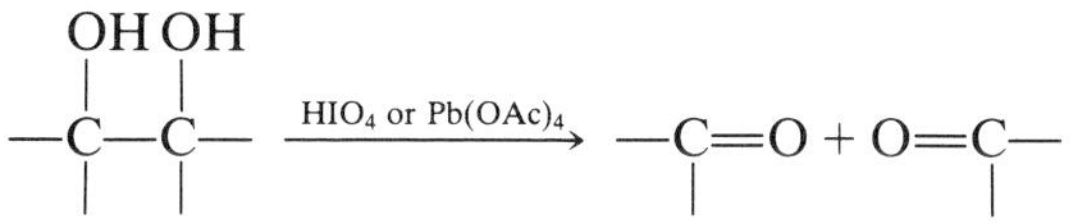

$$\underset{\underset{|}{\mid}}{\overset{\overset{|}{\mid}}{-}}\!\!C\!\!-\!\!C\!\!-\quad\xrightarrow{\text{HIO}_4\text{ or Pb(OAc)}_4}\quad-C{=}O + O{=}C-$$

Neither oxidizing agent reacts to any appreciable extent with poly(vinyl alcohol); however, both cause a reduction in molecular weight as measured by dilute solution viscometry. The results indicate that the polymer has a predominantly head-to-tail structure, but occasional head-to-head units are present and undergo chain scission with the subsequent decrease in molecular weight.

Ozonolysis is useful for determining structures of polydienes.[6] Ozonolysis of natural rubber, for example, followed by hydrolysis of the intermediate ozonide under reductive conditions, yields 4-ketopentanal, which establishes the structure as the head-to-tail 1,4-addition polymer of 2-methyl-1,3-butadiene (isoprene):

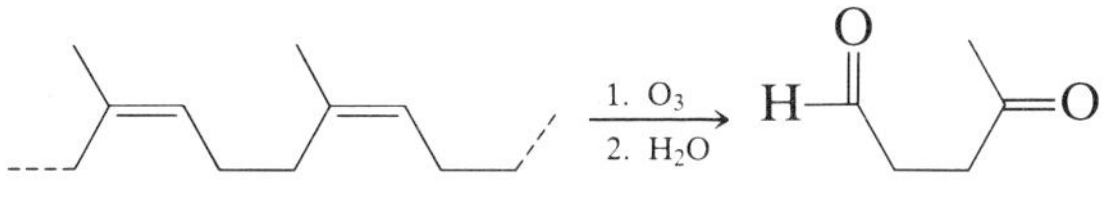

More recently[7] ozonolysis has been used to establish that polyacetylene, $-\!\!\!-\!\!(CH{=}CH)\!\!-\!\!\!-$, consists of more than 98% of the linear, conjugated polymer.

Other applications in which chemical methods are useful include the identification of polyesters and polyamides, or other polymers containing functional groups,[8] and the routine testing for polymer additives.[9]

5.3   Spectroscopic methods of analysis

Spectroscopic methods commonly employed for studying polymers are summarized in Table 5.1. Also listed are non-Raman scattering methods (discussed in the following section), which are not spectroscopic methods per se. Infrared (IR) and NMR spectroscopy are the most widely used, with recent emphasis on Fourier transform IR (FT-IR) and carbon-13 (^{13}C) NMR. Spectroscopic and other methods applied to *surface* analysis are covered in Section 5.5. The following discussion follows the order in Table 5.1.

5.3.1   Infrared[10,11a–16a]

The two instrumental variations of IR spectroscopy are the older dispersive method, in which prisms or gratings are used to disperse the IR radiation, and the more recent Fourier transform (FT) method, which uses the principle of interferometry. Advantages of the latter include small sample size requirements, rapid development of spectra, and—because the instrument has a dedicated computer—the ability to store and manipulate spectra. (Modern dispersive instruments are also being equipped with microcomputers for storage and manipulation of spectra.)

Dispersive spectra of most polymers of commercial import have been recorded,[4,17,18] hence qualitative identification of unknowns can frequently be accomplished by comparison. This includes polymers having varying stereochemistries or monomer sequence distribution, because such differences usually give rise to different spectra. Where comparative spectra are not available, insight into polymer structure may be gained by the usual consideration of functional group absorption bands, or by comparing spectra with

Table 5.1. Spectroscopic and scattering methods commonly used for studying polymers (excluding surfaces)

Vibrational
  Infrared (IR)
  Raman

Spin resonance
  Nuclear magnetic resonance (NMR)
  (proton and carbon-13)
  Electron spin resonance (ESR)

Electronic
  Ultraviolet (UV)-visible
  Fluorescence

Scattering
  X-ray
  Electron
  Neutron

those of readily characterizable low-molecular-weight *model compounds* of similar structure. An example of the latter is shown in Figure 5.1.[19] Apart from expected differences in the aromatic C-H bending region (650–900 cm^{-1}) arising from *para*-disubstituted versus monosubstituted benzene rings, the spectra compare favorably.

FT-IR[12–16a] brings greater versatility to polymer structural studies. Because spectra can be scanned, recorded, and transformed in a matter of seconds, the technique facilitates the study of such polymer reactions as degradation or crosslinking. The very small sample-size requirements facilitate coupling of the FT-IR instrument with a microscope for analysis of highly localized sections of a polymer sample.[20] And the capability for digital subtraction allows one to generate otherwise "hidden" spectra. An example of spectral subtraction is shown in Figure 5.2 for isotactic polystyrene.[21] The spectral region represented, with the scale greatly expanded, encompasses the carbon–hydrogen bending region of the pendant benzene rings. (Note that these are absorption spectra, rather than the more conventional transmittance spectra shown in Figure 5.1.) Spectrum A is that of semicrystalline polymer obtained by annealing; B is that of the same polymer heated above the T_m and quenched in the amorphous state. Subtraction of B from A reveals more well-defined absorbances characteristic of the crystalline regions where the benzene rings are "frozen" into relatively specific conformations.

FT-IR is particularly useful in studying polymer blends. Whereas the immiscible blend exhibits an IR spectrum that is the superposition of the two homopolymer spectra, the spectrum of a miscible blend is the superposition of *three* components—spectra of the homopolymers and an *interaction spectrum* arising from chemical or physical interaction between the homopolymers. For the commercial polyblend Noryl, for example, which consists of polystyrene and poly(phenylene oxide) (see Section 3.10), subtraction of the two homopolymer spectra from the spectrum of the blend reveals an interaction spectrum apparently arising from conformational changes brought about by dipole-dipole coupling of the benzene rings of the two homopolymers.[22] The interaction is greatest in blends of 30% polyether to 70% polystyrene, which is also the percentage concentration that offers the best mechanical properties.

5.3.2   Raman[16b,23a,24]

Like IR spectroscopy, Raman derives from vibrational transitions in molecules. When visible light impinges on molecules, the light is scattered; the frequency of the scattered light varies according to the vibrational modes of the scattering molecules. This is referred to as the *Raman effect*. Whereas IR absorption spectra are indicative of unsymmetric bond stretching and bending, the Raman effect responds to the *symmetric* vibrational modes. Thus IR and Raman spectroscopy are complementary.

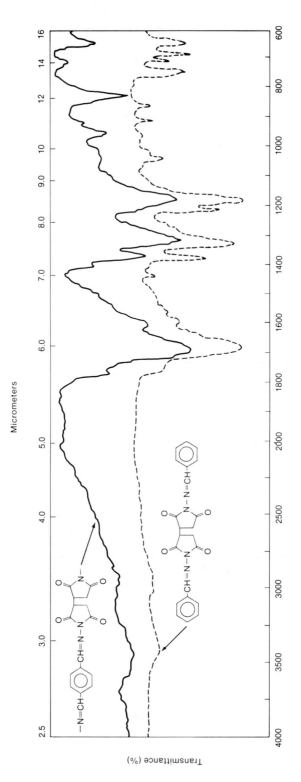

Figure 5.1. Infrared spectrum (KBr pellets) of polyimide (—) and model compound (--). [Reprinted from Troy, Sobanski, and Stevens,[19] by courtesy of Marcel Dekker, Inc.]

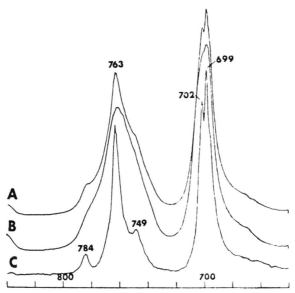

Figure 5.2. Fourier transform-infrared spectra of isotactic polystyrene in the 640 to 840 cm⁻¹ region: (A) semicrystalline; (B) amorphous; and (C) the difference spectrum obtained by subtracting B from A. [From Painter and Koenig,[21] copyright 1977. Reprinted by permission of John Wiley & Sons, Inc.]

A major difficulty with Raman is that the intensity of scattering is extremely low (10^{-8} to 10^{-5} of the incident intensity), hence powerful monochromatic light sources are needed, such as mercury-arc lamps or—in the more modern instruments—lasers. Also, because Raman is most responsive to symmetrical stretching in bonds such as C—C or C=C, the technique has been applied more to conformational studies of macromolecular carbon chains by comparison with long-chain (20–60 carbons) "model" alkanes. Where Raman is particularly useful is in conformational studies of biological macromolecules in aqueous solution as a function of pH. This is possible because Raman scattering by water is negligible compared with water's intense IR absorption.

5.3.3 Nuclear magnetic resonance[16c;23b,c;25a;26–29]

No single spectroscopic technique has provided so much valuable information to the polymer chemist as NMR. With the commercial development of NMR in the 1960s, initial applications dealt with proton (¹H) spectra of polymer solutions.[29] Thus structural units of polymers are identified from a combination of chemical shift data and spin-spin splitting. With the very high resolution afforded by modern NMR instruments, polymer scientists may now gain insight into polymer stereochemistry and monomer sequencing. A case in point, shown in Figure 5.3, is illustrated by the 500-MHz spectra of

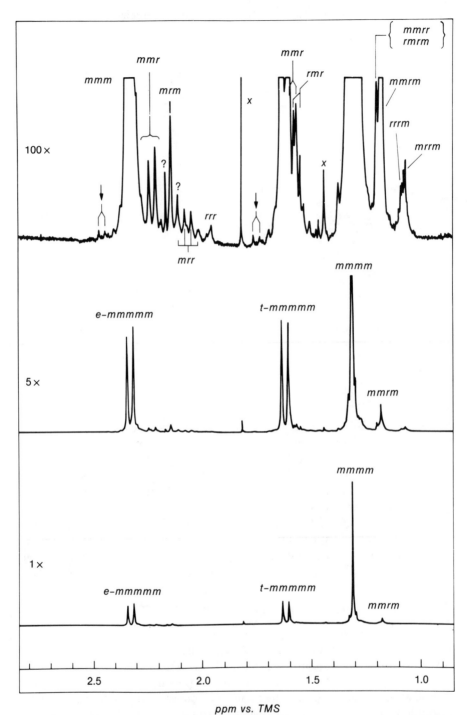

100×

5×

1×

mmm

mmr

mrm

? ?

mrr

rrr

x

mmr

rmr

x

{ *mmrr* \
rmrm }

mmrm

rrrm

mrrm

e–mmmmm

t–mmmmm

mmmm

mmrm

e–mmmmm

t–mmmmm

mmmm

mmrm

2.5　　2.0　　1.5　　1.0

ppm vs. TMS

Figure 5.3. 500-MHz ^{1}H nuclear magnetic resonance spectrum of isotactic poly(methyl methacrylate) (10% weight/volume solution in chlorobenzene-d_5 at 100°C) at three values of gain. *e, t: erythro* and *threo* placement of methylene protons with respect to ester groups; *m, r: meso* and *racemic* dyads. [Courtesy of Dr. Frank Bovey, Bell Laboratories. From Schilling, Bovey, Bruch, and Kozlowski,[30] copyright 1985. Reprinted by permission of the American Chemical Society.]

predominantly (95%) isotactic poly(methyl methacrylate)[30]:

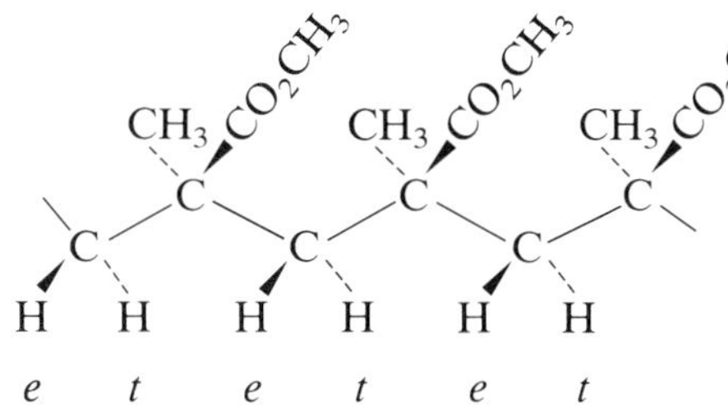

CH₃, CO₂CH₃   CH₃, CO₂CH₃   CH₃, CO₂CH₃

H   H   H   H   H   H
e   t   e   t   e   t

The spectral range shows the methyl resonance as a singlet at about 1.30 ppm and two methylene proton doublets at about 1.62 and 2.33 ppm. The coupling constant for the dissimilar methylene protons is about 15 Hz. (The methoxyl proton resonance occurs further downfield at about 3.6 ppm.) The designations *m* and *r* refer to the *meso* and *racemic* dyads, as discussed in Chapter 3; thus the methyl protons are resolved to the pentad (*mmmm*) level and the methylene protons to the hexad (*mmmmm*) level. The *e* and *t* notations refer to the *erythro* and *threo* placement, respectively, of the methylene protons with respect to the ester groups, the *erythro* resonance being further downfield because of the deshielding influence of the ester group. At the lowest sensitivity (1X) the spectrum resembles that of almost purely isotactic (all *m*) polymer with the exception of the small *mmrm* pentad at about 1.16 ppm. At higher gain (5X and 100X), deviations from purely isotactic are evident. Syndiotactic placement, by contrast, exhibits predominantly *racemic* (*r*) sequences at about 1.05 ppm for methyl and, because both methylene protons lie in identical magnetic environments, one singlet methylene resonance at 1.83 ppm. The methoxyl protons absorb at 3.42 ppm. As might be expected, atactic poly(methyl methacrylate) displays a broad range of sequences in its NMR spectrum, although the largest peaks correspond to those of syndiotactic placement.[30] A mechanistic rationale for this phenomenon is given in Chapter 6.

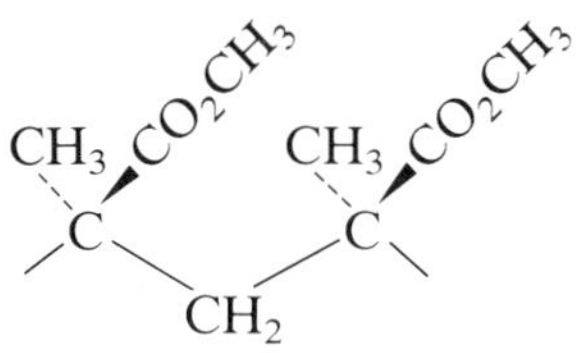

CH₃, CO₂CH₃   CH₃, CO₂CH₃   CH₃, CO₂CH₃   CO₂CH₃, CH₃

CH₂   CH₂

meso (*m*) dyad 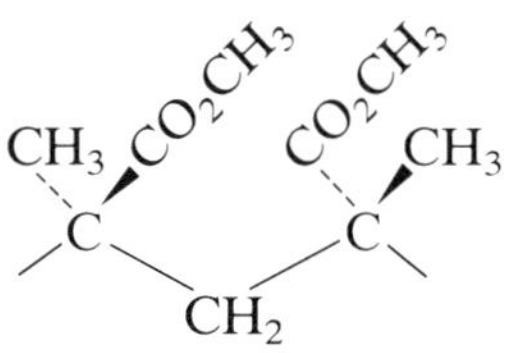   *racemic* (*r*) dyad

Introduction of FT techniques shifted the emphasis to ¹³C NMR. Major advantages of ¹³C NMR compared with ¹H NMR are the much larger chemical shifts for the former (which allow better signal resolution) and the absence of carbon–carbon coupling because of ¹³C's low natural abundance (1.11%). Pulse FT spectrometers that allow rapid scanning and accumulation of spectra to enhance the signal-to-noise ratio circumvent the low sensitivity arising from ^{13}C's low abundance. Spectra may be simplified further by using a second radiofrequency to decouple the spin interactions of ^{13}C with neighboring protons, thus yielding singlet resonances for each carbon. Figure 5.4 shows high-resolution proton-decoupled ^{13}C spectra of polypropylene of varying tacticity. Fine structure in the spectra may again be correlated with pentad sequences,[31,32] as discussed previously.

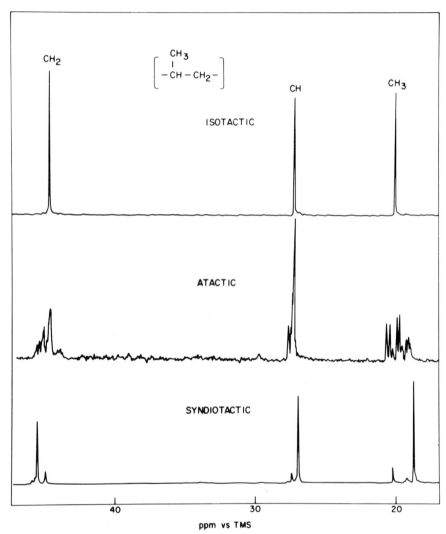

Figure 5.4. 25-MHz ^{13}C nuclear magnetic resonance spectrum of polypropylene (3.5% weight/volume in 1,2,4-trichlorobenzene at 135°C). [Courtesy of Dr. Frank Bovey, Bell Laboratories.]

Application of ^{13}C NMR to polymer solids gives rise to very broad resonances (of the order of a kilohertz) arising from dipolar interactions (mainly ^{13}C-^{1}H), long relaxation times in the solid state (of the order of minutes), and anisotropy of the chemical shift arising from the molecules being oriented in all possible directions, an effect that is averaged out in polymer solutions. Such difficulties may be resolved by a combination of instrumental refinements: proton decoupling to minimize dipolar broadening; application of radiofrequency pulses to make the ^{13}C and ^{1}H nuclei spin at the same rate, which effectively speeds up carbon's spin relaxation by transfer of magnetic polarization to the protons, a technique called *cross-polarization*; and rapid spinning of the sample at an angle of 54.7° with respect to the applied magnetic field (*magic angle spinning*), which has the effect of averaging out the spin vectors and minimizing the anisotropy effect. The combination of cross-polarization and magic angle spinning (CP-MAS) along with proton decoupling yields spectra of solids approaching the resolution of those obtained with solutions. As an example, Figure 5.5 shows the solid-state ^{13}C NMR spectrum of polycarbonate,

$$\left[\!\! - \!\! \bigcirc \!\! - \!\! \underset{\underset{CH_3}{|}}{\overset{\overset{CH_3}{|}}{C}} \!\! - \!\! \bigcirc \!\! - \!\! O\overset{\overset{O}{\|}}{C}O \!\! - \!\! \right]$$

with and without MAS, in the spectral region encompassing the carbonyl and ring carbon resonances.[33] Scientists thus have the capability for spectroscopic studies of crosslinked polymers and bulk polymers containing additives or fillers. CP-MAS also affords chemists the opportunity to study soluble polymers in the physical state in which they are actually used, which leads to a better understanding of solid-state morphology and chain conformation.

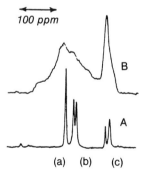

Figure 5.5. Solid-state proton-decoupled ^{13}C nuclear magnetic resonance spectrum of polycarbonate (A) with and (B) without cross-polarization–magic angle spinning. Peak assignments: (a) carbonyl carbon; (b) substituted ring carbons; and (c) unsubstituted ring carbons. [From Schaefer, Stejskal, and Buchdahl,[33] copyright 1977. Reprinted by permission of the American Chemical Society.]

5.3.4 Electron spin resonance[16d;23d,e;34–39]

Also called electron paramagnetic resonance (EPR), ESR works on the same principle as NMR except that microwave (rather than radiowave) frequencies are employed, and spin transitions of unpaired electrons rather than nuclei are recorded. Unlike NMR spectra, where absorption is recorded directly, ESR spectrometers plot the first derivative of the absorption curve.

Given the fact that the presence of unpaired (paramagnetic) electrons is prerequisite to obtaining an ESR spectrum, the utility of ESR in polymer chemistry is primarily for studying such free radical processes as polymerization, degradation, and oxidation. Figure 5.6, for example, represents the ESR spectrum of the free radical initially formed when poly(vinyl chloride) is irradiated with ultraviolet light[34]:

$$\left[\begin{array}{c} -CH_2CH- \\ | \\ Cl \end{array}\right] \xrightarrow{UV} -CH_2\dot{C}H- + Cl\cdot$$

The six-line signal arises from interaction of the unpaired electron with the five surrounding protons (4β and 1α), a phenomenon called *hyperfine splitting*. Information on the radical structure can be gained from a consideration of the line shape, intensity, position, and hyperfine splitting.

Another application of ESR is to attach a stable free radical (called a *spin label*) by covalent bonding to a polymer (for example, a transition metal atom or a nitroxide radical), or to mix a compound containing an unpaired electron

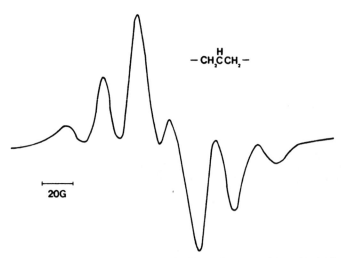

$$-CH_2\overset{H}{\underset{}{C}}CH_2-$$

20G

Figure 5.6. Electron spin resonance spectrum of UV-irradiated poly(vinyl chloride) at $-196°C$. [From Yang, Liutkas, and Haubenstock,[34] copyright 1980. Reprinted by permission of the American Chemical Society.]

(called a *spin probe*) with a polymer. Spin labels and spin probes are used primarily for studying molecular motion and relaxation phenomena in polymers.[37]

5.3.5   Ultraviolet (UV)-visible[16a,40,41]

Ultraviolet-visible absorption spectroscopy is used for the detection and quantitative measurement of chromophores that undergo $n \to \pi^*$ or $\pi \to \pi^*$ transitions. Because of its sensitivity, UV-visible spectroscopy has been particularly useful in identifying and analyzing "foreign" material in polymers— residual monomer, inhibitors, antioxidants, and so on. Styrene monomer in polystyrene, for example, may be determined quantitatively using styrene's λ_{max} at 292 nm.[9]

Copolymer composition also lends itself to UV-visible analysis if one of the repeating units has the requisite chromophore, as in the case of styrene copolymers. Beer-Lambert law interpretations of copolymer composition must be treated with some caution, for molar absorptivities vary with tacticity, sequence length, interactions between chromophores and functional groups, and solvent.[42] UV-visible spectroscopy may also be used to identify end groups, and—if one chain end or both ends are known to contain the chromophore—to determine number average molecular weights[43] (see Chapter 2, exercise 14).

Optical rotatory dispersion (ORD), which measures changes in optical rotation as a function of wavelength, is another technique employing UV-visible radiation. Although useful for studies of synthetic polymers containing optically active centers, ORD has been applied primarily to biopolymers.[44]

5.3.6   Fluorescence[16f,45–48]

When a molecule absorbs light energy to form an excited state, the excited state can lose its acquired energy by a variety of mechanisms. If energy loss occurs with emission of radiation, the process is termed *luminescence*. *Fluorescence* is a luminescence pathway involving energy loss from the lowest excited singlet state of the absorbing molecules. Fluorescence spectrophotometers measure wavelength of the absorbed or emitted light as a function of fluorescence intensity.

Fluorescence spectrometry has long been studied with biopolymers. Its application to synthetic polymers is just gaining momentum. The two major areas of interest with the latter are molecular motions and polymer-polymer compatibility. Fluorescent groups are incorporated into polymers either by covalent bonding or by dispersion of low-molecular-weight fluorescent molecules through the polymer matrix. Fluorescence yield—the amount of light emitted relative to that absorbed—varies according to molecular motion in solution or in the melt, and can thus be related to rotational freedom and chain conformation. Alternatively, a fluorescence donor and a fluorescence acceptor can be incorporated into the polymer. In such cases, charge transfer complexes (*exiplexes*) form, which emit at characteristic wavelengths. The ability of donor and acceptor to interact is also a function of molecular motion. In compatibility studies, the fluorescence donor is incorporated into one polymer and the acceptor into the other. Where the two polymers are completely miscible, energy transfer between donor and acceptor is very efficient, but efficiency drops off as miscibility decreases. Such studies have shown that even with incompatible polymers, there is a limiting energy transfer efficiency, which suggests a gradual rather than a sharp boundary separation between the immiscible phases.

5.4   X-ray, electron, and neutron scattering

Scattering methods already discussed are Raman (this chapter) and laser scattering for molecular weight determinations (Chapter 2). A vast amount of information on polymer morphology and molecular dimensions can be obtained by scattering of other energy sources, particularly x-rays, electrons, and neutrons.[49] Of the three, x-ray diffraction is most important. Indeed, it is possible with x-ray techniques to determine the spatial arrangements of all the atoms in such complex biopolymers as proteins.

5.4.1   *X-ray scattering*[50–53]

X-rays are generated in cathode-ray tubes when high-energy electrons impinge on metal targets. When x-rays are focused on a polymer sample (in pellet or cylinder form), two types of scattering occur. If the sample is crystalline, the x-rays are scattered *coherently*; that is, there is no change in wavelength or phase between the incident and scattered rays. Coherent scattering is commonly referred to as x-ray *diffraction*. If the sample has a nonhomogeneous (semicrystalline) morphology, the scattering is *incoherent*: there is a change in both wavelength and phase. Incoherent scattering (also called *Compton* scattering) is referred to as *diffuse diffraction* or simply as scattering.

Coherent scattering is determined by *wide-angle* measurements and incoherent scattering by *small-angle* measurements, as shown in Figure 5.7. (The acronyms WAXS and SAXS are used to denote wide-angle and small-angle x-ray scattering, respectively.) The wide-angle diffraction pattern consists of a series of concentric cones arising from scattering by the crystal planes; it is recorded as concentric rings on the x-ray plate superimposed on a diffuse background of incoherent scatter. A representative ring pattern is shown in Figure 5.7. As the degree of crystallinity increases, the rings become more sharply defined, and as the crystallites are oriented (as by drawing), the circles give way to a pattern of arcs and spots more nearly resembling diffraction patterns of low-molecular-weight crystalline compounds. Small-angle scatter patterns are very diffuse. For single crystal studies, the crystal is

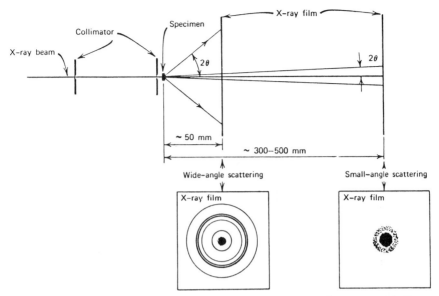

Figure 5.7. Wide-angle and small-angle x-ray diffraction techniques. [From E. A. Collins, J. Bares, and F. Billmeyer, *Experiments in Polymer Science*, Wiley-Interscience, copyright © 1973. Reprinted by permission of John Wiley & Sons, Inc.]

rotated at an angle perpendicular to the incident beam so that diffraction patterns at all possible angles are recorded.

Considerable insight into polymer morphology and structure may be gained from visual examination and mathematical interpretation of the pattern and intensity of scattered radiation, including degree of crystallinity, dimensions of crystalline domains, bond distances and angles, and type of conformation in the crystalline regions.

5.4.2   Electron scattering[13b]

Electron diffraction is usually accomplished using a transmission electron microscope set in the diffraction rather than the imaging mode. The diffraction pattern is projected on the microscope screen. Polymer samples must be very thin—in the range of several hundred angstrom units. As with x-ray diffraction, information gathered by electron diffraction has to do with morphology—crystal dimensions, degree of crystallinity, and so on. The main advantages of electron diffraction are that (1) both diffraction and transmission measurements are possible with a given sample, (2) sample size requirements are very small, and (3) diffraction intensities and the number of reflections are much higher. A disadvantage is that electrons may cause free radical reactions (e.g., chain scission, crosslinking) in the sample.

When the electron microscope is operated in the imaging mode (*transmission electron microscopy*), it is possible to resolve such morphological features as polymer single crystals with a resolution in the range of 2 to 5 Å at magnifications of 200,000 to 500,000.

5.4.3   *Neutron scattering*[23f,g;54;55]

Neutron scattering has the major disadvantage that a nuclear reactor is needed as a neutron source. Hence its applicability is limited. Furthermore, sophisticated neutron counters are required. A typical counter consists of a tube filled with ^{10}B-enriched boron trifluoride, which reacts with neutrons to form ionizing α-particles:

$$^{10}B + {}^1n \rightarrow {}^7Li + {}^4He$$

There are advantages to the method, however. It is the one scattering technique that provides much useful information from inelastic scattering, especially scattering by hydrogen atoms, where the scattering cross section is significantly higher than it is with other atoms. Thus, where elastic neutron scattering affords the same kind of information as x-ray and electron diffraction, inelastic neutron scattering may be used to study the movement of hydrogen atoms in the polymer. Neutrons are also more sensitive to observations of very-short-chain segments, for example in studies of chain folding in crystalline lamellae (see Chapter 3, Figure 3.17). A further advantage to neutron scattering is that the sample can be in virtually any shape or form, solid or liquid (including solutions).

5.5   Characterization and analysis of polymer surfaces[56-59]

Surface analysis has assumed great importance in recent years, particularly in studies of catalysts, semiconductors, and other electronics devices. In the polymer field, coatings, polymer-catalyst and polymer-filler interfaces, surface oxidation, and surface morphology are among the areas of interest. Basically, surface analysis involves irradiation of the surface with an energy source (photons, electrons, or ions) of sufficient energy to penetrate and cause some type of transition that results in the emission from the surface of an energy beam that can be analyzed. A great variety of techniques have been developed; those used most commonly for polymers are listed in Table 5.2. Where a method is known by more than one name, the alternative is given in parentheses. Not included is optical microscopy, which is limited to resolving surface characteristics of the order of 2000 Å.

Certain problems are inherent to surface analysis. If ions or electrons are the incident beam, or if electrons or ions are emitted from the surface, an electrical charge develops in the sample which may cause spectral distortion. This is most severe in the case of insulators (the great majority of polymers) in which the charge cannot be dispersed by conduction. Surface charge must be compensated for by calibration methods or, if possible, by charge neutralization. Surface contamination is another problem. Occasionally the contamination itself is the focus of the analysis. When the uncontaminated surface is of

Table 5.2. Methods of polymer surface analyses

Incident beam	Exit beam	Technique	Abbreviation
Electrons	Electrons	Scanning electron microscopy	SEM
Photons (IR)	Photons (IR)	Attenuated total reflectance spectroscopy (internal reflectance spectroscopy)	ATR (IRS)
Photons (IR or UV)	Acoustical signal	Photoacoustic spectroscopy	PAS
Photons (x-ray)	Electrons	Electron spectroscopy for chemical analysis (or applications) (x-ray photoelectron spectroscopy)	ESCA (XPS)
Photons (x-ray) or electrons	Electrons	Augér electron spectroscopy	AES
Ions	Ions	Secondary ion mass spectrometry	SIMS
Ions	Ions	Ion-scattering spectroscopy	ISS

interest, careful cleaning and the use of very low pressures (10^{-11}–10^{-7} torr) are often necessary. Low pressures are also necessary to avoid scattering of electrons or ions by air. Finally, the incident beam may cause surface damage. This is not a problem with IR or UV, nor to any great extent with x-rays; but electrons are particularly destructive to organic insulators, which limits the use of electron beam methods in the polymer field. Ion incident beams cause the loss of surface ions—a phenomenon called *sputtering*; measurement of the sputtered ions is, however, the basis of ion beam techniques.

5.5.1 Scanning electron microscopy (SEM)[60]

SEM differs from transmission electron microscopy (TEM) in that a very fine electron incident beam is scanned across the sample surface in synchronization with the beam in a cathode-ray tube. Scattered electrons are used to produce a signal that modulates the beam in the cathode-ray tube, producing an image with great depth of field and an almost three-dimensional appearance. SEM is of limited use in studying surface morphology, but it provides useful information on surface topology with a resolution of about 100 Å. Typical applications include studies of pigment dispersions in paints, blistering or cracking of coatings, phase boundaries in immiscible polyblends, cell structure of polymer foams, and adhesive failures. SEM is particularly valuable in evaluating how well polymeric surgical implants react to the host environment.

5.5.2 Attenuated total reflectance spectroscopy (ATR)[11b,58a]

Also called *internal reflection spectroscopy* (IRS), ATR is probably the most widely used method of analyzing polymer surfaces because it can be adapted to most IR instruments and is lower in cost. In the ATR method, a polymer sample is attached to a crystal such as ThBr or AgCl that is transparent to IR and has a higher refractive index than that of the polymer. An IR beam is directed into the crystal (sometimes called the *sample plate*) at an angle greater than the critical angle, as shown in Figure 5.8(a). After penetrating into the polymer surface a small distance (a few micrometers), the beam is reflected. Some of the IR frequencies are absorbed by the surface, and the attenuated reflected beam is directed into the spectrometer where it is plotted as a function of wavelength. Thus the ATR spectrum resembles a normal absorption spectrum. The amount of absorption can be increased by using an internal reflector plate, as shown in Figure 5.8(b), which allows the beam to be reflected a number of times from the surface. This is called *multiple internal reflection* (MIR). How many times the beam is reflected and how deep it penetrates into the polymer surface is a function of the relative refractive indexes, the incident angle, the geometry of the reflectance optics, and the wavelength.

ATR is particularly useful for studying surface coatings, surface oxidation, adhesive–substrate interfaces, and other aspects of polymer surfaces. Microscopy units are also available as adjuncts to FT-IR instruments; they allow movement of the sample under study so that different parts of the specimen surface may be analyzed. Because surface penetration is extremely small, ATR spectra should not be interpreted as being necessarily characteristic of the bulk polymer.

5.5.3 *Photoacoustic spectroscopy (PAS)*[61,62]

Although PAS encompasses both the IR and the UV-visible range and is applicable to bulk samples, its primary use with polymers is to characterize surfaces using FT-IR methodology. A solid polymer sample enclosed in an air-tight chamber is exposed to intensity-modulated IR radiation. The heat energy released following absorption of radiation by the sample causes a modulated pressure change in the chamber atmosphere, which is detected by a sensitive microphone. Figure 5.9 illustrates the process. The spectrum is

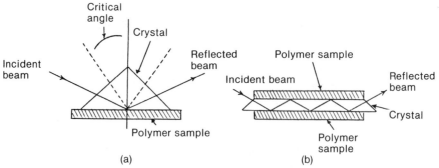

Figure 5.8. Attenuated total reflectance (ATR): (a) single reflection, and (b) multiple internal reflection (MIR).

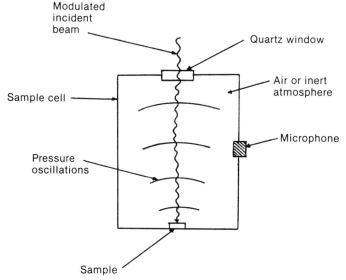

Modulated incident beam

Quartz window

Sample cell

Air or inert atmosphere

Microphone

Pressure oscillations

Sample

Figure 5.9. Schematic of the photoacoustic process.

plotted as frequency versus arbitrary units of the amplified photoacoustic signal, and as such resembles a normal IR spectrum. Because the signal-to-noise ratio of PAS is very low, several thousand scans are needed to obtain a spectrum. The major advantage of PAS is that sample preparation is virtually unnecessary; thus it is particularly useful with very air- or moisture-sensitive polymers such as polyacetylene for which conventional absorption spectra are difficult to obtain. Another feature of PAS is that subsurface layers may be characterized by changing the modulation range, thereby providing a depth profile of the sample.

5.5.4 *Electron spectroscopy for chemical analysis (or applications) (ESCA) and Augér electron spectroscopy (AES)*[23h,58b,63,64]

Less commonly called *x-ray photoelectron spectroscopy* (XPS), ESCA employs soft x-rays to detach electrons from film or powder surfaces. The difference in energy between the incident x-rays ($h\nu$) and the kinetic energy of the emitted election (E_{kin}) is equal to the binding energy (E_B) of the electron; that is,

$$E_B = h\nu - E_{kin}$$

Two types of electrons are emitted—inner core and the more loosely held valence electrons. When the photon ejects the inner core electron (photo-ionization), the effective increase in nuclear charge causes a reorganization of the valence electrons, resulting in excitation of a valence electron from an occupied to an unoccupied level (a process called *shake-up*) and the loss of a valence electron (*shake-off*). The emitted electrons are analyzed by an electrostatic analyzer. A schematic of an ESCA spectrometer is shown in Figure 5.10. The spectrometer is evacuated to 10^{-10} to 10^{-7} torr to preclude electron scattering by air. An ESCA spectrum plots intensity against binding energy of the inner core electron; thus it provides a surface "map" of the elements present. Peaks corresponding to photoionization are often accompanied by satellite peaks arising from the accompanying shake-up and shake-off, especially where pi bonding is present. Generally, ESCA analyzes the surface to a depth of about 50 Å. Argon ion sputtering may be used to expose greater depths for ESCA analysis.

An example of how ESCA may be used to characterize a film surface is shown in Figure 5.11.[65] Included are spectra of polystyrene and poly(ethylene oxide), and a block copolymer of the two containing 21 mol % polystyrene cast from two different solvents. Polystyrene exhibits a carbon peak at 285.0 electron volts (eV) and a low energy $\pi \rightarrow \pi^*$ shake-up peak at 291.6 eV. Poly(ethylene oxide) has peaks at 286.5 eV for carbon (shifted 1.5 eV from that in polystyrene because of the influence of neighboring oxygen atoms) and 533.3 eV for oxygen. Carbon and oxygen peaks both arise from photoionization of core 1s electrons. Both film surfaces are seen to be richer in the less abundant polystyrene, especially when ethylbenzene, a better solvent for polystyrene, is used. Interestingly, optical and electron microscopy show variations in surface morphology with the surface concentration of polystyrene.

AES complements ESCA and uses essentially the same instrumentation. When a 1s electron is ejected, a 2s electron can drop to the 1s level with simultaneous loss of a 2p electron (also called an Augér electron after the

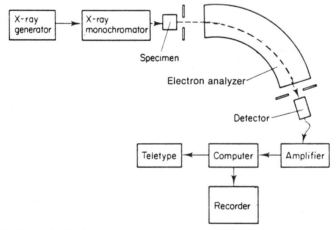

Figure 5.10. Schematic diagram of an electron spectroscopy for chemical analysis (ESCA) spectrometer. [From Rabek,[63] copyright 1980. Reprinted by permission of John Wiley & Sons, Ltd.]

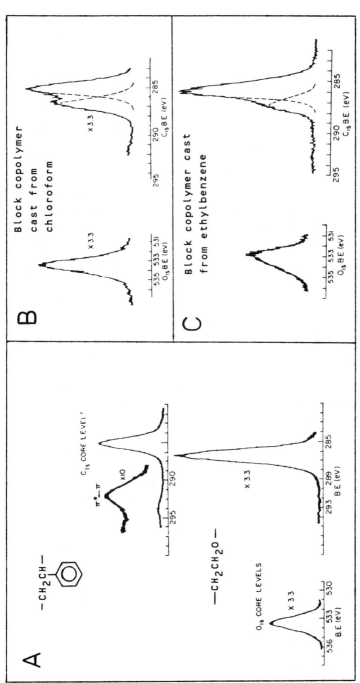

Figure 5.11. Electron spectroscopy for chemical analysis (ESCA) spectra of films: (A) Polystyrene and poly(ethylene oxide); and (B), (C) Polystyrene-*block*-poly(ethylene oxide) cast from chloroform and ethylbenzene, respectively. [From Thomas and O'Malley,[65] copyright 1979. Reprinted by permission of the American Chemical Society.]

French physicist Pierre Victor Augér, who laid the theoretical groundwork). An AES spectrum plots intensity against binding energy of the Augér electrons. Binding energy, E_b, given by

$$E_b = E_{kin}(1s) + E_b(2s) + E_b(2p)$$

Unlike ESCA, AES spectra are plotted as derivative peaks (as in ESR) to separate the signal from scattered electron background. AES has not been used extensively in polymer studies, but is useful for characterizing polymer-metal interfaces.

5.5.5 Secondary-ion mass spectrometry (SIMS) and ion-scattering spectroscopy (ISS)[56,57]

SIMS and ISS spectra are obtained by irradiating a surface with inert gas ions (called *primary ions*). Two phenomena occur, as shown in Figure 5.12. The primary ions are reflected in an elastic scattering process, with the energy of the scattered ions depending on the atoms causing the scattering. Detection and *kinetic energy* measurement of the scattered ions are the basis of ISS. Additionally, primary ions penetrate the surface and eject atomic and molecular fragments in a process that might be likened to sandblasting. This is the phenomenon of surface sputtering. Detection and mass measurement of the sputtered ions are the basis of SIMS.

The sputtered fragments are primarily neutral; however, enough positive and negative ions are formed to be focused and measured in a medium-resolution mass spectrometer. The SIMS spectrum, a plot of mass/charge versus intensity, thus resembles a typical mass spectrum. Because SIMS is very dependent on matrix effects, quantitative measurements are not practical. SIMS is very sensitive, however. Figure 5.13, for example, shows the SIMS-positive ion spectrum of polytetrafluoroethylene,[57] which clearly demonstrates surface contamination by sodium, aluminum, potassium, and saturated hydrocarbons, in addition to the fragmentation pattern of the polymer.

ISS complements SIMS. The great value of ISS lies in the fact that it

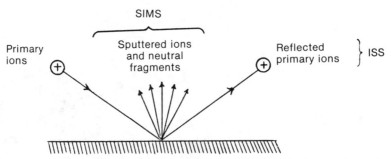

Figure 5.12. Representation of surface sputtering and primary ion elastic scattering as the basis of secondary-ion mass spectometry (SIMS) and ion-scattering spectroscopy (ISS), respectively

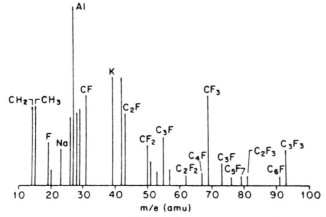

Figure 5.13. Secondary-ion mass spectometry-positive ion spectrum of polytetrafluoroethylene showing surface contamination. (^{20}Ne used as the sputtering gas.) [From Hercules and Hercules,[57] reprinted with permission.]

measures the surface layer composition to the exclusion of subsurface structure. Primary ions that penetrate the monolayer are scattered inelastically and are represented as background in the ISS spectrum. Another advantage is that ISS detects any atom heavier than the primary ion. Thus with helium as the primary ion, elements from lithium up through the periodic table are detectable.

5.6 Thermal analysis[66–68]

A number of thermal properties of polymers are discussed in Chapters 3 and 4, including crystalline melting point, glass transition temperature, flammability, and thermal stability. Flammability and thermal stability are concerned with chemical transformations, whereas melting and glass transition temperatures represent morphological changes.

Glass transition temperatures are most commonly measured by *differential scanning calorimetry* (DSC), *differential thermal analysis* (DTA), or *thermomechanical analysis* (TMA). Thermal degradation is determined by *thermogravimetric analysis* (TGA) and *pyrolysis-gas chromatography* (PGC), the latter frequently in combination with mass spectrometry. For measuring flammability, a variety of standardized procedures are used, from small-scale laboratory to large-scale room tests. These various techniques are described below.

5.6.1 Differential scanning calorimetry (DSC) and differential thermal analysis (DTA)[10–12a,69,70a,71a]

DTA is a very old technique, dating to Le Chatelier[72] in the last century, but it was not applied to polymers until the 1960s. DSC is of more recent vintage, and has become the method of choice for quantitative studies of thermal transitions in polymers. In both DSC and DTA, a polymer sample and an inert reference are heated, usually in a nitrogen atmosphere, and thermal transitions in the sample are detected and measured. The sample holder most commonly used is a very small aluminum cup (gold or graphite is used for analyses above 800°C), and the reference is either an empty cup or a cup containing an inert material in the temperature range of interest, such as anhydrous alumina. Sample sizes vary from about 0.5 to about 10 mg. Although the two methods provide the same type of information, there are significant differences in instrumentation. With DTA, both sample and reference are heated by the same heat source, and the difference in temperature (ΔT) between the two is recorded. When a transition occurs in the sample—for example, the glass transition or a crosslinking reaction—the temperature of the sample will lag behind that of the reference if the transition is endothermic, and will surge ahead if the transition is exothermic. With DSC, sample and reference are provided with individual heaters, and energy is supplied to keep the sample and reference temperatures constant. In this case, the electrical power difference between sample and reference ($d\Delta Q/dt$) is recorded. Schematic representations of DTA and DSC cells are given in Figure 5.14.

Data are plotted as ΔT (for DTA) or $d\Delta Q/dt$ (for DSC) on the ordinate against temperature on the abscissa. Such plots are called *thermograms*. Although ΔT and $d\Delta Q/dt$ are not linearly proportional, they are both related to heat capacity. Thus DSC and DTA thermograms have the same form. The major advantage of DSC is that peak areas of thermograms are related directly to enthalpy changes in the sample, hence may be used for measurements of heat capacities, heats of fusion, enthalpies of reactions, and the like.

An idealized DSC or DTA thermogram for a hypothetical crystallizable polymer is depicted in Figure 5.15, which illustrates the types of transitions of interest to polymer chemists. An actual thermogram[73] of poly (vinyl chloride) is shown for comparison in Figure 5.16. The glass transition causes an endothermic shift in the initial baseline because of the sample's increased heat capacity. Endothermic transitions lie below the baseline in these thermograms, and exothermic transitions lie above, although this varies with manufacturer. In reporting transition temperatures, it is important to indicate whether one is referring to the onset of the transition or to the inflection point or peak maximum, as shown in Figure 5.17. Both conventions are used.

5.6.2   *Thermomechanical analysis (TMA)*[70b,74a]

TMA employs a sensitive probe in contact with the surface of a polymer sample. As the sample is heated the probe senses thermal transitions such as T_g or T_m, by detecting either a change in volume (dilatometry) or a change in modulus. In the former case a flat probe tip is used, and in the latter a tapered tip, to penetrate the surface. Movement of the probe is detected by means of

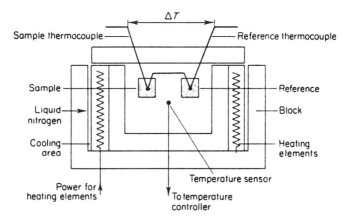

DTA

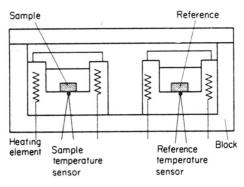

DSC

Figure 5.14. Schematic representations of differential thermal analysis (DTA) and differential scanning calorimetry (DSC) measuring cells. [From Rabek,[63] copyright 1980. Reprinted by permission of John Wiley & Sons, Ltd.]

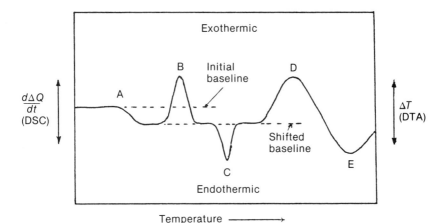

Figure 5.15. Idealized differential scanning calorimetry (DSC) or differential thermal analysis (DTA) thermogram: (A) temperature of glass transition, T_g; (B) crystalline melting point, T_m; (C) crystallization; (D) crosslinking; and (E) vaporization. $d\Delta Q/dt$ = electrical power difference between sample and reference; ΔT = difference in temperature between sample and reference.

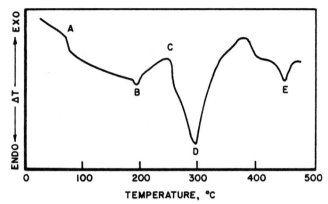

Figure 5.16. Differential thermal analysis (DTA) thermogram of poly(vinyl chloride): (A) temperature of glass transition, T_g; (B) melting point; (C) oxidative attack; (D) dehydrochlorination; and (E) probable depolymerization. [From Matlack and Metzger,[73] copyright © 1966. Reprinted by permission of John Wiley & Sons, Inc.]

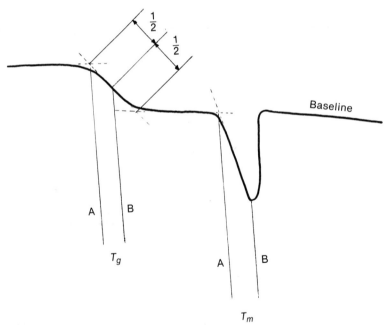

Figure 5.17. Methods of reporting transition temperatures: (A) at the onset, and (B) at the inflection point or maximum. T_g = glass transition temperature; T_m = crystalline melting point.

a variable transformer that records voltage output proportional to the displacement. TMA is generally more sensitive than DSC or DTA for detecting thermal transitions.

Another mechanical method less commonly used is *torsional braid analysis* (TBA). In TBA a glass braid or thread is impregnated with the polymer sample. The sample is then heated as the braid or thread is subjected to torsional oscillations. Variations in oscillation behavior are related to thermomechanical transitions.

5.6.3   Thermogravimetric analysis (TGA)[70b,75]

TGA is used primarily for determining thermal stability of polymers. Like DTA, TGA is an old techique[76] but has been applied to polymers only since the 1960s. The most widely used TGA method is based on continuous measurement of weight on a sensitive balance (called a *thermobalance*) as sample temperature is increased in air or in an inert atmosphere. This is referred to as *nonisothermal TGA*. Data are recorded as a thermogram of weight versus temperature. Weight loss may arise from evaporation of residual moisture or solvent, but at higher temperatures it results from polymer decomposition. Besides providing information on thermal stability, TGA may be used to characterize polymers through loss of a known entity, such as HCl from poly(vinyl chloride). Thus weight loss can be correlated with percent vinyl chloride in a copolymer. TGA is also useful for determining volatilities of plasticizers and other additives. Thermal stability studies are the major application of TGA, however. A typical thermogram illustrating the difference in thermal stability between a wholly aromatic polymer and a partially aliphatic polymer of analogous structure is shown in Figure 5.18.[77] Residual weight is frequently an accurate reflection of char formation, which is of interest in flammability testing.

A variation of the method is to record weight loss with time at a constant temperature. Called *isothermal TGA*, this is less commonly used than nonisothermal TGA. Modern TGA instruments allow thermograms to be recorded on microgram quantities of material. Some instruments are designed to record and process DSC and TGA data simultaneously, and may also be adapted for gas chromatographic and/or mass spectrometric analysis of effluent degradation products.

5.6.4   Pyrolysis-gas chromatography (PGC)[78,79]

Scientists have been studying polymer pyrolysis almost as long as they have been studying polymers, but it was not until 1954,[80] shortly after the technique of gas chromatography was first reported, that products of polymer pyrolysis were separated by this method. In 1959 the first report of the direct coupling of a pyrolysis device to the column inlet of a gas chromatograph appeared.[81] Since then the combination of pyrolysis and gas chromatography has proved to be a valuable tool for polymer characterization.

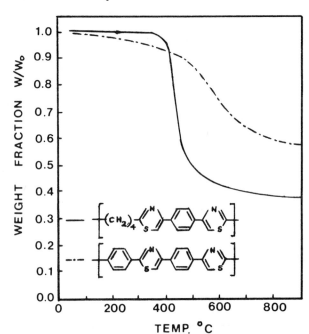

Figure 5.18. Thermogravimetric analysis (TGA) thermograms of polythiazoles. [From Marvel,[77] reprinted by courtesy of Marcel Dekker, Inc.]

Pyrolysis is accomplished in one of three ways: *furnace chamber pyrolysis, flash pyrolysis*, or *laser pyrolysis*. In each case the chromatogram of pyrolysis products is called a *pyrogram*. In the furnace chamber method, a polymer sample is either moved into a preheated pyrolysis chamber in a sample "boat" or heated rapidly from ambient to pyrolysis temperature. Gaseous pyrolysis products are swept directly into the gas chromatograph by carrier gas. The flash pyrolysis method employs a high-resistance coil or filament that is coated with polymer sample and attached directly to the sample inlet of the gas chromatograph. The coil is then heated by the application of electric current. Some flash pyrolyzers use ferromagnetic filaments that are heated by high-frequency induction to the Curie point, which brings the sample rapidly to a constant and reproducible temperature. Flash pyrolysis has certain disadvantages compared with furnace chamber pyrolysis: The metal coil may have a catalytic effect on the degradation; the coil may become fouled with char products, thus causing fluctuations in pyrolysis temperature; and quantitative analyses are much more difficult because the sample cannot be weighed accurately.

In the laser pyrolysis method, pyrolysis is accomplished with a focused laser beam. Pyrograms thus obtained are usually much simpler than those obtained by the other two methods. Because not all polymers absorb laser light, carbon black is often mixed with the sample to facilitate pyrolysis.

PGC of polymers is useful for both qualitative and quantitative analysis. With the former, polymers are often identifiable from the pattern of peaks (the "fingerprint") in the pyrogram. Even homopolymers of different tacticity, or random, block, and graft copolymers prepared from the same monomers invariably exhibit characteristic pyrograms. An example of quantitative analysis is the determination of vinyl acetate in poly[ethylene-*co*-(vinyl acetate)] by measuring the acetic acid evolved upon pyrolysis[82]:

$$\left[\begin{array}{c} CH_2-CH \\ | \quad | \\ H \quad O \\ | \\ O=C-CH_3 \end{array}\right] \longrightarrow -[CH=CH]- + HO-\overset{O}{\overset{\|}{C}}-CH_3$$

PGC is also useful in elucidating thermal degradation pathways. The pyrogram of polyethylene[83] (Figure 5.19) for example, consists of regularly spaced triplets corresponding to *n*-alkane, 1-alkene, and α, ω-alkadiene (each having the same number of carbons), which indicates that degradation occurs by random scission of the polymer backbone (discussed in Chapter 9). Most pyrograms are much more complex than that in Figure 5.19 because of further degradation of the initially formed pyrolysis products. Some polymers, such as polystyrene and poly (methyl methacrylate), exhibit large monomer peaks in their pyrograms arising from depolymerization reactions. Mass spectrometry (or the combination of GC and mass spectrometry) is also used to identify pyrolysis products.[25b,70d]

5.6.5 *Flammability testing*[84a–90]

Flammability is a difficult property to measure in a meaningful way because small-scale laboratory tests do not, in general, reflect burning behavior in true fire conditions. A polymer such as polyurethane, for example, which is widely used in carpeting and furniture upholstery, does not burn well if a match is applied to a small test sample; but in a burning room where temperatures are much higher and combustible gases have accumulated, the same sample may burn furiously. Room-size tests have demonstrated that when the accumulated combustible gases reach a certain temperature, they spontaneously ignite—a phenomenon called *flashover*—and the fire then spreads almost instantaneously to all other combustible objects in the room. It is not surprising, then, that a large number of tests have been developed.[85]

The most versatile small-scale test in widespread laboratory use is the *limiting oxygen index* (LOI) test.[86] The LOI is the minimum percentage of oxygen in an oxygen-nitrogen mixture that will initiate and support for three minutes the candlelike burning of a polymer sample; that is,

$$LOI = \frac{\text{vol. } O_2}{\text{vol. } O_2 + \text{vol. } N_2} \times 100$$

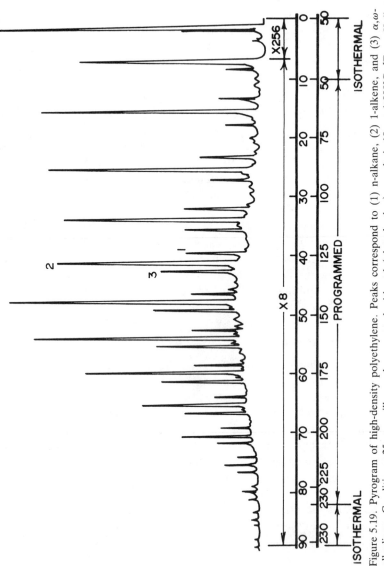

Figure 5.19. Pyrogram of high-density polyethylene. Peaks correspond to (1) n-alkane, (2) 1-alkene, and (3) α,ω-alkadiene. Conditions: 25-m capillary column coated with poly(phenyl ether); pyrolysis 10s at 1,000°C. [From Kolb et al.,[83] courtesy of Springer-Verlag.]

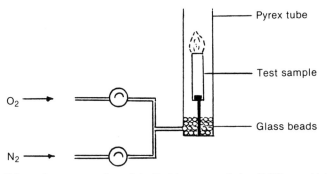

Figure 5.20. Schematic representation of the limiting oxygen index (LOI) test. [Adapted from Fenimore and Martin.[86]]

The procedure involves clamping a test sample of specified dimensions in a Pyrex cylinder, as shown in Figure 5.20. A metered mixture of O_2 and N_2 is fed in at the bottom, and the sample is ignited with a gas flame. Representative LOI values for some common polymers are given in Table 5.3. Noteworthy is the contrast between the "isomeric" poly(ethylene oxide) and poly(vinyl alcohol) (see exercise 6, Chapter 4). Dehydration of the latter results in cooling of the pyrolysis zone with resultant decrease in flammability. The flame-retardant properties of halogen are also apparent.

Table 5.3. Limiting oxygen indexes (LOI) of some common polymers[a]

Polymer	LOI
Polyoxymethylene	15
Poly(ethylene oxide)	15
Poly(methyl methacrylate)	17
Polypropylene	17
Polyethylene	17
Polystyrene	18
Poly(1,3-butadiene)	18
Poly(vinyl alcohol)	22
Polycarbonate	27
Poly(phenylene oxide)	28
Polysiloxane	30
Poly(vinyl chloride)	45
Poly(vinylidene chloride)	60
Polytetrafluoroethylene	95

[a] Values taken from Fenimore and Martin.[86]

Another widely used test procedure is to clamp a sample horizontally in air and to apply a flame at one end for 30 seconds. If, after removal of the flame, the sample fails to burn to a mark 4 inches from the point of ignition, it is said to be "self-extinguishing." In the LOI test, self-extinguishing corresponds to an LOI value of about 27.[86]

5.7 Measurement of mechanical properties[91]

To measure tensile strength, modulus, and elongation, a test specimen is clamped at each end. One end is fixed, and a gradually increasing load is applied to the other until the sample breaks. Plastics test specimens are usually of the dimensions shown in Figure 5.21(a). Fiber and elastomer specimens differ in form, but are tested in essentially the same manner. A typical testing instrument that measures stress and strain automatically with full scale loads from less than a gram to as high as 20,000 pounds is shown in Figure 5.22. Strength and elongation at the yield point, as well as at break, are usually recorded. The same instrument is used to measure compressive

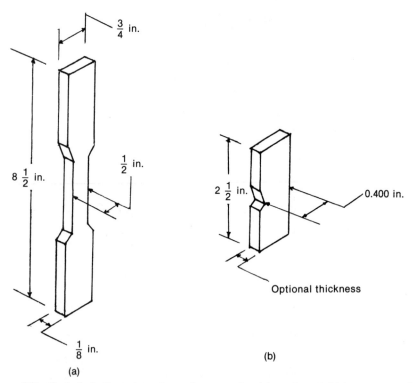

Figure 5.21. Typical plastics test specimens for measuring (a) tensile and (b) impact strength.

Figure 5.22. Instrument for performing tensile, compression, shear, and flexural testing. Shown with tensile testing grips and sample. [Courtesy of Instron Corp.]

and flexural strengths. Repeated flexing of a sample through a given distance often causes the sample to fail at lower stress than it would for a single flexure, a phenomenon known as *fatigue*.[84b] Alternating cycles of tensile and compressive stress are also used to evaluate fatigue. In general, fatigue resistance decreases as polymer rigidity or crosslinking increases.

A number of methods are used to measure impact strength; the choice depends on the type of sample. The most widely used test (the Izod test) involves clamping a test specimen at one end and striking the specimen with a weighted pendulum. The distance the pendulum travels after the sample breaks is taken as a measure of impact strength. Notched samples, as shown in Figure 5.21(b), are usually used to give more reproducible results. In such cases the pendulum strikes on the same edge as, and above, the notch, and the strength is calculated on the basis of 1 in. (or 1 m) of thickness. In all mechanical testing, a minimum of ten samples is used because a fairly wide spread of measured values is common with the same polymer batch. Representative mechanical properties are given in Chapter 4 (Tables 4.1 and 4.2).

Other commonly measured mechanical properties are the following:

Hardness. This is a measure of a polymer's resistance to surface indentation. The indenting device may be a spring-loaded needle-type indenter (Barcol test) or a weighted steel ball (Rockwell test).

Abrasion resistance. This is normally determined by measuring weight loss of a sample when the sample is subjected to some type of mechanical abrader or finely divided abrasive.

Tear resistance. A weighted pendulum-mounted blade is used to tear apart a test sample, and the distance the blade travels after tearing is correlated with tear resistance.

5.8   Evaluation of chemical resistance[74b]

Chemical resistance is usually determined by immersing test samples in the reagents of interest, often at elevated temperatures, then measuring some mechanical property such as tensile or flexural strength after a certain length of time. Exposure time may vary from days to months, depending on the polymer and test reagent. Chemical resistance is expressed in terms of the percent retention of properties. Visual effects of failure such as swelling, surface erosion, or *crazing* (the development of stress fractures) are also taken into account. Change in viscosity is a measurement also employed with thermoplastic polymers, particularly where immersion in a corrosive medium is not a consideration, as in photodegradation studies. Resistance to moisture is often measured simply as gain in weight after long-term exposure to water.

Weatherability testing is widely used in the plastics and coatings industries.[92] Test panels set facing south at 45° are commonly used for testing resistance to sunlight. Specially designed weathering chambers equipped with controlled humidity, water sprays, and UV lamps are also employed to duplicate the effects of weather cycles.

5.9   Evaluation of electrical properties[93–95]

Because of the widespread use of polymers in the electronics industry, as well as the traditional use of polymers as electrical insulators, measurement of electrical properties is of special significance, more so with the recent discovery of highly conducting polymers. Most important from the standpoint of evaluating polymers as resistors or conductors is the *volume resistivity*, which is determined by attaching electrodes to the polymer specimen, usually a disk as shown in Figure 5.23(a), and applying a potential of about 500 V. The volume resistivity in ohms, ρ_v, which is the inverse of conductivity, is given by

$$\rho_v = \frac{E}{I_v} \times \frac{A}{t} = \frac{R_v A}{t}$$

where E is the potential in volts, I_v the measured dc current in amps, A the area of the small electrode in square centimeters, t the thickness in centimeters, and R_v the resistance of the sample in ohms. By analogy with our earlier discussion in Section 4.8, samples having resistivities of about 10^8 ohm-cm or higher are considered good insulators, and those with resistivities below about 10^{-3} ohm-cm are good conductors. In between lie semiconductors.

To measure *surface resistivity*, ρ_s, the circuitry shown in Figure 5.23(b) is used. In this case,

$$\rho_s = \frac{E}{I_s} \frac{\pi D_m}{g} = \frac{R_s \pi D_m}{g}$$

where I_s is the current (amp), D_m the mean diameter of the gap (cm), g the width of the gap (cm), and R_s the resistance (ohm). *Insulation resistance* refers to the resistance of the sample to direct current both through the sample *and* across the surface.

Other electrical properties of interest are *dielectric strength* (a measure of the maximum voltage a sample can withstand before the sample undergoes physical breakdown); *dielectric constant* (a measure of a sample's contribution to the circuit capacitance); and *arc resistance* (the length of time before a high-voltage discharge arcs between two electrodes attached to the surface.)

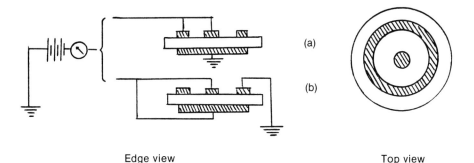

Edge view   Top view

Figure 5.23. Circuitry for measuring (a) volume resistivity and (b) surface resistivity. Cross-hatched areas are electrodes. [Adapted from Agranoff.[94]]

References

1. J. R. White, *Polymer Characterization*, Chapman and Hall, New York, 1989.
2. G. M. Kline (ed.), *Analytical Chemistry of Polymers*, Wiley-Interscience, New York, Part 1, 1959; Parts 2 and 3, 1962.
3. T. R. Crompton, *Analysis of Polymers: An Introduction*, Pergamon Press, New York, 1989.
4. J. Haslam, H. A. Willis, and D. C. M. Squirrell, *Identification and Analysis of Plastics*, 2nd ed., Butterworth, London, 1972, pp. 605ff.
5. P. J. Flory and F. S. Leutner, *J. Polym. Sci.*, **3**, 880 (1948); **5**, 267 (1950).
6. C. Marvel and R. E. Light, Jr., *J. Am. Chem. Soc.*, **72**, 3887 (1950).
7. G. C. G. Li, D. A, Kidwell, D. W. Brown, E. E. Hall, and G. E. Wnek, *J. Polym. Sci., Polym. Chem. Ed.*, **21**, 301 (1983).
8. D. Nissen, V. Rossbach, and H. Zahn, *Angew. Chem. Int. Ed. English*, **12**, 602 (1973).
9. T. R. Crompton, *Chemical Analysis of Additives in Plastics*, 2nd ed., Pergamon Press, Oxford, 1971.
10. G. J. Mol, R. J. Gritter, and G. E. Adams, in *Applications of Polymer Spectroscopy* (E. G. Brame, Jr., ed.), Academic Press, New York, 1978.
11. E. G. Brame, Jr., and J. G. Grasselli (eds.), *Infrared and Raman Spectroscopy*, Part C, Dekker, New York, 1977: (a) S. C. Brown and A. B. Harvey, Chap. 12; (b) C. D. Craver, Chap. 13.
12. J. L. Koenig, *Adv. Polym. Sci.*, **54**, 87 (1984).
13. B. Jassie, in *Developments in Polymer Characterization* (J. V. Dawkins, ed.), Vol. 4, Applied Science, London, 1983, p. 91.
14. L. D'Esposito and J. L. Koenig, in *Fourier Transform Infrared Spectroscopy*, Vol. 1 (J. R. Ferraro and L. J. Basile, eds.), Academic Press, New York, 1978, Chap. 2.
15. H. Ishida (ed.), *Fourier Transform Infrared Characterization of Polymers*, Plenum Press, New York, 1987.
16. W. Klöpffer, *Introduction to Polymer Spectroscopy*, Springer-Verlag, New York, 1984: (a) Chap. 7; (b) Chap. 6; (c) Chap. 10; (d) Chap. 9; (e) Chap. 3; (f) Chap. 4.
17. D. O. Hummel, *Infrared Spectra of Polymers*, Wiley-Interscience, New York, 1966.
18. R. Zbinden, *Infrared Spectroscopy of High Polymers*, Academic Press, New York, 1967.
19. R. C. Troÿ, D. B. Sobanski, and M. P. Stevens, *J. Macromol. Sci.-Chem.*, **A23**, 189 (1986).
20. R. G. Messerschmidt and M. A. Harthcock (eds.), *Infrared Microscopy*, Dekker, New York, 1988.
21. P. C. Painter and J. L. Koenig, *J. Polym. Sci., Polym. Phys. Ed.*, **15**, 1885 (1977).
22. J. L. Koenig and M. J. M. Tovar Rodriguez, *Appl. Spectrosc.*, **35**, 543 (1981).
23. K. J. Ivin (ed.), *Structural Studies of Macromolecules by Spectroscopic Methods*, Wiley-Interscience, New York, 1976: (a) P. J. Hendra, Chap. 6; (b) F. A. Bovey, Chap. 10; (c) J. Schaefer, Chap. 11; (d) A. T. Bullock and G. G. Cameron, Chap. 15; (e) K. Shimada and M. Szwarc, Chap. 16; (f) G. Allen, Chap. 1; (g) J. S. Higgins, Chap. 2; (h) D. T. Clark, Chap. 9.

24. J. F. Rabek, *Experimental Methods in Polymer Chemistry*, Wiley-Interscience, New York, 1980, Chap. 17.
25. L. S. Bark and N. S. Allen (eds.), *Analysis of Polymer Systems*, Applied Science, London, 1982: (a) J. R. Ebdon, Chap. 2; (b) S. Foti and G. Montaudo, Chap. 51.
26. V. C. McBrierty and D. C. Douglass, *Macromol. Rev.*, **16**, 291 (1981).
27. J. C. Randall, *Polymer Sequence Determination—Carbon-13 Method*, Academic Press, New York, 1977.
28. R. Voelkel, *Angew. Chem. Int. Ed. English*, **27**, 1468 (1988).
29. I. Ya Slonim and A. N. Lyubimov, *The NMR of Polymers*, Plenum Press, New York, 1970.
30. F. C. Schilling, F. A. Bovey, M. D. Bruch, and S. A. Kozlowski, *Macromolecules*, **18**, 1418 (1985).
31. A. Zambelli, D. E. Dorman, A. I. R. Brewster, and F. A. Bovey, *Macromolecules*, **6**, 925 (1973).
32. F. A. Bovey, *Chain Structure and Conformation of Macromolecules*, Academic Press, New York, 1982, Chap. 3.
33. J. Schaefer, E. O. Stejskal, and R. Buchdahl, *Macromolecules*, **10**, 384 (1977).
34. N.-L. Yang, J. Liutkas, and H. Haubenstock, in *Polymer Characterization by ESR and NMR* (A. E. Woodward and F. A. Bovey, eds.), ACS Symp. Ser. 142, American Chemical Society, Washington, D.C., 1980, p. 35.
35. P.-O. Kinell, B. Rånby, and V. Runnstrom-Reio (eds.), *ESR Applications to Polymer Research*, Wiley-Interscience, New York, 1973.
36. B. Rånby and J. F. Rabek, *ESR Spectroscopy in Polymer Research*, Springer-Verlag, Berlin, 1977.
37. R. F. Boyer and S. E. Keinath (eds.), *Molecular Motion in Polymers by ESR*, Harwood Academic, New York, 1980.
38. J. Sohma and M. Sakaguchi, *Adv. Polym. Sci.*, **20**, 109 (1976).
39. J. F. Rabek, *Experimental Methods in Polymer Chemistry*, Wiley-Interscience, New York, 1980, Chap. 21.
40. M. Tryon and E. Horowitz, in *Analytical Chemistry of Polymers* (G. M. Kline, ed.), Wiley-Interscience, New York, Part 2, 1962, Chap. 7.
41. J. F. Rabek, *Experimental Methods in Polymer Chemistry*, Wiley-Interscience, New York, 1980, Chap. 14.
42. M. Nencioni and S. Russo, *J. Macromol. Sci.-Chem.*, **A17**, 1255 (1982).
43. L. H. Garci Rubio, N. Ro, and R. D. Patel, *Macromolecules*, **17**, 1998 (1984).
44. R. W. Woody, *Macromol. Rev.*, **12**, 181 (1977).
45. J. F. Rabek, *Experimental Methods in Polymer Chemistry*, Wiley-Interscience, New York, 1980, Chap. 16.
46. H. Morawetz, *Science*, **203**, 405 (1979).
47. E. V. Anufrieva and Y. Ya. Gotlib, *Adv. Polym. Sci.*, **40**, 1 (1981).
48. K. P. Ghiggino, A. J. Roberts, and D. Phillips, *Adv. Polym. Sci.*, **40**, 69 (1981).
49. J. F. Rabek, *Experimental Methods in Polymer Chemistry*, Wiley-Interscience, New York, 1980, Chaps. 18, 28, 29.
50. L. E. Alexander, *X-Ray Diffraction Methods in Polymer Science*, Wiley-Interscience, New York, 1969.
51. P. Corradini, in *The Stereochemistry of Macromolecules* (A. D. Ketley, ed.), Dekker, New York, 1968, Chap. 1.
52. M. Kakudo and N. Kasai, *X-Ray Diffraction by Polymers*, Elsevier, Amsterdam, 1972.

53. R. S. Stein and G. L. Wilkes, in *Structure and Properties of Oriented Polymers* (I. M. Ward, eds.), Applied Sciences, London, 1975, Chap. 3.
54. E. W. Fischer, *Polym. J.*, **17**, 307 (1985).
55. R. W. Richards, *Dev. Polym. Charact.*, **5**, 1 (1986).
56. A. Czanderna (ed.), *Methods of Surface Analysis*, Elsevier, New York, 1975.
57. D. M. Hercules and S. H. Hercules, *J. Chem. Educ.*, **61**, 402, 483, 592 (1984).
58. D. T. Clark and W. J. Feast (eds.), *Polymer Surfaces*, Wiley-Interscience, New York, 1978: (a) H. A. Willis and V. J. I. Zichy, Chap. 15; (b) D. T. Clark, Chap. 16.
59. W. J. Feast and H. S. Munro (eds.), *Polymer Surfaces and Interfaces*, Wiley-Interscience, New York, 1987.
60. J. F. Rabek, *Experimental Methods in Polymer Chemistry*, Wiley-Interscience, New York, 1980, Chap. 27.
61. A. Rosencwaig, *Photoacoustics and Photoacoustic Spectroscopy*, Wiley-Interscience, New York, 1980.
62. J. F. McClelland, *Anal. Chem.*, **55**, 89A (1983).
63. J. F. Rabek, *Experimental Methods in Polymer Chemistry*, Wiley-Interscience, New York, 1980, Chap. 30.
64. D. T. Clark, *Adv. Polym. Sci.*, **24**, 125 (1977).
65. H. R. Thomas and J. J. O'Malley, *Macromolecules*, **12**, 323, (1979).
66. E. A. Turi (ed.), *Thermal Characterization of Polymeric Materials*, Academic Press, New York, 1981.
67. W. W. Wendlandt and L. W. Collins, *Thermal Analysis*, Wiley-Interscience, New York, 1977.
68. J. F. Rabek, *Experimental Methods in Polymer Chemistry*, Wiley-Interscience, New York, 1980, Chap. 34.
69. R. C. Mackenzie, *Differential Thermal Analysis*, Vols. 1 and 2, Academic Press, New York, 1970–72.
70. W. W. Wendlandt, *Thermal Methods of Analysis*, 2nd ed., Wiley-Interscience, New York, 1974: (a) Chaps. 5 and 6, (b) Chap. 11, (c) Chaps. 2–4, (d) Chap. 8.
71. B. Ke (ed.), *Newer Methods of Polymer Characterization*, Wiley-Interscience, New York, 1964: (a) B. Ke, Chap. 9, (b) E. Fisher, Chap. 7.
72. H. LeChatelier, *Bull. Soc. Fr. Mineral.*, **10**, 204 (1877).
73. J. D. Matlack and A. P. Metzger, *J. Polym. Sci.*, **B4**, 875 (1966).
74. I. M. Kolthoff, P. J. Elving, and F. H. Stross, *Treatise on Analytical Chemistry*, Part 3, Vol. 3, Wiley-Interscience, New York, 1976: (a) H. Friedman, pp. 542–551; (b) R. B. Seymour, pp. 341–392.
75. C. J. Keattch, *An Introduction to Thermogravimetry*, Heyden, London, 1969.
76. K. Honda, *Sci. Repts. Imp. Tohoku Univ.*, **4**, 97 (1915).
77. C. S. Marvel, in *High Temperature Polymers* (C. L. Segal, ed.) Dekker, New York, 1967, p. 9.
78. S. A. Liebman and E. J. Levy (eds.), *Pyrolysis and G. C. in Polymer Analysis*, Chromat. Sci. Ser. No. 29, Dekker, New York, 1985.
79. M. P. Stevens, *Characterization and Analysis of Polymers by Gas Chromatography*, Dekker, New York, 1969, Chap. 4.
80. W. H. T. Davison, S. Slaney, and A. L. Wragg, *Chem. Ind. (London)*, 1356 (1954).
81. R. S. Lehrle and J. C. Robb, *Nature*, **183**, 1671 (1959).
82. E. M. Barrall II, R. S. Porter, and J. F. Johnson, *Anal. Chem.*, **35**, 73 (1963).

83. B. Kolb, G. Kemmner, K. H. Kaiser, E. W. Cieplinski, and L. S. Ettre, *Z. Anal. Chem.*, **209**, 302 (1965).
84. I. M. Kolthoff, P. J. Elving, and F. H. Stross, *Treatise on Analytical Chemistry*, Part 3, Vol. 4, Wiley-Interscience, New York, 1977: (a) H. W. Eikner, pp. 1–56; (b) G. M. Sinclair, pp. 419–462.
85. K. Kishore and K. Mohandes, *J. Macromol. Sci. -Chem.*, **A18**, 379 (1982).
86. C. P. Fenimore and F. J. Martin, *Modern Plastics*, **44** (11), 141 (1966).
87. R. C. Nametz, *Ind. Eng. Chem.*, **59**, 99 (1967).
88. P. R. Johnson, *J. Polym. Sci.*, **18**, 491 (1974).
89. Y.-L. Hsieh and K.-N Yeh, *J. Appl. Polym. Sci.*, **28**, 1389 (1983).
90. R. M. Aseeva and G. E. Zaikov, *Combustion of Polymer Materials*, Hanser, Munich, 1986, Chap. 5.
91. N. M. Bikales (ed.), *The Mechanical Properties of Polymers*, Wiley-Interscience, New York, 1980.
92. A. Davis and D. Sims, *Weathering of Polymers*, Applied Science, London, 1983.
93. H. F. Mark, N. M. Bikales, C. G. Overberger, G. Menges, and J. I. Kroschwitz (eds.), *Encyclopedia of Polymer Science and Engineering*, Vol. 5, Wiley-Interscience, New York, 1986: (a) J. R. Perkins Jr., pp. 431–462; (b) K. N. Mathes, pp. 507–587.
94. J. Agranoff (ed.), *Modern Plastics Encyclopedia, 1984–85*, McGraw-Hill, New York, 1984, pp. 437ff.
95. C. C. Ku and R. Liepins, *Electrical Properties of Polymers*, Hanser, Munich, 1987.

Review exercises

1. Elemental analysis of a copolymer of propylene and vinyl chloride showed that the polymer contained 25.9% chlorine. What is the ratio of vinyl chloride to propylene in the copolymer?

2. Predict the ozonolysis $(1.0_3 ; 2.H_2O/Zn)$ products of (a) 3,4-polyisoprene; (b) 1,2-polyisoprene; (c) polyacetylene.

3. When a random copolymer of 1,3-butadiene (70%) and *o*-chlorostyrene (30%) was subjected to ozonolysis followed by hydrolysis (under oxidative conditions), the major organic products were 2-(*o*-chlorophenyl)hexanedioic acid, 2-carboxyhexanedioic acid, and butanedioic acid. Show the structural units in the copolymer that gave rise to the products. (ref. 6)

4. Justify the spectral differences in the 700 to 900-cm^{-1} region between the polymer and model compound shown in Figure 5.1.

5. Suppose you are an employee of a plastics firm and you are assigned the task of identifying a polymer used by a competitor to make certain automotive parts. You prepare a film of the polymer and determine its IR spectrum, as shown at the top of the next page. What information can you deduce from the spectrum? (A table of IR group frequencies might prove helpful.)

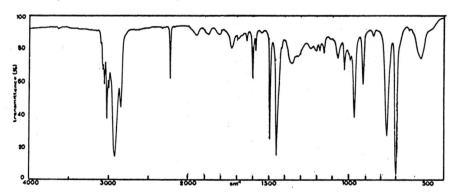

6. One stereoregular form of polystyrene (A) has a ^{1}H NMR spectrum containing a triplet centered at about 1.4 ppm and a quintet at about 1.9 ppm. Another stereoregular form (B) has an octet at about 1.6 ppm and a quintet at about 2.3 ppm. (Both spectra also exhibit phenyl proton resonances further downfield.) Interpret the spectra and determine which corresponds to isotactic and which to syndiotactic. (Syndiotactic polystyrene was first reported by N. Ishihara and coworkers in a 1986 paper. You may wish to locate the paper and check your answer.)

7. Figure 5.5 shows a partial ^{13}C (CP-MAS) NMR spectrum of polycarbonate. (a) How many additional peaks would appear in the full ^{13}C spectrum? (b) Describe or sketch the ^{1}H NMR spectrum expected for polycarbonate. (Assume high resolution.)

8. Two linear polybenzyls have been reported,

the latter by Lim and Choi in 1986. How could you distinguish between the two polymers spectroscopically (IR and NMR)?

9. In addition to giving information on stereochemical sequences, NMR can shed light on copolymer structural sequences. Poly(methyl methacrylate) exhibits a single resonance for the -OCH$_3$ protons at 3.74 ppm downfield from hexamethyl-disiloxane (HMDS) in benzonitrile solvent. Copolymers of methyl methacrylate (M_1) and acrylonitrile (M_2) exhibit three methoxyl proton peaks at 3.74, 3.82, and 3.86 ppm. The peaks at 3.82 and 3.86 get progressively larger at the expense of the 3.74 peak as the concentration of M_2 in the copolymer is increased. The peak at 3.86 ppm is the dominant methoxyl resonance at high (>75%) levels of M_2. To what triad sequences of structural units (not tacticity) do these resonances most likely correspond? (See R. Chujo, H. Ubara, and A. Nishioka, *Polym. J.*, **3**, 670, 1972.)

10. How would you expect the x-ray diffraction pattern of a rubber band to change as the band is stretched?

11. In your own words, explain the difference between (a) SEM and TEM, (b) SAXS and WAXS, (c) AES and ESCA, (d) ISS and SIMS, (e) DSC and DTA, (f) isothermal and nonisothermal TGA, (g) self-extinguishing and nonflammable.

12. Suggest a spectroscopic method for determining each of the following and describe the special feature(s) pertaining to each analysis: (a) The amount of styrene in a styrene-butadiene copolymer. (b) The amount of 1,2 polymer in poly(1,3-butadiene). (c) The stereochemistry of double bonds in 1,4-poly(1,3-butadiene). (d) Whether any unreacted polyacrylonitrile (PAN) is present in a sample of carbon fiber prepared by pyrolysis of PAN. (e) Whether a sample of poly(vinyl acetate) has undergone hydrolysis on exposure to moisture. (f) The identity of metallic driers used in a sample of painted wooden fence. (g) How completely a polyester surgical implant has been coated with polytetrafluoroethylene to improve its chemical resistance.

13. From the data in Table 3.5 (Chapter 3), sketch the anticipated DSC thermogram of poly(ethylene adipate) between -100 and $100°C$ (a) if the sample is highly crystalline, and (b) if the sample was initially quenched in the amorphous state. If the sample were to be cycled back to $-100°C$ at the same rate, would you expect the cooling and heating thermograms to appear the same? Explain.

14. A method of determining percent vinyl acetate in poly[ethylene-*co*-(vinyl acetate)] by pyrolysis GC is described in this chapter. Suggest a way to determine ethyl acrylate ($CH_2 = CHCO_2C_2H_5$) in poly[ethylene-*co*-(ethyl acrylate)]. (ref. 82)

15. A method employing pyrolysis and hydrogenation has been used to determine which products are unsaturated in a pyrogram (ref. 83). How would the chromatogram shown in Figure 5.19 change if this procedure were used? Suggest a reason why the largest peak in each triplet is 1-alkene. (You can check the answer in Chapter 9.)

16. An empirical method of determining percent crystallinity in high- and low-density polyethylene using PGC has been reported (D. Deur-Siftar, *J. Gas Chromatog.*, **5**, 72, 1967). If the journal is in your library, read the paper and summarize the method.

17. Given the data in Table 5.3, predict the approximate LOI of polybenzimidazole (structure 1, Chapter 4). Justify your prediction.

Part II
VINYL POLYMERS

6
Free radical vinyl polymerization

6.1 Introduction

From the commercial standpoint, vinyl polymers are the most important of all polymer types. As discussed in Chapter 1, polyethylene (high and low density), poly(vinyl chloride), polystyrene, and polypropylene lead the market in total volume output and should continue to do so in the foreseeable future. For this reason, a lot more research has been accomplished in the area of vinyl polymerization than with most other polymer types. It is not the purpose of this part of the book to discuss individual vinyl polymers, but rather to emphasize in general terms the differences among the various modes of polymerization and those factors in the polymerization processes responsible for the structure of the final product. For convenience we are classifying all polymers prepared by addition reactions of carbon-carbon double bonds as vinyl polymers. Some polymer scientists prefer to classify diene and vinylidene polymers separately.

It will be recalled from Chapter 1 that vinyl polymerization reactions are chain-reaction processes, which is the case with most (but certainly not all) addition polymerizations, and they require an initiator to begin the reaction. Polymerization occurs only at the reactive end of a growing chain; hence high molecular weights are reached rapidly at relatively low percentages of conversion, and monomer is always present in significant quantities during the process. While these generalizations apply to all chain polymerizations, the differences among the various types of initiation and propagation are significant enough to treat each separately. This chapter is concerned only with free radical polymerization.

A wide variety of monomer types lend themselves to free radical polymerization, which, as we shall see later, is not the case with ionic polymerization. Commercially important vinyl polymers, including polydienes, prepared by free radical processes are given in Table 6.1. Not included are *derivatives* of vinyl polymers, such as chlorinated polyethylene or poly(vinyl alcohol); these are discussed in Chapter 9. It should be emphasized here that

Table 6.1. Commercially important vinyl polymers prepared by free radical polymerization[a]

Polymer or copolymer	Method of manufacture[b]	Major uses
Polyethylene, low density (LDPE)	HP, B, Sol	Packaging film, wire and cable insulation, toys, flexible bottles, housewares, coatings
Poly(vinyl chloride)	B, S, Sol, E	Construction, rigid pipe, flooring, wire and cable insulation, film and sheet
Polystyrene[c]	B, S, Sol, E	Packaging (foam and film), foam insulation, appliances, housewares, toys
Styrene–butadiene rubber (SBR)	E	Tires, belting, flooring, molded goods, shoe soles, electrical insulation
Butadiene–acrylonitrile copolymer (nitrile rubber)	E	Fuel tanks, gasoline, hoses, adhesives, impregnated paper, leather and textiles
Acrylonitrile–butadiene–styrene copolymer (ABS)	E	Engineering plastics, household appliances, business machines, telephones, electronics, automotive parts, luggage
Polychloroprene (neoprene rubber)	E	Tires, wire coatings, belting, hoses, shoe heels, coated fabrics
Poly(methyl methacrylate)	B, S	Automotive parts, molding, compositions, decorative panels, skylights, glazing
Polyacrylonitrile[d]	S	Textile fibers, food packaging
Poly(vinyl acetate)	B, S, Sol, E	Water-based paints, adhesives, conversion to poly(vinyl alcohol)
Poly(vinylidene chloride)[e]	E, S	Food packaging
Poly(acrylic acid) and poly(methacrylic acid)	Sol, E	Adhesives, thickening agents, ionomers
Polyacrylamide	E	Thickening agent, flocculent
Polytetrafluoroethylene	HP, S, Sol	Electrical insulation, gaskets, bearings, bushings, valves, nonstick cooking utensils
Polytrichlorofluoroethylene	S, Sol	Gaskets, tubing, wire insulation
Poly(vinylidene fluoride)	S, Sol	Protective coatings, gaskets, pipe
Poly(vinyl fluoride)	S, Sol	Protective coatings
Allyl resins[f]	B, Sol	Lenses, electronics parts

[a] A number of copolymers are manufactured besides the high-volume ones listed.

[b] HP = high pressure; B = bulk; E = emulsion; S = suspension; Sol = solution.

[c] Poly(*p*-methylstyrene) also available.

[d] See Chapter 1 for a discussion of acrylonitrile-based textile fibers.

[e] Including copolymers, principally with vinyl chloride.

[f] Crosslinked diallyl and triallyl esters and ethers.

despite the very large volume of work published in the polymer field, most polymerization processes are still not completely understood. New information or new theories appear in the polymer literature with remarkable consistency, which makes it imperative that polymer scientists stay abreast of new developments.

6.2 Free radical initiators

Certain monomers, notably styrene and methyl methacrylate and some strained-ring cycloalkenes, undergo polymerization on heating in the absence of any added free radical initiator. Most monomers, however, require some kind of initiator. A large number of free radical initiators[1,2] are available; they may be classified into four major types: peroxides and hydroperoxides, azo compounds, redox initiators, and certain compounds that form radicals under the influence of light (*photoinitiators*). High-energy radiation (α- and β-particles, γ- and x-rays) can also promote free radical polymerization, although such radiation is used less commonly. *Plasma polymerization*[3,4]— polymerization initiated by a partially ionized gas generated by a radio-frequency glow discharge—is currently receiving much attention for preparing coatings and insulating layers, especially for microelectronics and medical applications. The polymerization mechanism, however, is complex and appears to involve both free radical and ionic species, and the polymers thus formed are highly crosslinked and bear little resemblance to polymers prepared by more conventional chain polymerization. Indeed, just about *any* organic compound will undergo plasma polymerization.

6.2.1 Peroxides and hydroperoxides

Of the various types of initiators, peroxides (ROOR) and hydroperoxides (ROOH) are most widely used. They are thermally unstable and decompose into radicals at a temperature and rate dependent on structure. The most commonly used peroxide is benzoyl peroxide (**1**), which undergoes thermal homolysis to form benzoyloxy radicals (reaction 6.1).* The benzoyloxy radicals may undergo a variety of reactions besides adding to monomer, including recombination (reverse of 6.1), decomposition to phenyl radicals and

$$1 \qquad (6.1)$$

---

* For radical structures such as the benzoyloxy radical, only the unpaired electron is shown. A more complete representation would include lone pair electrons on oxygen and contributing resonance structures.

carbon dioxide (6.2), and radical combination (6.3 and 6.4). These secondary reactions occur because of the confining effect of solvent molecules (the "cage" effect), and as a result, the concentration of initiator radicals is depleted. Induced decomposition (6.5) is another "wastage" reaction.

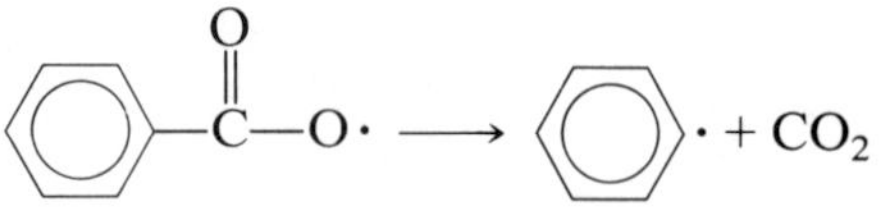

(6.2)

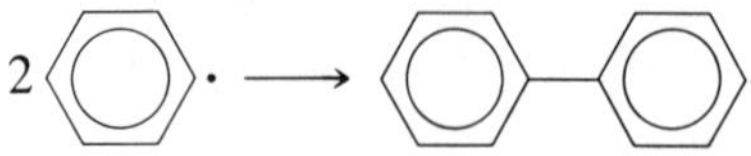

(6.3)

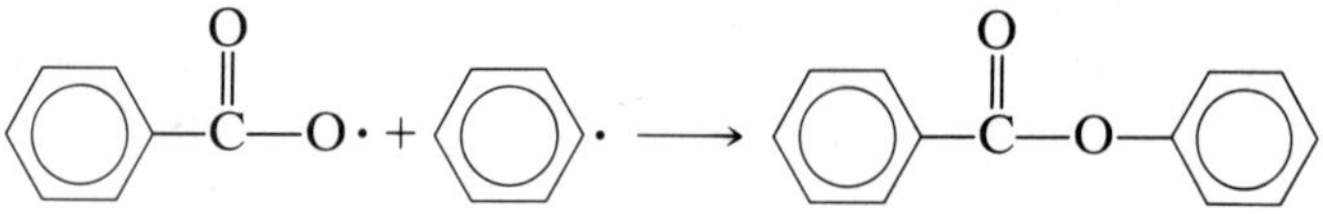

(6.4)

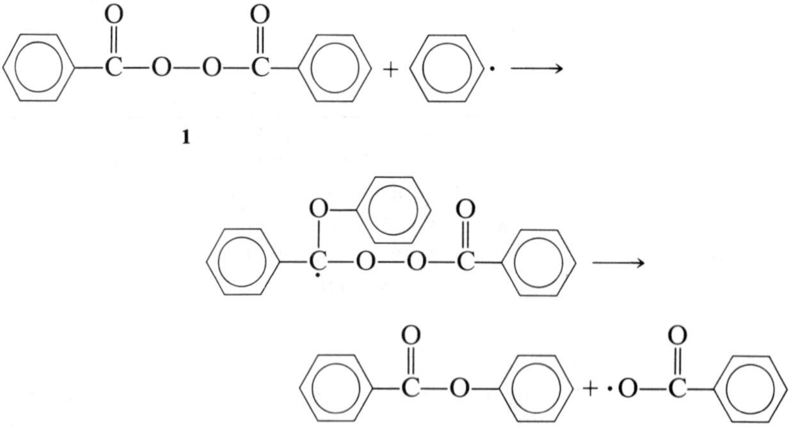

1

(6.5)

Two other common initiators are diacetyl peroxide (**2**) and di-*t*-butyl peroxide (**3**). Hydroperoxides such as cumyl hydroperoxide (**4**) decompose to form alkoxy and hydroxy radicals (6.6). Because hydroperoxides contain an active hydrogen atom, induced decomposition occurs readily, for example by a chain-end radical (6.7). Peroxy radicals may also combine with subsequent formation of oxygen (6.8).

$CH_3\overset{O}{\overset{\|}{C}}OO\overset{O}{\overset{\|}{C}}CH_3$ (**2**)

$(CH_3)_3\overset{O}{\overset{\|}{C}}OO\overset{O}{\overset{\|}{C}}(CH_3)_3$ (**3**)

4

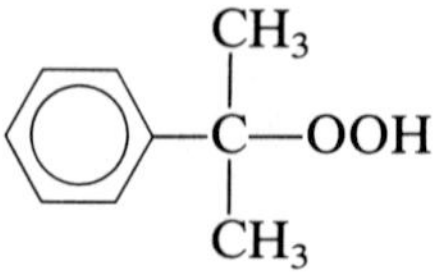

$$ROOH \longrightarrow RO\cdot + \cdot OH \qquad (6.6)$$

$$RO \left[CH_2CH \right]_n CH_2CH \cdot + ROOH \longrightarrow$$
$$\qquad\qquad R' \qquad\quad R'$$

$$RO \left[CH_2CH \right]_n CH_2CH_2 + ROO \cdot \quad (6.7)$$
$$\qquad\qquad R' \qquad\quad R'$$

$$2ROO \cdot \longrightarrow [ROO\!-\!OOR] \longrightarrow 2RO \cdot + O_2 \qquad (6.8)$$

The extent to which side reactions occur depends on the structure of the peroxide, the stability of the initially formed radicals, and the reactivity of the monomer. Ideally, a peroxide initiator should be relatively stable at room temperature but should decompose rapidly enough at polymer-processing temperatures to ensure a practical reaction rate. Benzoyl peroxide (half-life 30 minutes at 100°C) has the advantage that benzoyloxy radicals are stable enough that they tend to react with more reactive monomer molecules before eliminating carbon dioxide, thus reducing initiator wastage. Acetoxy radicals formed from **2**, on the other hand, are much less stable, and initiator wastage occurs more readily. More stable than benzoyloxy radicals are *t*-butoxy radicals, which are almost entirely captured by monomer; however, the decomposition temperature is relatively high (half-life ten hours at 126°C).

Decomposition of peroxides can frequently be induced at lower temperatures by the addition of *promoters*. For example, addition of N,N-dimethylaniline to benzoyl peroxide causes the latter to decompose rapidly at room temperature. Kinetics studies[5,6] indicate that the decomposition involves formation of an unstable ionic intermediate (6.9) that reacts further to give benzoyloxy radical and a radical cation (6.10). The radical cation apparently undergoes reactions other than addition to monomer, since polymers formed by this method contain no nitrogen.

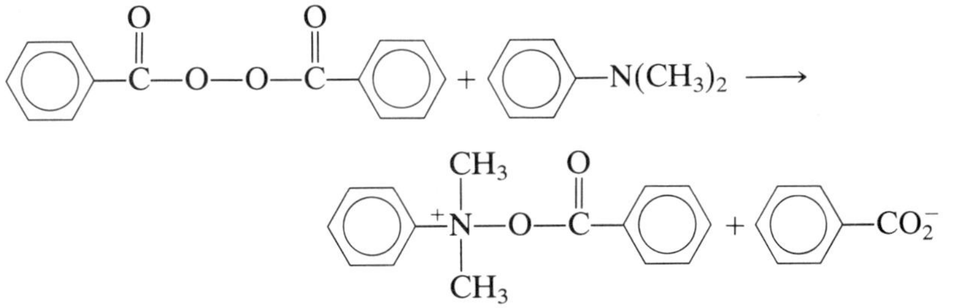

$$\qquad\qquad\qquad (6.9)$$

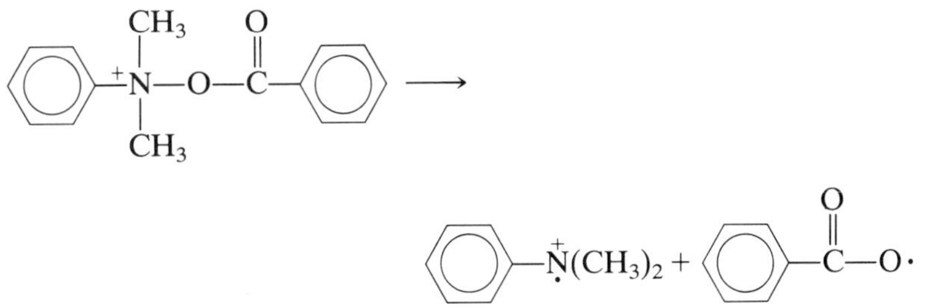

$$\qquad\qquad\qquad (6.10)$$

6.2.2 Azo compounds

The most commonly used azo compounds are those having cyano groups on the carbons attached to the azo linkage—for example, α,α'-azobis(isobutyronitrile) (**5**), which decomposes at relatively low temperatures (half-life 1.3 hours at 80°C). The driving force for decomposition is formation of nitrogen and the resonance-stabilized cyanopropyl radical (6.11). As with peroxide decomposition, the initially formed radicals can combine in the solvent cage to deplete initiator concentration. Combination leads to both tetramethylsuccinonitrile (**6**) and the ketenimine (**7**) (reactions 6.12 and 6.13, respectively).

$$(CH_3)_2\overset{\overset{\displaystyle CN}{|}}{C}-N{=}N-\overset{\overset{\displaystyle CN}{|}}{C}(CH_3)_2 \longrightarrow 2(CH_3)_2\overset{\overset{\displaystyle CN}{|}}{C}\cdot + N_2 \qquad (6.11)$$

$$\textbf{5}$$

$$2(CH_3)_2\overset{\overset{\displaystyle CN}{|}}{C}\cdot \quad \begin{cases} \longrightarrow (CH_3)_2\overset{\overset{\displaystyle CN}{|}}{C}-\overset{\overset{\displaystyle CN}{|}}{C}(CH_3)_2 \qquad (6.12) \\ \qquad\qquad \textbf{6} \\ \\ \longrightarrow (CH_3)_2C{=}C{=}N-\overset{\overset{\displaystyle CN}{|}}{C}(CH_3)_2 \qquad (6.13) \\ \qquad\qquad\qquad \textbf{7} \end{cases}$$

6.2.3 Redox initiators

Production of free radicals by one-electron transfer reactions is particularly useful in initiation of low-temperature polymerization and emulsion polymerization. Some typical examples are given in equations (6.14) to (6.16). Apart from the fact that low temperatures can be employed with redox systems, reaction rates are easy to control by varying the concentration of metal ion or peroxide. The decomposition of cumyl hydroperoxide (6.14) is used commercially for the low-temperature emulsion copolymerization of styrene and butadiene to form styrene–butadiene rubber. Hydrogen peroxide or persulfate systems, (6.15) and (6.16), are used in emulsion polymerization to produce radicals in the aqueous phase. For nonaqueous polymerization, metal ions are generally introduced as the naphthenates. Numerous combinations of possible redox initiator systems are available.

$$\text{C}_6\text{H}_5-\overset{\overset{\displaystyle CH_3}{|}}{\underset{\underset{\displaystyle CH_3}{|}}{C}}-OOH + Fe^{++} \longrightarrow C_6\text{H}_5-\overset{\overset{\displaystyle CH_3}{|}}{\underset{\underset{\displaystyle CH_3}{|}}{C}}-O\cdot + OH^- + Fe^{3+} \quad (6.14)$$

$$\textbf{4}$$

$$HOOH + Fe^{++} \longrightarrow HO\cdot + OH^- + Fe^{3+} \qquad (6.15)$$

$$^-O_3SOOSO_3^- + S_2O_3^{--} \longrightarrow SO_4^-\cdot + SO_4^{--} + \cdot S_2O_3^- \qquad (6.16)$$

6.2.4   Photoinitiators

Peroxides and azo compounds dissociate photolytically as well as thermally. The major advantage of photoinitiation is that the reaction is essentially independent of temperature; thus polymerizations may be conducted even at very low temperatures. Furthermore, better control of the polymerization reaction is generally possible because narrow wavelength bands may be used to initiate decomposition, and the reaction can be stopped simply by removing the light source. A wide variety of photolabile compounds are available, including disulfides (6.17), benzoin (**8**) (6.18), and benzil (**9**) (6.19).

$$RSSR \xrightarrow{h\nu} 2RS\cdot \qquad (6.17)$$

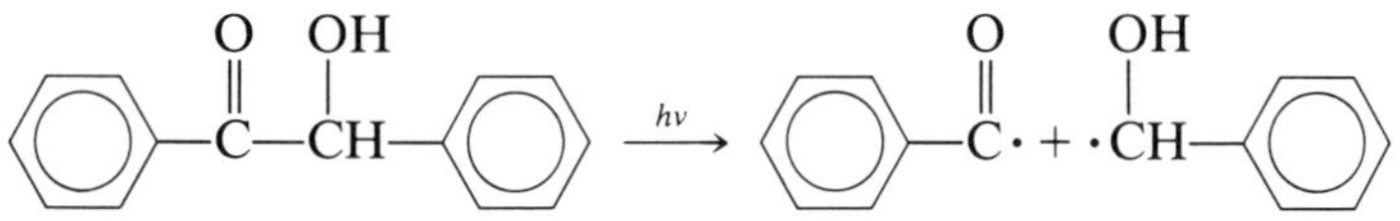

(6.18)

8

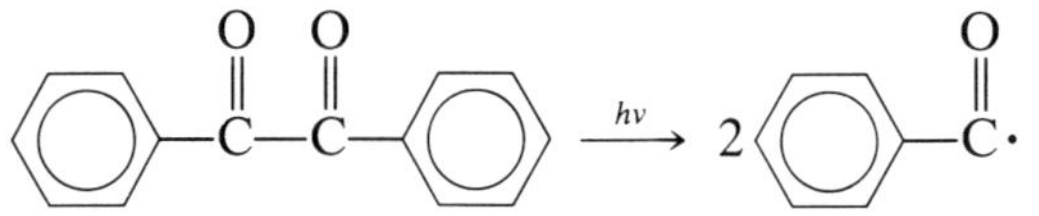

(6.19)

9

6.2.5   Thermal polymerization

Some monomers polymerize slowly on heating in the absence of added initiator.[7a] In such cases, free radical initiating species are generated in situ by mechanisms that are not, in most cases, well understood. Of the commercially important vinyl monomers, styrene undergoes thermal polymerization most rapidly. Several pathways have been suggested for thermal polymerization of styrene[7b]; recent work[8-10] suggests that an initial Diels–Alder dimer (**10**) forms (6.20), which transfers a hydrogen atom to monomer (6.21) to yield an initiator styryl radical (**11**) and a benzylic radical (**12**). This is an example of what is referred to as *molecule-induced homolysis*—the rapid formation of radicals by reaction of nonradical species.[11] Photopolymerization of styrene with no added initiator appears to follow a similar pathway.[12]

$$2CH_2{=}CH\phi \longrightarrow \qquad (6.20)$$

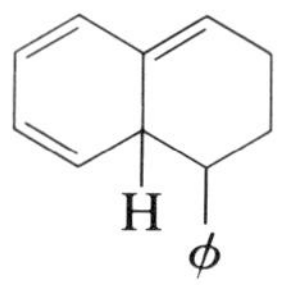

10

$$10 + CH_2{=}CH\phi \longrightarrow CH_3\overset{\cdot}{C}H\phi +$$

(6.21)

11   12

Apart from some limited applications with styrene, thermal polymerization has been used commercially only with certain polymers or oligomers end-capped with thermally labile moieties such as maleimide or nadimide for manufacture of composites (see Chapter 4, Table 4.4).

6.3   Techniques of free radical polymerization

Free radical polymerization can be accomplished in bulk, suspension, solution, or emulsion.[13] Ionic and other nonradical polymerizations are usually confined to solution techniques. Each of the methods has advantages and disadvantages, as outlined in Table 6.2. In addition, work has also been done on solid- and gas-phase polymerizations of vinyl monomers, but these are of lesser importance. Because polymers are not volatile, the term *gas-phase polymerization* means, in effect, bulk polymerization in which monomer vapors diffuse to the polymerization site. A special case of solid-phase polymerization involving inclusion complexes is discussed in Section 6.6.

6.3.1   Bulk

Bulk polymerization is simplest from the standpoint of formulation and equipment, but it is also the most difficult to control, particularly when the polymerization reaction is very exothermic. This, coupled with problems of heat transfer as the monomer–polymer solution increases in viscosity, limits the use of bulk methods in commercial production, although more efficient bulk processes have been developed in recent years.

In cases where polymer is insoluble in monomer, polymer precipitates and the viscosity of the medium does not change appreciably. Problems still arise, however, as a result of free radicals (detectable by ESR) being occluded in the polymer droplet, which can lead to *autoacceleration*, that is, a rapid increase in the polymerization rate.[7c] In some instances, particularly with diene monomers, this occlusion effect may lead to formation of insoluble crosslinked polymer nodules, a phenomenon referred to as *popcorn polymerization*.[14a] The crosslinked nodules are usually of light weight and occupy considerably more volume than the monomers from which they are derived, which may cause fouling and even fracture of the polymerization apparatus.

Table 6.2. Free radical polymerization techniques

Method	Advantages	Disadvantages
Bulk	Simple; no contaminants added	Reaction exotherm difficult to control; high viscosity
Suspension	Heat readily dispersed; low viscosity; polymer obtained in granular form and may be used directly	Washing and/or drying required; agglomeration may occur; contamination by stabilizer
Solution	Heat readily dispersed; low viscosity; may be used directly as solution	Added cost of solvent; solvent difficult to remove; possible chain transfer with solvent; possible environmental pollution
Emulsion	Heat readily dispersed; low viscosity; high molecular weight obtainable; may be used directly as emulsion; works on tacky polymers	Contamination by emulsifier and other ingredients; chain transfer agents often needed to control DP; washing and drying necessary for bulk polymer

The major commercial uses of bulk vinyl polymerization are in casting formulations and low-molecular-weight polymers for use as adhesives, plasticizers, tackifiers, and lubricant additives.

6.3.2   Suspension

Suspension polymerization involves mechanically dispersing monomer in a noncompatible liquid, usually water, and polymerizing the resultant monomer droplets by use of a monomer-soluble initiator. Monomer is kept in suspension by continuous agitation and the use of *stabilizers* such as poly-(vinyl alcohol) or methyl cellulose. If the process is carefully controlled, polymer is obtained in the form of granular beads, which are easy to handle and can be isolated by filtration or by spraying into a heated chamber (*spray drying*). A major advantage is that heat transfer is very efficient and the reaction is therefore easily controlled. Suspension polymerization cannot be used for tacky polymers such as elastomers because of the tendency for agglomeration of polymer particles. From the standpoint of kinetics and mechanism, suspension polymerization is identical to bulk polymerization.

Suspension methods are used to prepare a number of granular polymers, including polystyrene, poly(vinyl chloride), and poly(methyl methacrylate).

6.3.3   Solution

Like suspension, solution polymerization allows efficient heat transfer. Solvent must be chosen carefully, otherwise chain transfer reactions (discussed in the next section) may severely limit the molecular weight. Because of problems in removing solvent completely from the resultant polymer, the method is best suited to applications where the solution may be used directly, as with certain adhesives or solvent-based paints.

6.3.4   Emulsion

Developed at Goodyear Tire and Rubber Company in the 1920s, emulsion polymerization resembles suspension polymerization in that water is used as a dispersing medium and heat transfer is very efficient; but there the similarity ends.[15-17a] Monomer is dispersed in the aqueous phase by an emulsifying agent such as a soap or detergent. Initiator radicals, usually of the redox type, are generated in the aqueous phase and diffuse into soap micelles swollen with monomer molecules. As monomer is used up in the polymerization reaction, more monomer migrates into the micelles, and thus the reaction continues. Termination of polymerization occurs by radical combination when a new radical diffuses into the micelle. Because only one radical is present in the micelle prior to termination, extremely high molecular weights are obtainable, generally too high to be of practical value unless compounds called *chain transfer agents* are added that control the degree of polymerization. (How chain transfer agents work is discussed later in the chapter.) The

Table 6.3. Typical emulsion polymerization recipes[a]

Ingredients, conditions	Styrene–butadiene copolymer	Polyacrylate latex
Ingredients (parts by weight)		
Water	190	133
Butadiene	70	—
Styrene	30	—
Ethyl acrylate	—	93
2-Chloroethyl vinyl ether	—	5
p-Divinylbenzene	—	2
Soap	5	3[b]
Potassium persulfate	0.3	1
n-Dodecyl mercaptan	0.5	—
Sodium pyrophosphate	—	0.7
Conditions		
Time	12 hr	8 hr
Temperature	50°C	60°C
Yield	65%	~100%

[a] Recipes from Cooper.[17a]

[b] Sodium lauryl sulfate.

overall process is complex, with reaction kinetics differing significantly from that of bulk or solution processes.

Emulsion polymerization is widely used in industry for large-scale preparations, and is particularly useful for manufacturing water-based (latex) paints or adhesives in which the emulsified product is used directly. Emulsion polymerization is also suitable for preparing tacky polymers because the very small particles are stable and resist agglomeration. Two typical emulsion recipes are given in Table 6.3.

A much less commonly used emulsion technique involves dispersing an aqueous solution of monomer in a nonaqueous phase. This is referred to as an *inverse* or *water-in-oil* emulsion (as opposed to the more conventional *oil-in-water* type). The mechanism of polymerization is similar for the two techniques, but inverse emulsions tend to be less stable.

6.4 Kinetics and mechanism of polymerization

Initiation of free radical chain polymerization involves two reactions: formation of the initiator radical (6.22), and addition of the initiator radical to monomer (6.23). Evidence for the incorporation of initiator radicals arises from spectroscopic and chemical analysis of end groups. Addition of monomer radical to another monomer molecule, followed by successive additions of oligomer and polymer radicals to available monomer (6.24), comprise the propagation reactions.

$$\text{Initiator} \longrightarrow \text{R} \cdot \tag{6.22}$$

$$\text{R} \cdot + \text{CH}_2{=}\underset{\underset{\text{Y}}{|}}{\text{CH}} \longrightarrow \text{RCH}_2\underset{\underset{\text{Y}}{|}}{\text{CH}} \cdot \tag{6.23}$$

$$\text{RCH}_2\underset{\underset{\text{Y}}{|}}{\text{CH}} \cdot + \text{CH}_2{=}\underset{\underset{\text{Y}}{|}}{\text{CH}} \longrightarrow \text{RCH}_2\underset{\underset{\text{Y}}{|}}{\text{CH}}\text{CH}_2\underset{\underset{\text{Y}}{|}}{\text{CH}} \cdot \xrightarrow{n\text{CH}_2{=}\text{CHY}}$$

$$\text{RCH}_2\underset{\underset{\text{Y}}{|}}{\text{CH}}{-}\left[\text{CH}_2\underset{\underset{\text{Y}}{|}}{\text{CH}}\right]_n{-}\text{CH}_2\underset{\underset{\text{Y}}{|}}{\text{CH}} \cdot \tag{6.24}$$

It was mentioned in Chapter 3 (Section 3.5) that each addition step follows the predominant *head-to-tail* orientation shown in reaction (6.24). This is due to a combination of steric and electronic effects. Steric repulsion favors attack by the radical at the least hindered carbon of the double bond; and resonance stabilization favors formation of the more stable free radical. Head-to-tail polymerization does not occur exclusively. Significant amounts of *head-to-head* structures have been found (by ^{19}F NMR spectroscopy) to occur in certain fluorine-containing polymers,[18] notably poly(vinyl fluoride) (**13**) and poly(vinylidene fluoride) (**14**) (13–17% and 5–6%, respectively), and in poly(allyl acetate) (**15**) (up to 19%, depending on polymerization temperature.[19]) Where the intermediate radical is significantly resonance stabilized, however, head-to-tail structures appear to form almost exclusively.

$$-\!\!\{CH_2CHF\}\!\!- \qquad -\!\{CH_2CF_2\}\!- \qquad -\!\{CH_2CH\}\!-$$
$$CH_2OCOCH_3$$

           13   14   15

    Propagation continues until some reaction occurs to terminate it. The two principal ways that termination may occur in free radical polymerization are radical *coupling*, or *combination* (6.25), and *disproportionation* (6.26), which involves transfer of an atom, usually hydrogen, from one chain end to another.[17b,20a] Combination leads to initiator fragments at both ends of the polymer chain, whereas disproportionation results in an initiator fragment at one end. Also, combination leads to a head-to-head linkage. Both termination reactions are diffusion controlled and require pairing of electron spins.

$$\text{\textasciitilde CH}\!-\!\text{CH}\text{\textasciitilde} \qquad\qquad (6.25)$$
$$Y\quad Y$$

$$\text{\textasciitilde CH}_2\text{CH}\cdot + \cdot\text{CHCH}_2\text{\textasciitilde}$$
$$Y\qquad Y$$

$$\text{\textasciitilde CH}\!=\!\text{CH} + \text{CH}_2\text{CH}_2\text{\textasciitilde} \qquad (6.26)$$
$$Y\qquad\quad Y$$

    Whether termination occurs by coupling or by disproportionation depends in large measure on monomer structure or, more exactly, on the structure of the chain-end radical. Polystyryl radicals undergo coupling (6.27) almost exclusively at low temperatures, whereas poly(methyl methacrylate) radicals undergo mainly disproportionation (6.28); but generally both processes occur. What favors one termination reaction over the other? Steric repulsion is one factor. Coupling of poly(methyl methacrylate) radicals would lead to four bulky groups on adjacent carbons. Electrostatic repulsion of polar groups (such as ester) may also raise the activation energy for coupling. Availability of alpha hydrogens for hydrogen transfer is a third factor: poly(methyl methacrylate) radicals have five, compared with two for polystyryl radicals. It is no easy task to predict how termination will occur, however, as shown by the fact that polyacrylonitrile undergoes coupling virtually exclusively at 60°C whereas poly(vinyl acetate) undergoes disproportionation.

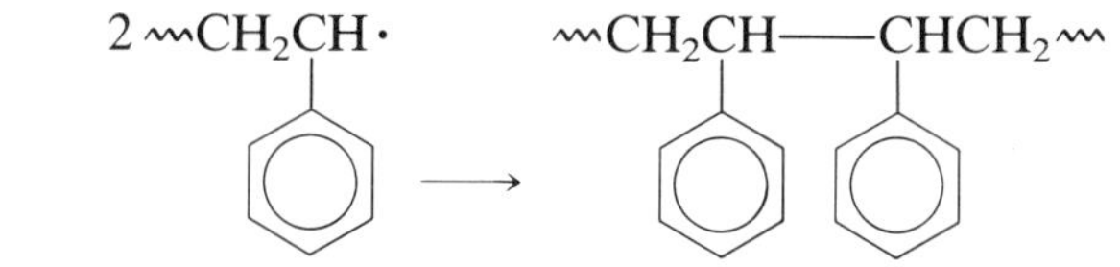

$$2 \sim CH_2CH\cdot \quad \sim CH_2CH\!\!-\!\!CHCH_2\sim$$

(6.27)

$$2 \sim CH_2\underset{CO_2CH_3}{\overset{CH_3}{C\cdot}} \longrightarrow \sim CH_2\underset{CO_2CH_3}{\overset{CH_3}{CH}} + \sim CH=\underset{CO_2CH_3}{\overset{CH_3}{C}}$$

$$\text{or} \quad \sim CH_2\!\!-\!\!\underset{CO_2CH_3}{\overset{\overset{\displaystyle CH_2}{\|}}{C}}$$

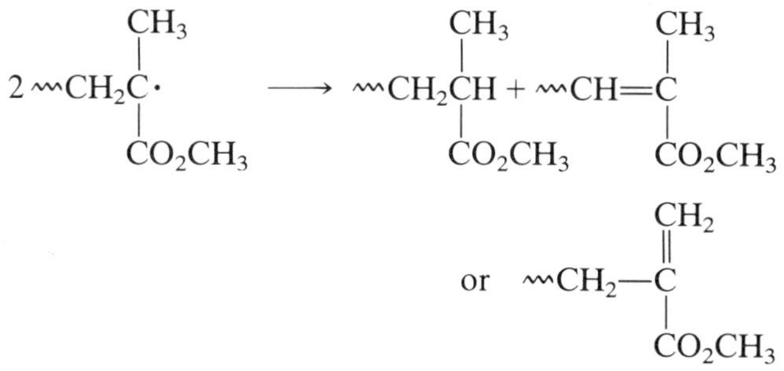

(6.28)

Another possible termination reaction involves combination of initiator radicals with chain end radicals (6.29). Called *primary radical termination*, this process is significant only at relatively high initiator levels or when very high viscosities limit the diffusion of high-molecular-weight chain end radicals.

$$\sim\!\!\sim\!\!CH_2\underset{Y}{\overset{}{CH}}\cdot + R\cdot \longrightarrow \sim\!\!\sim\!\!CH_2\underset{Y}{\overset{}{CHR}}$$

(6.29)

A simplified picture of the kinetics of free radical polymerization[7d,17c,21] assumes that the rates of initiation, propagation, and termination are all different, but that each propagation step is exclusively head-to-tail and goes at the same rate, independent of chain length. This last assumption is basically valid once the chain has grown to about four monomer units, at which point the effect of the end group becomes negligible. Similarly, it is assumed that the rates of termination by combination or disproportionation are independent of chain length. (In cases where there is a significant degree of head-to-head orientation, one must take into account four separate propagation reactions: head-radical-to-head, head-radical-to-tail, tail-radical-to-head, and tail-radical-to-tail.)

Initiation proceeds in two steps: decomposition of initiator to yield initiator radicals, R·, followed by addition of R· to monomer, M, to give a new radical, $M_1\cdot$.

Initiation

$$\text{Initiator} \xrightarrow{k_d} R\cdot$$

$$R\cdot + M \xrightarrow{k_i} M_1\cdot$$

Rate constants for the two reactions are k_d and k_i.

In the initial propagation step (rate constant k_p), $M_1\cdot$ adds to another monomer molecule to form a new radical, $M_2\cdot$, which, in turn, adds to M to form $M_3\cdot$, and so on.

Propagation

$$M_1\cdot + M \xrightarrow{k_p} M_2\cdot$$

$$M_2\cdot + M \xrightarrow{k_p} M_3\cdot$$

$$M_x\cdot + M \xrightarrow{k_p} M\cdot_{(x+1)}$$

As was mentioned above, termination occurs principally by radical coupling or disproportionation, where k_{tc} and K_{td} are the respective rate constants.

Termination

$$M_x\cdot + M_y\cdot \xrightarrow{k_{tc}} M\cdot_{(x+y)} \qquad \text{Coupling}$$

$$M_x\cdot + M_y\cdot \xrightarrow{k_{td}} M_x + M_y \qquad \text{Disproportionation}$$

Assuming the rate of initiator decomposition is very slow relative to that of addition of initiator radical to monomer (reasonable, given the high reactivity of free radicals), and taking into account the fact that two initiator radicals are formed with each decomposition, the expression for initiation rate, R_i, is

$$R_i = \frac{-d[M\cdot]}{dt} = 2fk_d\,[I]$$

where $[M\cdot]$ is the total concentration of chain radicals, $[I]$ is the molar concentration of initiator, and f is the *initiator efficiency*, that is, the fraction of initiator radicals that actually start a polymer chain and are not consumed in the wastage reactions discussed in Section 6.2. Thus,

$$f = \frac{\text{radicals that initiate a polymer chain}}{\text{radicals formed from initiator}}$$

Efficiency is determined most conveniently by measuring the end groups arising from the addition of initiator radicals (by using isotopically labeled initiator, for example) and comparing with the amount of initiator reacted. For most common free radical polymerization processes, f falls in the range 0.3 to 0.8.

For termination the rate expression is

$$R_t = \frac{-d[M\cdot]}{dt} = 2k_t[M\cdot]^2$$

The factor 2 takes into account that in any termination reaction two radicals are consumed. The rate constant, k_t, represents the summation of k_{tc} and k_{td}.

Because the rate constants for termination are very much greater than those of initiation, we may assume that shortly after the reaction begins, the formation and destruction of radicals occur at the same rates, hence the concentration of radicals, $[M\cdot]$, remains constant. This is referred to as the *steady-state assumption*. Thus, $R_i = R_t$, or

$$2fk_d[I] = 2k_t[M\cdot]^2$$

Solving for $[M\cdot]$ gives

$$[M\cdot] = \sqrt{\frac{fk_d[I]}{k_t}}$$

The rate expression for propagation is

$$R_p = \frac{-d[M]}{dt} = k_p[M][M\cdot]$$

Substituting the expression for $[M\cdot]$ above, we obtain

$$R_p = \frac{-d[M]}{dt} = k_p[M]\sqrt{\frac{fk_d[I]}{k_t}}$$

Because propagation involves large numbers of monomer molecules per chain whereas initiation consumes only one, the rate of polymerization is, for all practical purposes, equivalent to the rate of propagation. The polymerization rate, therefore, is proportional to the square root of initiator concentration and to the first power of monomer concentration. Thus, doubling the initiator concentration causes the rate to increase by a factor of about 1.4. This relationship has been confirmed experimentally for a variety of free radical polymerizations. Propagation and termination rate constants, as well as the corresponding activation energies, for several commercially important monomers are given in Table 6.4.[22a]

Table 6.4. Representative propagation and termination rate constants, k_p and k_t and activation energies, E_p and E_t[a]

Monomer	Temperature (°C)	$k_p \times 10^{-3}$ (L/mol-s)	E_p^* (kJ/mol)	$k_t \times 10^{-7}$ (L/mol-s)	E_t^* (kJ/mol)
Acrylonitrile	60	1.96	16.2	78.2	15.5
Chloroprene	40	0.220	—	9.7	—
Ethylene	83	0.242	18.4	54.0	1.3
Methyl acrylate	60	2.09	29.7	0.95	22.2
Methyl methacrylate	60	0.515	26.4	2.55	11.9
Styrene	60	0.176	26.0	7.2	8.0
Vinyl acetate	50	2.64	30.6	11.7	21.9
Vinyl chloride	50	1.10	16	21.0	17.6
Tetrafluoroethylene	40	7.40	17.4	7.4	13.6

[a] Data from Korus and O'Driscoll.[22a]

Another important parameter related to polymerization rate is the *average kinetic chain length*, $\bar{v}$, which is defined as the average number of monomer units polymerized per chain initiated, which is equal to the rate of polymerization per rate of initiation. Since $R_i = R_t$ under steady-state conditions,

$$\bar{v} = \frac{R_p}{R_i} = \frac{R_p}{R_t}$$

Substituting for R_p and R_t from the above expressions:

$$\bar{v} = \frac{k_p[M][M\cdot]}{2k_t[M\cdot]^2} = \frac{k_p[M]}{2k_t[M\cdot]}$$

Substituting the expression for $[M\cdot]$ derived previously,

$$\bar{v} = \frac{k_p[M]}{2(fk_tk_d[I])^{1/2}}$$

Kinetic chain length is thus seen to be related to a variety of rate and concentration parameters. It should be noted that $\bar{v}$ will decrease as both the initiator concentration and initiator efficiency increase. This is entirely reasonable because increasing the number of growing chains increases the probability of termination. Thus, varying initiator concentration is one way of controlling molecular weight.

✗ In the absence of any side reactions, kinetic chain length is related directly to degree of polymerization (and hence number average molecular weight) according to the mode of termination. If termination occurs exclusively by disproportionation, $\overline{DP} = \bar{v}$; if it occurs by coupling, $\overline{DP} = 2\bar{v}$.

While these kinetics expressions apply to a great many vinyl polymerization reactions, deviations are common. One type of deviation, variously referred to as the *gel effect*, the *Trommsdorff effect*, or the *Norris–Smith effect*, occurs in bulk or concentrated solution polymerizations when the medium viscosity gets very high, or in solution polymerizations when the polymer precipitates. In the more viscous medium, chain mobility is reduced, with the result that chain-end radicals have a lower probability of being in position to effect termination. Because the small monomer molecules can still diffuse to the active chain ends even as the termination rate decreases, there is a marked increase in polymerization rate, or autoacceleration. Propagation rate may also increase because of oriented monomer–polymer association in precipitated polymer.[23] Because the increase in rate is usually accompanied by an increase in reaction exotherm, autoacceleration may cause processing difficulties, particularly in bulk polymerizations. Higher molecular weights also result from the decrease in termination rate.

Deviations in predicted kinetic behavior also arise from *chain transfer reactions*—the transfer of reactivity from the growing polymer chain to another species. Because this process terminates the chain, but at the same time generates a new radical, chain transfer reactions result in lower molecular weight product. Since chain transfer reactions invariably arise in most free radical vinyl polymerizations, relatively broad molecular weight distributions are common. Let us examine some of the known chain transfer reactions.

A chain-end radical may abstract a hydrogen atom from a chain, leading to a reactive site for chain branching (6.30). Hydrogen abstraction may also

$$
\text{wCH} \cdot + \text{wCH}_2\text{CHw} \longrightarrow \text{wCH}_2 + \text{wCH}_2\dot{\text{C}}\text{w}
$$

with the Y substituents shown, leading to

$$
\underset{\text{Y}}{\text{wCH}_2}\dot{\text{C}} \underset{\text{Y}}{\text{wCH}_2\dot{\text{CHY}}} \xleftarrow{\text{CH}_2=\text{CHY}} \qquad (6.30)
$$

occur intramolecularly, a process referred to as *backbiting*. Polyethylene, produced by high-pressure free radical polymerization of ethylene, is highly branched, and most of the branches contain a small number of carbon atoms. This has been attributed to backbiting involving five- or six-membered cyclic transition states (6.31). These two types of chain transfer are very important because they lead to branched polymers whose properties may differ markedly from those of the corresponding linear polymers.

$$
\begin{array}{c}
\text{CH}_2\!-\!\text{CH}_2 \\
/ \qquad \backslash \\
\text{wCH} \qquad \text{CH}_2 \\
| \qquad\quad / \\
\text{H----CH}_2
\end{array}
\longrightarrow
\begin{array}{c}
\text{CH}_2\!-\!\text{CH}_2 \\
/ \qquad \backslash \\
\text{w}\dot{\text{C}}\text{H} \qquad \text{CH}_2 \\
\qquad\quad / \\
\text{CH}_3
\end{array}
\qquad (6.31)
$$

Chain transfer may also occur with initiator (6.32) or monomer (6.33), or it may take place with solvent. Polystyrene prepared in carbon tetrachloride,

$$
\text{wwCH}_2\dot{\text{C}}\text{H} \underset{\text{Y}}{\Big|} \xrightarrow{\text{ROOR}} \text{wCH}_2\underset{\text{Y}}{\text{CHOR}} + \text{RO}\cdot \qquad (6.32)
$$

$$
\xrightarrow{\text{CH}_2=\text{CHY}} \text{wCH}_2\text{CH}_2 + \text{CH}_2\!=\!\underset{\text{Y}}{\dot{\text{C}}}\text{Y} \qquad (6.33)
$$

for example, contains chlorine at the chain ends as a result of chlorine transfer (6.34) and initiation by the resultant $\cdot\text{CCl}_3$ radicals (6.35). Transfer to monomer is particularly important with monomers containing allylic hydrogen, such as propylene, because the formation of resonance-stabilized allylic radicals (6.36) is highly favorable. Thus, high-molecular-weight polypropylene cannot be prepared under the usual conditions of free radical polymerization. This is not a problem with methyl methacrylate, which also contains allylic

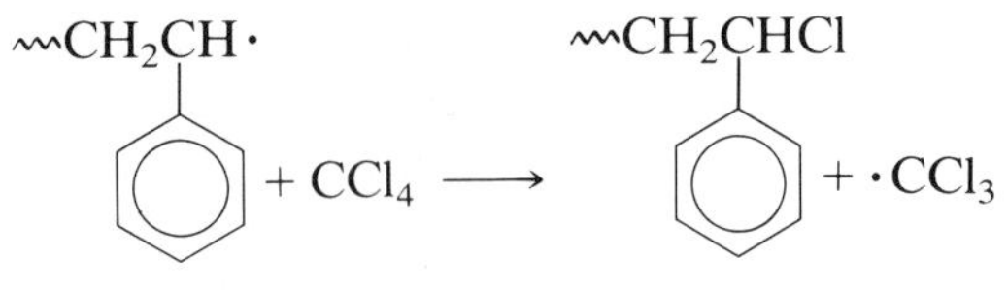

$$\text{wCH}_2\text{CH}\cdot \quad (\text{C}_6\text{H}_5) + \text{CCl}_4 \longrightarrow \text{wCH}_2\text{CHCl} \quad (\text{C}_6\text{H}_5) + \cdot\text{CCl}_3 \tag{6.34}$$

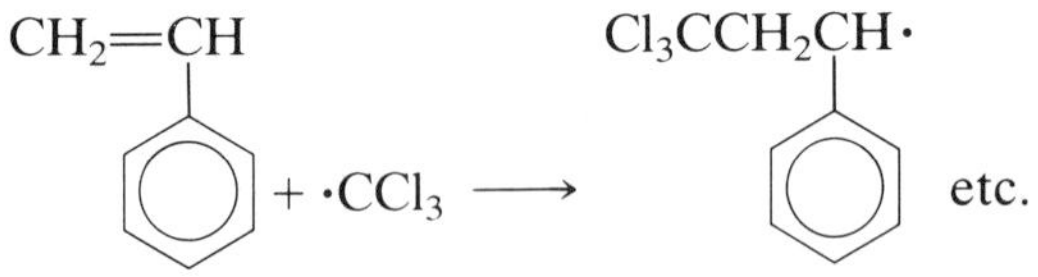

$$\text{CH}_2\!=\!\text{CH} \quad (\text{C}_6\text{H}_5) + \cdot\text{CCl}_3 \longrightarrow \text{Cl}_3\text{CCH}_2\text{CH}\cdot \quad (\text{C}_6\text{H}_5) \quad \text{etc.} \tag{6.35}$$

$$\underset{\text{CH}_3}{\text{wCH}_2\text{CH}\cdot} + \text{CH}_3\text{CH}\!=\!\text{CH}_2 \longrightarrow \underset{\text{CH}_3}{\text{wCH}_2\text{CH}_2} + [\text{CH}_2\!=\!\!=\!\!\text{CH}\!=\!\!=\!\!\text{CH}_2]\cdot \tag{6.36}$$

hydrogens, because addition to monomer to form a resonance-stabilized radical is even more favorable.

Most important from the standpoint of molecular weight control is transfer to a modifier, such as a mercaptan (6.37), a compound with a high affinity for

$$\underset{\text{Y}}{\text{wCH}_2\text{CH}\cdot} + \text{RSH} \longrightarrow \underset{\text{Y}}{\text{wCH}_2\text{CH}_2} + \text{RS}\cdot \tag{6.37}$$

hydrogen transfer. In such cases it is necessary to redefine the kinetic chain length as being the ratio of propagation rate to the combined rates of termination and transfer; that is,

$$\bar{\nu}_{tr} = \frac{R_p}{R_t + R_{tr}}$$

Since transfer reactions are second order, with

$$R_{tr} = k_{tr}[M\cdot][T]$$

where T is the transfer agent, we can rewrite the expression for $\bar{\nu}_{tr}$, taking into account all possible transfer reactions, as

$$\bar{\nu}_{tr} = \frac{k_p[M][M\cdot]}{2k_t[M\cdot]^2 + \Sigma k_{tr}[M\cdot][T]} = \frac{k_p[M]}{2k_t[M\cdot] + \Sigma k_{tr}[T]}$$

Remembering that

$$\bar{\nu} = \frac{k_p[M]}{2k_t[M\cdot]}$$

we can write the reciprocal of the above expression for $\bar{\nu}_{tr}$ as

$$\frac{1}{\bar{\nu}_{tr}} = \frac{1}{\bar{\nu}} + \frac{\Sigma k_{tr}[T]}{k_p[M]}$$

The ratio of the transfer rate constant to that of propagation is commonly defined as the chain transfer constant, C_T, for a particular monomer:

$$\frac{k_{tr}}{k_p} = C_T$$

Substituting, we have

$$\frac{1}{\bar{\nu}_{tr}} = \frac{1}{\bar{\nu}} + \frac{\Sigma C_T[T]}{[M]}$$

Clearly, as the rate of transfer and concentration of the transfer agent increase, the kinetic chain length becomes progressively smaller. Chain transfer constants for a number of compounds and monomers are available in the polymer literature,[22b] several of which are given in Table 6.5. In cases where the highly effective mercaptans are used to control the kinetic chain length, one can, to a first approximation, ignore other transfer processes to determine the amount of transfer agent needed to obtain a given molecular weight. When the concentration of transfer agent is high and k_{tr} is much greater than k_p, very-low-molecular-weight polymers called *telomers* are obtained. Then the process is called *telomerization*.

Chain transfer reactions can also be used to *prevent* free radical polymerizations. One type of compound, added as a stabilizer to vinyl monomers, is an alkylated phenol (16), which can transfer its phenolic hydrogen to form a new radical (6.38) that undergoes coupling reactions (6.39) rather than initiating polymerization. Such compounds, called *inhibitors*, are commonly added to monomers to prevent premature polymerization during shipment or storage. Before the monomer can be used it must be distilled, or the inhibitor must be removed by extraction. Alternatively, sufficient initiator must be added to "use up" the inhibitor before polymerization can occur. The period during

Table 6.5. Representative chain transfer constants, C_T, for styrene and methyl methacrylate[a]

Transfer agent	$C_T \times 10^4$	
	Styrene	*Methyl methacrylate*
Benzene	0.023	0.040
Cyclohexane	0.031	0.10[b]
Toluene	0.125	0.20
Chloroform	0.5	1.77
Carbon tetrachloride	90	2.40
Carbon tetrabromide	22,000	2700
n-Butyl mercaptan	210,000	6600

[a] Reaction temperature 60°C unless noted otherwise. Data from Young.[22b]
[b] 80°C.

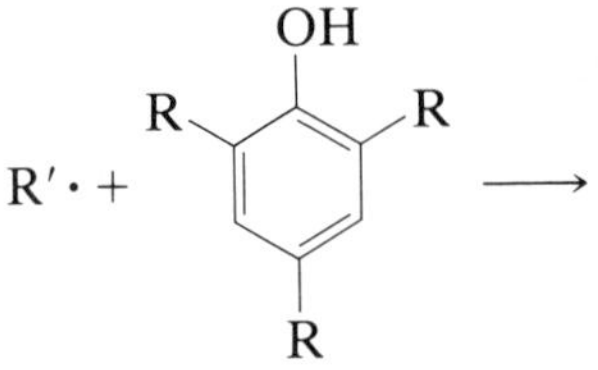

16

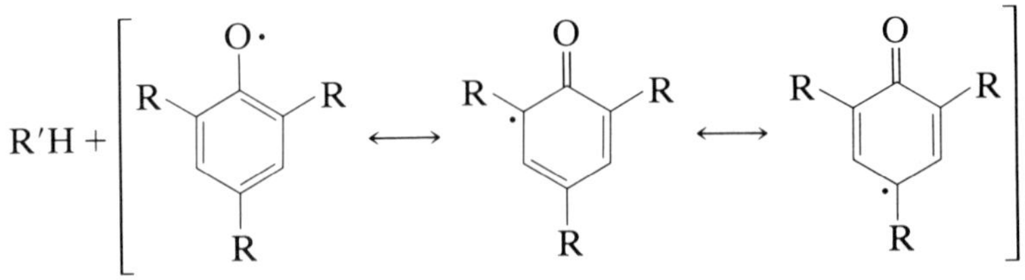

$$R'H + \left[\text{(6.38)} \right]$$

(6.38)

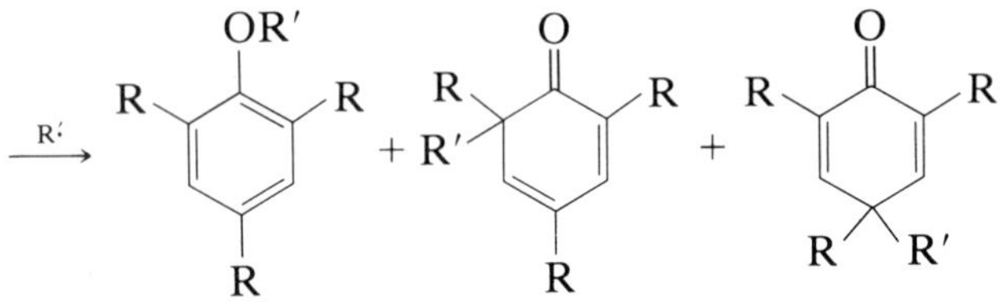

(6.39)

which inhibitor is consumed is known as the *induction period*. Introduction periods are common even with purified monomer because of the presence of oxygen, which is, itself, an inhibitor. A variety of compounds are used commercially as inhibitors.[14b]

Before leaving the topic of kinetics and mechanism, let us reconsider the case of emulsion polymerization. Unlike the case with free radical polymerization in bulk or solution, increasing rates of emulsion polymerization usually lead to *higher* molecular weights, especially when emulsifier concentration is increased. As was discussed previously under bulk polymerization, increasing the rate of formation of initiator radicals—which increases polymerization rate—also increases the rate of termination. With emulsion polymerization, on the other hand, polymerization rate competes with the rate of diffusion of terminator radicals into the micelles where polymerization is taking place. Once termination occurs, a new radical enters the micelle and polymerization begins again and continues until another terminator radical enters. And so the reaction continues. Assuming a constant rate of diffusion, therefore, each particle contains a growing radical half the time. The overall polymerization rate depends on the polymerization rate within each particle, as well as the number of particles. According to the kinetics treatment originally developed by Smith and Ewart (appropriately called the Smith–Ewart kinetics),[24]

$$R_p = k_p[M]\left(\frac{N}{2}\right)$$

where k_p is the overall polymerization rate constant, $[M]$ is the concentration of monomer in the micelle particles, and N is the number of particles. The factor, $N/2$, which takes into account that only half the particles contain a growing chain at any given time, is equal to the concentration of radicals, $[M\cdot]$. Because the rate of polymerization within each particle is equal to $k_p[M]$, and the rate of termination is equal to the rate of terminator radical entry into the particle, ρ, the average kinetic chain length is given by,

$$\bar{\nu} = \frac{k_p[M]}{\rho}$$

Since ρ is equal to the rate of initiator radical formation divided by the number of particles, the above expression may be rewritten:

$$\bar{\nu} = \frac{k_p[M]}{R_i/N} = \frac{k_p[M]N}{2fk_d[I]}$$

The terminator radical is derived from the initiator; therefore the molecular weight is not significantly altered by termination, and $\bar{\nu} = \overline{\mathrm{DP}}$.

From the above relationship we can see why the molecular weight increases with polymerization rate. If the initiator concentration remains constant, the polymerization rate can be enhanced by adding more emulsifier, which increases the number of micelles, which in turn increases both R_p and $\bar{\nu}$.

6.5 Stereochemistry of polymerization

Addition of radicals to monomers is not as simple a process as is implied in the previous section, because the stereochemistry of addition can profoundly affect the ultimate properties of the polymer. It will be recalled that for monosubstituted and 1,1-disubstituted monomers (the two substituents of the latter being different), two stereoregular arrangements are possible in the heat-to-tail polymer: *isotactic*, with like substituents all on the same side, and *syndiotactic*, with like substituents of alternate repeating units on the same side. A random arrangement is referred to as *atactic*. (See Figure 3.10, Chapter 3). There are two factors to consider in describing the stereochemistry of addition of a monomer molecule to a radical chain end: (1) interaction between the terminal chain carbon and the approaching monomer molecule, and (2) the configuration of the penultimate repeating unit in the polymer chain.

Consider first the approach of monomer. The terminal carbon atom presumably has sp^2 hybridization and is planar. For polymerization of monomer $CH_2 = CXY$, there are two ways that a monomer molecule can approach the terminal carbon: the mirror approach, with like substituents on the same side

(6.40), or the nonmirror approach, with like substituents on opposite sides (6.41). As long as stereochemistry is maintained—that is, free rotation does not occur before the next monomer molecule adds—then the mirror approach (6.40) will always lead to isotactic polymer and the nonmirror approach (6.41) to syndiotactic polymer. If, on the other hand, interactions between the substituents of the penultimate repeating unit and the terminal carbon atom are significant, then conformational factors could also have an influence and the mode of addition could be governed by the conformation that allows the least steric or electrostatic interaction.

$$(6.40)$$

$$(6.41)$$

It was believed for many years that with free radical polymerization these effects were not significant and that only atactic polymer was formed. This was not unreasonable because free radical processes lack the significant steric and electrostatic factors associated with complexes in Ziegler–Natta polymerization or with counterions in ionic polymerization. In 1958, however, it was reported that free radical polymerization of methyl methacrylate at temperatures below 0°C gave crystalline polymer,[26] and this was subsequently shown to be predominantly syndiotactic by the use of high-resolution NMR.[27] Since this initial work, considerable interest has developed in stereoregular free radical polymerization.

One model[28] for the methyl methacrylate polymerization that takes into account penultimate unit interactions and that allows for the least steric hindrance between substituent groups during reaction with monomer is shown in reaction (6.42). (The polymer chain, P, is the bulkiest group.)

$$(6.42)$$

In such an arrangement the radical receives maximum exposure for reaction with monomer. The newly formed radical then assumes the preferred conformation with another monomer molecule, and so the process repeats itself to give the syndiotactic structure. This model assumes an energy barrier to rotation that is considerably higher than that of propagation; otherwise a random distribution would result. This, of course, would be temperature dependent, and as expected, the degree of stereoregularity decreases at higher temperatures.

While the above model satisfactorily explains the formation of syndiotactic poly(methyl methacrylate), it does not take into account possible interaction between chain-end radical and monomer. Using calculations of interaction energies that take into account such factors as distance of separation, angle of approach, and point of radical addition (at the carbon atom or at the π orbital of the double bond), other researchers have concluded that interaction between monomer and terminal unit is the predominant factor resulting in syndiotactic placement as long as the terminal carbon has the sp^2 planar structure.[29] This is the nonmirror arrangement shown in reaction (6.41). If, however, the terminal radical assumes a substantial amount of sp^3 hybridization, which it might do as it enters the transition state, then the resultant change in configuration could cause interactions with the penultimate unit to be of greater significance. Whichever is the case, syndiotactic placement is favored.

Apart from poly(methyl methacrylate), numerous other polymers prepared by free radical initiation have been shown to have varying degrees of syndiotacticity, but no clear-cut trends can be observed.[25] It would seem that increasing the size of substituent groups would increase the degree of stereoregularity, yet poly(2,4,6-triphenylbenzyl methacrylate) (17) appears to be *less* syndiotactic than poly(methyl methacrylate) prepared under the same conditions. No doubt a combination of steric and polar effects is responsible for the stereochemistry of free radical polymerization, but more work needs to be done before a definitive mechanism can be written. As will be seen in

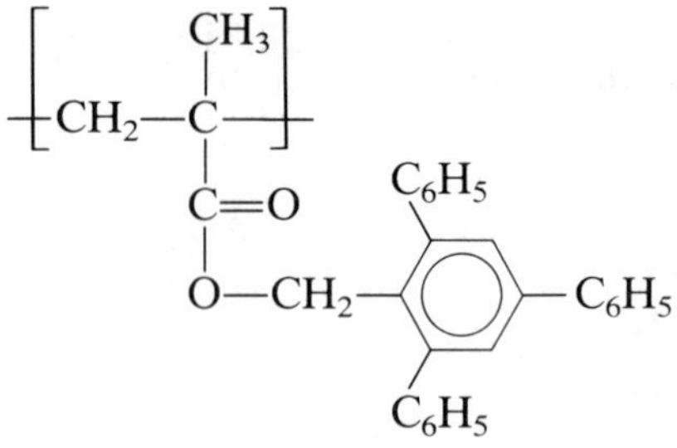

17

the following chapters, ionic and coordination polymerization can lead to much greater stereochemical control.

6.6   Polymerization of dienes

6.6.1   Isolated dienes

Polymerization of isolated dienes leads to crosslinked polymers when the double bonds react independently of one another. In certain cases, cooperative addition reactions of one double bond to the other of the monomer lead to cyclic polymers.[7d,30] This is an example of *cyclopolymerization*—the formation of a cyclic structural unit in a propagation step. Cyclopolymerization of divinylformal (**18**) (6.43) illustrates this process.

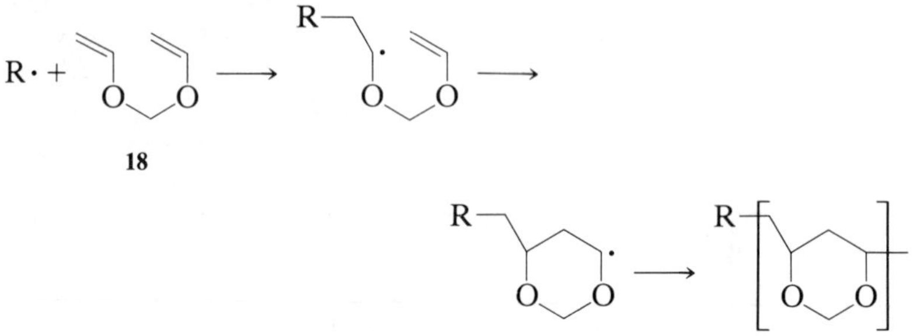

18

$$(6.43)$$

The normal (head-to-tail) mode of addition yields the six-membered ring, whereas abnormal (head-to-head) addition (6.44) gives the five-membered ring. The former is preferred, particularly when substituent groups are on the second carbon, but both are formed in significant amounts.

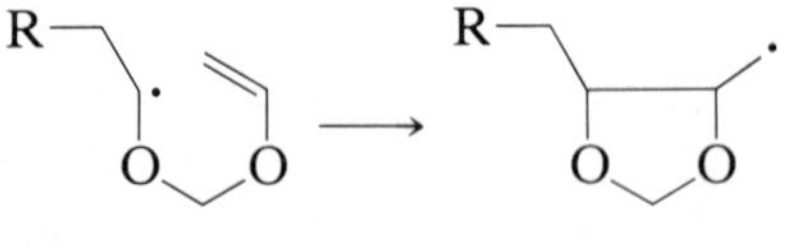

$$(6.44)$$

Such cyclopolymerization reactions usually proceed concomitantly with ordinary vinyl polymerization so that some pendant unsaturated groups are formed (**19**). Further polymerization leads to chain branching and, eventually, to insoluble crosslinked products. It is of interest that divinyl ketals (**20**) give much greater amounts of cyclopolymer than divinyl formal. This has been attributed to a conformational effect. The *s-cis* (*cisoid*) conformation (**18** or **20**) is preferred for cyclopolymerization, and the *s-trans* (*transoid*) (**21**) for ordinary vinyl polymerization. Presumably, for steric reasons, substituents on the central carbon atom favor the *s-cis* conformation.

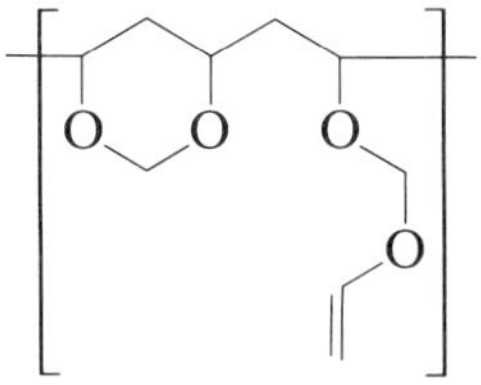

19 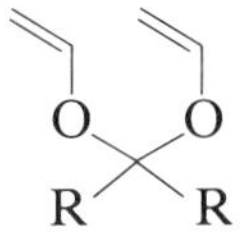   20 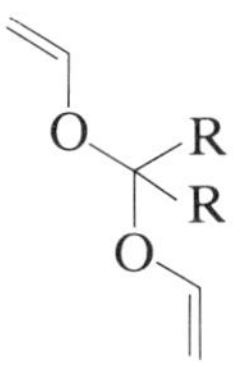   21

Several diallyl monomers are used commercially to prepare highly cross-linked thermosetting *allyl resins* or for crosslinking other polymers.[31] Among the most important are diallyl phthalate (**22**) and diethylene glycol bis(allyl carbonate) (**23**). Monomer **22** is used primarily for manufacturing electrical or electronics items (circuit boards, insulators, television components, etc.) or for preimpregnating glass cloth or fiber for fiber-reinforced plastics. Monomer **23** is used to make polymers for applications requiring good optical clarity, such as eyeware lenses, camera filters, panel covers, and the like. Interestingly, monomer **22** has been shown to undergo about 40% intramolecular cyclization (6.45) to yield the 11-membered ring,[32] no doubt a consequence of the proximity of the reacting groups.

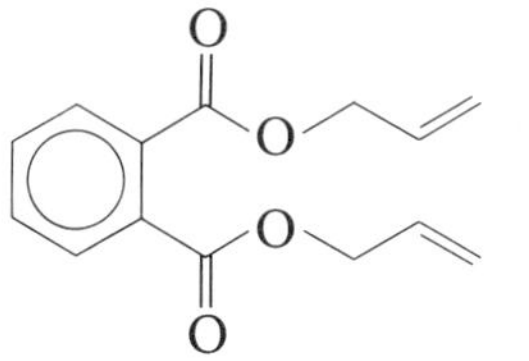

22 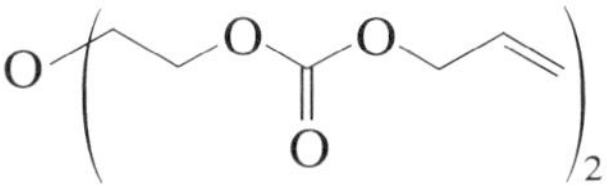   23

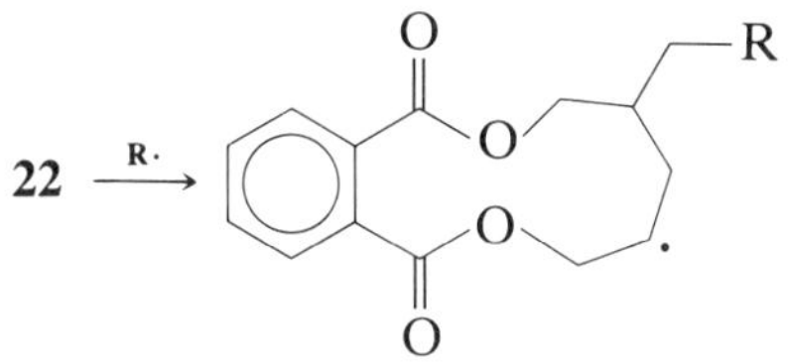

22 $\xrightarrow{\text{R·}}$ etc.   (6.45)

6.6.2 Conjugated dienes

Conjugated dienes such as 1,3-butadiene (**24**) undergo both 1,2- and 1,4-addition. Addition polymerization thus leads by analogy to 1,2- and 1,4-addition polymers via the delocalized radical intermediate **25** (6.46). Thus,

1,2-addition gives polymer (**26**) with pendant vinyl groups, while 1,4-addition leads to polymer with unsaturation in the chain. In the latter case, both *cis* (**27**) and *trans* (**28**) configurations are possible.

$$CH_2{=}CH{-}CH{=}CH_2 \xrightarrow{\text{R·}} [RCH_2{-}CH{=\!=\!=}CH{=\!=\!=}CH_2]\cdot \qquad (6.46)$$

24   25

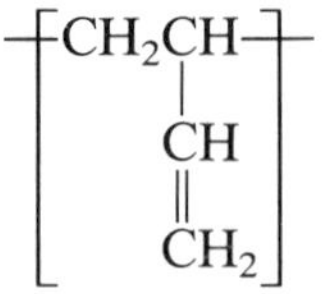

26 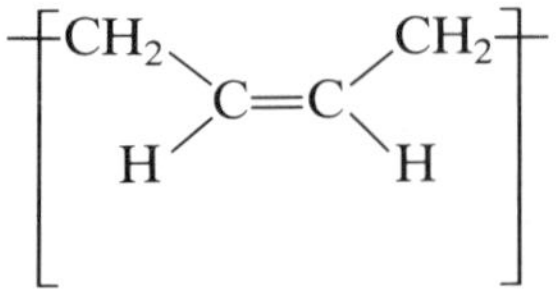   27 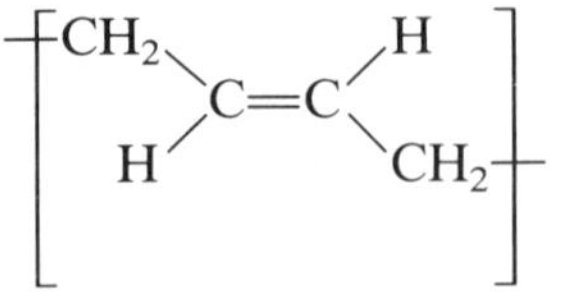   28

In general free radical polymerization of 1,3-butadiene results in approximately 20% of the 1,2 structure, while most of the 1,4 polymer has the *trans* structure. As the polymerization temperature is increased, the amount of *cis*-1,4 increases, but the yield of 1,2 polymer remains about the same.

With substituted dienes like isoprene (2-methyl-1,3-butadiene) the situation is more complicated; 1,2 and 3,4 structures (**29** and **30**, respectively) as well as *cis*- and *trans*-1,4 (**31** and **32**), are possible. All are formed in free radical polymerization, but the head-to-tail *trans*-1,4 (**32**) predominates. Natural rubber (discussed in Chapter 17) is a polyterpene consisting of head-to-tail 1,4-polyisoprene units, with the most important type, *Hevea* rubber, having the *cis* structure (**31**).

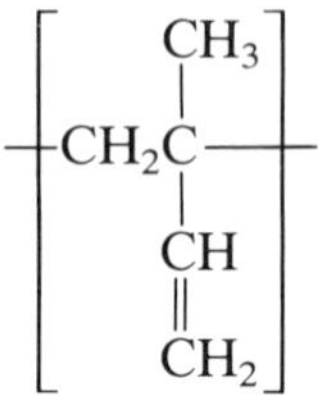

29 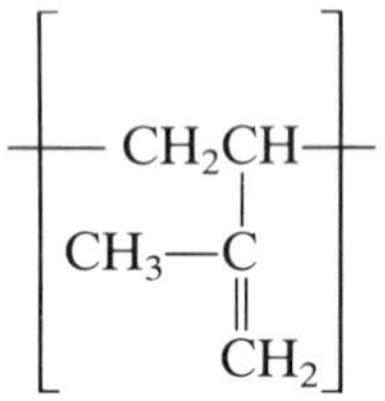   30

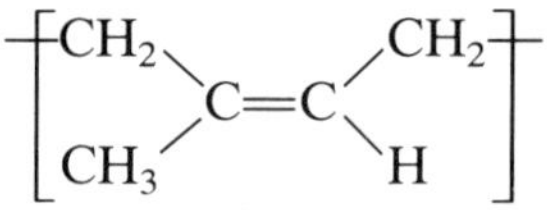

31 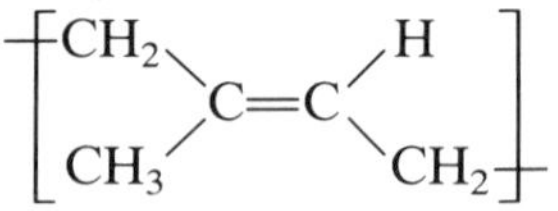   32

Table 6.6 lists the percentages of each structural unit at different polymerization temperatures for the three most important diene monomers—butadiene, isoprene, and chloroprene (2-chloro-1,3-butadiene). Several points are worth noting. In the case of isoprene, the amount of *cis*-1,4 structure increases with increasing temperature up to about 100°C, then it decreases. The total amount of 1,2 and 3,4 polymer (approximately 10%) does not change appreciably with temperature; however, the 3,4 structure is

Table 6.6. Structure of free radical-initiated diene polymers[a]

Monomer	Polymerization temperature (°C)	Percent			
		cis-1,4	*trans*-1,4	1,2	3,4
Butadiene	−20	6	77	17	—
	−10	9	74	17	—
	5	15	68	17	—
	20	22	58	20	—
	100	28	51	21	—
	175	37	43	20	—
	233	43	39	18	—
Isoprene	−20	1	90	5	4
	−5	7	82	5	5
	10	11	79	5	5
	50	18	72	5	5
	100	23	66	5	6
	150	17	72	5	6
	203	19	69	3	9
	257	12	77	2	9
Chloroprene	−46	5	94	1	0.3
	10	9	84	1	1
	46	10	81–86	2	1
	100	13	71	2.4	2.4

[a] Data from Cooper,[25] p. 275. Reprinted by courtesy of Marcel Dekker, Inc.

favored above 150°C. With chloroprene, the *trans*-1,4 structure also decreases with increasing temperature, but the amount of *trans*-1,4 is slightly higher than that of isoprene over the temperature range measured.

Simple diene monomers exist mainly in the *s-trans* conformation. It is believed that stereochemistry is retained during the polymerization process, which accounts for the predominance of *trans* structure.[25] Thus at higher temperatures, as the proportion of *s-cis* conformation increases, the yield of *cis* polymer increases proportionately. Support for this is gained from the fact that 2-(tri-*n*-butyltin)butadiene (**33**), which, because of the large size of the substituent group, exists almost entirely in the *s-cis* conformation, gives almost 100% *cis*-1,4 polymer at 60°C.

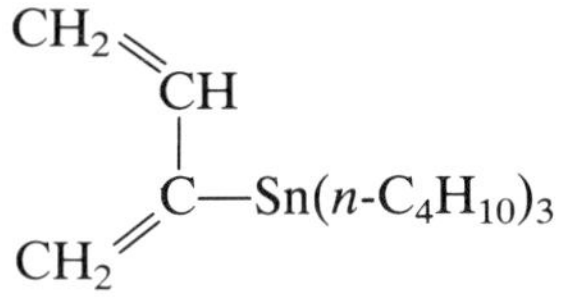

$$CH_2=CH$$
$$| $$
$$CH_2=C-Sn(n\text{-}C_4H_{10})_3$$

33

A miscellaneous method of forming highly stereoregular polymers involves the use of *inclusion compounds* (also called *clathrates*), formed between monomer and a crystalline host material such as urea or thiourea.[33,34] Under the influence of gamma radiation the monomers undergo solid-state free radical polymerization within the inclusion canals. Diene monomers form 1,4-*trans* polymer exclusively because of the steric confinements of the clathrate. Nondiene monomers such as vinyl chloride or methyl methacrylate also undergo free radical inclusion polymerization, but the polymers do not exhibit such a high degree of stereoregularity. For monomers such as *trans*-1,3-pentadiene (**34**), the inclusion polymer (**35**) contains a chiral center and is highly isotactic; if the host compound is optically active, asymmetric inclusion polymerization occurs to yield optically active polymer. Such stereochemical control has not been observed in conventional free radical polymerization of 1-substituted dienes. Another feature of inclusion polymerization is that block copolymers can be formed by polymerizing one monomer, then adding a second monomer and reirradiating. Random copolymers are formed if both monomers are introduced simultaneously.

34   **35**

6.7   Monomer reactivity

Consideration of free energies of polymerization,[7e,17d] that is,

$$\Delta G_p = \Delta H_p - T \Delta S_p$$

suggests that almost any compound containing a double bond is capable of being transformed into polymer. The enthalpies and entropies of polymerization for several common monomers (i.e., the difference in entropy and enthalpy between monomer and polymer) are listed in Table 6.7.[22c] Clearly the enthalpy of reaction is very favorable in each case, more than enough to offset the unfavorable entropy effect. A favorable free energy of polymerization, however, does not guarantee a favorable polymerization reaction under any set of conditions. Neither propylene nor isobutylene, for example, undergoes normal free radical polymerization to any extent. Polarity and size of the substituent groups on the double bond as well as the propensity for chain transfer determine the kinetic feasibility of polymerizing a particular monomer.

In free radical polymerizations two factors need to be considered in any discussion of monomer reactivity: the stability of the monomer toward addition of a free radical, and the stability of the monomer radical thus formed. Of the two, the latter is of overriding importance, as shown by the fact that styrene is particularly susceptible to polymerization despite the fact that the

Table 6.7. Representative enthalpies, ΔH, and
entropies, ΔS, of polymerization[a]

Monomer	$-\Delta H$ (kJ/mol)	$-\Delta S$ (J/mol)
Acrylonitrile[b]	77	109
1,3-Butadiene[c]	78	89
Ethylene[d]	109	155
Isoprene	75	101
Methyl methacrylate	65	117
Propylene	84	116
Styrene	70	104
Tetrafluoroethylene	163	—
Vinyl acetate	90	—
Vinyl chloride	71	—

[a] Values selected from Ivin.[22c]. Data refer to conversion
of liquid monomer to amorphous or slightly crystalline
polymer at 25°C unless otherwise indicated.

[b] 74.5°C.

[c] Refers to 1,4-polymer.

[d] Gaseous monomer to semicrystalline polymer.

vinyl group is conjugated with a benzene ring. Stabilization of the resultant
radical (**36**) thus reduces the activation energy for attack by initiator radicals.

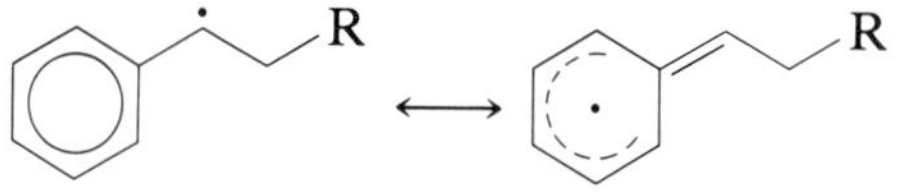

36

Nitrile groups are also very effective in stabilizing radicals, as shown by the
addition of ethyl radicals to monomers in the gas phase[35] where the rate order
is acrylonitrile > styrene > vinyl acetate. This does not necessarily mean the
same order applies in initiation of polymerization, however. Reaction of
monomers with ^{14}C-labeled benzoyloxy radicals has been studied[36] by comparing the rates of decarboxylation (6.47) and addition to monomer (6.48).

$$\phi^{14}CO_2 \cdot \rightarrow \phi \cdot + {}^{14}CO_2 \tag{6.47}$$

$$\phi^{14}CO_2 \cdot + M \rightarrow \phi^{14}CO_2M \cdot \tag{6.48}$$

The more reactive the monomer, the greater the amount of initiation by
$\phi^{14}CO_2 \cdot$ relative to initiation by $\phi \cdot$, hence the more ^{14}C incorporated into the
polymer. Such studies indicate an order: styrene > vinyl acetate >
acrylonitrile. Clearly polar effects are important in reducing the reactivity of
acrylonitrile toward benzoyloxy radicals.

Propagation rate data (Table 6.4) also reflect, to a degree, the inverse relationship between monomer stability and polymerization rate. Vinyl acetate, a very stable monomer, has one of the highest rate constants, the reverse of the situation with styrene. Both steric and polar effects play such an important role in monomer reactivity, however, that it is impossible to make clear-cut generalizations.[17c] Acrylonitrile, for example, has a relatively high rate of polymerization despite the stabilizing effect of the nitrile group. The lower rate constant of methyl methacrylate compared with methyl acrylate may be attributed to additional resonance stabilization of the intermediate radical by hyperconjugation or increased steric hindrance, or both.

Steric effects are of particular significance with 1,2-disubstituted monomers,[20b] which are difficult to polymerize under ordinary free radical conditions. (Exceptions are fluorine-containing monomers.) This is not because of any increased steric interaction in the polymer. On the contrary, polymers derived from 1,2-disubstituted monomers are generally more thermodynamically stable than those derived from 1,1-disubstituted monomers of analogous structure. Rather, the reduced activity appears to arise from steric interactions between substituent groups of the approaching monomer molecule and the substituent on the carbon adjacent to the chain end radical (37).

37

Evidence to support this hypothesis arises from copolymerization studies. Monosubstituted or 1,1-disubstituted monomers, which have only hydrogen atoms attached to the second carbon of the double bond, usually copolymerize readily with 1,2-disubstituted monomers.

Steric effects are also evident when comparing reactivities of *cis* and *trans* isomers. Fumarate esters (38) are more reactive in free radical polymerizations than maleate esters (39), even though the latter are thermodynamically

38   39

less stable. Because addition of initiator radical, $R\cdot$, leads to the same free radical in both instances, the greater reactivity of the *trans* isomer is best explained in terms of coplanarity requirements of the transition state leading to radical formation (40). Steric interference between the ester groups of the *cis* isomer would reduce coplanarity.

40

That polar effects also play a role is shown by the fact that *para*-substituted styrene derivatives follow the Hammett relationship,*

$$\log\left(\frac{k}{k_0}\right) = \rho\sigma$$

in polymerization reactions, with electron-attracting groups increasing the rate and electron-donating groups retarding it. As would be expected for a free radical process, the substituent effect is not large ($\rho = +0.6$). Polarity is of much greater significance in copolymerization.

One other consideration regarding monomer reactivity involves the reversibility of vinyl polymerization reactions; that is,

$$M_x \cdot + M \underset{k_{dp}}{\overset{k_p}{\rightleftharpoons}} M \cdot_{(x+1)}$$

where k_{dp} is the depropagation rate constant. Just as radical stability retards propagation, it enhances depropagation. As the polymerization temperature is raised, the depropagation rate increases until a point is reached where the forward and back reactions are equal. The temperature at which this occurs is called the *ceiling temperature* (T_c), and at that temperature, ΔG of polymerization is zero, or

$$T_c = \frac{\Delta H}{\Delta S}$$

---

*The $\rho\sigma$ relationship was originally established by Hammett to correlate the electron-donating or electron-attracting properties of substituent groups on reaction rates of *meta*- and *para*-substituted benzene derivatives. In this equation, k and k_0 are the rate constants of substituted and unsubstituted benzene compounds, respectively; σ is a substituent constant having a single value for each substituent that reflects its ability at its particular ring position to release electrons to or withdraw electrons from the reaction center; and ρ is a reaction constant reflecting the effect of the availability of electrons on a particular reaction. Sigma values are normally determined from the ionization of substituted benzoic acids in water at 25°C. For this reaction, the value of

ρ is defined at $+1.000$. Groups that release electrons relative to hydrogen have negative σ values, whereas groups that attract electrons have positive values. When electron-attracting substituents facilitate a reaction, ρ has a positive value.

A discussion of the Hammett equation can be found in most up-to-date advanced texts in organic chemistry.

It follows that whatever affects ΔH or ΔS for a given monomer (e.g., monomer concentration, pressure) will also affect T_c. Table 6.8 lists ceiling temperatures for several pure liquid monomers.[37]

Because propagation and depropagation are in equilibrium, there will always be some residual monomer in the polymer. With most common vinyl monomers, however, ceiling temperatures are sufficiently high that the amount of residual monomer is extremely small. A notable exception is α-methylstyrene (**41**), which has a T_c of about 66°C in bulk polymerization.

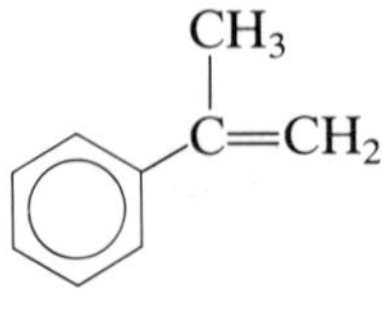

$$\underset{\text{(benzene ring)}}{}\overset{\displaystyle CH_3}{\underset{\displaystyle}{\overset{|}{C}}}=CH_2$$

41

Steric effects arising from having bulky methyl and phenyl groups located together on alternate carbon atoms of the polymer, and the relative stability of the tertiary benzylic α-methylstyryl radical, combine to give α-methylstyrene its low T_c. As a result, poly(α-methylstyrene) is difficult to obtain by conventional radical polymerization. In cases where the T_c is very high (polythylene or polytetrafluoroethylene, for example), polymers degrade by bond cleavage reactions other than depropagation at temperatures well below the T_c.

Before we leave the subject of polymer reactivity, it should be pointed out that monomers do not necessarily have to be low-molecular-weight compounds. The reactive double bond could be at the end of an oligomeric or even a polymeric chain. Monomers of this type are called *macromonomers*, or *macromers*.[38–40] Polymerization of a macromonomer (6.49) results in a comb

Table 6.8. Representative ceiling temperatures, T_c, of pure liquid monomers[a]

Monomer	T_c (° C)
1,3-Butadiene	585
Ethylene	610
Isobutylene	175
Isoprene	466
Methyl methacrylate	198
α-Methylstyrene	66
Styrene	395
Tetrafluoroethylene	1100

[a] Data from Allcock and Lampe.[37]

polymer, in which long-chain units reside on alternate carbon atoms of the polymer backbone. Comb polymers, which are structurally related to graft copolymers (discussed in Chapter 9), may exhibit crystallinity or liquid crystallinity if the macromonomer has a regular structure.[41,42]

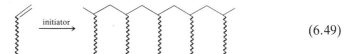

$$(6.49)$$

6.8 Copolymerization

The mechanism of copolymerization[43a] is analogous to that of homopolymerization, but reactivities of the various monomers toward other monomers can vary considerably. Already mentioned in the previous section is the case of 1,2-disubstituted ethylenes which, in most instances, fail to homopolymerize, yet enter into copolymerization reactions readily with a variety of mono- or 1,1-disubstituted monomers.

Consider the case where two monomers, M_1 and M_2, copolymerize. The monomers can undergo either *self-propagation* ($M_1 \cdot$ reacting with M_1, or $M_2 \cdot$ reacting with M_2) or *cross-propagation* ($M_1 \cdot$ reacting with M_2, or $M_1 \cdot$ reacting with M_1). This results in four separate reactions, each with its own rate constant:

$$M_1 \cdot + M_1 \xrightarrow{k_{11}} M_1 \cdot$$
$$M_1 \cdot + M_2 \xrightarrow{k_{12}} M_2 \cdot$$
$$M_2 \cdot + M_1 \xrightarrow{k_{21}} M_1 \cdot$$
$$M_2 \cdot + M_2 \xrightarrow{k_{22}} M_2 \cdot$$

If we assume that the concentrations of $M_1 \cdot$ and $M_2 \cdot$ remain constant (steady-state treatment), then the rate of addition of $M_1 \cdot$ to M_2 will equal the rate of addition of $M_2 \cdot$ to M_1; that is,

$$k_{12}[M_1 \cdot][M_2] = k_{21}[M_2 \cdot][M_1]$$

The rate of disappearance of M_1 and M_2 may be expressed by the following:

$$\frac{-d[M_1]}{dt} = k_{11}[M_1 \cdot][M_1] + k_{21}[M_2 \cdot][M_1]$$

$$\frac{-d[M_2]}{dt} = k_{12}[M_1 \cdot][M_2] + k_{22}[M_2 \cdot][M_2]$$

The ratio of the two rates is then

$$\frac{d[M_1]}{d[M_2]} = \frac{[M_1]}{[M_2]} \left(\frac{k_{11}[M_1 \cdot] + k_{21}[M_2 \cdot]}{k_{12}[M_1 \cdot] + k_{22}[M_2 \cdot]} \right)$$

If we combine this equation with the steady-state expression above and define

$$\frac{k_{11}}{k_{12}} = r_1 \quad \text{and} \quad \frac{k_{22}}{k_{21}} = r_2$$

we derive the *copolymer equation* (also called the *copolymer composition equation*).

$$\frac{d[M_1]}{d[M_2]} = \frac{[M_1]}{[M_2]} \left(\frac{r_1[M_1] + [M_2]}{[M_1] + r_2[M_2]} \right)$$

The ratio $d[M_1]/d[M_2]$ represents the molar ratio of the two monomers in the copolymer, while $[M_1]$ and $[M_2]$ are the initial molar concentration of monomers in the reaction mixture (or *feed*, as the mixture is frequently called.) The terms r_1 and r_2, called the *reactivity ratios*, define the relative tendencies of the monomers to self-propagate or cross-propagate. If $r_1 > 1$, then M_1 tends to self-propagate, whereas if $R_1 < 1$, copolymerization is preferred. Thus if we know the reactivity ratios, we can calculate the initial molar ratio of monomers needed to give a desired ratio of monomeric units in the copolymer, at least in the initial stages of copolymerization (up to about 10%). In most cases, one monomer is incorporated in preference to the other; therefore the monomer composition in the feed changes as the copolymerization reaction proceeds, and the polymer composition changes with increasing conversion. Not infrequently it may be necessary to add one monomer continuously in order to maintain the desired copolymer composition. The validity of the copolymer equation has been confirmed experimentally.

An alternative method of expressing the copolymer equation is in terms of mole fractions. Thus f_1 and f_2 are the mole fractions of M_1 and M_2 in the feed, respectively, and F_1 and F_2 are the mole fractions of M_1 and M_2 in the copolymer. Because

$$f_1 = 1 - f_2 = \frac{[M_1]}{[M_1] + [M_2]}$$

and

$$F_1 = 1 - F_2 = \frac{d[M_1]}{d[M_1] + d[M_2]}$$

the copolymer equation may be rewritten

$$F_1 = \frac{r_1 f_1^2 + f_1 f_2}{r_1 f_1^2 + 2f_1 f_2 + r_2 f_2^2}$$

Determination of reactivity ratios[44-46] is no easy task, requiring analysis of copolymer composition in the early stages of reaction for a variety of monomer feed compositions. More recent methods[47,48] have stressed using NMR triad data. Fortunately, large numbers of reactivity ratios have been compiled[22d,43b]; several of them are listed in Table 6.9.

Table 6.9. Representative free radical monomer (*M*) reactivity ratios (*r*)[a]

M_1	M_2	r_1	r_2	Temperature (°C)
Styrene	Methyl methacrylate	0.52	0.46	60
Styrene	Acrylonitrile	0.40	0.04	60
Styrene	Vinyl acetate	55	0.01	60
Styrene	Maleic anhydride	0.041	0.01	60
Styrene	Vinyl chloride	17	0.02	60
Styrene	1,3-Butadiene	0.58	1.35	50
Styrene	Isoprene	0.54	1.92	80
Methyl methacrylate	Vinyl chloride	10	0.1	68
Methyl methacrylate	Vinyl acetate	20	0.015	60
Methyl methacrylate	Acrylonitrile	1.20	0.15	60
Methyl methacrylate	1,3-Butadiene	0.25	0.75	90
Ethylene	Tetrafluoroethylene	0.38	0.1	25
Ethylene	Acrylonitrile	0	7	20
Ethylene	Vinyl acetate	0.97	1.02	130

[a] Data selected from Young[22d] (see this reference for polymerization conditions).

Clearly, a knowledge of reactivity ratios is of immense benefit in formulating feed ratios for copolymers. Let us examine several hypothetical situations involving different values of r_1 and r_2. First we consider the situation where

$$r_1 = r_2 = 1.0$$

In this case the monomers would exhibit no preference for homopolymerization or copolymerization, and a truly random copolymer would result. Under these conditions the copolymer equation reduces to

$$F_1 = f_1$$

and a plot of f_1 versus F_1 (Figure 6.1) would show the same composition in both feed and copolymer. A situation approaching this is the copolymerization of ethylene (M_1) and vinyl acetate (M_2) ($r_1 = 0.97$, $r_2 = 1.02$). This is shown by curve A in Figure 6.1, which is very close to the diagonal.

Next we consider the case where

$$r_1 = r_2 = 0$$

Here, neither monomer would exhibit a tendency to homopolymerize, and a truly *alternating* copolymer would result. Then $F_1 = 0.5$. Approaching this situation is the styrene (M_1)-maleic anhydride (M_2) system where $r_1 = 0.041$ and $r_2 = 0.01$. This is shown by curve B in Figure 6.1.

A far more common situation is where both r_1 and r_2 lie between 0 and 1, such as with the styrene (M_1)-methyl methacrylate (M_2) system ($r_1 = 0.52$,

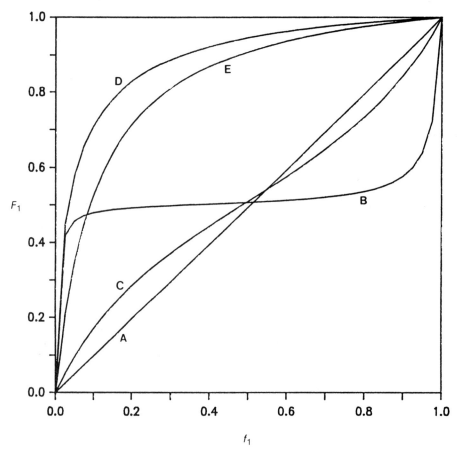

Figure 6.1. Variation of F_1 with f_1 for copolymerizations (F_1 = mole fraction of monomer in copolymer; f_1 = mole fraction of monomer in feed): (A) $r_1 = 0.97$, $r_2 = 1.02$ (r = reactivity ratio); (B) $r_1 = 0.041$, $r_2 = 0.01$; (C) $r_1 = 0.52$, $r_2 = 0.46$; (D) $r_1 = 55$, $r_2 = 0.01$; (E) $r_1 = 10$, $r_2 = 0.1$.

$r_2 = 0.46$), shown by curve C in Figure 6.1. In such cases there is some tendency toward alternation. (If $r_1 = 0.5$, M_1 reacts twice as fact with M_2 as with M_1.) The point at which curve C crosses the diagonal represents a composition where polymerization proceeds to relatively high conversion with no change in either feed ratio or copolymer composition. Such cases are called *azeotropic polymerization*. Since with azeotropic polymerization

$$\frac{d[M_1]}{d[M_2]} = \frac{[M_1]}{[M_2]}$$

the copolymer equation reduces to

$$\frac{[M_1]}{[M_2]} = \frac{1 - r_2}{1 - r_1}$$

under azeotropic conditions. Azeotropic polymerization is also possible, in principle, when both r_1 and r_2 are greater than 1; however, this situation has not been observed in free radical copolymerization.

Now let us consider the case where r_1 is much greater than 1 and r_2 much less than 1; for example, styrene (M_1)-vinyl acetate (M_2) $(r_1 = 55, r_2 = 0.01)$ or styrene (M_1)-vinyl chloride (M_2) $(r_1 = 17, r_2 = 0.02)$. Curve D in Figure 6.1 illustrates the former. Such cases lead to incorporation of monomer 1 almost exclusively in the early stages of polymerization. When r_1 is very high and r_2 is close to 0, one obtains essentially a homopolymer of M_1.

It is common among polymer chemists to refer to the situation where $r_1 r_2 = 1$ as *ideal copolymerization* following a parallel with vapor-liquid equilibria in ideal liquid mixtures. One should keep in mind, however, that "ideal" copolymerization in the sense of a random distribution of monomer units in the copolymer occurs only when both r_1 and r_2 are close to 1. The more r_1 and r_2 diverge, the less random will be the distribution. A case in point is the "ideal copolymerization" of methyl methacrylate (M_1) and vinyl chloride (M_2) where $r_1 = 10$ and $r_2 = 0.1$ and $r_1 r_2 = 1$. This is shown with curve E in Figure 6.1.

To simplify the process of determining reactivity ratios, a semiempirical relationship has been developed that expresses the ratios in terms of constants that are characteristic of each monomer but independent of comonomer, thus eliminating the need to determine experimentally the ratios for each pair of monomers in a particular copolymerization. This relationship is called the *Q-e scheme*[43c]:

$$r_1 = \left(\frac{Q_1}{Q_2}\right) \exp[-e_1(e_1 - e_2)]$$

$$r_2 = \left(\frac{Q_2}{Q_1}\right) \exp[-e_2(e_2 - e_1)]$$

where Q_1 and Q_2 are measures of the reactivity of monomers M_1 and M_2 and are related to resonance stabilization of monomer. The constants e_1 and e_2 are measures of the polarity of the monomers. Styrene was chosen as the standard and assigned values of $Q = 1.00$ and $e = -0.80$. Values of Q increase with increasing resonance stabilization, while e values become less negative as groups attached to the double bond become more electron-attracting. A few representative values of Q and e are given in Table 6.10.[43d] While the trends in Q and e values are obvious, it should be pointed out that these values are only approximate and provide, at best, a semiquantitative but nevertheless extremely useful means of predicting reactivity ratios where other experimental data are not available.

Reactivity ratios are the result of a combination of steric, resonance, and polar effects. Steric effects have already been discussed as they apply to 1,2-disubstituted ethylenes, and steric inhibition of resonance was cited in the reduced reactivity of maleate esters relative to the *trans* isomers. As was

Table 6.10. Representative values for
reactivity (*Q*) and polarity (*e*) of monomers[a]

Monomer	*Q*	*e*
1-Vinylnaphthalene	1.94	−1.12
p-Nitrostyrene	1.63	0.39
p-Methoxystyrene	1.36	−1.11
Styrene	1.00	−0.80
Methyl methacrylate	0.74	0.40
Acrylonitrile	0.60	1.20
Methyl acrylate	0.42	0.60
Vinyl chloride	0.044	0.20
Vinyl acetate	0.026	−0.22

[a] Data from Young.[43d]

mentioned earlier, resonance affects reactivity by stabilization, of the in-termediate radical—the more resonance stabilization, the less reactive is the monomer toward propagation. Thus, resonance-stablized styryl radicals ex-hibit little tendency to add to vinyl acetate because the resultant radical would not be stabilized. Styrene does, however, copolymerize with methyl methacrylate because the latter, like styrene, forms a delocalized radical (**42**).

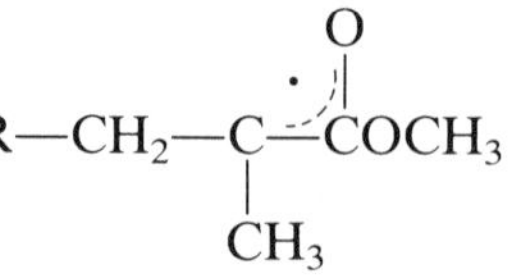

$$R-CH_2-\overset{\displaystyle CH_3}{\underset{\displaystyle CH_3}{C}}\overset{\displaystyle O}{\underset{}{\overset{\|}{-COCH_3}}}$$

42

From the reactivity ratios of the styrene–methyl methacrylate system, it can be seen that each radical has about twice as much tendency to react with its opposite number, a phenomenon attributable to polar effects. The differ-ence in electron-withdrawing or electron-donating effects of the substituent groups lowers the activation energy for addition of styryl radicals to methyl methacrylate relative to addition to styrene. Support for this is gained from the fact that appropriately substituted styrenes follow the Hammett rela-tionship in copolymerization as well as in homopolymerization.

Two theories have been proposed to explain those cases where there is a strong tendency toward alternation (r_1 and $r_2 = \sim 0$), one involving polar effects and the other charge-transfer interactions. First, let's consider polar effects as applied to the copolymerization of styrene and maleic anhydride.[43a] Addition of maleyl radical to styrene (6.50) leads to a transition state that can

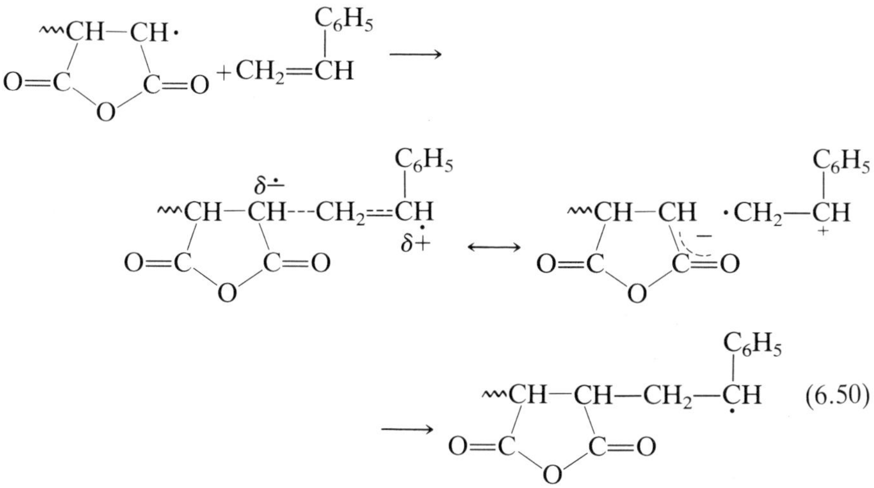

$$\text{(scheme 6.50)}$$

be stabilized by resonance, including one nonbonded resonance form involving electron transfer. Such structures may be invoked in cases where the monomer contains an abundance of electronegative substituents, as with maleic anhydride. Addition of styryl radical to maleic anhydride (6.51) would give rise to a similarly stabilized transition state. The net effect is that styryl radicals prefer to add to maleic anhydride and maleyl radicals prefer to add to styrene.

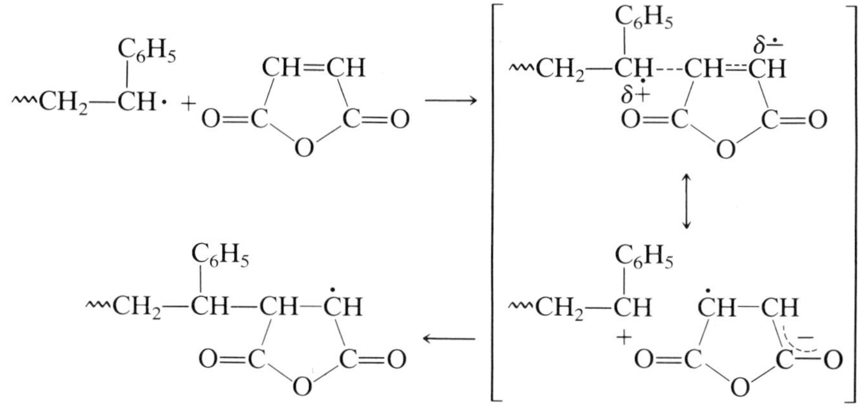

$$\text{(scheme 6.51)}$$

The other interpretation of alternation is based on formation of charge-transfer complexes between comonomers.[49] Such complexes can be detected by observation of characteristic ultraviolet absorption maxima in solutions of monomers. According to this theory, an alternating copolymerization reaction is, in effect, a *homopolymerization of the charge-transfer complex*. Thus the propagating chain end is not a conventional radical, nor is it an ionic species, but rather it is a donor (D)-acceptor (A) complex that reacts preferentially with other charge-transfer complexes (6.52). There is considerable evidence favoring this theory:

$$\text{ww}(DA)_n \overset{+}{D}..\bar{A} + \overset{+}{D}..\bar{A} \longrightarrow \text{ww}(DA)_{n+1}\overset{+}{D}..\bar{A} \qquad (6.52)$$

1. The rate of polymerization reaches a maximum when the monomer composition ratio is about 1:1, which is where the donor-acceptor complex is at maximum concentration.
2. Alternation is independent of monomer feed ratios, and other reactive monomers included with the feed fail to react while alternating copolymer is forming.
3. The rate is enhanced by addition of Lewis acids, which increase the acceptor properties of one of the monomers.[50,51]
4. Chain transfer agents have little effect on the molecular weight of the copolymer.
5. Alternating copolymers of *N*-phenylmaleimide (**43**) and 2-chloroethyl vinyl ether (**44**) contain predominantly *cis* ring structures arising from addition of chain-end radical (6.53) to the stereochemically fixed charge-transfer complex (**45**) on the side of the complex syn to the vinyl ether (6.53).[52] (The charge-transfer complex geometry is based on maximum overlap between the highest occupied molecular orbital of the donor with the lowest unoccupied molecular orbital of the acceptor.)

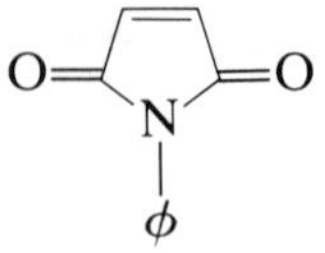

43

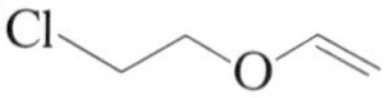

44

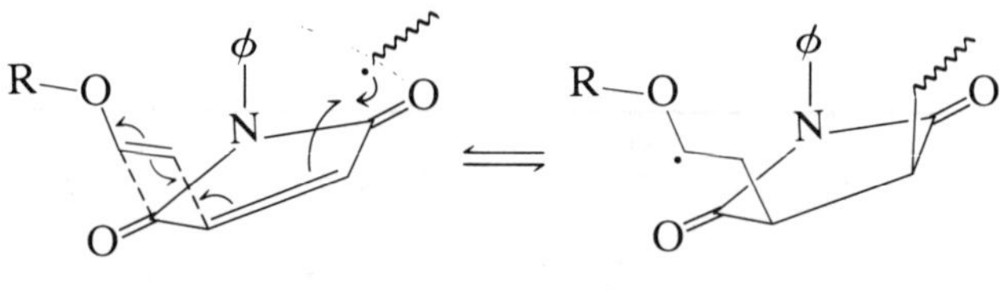

45

(6.53)

Examples of other compounds that undergo free radical-initiated copolymerization with vinyl monomers are carbon monoxide and sulfur dioxide, the former to yield polyketones (6.54) and the latter polysulfones (6.55). These reactions also tend toward alternation, and sulfur dioxide is known to form 1:1 complexes with alkenes.

$$\underset{\substack{| \\ CH_2=CH}}{\overset{R}{}} + CO \longrightarrow \underset{\substack{| \quad || \\ CH_2-CH-C}}{\overset{R \quad O}{\Big[\Big]}} \qquad (6.54)$$

$$CH_2{=}CH + SO_2 \longrightarrow \left[CH_2{-}\overset{\displaystyle R}{\underset{\displaystyle |}{CH}}{-}SO_2\right] \qquad (6.55)$$

Charge-transfer mechanisms have also been proposed to explain certain cyclocopolymerization reactions.[53] Examples are the free radical-initiated reaction of divinyl ether with maleic anhydride[54] (6.56) and the "spontaneous" cyclopolymerization of divinyl ether and divinyl sulfone[55] (6.57).

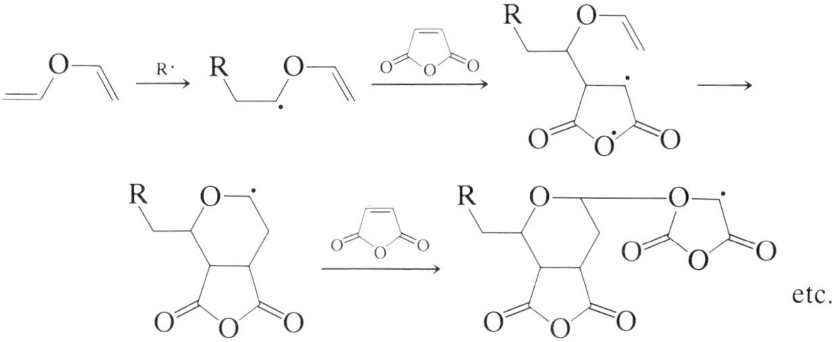

$$(6.56)$$
etc.

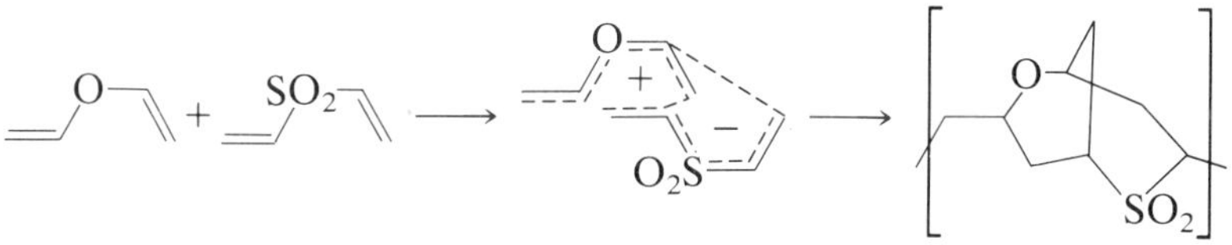

$$(6.57)$$

Which is responsible for alternation in free radical copolymerization—polar effects in the transition state or charge-transfer complexes, or both—is still open to question,[56] despite the volume of literature published on the subject.

References

1. W. A. Pryor, *Free Radicals*, McGraw-Hill, New York, 1966.
2. S. Patai (ed.), *The Chemistry of Peroxides*, Wiley-Interscience, New York, 1983.
3. M. Shen and A. T. Bell (eds.), *Plasma Polymerization*, ACS Symp. Ser. 108, American Chemical Society, Washington, D.C., 1979.
4. M. Yasuda, *Plasma Polymerization*, Academic Press, San Diego, 1985.
5. C. Walling, *Free Radicals in Solution*, Wiley-Interscience, New York, 1957.
6. W. A. Pryor and W. H. Hendrickson Jr., *Tetrahedron Lett.*, **24**, 1459 (1983).
7. G. E. Ham (ed.), *Vinyl Polymerization*, Part 1, Dekker, New York, 1967: (a) G. E. Ham, Chap. 1; (b) M. H. George, pp. 147–152; (c) A. D. Jenkins, Chap. 6; (d) W. E. Gibbs and T. M. Barton, Chap. 2; (e) R. M. Joshi and B. J. Zwolinski, Chap. 8.
8. W. D. Graham, J. G. Green, and W. A. Pryor, *J. Org. Chem.*, **44**, 907 (1979).
9. A. Husain and A. H. Hamielec, *J. Appl. Polym. Sci.*, **22**, 1207 (1978).
10. O. F. Olaj, H. F. Kauffmann, and J. W. Breitenbach, *Makromol. Chem.*, **177**, 3065 (1976).

11. W. A. Pryor, *Free Radicals*, McGraw-Hill, New York, 1966, pp. 119–126.
12. H. F. Kauffmann, *Makromol. Chem.*, **180**, 2649, 2665, 2681 (1981).
13. C. E. Schildknecht, *Polymer Processes*, Wiley-Interscience, New York, 1956.
14. H. F. Mark, N. G. Gaylord, and N. M. Bikales (eds.), *Encyclopedia of Polymer Science and Technology*, Wiley-Interscience, New York, 1969: (a) J. W. Breitenbach, Vol. 11, pp. 587–597; (b) G. Goldfinger, W. Yee, and R. D. Gilbert, Vol. 7, pp. 644–664.
15. D. C. Blakely, *Emulsion Polymerization—Theory and Practice*, Applied Science, London, 1975.
16. D. R. Bassett and A. E. Hamielec (eds.), *Emulsion Polymers and Emulsion Polymerization*, ACS Symp. Ser. 165, American Chemical Society, Washington, DC., 1981.
17. A. D. Jenkins and A. Ledwith (eds.), *Reactivity, Mechanism and Structure in Polymer Chemistry*, Wiley-Interscience, New York, 1974: (a) W. Cooper, Chap. 7; (b) A. M. North, Chap. 5; (c) A. D. Jenkins, Chap. 4; (d) K. J. Ivin, Chap. 16.
18. C. W. Wilson III and E. R. Santee Jr., *J. Polym. Sci. C*, **8**, 97 (1965).
19. A. Matsumoto, K. Iwanami, and M. Oiwa, *J. Polym. Sci., Polym. Lett. Ed.*, **18**, 211 (1980).
20. T. Tsuruta and K. F. O'Driscoll (eds.), *Structure and Mechanism in Vinyl Polymerization*, Dekker, New York, 1969: (a) A. M. North and D. Postlethwaite, Chap. 4; (b) Y. Minouro, Chap. 7.
21. C. H. Bamford, W. G. Barb, A. D. Jenkins, and P. F. Onyan, *The Kinetics of Vinyl Polymerization by Free Radical Mechanisms*, Academic Press, New York (1958).
22. J. Brandrup, E. H. Immergut, and W. McDowell (eds.), *Polymer Handbook*, 2nd ed., Wiley-Interscience, New York, 1975: (a) R. Korus and K. F. O'Driscoll, pp. II-45ff and II-451ff; (b) L. J. Young, pp. II-57ff; (c) K. J. Ivin, pp. II-421ff; (d) L. J. Young, pp. II-105ff; (e) L. J. Young pp. II-387ff.
23. A. Chapiro, in *Polymer Science Overview* (G. A. Stahl, ed.), ACS Symp. Ser. 175, American Chemical Society, Washington, D.C., 1981, Chap. 16.
24. W. V. Smith and R. H. Ewart, *J. Chem. Phys.*, **16**, 592 (1948).
25. W. Cooper, in *The Stereochemistry of Macromolecules*, Vol. 2 (A. D. Ketley, ed.), Dekker, New York, 1967, Chap. 5.
26. T. G. Fox, B. S. Garrett, W. E. Goode, S. Gratch, J. F. Kincaid, A. Spell, and J. D. Stroupe, *J. Am. Chem. Soc.*, **80**, 1768 (1958).
27. F. A. Bovey, *J. Polym. Sci.*, **46**, 59 (1960).
28. D. J. Cram and K.R. Kopecky, *J. Am. Chem. Soc.*, **81**, 2748 (1959).
29. C. E. H. Bawn, W. H. Janes, and A. M. North, *J. Polym. Sci. C*, **4**, 427 (1963).
30. R. J. Cotter and M. Matzner, *Ring-Forming Polymerizations*, Part A, Academic Press, New York, 1969, Chap. 2.
31. C. E. Schildknecht, *Allyl Compounds and Their Polymers*, Wiley-Interscience, New York, 1973.
32. A. Matsumoto, K. Iwanami, and M. Oiwa, *J. Polym. Sci., Polym. Lett. Ed.*, **18**, 307 (1980).
33. M. Farina, G. Disilvestro, and P. Sozzani, in *Crystallographically Ordered Polymers* (D. J. Sandman, ed.), ACS Symp. Ser. 337, American Chemical Society, Washington, D.C., 1987, p. 79.
34. K. Takemoto and M. Miyata, *J. Macromol. Sci., Rev. Macromol. Chem.*, **C18**, 83 (1980).

35. D. G.L. James and T. Ogawa, *Can. J. Chem.*, **43**, 640 (1965).
36. C. A. Barson, J. C. Bevington, and D. E. Eaves, *Trans. Faraday Soc.*, **54**, 1678 (1958).
37. H. R. Allcock and F. W. Lampe, *Contemporary Polymer Chemistry*, Prentice-Hall, Englewood Cliffs, N. J., 1981, p. 225.
38. P. F. Rempp and E. Franta, *Adv. Polym. Sci.*, **58**, 1 (1984).
39. G. O. Schulz and R. Milkovich, *J. Polym. Sci., Polym. Chem. Ed.*, **22**, 1633 (1984).
40. Y. Yamashita, *J. Appl. Poly. Sci., Appl. Polym. Symp.*, **36**, 193 (1981).
41. N. A. Platé and V. P. Shibaev, *Comb-Shaped Polymers and Liquid Crystals*, Plenum Press, New York, 1987.
42. N .A. Platé and V. P. Shibaev, *Macromol. Rev.*, **8**, 117 (1974).
43. G. E. Ham (ed.), *Copolymerization*, Wiley-Interscience, New York, 1964: (a) G. E. Ham, Chap. 1; (b) H. Mark, B. Immergut, and E. H. Immergut, Appendix A; (c) T. Alfrey Jr. and L. J. Young, Chap. 2; (d) L. J. Young, Appendix B.
44. D. Braun, H. Cherdron, and W. Kern, *Practical Macromolecular Organic Chemistry*, 3rd ed., Harwood, New York, 1984, pp. 206–221.
45. P. W. Tidwell and G. A. Mortimer, *J. Macromol. Sci.-Rev. Macromol. Chem.*, **C4**, 281 (1970).
46. R. C. McFarlane, P. M. Reilly, and K. F. O'Driscoll, *J. Polym. Sci., Polym. Chem. Ed.*, **18**, 251 (1980).
47. A. Rudin, K. F. O'Driscoll, and M. S. Rumack, *Polymer*, **22**, 740 (1981).
48. J. J. Uebel and F. J. Dinan, *J. Polym. Sci., Polym. Chem. Ed.*, **21**, 917 (1983).
49. D. Bartlett and K. Nozaki, J. Am. Chem. Soc., **68**, 1495 (1946).
50. H. Hirai, *J. Macromol. Sci.-Rev. Macromol. Chem.*, **11**, 47 (1976).
51. C. H. Bamford, in *Alternating Copolymerization* (J. M. G. Cowie, ed.), Plenum Press, New York, 1985, Chap. 3.
52. G. B. Butler, K. G. Olson, and C.-L. Tu, *Macromolecules*, **17**, 1884 (1984).
53. G. B. Butler, *Pure Appl. Chem.*, **23**, 255 (1970).
54. G. B. Butler and A. F. Campus, *J. Polym. Sci., A-1*, **8**, 545 (1970).
55. G. B. Butler and A. J. Sharpe, *J. Polym. Sci., B*, **9**, 125 (1971).
56. J. R. Ebdon, C. R. Towns, and K. Dodgson,*J. Macromol. Sci., Rev. Macromol. Chem. Phys.*, C26, 523 (1986).

Review exercises

1. Give reasonable explanations for the following facts:
   (a) Free radical polymerization has received more widespread commercial development than ionic polymerization.
   (b) Almost all substituted ethylene monomers of commercial importance have the structure $CH_2 = CHR$ or $CH_2 = CRR'$ (where R or R' represent identical or different substituent groups).
   (c) Syndiotactic placement is favored at low temperatures in free radical polymerization.
   (d) In general, increasing polymerization rate decreases molecular weight in bulk polymerization, whereas the reverse is the case in emulsion polymerization.
   (e) Isobutylene and allyl acetate cannot be polymerized to high molecular weight under the usual free radical conditions.

2. Discuss the similarities and differences between suspension and emulsion free radical polymerization, including advantages and disadvantages of each.

3. Summarize the causes and ramifications of the gel (Trommsdorff or Norris–Smith) effect.

4. Wastage reaction products of diacetyl peroxide (structure **2**) include carbon dioxide and ethane. Those of di-*t*-butyl peroxide (**3**) include acetone, methane, *t*-butyl alcohol, and isobutylene oxide. Suggest plausible reaction pathways by which these various side products might form.

5. In normal free radical processes with added initiator, the polymerization rate is proportional to the first power of monomer concentration. What would be the effect of monomer concentration on $\overline{DP}$ in thermal polymerization? Derive a kinetics expression to support your conclusion.

6. If equal concentrations of acrylonitrile and methyl methacrylate were each polymerized at 60°C with equal concentrations of the same initiator, which polymer would have the higher $\overline{DP}$ and by how much? Assume polyacrylonitrile undergoes termination only by radical combination and poly(methyl methacrylate) by disproportionation, that no chain transfer occurs, and that initiator efficiencies are the same in both reactions. Which polymer would have the higher $\overline{M}_n$, and by how much?

7. What concentration of benzoyl peroxide ($k_d = 1.45 \times 10^{-6}$ L/mol-s at 60°C) would be needed to polymerize a 1.00 M solution of styrene to a molecular weight of 125,000? Assume that termination occurs only by radical combination, that no chain transfer occurs, and that initiator efficiency is 100%.

8. A certain monomer is polymerized by benzoyl peroxide labeled in the benzene ring with carbon-14. (a) Analysis by carbon counting shows that the polymer contains an average of 1.27 labeled phenyl groups per molecule. Assuming no chain transfer occurred, what does this indicate about the mechanism of termination? (b) Vigorous basic hydrolysis removed 60% of the carbon-14. What does this indicate about the mechanism of initiation?

9. Suppose you wanted to repeat the experiment given in question 7 (using the same initiator concentration), but wanted to limit the molecular weight to 15,000. What concentration of *n*-butyl mercaptan (1-butanethiol) would you add?

10. Predict the order of polymer molecular weight obtainable (highest to lowest) by solution polymerization of styrene in benzene, toluene, ethylbenzene, *i*-propylbenzene, and *t*-butylbenzene. Assume each polymerization is run under identical conditions other than choice of solvent. Justify your answer.

11. Explain how monomer concentration might affect the outcome of cyclopolymerization of an isolated diene.

12. Given the fact that propylene forms only oligomers under normal free radical con-

ditions, predict the results of free radical polymerization of propylene contained within the channels of an inclusion compound. (G. Di Silvestro, P. Sozzani, and M. Farina, *Polym. Preprints*, **27**(1), 92, 1986)

13. Inclusion polymerization of diene monomers invariably leads to highly crystalline stereoregular (*trans*-1,4) polymer. An exception has been noted (ref. 33) with isoprene polymerized in perhydrotriphenylene host. The polyisoprene thus obtained was indeed shown to be *trans*-1,4 by IR spectroscopy; nevertheless, the polymer was noncrystalline. An examination of the polymer's ^{13}C NMR spectrum revealed four signals in the CH_2 region. Suggest a reason for the polymer's failure to crystallize.

14. On the basis of chemical structure, predict the result of free radical copolymerization of the following monomer pairs: (a) acrylamide and acrylonitrile, (b) ethylene and methyl acrylate, (c) 1,3-butadiene and vinyl stearate (vinyl octadecanoate), (d) chlorotrifluoroethylene and ethyl vinyl ether. When you have made your predictions, consult a polymer handbook and check your answers on the basis of reactivity ratios.

15. Predict the ratio of butadiene to styrene repeating units initially produced by free radical copolymerization of an equimolar mixture of the two monomers at 50°C. (Consult Table 6.9.)

16. What weight ratio of the two monomers in the previous problem would be necessary to have twice as many butadiene as styrene units in the copolymer.

17. Compare the reactivity ratios determined for methyl methacrylate and vinyl acetate by the *Q-e* scheme with those determined experimentally (Table 6.9).

18. In the presence of a free radical initiator, furan and maleic anhydride form a 1:1 copolymer at 70°C in benzene solution. In the absence of the initiator, the two reactants form a 1:1 Diels–Alder adduct. The Diels–Alder adduct undergoes polymerization at temperatures higher than 70°C to give a polymer identical with the 1:1 copolymer. IR and ^{13}C NMR of the polymer exhibited olefinic unsaturation of the type found in the Diels–Alder adduct. What is the most likely structure of the polymer? What other structures might be feasible in the absence of the spectroscopic evidence?   References pertaining to this problem: N. G. Gaylord et al., *J. Macromol. Sci.-Chem.*, **A6**, 1459, 1972; B. Kamo et al., *Polym. J.*, **6**, 121, 1974; Y. A. Ragab and G. B. Butler, *J. Polym. Sci., Polym. Lett. Ed.*, **14**, 273, 1976; *idem., J. Polym. Sci., Polym. Chem. Ed.*, **19**, 1175, 1981.

7
Vinyl polymerization with ionic and group transfer initiators

7.1 Introduction

A fundamental difference between free radical and ionic polymerization reactions is that, with the latter, *counterions* (also called *gegenions*) are present in the reaction medium to preserve electrical neutrality. The mechanism of polymerization may be strongly influenced by the counterions, depending on whether they are strongly or weakly associated with the propagating chain. Such association depends in large measure on solvation effects. Ionic polymerizations are therefore considerably more complex than free radical polymerizations, but at the same time they are more versatile in their degree of steric control. They are also more versatile in their scope, having application in ring-opening polymerizations of cyclic ethers (to form polyethers), lactams (to form polyamides), and lactones (to form polyesters), and in the polymerization of aldehydes and ketones (to form polyethers). These applications are covered in later chapters dealing with the appropriate polymer type.

Despite the versatility of ionic polymerization, commercial processes based on ionic vinyl polymerizations are far fewer in number and reflect a much narrower choice of monomers than the corresponding free radical processes. This results from the polar nature of the propagating chain end in ionic polymerization; monomers must contain substituent groups capable of stabilizing carbocations or carbanions. A further limiting factor is the necessity for solution polymerization in most instances, although bulk processes are sometimes used. Examples of vinyl polymers prepared commercially by ionic polymerization are given in Table 7.1.

In the early 1980s the new method of *group transfer polymerization* was introduced. Group transfer polymerization is not an ionic process in the strictest sense—the propagating chain end is covalent in character. Anionic species or Lewis acids are used as polymerization catalysts, however, and the propagation reactions are not unlike those of anionic polymerization.

234

Table 7.1. Commercially important polymers prepared by ionic polymerization

Polymer or copolymer	Major uses
Cationic[a]	
Polyisobutylene and polybutenes[b] (low and high molecular weight)	Adhesives, sealants, insulating oils, lubricating oil and grease additives, moisture barriers
Isobutylene–isoprene copolymer[c] ("butyl rubber")	Inner tubes, engine mounts and springs, chemical tank linings, protective clothing, hoses, gaskets, electrical insulation
Isobutylene-cyclopentadiene copolymer	Ozone-resistant rubber
Hydrocarbon[d] and polyterpene resins	Inks, varnishes, paints, adhesives, sealants
Coumarone–indene resins[e]	Flooring, coatings, adhesives
Poly(vinyl ether)s	Polymer modifiers, tackifiers, adhesives
Anionic[f]	
cis-1,4-Polybutadiene	Tires
cis-1,4-Polyisoprene	Replacement for natural rubber
Styrene–butadiene rubber (SBR)[g]	Tire treads, belting, hose, shoe soles, flooring, coated fabrics
Styrene–butadiene block and star copolymers	Flooring, shoe soles, artificial leather, wire and cable insulation
ABA block copolymers (A = styrene, B = butadiene or isoprene)	Thermoplastic elastomers
Polycyanoacrylate[h]	Adhesives

[a] $AlCl_3$ and BF_3 most frequently used coinitiators.

[b] "Polybutenes" are copolymers based on C_4 alkenes and lesser amounts of propylene and C_5 and higher alkenes from refinery streams.

[c] Terpolymers of isobutylene, isoprene, and divinylbenzene are also used in sealant and adhesive formulations.

[d] Aliphatic and aromatic refinery products.

[e] Coumarone (benzofuran) and indene (benzocyclopentadiene) are products of coal tar.

[f] n-Butyllithium most common initiator.

[g] Contains higher *cis* content than SBR prepared by free radical polymerization.

[h] Monomer polymerized by adventitious water.

7.2 Cationic polymerization

7.2.1 Cationic initiators

In cationic chain polymerization [1–3a,4a] the propagating species is a carbocation.* Initiation is brought about by addition of an electrophile to a monomer molecule (7.1).

---

* The terms *carbonium ion* and *carbenium ion* are both used to denote a positively charged carbon species. Although the former is the traditional term, many chemists prefer that it be reserved for pentavalent carbocations, including nonclassical ions, and the latter for "classical" trivalent, trigonal carbocations. Because such terminology is not universally accepted, we will avoid the controversy and refer to the propagating species as a carbocation.

$$E^+ + CH_2{=}CR_2 \rightarrow ECH_2\overset{+}{C}R_2 \qquad (7.1)$$

Compounds used most frequently to effect cationic polymerization are mineral acids, particularly H_2SO_4 and H_3PO_4, and such Lewis acids as $AlCl_3$, BF_3, $TiCl_4$, and $SnCl_4$. Lewis acids are seldom effective alone; rather they require the presence of trace amounts of water or some other proton or cation source, which, on reaction with the Lewis acid, forms the electrophilic species that initiates polymerization. Examples are the reactions of BF_3 with water (7.2) and aluminum chloride with an alkyl chloride (7.3). Because it is the proton or carbocation that initiates the polymerization reaction, the compounds that give rise to them are correctly referred to as the *initiators*, and the

$$BF_3 + H_2O \rightleftharpoons HOBF_3^-H^+ \qquad (7.2)$$

$$AlCl_3 + RCl \rightleftharpoons AlCl_4^-R^+ \qquad (7.3)$$

Lewis acids are *coinitiators[1]* (not the other way around, as is commonly done in the polymer literature). The combination of Lewis acid and proton or cation source is the *initiating system*. With certain very active Lewis acids, *autoionization* (7.4) may occur. In such cases the initiator and coinitiator are the same.

$$2AlBr_3 \rightleftharpoons AlBr_4^-AlBr_2^+ \qquad (7.4)$$

A variety of other initiators can bring about cationic polymerization,[1,5] among them ionizable compounds such as triphenylmethyl halides (7.5) or tropylium halides (7.6), and iodine, which may react by in situ generation of HI (7.7) or by ion-pair formation (7.8). Radical cations formed in electron transfer processes (7.9) in which monomer (M) is the electron donor may also initiate cationic polymerization. The reaction between a nitro compound and N-vinylcarbazole, for example, gives rise to a radical cation (7.10) that initiates polymerization of the vinyl monomer.

$$(C_6H_5)_3CCl \rightleftharpoons (C_6H_5)_3C^+ + Cl^- \qquad (7.5)$$

$\rightleftharpoons$ $+ Cl^- \qquad (7.6)$

$$I_2 + CH_2{=}CR_2 \rightleftharpoons ICH_2CIR_2 \begin{cases} \rightleftharpoons ICH{=}CR_2 + HI & (7.7) \\ \rightleftharpoons ICH_2\overset{+}{C}R_2\ I^- & (7.8) \end{cases}$$

$$M + A \longrightarrow M{\cdot}^+ + A{\cdot}^- \qquad (7.9)$$

$+ RNO_2 \longrightarrow$ $+ R\dot{N}O_2^- \quad (7.10)$

Interestingly, mixtures of donor and acceptor monomers may undergo spontaneous reaction to yield homopolymer or copolymer. In such cases, polymerization appears to be initiated ionically by a zwitterion resulting from decay of an initially formed charge-transfer complex of donor (D) and acceptor (A) monomers (7.11).[5]

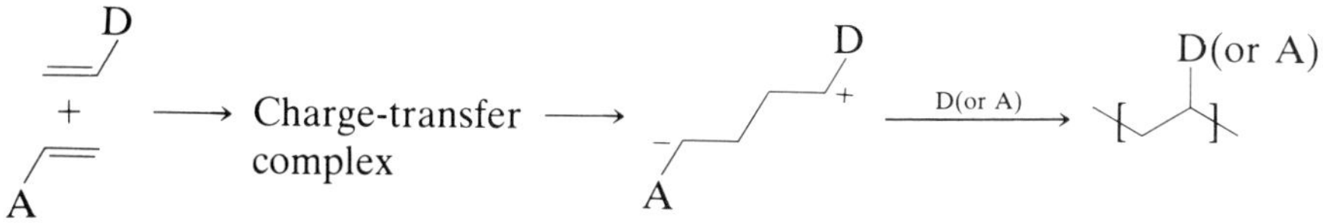

$$\text{(7.11)}$$

It should be kept in mind that not all initiating systems are equally effective. Relatively stable carbocations of the triphenylmethyl or tropylium type are only useful with very reactive monomers such as vinyl ethers; and mineral acid initiators seldom lead to very-high-molecular-weight polymer. Furthermore, in most cases of cationic polymerization, a clear understanding of the initiation mechanism is lacking.[6]

7.2.2 Mechanism, kinetics, and reactivity in cationic polymerization

In carbocationic initiation, addition of the electrophilic species follows Markovnikov's rule, that is, the more stable carbocation intermediate is formed. As would be expected from a consideration of carbocation stability, the rate of addition to aliphatic monomers is in the order

$$(CH_3)_2C{=}CH_2 > CH_3CH{=}CH_2 > CH_2{=}CH_2$$

Of these three monomers, only isobutylene provides the requisite carbocation stability for cationic polymerization. For a series of *para*-substituted styrenes, the reactivity for substituent groups in cationic initiation is in the order of ring activation, namely

$$OCH_3 > CH_3 > H > Cl$$

Ortho substituents retard the addition regardless of whether they activate or deactivate the ring, because of steric hindrance. Vinyl ethers are particularly reactive toward cationic initiators (7.12) because delocalization of the intermediate carbocation gives rise to one contributing resonance form having no electron-deficient atoms.

$$CH_2{=}CH\ddot{O}R \xrightarrow{R'+} R'CH_2{-}\overset{+}{C}H{-}\ddot{O}R \longleftrightarrow R'CH_2{-}CH{=}\overset{+}{\ddot{O}}R \quad (7.12)$$

In the propagating step, the reaction is favored by increasing carbocation stability. It may be recalled that in free radical propagation the *less* stable radicals react faster. Possibly cationic propagation involves two steps—a rate-limiting formation of a pi complex between the chain end and an approaching monomer molecule, followed by covalent bond formation (7.13), whereas covalent bond formation is rate limiting in free radical processes.

$$\text{wwww} + + CH_2{=}CR_2 \xrightarrow{\text{slow}} \text{wwww} \left\langle \begin{array}{c} CR_2 \\ + \| \\ CH_2 \end{array} \right. \xrightarrow{\text{fast}} \text{wwww} CH_2\overset{+}{C}R_2 \quad (7.13)$$

Solvent polarity influences polymerization rate, but it is extremely difficult to make concrete predictions as to how the effect is manifested. Predictably, increasing solvent polarity favors the initiation step where a charged species is being generated. The opposite is expected in propagation, however, because the charge is dispersed in the transition state. Another complicating factor is the degree of association between the cationic chain end and the anion (A^-). Between the extremes of pure covalent bonds and solvated ions are intimate ion pairs (also called contact ion pairs) and solvent-separated ion pairs (7.14). Which actually exists in any given solvent depends on both the

$$\text{wwww} A \rightleftharpoons \text{wwww} + A^- \rightleftharpoons \text{wwww} + \| A^- \rightleftharpoons \text{wwww} + + A^-$$

| Covalent | Intimate ion pair | Solvent-separated ion pair | Solvated ions |

$$(7.14)$$

dielectric strength of the solvent and the solvent's ability to coordinate the cation. (The "solvating power" of an ether solvent, for example, is enhanced by the lone-pair electrons on oxygen). In general, the more intimate the association, the lower the propagation rate. Thus one might reasonably expect that in poorly solvating solvents, increasing solvent polarity will enhance the propagation rate by shifting the association away from intimate ion pairs. As the solvating power of the solvent increases, however, the shift will be in the opposite direction, and propagation will be retarded. How solvent affects the stereochemisty of propagation is discussed in the next section.

Chain transfer reactions are common in cationic polymerization. Let us consider, for example, the polymerization of styrene with sulfuric acid. Possible transfer reactions include:

1. With monomer:

$$\text{ww}CH_2\overset{+}{C}H\,HSO_4^- + CH_2{=}CH \quad\quad \text{ww}CH{=}CH + CH_3\overset{+}{C}H\,HSO_4$$

$$(7.15)$$

2. By ring alkylation:

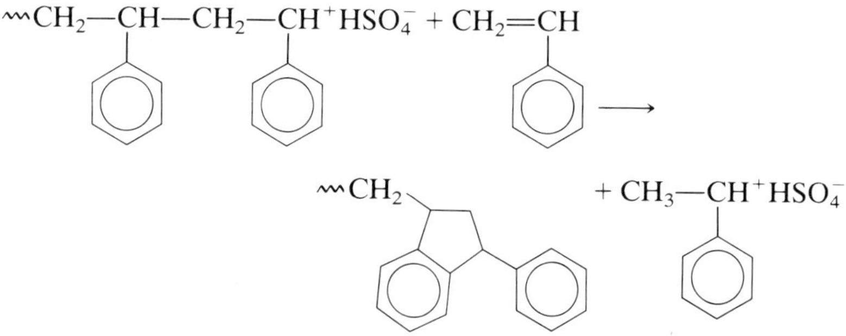

$$\text{$\sim$CH}_2\text{—CH—CH}_2\text{—CH}^+\text{HSO}_4^- + \text{CH}_2\text{=CH}$$

$$\text{$\sim$CH}_2 \qquad\qquad\qquad\qquad + \text{CH}_3\text{—CH}^+\text{HSO}_4^- \quad (7.16)$$

3. By hydride abstraction from the chain to form a more stable ion:

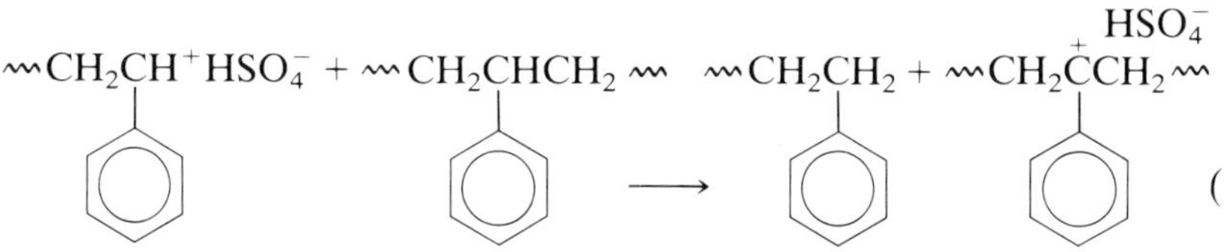

$$\text{$\sim$CH}_2\text{CH}^+\text{HSO}_4^- + \text{$\sim$CH}_2\text{CHCH}_2\text{$\sim$} \quad \text{$\sim$CH}_2\text{CH}_2 + \text{$\sim$CH}_2\overset{+}{\text{C}}\text{CH}_2\text{$\sim$} \qquad \overset{\text{HSO}_4^-}{}$$

$$(7.17)$$

4. With solvent—for example, benzene—by electrophilic substitution:

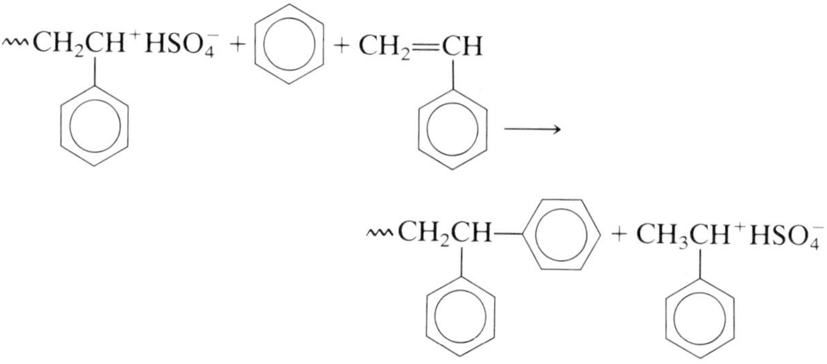

$$\text{$\sim$CH}_2\text{CH}^+\text{HSO}_4^- + \bigcirc + \text{CH}_2\text{=CH}$$

$$(7.18)$$

$$\text{$\sim$CH}_2\text{CH—}\bigcirc + \text{CH}_3\text{CH}^+\text{HSO}_4^-$$

Chain branching occurs via reaction (7.17) or by intermolecular ring alkylation.

Termination reactions are harder to define in cationic processes because they are easy to confuse with chain transfer. Examples of termination are the combination of chain end with counterion (i.e., a change from ionic to covalent bonding) as in the trifluoroacetic acid-initiated polymerization of styrene (7.19),[7] and chain-end chlorination in the BF_3/H_2O-initiated polymerization of isobutylene (7.20).[1]

$$\text{wwwCH}_2\text{CH}^+ \quad ^-\text{OCCF}_3 \longrightarrow \text{wwwCH}_2\text{CHOCCF}_3 \quad (7.19)$$

$$\underset{\underset{\text{CH}_3}{|}}{\overset{\overset{\text{CH}_3}{|}}{\text{wwwCH}_2\text{C}^+}} \quad \text{BCl}_3\text{OH}^- \longrightarrow \underset{\underset{\text{CH}_3}{|}}{\overset{\overset{\text{CH}_3}{|}}{\text{wwwCH}_2\text{C}}}\!\!-\!\text{Cl} + \text{BCl}_2\text{OH} \quad (7.20)$$

Because chain transfer to monomer is so common in cationic polymerization, various strategies have been devised to circumvent it.[1] One involves use of a *proton trap* such as 2,6-di-*t*-butylpyridine, which intercepts the proton (7.21) before it transfers to monomer.[8] The result is a lower overall yield but higher molecular weight and lower polydispersity index. The bulky *t*-butyl groups prevent reaction with electrophiles larger than the proton.

$$\text{wwwCH}_2\text{CR}_2^+ + \quad \underset{\text{N}}{\underset{\times \quad \times}{\bigcirc}} \quad \longrightarrow \quad \text{wwwCH}=\text{CR}_2 + \quad \underset{\underset{\text{H}}{\overset{|}{\text{N}^+}}}{\underset{\times \quad \times}{\bigcirc}} \quad (7.21)$$

Transfer reactions have also been used to advantage to prepare telechelic polymers.[1] A bifunctional compound called an *inifer*, which brings about both *ini*tiation and chain trans*fer*, is used. Steps in the inifer method, using α, α'-dichlorodiisopropylbenzene as the inifer, BCl_3 as cocatalyst, and isobutylene monomer include formation of the initiating cation (7.22), polymerization (7.23), and chain transfer (7.24). Transfer at the other labile chlorine

$$\underset{\underset{\text{CH}_3}{|}}{\overset{\overset{\text{CH}_3}{|}}{\text{Cl}-\text{C}}}\!\!-\!\!\bigcirc\!\!-\!\!\underset{\underset{\text{CH}_3}{|}}{\overset{\overset{\text{CH}_3}{|}}{\text{C}}}\!\!-\!\text{Cl} + \text{BCl}_3 \rightleftharpoons \underset{\underset{\text{CH}_3}{|}}{\overset{\overset{\text{CH}_3}{|}}{\text{Cl}-\text{C}}}\!\!-\!\!\bigcirc\!\!-\!\!\underset{\underset{\text{CH}_3}{|}}{\overset{\overset{\text{CH}_3}{|}}{\text{C}^+}} \quad \text{BCl}_4^- \quad (7.22)$$

$$\underset{\underset{\text{CH}_3}{|}}{\overset{\overset{\text{CH}_3}{|}}{\text{Cl}-\text{C}}}\!\!-\!\!\bigcirc\!\!-\!\!\underset{\underset{\text{CH}_3}{|}}{\overset{\overset{\text{CH}_3}{|}}{\text{C}^+}} + \text{CH}_2\!\!=\!\!\underset{\underset{\text{CH}_3}{|}}{\overset{\overset{\text{CH}_3}{|}}{\text{C}}} \longrightarrow$$

$$\underset{\underset{\text{CH}_3}{|}}{\overset{\overset{\text{CH}_3}{|}}{\text{Cl}-\text{C}}}\!\!-\!\!\bigcirc\!\!-\!\!\underset{\underset{\text{CH}_3}{|}}{\overset{\overset{\text{CH}_3}{|}}{\text{C}}}\text{wwwCH}_2\underset{\underset{\text{CH}_3}{|}}{\overset{\overset{\text{CH}_3}{|}}{\text{C}^+}} \quad (7.23)$$

$$\sim\!\!\sim\!\!\sim CH_2\overset{CH_3}{\underset{CH_3}{\overset{|}{\underset{|}{C}}}}{}^{+} + Cl\!-\!\overset{CH_3}{\underset{CH_3}{\overset{|}{\underset{|}{C}}}}\!\!-\!\!\bigcirc\!\!-\!\!\overset{CH_3}{\underset{CH_3}{\overset{|}{\underset{|}{C}}}}\!\!-\!\!Cl \longrightarrow$$

$$\sim\!\!\sim\!\!\sim CH_2\overset{CH_3}{\underset{CH_3}{\overset{|}{\underset{|}{C}}}}\!\!-\!\!Cl + {}^{+}\overset{CH_3}{\underset{CH_3}{\overset{|}{\underset{|}{C}}}}\!\!-\!\!\bigcirc\!\!-\!\!\overset{CH_3}{\underset{CH_3}{\overset{|}{\underset{|}{C}}}}\!\!-\!\!Cl \quad (7.24)$$

leads to polyisobutylene terminated with chloro groups,

$$Cl\!-\!\overset{CH_3}{\underset{CH_3}{\overset{|}{\underset{|}{C}}}}\!\!\sim\!\!\sim\!\!\sim\!\!\overset{CH_3}{\underset{CH_3}{\overset{|}{\underset{|}{C}}}}\!\!-\!\!\bigcirc\!\!-\!\!\overset{CH_3}{\underset{CH_3}{\overset{|}{\underset{|}{C}}}}\!\!\sim\!\!\sim\!\!\sim\!\!\overset{CH_3}{\underset{CH_3}{\overset{|}{\underset{|}{C}}}}\!\!-\!\!Cl$$

that are readily convertible to other functionalities.

Some polymerizations involving cationic initiators appear to involve covalent bonding with initiator at the growing chain end rather than free ions or ion pairs. An example is the perchloric acid-initiated polymerization of styrene at low initiator concentrations in chlorocarbon solvents, which proceeds at a very slow rate compared with other cationic processes. The chain end is believed to be the covalently bonded ester that undergoes propagation by inserting monomer between carbon-oxygen bonds (7.25). Such polymerizations, referred to as *pseudocationic*,[9,10] involve rapid ester formation followed by rate-determining propagation. Spectrophotometric and conductometric studies indicate an absence of ionic species during propagation. One might think of pseudocationic polymerization in terms of one extreme of the ion-pair dissociation spectrum (7.14).

$$-CH_2CH\!-\!OClO_3 + CH_2\!\!=\!\!CH \qquad -CH_2CHCH_2CH\!-\!OClO_3$$

$$\bigcirc \qquad\qquad \bigcirc \longrightarrow \bigcirc\bigcirc \qquad (7.25)$$

It was discovered in the 1980s that it is possible to achieve cationic polymerization in the complete *absence* of termination or chain transfer reactions. Polymerization of isobutylene with a tertiary ester and BCl_3,[11–14] for example, involves formation of a tertiary carbocation-initiating species (7.26) and polymerization (7.27) to yield polyisobutylene terminated with what appears to be a very tightly bound—but still active—ion pair. A similar situation occurs when I_2/HI or I_2/ZnI_2 is used as the initiating system.[15–18] In this case, shown for vinyl ether propagation (7.28), the mechanism apparently involves insertion of monomer into an activated carbon–iodine bond.[17]

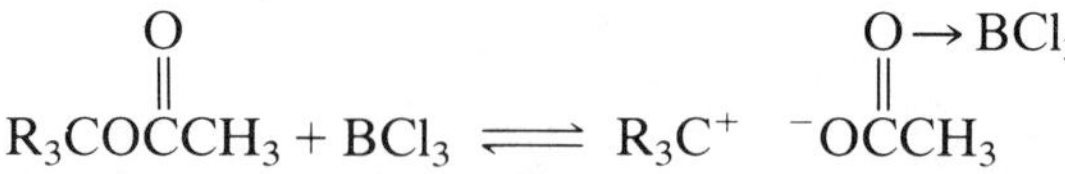

$$R_3COCCH_3 + BCl_3 \rightleftharpoons R_3C^+ \ {}^-OCCH_3 \qquad (7.26)$$

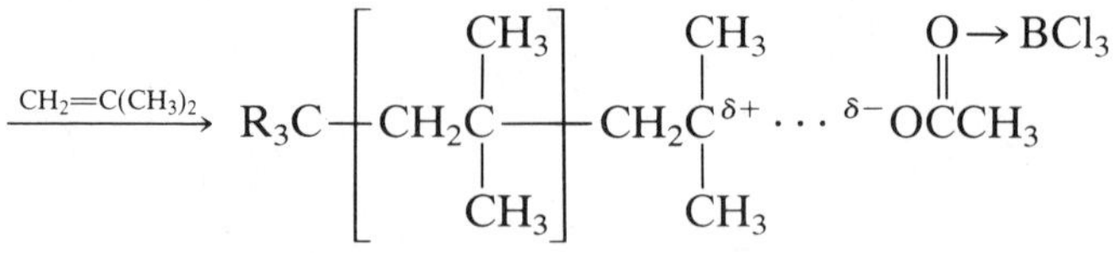

$$\xrightarrow{CH_2=C(CH_3)_2} R_3C\left[CH_2C\underset{CH_3}{\overset{CH_3}{|}}\right]CH_2C^{\delta+}\cdots{}^{\delta-}OCCH_3 \qquad (7.27)$$

$$\overset{\text{\tiny mmm}CH_2}{\underset{CH_2}{\diagdown}}\overset{OR}{\underset{\diagup}{CH}} \quad I\text{---}ZnI_2 \longrightarrow \overset{\text{\tiny mmm}CH_2}{\underset{CH_2}{\diagdown}}\overset{OR}{\underset{\diagup}{CH}} \quad I\text{---}ZnI_2 \qquad (7.28)$$

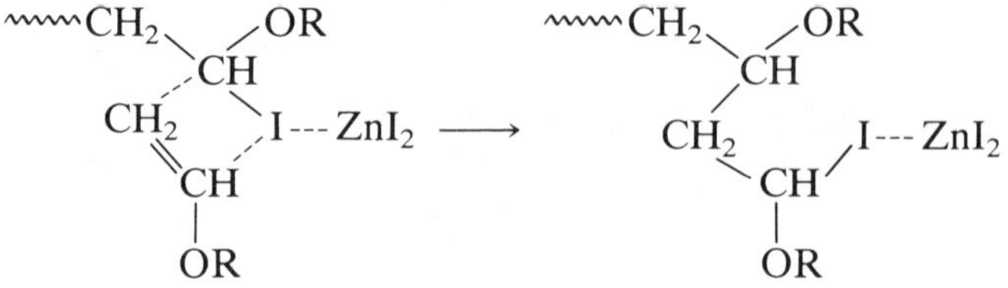

Where termination is totally suppressed, the polymers—which possess still-active chain ends—are called *living polymers*. Addition of more monomer results in further increase in molecular weight and, if the added monomer is different from the starting monomer, in formation of block copolymer. Living polymers have been known much longer and exploited much more extensively with anionic polymerization, hence the practical and kinetic consequences of this important concept are elaborated on in Section 7.3. Worth noting, however, is that living polymers of the type shown in reaction 7.28 may be prepared at room temperature; most living polymers require low-temperature synthesis.

Because of the complexity of cationic initiation and uncertainty concerning the extent of ion-pair formation at the propagating chain end, any kinetic treatments of cationic polymerization reflect approximations at best. Without specifying the nature of the initiation-coinitiation mechanism, we can write general initiation, propagation, termination, and transfer rates as follows:

$$R_i = k_i[I][M]$$

$$R_p = k_p[M][M^+]$$

$$R_t = k_t[M^+]$$

$$R_{tr} = k_{tr}[M][M^+]$$

where $[I]$, $[M]$ and $[M^+]$ represents molar concentrations of initiator, monomer, and cationic chain end, respectively. As with free radical polymerization (but with less certainty), we can approximate a steady state for the growing chain end; thus $R_i = R_t$, and

$$k_i[I][M] = k_t[M^+]$$

or

$$[M^+] = \frac{k_i[I][M]}{k_t}$$

Substituting for $[M^+]$ in R_p, one obtains

$$R_p = \frac{k_p k_i [I][M]^2}{k_t}$$

Some representative cationic propagation rate constants are given in Table 7.2.

In the absence of any chain transfer, the kinetic chain length, $\bar{\nu}$, is equal to $\overline{DP}$ and is expressed as

$$\bar{\nu} = \overline{DP} = \frac{R_p}{R_t} = \frac{k_p[M][M^+]}{k_t[M^+]} = \frac{k_p[M]}{k_t}$$

If transfer is the predominant mechanism controlling chain growth,

$$\bar{\nu} = \overline{DP} = \frac{R_p}{R_{tr}} = \frac{k_p[M][M^+]}{k_{tr}[M][M^+]} = \frac{k_p}{k_{tr}}$$

These kinetics expressions illustrate some basic differences between free radical and cationic processes. In the former, the propagation rate is proportional to the square root of initiator concentration, whereas cationic polymerization shows a first-order dependence. Furthermore, the molecular weight in cationic polymerization is independent of initiator concentration, unlike free radical polymerization where $\overline{DP}$ is inversely proportional to $[I]^{1/2}$ in the absence of chain transfer. The difference arises from radical disproportionation and combination reactions characteristic of free radical termination. Increasing initiator concentration thus increases the probability of radical termination, which is not the case in ionic polymerization.

Diene monomers also undergo cationic polymerization (primarily *trans-1,4*), but their use is limited mainly to copolymer synthesis. Nonconjugated dienes may undergo cationic cyclopolymerization[19,20] by a mechanism analogous to that of free radical cyclopolymerization (Chapter 6, Section 6.6). A

Table 7.2. Representative cationic propagation rate constants, R_p[a]

Monomer	Solvent	Temperature (°C)	Initiator	k_p (L/mol-s)
Styrene	None	15	Radiation	3.5×10^6
α-Methylstyrene	None	0	Radiation	4×10^6
i-Butyl vinyl ether	None	30	Radiation	3×10^5
i-Butyl vinyl ether	CH_2Cl_2	0	$C_7H_7{}^+SbCl_6{}^-$	5×10^3
t-Butyl vinyl ether	CH_2Cl_2	0	$C_7H_7{}^+SbCl_6{}^-$	3.5×10^3
Methyl vinyl ether	CH_2Cl_2	0	$C_7H_7{}^+SbCl_6{}^-$	1.4×10^2
2-Chloroethyl vinyl ether	CH_2Cl_2	0	$C_7H_7{}^+SbCl_6{}^-$	2×10^2

[a] Data from Ledwith and Sherrington,[3a] p. 278. Copyright 1974. Reprinted by permission of John Wiley & Sons, Ltd.

typical example is the formation of a six-membered ring-containing polymer from 2,6-diphenyl-1,6-heptadiene (7.29).

(7.29)

7.2.3   Stereochemistry of cationic polymerization

It has been recognized for a long time that cationic polymerization can lead to stereoregular structures. Most of the work in this area has been done on vinyl ethers.[21a] although other monomers, notably α-methylstyrene, have also been observed to form stereoregular polymer. Observations resulting from these studies are that (1) greater stereoregularity is achieved at lower temperatures, (2) the degree of stereoregularity can vary with initiator, and (3) the degree and type of stereoregularity (isotactic or syndiotactic) vary with solvent polarity. For example, t-butyl vinyl ether forms isotactic polymer in nonpolar solvents and mainly syndiotactic polymer in polar solvents. This behavior can be rationalized in terms of the extent to which the cationic chain end and the counterion are associated. In polar solvents both ions would be strongly solvated and the chain end would exist as a free carbocation surrounded by solvent molecules. The same conformational factors would thus apply here that are responsible for syndiotactic placement in free radical polymerization. In nonpolar solvents, however, association between carbocation chain end and counterion would be strong, and counterion could influence the course of steric control.

How isotactic polymer forms is the subject of considerable debate. Several models have been proposed, the major aspects of which are presented here, but the true course of stereoregular polymerization is still not clear. Two of the models proposed for vinyl ether polymerization[22,23] suggest a six-membered cyclic chain end formed by coordination of the carbocation of the terminal carbon with the oxygen of the alkoxy group attached to the fifth carbon atom (7.30). For this cyclic oxonium ion, the polymer chain, being the bulkiest group, would be equatorial, the alkoxy group on the third carbon would be axial, and the terminal alkoxy group would be equatorial. Attack of monomer at C_1 will be from the back side causing inversion; hence the alkoxy group at what used to be C_1 will have the same configuration as that at the original C_3 (7.31). A new cyclic oxonium ion would then form, and the process would repeat itself to give isotactic polymer. The major difference between the two models proposed lies in the degree of participation by the counterion in complexing the reacting monomer molecule.

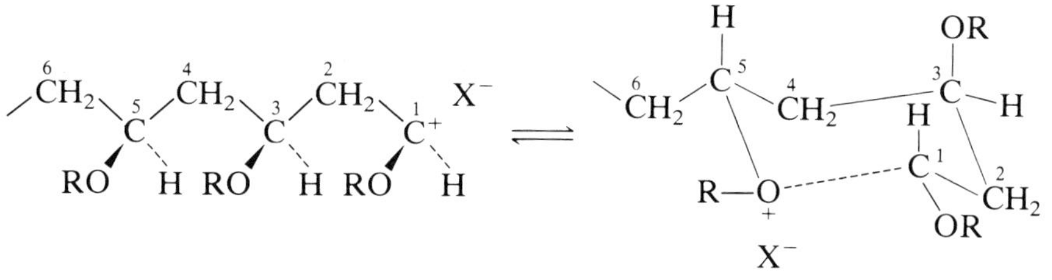

$$(7.30)$$

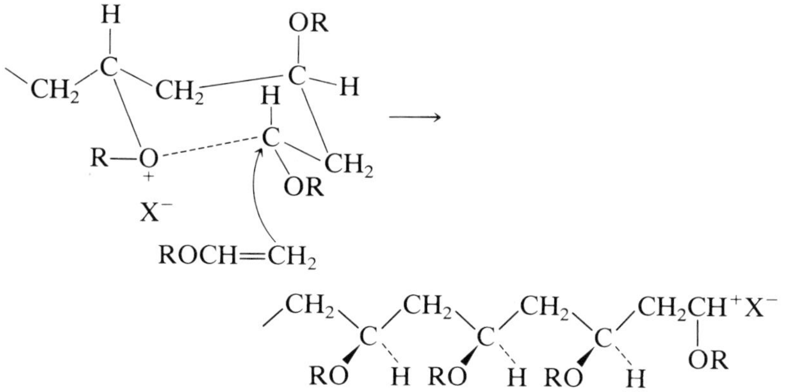

$$(7.31)$$

Another model[24] proposes a strong ion–counterion association at the chain end, with monomer inserting itself in such a way that the alkoxy groups are *anti* to one another (7.32). As long as propagation proceeds faster than rotation about carbon–carbon bonds in the chain, a regular repeating isotactic polymer results. Some coordination of monomer with counterion may also be assumed in this model.

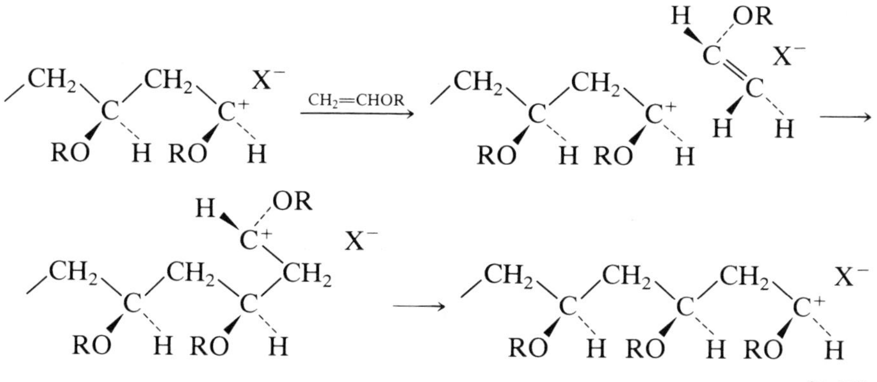

$$(7.32)$$

The major drawback to these mechanisms is that they all assume some tetrahedral character for the carbocation. With vinyl ethers this would diminish the considerable resonance stabilization afforded by delocalization of the unshared *p* electrons on oxygen, which requires a planar configuration. Furthermore, the first two, involving cyclic oxonium ions, are suitable only for vinyl ether polymerization but are not applicable to such monomers as α-methylstyrene which, in general, exhibit similar steric control in polymerization.

A fourth model has been proposed[25] that takes into account normal sp^2 hybridization of the terminal chain carbon atom and is based on accepted mechanistic theories of displacement reactions. According to this model, in *polar* solvents the cationic chain end may be considered essentially free, and monomer approaches from the front side in the preferred syndiotactic arrangement analogous to that of syndiotactic free radical polymerization. In *nonpolar* solvents, the chain end exists as an ion pair, the "tightness" of which depends on solvent polarity and counterion. Assuming that considerable energy is required to separate the ion pair, the incoming monomer would then prefer *back-side* attack, or an S_N2 type of reaction leading to isotactic placement.* These two possibilities are shown in equations (7.33) and (7.34).

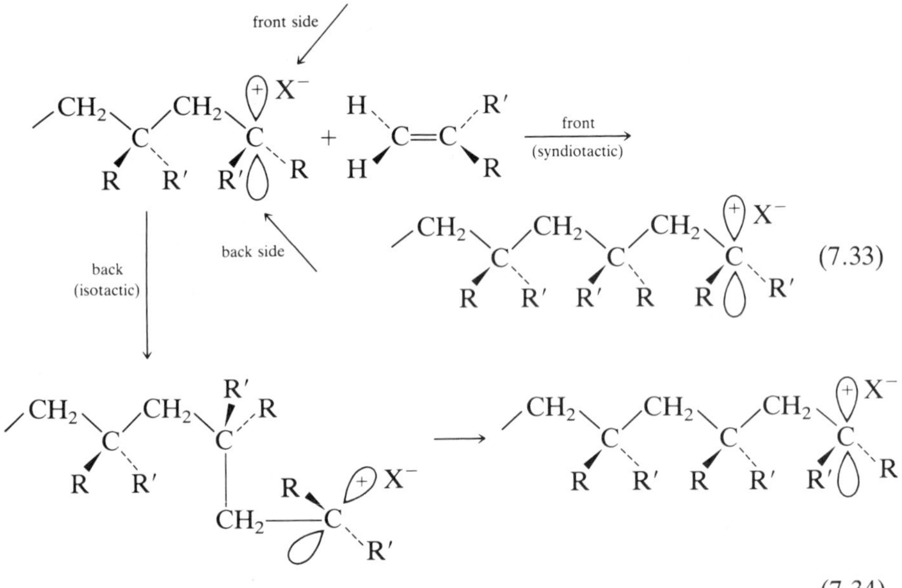

$$(7.33)$$

$$(7.34)$$

---

* This theory has some precedence in that one school of thought concerning nucleophilic displacement reactions is that they all involve ion-pair intermediates. According to this theory, pure S_N1 displacement involves ion-pair formation in the rate-determining step, whereas pure S_N2 displacement involves ion-pair destruction by nucleophilic displacement. See, for example, R. A. Sneen and J. W. Larsen, *J. Am. Chem. Soc.*, **91**, 362, (1969). (For a discussion of the pros and cons of this theory, see T. H. Lowry and K. S. Richardson, *Mechanism and Theory in Organic Chemistry*, 2nd ed., Harper and Row, New York, 1981, Chap. 4.)

With poly(α-methylstyrene), where two bulky groups are attached to the third carbon of the polymer backbone, back-side attack is hindered and syndiotactic placement is favored in both polar and nonpolar solvents, although the amount of isotacticity does increase as solvent polarity is reduced. Vinyl ethers, being less hindered, exhibit a more pronounced preference for isotactic placement in nonpolar solvents.

The fact that increasing temperature decreases stereoregularity can be interpreted in terms of small differences in activation energy for front- and back-side attack as well as of the effect on conformational mobility of the growing chain. Counterion–monomer interactions are assumed to be negligible, since both may be considered "electron rich."

While the last model (reactions 7.33 and 7.34) satisfactorily explains most of the experimental observations, a point to be gained from this discussion is that, in many polymerization processes, no single theory is necessarily adequate to explain all the facts. One should approach these subjects with an open mind and a realization that considerably more work may be necessary before a mechanism is completely clarified.

7.2.4 Cationic copolymerization

The copolymerization equation derived for free radical initiation (Chapter 6) may also be applied to cationic copolymerization to determine reactivity ratios, several of which are listed in Table 7.3. The situation is complicated, however, by counterion effects.[1,26a] Thus, unlike free radical copolymerization, different cationic initiators can cause variations in reactivity ratios. Furthermore, solvent polarity may also have an effect, because this governs the degree of chain-end ion-pair association. Another significant difference in ionic copolymerization is that there is no apparent tendency for alternating copolymers to form. On the contrary, block copolymers or homopolymer blends are more likely.

Variations in reactivity ratios are often, but not always, observed for different initiators where solvent is kept constant, but it is difficult to make clear-cut predictions, because such factors as carbocation reactivity or preferential solvation of counterion by one monomer may play a part. Similarly, such complications frequently result in unpredictable behavior when initiator is kept constant and solvent is varied.

Yet another unpredictable effect is that of temperature. In free radical processes, increasing temperature results in greater randomness, which is to be expected, since differences in activation energies between self-propagation and cross-propagation would be reduced. Cationic copolymerizations, on the other hand, show greater or lesser tendency toward randomness from one system to another, and no definite trend is observable.

Steric effects are also important in cationic copolymerization, but not always to the extent found in free radical processes. For example, *cis*-and

Table 7.3. Representative cationic reactivity ratios, $(r)^a$

Monomer 1	Monomer 2	Coinitiator[b]	Solvent[b]	Temperature (°C)	r_1	r_2
Isobutylene	1,3-Butadiene	AlEtCl$_2$	CH$_3$Cl	-100	43	0
	1,3-Butadiene	AlCl$_3$	CH$_3$Cl	-103	115	0
	Isoprene	AlCl$_3$	CH$_3$Cl	-103	2.5	0.4
	Cyclopentadiene	BF$_3 \cdot$OEt$_2$	ϕCH$_3$	-78	0.60	4.5
	Styrene	SnCl$_4$	EtCl	0	1.60	0.17
	Styrene	AlCl$_3$	CH$_3$Cl	-92	9.02	1.99
	α-Methylstyrene	TiCl$_4$	ϕCH$_3$	-78	1.2	5.5
Styrene	α-Methylstyrene	SnCl$_4$	EtCl	0	0.05	2.90
	p-Methylstyrene	SnCl$_4$	CCl$_4$	-78	0.33	1.74
	trans-β-Methylstyrene	SnCl$_4$	CH$_2$Cl$_2$	0	1.80	0.10
p-Chlorostyrene	cis-β-Methylstyrene	SnCl$_4$	CCl$_4$/ϕNO$_2$(1:1)	0	1.0	0.32
	trans-β-Methylstyrene	SnCl$_4$	CCl$_4$/ϕNO$_2$(1:1)	0	0.74	0.32
Ethyl vinyl ether	i-Butyl vinyl ether	BF$_3$	CH$_2$Cl$_2$	-78	1.30	0.92
2-Chloroethyl vinyl ether	α-Methylstyrene	BF$_3$	CH$_2$Cl$_2$	-23	6.02	0.42

[a] Data from Kennedy and Marechal.[1]

[b] Et = C$_2$H$_5$, ϕ = phenyl.

trans-β-methylstyrene exhibit similar reactivity ratios when copolymerized with *p*-chlorostyrene, which suggests that steric inhibition of resonance in the transition state of propagation may not be overly significant (see structure **40**, Chapter 6). This might not be unreasonable if formation of a pi complex rather than a carbocation is indeed rate limiting. That 1,2-disubstituted monomers often undergo cationic homopolymerization (not the case under free radical conditions) lends support to this hypothesis, for in pi complex formation, the chain end might be expected to approach the monomer molecule perpendicularly to the double bond (7.13) rather than in the sterically unfavorable orientation depicted previously (structure **37**, Chapter 6). (Interestingly, one of the best methods of preparing polymers having substituents on each backbone carbon is by polymerization of diazoalkanes, $RCHN_2$—shown for diazomethane in Chapter 1, reaction 1.22—which also happens to be a cationic process.[27])

Steric effects are also evident in copolymerization of styrene with alkyl-substituted styrenes where *p*- and α-methyls increase, but β-methyl reduces reactivity.

The most important of the commercial cationic copolymers is *butyl rubber*, prepared from isobutylene and isoprene. Because of its impermeability, butyl rubber finds wide use in protective clothing and tire inner tubes. It is manufactured by low-temperature ($-100°C$) copolymerization of about 97% isobutylene and 3% isoprene in chlorocarbon solvents (Tables 7.1 and 7.3). More recently an ozone-resistant copolymer of isobutylene and cyclopentadiene has been marketed. (Recall the discussion of ozone resistance in Section 4.6.)

7.2.5 *Isomerization in cationic polymerization*

That carbocations undergo 1,2 shifts to form more stable carbocations is well known. This phenomenon is also observed in cationic polymerization.[19,28] When 3-methyl-1-butene is polymerized using Lewis acid coinitiator at low temperature (7.35), a hydride shift occurs to yield what is, in effect, a 1,3-addition polymer. The driving force for rearrangement is formation of the more stable tertiary carbocation. Similarly, rearrangement during cationic polymerization of the terpene β-pinene (7.36) relieves strain in the bicyclic system. As in the case with cyclopentadiene, β-pinene also yields an ozone-resistant elastomer upon copolymerization with isobutylene. Numerous examples of isomerization polymerization have been reported.

$$CH_2{=}CH{-}\underset{\diagdown CH_3}{\overset{\diagup CH_3}{CH}} \xrightarrow{X^+} XCH_2{-}\overset{+}{CH}{-}\underset{\diagdown CH_3}{\overset{\diagup CH_3}{CH}} \xrightarrow{H:shift}$$

$$XCH_2{-}CH_2{-}\underset{\diagdown CH_3}{\overset{\diagup CH_3}{\overset{+}{C}}} \xrightarrow{monomer} \left[CH_2{-}CH_2{-}\underset{\underset{CH_3}{|}}{\overset{\overset{CH_3}{|}}{C}} \right] \quad (7.35)$$

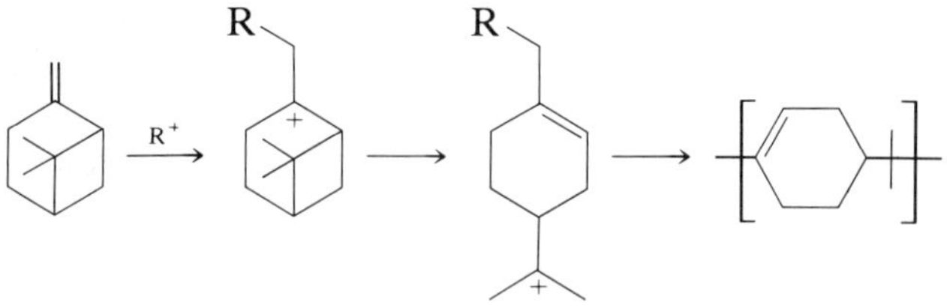

$$(7.36)$$

7.3 Anionic polymerization

7.3.1 *Anionic initiators*

In anionic vinyl polymerization,[3b,29-34] the propagating chain is a carbanion; hence initiation is brought about by species that undergo nucleophilic addition to monomer (7.37). Monomers having substituent groups capable of stabilizing a carbanion through resonance or induction are most susceptible to anionic polymerization. Examples of such groups are nitro, cyano, carboxyl, vinyl, and phenyl.

$$Nu^- + CH_2{=}CHR \longrightarrow NuCH_2{-}\underset{\underset{R}{|}}{C}H^- \qquad (7.37)$$

The strength of the base necessary to initiate polymerization depends in large measure on monomer structure. Thus potassium bicarbonate is strong enough to initiate polymerization of 2-nitropropene (7.38), whereas a base as weak as water causes vinylidene cyanide to polymerize violently (7.39). Such high reactivity has been exploited commercially with the development of *cyanoacrylate* adhesives—vinyl monomers substituted on the same carbon with CN and CO_2R groups that polymerize on exposure to atmospheric or surface moisture. Most applications of anionic polymerization, however, involve initiation by alkali metals or their compounds.

$$CH_2{=}\underset{\underset{NO_2}{|}}{\overset{\overset{CH_3}{|}}{C}} \xrightarrow{KHCO_3} \left[CH_2{-}\underset{\underset{NO_2}{|}}{\overset{\overset{CH_3}{|}}{C}} \right] \qquad (7.38)$$

$$CH_2{=}\underset{\underset{CN}{|}}{\overset{\overset{CN}{|}}{C}} \xrightarrow{H_2O} \left[CH_2{-}\underset{\underset{CN}{|}}{\overset{\overset{CN}{|}}{C}} \right] \qquad (7.39)$$

The most typically used anionic initiators may be classified into two basic types—those that react by addition of a negative ion and those that undergo electron transfer.[33] The most common initiators that react by addition of a negative ion are simple organometallic compounds of the alkali metals—for example, butyllithium. Organolithium compounds are generally low melting and soluble in inert organic solvents. Organometallic compounds of the higher alkali metals have more ionic character and are generally insoluble; hence they initiate polymerization by a heterogeneous process. Other initiators of this type include organic compounds of calcium and barium and Grignard reagents, but these are not used as commonly as the alkali metal compounds.

Initiation by charge transfer can be brought about by free alkali metals or by addition complexes of alkali metals and unsaturated or aromatic compounds. Free metals may be employed as solutions in liquid ammonia or certain ether solvents or as suspensions in inert solvents. The latter are prepared by heating the solvent and metal above the melting point of the metal, stirring vigorously to form an emulsion, then cooling to obtain a fine, solid dispersion. The metal may also be used as a free-flowing powder coated on an inert support such as alumina. The polymerization process is heterogeneous if the metal is used as a dispersion and homogeneous if it is in solution. (Not all ammonia solutions of alkali metals react by electron transfer; potassium is believed to form potassium amide, which initiates polymerization by addition of amide ion.) Electron transfer processes involving metal donor, $D\cdot$, and monomer, M, may be written in general form as follows (7.40):

$$D\cdot + M \rightarrow D^+ + M\cdot^- \tag{7.40}$$

Stable addition complexes of alkali metals that initiate polymerization by electron transfer are formed by reaction of metal and substrate in an inert solvent. Examples are reactions of sodium with naphthalene (7.41) or stilbene

$$Na\cdot + \text{(naphthalene)} \longrightarrow \text{(naphthalene anion radical)} Na^+ \tag{7.41}$$

$$Na\cdot + C_6H_5-CH{=}CH-C_6H_5 \rightarrow [C_6H_5-\overset{\bullet}{C}H-\overset{..}{C}HC_6H_5]\,Na^+ \tag{7.42}$$

(7.42). A general equation for the reaction of this type of donor, $D\cdot^-$ with monomer, M, may be written (7.43)

$$D\cdot^- + M \rightarrow D + M\cdot^- \tag{7.43}$$

Related to the foregoing are addition complexes (called *ketyls*) of alkali metals and nonenolizable ketones, such as benzophenone (7.44) The alkali metal ketyls normally exist in equilibrium with the dianion dimer (7.45),

$$Na\cdot + C_6H_5\overset{\overset{O}{\|}}{C}C_6H_5 \longrightarrow \left[C_6H_5\overset{\overset{O}{\|}}{C}C_6H_5\right]^{\cdot} Na^+ \tag{7.44}$$

and initiation can be brought about by electron transfer or by anion addition. Similar addition complexes are formed with compounds having nitrogen–oxygen, nitrogen–nitrogen, and carbon–nitrogen double bonds.

$$2\left[C_6H_5\overset{O}{\overset{\|}{C}}C_6H_5 \right]^{\bar{\cdot}} Na^+ \longrightarrow \underset{\displaystyle (C_6H_5)_2C-C(C_6H_5)_2}{\overset{\displaystyle NaO \quad ONa}{\overset{|}{}\quad\overset{|}{}}} \qquad (7.45)$$

7.3.2 *Mechanism, kinetics, and reactivity in anionic polymerization*

The mechanism of *initiation* depends on the type of initiator used. For those that react by ion addition, such as the alkali metal alkyls, essentially two different initiation processes can occur, depending on the degree of association between metal ion and alkyl group. Lithium compounds, for example, may have covalent carbon–metal bonds, whereas the higher alkali metals tend to form more ionic bonds. Between the extremes of pure covalent bonds and solvated ions are ion pairs of varying degrees of association, as is the case with cationic polymerization, with solvent polarity and solvating power playing an equally impotant role.

In polar solvents where free solvated ions predominate, the mechanism involves simple addition of anion to monomer (7.37). In nonpolar solvents where there is close association between ions, the mechanism is believed to involve initial pi-complex formation, as shown in equation (7.46) for an alkyllithium compound. Which mechanism prevails depends, therefore, on a variety of factors including solvent polarity, type of cation, and temperature.

$$R-Li + CH_2\!\!=\!\!\underset{\underset{R'}{|}}{CH} \rightleftharpoons R-Li\text{---}\underset{\displaystyle CH_2}{\overset{\displaystyle CHR'}{\|}} \longrightarrow RCH_2\underset{\underset{R'}{|}}{CHLi} \quad (7.46)$$

The rate of initiation depends on the structure of both initiator and monomer. In some cases initiation occurs much more rapidly than propagation because of the lower reactivity of the propagating carbanion relative to that of initiator.

Initiation by electron transfer is considerably different. Let's consider first initiation by alkali metal. When the metal is dispersed in an inert solvent, the process is heterogeneous. The reaction, shown here for sodium, involves, first, formation of a radical anion (7.47), followed by radical combination to form a dianion (7.48). Evidence that this occurs is based on a rapidly dis-

$$Na\cdot + CH_2\!\!=\!\!\underset{\underset{R}{|}}{CH} \longrightarrow \left[CH_2\text{===}\underset{\underset{R}{|}}{CH} \right]^{\bar{\cdot}} Na^+ \qquad (7.47)$$

$$2\left[CH_2\text{===}\underset{\underset{R}{|}}{CH} \right]^{\bar{\cdot}} Na^+ \longrightarrow Na^+ \overset{..}{\underset{\underset{R}{|}}{C}}HCH_2CH_2\overset{..}{\underset{\underset{R}{|}}{C}}HNa^+ \qquad (7.48)$$

appearing ESR signal on mixing initiator and monomer, and the quantitative formation of dicarboxylic acid on addition of carbon dioxide to an equimolar mixture of the two. For alkali metals in solution the process is essentially the same except that, in the case of ammonia, free solvated electrons probably add to monomer rather than being transferred directly from the metal. As was mentioned earlier, potassium in ammonia appears to initiate polymerization by addition of amide ion rather than by electron transfer. Addition complexes such as naphthalenesodium or the alkali metal ketyls also react by electron transfer to form dianions.

The kinetic and mechanistic aspects of anionic polymerization are better understood than those of cationic polymerization. In the case of the potassium amide-initiated polymerization in liquid ammonia, initiation involves dissociation (7.49) followed by addition of amide ion to monomer (7.50).

$$KNH_2 \rightleftharpoons K^+ + NH_2^- \tag{7.49}$$

$$NH_2^- + M \rightarrow H_2N-M^- \tag{7.50}$$

Because the second step is slow relative to the first,

$$R_i = k_i[NH_2^-][M]$$

Chain termination is known to result primarily by transfer to solvent (7.51).

$$H_2N(M)_nM^- + NH_3 \rightarrow H_2N(M)_nMH + NH_2^- \tag{7.51}$$

Rate expressions for propagation and transfer may be written in the conventional way:

$$R_p = k_p[M][M^-]$$

$$R_{tr} = k_{tr}[M^-][NH_3]$$

Assuming a steady state whereby $R_i = R_{tr}$,

$$k_i[NH_2^-][M] = k_{tr}[M^-][NH_3]$$

and

$$[M^-] = \frac{k_i[NH_2^-][M]}{k_{tr}[NH_3]}$$

Substituting in R_p we obtain

$$R_p = \frac{k_p k_i[NH_2^-][M]^2}{k_{tr}[NH_3]}$$

The average kinetic chain length, $\bar{v}$, is expressed as

$$\bar{v} = \frac{R_p}{R_{tr}} = \frac{k_p[M][M^-]}{k_{tr}[M^-][NH_3]} = \frac{k_p[M]}{k_{tr}[NH_3]}$$

Because $\bar{v} = \overline{DP}$ for this process, it is clear that both molecular weight and propagation rate are inversely proportional to the concentration of ammonia.

Other types of transfer reactions are recognized in anionic polymerization. Unsaturated end groups in polyacrylonitrile, for example, are believed to arise from transfer to monomer (7.52) followed by initiation by the resultant vinyl anion (7.53). In many instances, however, the rigorous exclusion of

$$\text{wwCH}_2\text{CH:}^- + \underset{\underset{\displaystyle\text{CN}}{|}}{\text{CH}_2\!\!=\!\!\text{CH}} \longrightarrow \text{wwCH}_2\text{CH}_2 + \underset{\underset{\displaystyle\text{CN}}{|}}{\text{CH}_2\!\!=\!\!\text{C:}^-} \qquad (7.52)$$

with the first CH:⁻ bearing a CN group.

$$\underset{\underset{\displaystyle\text{CN}}{|}}{\text{CH}_2\!\!=\!\!\text{C:}^-} + \underset{\underset{\displaystyle\text{CN}}{|}}{\text{CH}_2\!\!=\!\!\text{CH}} \longrightarrow \underset{\underset{\displaystyle\text{CN}}{|}}{\text{CH}_2\!\!=\!\!\text{C}}\;\;\underset{\underset{\displaystyle\text{CN}}{|}}{\text{CH}_2\text{CH:}^-} \qquad (7.53)$$

impurities leads to polymerization reactions having *no* inherent termination or chain transfer. When styrene is polymerized with naphthalenesodium, for example, the green color of initiator changes to red, the color of the styryl anion. The red color persists even at 100% conversion, and addition of more styrene results in further polymerization. It was experiments such as this that first demonstrated the existence of living polymers.[35,36] Addition of carbon dioxide to the living polymer results in quantitative conversion to carboxyl end groups, whereas addition of a second monomer leads to block co-polymers.

In situations where the rate of initiation is very high relative to that of propagation,

$$\frac{d[M]}{dt} = k_p[I]_0[M]$$

where $[I]_0$ is the initial concentration of initiator. This means, in effect, that all chains begin to grow simultaneously. Integration of the rate expression gives

$$[M] = [M]_0 e^{k[I]t}$$

If no termination or chain transfer reactions occur, the average kinetic chain length is simply equal to the number of monomer molecules reacted divided by the number of chains initiated; that is,

$$\bar{\nu} = \frac{[M]_0 - [M]}{[I]_0}$$

or, as monomer is completely consumed,

$$\bar{\nu} = \frac{[M]_0}{[I]_0}$$

For those cases where initiation is by a simple anionic species (butyllithium, for example), $\bar{\nu} = \overline{DP}$. In the case of electron-transfer initiators that combine according to equation (7.48), $\overline{DP} = 2\bar{\nu}$ because the polymer chain grows simultaneously from each end of the initiating species.

An important consequence of having no termination or chain transfer, coupled with the virtually simultaneous initiation of all polymer chains, is that the molecular weight distribution is very narrow. Indeed, anionic polymerization is used to prepare the low polydispersity standards for calibration in gel permeation chromatography (Chapter 2). Furthermore, where living polymer is formed at both chain ends, quenching with carbon dioxide results in carboxyl-terminated telechelic polymer.

Association between counterion and terminal carbanion is an important factor in propagation rate. In poorly solvating hydrocarbon solvents with Li^+ counterion, for example, association of the type shown in reaction (7.54) impairs reactivity. Larger alkali metal cations,

$$2 \text{$\sim$CH}_2\text{CHLi} \longrightarrow \text{$\sim$CH}_2\text{CH} \underset{\substack{| \\ R}}{\overset{\substack{\text{Li} \\ }}{\Big\langle}} \underset{\substack{\text{Li} \\ }}{\Big\rangle} \underset{\substack{| \\ R}}{\text{CHCH}_2\text{$\sim$}} \tag{7.54}$$

which do not coordinate as strongly as Li^+, generally allow faster propagation under these conditions. In a more polar, solvating medium such as an ether, however, Li^+ is the most strongly solvated of the alkali metal cations, and lithium-based initiators afford the fastest rates. The effect of ion-pair association and solvent polarity are evident in the representative propagation rate constants for polystyrene in Table 7.4, where the order of solvating power is seen to be

1,2-dimethoxyethane > tetrahydrofuran > benzene > cyclohexane

Monomer structure is another consideration. For initiation with amide ion, the following order of reactivity has been observed:

$$\text{CH}_2\text{=CHCN} > \text{CH}_2\text{=}\overset{\substack{\text{CH}_3 \\ |}}{\text{C}}\text{CO}_2\text{CH}_3 > \text{CH}_2\text{=CHC}_6\text{H}_5 > \text{CH}_2\text{=CH}\text{—}\text{CH}\text{=CH}_2$$

Table 7.4. Representative anionic propagation rate constants, k_p, for polystyrene[a]

Counterion	Solvent	k_p (L/mol-s)[b]
Na^+	Tetrahydrofuran	80
Na^+	1,2-Dimethoxyethane	3600
Li^+	Tetrahydrofuran	160
Li^+	Benzene	10^{-4}–10^{-1} [c]
Li^+	Cyclohexane	10^{-4}–10^{-4} [c]

[a] Data from Morton.[29]

[b] At 25°C unless otherwise noted.

[c] Variable temperature.

Methyl substitution on the α-carbon decreases the rate, with reactivity in the order

$$CH_2{=}CHCN > CH_2{=}\overset{\overset{\displaystyle CH_3}{|}}{C}CN > CH_2{=}CHCO_2CH_3 > CH_2{=}\overset{\overset{\displaystyle CH_3}{|}}{C}CO_2CH_3$$

This results from inductive destabilization of the carbanion and steric interference with both chain-end solvation and approach of monomer. Similar effects have been observed in the polymerization of α-methylstyrene and ring-substituted styrenes where alkyl or methoxyl groups decrease the rate.

7.3.3 Stereochemistry of anionic polymerization

For nondiene vinyl monomers, stereochemical control in anionic polymerization parallels that of cationic polymerization. With soluble anionic initiators (homogeneous conditions) at low temperatures, polar solvents favor syndiotactic placement, whereas nonpolar solvents favor isotactic placement.[21b] Thus, stereochemistry depends in large measure on the degree of association with counterion, as it does in cationic polymerization.

Syndiotactic placement may be rationalized in terms of the preferred stereochemical arrangement between the chain-end carbanion and approaching monomer under conditions where the chain end exists as a solvated ion not closely associated with counterion. This was discussed earlier in this chapter (Section 7.2.3) for cationic polymerization and also in Chapter 6 (Section 6.5) where an analogous situation exists under free radical conditions. The greatest success in obtaining highly syndiotactic or isotactic polymer under anionic conditions has been achieved with acrylic monomers, particularly methyl methacrylate; hence theories pertaining to isotactic placement have centered mainly on this monomer.

Early proposals to explain isotactic placement centered on conformational effects in the polymer backbone[37] or coordination of the anionic chain end with counterion to yield a cyclic unit.[22,23] One example of the latter[22] is shown in reactions (7.55) and (7.56). According to this model, the terminal delocalized carbanion coordinates the carbonyl group of the penultimate ester group via the oxygen to give a six-membered ring that is coordinated strongly with lithium counterion at both oxygens (7.55). Monomer then approaches from

$$(7.55)$$

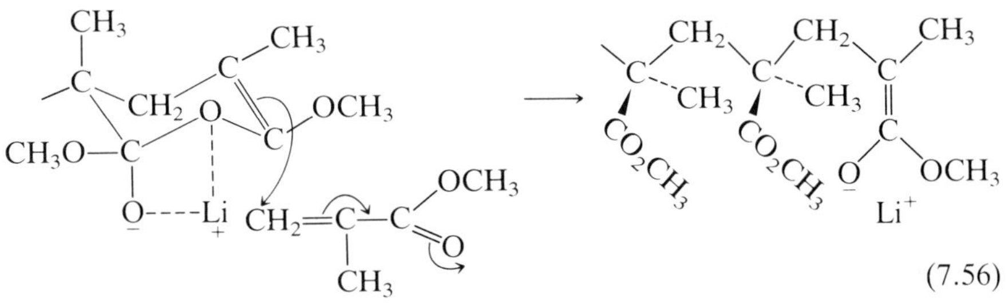

$$(7.56)$$

below (7.56) because of steric blocking of the top side by the axial methyl group, but also because this is the most favorable orientation to coordinate the counterion. Once addition occurs, the chain end again assumes the preferred cyclic orientation for approach of the next monomer molecule. The level of isotactic placement thus decreases as the solvent polarity is increased, or as lithium is replaced with the less strongly coordinating higher alkali metal ions. Another model, depicted in Scheme 7.1 (and somewhat reminiscent of the front-side/back-side model for cationic polymerization), suggests ion pairing regardless of stereochemistry, with counterion–monomer coordination assisting isotactic placement in poorly solvating solvents, and counterion–solvent interaction blocking isotactic placement in polar solvents.[38]

A number of mechanistic refinements to explain isotactic and syndiotactic placement have appeared over the years,[39] encompassing such monomers as vinyl ketones and vinylpyridines as well as acrylic esters. Invariably, coordination of chain-end with counterion and counterion solvation are key ingredients. As is the case with cationic polymerization, a definitive mechanism of stereochemical control is still lacking.

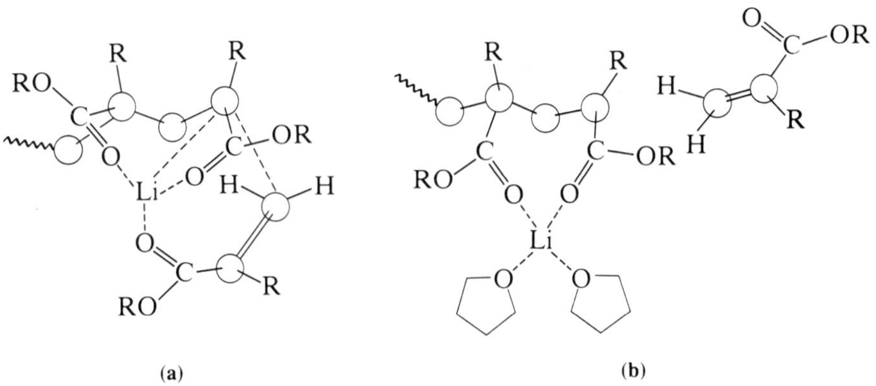

(a)   **(b)**

Scheme 7.1. (a) Isotactic approach of methyl methacrylate in a nonpolar solvent. (b) Syndiotactic approach of methyl methacrylate in tetrahydrofuran. (Circles represent backbone or incipient backbone carbons; R = methyl. Backbone hydrogens omitted.)

From the commercial standpoint, stereochemical control of diene polymerization[21b] is most important, in particular isoprene and 1,3-butadiene (Table 7.1). For both monomers, the level of *cis*-1,4 polymer is increased by using lithium-based initiators in nonpolar solvents. This is especially so with isoprene, where almost entirely *cis*-1,4 polymer is produced using butyl-lithium initiator in pentane or hexane. Temperature has little effect on stereochemistry. Thus a product virtually identical in structure to the most important form of natural rubber (*Hevea* rubber) is accessible by anionic polymerization. The synthetic product is, in fact, often referred to as "synthetic natural rubber."

Formation of *cis*-polyisoprene has been attributed to lithium's ability either to hold isoprene in its *s-cis* conformation by pi complexation (7.57) or to

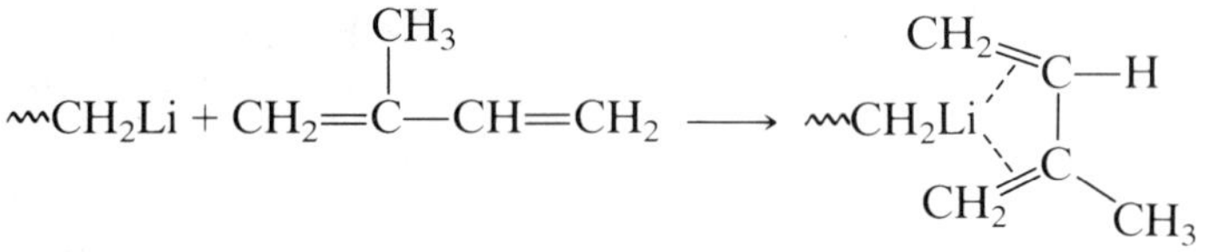

$$\text{w}CH_2Li + CH_2{=}\overset{\overset{\textstyle CH_3}{|}}{C}{-}CH{=}CH_2 \longrightarrow \text{w}CH_2Li \qquad (7.57)$$

"lock" the isoprene into a *cis* configuration by forming a six-membered ring transition state (7.58).

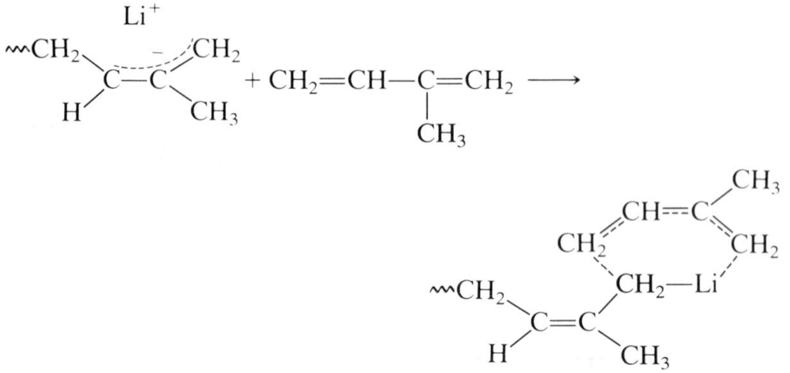

$$(7.58)$$

Predominantly *cis*-1,4-poly(1,3-butadiene) is also formed under the conditions given above for *cis*-polyisoprene, although at a somewhat lower stereospecificity. This may arise from butadiene's preference for the *s-trans* conformation, whereas the steric effect of the methyl group of isoprene favors *s-cis* (7.59).

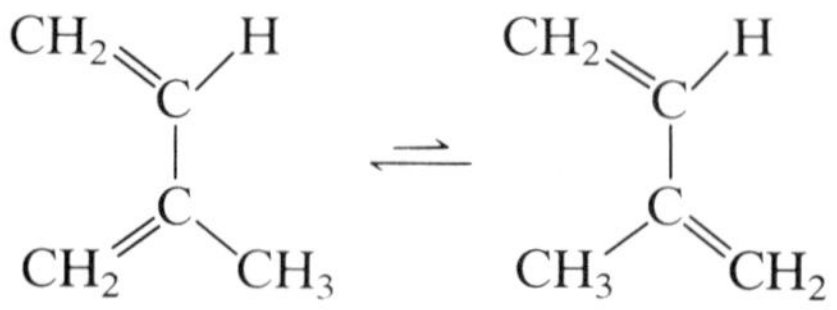

$$(7.59)$$

7.3.4 Anionic copolymerization

As is the case with cationic polymerization, relatively few reactivity ratios have been determined for anionic copolymerization processes because of the complicating effects of counterion.[29] Some representative values are given in Table 7.5.

Among the complicating factors are the solvating power of the solvent, the effect of temperature, possible competition between anionic and free radical polymerization when electron transfer initiators are used, and contrasts between homogeneous and heterogeneous polymerization systems. Solvent effects are shown most dramatically with styrene–butadiene copolymerization. A change from the nonpolar hexane to the highly solvating tetrahydrofuran reverses the order of reactivity when the soluble *n*-butyllithium is used as initiator. It should be noted that in homopolymerization, styrene is *more* reactive than butadiene. Clearly the lithium counterion plays a significant role in the preference polystyryl anion exhibits towards butadiene, a role that is effectively nullified in the presence of tetrahydrofuran.

The influence of temperature is also demonstrated by the even greater difference in reactivity ratios at $-78°C$. One might expect solvation to be felt more strongly at lower temperatures, which is the case in tetrahydrofuran but not to any appreciable extent in hexane.

Differences between homogeneous and heterogeneous processes have been demonstrated in copolymerization of styrene and methyl methacrylate.[40] With soluble *n*-butyllithium initiator, only poly(methyl methacrylate) is formed. With metallic lithium or methyllithium, both insoluble, block copolymers result, possibly by initial styrene polymerization on the initiator surface followed by detachment, and subsequent initiation of methyl methacrylate by the detached polystyryl anion. Whether or not free radical processes compete with anionic polymerization is difficult to verify experimentally, but it cannot be ruled out when alkali metals or addition complexes are used as initiators.

A variation on the anionic copolymerization process involves nucleophilic displacement of aliphatic dihalides by the dianion formed by electron transfer initiation (7.60).[41] While this reaction clearly does not fit the definition of vinyl polymerization, it is of interest because impurities need not be so rigorously excluded as they have to be in conventional anionic polymerization, and the reaction provides a route to copolymers having a regular unit.

$$\text{Li}^+\overset{-}{\text{C}}\text{HCH}_2\text{CH}_2\overset{-}{\text{C}}\text{HLi}^+ + \text{Cl—R—Cl} \longrightarrow \begin{bmatrix} \text{CHCH}_2\text{CH}_2\text{CH—R} \\ | \qquad\qquad | \\ \text{X} \qquad\qquad \text{X} \end{bmatrix}$$

$$\qquad\qquad\qquad\qquad\qquad | \qquad\qquad | \qquad\qquad\qquad\qquad\qquad\qquad\qquad\qquad\qquad$$
$$\qquad\text{X} \qquad\qquad \text{X}$$

$$(7.60)$$

Most interesting from the standpoint of commercial development is the formation of block copolymers by the living polymer method.[29,42] This is

Table 7.5. Representative anionic reactivity ratios $(r)^a$

Monomer 1	Monomer 2	Initiator[b]	Solvent[c]	Temperature (°C)	r_1	r_2
Styrene	Methyl methacrylate	Na	NH₃	d	0.12	6.4
		n-BuLi	None	d	e	e
	Butadiene	n-BuLi	None	25	0.04	11.2
		n-BuLi	Hexane	25	0.03	12.5
		n-BuLi	Hexane	50	0.04	11.8
		n-BuLi	THF	25	4.0	0.3
		n-BuLi	THF	-78	11.0	0.4
		EtNa	Benzene	d	0.96	1.6
	Isoprene	n-BuLi	Cyclohexane	40	0.046	16.6
	Acrylonitrile	RLi	None	d	0.12	12.5
	Vinyl acetate	Na	NH₃	d	0.01	0.01
Butadiene	Isoprene	n-BuLi	Hexane	50	3.38	0.47
Methyl methacrylate	Acrylonitrile	NaNH₂	NH₃	d	0.25	7.9
		RLi	None	d	0.34	6.7
	Vinyl acetate	NaNH₂	NH₃	d	3.2	0.4

[a] Data from Morton.[29]

[b] Bu = butyl, Et = ethyl, R = alkyl.

[c] THF = tetrahydrofuran.

[d] Temperature not specified.

[e] No detectable styrene in polymer.

illustrated in reaction (7.61) for polystyrene-*block*-poly(methyl methacrylate). The order of monomer addition is important in block copolymerization.

$$n \quad CH_2{=}CH \atop \qquad\quad C_6H_5 \xrightarrow{R:^-} \left[CH_2CH\atop C_6H_5\right]_{n-1}\!\!\!\!-CH_2CH:^-\atop \qquad\qquad C_6H_5 \xrightarrow{m \quad CH_2{=}CCO_2CH_3 \atop \qquad\qquad\quad CH_3}$$

$$\left[CH_2CH\atop \;\;C_6H_5\right]_n\!\!\!\left[CH_2C\atop CO_2CH_3\right]_{m-1}\!\!\!\!\!-CH_2C:^-\atop \qquad\qquad CO_2CH_3 \qquad (7.61)$$

In the example given, the copolymer would not form if methyl methacrylate were polymerized first, because living poly(methyl methacrylate) is not basic enough to add to styrene. Nevertheless, large numbers of block copolymers have been synthesized, including some consisting of vinyl blocks coupled with nonvinyl blocks, or consisting of only nonvinyl blocks.

Greatest commercial success has been achieved with ABA triblock polymers, which have proved to be versatile thermoplastic elastomers (Chapter 3, Section 3.9). They can be synthesized by a three-stage sequential formation of each block using a monofunctional initiator such as *n*-butyllithium, or in a two-stage process using a difunctional initiator (i.e., the type that undergoes radical combination) such as naphthalenesodium. The latter is illustrated schematically (7.62) for the commercially important styrene–butadiene–styrene (SBS) triblock polymer. Typically, thermoplastic elastomers of this

$$B \xrightarrow{\text{initiator}} B:^{\cdot-} \xrightarrow{\text{combination}} {}^-:BB:^- \xrightarrow{B} {}^-:BBBBBBBBBB:^- \xrightarrow{S}$$

$${}^-:SSSSSBBBBBBBBBBBSSSSS:^- \qquad (7.62)$$

type have molecular weights of about 50,000 to 70,000 for the polybutadiene blocks and 10,000 to 15,000 for the polystyrene blocks.

Other commercial block copolymers include styrene–butadiene (AB-type) and star-block (also called *radial*) copolymers.[42] The latter are formed by the reaction of living block copolymer with, for example, silicon tetrachloride (7.63). A major advantage of star-block copolymers is that they exhibit much lower melt viscosities, even at very high molecular weights, than their linear counterparts.

$$4\,\text{~~~}:^- + SiCl_4 \longrightarrow \qquad\qquad (7.63)$$

7.4   Group transfer polymerization

Of the various vinyl polymerization methodologies, that of group transfer polymerization (GTP) has arrived on the scene most recently, [43–45] Its commercial potential and basic chemistry are the subject of intense investigation.

The GTP method converts monomers frequently used in anionic polymerization into living polymers; however, the propagating chain is *covalent* in character. Typically, an organosilicon compound is used to initiate the polymerization in solution in the presence of an anionic or Lewis acid catalyst. Examples of each type of compound are given in Table 7.6. The process is illustrated for the polymerization of methyl methacrylate using the methyl trimethylsilyl acetal of dimethylketene* as initiator and bifluoride ion as catalyst (7.64). In each propagation step the SiR_3 group is transferred to

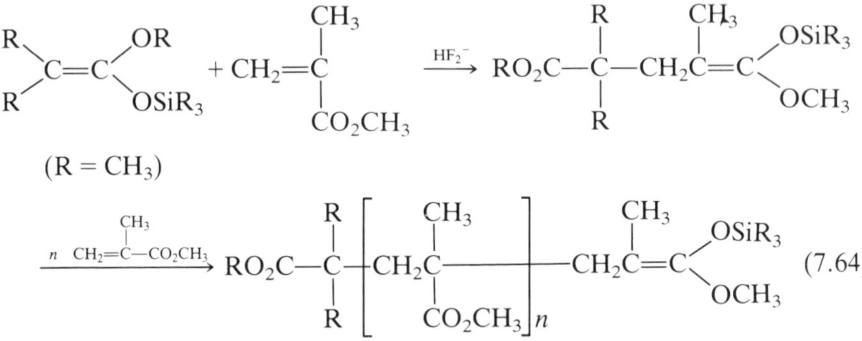

$$(R = CH_3)$$

$$(7.64)$$

the carbonyl oxygen of the incoming monomer, hence the name GTP. If a difunctional initiator is used, the chain propagates from each end[46] (7.65).

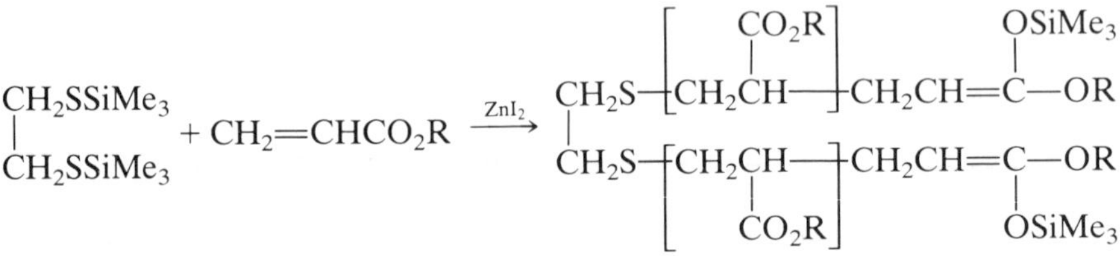

$$(7.65)$$

Once the monomer is consumed, a different monomer may be added (to form block copolymer), or the chain can be terminated by removal of catalyst or by such a desilylation reaction as protonation (7.66) or alkylation (7.67).

---

* The initiator is synthesized[47] by converting the appropriate ester to its anion with a strong base, followed by reaction with chlorotrimethylsilane, as follows ($R = CH_3$):

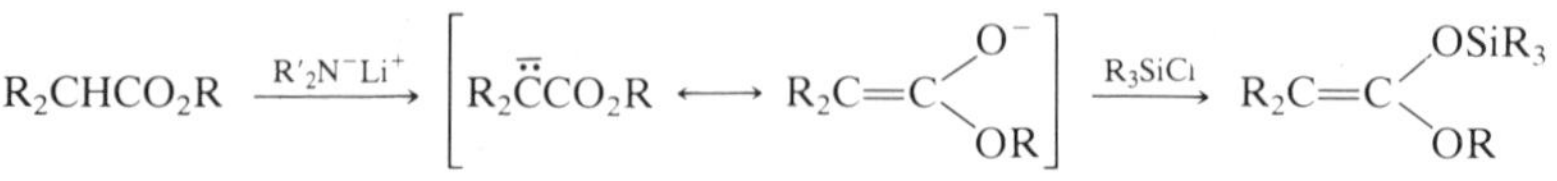

Table 7.6. Representative compounds used in group transfer polymerization

Monomers[a]	Initiators[a]	Catalysts[a]	Solvents
$CH_2{=}CHCO_2R$	$Me_2C{=}C\overset{OMe}{\underset{OSiMe_3}{\diagdown}}$	Anionic[b] HF_2^- CN^- N_3^-	Acetonitrile 1,2-Dichloroethane[d] Dichloromethane[d]
$CH_2{=}\overset{Me}{\underset{\vert}{C}}CO_2R$	$Me_3SiCH_2CO_2Me$	Me_3SiF_2 Lewis acid[c] ZnX_2 R_2AlCl $(R_2Al)_2O$	N,N-Dimethylacetamide N,N-Dimethylformamide Ethyl acetate Propylene carbonate Tetrahydrofuran
$CH_2{=}CHCONR_2$	Me_3SiCN		
$CH_2{=}CHCN$	$RSSiMe_3$		Toluene[d]
$CH_2{=}\overset{Me}{\underset{\vert}{C}}CN$	$ArSSiMe_3$		
$CH_2{=}CH\overset{O}{\overset{\Vert}{C}}R$			

[a] R = alkyl, Ar = aryl, Me = methyl, X = halogen.

[b] 0.1 mol % relative to initiator.

[c] 10–20 mol % ZnX_2 relative to monomer; 10–20 mol % of Al catalyst relative to initiator.

[d] Preferred with Lewis acid catalysts.

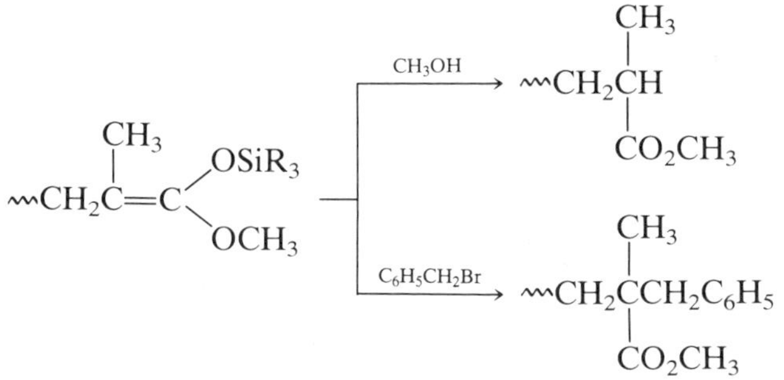

$$
\overset{\displaystyle CH_3}{\underset{\displaystyle OCH_3}{\text{www}CH_2C{=}C{\diagdown}^{OSiR_3}}}
\begin{cases}
\xrightarrow{CH_3OH} \text{www}CH_2\overset{CH_3}{\underset{CO_2CH_3}{\underset{\vert}{\overset{\vert}{C}H}}} & (7.66) \\[2em]
\xrightarrow{C_6H_5CH_2Br} \text{www}CH_2\overset{CH_3}{\underset{CO_2CH_3}{\underset{\vert}{\overset{\vert}{C}}}}CH_2C_6H_5 & (7.67)
\end{cases}
$$

The mechanism of GTP is not known with certainty, but in the case of anionic catalysis, it has been proposed[43] that the nucleophilic catalyst (Nu:⁻) activates the silyl group, possibly via a hypervalent silicon intermediate, which promotes a Michael-type addition of initiator to monomer with a concomitant transfer of the silyl group to the monomer's carbonyl oxygen (7.68). In the case of Lewis acid catalysis, it is believed that the catalyst coordinates with the carbonyl oxygen of monomer, thus rendering the monomer more susceptible to nucleophilic attack by initiator. In general, higher molecular weights and lower polydispersities are realized with anionic catalysts.

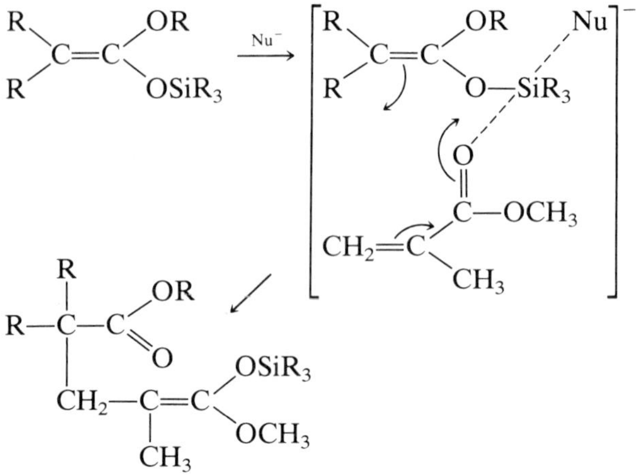

$$\underset{R}{\overset{R}{>}}C=C\underset{OSiR_3}{\overset{OR}{<}} \quad \xrightarrow{Nu^-} \quad \left[\underset{R}{\overset{R}{>}}C=C\underset{O-SiR_3}{\overset{OR\quad Nu}{<}}\right]^{-}$$

$$\underset{CH_2-C=C}{\overset{R}{\underset{|}{R-C-C}}}\overset{OR}{\underset{\overset{\|}{O}}{<}}$$

$$\underset{CH_3}{\overset{}{CH_2-C=C}}\overset{OSiR_3}{<}_{OCH_3}$$

(7.68)

Chain transfer in GTP can be brought about by moderately weak carbon acids such as α-phenylpropionitrile or methyl α-phenylacetate in the presence of anionic catalysts.[48] The reaction involves transfer of a trialkylsilyl group from the chain end to the transfer agent, shown in reaction (7.69) for

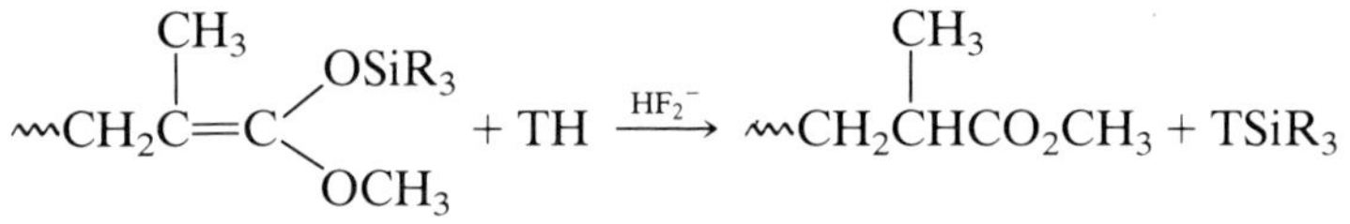

$$\underset{OCH_3}{\overset{\overset{CH_3}{\underset{|}{\quad}}}{\mathbf{\textit{mm}}CH_2C=C}}\overset{OSiR_3}{<} + TH \xrightarrow{HF_2^-} \overset{\overset{CH_3}{\underset{|}{\quad}}}{\mathbf{\textit{mm}}CH_2CHCO_2CH_3} + TSiR_3 \qquad (7.69)$$

poly(methyl methacrylate), with TH denoting the acidic transfer agent. The silylated transfer agent then initiates a new chain (7.70).

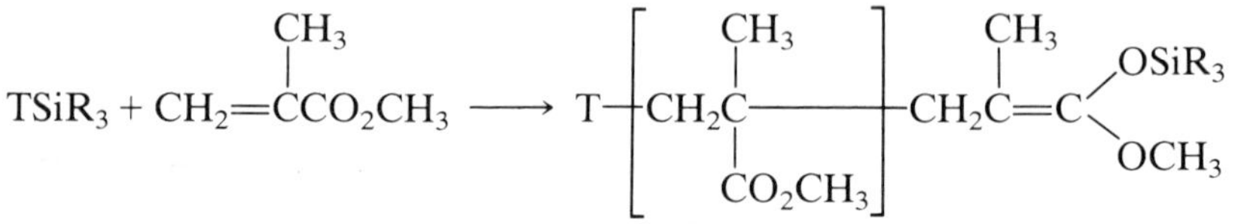

$$TSiR_3 + CH_2=\overset{\overset{CH_3}{\underset{|}{\quad}}}{C}CO_2CH_3 \longrightarrow T\left[\overset{\overset{CH_3}{\underset{|}{\quad}}}{\underset{\overset{|}{CO_2CH_3}}{CH_2C}}\right]CH_2\overset{\overset{CH_3}{\underset{|}{\quad}}}{C}=C\overset{OSiR_3}{<}_{OCH_3}$$

(7.70)

There are distinct advantages of GTP:

1. The reaction proceeds rapidly at room temperature to give quantitative yields of living polymer. (Temperature ranges of about -100 to $100°C$ have been employed.)
2. Polydispersities close to unity are obtainable.
3. Mixtures of monomers give essentially random copolymer.
4. The method is very tolerant of other functionalities in the monomer.
   For example, the monomer *p*-vinylbenzyl methacrylate is readily con-

verted to polymer (7.71) without affecting the *p*-vinylbenzyl group, a vinyl function that is very sensitive to free radical polymerization.[49]

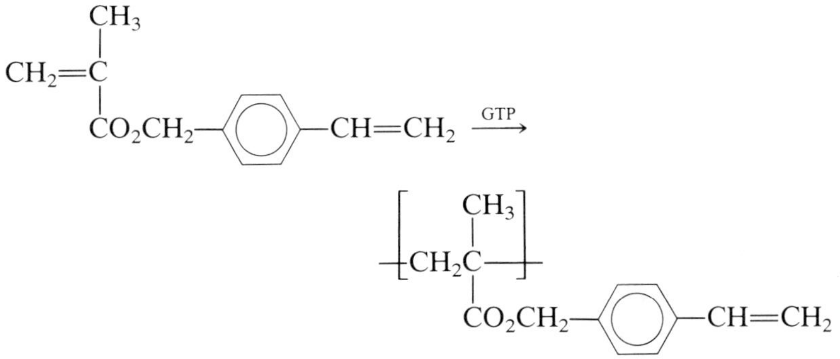

$$
CH_2=\overset{\overset{\displaystyle CH_3}{|}}{\underset{\underset{\displaystyle CO_2CH_2-}{|}}{C}} \!\!\!-\!\!\! \bigcirc \!\!\! -CH=CH_2 \xrightarrow{\text{GTP}}
$$

$$
\left[\!\! \begin{array}{c} CH_3 \\ | \\ CH_2C \\ | \\ CO_2CH_2- \end{array} \!\! \right] \!\!\! \bigcirc \!\!\! -CH=CH_2 \quad (7.71)
$$

5. Liquid crystalline polymers are obtainable by polymerizing acrylic esters having rigid structures in the ester group.[50]
6. Predominantly syndiotactic polymer can be prepared independent of solvent polarity. Typically, anionic catalysts at $-78°C$ or Lewis acid catalysts at room temperature yield polymers having a syndiotactic/heterotactic ratio of about $2:1$. Roughly equal amounts of syndiotactic and heterotactic sequences are obtained with anionic catalysts at room temperature.
7. By using initiators containing "protected" functional groups, it is possible to prepare functionally terminated polymers.[51] If the OR group of the initiator is replaced with $OSiR_3$ in reaction (7.64), for example, one obtains a polymer that can be hydrolyzed quantitatively to give carboxyl-terminated polymer (7.72).

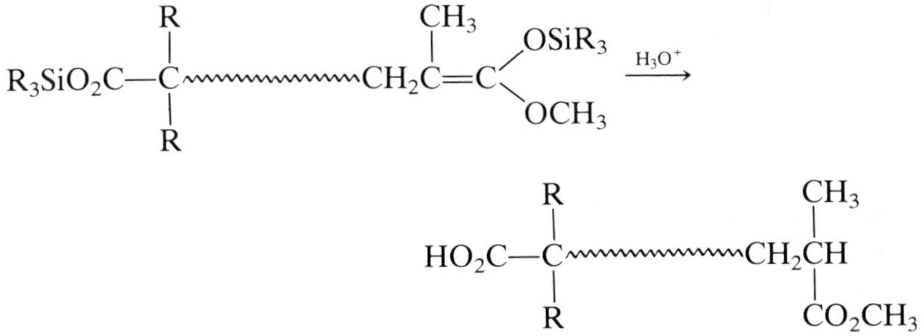

$$
R_3SiO_2C-\overset{\overset{\displaystyle R}{|}}{\underset{\underset{\displaystyle R}{|}}{C}}\!\!\!\sim\!\!\!\!\sim\!\!\!\!\sim\!\!\!CH_2\overset{\overset{\displaystyle CH_3}{|}}{C}\!\!=\!\!C\!\!\begin{array}{c} \nearrow OSiR_3 \\ \searrow OCH_3 \end{array} \xrightarrow{H_3O^+}
$$

$$
HO_2C-\overset{\overset{\displaystyle R}{|}}{\underset{\underset{\displaystyle R}{|}}{C}}\!\!\!\sim\!\!\!\!\sim\!\!\!\!\sim\!\!\!CH_2\overset{\overset{\displaystyle CH_3}{|}}{\underset{\underset{\displaystyle CO_2CH_3}{|}}{CH}} \quad (7.72)
$$

8. Telechelic polymers can be prepared by coupling end-protected living polymer, then removing the protecting group.[51] Coupling may be accomplished with a difunctional alkylating agent such as 1,4-bis(bromomethyl)benzene. Thus, the polymer shown in reaction (7.72) can be converted to carboxyl-terminated telechelic poly(methyl methacrylate) (7.73).

$$\underset{\underset{R}{\overset{\overset{R}{|}}{\overset{|}{|}}}{R_3SiO_2-C}\text{〜〜〜〜〜〜}CH_2\underset{}{C}=C\overset{OSiR_3}{\underset{OCH_3}{\diagup}}\quad \xrightarrow[\text{2. } H_3O^+]{\text{1. } BrCH_2-\!\!\!\!\bigcirc\!\!\!\!-CH_2Br}$$

$$\underset{\underset{R}{\overset{\overset{R}{|}}{|}}}{HO_2C-C}\text{〜〜〜〜〜}\underset{\underset{CO_2CH_3}{\overset{\overset{CH_3}{|}}{|}}}{C}-CH_2-\!\!\!\bigcirc\!\!\!-CH_2-\underset{\underset{CO_2CH_3}{\overset{\overset{CH_3}{|}}{|}}}{C}\text{〜〜〜〜〜}\underset{\underset{R}{\overset{\overset{R}{|}}{|}}}{C}-CO_2H \qquad (7.73)$$

It seems likely, on the basis of what has already been learned, that GTP may eventually rank in importance with the more traditional methods of vinyl polymerization.

References

1. J. P. Kennedy and E. Marechal, *Carbocationic Polymerization*, Wiley-Interscience, New York, 1982.
2. J. P. Kennedy, *Cationic Polymerization of Olefins: A Critical Inventory*, Wiley-Interscience, New York, 1975.
3. A. D. Jenkins and A. Ledwith (eds.), *Reactivity, Mechanism and Structure in Polymer Chemistry*, Wiley-Interscience, New York, 1974: (a) A. Ledwith and D. C. Sherrington, Chap. 9; (b) A. Parry, Chap. 11.
4. G. E. Ham (ed.), *Vinyl Polymerization*, Part 2, Dekker, New York, 1969: (a) Z. Zlamal, Chap. 6; (b) M. Morton, Chap. 5.
5. H. K. Hall Jr. and T. Gotoh, *Polym. Preprints*, **26** (1), 34 (1985).
6. A. Gandini and H. Cheradame, *Adv. Polym. Sci.*, **34/35**, 1 (1980).
7. J. J. Throssel, S. P. Good, M. Szwarc, and V. Stannett, *J. Am. Chem. Soc.*, **78**, 1122 (1956).
8. J. P. Kennedy, T. Kelen, S. C. Guhaniyogi, and R. T. Chou, *J. Macromol. Sci.-Chem.*, **A18**, 129 (1982), and preceding papers.
9. P. H. Plesch, *Pure Appl. Chem.* **12**, 117 (1966).
10. P. H. Plesch, *Adv. Polym. Sci.*, **8**, 137 (1971).
11. M. K. Mishra and J. P. Kennedy, *J. Macromol. Sci.-Chem.*, **A24**, 999 (1987).
12. M. K. Mishra and J. P. Kennedy, *Polym. Bull.*, **17**, 7 (1987).
13. R. Faust and J. P. Kennedy, *J. Polym. Sci. A. Polym. Chem.*, **25**, 1847 (1987).
14. R. Faust and J. P. Kennedy, *Polym. Bull.*, **19**, 21, 29, 35 (1988).
15. M. Sawamoto and T. Higashimura, *Makromol. Chem., Makromol. Symp.*, **3**, 83 (1986).
16. T. Higashimura, S. Aoshima, and M. Sawamoto, *Makromol. Chem., Makromol. Symp.*, **3**, 99 (1986).
17. M. Sawamoto, C. Okamoto, and T. Higashimura, *Macromolecules*, **20**, 2693 (1987).
18. T. Higashimura, K. Kojima, and M. Sawamoto, *Polym. Bull.*, **19**, 7 (1988).
19. J. P. Kennedy and A. P. Langer Jr., *Adv. Polym. Sci.*, **3**, 508 (1964).

20. R. J. Cotter and M. Matzner, *Ring-Forming Polymerizations*, Part A, Academic Press, New York, 1969, Chap. 2.
21. A. D. Ketley (ed.), *The Stereochemistry of Macromolecules*, Vol. 2, Dekker, New York, 1967: (a) A. D. Ketley, Chap. 2; (b) D. Braun, Chap. 1.
22. D. J. Cram and K. R. Kopecky, *J. Am. Chem. Soc.*, **81**, 2748 (1959).
23. C. E. H. Bawn and A. Ledwith, *Quart. Rev.*, **16**, 361 (1962).
24. T. Higashimura, T. Yonezawa, S. Okamura, and K. Fukui, *J. Polym. Sci.*, **39**, 487 (1959).
25. T. Kunitake and C. Aso, *J. Polym. Sci. A-1*, **8**, 665 (1970).
26. G. E. Ham (ed.), *Copolymerization*, Wiley-Interscience, New York, 1964: (a) J. P. Kennedy, Chap. 5; (b) M. Morton, Chap. 7; (c) A. S. Hoffman and R. Bacskai, Chap. 6.
27. G. W. Cowell and A. Ledwith, *Quart. Rev.*, **24**, 119 (1970).
28. A. D. Ketley and L. P. Fisher, in *Structure and Mechanism in Vinyl Polymerization*, (T. Tsuruta and K. F. O'Driscoll, eds.), Dekker, New York, 1969, Chap. 12.
29. M. Morton, *Anionic Polymerization: Principles and Practice*, Academic Press, New York, 1983.
30. M. Morton and L. J. Fetters, in *Polymerization Processes* (C. E. Schildknecht, ed.), Wiley-Interscience, New York, 1977, Chap. 9.
31. J. E. McGrath (ed.), *Anionic Polymerization: Kinetics, Mechanism and Synthesis*, ACS Symp. Ser. 166, American Chemical Society, Washington, D.C., 1981.
32. T. E. Hogen-Esch and J. Smid (eds.), *Recent Advances in Anionic Polymerization*, Elsevier, New York, 1987.
33. L. Reich and A. Schindler, *Polymerization by Organometallic Compounds*, Wiley-Interscience, New York, 1966, Chap. 7.
34. M. Szwarc, *Carbanions, Living Polymers, and Electron-Transfer* Processes, Wiley-Interscience, New York, 1968.
35. M. Szwarc, *Nature*, **178**, 1168 (1956).
36. M. Szwarc, M Levy, and R. Milkovich, *J. Am. Chem. Soc.*, **78**, 2656 (1956).
37. M. Szwarc, *Chem. Ind.* (London), 1589 (1958).
38. W. Fowells, C. Schuerch, F. A. Bovey, and F. P. Hood, *J. Am. Chem. Soc.*, **89**, 1396 (1967).
39. T. E. Hogen-Esch and J. Smid (eds.), *Recent Advances in Anionic Polymerization*, Elsevier, New York, 1987, pp. 187–272.
40. C. G. Overberger and N. Yamamoto, *J. Polym. Sci. A-1, Polym. Chem.*, **4**, 3101 (1966).
41. D. H. Richards, N. F. Scilly, and F. Williams, *Polymer*, **10**, 603 (1969).
42. A. Noshay and J. E. McGrath, *Block Copolymers: Overview and Critical Survey*, Academic Press, New York, 1977.
43. O. W. Webster, W. R. Hertler, D. Y. Sogah, W. B. Farnham, and T. V. Rajan-Babu, *J. Am. Chem. Soc.*, **105**, 5706 (1983).
44. W. R. Hertler, D. Y. Sogah, O. W. Webster, and B. M. Trost, *Macromolecules*, **17**, 1415 (1984).
45. D. Y. Sogah, W. R. Hertler, O. W. Webster, and G. M. Cohen, *Macromolecules*, **20**, 1473 (1987).
46. M. T. Reetz, R. Ostarek, K.-E. Piejko, D. Arlt, and D. Bomer, *Angew. Chem. Int. Ed. Engl.*, **25**, 1108 (1986).
47. C. Ainsworth, F. Chen, and Y.-N. Kuo, *J. Organomet. Chem.*, **46**, 59 (1972).
48. W. R. Hertler, *Macromolecules*, **20**, 2976 (1987).

49. C. Pugh and V. Percec, *Polym. Bull.* **14**, 109 (1985).
50. W. Kreuder and O. W. Webster, *Makromol. Chem., Rapid Commun.*, **7**, 5 (1986).
51. O. W. Webster, W. R. Hertler, D. Y. Sogah, W. B. Farnham, and T. V. Rajan-Babu, *J. Macromol. Sci.-Chem.*, **A21**, 943 (1984).

Review exercises

1. Give plausible explanations for the following facts:
   (a) Molecular weight does not depend on initiator concentration in ionic polymerization as it does in free radical polymerization.
   (b) Polymerization rate and polymer stereochemistry are more sensitive to solvent effects in ionic polymerization than in free radical polymerization.
   (c) $\overline{DP} = \overline{\nu}$ in cationic polymerization, but this is not always the case in free radical or anionic polymerization.
   (d) Ethyl vinyl ether undergoes cationic polymerization faster than β-chloroethyl vinyl ether under the same conditions.
   (e) Cationic polymerization of methyl α-methylvinyl ether gives predominantly syndiotactic polymer in toluene, whereas methyl vinyl ether yields mainly isotactic polymer.
   (f) Alkyllithium initiators favor the *cis* stereochemistry in polymerization of conjugated dienes.
   (g) If methyl methacrylate is polymerized with *n*-butyllithium in toluene at $-30°C$, the resultant polymer is almost 60% isotactic. If 10% of the toluene is replaced with pyridine, polymerization yields polymer that is nearly 60% syndiotactic. (R. L. Miller, *SPE Trans.*, **3**, 1, 1963)

2. Predict the order of reactivity (and justify your prediction): (a) in cationic polymerization: styrene, *p*-methoxystyrene, *p*-chlorostyrene, *p*-methylstyrene; (b) in anionic polymerization: styrene, 2-vinylpyridine, 3-vinylpyridine, 4-vinylpyridine.

3. Write reactions illustrating transfer to monomer and transfer to polymer in the cationic polymerization of propylene and isobutylene. Which type of transfer would you expect to predominate in each case. Suggest a reason why propylene does not form high-molecular-weight polymer under cationic conditions.

4. If trifluoroacetic acid is added dropwise to styrene, no polymerization occurs. If styrene is added to the acid, however, high-molecular-weight polymer forms rapidly. Suggest an explanation. (See reaction 7.19 and ref. 7.)

5. Rearrangement in cationic polymerization is temperature dependent, with rearranged polymer favored at low temperatures. (a) Explain how NMR might be used to evaluate the degree of rearrangement in the cationic polymerization of 3-methyl-1-butene (reaction 7.35). (b) Predict the principal repeating units for the polymers obtained by low-temperature cationic polymerization of 3,3-dimethyl-1-butene and vinylcyclohexane. (J. P. Kennedy et al., *J. Polym. Sci. A*, **2**, 5029, 1964)

6. Proton and ^{13}NMR of poly(4-methyl-1-pentene) formed under cationic conditions show the presence of five different repeating units. Predict their structure and explain their formation. (G. Ferraris et al., *Macromolecules*, **10**, 188, 1977)

7. What number average molecular weight polystyrene would be formed by polymerization of 2.0 M styrene with 1.0×10^{-3} *M* naphthalenesodium in tetrahydrofuran? If the reaction were run at 25°C, how long would it take to reach 90% conversion?

8. Write reactions illustrating the synthesis of poly(ethyl vinyl ether)-*block*-poly-(*p*-methoxystyrene).

9. Write reactions illustrating the synthesis of poly(methyl methacrylate)-*block*-polyacrylonitrile (a) by anionic polymerization, (b) by group transfer polymerization. (c) How could you maximize isotactic placement in the poly(methyl methacrylate) block? (d) Show how you could synthesize an ABA block copolymer of the two monomers.

10. Predict the structure of the polymer formed from 2,6-diphenyl-1,6-heptadiene under anionic conditions. Write a mechanism for its formation.

11. The predominant repeating unit obtained in the following free radical polymerization is as shown:

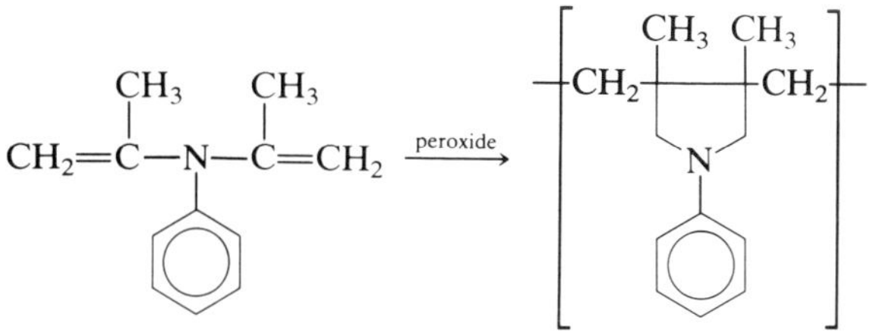

Would you expect the same result if GTP were used? Explain. (See J. J. Kozakiewicz et al., *Polym. Preprints*, **28**(2), 347, 1987.)

12. Apart from lower viscosity, what other advantage might AB star-block copolymers have over linear AB block copolymers?

13. Silyl vinyl ethers are polymerized in the presence of aldehyde initiators and Lewis acid catalysts to give silylated poly(vinyl alcohol):

$$\mathrm{R'CHO + CH_2{=}CHOSiR_3 \xrightarrow{ZnBr_2} R'\underset{\displaystyle OSiR_3}{\underset{|}{C}H}\!-\!\!\left[CH_2\underset{\displaystyle OSiR_3}{\underset{|}{C}H}\right]\!\!-\!CH_2CHO}$$

This reaction has been dubbed "aldol-GTP." Propose a mechanism for the polymerization and explain the terminology. (D. Y. Sogah, *Polym. Preprints*, **27**(1), 163, 1986)

14. GTP has been referred to as the "mirror image" of pseudocationic polymerization and could be appropriately called "pseudoanionic" (K. Matyjaszewski, *J. Polym.*

Sci. A., Polym. Chem., **25**, 765, 1987). Suggest a mechanism for pseudocationic polymerization consistent with this description.

15. Reaction of 2,3-dihydrofuran with BF_3 at dry ice temperatures yields a polymer that, on exposure to air, develops strong IR absorptions at 3435, 1771, and 1726 cm^{-1}. Give the polymer structure and explain the probable cause of the IR absorptions. If the air-exposed polymer is dissolved in $CHCl_3$ or $Cl_2CHCHCl_2$, a yellow solution is formed and, in the case of the latter (but not the former), the solution gels rapidly. Speculate on what might have happened. What would you expect an elemental analysis to reveal? (J. A. Moore and R. Wille, *Polym. Preprints*, **24**(2), 317, 1983; *Macromolecules*, **19**, 3004, 1986)

16. Grignard reagents are effective anionic initiators. Show with equations how vinyl-benzylmagnesium chloride (*m*- and *p*-CH_2=$CHC_6H_4CH_2MgCl$) might be used to prepare a comb polymer. (See K. Hadada et al., *Polym. J.*, **18**, 581, 1986.) (Note: The initiator undergoes spontaneous polymerization in tetrahydrofuran but is stable in dry ether.)

17. The following sequence of reactions has been used to prepare comb polymers (R. Asami et al., *Polym. Bull.*, **16**, 125, 1986). Give the structures of A through D.

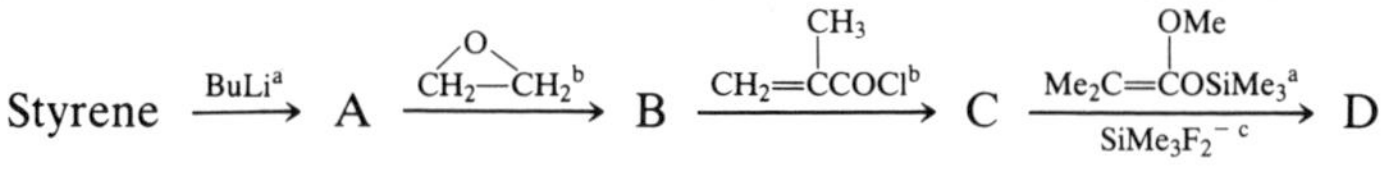

$$\text{Styrene} \xrightarrow{\text{BuLi}^a} \text{A} \xrightarrow{\overset{O}{\overset{\frown}{CH_2-CH_2}}\,^b} \text{B} \xrightarrow{CH_2=\overset{\overset{\displaystyle CH_3}{|}}{C}COCl^b} \text{C} \xrightarrow[\text{SiMe}_3F_2^{-\,c}]{Me_2C=\overset{\overset{\displaystyle OMe}{|}}{C}OSiMe_3\,^a} \text{D}$$

^{a} Initiator, ^{b} Equimolar, ^{c} Catalyst.

8
Vinyl polymerization with complex coordination catalysts

8.1 Introduction

In the early 1950s, Karl Ziegler in Germany discovered that certain combinations of transition metal compounds and organometallic compounds polymerize ethylene at low temperatures and pressures to give polyethylene that has an essentially linear structure.[1] Now referred to as high-density polyethylene (HDPE), the product is denser, tougher, and higher melting than the branched low-density polyethylene (LDPE) prepared at high pressures with free radical initiators, because the more regular structure allows closer chain packing and a high degree of crystallinity. Ziegler's remarkable discovery, the culmination of work begun before the Second World War, led to speculation that the high-pressure method of making polyethylene was doomed. In fact, substantial differences in properties between HDPE and LDPE assured ample markets for both, as may be recalled from our discussion of commodity plastics in Chapter 1. A more recent threat to LDPE is what is called *linear low-density polyethylene* (LLDPE), which is actually a copolymer of ethylene and 1-butene (with lesser amounts of 1-hexene and higher 1-alkenes to vary the density) manufactured with Ziegler-type catalysts. First marketed in the late 1970s, LLDPE is attractive because it requires considerably less energy to produce than LDPE.

Following close on the heels of Ziegler's discovery was the recognition by Giulio Natta in Italy that catalysts of the type described by Ziegler were capable of polymerizing 1-alkenes (or *alpha olefins*, as they are called in the chemical industry) to yield stereoregular polymers.[2] Natta and his coworkers, as well as scientists in other laboratories, subsequently extended the range of catalysts, both soluble and insoluble, to yield polymers exhibiting a wide range of stereoregular structures, including those derived from dienes and cycloalkenes.[3,4] Many are now manufactured on a commercial scale, as outlined in Table 8.1. Indeed, stereoregular (isotactic) polypropylene, which cannot be made with sufficiently high molecular weight by free radical or ionic polymerization, has already achieved the status of a commodity plastic (see Chapter 1). Amorphous polypropylene, previously an unwanted by-product

271

Table 8.1. Commercially available polymers synthesized with complex coordination catalysts

Polymer	Principal stereochemistry	Typical uses
Plastics		
Polyethylene, high density (HDPE)	—	Bottles, drums, pipe, conduit, sheet, film, wire and cable insulation
Polypropylene	Isotactic	Automobile and appliance parts, rope, cordage, webbing, carpeting, film
Poly(1-butene)	Isotactic	Film, pipe
Poly(4-methyl-1-pentene)[a]	Isotactic	Packaging, medical supplies, lighting
1,4-Polyisoprene	*trans*	Golf ball covers, orthopedic devices
Ethylene–1-butene[b] copolymer (linear low-density polyethylene, LLDPE)	—	Blending with LDPE, film packaging, bottles
Ethylene–propylene block copolymers (polyallomers)	Isotactic	Food packaging, automotive trim, toys, bottles, film, heat-sterilizable containers
Polydicyclopentadiene[c]	—	Reaction injection molding (RIM) structural plastics
Elastomers		
1,4-Polybutadiene	*cis*	Tires, conveyer belts, wire and cable insulation, footware
1,4-Polyisoprene	*cis*	Tires, footware, adhesives, coated fabrics
Poly(1-pentenylene) (polypentenamer)[c]	*trans*	Tires
Poly(1-octenylene) (polyoctenamer)[c]	*trans*	Blending with other elastomers
Poly(1,3-cyclo-pentenylenevinylene) (norbornene polymer)[c]	*trans*	Molding compounds, engine mounts, car bumper guards
Polypropylene (amorphous)	—	Asphalt blends, sealants, adhesives, cable coatings
Ethylene–propylene copolymer (EPM, EPR)	—	Impact modifier for polypropylene, car bumper guards
Ethylene–propylene–diene copolymer (EPDM)	—	Wire and cable insulation, weather stripping, tire side walls, hose, seals

[a] Usually copolymerized with small amounts of 1-pentene.

[b] 1-Hexene and 1-octene used in smaller amounts.

[c] Synthesized by metathesis polymerization of the corresponding cycloalkene.

of isotactic polypropylene manufacture, is also becoming increasingly important, particularly for modifying asphalt roofing compounds. Not included in Table 8.1 is polyacetylene, which is still in the developmental stage but of interest because of its aforementioned conducting properties.

Catalysts developed by Ziegler and Natta have come to be known, appropriately enough, as *Ziegler–Natta catalysts*. (Because they are not consumed in the polymerization reaction, as was the case with free radical and ionic initiators, it is appropriate to designate them catalysts.) Other types of complex catalysts that have received attention are the *reduced metal oxides* and the *alfin catalysts*, the latter prepared from compounds of sodium. All three types are used principally in heterogeneous polymerization, although some homogeneous processes are also commercially important.

Before we proceed with the fundamentals of complex coordination polymerization, it is worth recalling that Ziegler and Natta shared the Nobel Prize in Chemistry in 1963.

8.2 Ziegler–Natta catalysts

The number of compounds and combinations of compounds that fit into the category of Ziegler–Natta catalysts[3,5a,6a] are far too numerous to describe here, but a Ziegler–Natta catalyst may be defined as a combination of (1) a transition metal compound of an element from groups IV to VIII, and (2) an organometallic compound of a metal from groups I to III of the periodic table. The transition metal compound is referred to as the catalyst and the organometallic compound as the cocatalyst.

Most commonly the catalyst component consists of halides or oxyhalides of titanium, vanadium, chromium, molybdenum, or zirconium. Compounds of iron and cobalt have, in some instances, also been found to be effective. Some ligands other than halide or oxyhalide that have been investigated include alkoxy, acetylacetonyl, cyclopentadienyl, and phenyl. Cocatalysts are usually hydrides, alkyls, or aryls of metals, such as aluminum, lithium, zinc, tin, cadmium, beryllium, and magnesium. By far the most important and most thoroughly studied Ziegler–Natta systems are combinations of titanium trihalides and tetrahalides with trialkylaluminum compounds.

Catalysts are prepared by mixing the components in a dry, inert solvent in the absence of oxygen, usually at a low temperature. They are characterized by having high reactivity toward many nonpolar monomers and are usually capable of giving polymers having a high degree of stereoregularity. Catalyst activity usually changes with time, and it is not uncommon for maximum activity to be reached after aging periods of one to two hours. The nature of the insoluble Ziegler–Natta catalysts is not well understood, and the trial-and-error method is frequently employed in developing new catalysts. For the $TiCl_4$–AlR_3 (R = alkyl) system it is recognized that initially exchange reactions take place (8.1 to 8.3). Organotitanium compounds are then believed to undergo reduction via homolytic bond cleavage (8.4 and 8.5).

$$AlR_3 + TiCl_4 \rightarrow AlR_2Cl + TiRCl_3 \tag{8.1}$$

$$AlR_2Cl + TiCl_4 \rightarrow AlRCl_2 + TiRCl_3 \tag{8.2}$$

$$AlR_3 + TiRCl_3 \rightarrow AlR_2Cl + TiR_2Cl_2 \tag{8.3}$$

$$TiRCl_3 \rightarrow TiCl_3 + R\cdot \tag{8.4}$$

$$TiR_2Cl_2 \rightarrow TiRCl_2 + R\cdot \tag{8.5}$$

Further reduction may also occur:

$$TiRCl_2 \rightarrow TiCl_2 + R\cdot \tag{8.6}$$

$$TiRCl_3 \rightarrow TiCl_2 + RCl \tag{8.7}$$

In addition, $TiCl_3$ may be formed by the equilibrium

$$TiCl_4 + TiCl_2 \rightleftharpoons 2TiCl_3 \tag{8.8}$$

Radicals formed in these reactions may be removed by combination, disproportionation, or reaction with solvent. While such reactions undoubtedly occur in catalyst formation, it is not known to what extent, and the aging process certainly requires more clarification. Because of the heterogeneous nature of the catalyst it is certain that reduction of Ti(IV) is not complete. Soluble catalysts, on the other hand, appear to form well-defined complexes. As an example, the soluble complex formed from triethylaluminum and bis(cyclopentadienyl)titanium dichloride has been shown by elemental and x-ray analysis to have structure **1** (C_5H_5 = cyclopentadienyl) in which both metals have the tetrahedral configuration.

1

Although $TiCl_4$ is frequently used in Ziegler–Natta catalysts, better activity is achieved by using $TiCl_3$ directly rather than by forming it by reduction. It is of interest that $TiCl_3$ exists in four different crystalline forms, referred to as α, β, γ, and δ. The α, γ, and δ forms give polymers having a high degree of stereoregularity. All three forms also have similar close-packed layered crystal structures. The β form, however, which has a linear structure, gives much more atactic polymer, indicating that stereoregularity is very much dependent on surface characteristics of the catalyst.

The ratio of transition metal to metal from groups I to III and the valency of the transition metal can affect rate of polymerization as well as yield, molecular weight, and stereoregularity of the polymer. It is possible, therefore, to vary the molecular weight by changing the ratio. The polymerization process can also be affected by the use of certain additives, particularly electron donors such as amines. For example, the $TiCl_3$–$Al(C_2H_5)Cl_2$ system fails to give stereoregular poly(1-alkenes). Addition of amine, however, which interacts strongly with both catalyst and cocatalyst, results in highly stereoregular polymer. Tables 8.2, 8.3, and 8.4 show how some of the structural variables of Ziegler–Natta catalysts affect the stereoregularity of polypropylene. Table 8.2 illustrates the influence of transition metal compound,[7] Table 8.3 the influence of the organic group of the organoaluminum cocatalyst, and Table 8.4 the effect of different catalyst–cocatalyst combinations, many with additives.

A problem with many of the earlier catalyst systems was their low efficiency. Substantial amounts of catalyst were needed to achieve acceptable yields of polymer, and the spent catalyst had to be removed from the finished product. Efficiency of heterogeneous catalysts can be improved significantly by impregnating the catalyst on a solid support such as $MgCl_2$ or MgO. As an example, a typical $TiCl_3$–AlR_3 catalyst yields about 50 to 200 g of polyethylene per gram of catalyst per hour per atmosphere of ethylene, whereas as much as 7000 g may be produced using a $MgCl_2$-supported catalyst, and without the need for the costly step of removing catalyst from the product. Presumably the support maximizes the number of active sites on the catalyst surface. Such catalyst systems are often referred to as *high-mileage catalysts*.

Soluble catalysts have much higher efficiencies. Remarkably, bis(cyclopentadienyl)zirconium dichloride (2) complexed with an aluminoxane (alternating aluminum and oxygen atoms in ring or chain form) has been reported[8] to have efficiencies as high as 5×10^6 g of polyethylene per gram of

Table 8.2. Influence of transition metal on stereoregularity of polypropylene [cocatalyst, $Al(C_2H_5)_3$][a]

Transition metal compound	Stereoregularity (%)
$TiCl_4$	48
$TiBr_4$	42
$TiCl_3$ (α, γ, or δ)	80–92
$TiCl_3$ (β)	40–50
$ZrCl_4$	55
VCl_3	73
$CrCl_3$	36
VCl_4	48
$VOCl_3$	32

[a] From Dawans and Teyssie,[7] courtesy of La Société Chimique de France.

Table 8.3. Influence of the R group of AlR_3 on the stereoregularity of polypropylene[a]

R	Stereoregularity (%)	
	$TiCl_3$	$TiCl_4$
C_2H_5	79.4	47.8
n-C_3H_7	71.8	50.9
i-C_4H_9	74.5	30.0
C_6H_5	65.4	—
C_6H_{13}	64.0	26.2
$C_{16}H_{33}$	59.0	16.2

[a] Reprinted from Jordan,[5a] p. 30, courtesy of Marcel Dekker, Inc.

Table 8.4. Stereoregularity of polypropylene with different catalyst systems[a]

Catalyst[b]	Stereoregularity (%)
$R_3Al + TiCl_4$	35.2
$R_3Al + \alpha\text{-}TiCl_3$	84.7
$R_3Al + \beta\text{-}TiCl_3$	45
$R_3Al + TiCl_4 + NaF$	97
$R_3Al + TiCl_4 +$ compounds of P, As, or Sb	98
$R_3Al + TiCl_3 +$ amine	81
$R_3Al + Ti(O\text{-}i\text{-}butyl)_4$	20
$R_3Al + V(acac)_3$	0
$R_3Al + Ti(C_5H_5)Cl_2$	70–90
$R_3Al + Tl(C_2H_5)_2Cl_2$	85
$R_2AlX + TiCl_3$	90–99
$RAlX_2 + \gamma\text{-}TiCl_3 +$ amine	99
$RAlX_2 + TiCl_3 + HPT$	97
$RNa + TiCl_3$	90
$RNa + TiCl_4$	90
$RLi + TiCl_4$	90
$R_2Zn + TiCl_3$	65
$R_2Zn + TiCl_3 +$ amine	93

[a] Reprinted from Jordan,[5a] p. 28, courtesy of Marcel Dekker, Inc.

[b] R = alkyl, (acac) = acetylacetonate, X = halogen, HPT = hexamethyl phosphoric triamide, C_5H_5 = cyclopentadienyl.

zirconium per hour and, unlike most soluble catalysts, the complex is stable indefinitely in solid form. Most soluble catalysts have efficiencies of the order of 10^4 to 10^5 g.

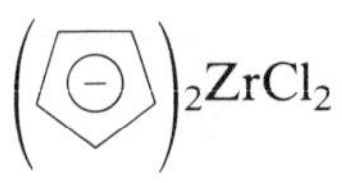

2

8.3   Mechanism and reactivity in Ziegler–Natta polymerization

Despite the tremendous amount of research that has been conducted in this area,[3,4,9,10] the true mechanism of Ziegler–Natta polymerization is still not entirely clear. It is generally agreed, however, that heterogeneous polymerization occurs at localized active sites on the catalyst surface. The organometallic component is believed to activate the site by alkylation of a transition metal atom at the surface. Of the various mechanisms that have been proposed, the two that are most generally accepted are the so-called *monometallic* and *bimetallic* mechanisms,[3,5b,c] the former being favored in heterogeneous processes. In both mechanisms, monomer is pictured as being incorporated into the polymer by an insertion reaction between a transition metal atom and a terminal carbon of a coordinated polymer chain.

The monometallic mechanism, with titanium as the transition metal, is shown in Scheme 8.1. According to this mechanism, monomer is complexed at a titanium atom exposed on the catalyst surface by a missing chlorine atom. An insertion reaction takes place, shifting the vacant octahedral position. Then migration of the chain occurs to reestablish the vacant site on the surface.

Either the monometallic or the bimetallic mechanism might occur with soluble catalysts. The bimetallic mechanism, illustrated in Scheme 8.2 for a titanium–aluminum system, involves initial pi complexation of monomer to the transition metal (Ti) that is bridged through alkyl groups to an aluminum atom. This is followed by ionization of the transition metal–alkyl bond, formation of a six-membered cyclic transition state, then insertion. In both mechanisms, the polymer chain grows from the catalyst surface by successive insertion reactions of complexed monomer, and the R group originally present in the organometallic cocatalyst ends up as the terminal group of the chain.

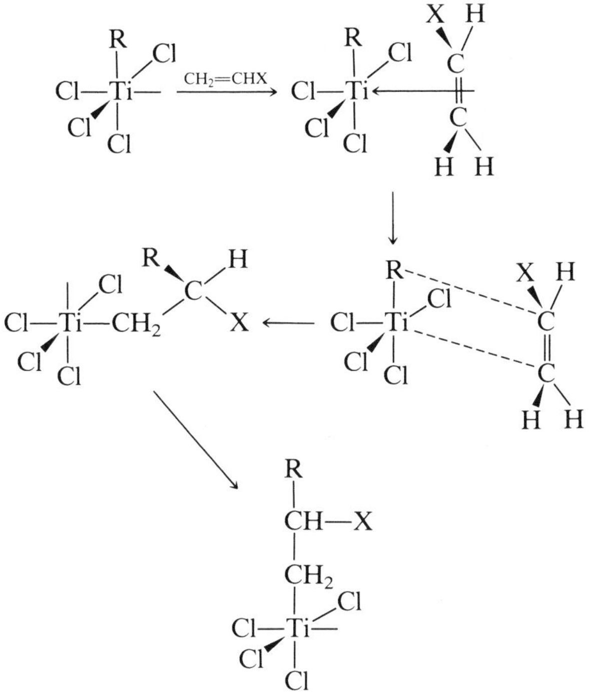

Scheme 8.1. Monometallic mechanism of Ziegler–Natta polymerization.

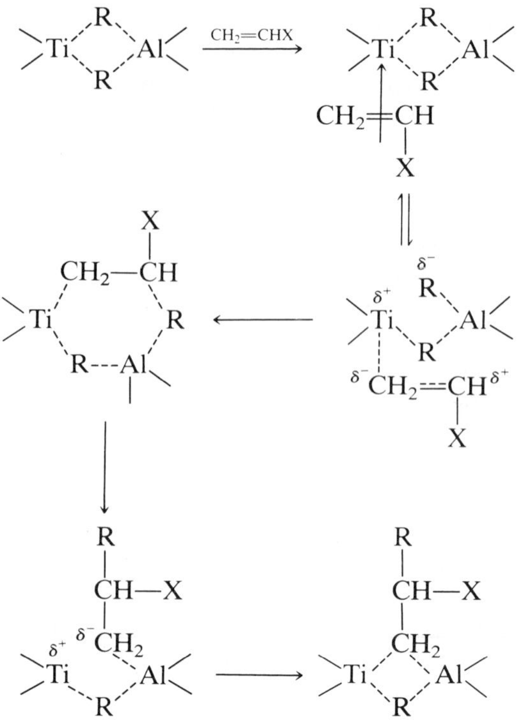

Scheme 8.2. Bimetallic mechanism of Ziegler–Natta polymerization.

That insertion is the key step has been demonstrated[11] by isolation and x-ray characterization of the initial insertion product (**5**) formed by reaction (8.9) between the "monomer," trimethyl(phenylethynyl)silane (**3**), and the soluble complex (**4**) (C_5H_5 = cyclopentadienyl). It has been proposed that the active component is an electrophilic titanium species (**6**) formed in the equilibrium (8.10).

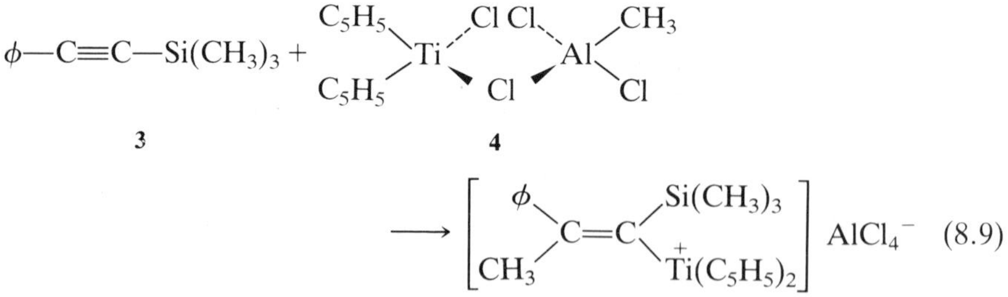

$$\phi-C\equiv C-Si(CH_3)_3 \; + \quad \text{(4)}$$

3   **4**

$$\longrightarrow \left[\begin{array}{c} \phi \\ CH_3 \end{array} C=C \begin{array}{c} Si(CH_3)_3 \\ \overset{+}{Ti}(C_5H_5)_2 \end{array} \right] AlCl_4^- \quad (8.9)$$

5

$$4 \; \rightleftharpoons \; (C_5H_5)_2\overset{+}{Ti}CH_3 \quad AlCl_4^- \qquad (8.10)$$

6

As might be expected from a consideration of the metal–alkyl bond at the active site of the catalyst, Ziegler–Natta polymerization bears a superficial resemblance to anionic polymerization. Thus, active hydrogen compounds bring about termination just as they do in anionic processes. Furthermore, the propagating chains have relatively long lifetimes once monomer is consumed. More important, however, anionic chain ends are not a characteristic of Ziegler–Natta polymerizations.

A wide variety of nonpolar monomers are polymerizable with Ziegler–Natta catalysts and, predictably, monomer activity decreases with increasing steric hindrance about the double bond. Following is the observed order of reactivity for several 1-alkenes:

$$CH_2{=}CH_2 > CH_2{=}CHCH_3 > CH_2{=}CHCH_2CH_3 >$$
$$CH_2{=}CHCH_2CH(CH_3)_2 > CH_2{=}CHCH(CH_3)_2 >$$
$$CH_2{=}CHCH(CH_2CH_3)_2 > CH_2{=}CHC(CH_3)_3$$

Alkynes are also subject to polymerization with Ziegler–Natta catalysts, as mentioned earlier for polyacetylene.

Termination of chain growth may occur a number of ways. Chain transfer to monomer (8.11) or internal hydride transfer (8.12) results in unsaturated

$$\text{Cat}{-}\text{CH}_2\overset{X}{\text{CH}}{+}\text{CH}_2\overset{X}{\text{CH}}{+}_n\text{R}$$
7

$$\xrightarrow{CH_2=CHX} \text{Cat}{-}\text{CH}_2\text{CH}_2\text{X}$$

$$+\text{CH}_2{=}\overset{X}{\text{C}}{+}\text{CH}_2\overset{X}{\text{CH}}{+}_n\text{R} \quad (8.11)$$

$$\longrightarrow \text{Cat}{-}\text{H} + \text{CH}_2{=}\overset{X}{\text{C}}{+}\text{CH}_2\overset{X}{\text{CH}}{+}_n\text{R} \quad (8.12)$$

chain ends. ("Cat" in polymer **7** refers to the catalyst surface.) Although unsaturation has been detected spectroscopically, saturated chain ends predominate in Ziegler–Natta polymerizations. Chain transfer involving the cocatalyst (8.13) may also occur. Indeed, $Zn(C_2H_5)_2$ has been used commer-

$$\textbf{7} + \text{AlR}_3 \longrightarrow \text{Cat}{-}\text{R} + \text{R}_2\text{Al}{-}\text{CH}_2\overset{X}{\text{CH}}{+}\text{CH}_2\overset{X}{\text{CH}}{+}_n\text{R} \quad (8.13)$$

cially to control molecular weight. The preferred transfer agent for molecular weight control, however, is molecular hydrogen (8.14) because it reacts

$$\textbf{7} + \text{H}_2 \longrightarrow \text{Cat}{-}\text{H} + \text{CH}_3\overset{X}{\text{CH}}{+}\text{CH}_2\overset{X}{\text{CH}}{+}_n\text{R} \quad (8.14)$$

cleanly, leaves no residue, and is low in cost. Without the application of such transfer reagents, molecular weight would, in most instances, be too high for commercial use. A true termination reaction is brought about by compounds containing active hydrogen (8.15).

$$\mathbf{7} + HX \longrightarrow Cat—X + \underset{\underset{X}{|}}{CH_3CH}\left[\underset{\underset{X}{|}}{CH_2CH}\right]R \qquad (8.15)$$

Because of the complexities of Ziegler–Natta polymerization, no comprehensive or unified kinetics scheme has emerged that adequately takes into account surface adsorption, catalyst–cocatalyst interaction, decay of catalyst activity, catalyst morphology, particle size, and so on.[9] Suffice it to say that three types of rate curves have been identified: a constant rate type, a decaying rate type, and one that combines elements of the other two. All three, which are shown schematically in Figure 8.1, exhibit what is referred to as an induction period, that is, a time period leading up to maximum rate. Of the three, the decaying rate type is most common. Rate decay has been attributed to such factors as structural changes that reduce the number or

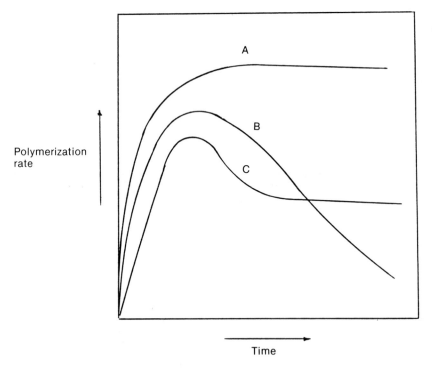

Figure 8.1. Types of rate curves observed in Ziegler–Natta polymerization: (A) constant; (B) decaying; and (C) decaying to constant.

activity of active centers, and encapsulation of active centers by polymer, which prevents approach by monomer.

Molecular weight distributions are generally broad when insoluble catalysts are used, and much narrower with soluble catalysts.[12] In the former case, the broad distribution may arise from the decay of catalyst activity or from the presence of sites of variable activity.

8.4 Stereochemistry of Ziegler–Natta polymerization

Without doubt, control of stereochemistry is the most fascinating and potentially useful aspect of Ziegler–Natta polymerization. Isotactic polymers are obtained with insoluble catalysts, with the level of stereoregularity probably depending, to a degree, on how exposed the active site is on the catalyst surface. Presumably, the more exposed the site, the less the stereoregularity. With 1-alkenes, isotactic polymer is formed regardless of the size of the alkyl substituent on the double bond; hence it appears that when monomer is complexed the smaller CH_2 group is inserted into the catalyst surface with the bulkier end pointing out. When the polymer chain migrates to the carbon bearing the substituent (Schemes 8.1 and 8.2), it thus always approaches from the same side, giving rise to isotactic placement. It is also apparent that the double bond of the monomer undergoes *cis* opening exclusively, that is, *cis* addition to the double bond occurs. This was shown by polymerization of *cis*- and *trans*-1-deuteropropene to give *erythro*-diisotactic and *threo*-diisotactic poly(1-deuteropropylene), respectively (8.16 and 8.17). Similar behavior, has been observed with other 1,2-disubstituted olefins. *Trans-i*-butyl propenyl ether, for example, gives *threo*-diisotactic polymer with Ziegler–Natta catalysts. *Cis* addition is also evident in the previously described reaction (8.9) involving the methyltitanium complex **6** and the alkyne **3**. Polymerization of acetylene using an $Al(C_2H_5)_3$–$Ti(OC_4H_9\text{-}n)_4$ catalyst yields *cis* polymer at temperatures below $-78°C.$[13] At 150°C, however, *trans*-polyacetylene is formed, probably by thermal isomerization of the initially formed *cis* polymer. Other Ziegler–Natta catalysts may be used, but this system is the one of choice for achieving good electrical conductivity.[13]

$$(8.16)$$

erythro-Diisotactic

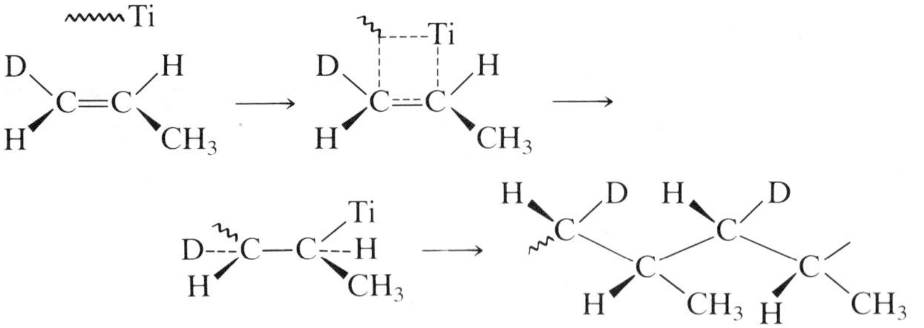

$$\text{(8.17)}$$

threo-Diisotactic

Syndiotactic polypropylene has been prepared under both heterogeneous and homogeneous conditions, but only under homogeneous conditions has a high degree of syndiotactic placement been achieved, and then only at temperatures below about $-45°C$,[3,14] Soluble catalysts based on vanadium are particularly effective. Presumably syndiotactic placement occurs for essentially the same reasons that it occurs in low-temperature free radical polymerization; that is, the approaching monomer and the propagating chain end assume the preferred conformation that minimizes steric or electronic interactions. This would be the case if steric effects at the active catalyst site prevented rotation of the bound polymer chain. Alternatively, one might invoke a model similar to that shown in Scheme 8.1; in the case of soluble catalysts, however, migration of the active site back to the surface is precluded, and a second monomer molecule can approach the vacant site with its substituent group directed away from the substituent of the monomer already inserted. This is illustrated in Scheme 8.3 for a vanadium complex. At

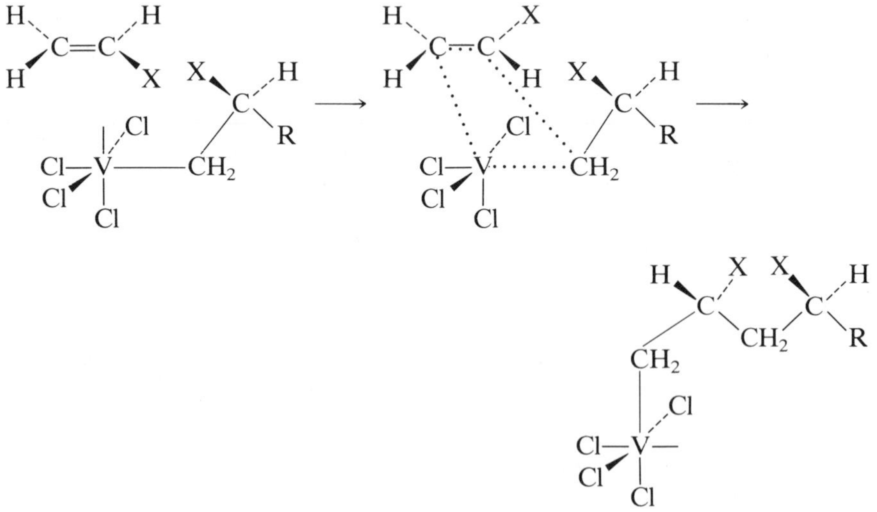

Scheme 8.3. Syndiotactic placement at the active site of a soluble vanadium complex.

higher temperatures, rotational barriers are overcome, and less syndiotactic placement occurs.

8.5 Polymerization of dienes with Ziegler–Natta catalysts

The versatility of Ziegler–Natta catalysts is clearly demonstrated in the polymerization of conjugated dienes where several stereochemical arrangements are possible. For 1,3-butadiene, the four possible structures—*cis*-1,4; *trans*-1,4; isotactic 1,2; and syndiotactic 1,2—can all be obtained in relatively pure form with the appropriate choice of catalyst. Some of these catalyst systems, gathered from different literature sources,[2,15–20] are shown in Table 8.5. Not all catalysts, of course, give such a high degree of stereoregularity; most of the vast number that have been studied give mixtures of stereoisomers. The ones in Table 8.5 were chosen from those reported to give greater than 90% stereospecificity. Included in the table are some catalysts, such as the nickel pi complexes, that are not strictly Ziegler–Natta catalysts but that react in analogous fashion by complexation with monomer.

Table 8.6 shows the effect of different catalysts on the polymerization of isoprene.[17,20–22] With this monomer it is possible, in principle, to have *cis*- and *trans*-1,4, 1,2, and 3,4 polymerization. No 1,2 polymer has yet been

Table 8.5. Catalysts for the stereospecific polymerization of butadiene

Catalyst[a]	Yield (%)	Polymer structure[b]	Ref. no.
$R_3Al + VCl_4$	97–98	*trans*-1,4	2
$R_3Al + VCl_3$	99	*trans*-1,4	2
$R_3Al + VOCl_3$	97–98	*trans*-1,4	2
Bis(π-crotyl NiI)	94	*trans*-1,4	15
(π-Cyclooctadiene)$_2$Ni, HI	~100	*trans*-1,4	16
$R_3Al + TiI_4$	93–94	*cis*-1,4	2
$R_3Al + CoCl_2 \cdot$ pyridine	90–97	*cis*-1,4	17
$R_2AlCl + CoCl_2$	96–97	*cis*-1,4	2
Bis(π-crotyl NiCl)	92	*cis*-1,4	15
(π-Cyclooctadiene)$_2$Ni, CF_3CO_2H (CF_3CO_2H : Ni = 1:1)	~100	*cis*-1,4	18
$R_3Al + Ti(OC_6H_9)_4$	90–100	1,2	19
$Et_3Al + V(acac)_3$	~90	st-1,2	2
$Et_3Al + Cr(C_6H_5CN)_6$			
Al/Cr = 2	~100	st-1,2	20
Al/Cr = 10	~100	it-1,2	20

[a] Et = ethyl, (acac) = acetylacetonate.
[b] st = syndiotactic, it = isotactic.

Table 8.6. Catalysts for the stereospecific polymerization of isoprene

Catalyst[a]	Yield (%)	Polymer structure	Ref. no.
$R_3Al + \alpha$-$TiCl_3$	91	trans-1,4	21
$Et_3Al + VCl_3$	99	trans-1,4	20
$Et_3Al + TiCl_4$			
Al/Ti < 1	95	trans-1,4	20
Al/Ti > 1	96	cis-1,4	20
$R_3Al + TiCl_4$ + amine	~100	cis-1,4	22
$R_2AlCl + CoCl_2 \cdot$ pyridine	95.5	cis-1,4	17
$R_3Al + V(acac)_3$	90	3,4	21
$Et_3Al + Ti(OR)_4$	95	3,4	20

[a] Et = ethyl, (acac) = acetylacetonate.

observed, however, which is not surprising, since the 1,2 bond is more hindered. Butadienes substituted on terminal carbons can be polymerized to give polymers having *cis* or *trans* double bonds and chiral carbons in the same structure. *Cis*- or *trans*-1,3-pentadiene, for example, yields isotactic *trans*-1,4 polymer in the presence of AlR_3–$TiCl_3$ or VCl_3 catalyst and mainly isotactic *cis*-1,4 with AlR_3–$Ti(OR)_4$.[20]

It is not possible at the present time to give a definitive mechanism for why any particular stereochemical arrangement is preferred. In addition to the catalyst structure, transition metal valency, crystalline form, and so on, as well as the conformation (*s-cis* or *s-trans*) of monomer must also play a part. One theory of whether 1,4 or 1,2 (or 3,4) polymerization occurs is based on whether the catalyst coordinates one or both double bonds of the diene. Coordination of one would thus lead to 1,2 polymerization and coordination of both to 1,4. Another theory is that coordination of a π-allylic structure occurs and that the direction of approach of monomer determines structure. Equations (8.18) and (8.19) illustrate this argument using butadiene as an example. If monomer approaches the CH_2–metal (M) bond of the complex, 1,4 polymer forms (8.18). If it approaches the CH–metal bond (8.19), 1,2 polymerization results. This is, of course, a simplified picture that provides no information on the geometric arrangement of the double bond or the tacticity at a chiral carbon.

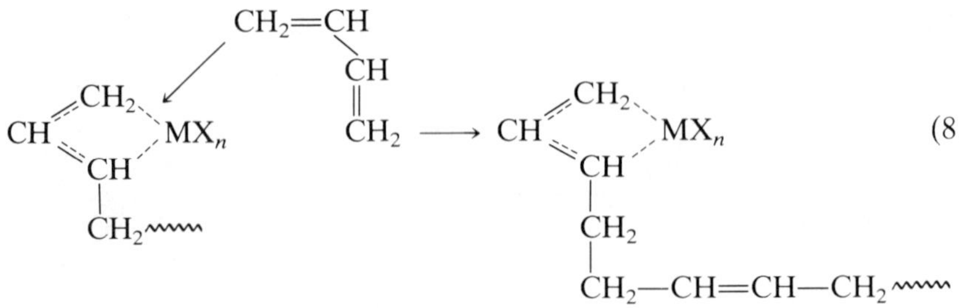

$$(8.18)$$

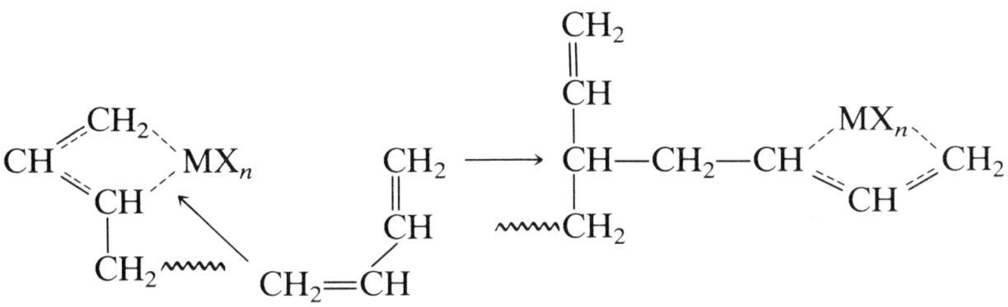

$$\text{(8.19)}$$

Conjugated cyclic dienes such as 1,3-cyclohexadiene (**8**) also undergo Ziegler–Natta polymerization (8.20). Also, as will be seen in the next section, polydienes can be synthesized by ring-opening reactions of cyclobutene and other cycloalkenes.

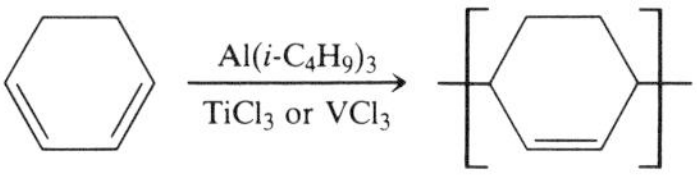

$$\text{(8.20)}$$

8

Nonconjugated dienes can be polymerized with coordination catalyst[5d,23] to give cyclic repeating units analogous to those already described in free radical and ionic cyclopolymerizations. Typically, 1,5-hexadiene (**9**) is polymerized with $TiCl_4$–$Al(C_2H_5)_3$ to polymer having repeating cyclopentane rings bridged with methylene groups *cis* to each other at positions 1 and 3

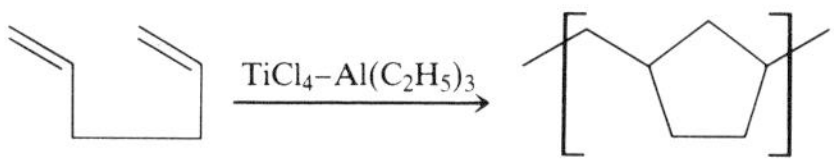

$$\text{(8.21)}$$

9

(8.21). Soluble catalysts, however, tend to give ordinary 1,2 polymerization.

8.6 Metathesis polymerization

Polymerization of cyclobutene[5d] (**10**) in the presence of $TiCl_4$–$Al(C_2H_5)_3$ gives approximately 95% of a mixture of *cis*- and *trans*-1,4-polybutadiene (8.22). Molybdenum and tungsten catalysts also favor ring-opening. Vanadium and chromium catalysts, on the other hand, result in formation of polycyclobutene (8.23) of which two stereochemical forms have been

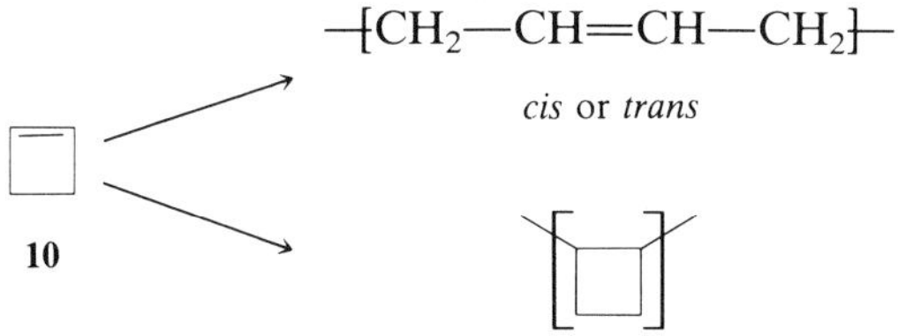

$$-\!\{CH_2\!-\!CH\!=\!CH\!-\!CH_2\}\!- \qquad \text{(8.22)}$$

cis or *trans*

10

$$\text{(8.23)}$$

erythro-Diisotactic or
erythro-disyndiotactic

observed, depending on the catalyst used. Heterogeneous catalysts, such as VCl_4–$Al(C_6H_{13})_3$, give *erythro*-diisotactic polycyclobutene, whereas soluble catalysts such as $V(acetylacetonate)_3$–$Al(C_2H_5)_2Cl$, give the *erythro*-disyndiotactic arrangement. (See Chapter 3, Figure 3.14, for these polymer structures.) Polymerization of 3-methylcyclobutene (11) results in ring-opening to give a polymer identical to 1,4-poly(1,3-pentadiene) (8.24).

$$\text{(11, structure)} \longrightarrow \left[\text{CH}-\text{CH}=\text{CH}-\text{CH}_2 \right] \quad (8.24)$$

11

Ring-opening reactions have been studied extensively in recent years. Some have led to commercial development (Table 8.1). Ring-opening reactions initiated with organometallic catalysts are of a general redistribution type referred to as *olefin metathesis*.[24-27] * (The acronym ROMP is often used to designate ring-opening metathesis polymerization.) Particularly effective catalysts for this reaction are prepared from the reaction product of tungsten hexachloride with alcohol together with ethylaluminum dichloride. Other transition metals, notably titanium, molybdenum, and rhenium, have also been used. Redistribution, shown in reaction (8.25), involves cleavage of double bonds (transalkylidenation). The alternate possibility, scission of bonds adjacent to double bonds (transalkylation), has been shown not to occur by experiments with deuterated olefins. An example of the metathesis reaction for simple alkenes is the formation of a 1:2:1 mixture of 2-butene, 2-pentene, and 3-hexene from 2-pentene.

$$\begin{array}{ccc} R_1CH{=}CHR_2 & R_1CH & CHR_2 \\ & \rightleftharpoons \; \| & \| \\ R_1CH{=}CHR_2 & R_1CH & CHR_2 \end{array} \quad (8.25)$$

There is ample evidence to suggest that this remarkable bond redistribution is brought about by an intermediate *metal carbene* species (12) that reacts with alkene to form a transient metallacyclobutane (13) leading subsequently to products (8.26).[25,28] The metal (Mt) is bracketed in reaction (8.26) to indicate that other ligands are present. Although it is not clear at the present time how the metal carbene is formed from monomer and catalyst, it is known that stable metal carbenes such as $(C_6H_5)_2C{=}W(CO)_5$ initiate metathesis polymerization[29]; in some instances metal carbenes have been identified in polymerization reactions by NMR.[30,31]

$$\begin{array}{ccc} \text{C} + \text{C} & \text{C}-\text{C} & \text{C}=\text{C} \\ \| \quad \| \longrightarrow & | \quad | \longrightarrow & \\ [\text{Mt}] \quad \text{C} & [\text{Mt}]-\text{C} & [\text{Mt}]{=}\text{C} \end{array} \quad (8.26)$$

12   13

* From the Greek *meta* (change) and *tithemi* (place).

One of the propagation steps in metathesis polymerization of a cycloalkene is shown in reaction (8.27). High molecular weights (in excess of 10^5) are usually obtained, and polymerization is accompanied by formation of low-molecular-weight oligomers. Rate of polymerization appears to be a function of the metal carbene reactivity rather than of ring strain in the cycloalkene.[32] Even relatively unstrained monomers such as cyclohexene will undergo ring-opening polymerization under certain conditions.[33]

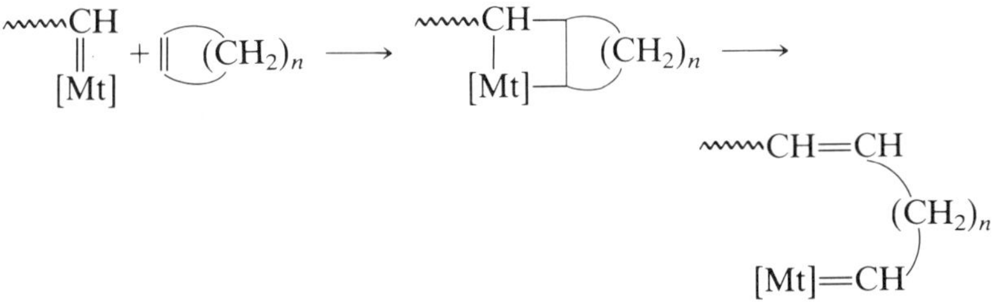

$$ (8.27) $$

One of the most unexpected findings in metathesis polymerization is that certain group VIII metal compounds, for example $RuCl_3$, catalyze the ring-opening polymerization of monomers such as 7-oxanorbornene derivatives (**14**) to molecular weights in excess of 10^6 *in aqueous solution* (8.28).[34]

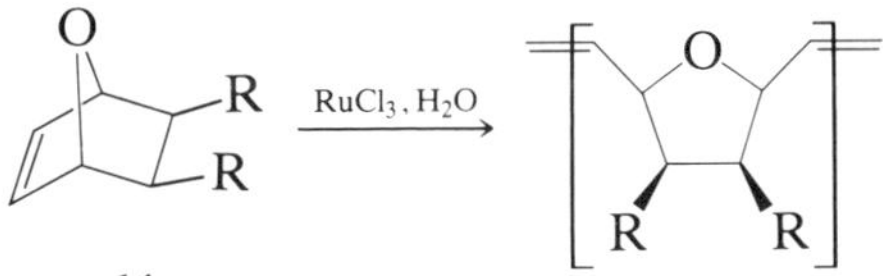

$$ (8.28) $$

14

This is contrary to the usual polymerization conditions, in which moisture must be rigorously excluded. Furthermore, the catalyst, which appears to owe its activity to the formation of a Ru^{2+}–monomer complex, can be recycled, and the activity actually increases with successive use. Considerably more work in this area may be anticipated in view of the advantages inherent in the use of aqueous polymerization conditions.

Examples of three metathesis polymerizations that have been developed commercially are shown in reactions (8.29) through (8.31). In all cases, the double-bond configuration is primarily *trans*. Synthesis of polypentenamer (8.29) was developed in the United States in the early 1970s, ostensibly for the tire market, but problems have delayed large-scale production. Polyoctenamer (8.30) and norbornene polymer (8.31) are manufactured in Europe.

$$ -[CH_2=CH(CH_2)_3]- \qquad (8.29) $$

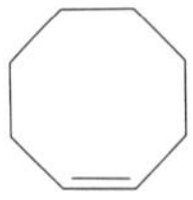

$$ -[CH_2=CH(CH_2)_6]- \qquad (8.30) $$

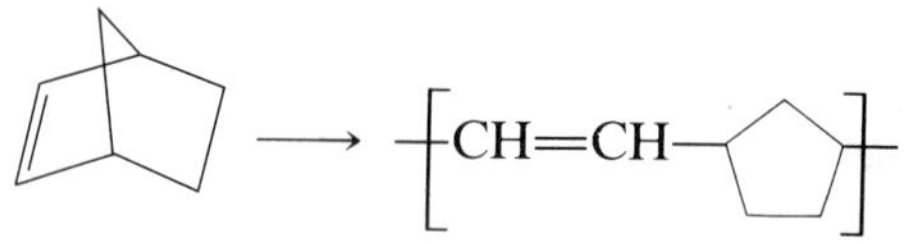

$$(8.31)$$

Another commercial product, polydicyclopentadiene, is prepared by metathesis polymerization of the strained tricyclic monomer, as shown in Scheme 8.4. Two steps are involved: initial opening of the more strained bicyclic system, followed by opening of the less strained cyclopentene ring with subsequent crosslinking. Because of the monomer's high reactivity, the process is adaptable to reaction injection molding (see Chapter 4) with monomer plus catalyst constituting one component of the RIM system and monomer plus activator constituting the other.[35]

Metathesis polymerization works equally as well with cyclic polyenes. Polymerization of 1-methyl-1,5-cyclooctadiene (**15**) (8.32) yields an elastomer having the structure of a regularly alternating copolymer of isoprene and

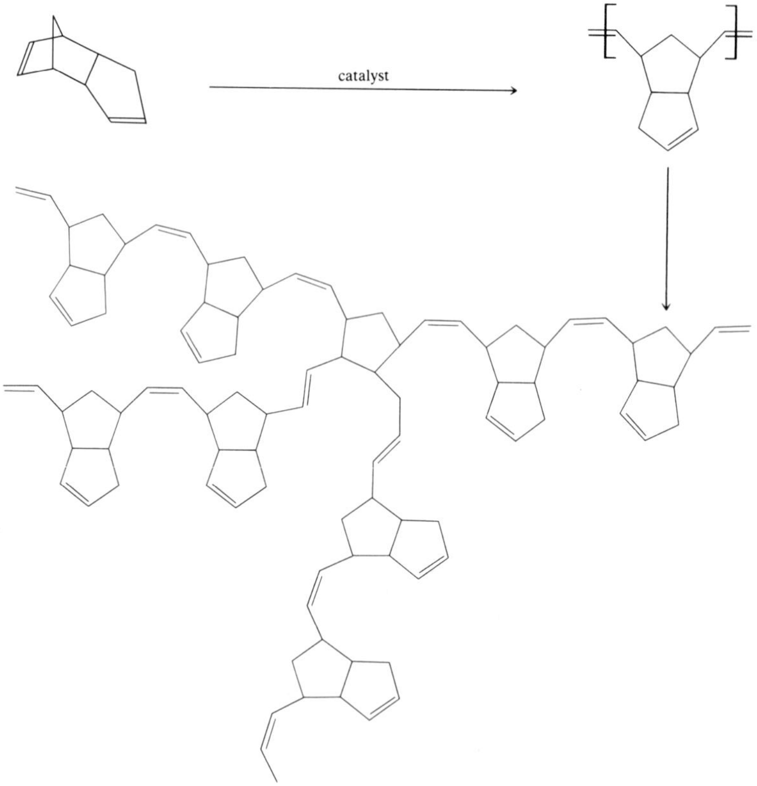

Scheme 8.4. Metathesis polymerization of dicyclopentadiene. [Adapted from R. P. Geer. *Proc. 1983 Natl. Tech. Conf., SPE*, Detroit.]

butadiene. Similarly, *cis,trans*-cyclodeca-1,5-diene (**16**) forms a polymer having the structure of a regular 2:1 polymer of butadiene and ethylene (8.33). The synthesis of polyacetylene (primarily *trans*) based on metathesis polymerization of 1,3,5,7-cyclooctatetraene (**17**) (8.33) has also been reported.[36,37]

$$
\underset{\textbf{15}}{\text{(structure)}} \xrightarrow[\text{C}_2\text{H}_5\text{OH}]{\text{WCl}_6,\text{C}_2\text{H}_5\text{AlCl}_2} \left[-\text{CH}_2\overset{\overset{\text{CH}_3}{|}}{\text{C}}=\text{CHCH}_2\text{CH}_2\text{CH}=\text{CHCH}_2- \right] \quad (8.32)
$$

$$
\underset{\textbf{16}}{\text{(structure)}} \xrightarrow[\text{C}_2\text{H}_5\ \text{OH}]{\text{WCl}_6,\text{C}_2\text{H}_5\text{AlCl}_2} -(\text{CH}_2\text{CH}_2)(\text{CH}_2\text{CH}=\text{CHCH}_2)_2- \quad (8.33)
$$

$$
\underset{\textbf{17}}{\text{(structure)}} \xrightarrow[\text{Al}(\text{C}_2\text{H}_5)_2\text{Cl}]{\text{W}[\text{OCH}(\text{CH}_2\text{Cl})_2]_{2\ \text{or}\ 3}} -(\text{CH}=\text{CH})_4- \quad (8.34)
$$

Less commonly, metathesis polymerization has been applied to open-chain unsaturated monomers. Acetylene, for example, forms polyacetylene under metathesis conditions[38]; isolated dienes such as 1,5-hexadiene (**18**) undergo step-growth polycondensation (8.35) with loss of ethylene.[39]

$$
\underset{\textbf{18}}{\text{(structure)}} \xrightarrow[\text{WCl}_6]{\text{C}_2\text{H}_5\text{AlCl}_2} -(\text{CH}=\text{CHCH}_2\text{CH}_2)- + \text{CH}_2=\text{CH}_2 \quad (8.35)
$$

8.7   Ziegler–Natta copolymerization

Random copolymers of ethylene and 1-alkenes (α-olefins) are obtainable with Ziegler–Natta catalysts, the most important being those of ethylene and 1-butene (LLDPE) and ethylene with propylene (EPM or EPR and EPDM).[5e,40a,b] From the representative reactivity ratios presented in Table 8.7, it can be seen that ethylene is much more reactive than higher alkenes, and the ratios vary with the nature and physical state of the catalyst. In most instances, $r_1 r_2$ is close to unity. With heterogeneous catalysts, a wide range of compositions is generally obtained, possibly because different active sites may give rise to different reactivity ratios, or because of the previously mentioned encapsulation of active sites leading to decay of activity.

Table 8.7. Representative reactivity ratios (*r*) in Ziegler–Natta polymerization[a]

Monomer 1	Monomer 2	Catalyst[b]	Reaction type[c]	r_1	r_2
Ethylene	Propylene	$TiCl_3/AlR_3$	H	15.72	0.110
		$TiCl_4/AlR_3$	C	33.36	0.032
		VCl_3/AlR_3	H	5.61	0.145
		VCl_4/AlR_3	C	7.08	0.088
Ethylene	1-Butene	VCl_3/AlR_3	H	26.96	0.043
		VCl_4/AlR_3	C	29.60	0.019
Propylene	1-Butene	VCl_3/AlR_3	H	4.04	0.252
		VCl_4/AlR_3	C	4.32	0.227

[a] Data from Boor.[3]
[b] $R = C_6H_{13}$.
[c] H = heterogeneous; C = colloidal.

A more homogeneous polymer composition is obtained with soluble catalysts, particularly if monomer composition is carefully controlled to remain relatively constant during polymerization. Such is the case with production of EPM (EPR), which contains about 60 parts ethylene to 40 parts propylene. Crosslinking of such elastomers is accomplished with peroxides.

The other commercially important type of ethylene–propylene copolymer (EPDM) is prepared with small amounts of a nonconjugated diene to facilitate crosslinking.[41] Typical dienes are ethylidenenorbornene (**19**), dicyclopentadiene (**20**), and 1,4 hexadiene (**21**), with **19** being the diene of choice. Both types of elastomers have excellent ozone resistance by virtue of having no unsaturation as an integral part of the backbone. A variety of other copolymer of ethylene and 1-alkenes have been investigated.[40c]

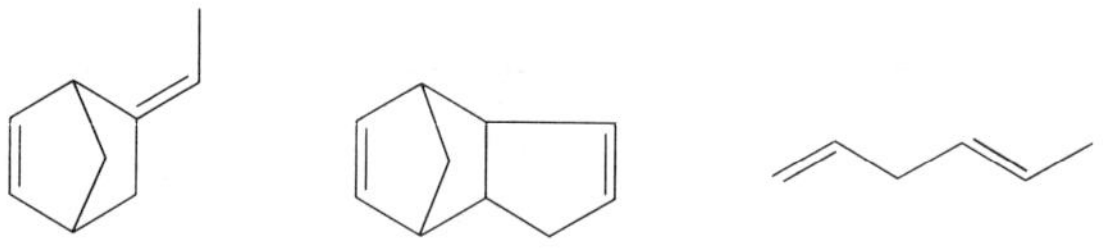

19   **20**   **21**

As was the case with free radical initiation, 1,2-disubstituted monomers display little tendency to homopolymerize under Ziegler–Natta conditions. They will copolymerize with ethylene or 1-alkenes[39a]; however, it is usually necessary to use a large excess of the 1,2 monomer because of its lower reactivity. It is of interest that there is some tendency toward alternation, with *erythro*-diisotactic stereochemistry (see Chapter 3, Figure 3.13), when sufficient 1,2-disubstituted monomer is used.

A number of block copolymers prepared with Ziegler–Natta catalysts have been reported; in most cases, however, there is serious question as to whether the compositions also include significant amounts of homopolymer.[3] The problem arises with uncertainties over the lifetimes of propagating chains, which may become detached from catalyst by chain transfer, thereby leaving the catalyst still active to initiate homopolymerization of a second monomer. At the present time, therefore, the Ziegler–Natta method is inferior to anionic polymerization for synthesizing carefully tailored block copolymers. Nevertheless, block copolymers of ethylene and propylene (Eastman Kodak's *Polyallomers*) have been commercialized. Unlike the elastomeric random copolymers of ethylene and propylene, these are high-impact plastics exhibiting crystallinity characteristics of both isotactic polypropylene and linear polyethylene, and, no doubt, they contain homopolymers in addition to block copolymers. (Ethylene–propylene block copolymers, designated TPR, have also been marketed by Uniroyal.)

8.8 Supported metal oxide catalysts

Oxides of a variety of metals on finely divided inert support materials initiate polymerization of ethylene and other vinyl monomers by a mechanism that is assumed to be similar to that of heterogeneous Ziegler–Natta polymerization; that is, initiation probably occurs at active sites on the catalyst surface.[3,42] Among the metals that have been investigated are chromium, vanadium, molybdenum, nickel, cobalt, niobium, tantalum, tungsten, and titanium. Typical supports include alumina, silica, and charcoal. Chromium-based catalysts originally developed by the Phillips Petroleum Company are often referred to as *Phillips catalyst*. Other metal oxide catalysts were developed primarily at Standard Oil of Indiana.

Catalysts are prepared by one of two methods. The support material is impregnated with the metal ion, then heated in air at a high temperature to form the metal oxide. Alternatively, when the support material is an oxide such as alumina, the two oxides are coprecipitated and dried in air. In each case the the catalyst is activated by treatment with a reducing agent such as hydrogen, metal hydride, or carbon monoxide. Poisoning of the catalyst occurs in the presence of such materials as water, oxygen, or acetylene. The function of the support appears to be more than simply providing a large surface area. Some type of interaction must occur between oxide and support because the oxide alone behaves differently.

Ethylene is the most important monomer used with supported metal oxide catalysts. Much of the high-density polyethylene is, in fact, now manufactured this way. Unlike Ziegler–Natta catalysts, which give rise to polymer having primarily saturated end groups, the metal oxides yield polyethylene with approximately equal amounts of saturated and unsaturated chain ends.

Initiation is presumed to involve initial formation of a pi complex which disproportionates to yield both metal–CH=CH$_2$ and metal–CH$_2$CH$_3$ coordination (8.36). Propagation then proceeds via an insertion mechanism.

$$[Mt] \xrightarrow{\text{C}_2\text{H}_4} [Mt] \overset{\text{CH}_2}{\underset{\text{CH}_2}{+\!\!\!\mid}} \longrightarrow [Mt]—CH=CH_2 + [Mt]—CH_2CH_3 \quad (8.36)$$

The supported metal oxides are not as active as Ziegler–Natta catalysts, and they do not give rise to a high degree of stereoregularity. Propylene forms partially crystalline polymer, but higher 1-alkenes give amorphous product. Styrene is unreactive. Dienes polymerize to yield primarily *trans*-1,4 polymer. Copolymers of ethylene and 1-alkenes have been prepared with chromia and molybdena catalysts.

8.9   Alfin catalysts[6b]

Alfin catalysts are a class of heterogeneous catalysts consisting of an alkenylsodium compound, an alkoxide, and an alkali metal halide. The name is derived from *al*cohol and ole*fin*, which are used to prepare the catalyst. Alfin catalysts are particularly effective in polymerizing butadiene and isoprene to very-high-molecular-weight polymer, although they will polymerize certain other monomers, notably styrene.

The most effective catalyst for diene polymerization consists of allylsodium, sodium isopropoxide, and sodium chloride, a combination typically formed by a reaction sequence such as that shown in reactions (8.37), (8.38), and (8.39).

$$1.5\ C_5H_{11}Cl + 3Na \rightarrow 1.5\ C_5H_{11}Na + 1.5\ NaCl \qquad (8.37)$$

$$1.5\ C_5H_{11}Na + (CH_3)_2CHOH \rightarrow$$
$$0.5\ C_5H_{11}Na + (CH_3)_2CHONa + C_5H_{12} \quad (8.38)$$

$$0.5\ C_5H_{11}Na + 0.5\ CH_2\!\!=\!\!CHCH_3 \rightarrow 0.5\ CH_2\!\!=\!\!CHCH_2Na + 0.5\ C_5H_{12}$$
$$(8.39)$$

Alfin catalysts polymerize butadiene within minutes to polymer having a molecular weight of several million, consisting of about 68% *trans*-1,4 and the remainder divided equally between *cis*-1,4 and 1,2. The process resembles Ziegler–Natta polymerization in that one allyl group is incorporated into each polymer chain. How this comes about, however, is open to speculation; both free radical and anionic mechanisms have been proposed.

Butadiene polymerizes about 40 times more rapidly than isoprene with alfin catalyst, and 2,3-dimethylbutadiene is unreactive.

The major drawback to the alfin system is that molecular weights are too high to be practicable, hence they have had little commercial success.

References

1. K. Ziegler, E. Holzkamp, H. Breil, and H. Martin, *Angew. Chem.*, **67**, 541 (1955).
2. G. Natta, *J. Polym. Sci.* **48**, 219 (1960).
3. J. Boor Jr., *Ziegler-Natta Catalysts and Polymerizations*, Academic Press, New York, 1979.
4. G. Natta and F. Danusso (eds.), *Stereoregular Polymers and Stereospecific Polymerizations*, 2 vols., Pergamon Press, New York, 1967.
5. A. D. Ketley (ed.), *The Stereochemistry of Macromolecules*, Vol. 1, Dekker, New York, 1967: (a) D. O. Jordan, Chap. 1; (b) D. F. Hoeg, Chap. 2; (c) P. Cossee, Chap. 3; (d) W. Marconi, Chap. 5; (e) I. Pasquon, A. Valvassori, and G. Sartori, Chap. 4.
6. L. Reich and A. Schindler, *Polymerization by Organometallic Compounds*, Wiley-Interscience, New York, 1966: (a) Chap. 4; (b) Chap. 5.
7. F. Dawans and P. Teyssié, *Bull. Soc. Chim. Fr.*, 2376 (1963).
8. D. O'Sullivan, *Chem. Eng. News*, July 4, 1983, p. 29.
9. T. Keii, *Kinetics of Ziegler-Natta Polymerization*, Halsted Press, New York, 1973.
10. V. A. Zakharov, G. D. Bukatov, and Y. I. Yermakov, *Adv. Polym. Sci.*, **51**, 61 (1983).
11. J. J. Eisch, A. M. Piotrowski, S. K. Brownstein, E. J. Gabe, and F. L. Lee, *J. Am. Chem. Soc.*, **107**, 7219 (1985).
12. U. Zucchini and G. Cecchin, *Adv. Polym. Sci.*, **51**, 101 (1983).
13. T. Ito, H. Shirakawa, and S. Ikeda, *J. Polym. Sci., Polym. Chem. Ed.*, **12**, 11 (1974).
14. E. A. Youngman and J. Boor Jr., *Macromol. Rev.*, **2**, 33 (1967).
15. T. Matsumoto and J. Furukawa, *J. Polym. Sci. B*, **5**, 935 (1967).
16. J. P. Durand, F. Dawans, and P. Teyssié, *J. Polym. Sci. B.*, **5**, 785 (1967).
17. B. A. Dologoplosk, E. N. Kropacheva, E. K. Khrennikova, E. I. Kuznetzova, and K. G. Golodova, *Dokl. Akad. Nauk SSSR*, **135**, 847 (1960); *Chem. Abstr.*, **55**, 13292f (1960).
18. J. P. Durand and P. Teyssié, *J. Polym. Sci. B*, **6**, 229 (1968).
19. G. Wilke, *J. Polym. Sci.*, **38**, 45 (1959).
20. G. Natta, L. Porri, A. Carbonaro, and G. Stoppa, *J. Makromol. Chem.*, **77**, 114 (1964); G. Natta, L. Porri, and A. Carbonaro, ibid, p. 126.
21. C. E. H. Bawn, A. M. North, and J. S. Walker, *Polymer*, **5**, 419 (1964).
22. G. A. Razuraev, K. S. Minsker, G. T. Fedoseeva, and L. A. Savel'ev, *Vysoko-molekul. Soedin.*, **1**, 1691 (1959); *Chem. Abstr.*, **54**, 19015g (1960).
23. R. J. Cotter and M. Matzner, *Ring-Forming Polymerizations*, Part A, Academic Press, New York, 1969, pp. 42–47.
24. V. Dragutan, A. T. Balaban, and M. Dimonii, *Olefin Metathesis and Ring-Opening Polymerization of Cyclo-olefins*, Wiley-Interscience, New York, 1984.
25. K. J. Ivin, *Olefin Metathesis*, Academic Press, New York, 1983.
26. N. Calderon, E. A. Ofstead, and W. A. Judy, *Angew. Chem. Intern. Ed. English*, **15**, 401 (1976).
27. N. Calderon, *J. Macromol. Sci., Rev. Macromol. Chem.*, **C7**, 105 (1972).
28. B. A. Dolgoplosk, *J. Polym. Sci., Polym. Symp.*, **67**, 99 (1978).
29. T. J. Katz, S. J. Lee, and N. Acton, *Tetrahedron Lett.*, 4247 (1976).

30. J. Kress, M. Wesolek, and J. A. Osborn, *J. Chem. Soc., Chem. Commun.*, 514 (1982).
31. J. Kress, J. A. Osborn, R. M. E. Green, K. J. Ivin, and J. J. Rooney, *J. Am. Chem. Soc.*, **109**, 899 (1987).
32. P. A. Patton and T. J. McCarthy, *Macromolecules*, **17**, 2939 (1984).
33. P. A. Patton, C. P. Lillya, and T. J. McCarthy, *Macromolecules,* **19**, 1266 (1986).
34. B. M. Novak and R. H. Grubbs, *J. Am. Chem. Soc.*, **110**, 7542 (1988).
35. R. P. Geer and R. D. Stoutland, *Plastics 85*, Proceedings of the SPE 43rd Annual Technical Conference, 1233 (1985).
36. Y. V. Korshak, V. V. Korshak, G. Kanischka, and H. Hocker, *Makromol. Chem., Rapid. Commun.*, **6**, 685 (1985).
37. F. L. Klavetter and R. H. Grubbs, *Polym. Preprints*, **28** (2), 425 (1987).
38. A. J. Amass, M. S. Beevers, T. R. Farren, and J. A. Stowell, *Makromol. Chem., Rapid Commun.*, **8**, 119 (1987).
39. M. Lindmark-Hamburg and K. B. Wagener, *Polym. Preprints*, **28** (1), 234 (1987).
40. G. E. Ham (ed.), *Copolymerization*, Wiley-Interscience, New York, 1964: (a) C. A. Lukach and H. M. Spurlin, Chap. 4A; (b) G. Crespi, A. Valvassori, and G. Sartori, Chap. 4C; (c) idem, Chap. 4D.
41. S. Cesca, *Macromol. Rev.*, **10**, 1 (1975).
42. D. R. Witt, in *Reactivity, Mechanism and Structure in Polymer Chemistry* (A. D. Jenkins and A. Ledwith, eds.), Wiley-Interscience, New York, 1974, Chap. 13.

Review exercises

1. Set up a table outlining the likelihood of converting the following monomers to high-molecular-weight polymer under free radical, cationic, anionic, group transfer, or coordination conditions: methyl acrylate, 1-pentene, methacrylonitrile, ethylene, propylene, vinyl acetate, allyl propionate, isoprene, *i*-propyl vinyl ether, 1,1-dichloroethylene. (List the monomers by structure in the table.)

2. Explain (a) how sites of variable activity in a heterogeneous catalyst might result in a polymer of high polydispersity, (b) why molecular hydrogen is useful for controlling molecular weight in Ziegler–Natta polymerization, and (c) why only coordination catalysts are effective for polymerizing 1-alkenes.

3. Give the structure and stereochemical designation of all possible polycyclohexenes that could, in principle, be obtained by stereospecific polymerization (*Pure Appl. Chem.*, **12**, 645, 1966). Which are unlikely to be obtainable in practice?

4. It was mentioned in Chapter 6 that 1,3-pentadiene undergoes free radical inclusion polymerization to form isotactic *trans*-1,4 polymer. In this chapter it was noted that both *cis*- and *trans*-isotactic-1,4-poly(1,3-pentadiene) can be prepared with Ziegler–Natta catalysts. There are *nine* other possible stereoregular forms for poly(1,3-pentadiene). What are they?

5. Assuming only head-to-tail polymerization and *cis* double-bond opening occurs, give the structure and stereochemical designation of all polymers obtainable in principle by stereospecific polymerization of (a) 2-methyl-1-butene, (b) *trans*-2-pentene, (c) isoprene, and (d) propyne.

6. Draw the structure of (a) *erythro*-diisotactic-*trans*-1,4-poly(2,4-heptadiene) and (b) *erythro*-diisotactic-*cis*-1,4-poly(1,3-cyclohexadiene).

7. Ignoring stereochemistry, predict the repeating units obtained by conventional (not metathesis) Zeigler–Natta polymerization of (a) 2,6-dimethyl-1,6-heptadiene and (b) 1,5-cyclooctadiene.

8. Predict the structure of the linear polymer formed by Ziegler–Natta polymerization of 1,6-heptadiyne. What type of properties might you expect for such a polymer? (J. K. Stille and D. A. Frey, *J. Am. Chem. Soc.*, **83**, 1697, 1961)

9. Predict the structure of the polymers formed by head-to-tail metathesis polymerization of the following monomers (ref. 26, and T. M. Swager et al., *J. Am. Chem. Soc.*, **110**, 2973, 1988):

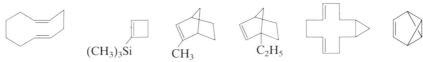

$(CH_3)_3Si$ CH_3 C_2H_5

10. When 1,3,5,7-cyclooctatetraene undergoes ring-opening metathesis polymerization (8.34) in dilute solution, benzene is formed irreversibly in 75% yield. Benzene formation is insignificant in bulk metathesis polymerization. Considering the mechanism of metathesis polymerization, suggest a reason for the different behavior. (F. L. Klavetter and R. H. Grubbs, *J. Am. Chem. Soc.*, **110**, 7807, 1988)

9
Reactions of vinyl polymers

9.1 Introduction

The subject of polymer reactions is so broad and the literature so extensive that it is not possible to present more than a survey of the field in a book of this type. Chemical modifications encompass a wide range of applications: ion-exchange resins; polymeric reagents and polymer-bound catalysts; polymeric supports for chemical reactions; degradable polymers to address medical, agricultural, or environmental concerns; flame-retardant polymers; and surface treatments to improve such properties as biocompatibility or adhesion, to name a few. Books and symposia proceedings[1–5] covering reactions of polymers treat individual topics in depth. The purpose of this chapter is to summarize and illustrate chemical modifications of vinyl polymers. For convenience these are grouped into five general categories: (1) reactions that involve the introduction or modification of functional groups, (2) reactions that introduce cyclic units into the polymer backbone, (3) reactions leading to block and graft copolymers, (4) crosslinking reactions, and (5) degradation reactions. Only with polymers in the first category is there no overall change in the gross structure of the polymer.

Before we discuss each type, it should be pointed out that while polymers will, in principle, undergo all the reactions common to low-molecular-weight compounds, contrasting behavior may arise from the polymer's molecular size, shape, or morphology. If a polymer is semicrystalline, for example, the crystalline regions, because of their impermeability, are usually inaccessible to chemical reactants, and reaction may be limited to the amorphous regions. Even in a completely amorphous polymer, not all the functional groups of interest are necessarily equally accessible. Conformational or localized steric effects might well influence the rate or extent of the reaction in different parts of the same molecule. At the same time, the proximity of functional groups might enhance a reaction by the *neighboring group effect*. Copolymers of acrylic acid and *p*-nitrophenyl methacrylate, for example, undergo base-catalyzed hydrolysis faster than simple *p*-nitrophenyl esters because of participation by neighboring carboxylate anions (9.1).[6] That neighboring group and conformational effects in polymers are important in influencing chemical reactions is demonstrated in convincing fashion in the area of enzyme chemistry.

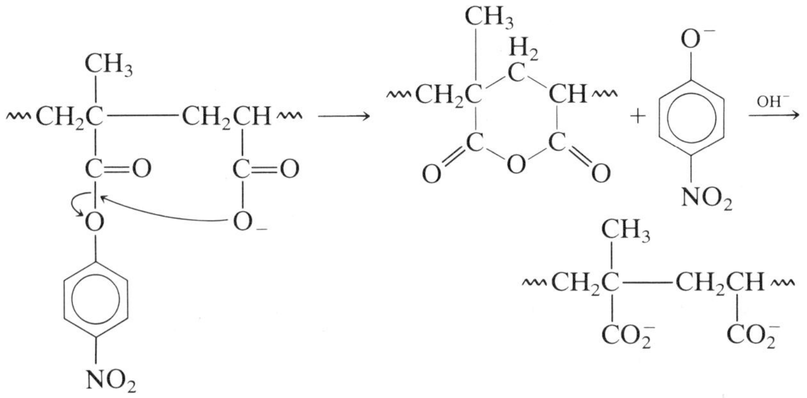

$$\quad (9.1)$$

Another complicating factor in polymer reactions is the possibility that the reaction itself might change the physical form of the polymer. Even at low levels of conversion a reaction could alter the polymer's conformation, thereby influencing the reaction rate at unreacted sites; or it might even cause the polymer to precipitate, which would effectively preclude further reaction in the precipitated regions. The possibility that such variables might play a role should be taken into account in effecting any polymer modification.

9.2   Functional group reactions

9.2.1   Introduction of new functional groups

Among the oldest commercial processes are the chlorination (9.2) and chlorosulfonation (9.3) of polyethylene.[7,8] The properties of polyethylene are

$$+CH_2CH_2+ \xrightarrow[(-HCl)]{Cl_2} \left[\begin{matrix} CHCH_2 \\ | \\ Cl \end{matrix} \right] \quad (9.2)$$

$$+CH_2CH_2+ \xrightarrow[(-HCl)]{Cl_2,SO_2} \left[\begin{matrix} CH_2CH \\ | \\ SO_2Cl \end{matrix} \right] \quad (9.3)$$

substantially altered by chlorination. Flammability is decreased. Solubility is increased or decreased depending on the level of substitution. If the chlorination is run under heterogeneous conditions (polyethylene suspended in an inert medium), the resultant polymer is more crystalline at comparable levels of chlorination than a similar polymer prepared using polyethylene solution. This is not unexpected, because the homogeneous process would distribute the chlorine more randomly. Chlorosulfonation provides sites for subsequent crosslinking reactions. Chlorination of poly(vinyl chloride) is also used commercially to increase the glass transition temperature. Free radical substitution reactions of these types cause crosslinking by radical combination reactions, hence the reaction conditions must be carefully controlled. Both chlorinated and chlorosulfonated polyethylene are, however, available as commercial products of long standing,[9] the latter under the well-known Hypalon (du Pont) trademark.

Fluorination may be accomplished by direct reaction between solid polymer and fluorine.[10] Special techniques such as use of mixtures of nitrogen and fluorine are needed to avoid degradation reactions, but polyethylene can be converted almost quantitatively to the fully fluorinated polymer by this technique. Where double bonds or benzene rings are present in the polymer, both substitution and addition occur. Polystyrene, for example, yields the fully fluorinated saturated polymer on reaction with fluorine (9.4). Where the reaction is particularly useful is in surface fluorination to enhance chemical inertness and—in the case of plastic bottles—to improve solvent barrier properties.[11]

$$
\left[\begin{array}{c} CH_2CH \\ | \\ C_6H_5 \end{array}\right] \xrightarrow[(-HF)]{F_2} \left[\begin{array}{c} CF_2CF \\ | \\ C_6F_{11} \end{array}\right] \tag{9.4}
$$

Aromatic substitution reactions (nitration, sulfonation, chlorosulfonation, etc.) occur readily on polystyrene[12,13] and are useful for manufacturing ion-exchange resins or for introducing sites for crosslinking or grafting. Chloromethylation (9.5) of polystyrene is particularly important for introducing new functionalities because the resultant benzylic chloride undergoes nucleophilic displacement so readily. Vinylbenzyl chloride (a mixture of *meta* and *para*) and its polymers are, in fact, available commercially for just such a purpose. The functionalized polymers, crosslinked by including small amounts of divinylbenzene during polymerization, are used for *solid phase synthesis*, in which the insoluble polymer is used as a "handle" for maintaining a catalyst or reaction product in a solid phase for easy separation from the reaction medium.[14–19] A very important example of this technique for synthesizing proteins is described in Chapter 17.

$$
\left[\begin{array}{c} CH_2CH \\ | \\ \bigcirc \end{array}\right] \xrightarrow[AlCl_3]{CH_3OCH_2Cl} \left[\begin{array}{c} CH_2CH \\ | \\ \bigcirc \\ CH_2Cl \end{array}\right] + CH_3OH \tag{9.5}
$$

Another example that illustrates the value of adding new functionalities is the introduction of ketone groups via the intermediate oxime (9.6) to render polyethylene photodegradable,[20] a process discussed later in the chapter.

$$
-[CH_2CH_2]- \xrightarrow{NOCl} \left[\begin{array}{c} NOH \\ || \\ CCH_2 \end{array}\right] \xrightarrow[H^+]{H_2O} \left[\begin{array}{c} O \\ || \\ CCH_2 \end{array}\right] \tag{9.6}
$$

9.2.2 Conversion of functional groups

Conversion of one functional group to another is used primarily to obtain polymers difficult or impossible to prepare by direct polymerization. An example is the commercially important synthesis of poly(vinyl alcohol)[21,22] by hydrolysis or alcoholysis of poly(vinyl acetate) (9.7). Poly(vinyl alcohol) cannot be made directly because vinyl alcohol is the unstable enol form of acetaldehyde (see exercise 11, Chapter 1). Isotactic poly(vinyl alcohol) can be prepared by acid cleavage of isotactic poly(vinyl *t*-butyl ether) (9.8) (the latter made by cationic polymerization of monomer at $-78°C$).[23]

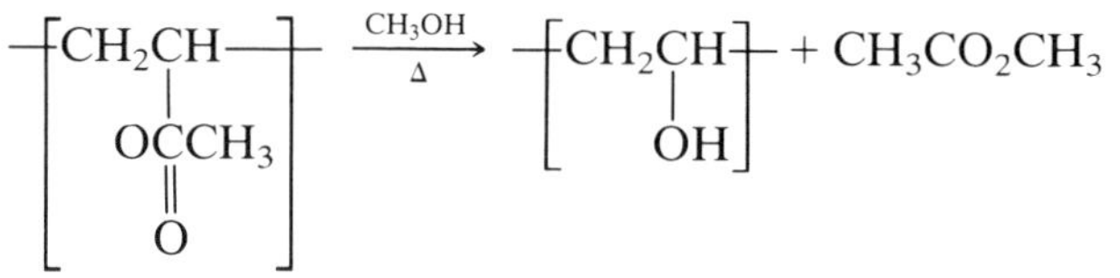

$$\left[\begin{array}{c} CH_2CH \\ | \\ OCCH_3 \\ \| \\ O \end{array}\right] \xrightarrow[\Delta]{CH_3OH} \left[\begin{array}{c} CH_2CH \\ | \\ OH \end{array}\right] + CH_3CO_2CH_3 \qquad (9.7)$$

$$\left[\begin{array}{c} CH_2CH \\ | \\ OC(CH_3)_3 \end{array}\right] \xrightarrow{HBr} \left[\begin{array}{c} CH_2CH \\ | \\ OH \end{array}\right] + BrC(CH_3)_3 \qquad (9.8)$$

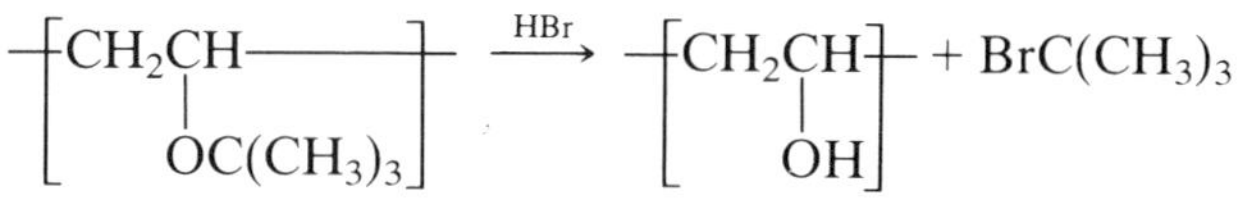

Examples of syntheses of other polymers difficult or impossible to obtain directly are the following:

1. Saponification of isotactic or syndiotactic poly(trimethylsilyl methacrylate) (both prepared under anionic conditions) to yield isotactic or syndiotactic poly(methacrylic acid), respectively (9.9).[24]

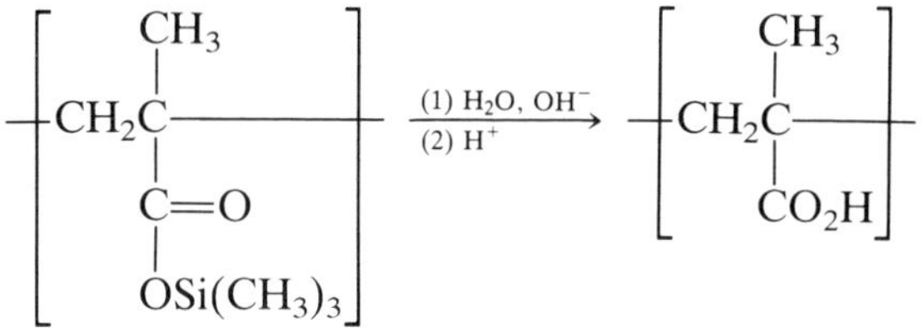

$$\left[\begin{array}{c} CH_3 \\ | \\ CH_2C \\ | \\ C=O \\ | \\ OSi(CH_3)_3 \end{array}\right] \xrightarrow[\text{(2) } H^+]{\text{(1) } H_2O,\ OH^-} \left[\begin{array}{c} CH_3 \\ | \\ CH_2C \\ | \\ CO_2H \end{array}\right] \qquad (9.9)$$

2. Hofmann degradation of polyacrylamide to give poly(vinyl amine) (9.10),[25] a reaction complicated by side reactions leading to crosslinking.

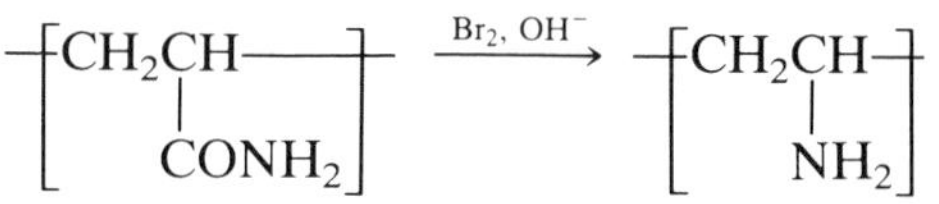

$$\left[\begin{array}{c} CH_2CH \\ | \\ CONH_2 \end{array}\right] \xrightarrow{Br_2,\ OH^-} \left[\begin{array}{c} CH_2CH \\ | \\ NH_2 \end{array}\right] \qquad (9.10)$$

3. Synthesis of "head-to-head poly(vinyl bromide)" by controlled bromination of 1,4-polybutadiene (9.11).[26]

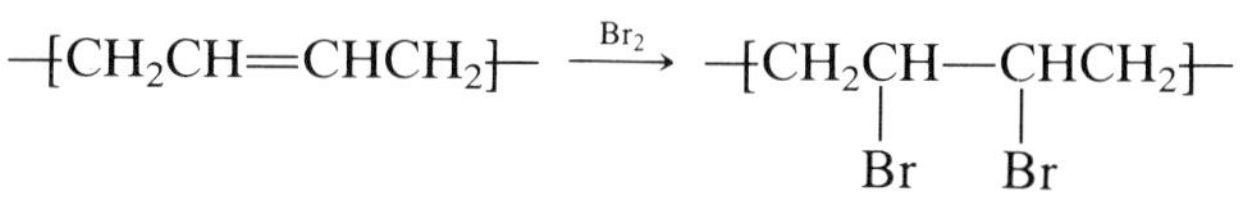

$$\left[CH_2CH=CHCH_2\right] \xrightarrow{Br_2} \left[\begin{array}{cc} CH_2CH-CHCH_2 \\ | \quad\quad | \\ Br \quad\quad Br \end{array}\right] \qquad (9.11)$$

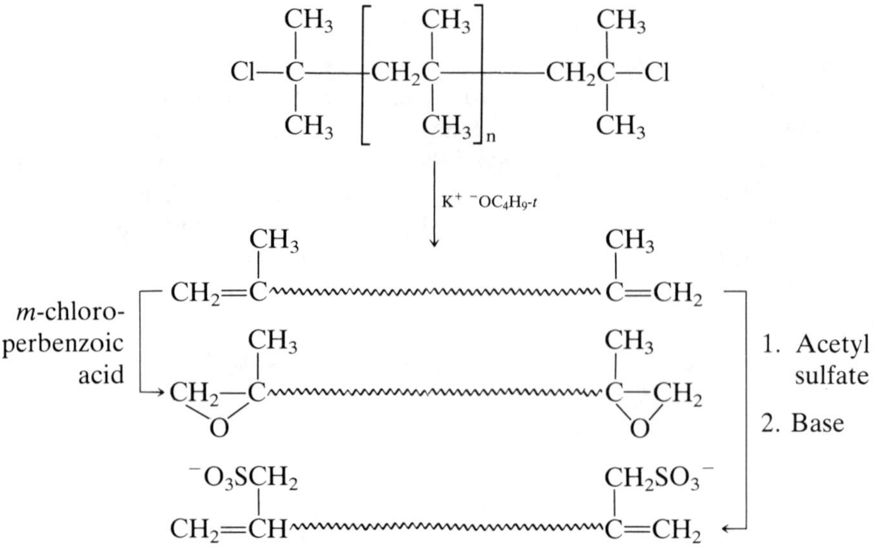

Scheme 9.1. Some functional group conversions of telechelic polyisobutylene.[32]

Other types of "classical" functional group conversions include dehydro-chlorination of poly(vinyl chloride) (9.12)[27] (an early synthesis of polyacetylene), hydroformylation of polypentenamer (9.13),[28] and hydroboration of 1,4-polyisoprene (9.14).[29]

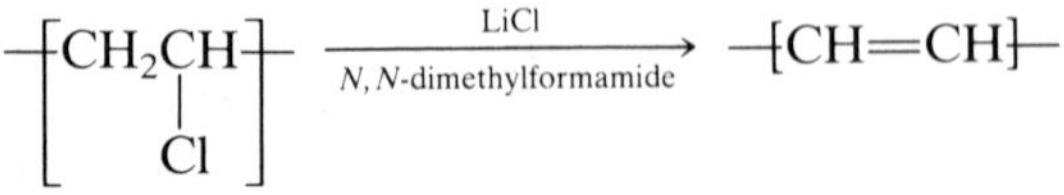

$$\left[\!\!\begin{array}{c} CH_2CH \\ | \\ Cl \end{array}\!\!\right] \xrightarrow[\text{N, N-dimethylformamide}]{\text{LiCl}} [CH=CH] \qquad (9.12)$$

$$[CH=CH(CH_2)_3] \xrightarrow[\text{catalyst}]{\text{CO, H}_2} \begin{array}{c} CH_2CH(CH_2)_3 \\ | \\ CHO \end{array} \qquad (9.13)$$

$$\left[\!\!\begin{array}{c} CH_3 \\ | \\ CH_2C=CHCH_2 \end{array}\!\!\right] \xrightarrow[\text{(2) NaOH, H}_2\text{O}_2]{\text{(1) B}_2\text{H}_6} \left[\!\!\begin{array}{c} CH_3 \\ | \\ CH_2CHCHCH_2 \\ | \\ OH \end{array}\!\!\right] \qquad (9.14)$$

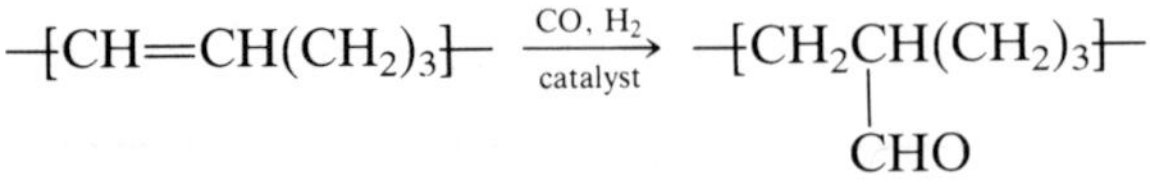

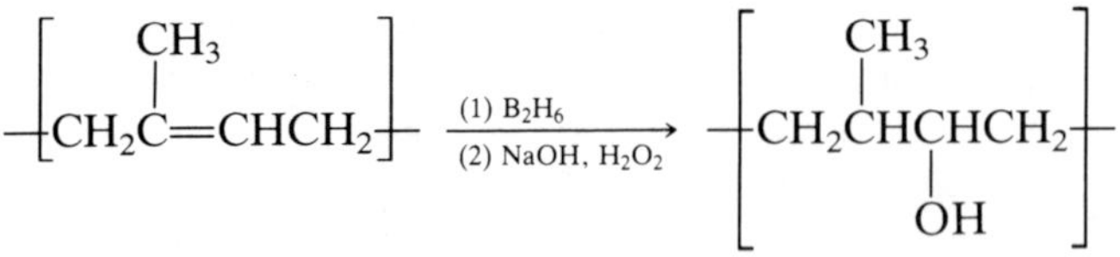

A less familiar reaction is the conversion of a fraction of the chloro groups of poly(vinyl chloride) to cyclopentadienyl (9.15) for purposes of cross-linking[30,31] (discussed in Section 9.4).

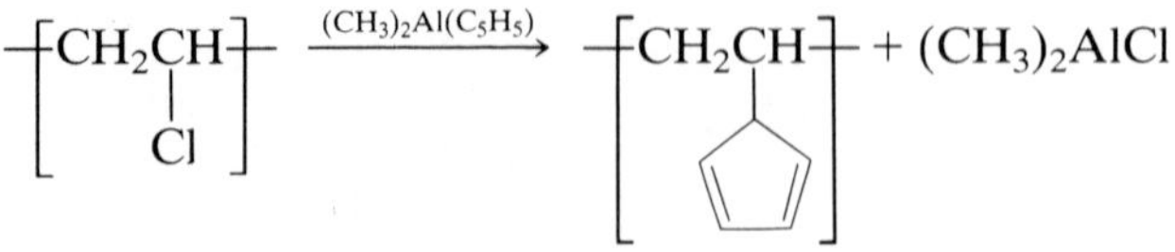

$$\left[\!\!\begin{array}{c} CH_2CH \\ | \\ Cl \end{array}\!\!\right] \xrightarrow{\text{(CH}_3)_2\text{Al(C}_5\text{H}_5)} \left[\!\!\begin{array}{c} CH_2CH \\ | \\ \end{array}\!\!\right] + (CH_3)_2AlCl \qquad (9.15)$$

Functional group conversions are not limited to pendant groups on the backbone; in some instances conversion of end groups of telechelic polymers leads to useful products. An example, illustrated in Scheme 9.1, involves dehydrochlorination of chloride-terminated polyisobutylene prepared by the inifer method (Chapter 7).[32] Epoxidation of the resultant vinyl groups with *m*-chloroperbenzoic acid leads to epoxide-terminated polymer that can be cured by reactions analogous to those used with epoxy resins (discussed in Chapter 11). Sulfonation followed by neutralization with base yields a polymer with properties of a thermoplastic elastomer. The elastomeric behavior arises from coulombic attraction of the ionic end groups leading to microdomains similar to those exhibited by ABA block copolymers.

9.3 Ring-forming reactions

As might be expected from our earlier consideration of restricted rotation (Chapter 3), the introduction of cyclic units into polymers generally results in greater rigidity, higher glass transition temperatures, and, frequently, improved thermal stability. One of the most important cyclization reactions leading to high thermal stability is the manufacture of *carbon fiber* (also called *graphite fiber*).* It is not a new process; in the 1870s Edison prepared carbon fiber filaments for his first electric light bulbs by pyrolyzing natural cellulose fibers. Most carbon fiber today is made by controlled pyrolysis of polyacrylonitrile fibers.[33-35] This is a complex free radical process involving a series of reactions leading eventually to a highly crosslinked graphitelike polymer (Scheme 9.2) accompanied by loss of HCN and N_2. Quinone-type structures

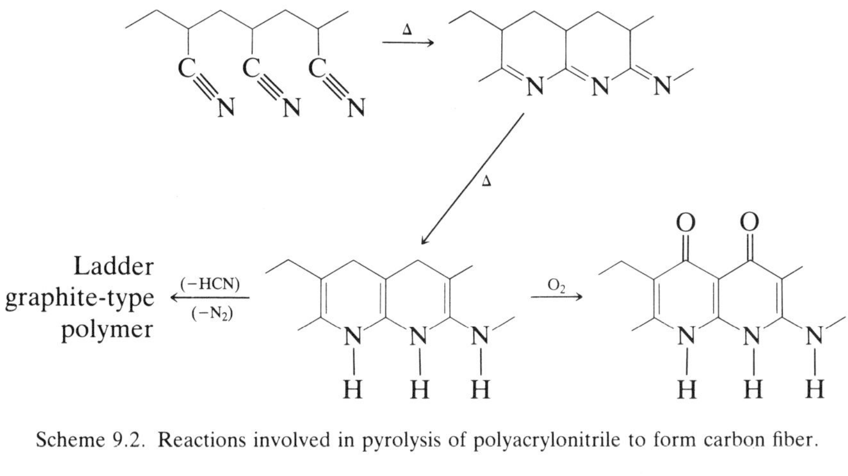

Scheme 9.2. Reactions involved in pyrolysis of polyacrylonitrile to form carbon fiber.

* While the terms *carbon fiber* and *graphite fiber* are often used synonymously, many workers in the field prefer to reserve the latter for fibers that have been treated at temperatures in excess of 2000°C, at which point the fibers assume a more truly graphitelike structure.

are formed if oxygen is present. If graphitization is applied to fibers under stress, longitudinal order results and the graphite fibers exhibit very high modulus. Such fibers, which may still contain residual nitrogen, are used primarily in high-strength composites and in applications requiring good heat resistance.

Ladder structures are also formed from poly(methyl vinyl ketone) by intramolecular aldol condensation (9.16).[36] Nonladder structures, on the other hand, result from such reactions as dechlorination of poly(vinyl chloride) (9.17) or intramolecular Friedel–Crafts acylation of styrene–methacryloyl chloride copolymer (9.18).[37] These reactions do not go to completion because cyclization occurs randomly, leaving some functional groups

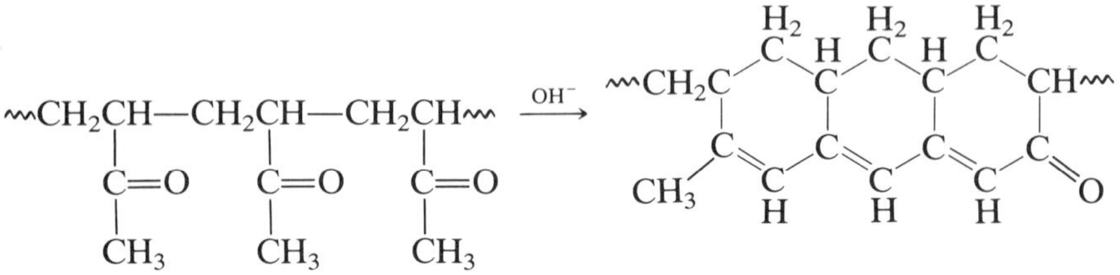

$$\text{(9.16)}$$

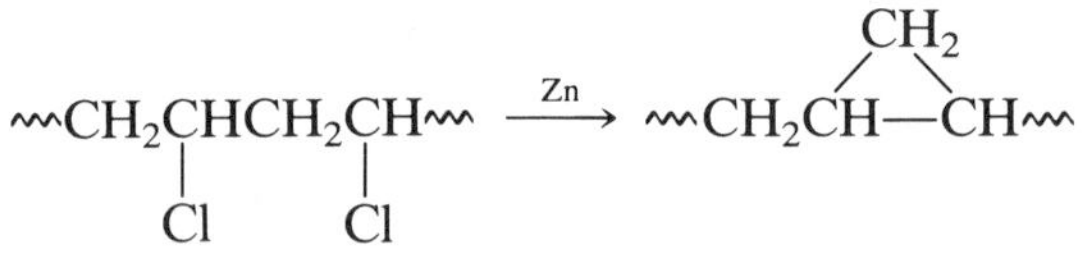

$$\text{(9.17)}$$

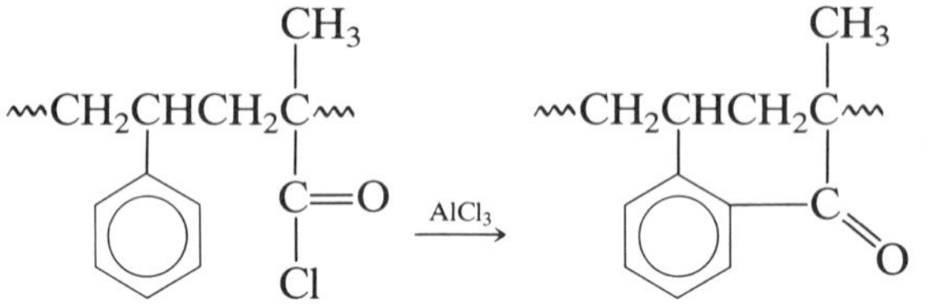

$$\text{(9.18)}$$

isolated between ring units. Only when the process is reversible can a cyclization reaction be made to approach its theoretical limits. Such a case is the formation of cyclic acetal groups by reaction of aldehydes with poly-(vinyl alcohol) (9.19) in which more than 90% of the hydroxyl groups can be

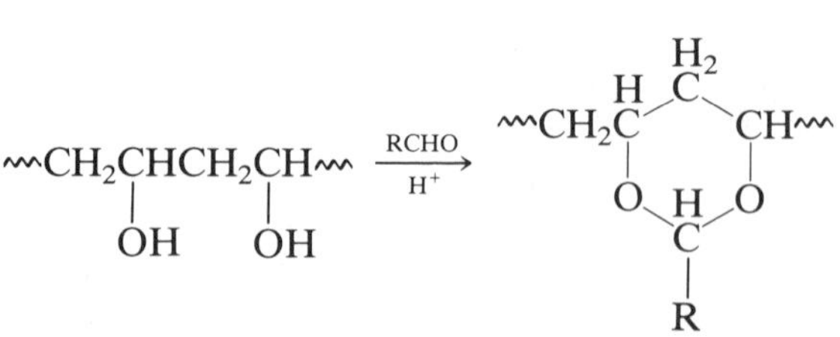

$$\text{(9.19)}$$

converted. This particular reaction is important commercially; the product when $R = C_3H_7$, commonly called *poly(vinyl butyral)*, is used as a plastic film in laminated safety glass.[22] Another commercially important cyclization is the epoxidation of natural rubber[38] (9.20) to increase oil resistance and decrease gas permeability.

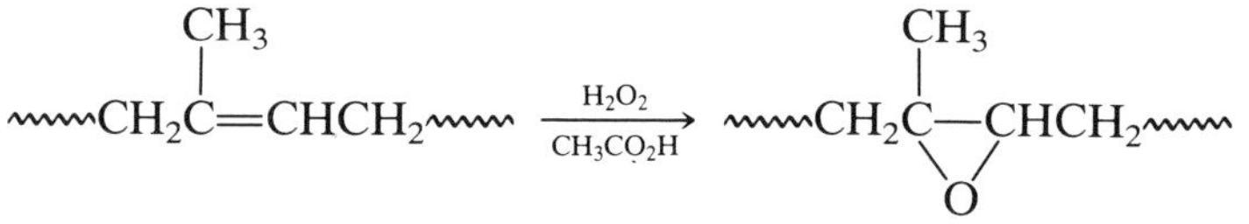

$$\text{wwwCH}_2\overset{\overset{\displaystyle CH_3}{|}}{C}=CHCH_2\text{www} \xrightarrow[CH_3CO_2H]{H_2O_2} \text{wwwCH}_2\overset{\overset{\displaystyle CH_3}{|}}{C}\overset{O}{\underset{\diagdown\diagup}{-}}CHCH_2\text{www} \quad (9.20)$$

Rubber and other diene polymers undergo cyclization in the presence of acid.[39] A probable pathway is shown for *cis*-1,4-polyisoprene (9.21); however, because of the high probability of transfer reactions occurring, only about one to four rings are fused together in any fused-ring sequence. As might be expected, the polymer loses its elastomeric properties with the introduction of cyclic units.

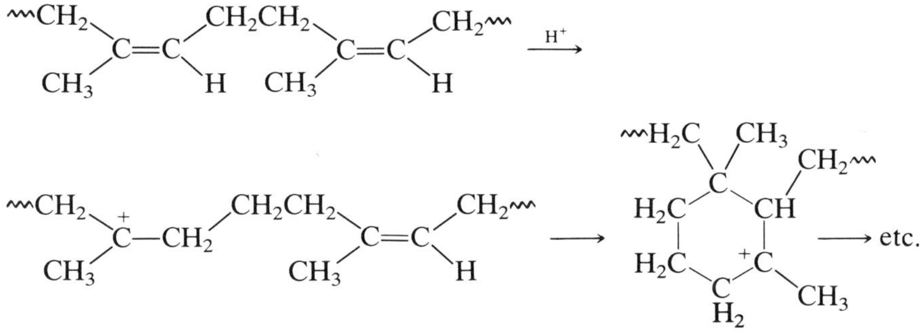

$$(9.21)$$

9.4 Crosslinking

From the commercial standpoint, crosslinking is the most important reaction of vinyl polymers and is fundamental to the rubber and elastomer industries. It can be brought about by (1) *vulcanization*, using peroxides, sulfur, or sulfur-containing compounds; (2) free radical reactions caused by ionizing radiation; (3) photolysis involving photosensitive functional groups; (4) chemical reactions of labile functional groups; or (5) coulombic interactions of ionic species. The physical and morphological consequences of chemical and physical crosslinking are discussed in Chapter 3. Here we are concerned with the reactions leading to chemical crosslinking.

9.4.1 Vulcanization

Vulcanization is a general term applied to crosslinking of polymers, particularly elastomers.[40–43] Peroxide-initiated crosslinking of saturated polymers such as polyethylene proceeds by hydrogen abstraction by radicals resulting from homolytic cleavage of peroxide (9.22) followed by radical combination (9.23). With unsaturated polymers, hydrogen abstraction probably occurs at the allylic position (9.24) with subsequent crosslinking again resulting from radical combination (9.25). It is apparent that addition-transfer processes

(9.26, 9.27) can also cause crosslinking, because in many instances considerably more crosslinks are formed than would be expected on the basis of only abstraction-combination reactions. Not all vinyl polymers are crosslinkable with peroxides; polypropylene and poly(vinyl chloride), for example, undergo degradation in preference to crosslinking.

$$RO\cdot + \text{Ɱ}CH_2CH_2\text{Ɯ} \longrightarrow \text{Ɯ}\dot{C}HCH_2\text{Ɯ} + ROH \tag{9.22}$$

$$
\begin{array}{c}
\text{Ɯ}\dot{C}HCH_2\text{Ɯ} \\
+ \\
\text{Ɯ}\dot{C}HCH_2\text{Ɯ}
\end{array}
\longrightarrow
\begin{array}{c}
\text{Ɯ}CHCH_2\text{Ɯ} \\
| \\
\text{Ɯ}CHCH_2\text{Ɯ}
\end{array}
\tag{9.23}
$$

$$\text{Ɯ}CH_2CH{=}CHCH_2\text{Ɯ} + RO\cdot \longrightarrow \text{Ɯ}\dot{C}HCH{=}CHCH_2\text{Ɯ} + ROH \tag{9.24}$$

$$
\begin{array}{c}
\text{Ɯ}\dot{C}HCH{=}CHCH_2\text{Ɯ} \\
+ \\
\text{Ɯ}\dot{C}HCH{=}CHCH_2\text{Ɯ}
\end{array}
\longrightarrow
\begin{array}{c}
\text{Ɯ}CHCH{=}CHCH_2\text{Ɯ} \\
| \\
\text{Ɯ}CHCH{=}CHCH_2\text{Ɯ}
\end{array}
\tag{9.25}
$$

$$
\begin{array}{c}
\text{Ɯ}\dot{C}HCH{=}CHCH_2\text{Ɯ} \\
+ \\
\text{Ɯ}CH_2CH{=}CHCH_2\text{Ɯ}
\end{array}
\longrightarrow
\begin{array}{c}
\text{Ɯ}CHCH{=}CHCH_2\text{Ɯ} \\
\diagdown \\
\text{Ɯ}CH_2CH{-}\dot{C}HCH_2\text{Ɯ}
\end{array}
\tag{9.26}
$$

$$\downarrow \text{Ɯ}CH_2CH{=}CHCH_2\text{Ɯ}$$

$$
\text{Ɯ}\dot{C}HCH{=}CHCH_2\text{Ɯ} \quad + \quad
\begin{array}{c}
\text{Ɯ}CHCH{=}CHCH_2\text{Ɯ} \\
\diagdown \\
\text{Ɯ}CH_2CH{-}CH_2CH_2\text{Ɯ}
\end{array}
\tag{9.27}
$$

The oldest method of vulcanization, discovered independently in 1839 by Goodyear in the United States and MacIntosh and Hancock in Great Britain, uses elemental sulfur. The mechanism appears to be ionic in nature, involving addition to a double bond to form an intermediate sulfonium ion (9.28) which then abstracts a hydride ion (9.29) or donates a proton (9.30) to form new

$$\text{Ɯ}CH_2CH{=}CHCH_2\text{Ɯ} + \text{Ɯ}\overset{\delta+}{S}{-}\overset{\delta-}{S}\text{Ɯ} \longrightarrow \text{Ɯ}CH_2CH{\cdots}CHCH_2\text{Ɯ} \tag{9.28}$$

$$\underset{\underset{\text{Ɯ}}{|}}{\overset{+}{S}} \; + \text{Ɯ}S^-$$

$$
\begin{array}{c}
\text{Ɯ}CH_2CH{\cdots}CHCH_2\text{Ɯ} \\
\underset{\underset{\text{Ɯ}}{|}}{\overset{+}{S}} \\
\\
+ \\
\\
\text{Ɯ}CH_2CH{=}CHCH_2\text{Ɯ}
\end{array}
\left[
\begin{array}{l}
\rightarrow \text{Ɯ}CH_2CH{-}CH_2CH_2\text{Ɯ} + \text{Ɯ}\overset{+}{C}HCH{=}CHCH_2\text{Ɯ} \\
\qquad\quad \underset{\underset{}{|}}{S} \\
\qquad\quad \S \\
\\
\\
\rightarrow \text{Ɯ}CH_2CH{-}CH{=}CH\text{Ɯ} + \text{Ɯ}CH_2CH_2{-}\overset{+}{C}HCH_2\text{Ɯ} \\
\qquad\quad \underset{\underset{}{|}}{S} \\
\qquad\quad \S
\end{array}
\right.
\begin{array}{c}
(9.29) \\
\\
\\
\\
(9.30)
\end{array}
$$

cations for propagating the reaction. Termination occurs by reaction between sulfenyl anions and carbocations. Studies on model compounds indicate that, in addition to simple monosulfide or disulfide linkages, some polysulfide groups and cyclic monosulfide groups resulting from intramolecular addition of pendant sulfur to a double bond are also present.

The rate of vulcanization with sulfur can be, and normally is, increased by addition of accelerators such as zinc salts of dithiocarbamic acids (**1**) or organosulfur compounds such as tetramethylthiuram disulfide (**2**). Other compounds, notably zinc oxide and stearic acid, are also added as activators.

$$\underset{\textbf{1}}{[(CH_3)_2\overset{\overset{\displaystyle S}{\|}}{N}CS^-]_2Zn^{2+}} \qquad\qquad \underset{\textbf{2}}{(CH_3)_2\overset{\overset{\displaystyle S}{\|}}{N}C\overset{\overset{\displaystyle S}{\|}}{S}SCN(CH_3)_2}$$

Although the mechanism of acceleration is not well understood, acceleration is known to decrease the number of cyclic monosulfide groups and increase the number of monosulfide and disulfide crosslinks.

9.4.2 Radiation crosslinking

When vinyl polymers are subjected to ionizing radiation, whether it be photons, electrons, neutrons, or protons, two main types of reaction occur—crosslinking and degradation.[44-48] Generally, both occur simultaneously, although degradation predominates with high doses of radiation. With low doses the polymer structure determines which will be the major reaction. Geminally disubstituted polymers tend to undergo chain scission, with monomer being a major degradation product. Thus, such polymers as poly(α-methylstyrene), poly(methyl methacrylate), or polyisobutylene will decrease in molecular weight on exposure to radiation. Halogen-substituted polymers, such as poly(vinyl chloride), break down with loss of halogen. With most other vinyl polymers, crosslinking predominates. A limitation of radiation crosslinking is that radiation does not penetrate very far into the polymer matrix; hence the method is primarily used with films.

The mechanism of crosslinking is free radical in nature and probably involves initial ejection of a hydrogen atom (9.31), which, in turn, removes another hydrogen atom from an adjacent site on a neighboring chain (9.32). Crosslinking then occurs by radical combination (9.33). This is a reasonable

$$-CH_2CH_2- \xrightarrow{h\nu} -\overset{\displaystyle\cdot}{C}HCH_2- + H\cdot \qquad (9.31)$$

$$-CH_2CH_2- + H\cdot \longrightarrow -\overset{\displaystyle\cdot}{C}HCH_2- + H_2 \qquad (9.32)$$

(neighboring chain)

$$-\overset{\displaystyle\cdot}{C}HCH_2- + -\overset{\displaystyle\cdot}{C}HCH_2- \longrightarrow \begin{matrix} -CHCH_2- \\ | \\ -CHCH_2- \end{matrix} \qquad (9.33)$$

assumption because hydrogen is a major side product, and random formation of radicals would not give the efficiency of crosslinking that is generally observed.

Chain branches are also ejected. Low-density polyethylene, for example, gives larger amounts of gaseous hydrocarbons on irradiation than does linear high-density polyethylene. Fragmentation reactions of this type, as well as ejection of hydrogen, also lead to double bonds in the polymer chains (9.34) in addition to crosslinking.

$$
\begin{array}{ccc}
-CH_2CH- & \longrightarrow & -\overset{\bullet}{C}HCH- + H\cdot \longrightarrow -CH{=}CH- + RH \quad (9.34)\\
\mid & & \mid \\
R & & R
\end{array}
$$

Radiolysis effects on numerous vinyl polymers have been studied, but polyethylene is most important. Irradiated polyethylene film is used commercially because of its improved tensile and thermal properties. Polystyrene is quite resistant to radiation, a characteristic of aromatic polymers in general, but it can be crosslinked with higher doses.

9.4.3 Photochemical crosslinking

Ultraviolet or visible light-induced crosslinking[49,50] (*photocrosslinking*) has taken on increasing importance in recent years. Among the numerous applications are printed circuits for electronic equipment; printing inks; coatings for optical fibers; varnishes for paper and carton board; finishes for vinyl flooring, wood, paper, and metal; and curing of dental materials. Photocrosslinking applied to photoresist technology is described in Chapter 4. There are two basic methods for bringing about photocrosslinking: (1) incorporating photosensitizers into the polymer, which absorb light energy and thereby induce formation of free radicals, and (2) incorporating groups that undergo either *photocycloaddition* reactions or light-initiated polymerization.

When triplet sensitizers such as benzophenone are added to polymer, absorption of ultraviolet results in $n \rightarrow \pi^*$ excitation of the sensitizer followed by hydrogen abstraction from the polymer to yield radical sites available for crosslinking by combination reactions. If the chromophore is built into the polymer backbone, as with polymers or copolymers of vinyl ketones, then degradation occurs, either by α-cleavage of the excited polymer (9.35) or by abstraction of a γ-hydrogen (9.36). The former leads to active sites for crosslinking, while the latter results in chain cleavage. (Recall the discussion of Norrish type II reactions in Chapter 4, Section 4.7.) Poly(vinyl ester)s undergo analogous α-cleavage reactions (9.37) with subsequent crosslinking. Copolymers of vinyl esters and fluorinated monomers that can be crosslinked by ultraviolet have been developed for use as weather-resistant wood coatings. In this application, the vinyl ester constitutes about 10% of the copolymer and benzophenone is added as a sensitizer.

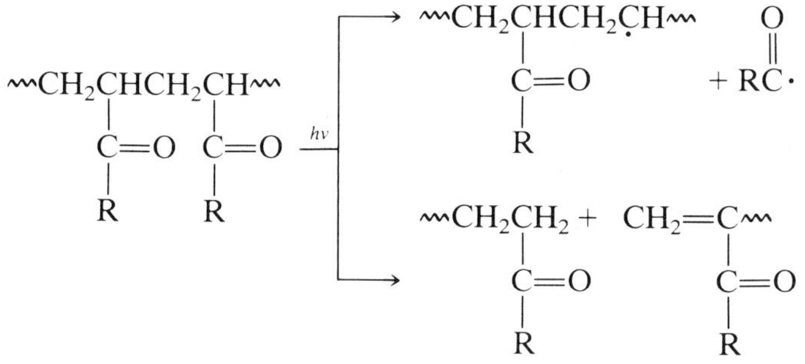

(9.35)

(9.36)

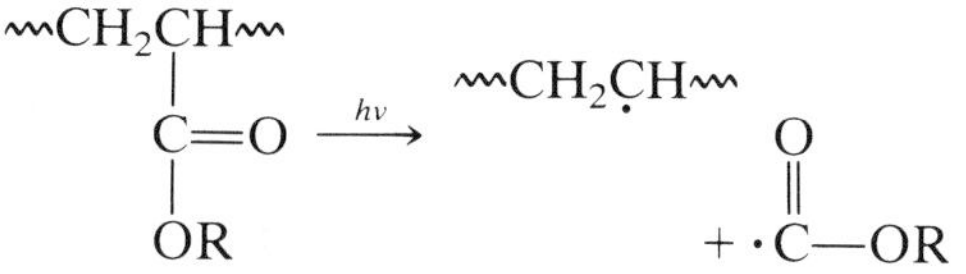

(9.37)

As is the case with radiolysis, polymers prepared from 1,1-disubstituted monomers undergo more extensive degradation, particularly to monomer, on photolysis.

A wide variety of functional groups has been used to effect photocycloaddition or light-induced polymerization crosslinking.[49–56] Many of them are listed in Table 9.1. The groups may be present as part of the backbone or,

Table 9.1. Groups used to effect photocrosslinking[49–56]

Type	*Structure*
Alkyne	R—C≡C—R
Anthracene	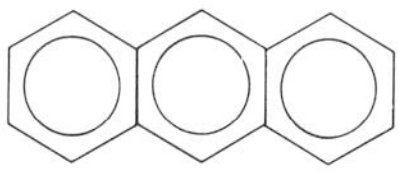
Benzothiophene dioxide	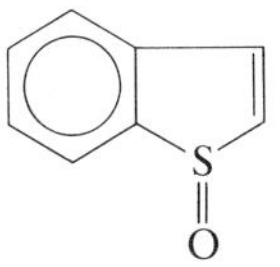
Chalcone	$ArCH{=}CHC({=}O)Ar$
Cinnamate	$ArCH{=}CHCO_2R$
Coumarin	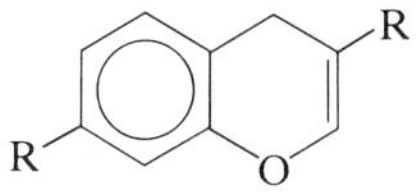

(*continues*)

Table 9.1. (*Continued*)

Type	Structure
Dibenz[b, f]azepine	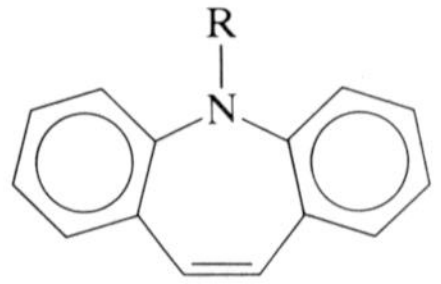
Diphenylcyclopropenecarboxylate	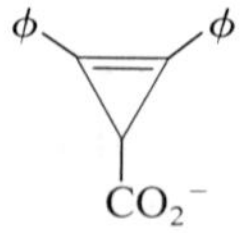
Episulfide	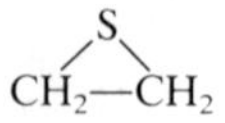
Maleimide (R = H, CH₃, Cl)	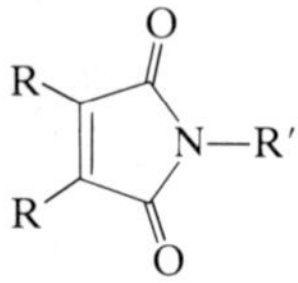
Stilbazole	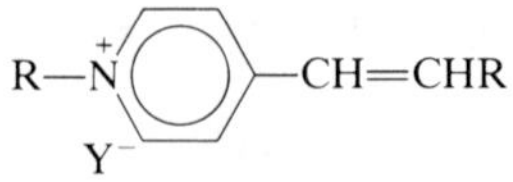 $R—\overset{+}{N}$ ⬡ $—CH=CHR$ Y⁻
Stilbene	ArCH=CHAr
Styrene	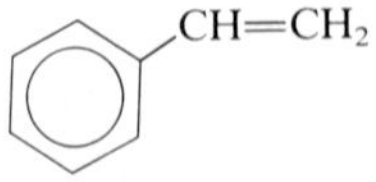
1,2,3-Thiadiazole	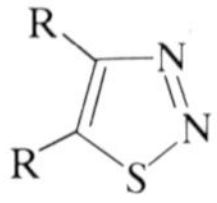
Thymine	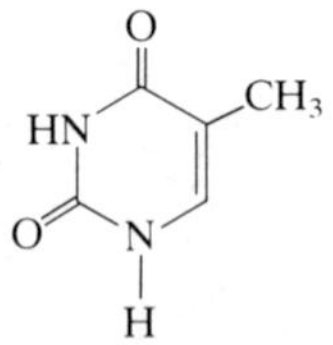

more commonly, as pendant groups. In most cases $2\pi + 2\pi$ cycloaddition occurs to give cyclobutane crosslinks, as shown for cinnamate ester in Scheme 9.3(a). With anthracene, Scheme 9.3(b), cycloaddition is $4\pi + 4\pi$. Free radical vinyl-type polymerization may accompany cycloaddition, as is the case with maleimide[55]; or vinyl polymerization may predominate, as with

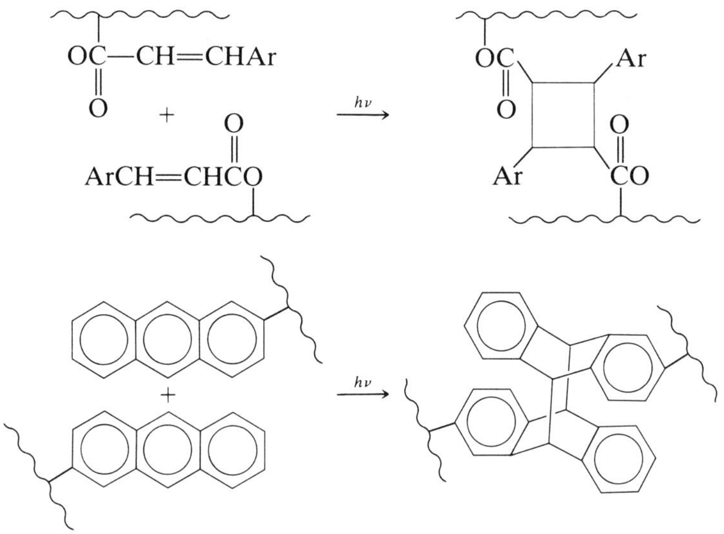

Scheme 9.3. Photocrosslinking (a) by $2\pi + 2\pi$ cycloaddition and (b) by $4\pi + 4\pi$ cycloaddition.

styrene groups.[51] The reactive groups may be incorporated into polymer during the polymerization reaction; for example, β-vinyloxyethyl cinnamate (**3**) undergoes cationic polymerization through the vinyl ether (9.38) to yield linear polymer containing pendant cinnamate ester.[56] Alternatively, the group may be added to preformed polymer, as in the Friedel–Crafts alkylation of polystyrene with N-chloromethylmaleimide (**4**) (9.39).[55]

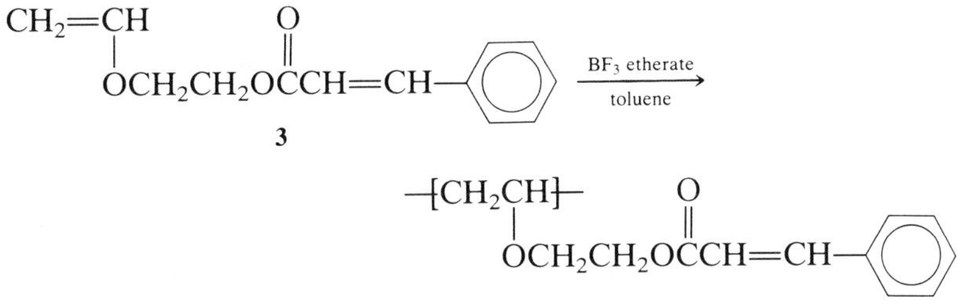

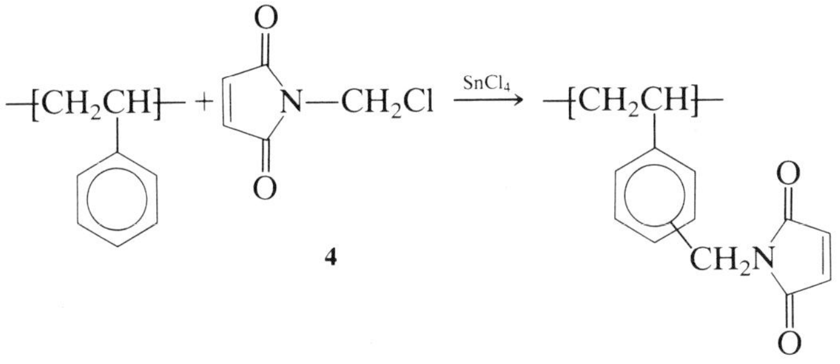

9.4.4 *Crosslinking through labile functional groups*

Reaction between appropriate difunctional or polyfunctional reagents with labile groups on the polymer chains can bring about crosslinking, but such reactions have not been exploited nearly as much as the more convenient vulcanization methods.

Polymers containing acid chloride groups, such as the previously described chlorosulfonated polyethylene (9.3), react with diamines (9.40) or diols (9.41) to yield sulfonamide and sulfonate crosslinks, respectively.[57] Dihalogen compounds can be used to crosslink polystyrene by the Friedel–Crafts reaction (9.42).[58]

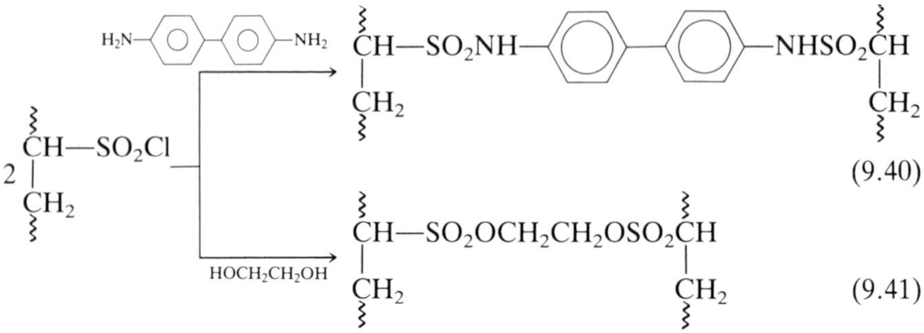

$$(9.40)$$

$$(9.41)$$

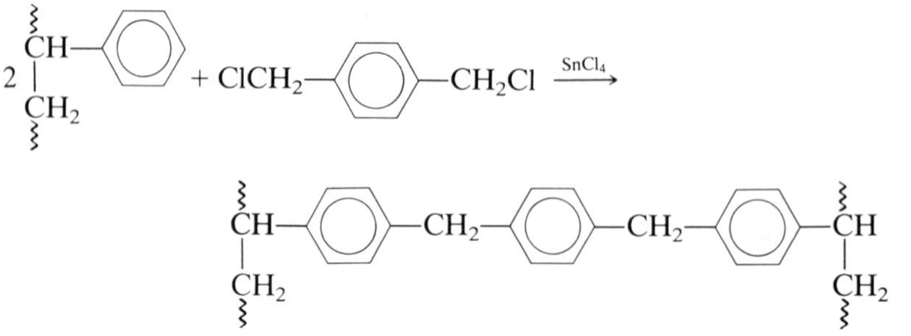

$$(9.42)$$

An interesting cycloaddition process has been used to crosslink copolymers of 2-phenyl-5-(4′-vinyl)phenyltetrazole (**5**) and acrylonitrile. The functional groups remain intact during free radical copolymerization, but on heating the tetrazole ring decomposes and cycloaddition with nitrile occurs to give triazole crosslinks (9.43).[59]

A particularly interesting example of cycloaddition crosslinking involves the Diels–Alder reaction of cyclopentadiene-substituted polymer (see reaction 9.15). It is well known that cyclopentadiene dimerizes by Diels–Alder cycloaddition, one ring acting as diene, the other as dienophile. At about 180°C the dimer reverts to monomer. Cyclopentadiene substituent groups on polymers undergo an analogous reaction, yielding crosslinked polymer by cycloaddition and linear polymer by retrograde Diels–Alder reaction of the

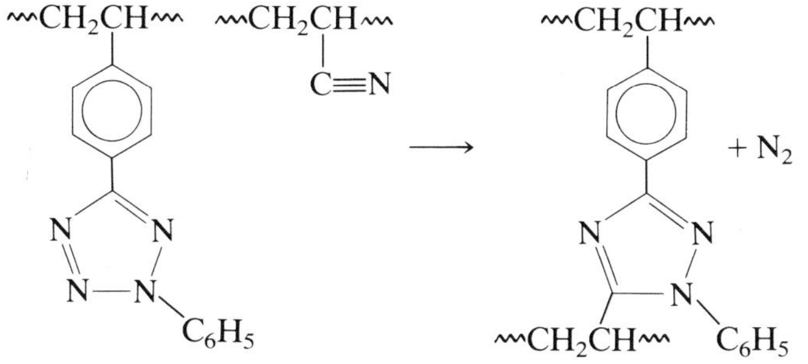

$$\text{(9.43)}$$

5

crosslinked polymer at elevated temperatures (9.44).[30] Such reversible systems have potential as thermoplastic elastomers.

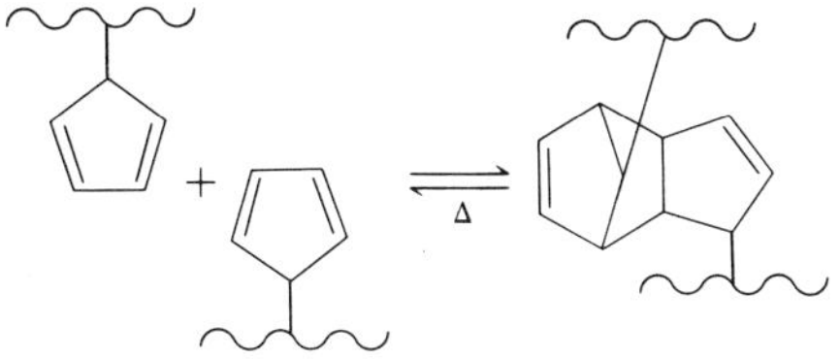

$$\text{(9.44)}$$

9.4.5 Ionic crosslinking

Examples of ionic crosslinking are the hydrolysis of chlorosulfonated polyethylene with aqueous lead oxide (9.45) to yield lead sulfonate crosslinks, and the partial conversion of poly(ethylene-*co*-methacrylic acid) (**6**) to salts of divalent metals. The latter type, marketed under the du Pont trade name Surlyn, are called *ionomers*.[60,61] (Commercial products may contain univalent as well as divalent metal salts.)

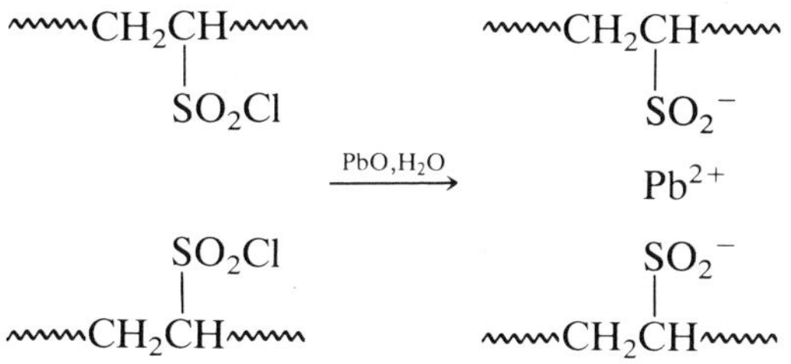

$$\text{(9.45)}$$

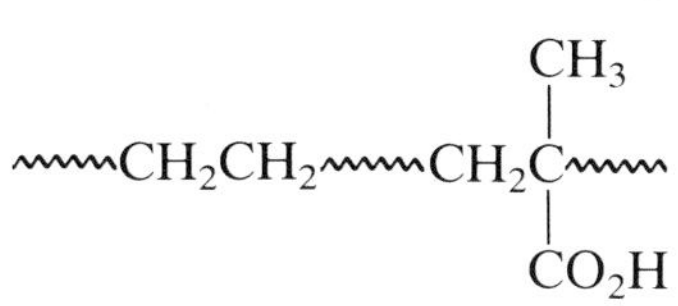

6

Ionomers have interesting properties compared with the nonionized copolymer. Introduction of ions causes disordering of the semicrystalline structure, which makes the polymer transparent. Crosslinking gives the polymer elastomeric properties, but it can still be molded at elevated temperatures. In fact, molding is generally facilitated by a broader melt range than is the case with the nonionic polymer. Ionic polymers, therefore, also qualify as thermoplastic elastomers. Increased polarity also improves adhesion. Ionomers are used as coatings, as adhesive layers for bonding wood to metal, as blow-molded and injection-molded containers, as golf ball covers, and as blister packaging material. Because of its adhesive qualities, an ionic acrylic copolymer is used as a binder for aluminosilicate dental fillings.

9.5 Block and graft copolymer formation

In Chapters 7 and 8 we encountered the living polymer method of forming block copolymers. In this section we explore some other methods of preparing block copolymers and, for the first time, a few of the numerous procedures developed for preparing graft copolymers.[62-65]

9.5.1 Block copolymers

Polymers containing functional end groups are generally amenable to forming block copolymers.[66] Styrene polymerized with the H_2O_2–$FeSO_4$ redox system, for example, forms hydroxy-terminated polystyrene. Subsequent reaction with an isocyanate-terminated polymer yields an AB block copolymer via urethane linkages (9.46). Similarly, telechelic polymers (functional groups at both ends) can be converted to ABA block copolymer or, if both homopolymers are telechelic, to the $+AB+$ type.

$$\text{wwwCHCH}_2\text{OH} + \text{OCNwww} \longrightarrow \text{www CHCH}_2-\overset{\overset{\textstyle O}{\|}}{\text{OC}}\text{NHwww} \quad (9.46)$$

It is worth noting that polymers containing reactive functional groups are, in a sense, "living," and their conversion to block copolymers might be considered in the same light as the living polymer method of creating block copolymers discussed in the previous two chapters.

More recently, block copolymers have yielded to mixed mechanism techniques[67]; that is, the first block is formed by one mechanism (e.g., anionic), and the second by a different mechanism (cationic, free radical, etc.) A schematic representation of an anionic-cationic mixed mechanism synthesis is given in reaction (9.47).

$$xM_1 \xrightarrow{\text{BuLi}} \text{Bu}+M_1\underset{x-1}{\underbrace{}}M_1:^-\text{Li}^+ \xrightarrow{\text{Br}_2} \text{Bu}+M_1\underset{x-1}{\underbrace{}}M_1\text{Br} \xrightarrow{\text{AgClO}_4}$$

$$\text{Bu}+M_1\underset{x-1}{\underbrace{}}M_1^+\text{ClO}_4^- \xrightarrow{yM_2} \text{Bu}+M_1\underset{x}{\underbrace{}}[M_2\underset{y-1}{\underbrace{}}M_2^+\text{ClO}_4^- \quad (9.47)$$

Peroxide groups introduced to polymer chain ends can be used to initiate a new chain.[65a] Typically an initiator such as the monohydroperoxide of diisopropylbenzene is used, and the resultant polymer (**7**), which contains an isopropylbenzene end group, can itself be converted to hydroperoxide (9.48).

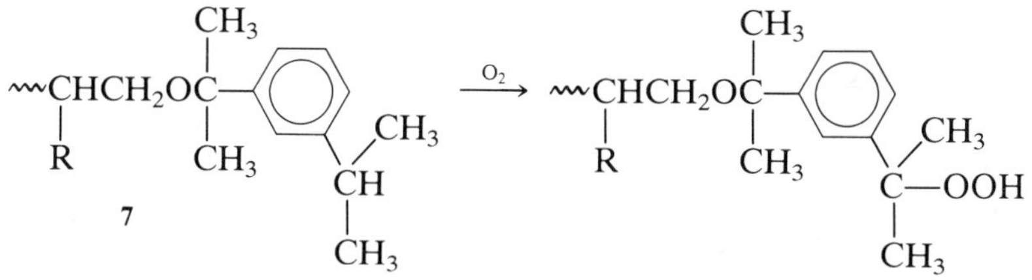

(9.48)

Addition of a second monomer followed by redox or thermal decomposition of the hydroperoxide initiates polymerization at the chain end. Similarly peroxide units can be formed in a polymer backbone by polymerizing a monomer in the presence of small amounts of oxygen (9.49). Thermal cleavage of the peroxide leads to radical-terminated chains capable of initiating a second monomer (9.50).

$$2x\,CH_2{=}\underset{\displaystyle X}{\underset{|}{C}H} + O_2 \xrightarrow{\text{initiator}} \left[CH_2\underset{\displaystyle X}{\underset{|}{C}H}\right]_x O{-}O \left[CH_2\underset{\displaystyle X}{\underset{|}{C}H}\right]_x \qquad (9.49)$$

$$\left[CH_2\underset{\displaystyle X}{\underset{|}{C}H}\right]_x O{-}O \left[CH_2\underset{\displaystyle X}{\underset{|}{C}H}\right]_x + 2y\,CH_2{=}CHY \longrightarrow$$

$$2\left[CH_2\underset{\displaystyle X}{\underset{|}{C}H}\right]_x O \left[CH_2\underset{\displaystyle Y}{\underset{|}{C}H}\right]_y \qquad (9.50)$$

Another way to form chain-end radicals is by mechanical degradation of homopolymers using ultrasonic radiation or high-speed stirring.[63] Polyethylene-*block*-polystyrene, for example, has been made by masticating polystyrene in the presence of ethylene. Mastication of mixtures of homopolymers yields similar results. Ionizing radiation may also be used to effect chain cleavage; however, this is more readily applicable to formation of graft copolymers.

9.5.2 *Graft copolymers*

There are three general methods of preparing graft copolymers: (1) A monomer is polymerized in the presence of a polymer with branching resulting from chain transfer. (2) A monomer is polymerized in the presence of a polymer having reactive functional groups or positions that are capable of being activated, for example, by irradiation. (3) Two polymers having reactive functional groups are coreacted. Chain transfer grafting will be considered first.

Three components are necessary for grafting by chain transfer: polymer, monomer, and initiator. The function of the initiator is either to polymerize monomer to form a polymeric radical, ion, or coordination complex which can then attack the original polymer, or to react with the original polymer to form such species on the backbone, which initiate polymerization of monomer. As with ordinary copolymerization, it is necessary to consider reactivity ratios of monomers to ensure that grafting will occur. It is also necessary to take into account the frequency of transfer to determine the number of grafts. Normally, mixtures of homopolymers result along with graft copolymer.

Grafting generally occurs at sites that are susceptible to transfer reactions, such as at carbons adjacent to double bonds in polydienes or on carbons adjacent to carbonyl groups. An example of the latter is the reaction of poly(vinyl acetate) with ethylene using a free radical initiator.[57] This particular reaction occurs both on the pendant methyl group and on the polymer backbone (9.51). This has been shown to be the case by hydrolysis of the product to give a mixture of poly(vinyl alcohol)-*graft*-polyethylene and long-chain carboxylic acids. Grafting efficiency is improved if a group that undergoes radical transfer readily, such as a mercaptan, is incorporated into the polymer backbone.

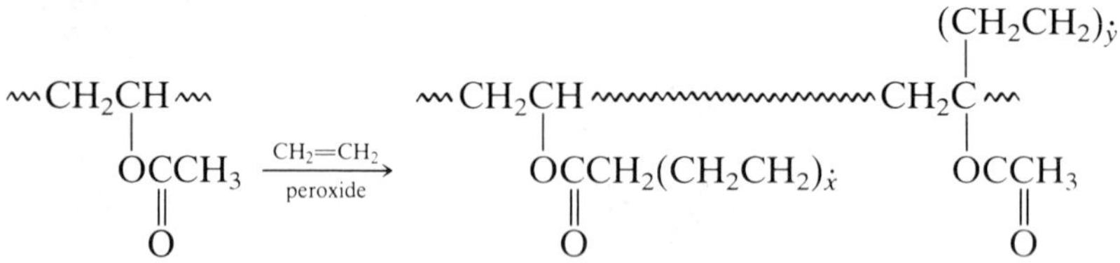

$$\text{(9.51)}$$

Cationic chain transfer grafting occurs when styrene is polymerized with BF_3 in the presence of poly(*p*-methoxystyrene) (9.52).[68,69] In this case the activated benzene rings undergo Friedel–Crafts attack by the carbocationic chain end of polystyrene. Isobutylene, which might be expected to behave similarly, fails to form graft copolymer under these conditions.

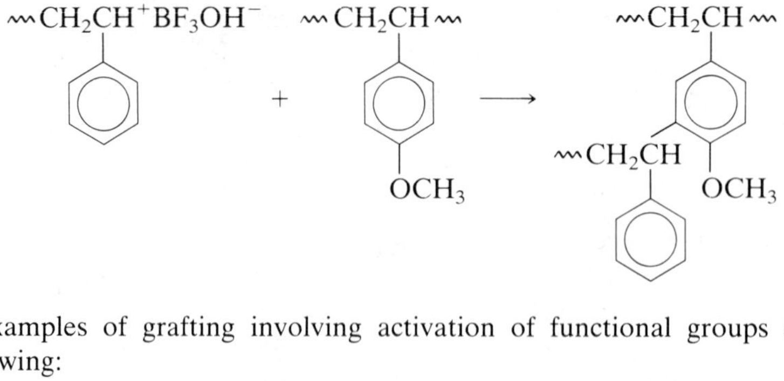

$$\text{(9.52)}$$

Examples of grafting involving activation of functional groups are the following:

1. Synthesis of poly(vinyl chloride)-*graft*-(*cis*-1,4-polybutadiene) by cationic initiation of butadiene with a diethylaluminum chloride-cobalt complex (9.53).[70] A butadiene-cobalt complex is presumably responsible for the *cis* stereochemistry.

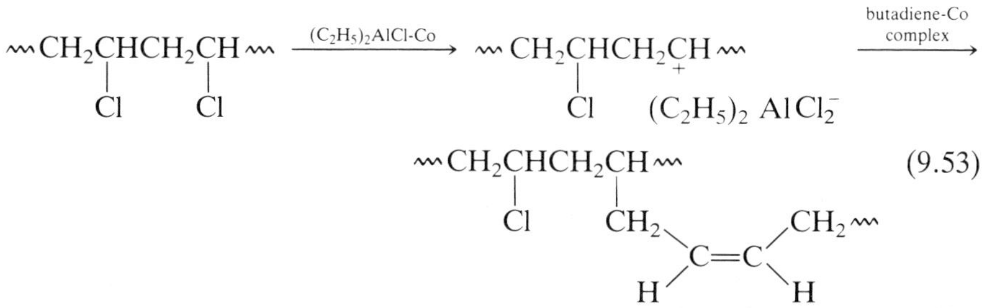

$$\text{butadiene-Co complex}$$

$$\sim\!\!CH_2CHCH_2CH\!\!\sim \xrightarrow{(C_2H_5)_2AlCl\text{-}Co} \sim\!\!CH_2CHCH_2CH\!\!\sim \xrightarrow{\text{butadiene-Co complex}}$$

with Cl, Cl substituents on the left and Cl, $(C_2H_5)_2\,AlCl_2^-$ on the right

$$\sim\!\!CH_2CHCH_2CH\!\!\sim \quad (9.53)$$

$$Cl \qquad CH_2\!\!\diagdown_{C=C}\diagup^{CH_2\sim}$$
$$\text{H}\diagup \qquad \diagdown\text{H}$$

2. Synthesis of poly(p-chlorostyrene)-*graft*-polyacrylonitrile by anionic initiation of acrylonitrile using naphthalenesodium in the presence of poly(p-chlorostyrene) (9.54).[71,72] A polymeric anion radical is the initiating species.

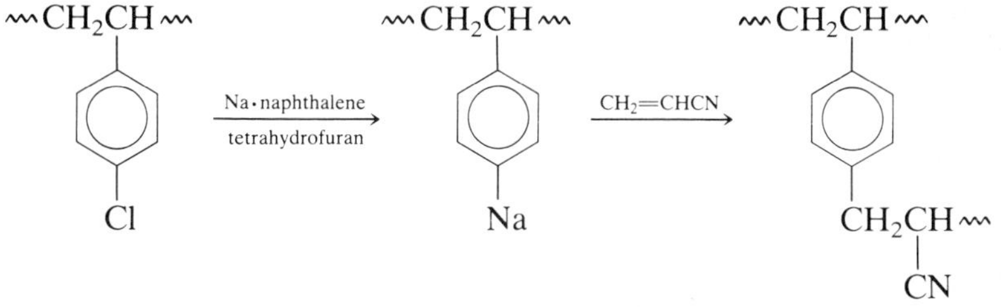

$$\sim\!\!CH_2CH\!\!\sim \xrightarrow[\text{tetrahydrofuran}]{Na \cdot naphthalene} \sim\!\!CH_2CH\!\!\sim \xrightarrow{CH_2=CHCN} \sim\!\!CH_2CH\!\!\sim \quad (9.54)$$

with Cl, Na, and $CH_2CH\!\!\sim$ / CN substituents respectively

3. Synthesis of 1,4-polybutadiene-*graft*-(isotactic polypropylene) by activation of the polybutadiene with diethylaluminum hydride and $TiCl_3$ followed by Ziegler–Natta polymerization of propylene (9.55).[73]

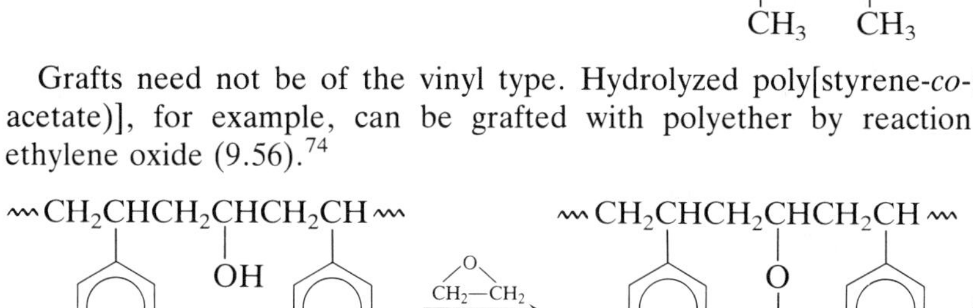

$$\sim\!\!CH_2CH=CHCH_2\!\!\sim \xrightarrow{(C_2H_5)_2AlH} \sim\!\!CH_2CHCH_2CH_2\!\!\sim \xrightarrow[(2)\ CH_2=CHCH_3]{(1)\ TiCl_3}$$
$$Al(C_2H_5)_2$$

$$\sim\!\!CH_2CHCH_2CH_2\!\!\sim$$
$$CH_2CHCH_2CH\!\!\sim \quad (9.55)$$
$$CH_3 \qquad CH_3$$

Grafts need not be of the vinyl type. Hydrolyzed poly[styrene-*co*-(vinyl acetate)], for example, can be grafted with polyether by reaction with ethylene oxide (9.56).[74]

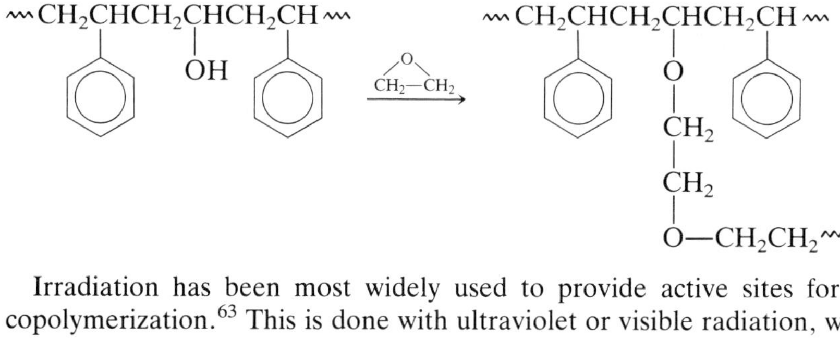

$$\sim\!\!CH_2CHCH_2CHCH_2CH\!\!\sim \xrightarrow{CH_2-CH_2 \ (O)} \sim\!\!CH_2CHCH_2CHCH_2CH\!\!\sim \quad (9.56)$$

with OH substituent on left; O—CH_2—CH_2—O—$CH_2CH_2\!\!\sim$ on right

Irradiation has been most widely used to provide active sites for graft copolymerization.[63] This is done with ultraviolet or visible radiation, with or without added photosensitizer, or with ionizing radiation, particularly the latter. Free radical reactions are involved in all cases. A major difficulty is that irradiation causes substantial amounts of homopolymerization along with grafting. This has been obviated to some extent by preirradiating the polymer prior to addition of the new monomer. One method is to preirradiate the polymer in the presence of air or oxygen to form hydroperoxide groups on the backbone. Subsequent addition of monomer and heating results in radical polymerization at the peroxide sites accompanied by some homopolymerization initiated by the hydroxy radicals formed on homolysis of hydroperoxide. Preirradiation can also be performed in the absence of air to form free radicals trapped in the viscous polymer matrix. Monomer is then added. The method is not very efficient because of the low concentration of radicals that can be trapped; and homopolymerization can still occur by chain transfer reactions.

Direct irradiation of monomer and polymer together has been most extensively used. Because homopolymerization can occur, monomer and polymer must be chosen carefully. Generally, the best combination is a polymer that is very sensitive to radiation—that is, one that forms a high concentration of radicals—and a monomer that is not very sensitive. Sensitivity is normally measured in terms of G values, which represent the number of free radicals formed per 100 eV of energy absorbed per gram. Table 9.2 lists some G values for a few common monomers and polymers.[47] According to these data, a good combination would be poly(vinyl chloride) and butadiene. Homopolymerization may be reduced by providing radiation in bursts while monomer is allowed to diffuse through the polymer. Irradiation grafting of polymer emulsions is also an effective way to minimize homopolymerization, since the reaction medium remains fluid even at high conversions.

Another method of irradiation grafting involves irradiating an intimate mixture of homopolymers. Apart from the fact that most polymers are

Table 9.2. Approximate G values of monomers and polymers[a]

Monomer	G	Polymer	G
Butadiene	Very low	Polybutadiene	2.0
Styrene	0.70	Polystyrene	1.5–3
Ethylene	4.0	Polyethylene	6–8
Acrylonitrile	5.0–5.6	—	—
Methyl methacrylate	5.5–11.5	Poly(methyl methacrylate)	6–12
Methyl acrylate	6.3	Poly(methyl acrylate)	6–12
Vinyl acetate	9.6–12.0	Poly(vinyl acetate)	6–12
Vinyl chloride	10.0	Poly(vinyl chloride)	10–15

[a] Data from Marvel, Sample, and Roy.[27] G values refer to number of free radicals formed per 100 eV of energy absorbed per gram of material.

incompatible, this technique is of limited use, since crosslinking between like polymer chains can occur with equal probability.

The third type of grafting process is exemplified by reactions of oxazoline-substituted polystyrene (a commercial product). The oxazoline group forms addition compounds with a variety of other functional groups including carboxylic acid, anhydride, alcohol, amine, epoxide, mercaptan, and phenol. Grafting of an oxazoline-substituted polymer with a carboxyl-terminated polymer is shown in reaction (9.57). Such grafting reactions can be used, for example, for compatibilizing polymer blends, or for improving surface adhesion between polystyrene molded parts and appropriately functionalized surface coatings.

$$
\begin{array}{c}
\text{oxazoline group} + HO_2C\text{~~~} \longrightarrow \underset{\overset{\|}{O}}{\overset{}{-C}}NHCH_2CH_2O\underset{\overset{\|}{O}}{\overset{}{C}}\text{~~~}
\end{array}
\qquad (9.57)
$$

9.6   Polymer degradation

Stability of polymers is of critical interest to manufacturers, and, understandably, polymer degradation reactions have received considerable attention.[75-78] There are three principle methods of degrading polymers: (1) chemical, (2) thermal, and (3) radiative. In addition one can use ultrasonic radiation or mechanochemical techniques (for example, mastication) as described in the previous section.

9.6.1   Chemical degradation

This discussion is limited to reactions that cause breakdown of the polymer backbone; it does not cover reactions involving pendant groups. Because the backbones of vinyl polymers are made up of carbon chains containing no functional groups other than the double bonds of diene polymers, chemical degradation is essentially limited to oxidation. Most important is oxidation with oxygen because this has a direct bearing on polymer durability.

Saturated polymers are degraded very slowly by oxygen, and the reaction is autocatalytic. It can be speeded up considerably by application of heat or light or by the presence of certain impurities that catalyze the oxidation process. Tertiary carbon atoms are most susceptible to attack, and this is reflected in the following order of resistance to oxidation of three common polymers: polyisobutylene > polyethylene > polypropylene. Reaction products are numerous and include water, carbon dioxide, carbon monoxide, hydrogen, and alcohols. Crosslinking always accompanies degradation. It is believed that decomposition of initially formed hydroperoxide groups is mainly responsible for chain scission (9.58).

$$\underset{\underset{R}{|}\,\,\underset{R}{|}\,\,\underset{R}{|}}{\text{www}\,CH_2CHCH_2CCH_2CH\,\text{www}} \overset{OOH}{} \longrightarrow \underset{\underset{R}{|}}{\text{www}\,CH_2CHCH_2\cdot} + \underset{\underset{R}{|}\,\,\underset{R}{|}}{\overset{\overset{O}{\|}}{C}CH_2CH\,\text{www}} + \cdot OH$$

$$(9.58)$$

Unsaturated polymers undergo oxidative degradation much more rapidly by complex free radical processes involving peroxide and hydroperoxide intermediates. Allylic carbon atoms are most sensitive to attack, because resonance-stabilized radicals are formed. Oxidation is inhibited in commercial polymers by addition of antioxidants of the type discussed in Chapter 4 (Section 4.9).

Unsaturated polymers are also very susceptible to attack by ozone. Degradation by ozonolysis is discussed in Chapter 4 (Section 4.6). The discussion will not be repeated here except to remind the reader that an effective strategy for improving ozone resistance is to position the alkene moiety that is necessary for crosslinking such that oxidative bond cleavage causes no reduction in molecular weight. Also discussed in Chapter 4 (Section 4.7) are approaches to *enhancing* the oxidative degradability of vinyl polymers by incorporating enzyme-sensitive polymers such as starch or cellulose to promote oxidation by soil microorganisms.

9.6.2 Thermal degradation

There are basically three types of thermal degradation reactions for vinyl polymers[77-79]: (1) *nonchain scission*, (2) *random chain scission*, and (3) *depropagation*. Nonchain scission refers to reactions involving pendant groups that do not break the polymer backbone. Typical of such reactions are dehydrochlorination of poly(vinyl chloride) (9.12), elimination of acid from poly(vinyl esters)—for example, poly(vinyl acetate) (9.59)—and elimination of alkene from poly(alkyl acrylate)s (9.60). The first two reactions lead to highly colored residues, indicating that the double bonds formed in the

$$\underset{\underset{\underset{O}{\|}}{\overset{|}{OCCH_3}}}{-\!\!\!\!\!\left[CH_2CH\right]\!\!\!-} \longrightarrow -\!\!\!\!\!\left[CH=\!CH\right]\!\!\!- + \underset{\underset{O}{\|}}{HOCCH_3} \qquad (9.59)$$

$$\underset{\underset{\underset{OCH_2CH_2R}{|}}{\overset{|}{O=\!C}}}{-\!\!\!\!\!\left[CH_2CH\right]\!\!\!-} \longrightarrow -\!\!\!\!\!\left[CH_2CH\right]\!\!\!- + CH_2=\!CHR \atop \underset{\underset{OH}{|}}{\overset{|}{O=\!C}} \qquad (9.60)$$

polymer backbone are primarily conjugated. Enough side reactions occur, however, that such elimination reactions are not satisfactory for synthesizing polyacetylene. Nonchain scission has, however, been used as one approach to solving the problems of polyacetylene's intractability.[80-83] The method involves synthesis of a reasonably stable, tractable precursor polymer that can be purified and fabricated, then converted thermally to polyacetylene. An example involves the tricyclic monomer (**8**), which undergoes metathesis polymerization to precursor polymer (**9**) (9.61). Thermal degradation of films of the latter (9.62) yields coherent films of polyacetylene. Developed at the University of Durham, this particular approach to making polyacetylene is referred to as the "Durham route."

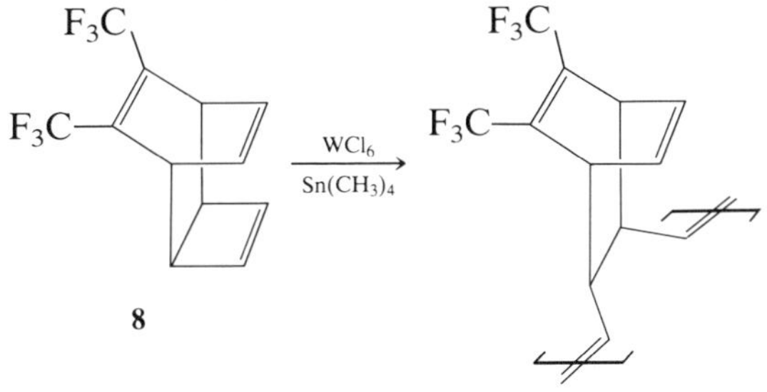

$$(9.61)$$

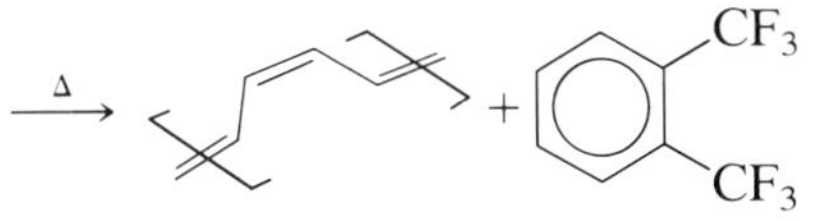

$$(9.62)$$

Nonchain scission reactions are useful for characterizing copolymers when the amount of a volatile degradation product can be correlated with the concentration of a given repeating unit.[84,85] (see Chapter 5, Section 5.5).

Random chain scission results from homolytic bond-cleavage reactions at weak points in the polymer chains. Complex mixtures of degradation products are formed, the origin of which may be explained in terms of radical transfer reactions such as (9.63). It may be recalled (Chapter 5, Figure 5.19) that a gas chromatographic pyrogram of polyethylene exhibits a homologous series of triplet peaks corresponding to alkane, 1-alkene, and α, ω-dialkene, with the largest peak in each triplet being that of 1-alkene. This follows from the breakdown pattern shown in Scheme 9.4. Random chain scission occurs with all vinyl polymers to varying degrees, but it occurs less with increasing substitution on the polymer backbone.

$$\sim\sim\sim CH_2CH_2CH_2CH_2 \longrightarrow \sim\sim\sim CH_2CH_2\cdot + \cdot CH_2CH_2\sim\sim\sim \longrightarrow$$

$$\sim\sim\sim CH=CH_2 + CH_3CH_2\sim\sim\sim \quad (9.63)$$

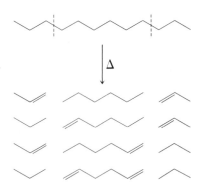

Scheme 9.4. Random chain scission of polyethylene.

Depropagation, or depolymerization (unzipping), to give monomer occurs mainly with polymers prepared from 1,1-disubstituted monomers. Initiation may be at a chain end or at a random site along the backbone. Poly(methyl methacrylate) appears to begin unzipping primarily at the chain ends, whereas poly(α-methylstyrene) does so at random sites along the chain. In both cases tertiary radicals are formed with each depropagation step (9.64). Polymers having single substituents on alternate carbons degrade by both depropagation and random chain scission, with the amount of monomer formed varying with temperature.[86]

$$\text{wwwCH}_2\underset{\underset{R}{|}}{\overset{\overset{R}{|}}{C}}\text{CH}_2\underset{\underset{R}{|}}{\overset{\overset{R}{|}}{C}}\cdot \longrightarrow \text{wwwCH}_2\underset{\underset{R}{|}}{\overset{\overset{R}{|}}{C}}\cdot + \text{CH}_2{=}\text{C}\overset{R}{\underset{R}{\diagdown}} \qquad (9.64)$$

9.6.3 Degradation by radiation

It was mentioned earlier that radiation may cause both crosslinking or degradation. Which predominates depends on radiation dosage, polymer structure, and temperature.[40,47] Ultraviolet or visible light causes 1,1-disubstituted polymers to degrade to monomer almost exclusively at elevated temperatures, whereas crosslinking and chain scission reactions predominate at room temperature. Other vinyl polymers undergo crosslinking primarily, regardless of temperature. Rearrangements may also occur as a result of homolysis and recombination reactions, especially with diene polymers.[87,88]

Ionizing radiation leads to much higher yields of monomer from 1,1-disubstituted polymers at room temperature. Degradations of this type initiated with electron beams are put to advantage in the manufacture of microcircuits using resist technology. At comparable levels of radiation, polyethylene and monosubstituted polymers undergo mainly crosslinking. All vinyl polymers tend to degrade under very high dosages of radiation.

References

1. E. M. Fettes (ed.), *Chemical Reactions of Polymers*, Wiley-Interscience, New York, 1964.
2. J. A. Moore (ed.), *Reactions of Polymers*, Reidel, Boston, 1973.
3. C. E. Carraher Jr., and M. Tsuda (eds.), *Modification of Polymers*, ACS Symp. Ser. 121, American Chemical Society, Washington, D.C., 1980.
4. C. E. Carraher Jr., and J. A. Moore (eds.), *Modification of Polymers*, Plenum Press, New York, 1983.
5. J. L. Benham and J. F. Kinstle (eds.), *Chemical Reactions on Polymers*, ACS Symp. Ser. 364, American Chemical Society, Washington, D.C., 1988.
6. H. Morawetz and P. E. Zimmering, *J. Phys. Chem.*, **56**, 753 (1954).
7. E. M. Fettes, in *Crystalline Olefin Polymers*, Part 2 (R. A. V. Raff and K. W. Doak, eds.), Wiley-Interscience, New York, 1965, Chap. 6.
8. M. A. Smook, W. J. Remington, and D. E. Strain, in *Polythene* (A. Renfrew and P. Morgan, eds.), Wiley-Interscience, New York, 1960, Chap. 14.
9. F. W. Keeley, in *Introduction to Rubber Technology* (M. Morton, ed.), Reinhold, New York, 1959, Chap. 14.
10. R. J. Lagow and J. L. Margrave, *J. Polym. Sci., Polym. Lett. Ed.*, **12**, 177 (1974).
11. D. M. Buck, P. D. Marsh, and K. J. Kallish, *Polym.-Plast. Technol. Engl.*, **26**, 71 (1987).
12. G. J. Jones, in *Chemical Reactions of Polymers* (E. M. Fettes, ed.), Wiley-Interscience, New York, 1964, Chap. 3.
13. M. Camps, M. Chatzopoulos, J.-M. Camps, and J.-P. Montheard, *J. Macromol. Sci.-Rev. Macromol. Chem. Phys.*, **C27**, 505 (1987–88).
14. N. K. Mathur, C. K. Narang, and R. E. Williams, *Polymers as Aids in Organic Chemistry*, Academic Press, New York, 1980.
15. W. T. Ford (ed.), *Polymeric Reagents and Catalysts*, ACS Symp. Ser. 308, American Chemical Society, Washington, D.C., 1986.
16. P. Hodge and D. C. Sherrington (eds.), *Polymer-Supported Reactions in Organic Synthesis*, Wiley-Interscience, New York, 1980.
17. G. Maneke and W. Storek, *Angew. Chem. Intern. Ed. English*, **17**, 657 (1978).
18. C. G. Overberger and K. N. Sannes, *Angew. Chem. Intern. Ed. English*, **13**, 99 (1974).
19. W. Heitz, *Adv. Polym. Sci.*, **23**, 1 (1977).
20. V. Pozzi, A. E. Silvers, L. Giuffre, and E. Cernia, *J. Appl. Polym. Sci.*, **19**, 923 (1975).
21. C. A. Finch, *Polyvinyl Alcohol: Properties and Applications*, Wiley-Interscience, New York, 1973.
22. J. G. Pritchard, *Poly(vinyl alcohol)*, Gordon and Breach, New York, 1970.
23. W. R. Sorensen and T. W. Campbell, *Preparative Methods of Polymer Chemistry*, 2nd ed., Wiley-Interscience, New York, 1968, pp. 239ff.
24. N. N. Aylward, *J. Polym. Sci. A-1*, **8**, 319 (1970).
25. M. Sugiura, M. Ochi, Y. Tani, and Y. Nagai, *Kogyo Kagaku Zasshi*, **72**, 1926 (1969); *Chem. Abstr.*, **72**, 22105b (1970).
26. H. Kawaguchi, P. Loeffler, and O. Vogl, *Polymer*, **26**, 1257 (1985).
27. C. S. Marvel, J. H. Sample, and M. F. Roy, *J. Am. Chem. Soc.*, **61**, 3241 (1939).
28. K. Sanui, W. J. MacKnight, and R. W. Lenz, *Macromolecules*, **7**, 952 (1974).

29. Y. Minoura and H. Ikeda, *J. Appl. Polym. Sci.*, **15**, 2219 (1971).
30. J. P. Kennedy and K. F. Castner, *J. Polym. Sci., Polym. Chem. Ed.*, **17**, 2039, 2055 (1979).
31. B. Iván, J. P. Kennedy, T. Kelen, and F. Tüdös, *Polym. Bull.*, **1**, 415 (1979).
32. J. P. Kennedy, *J. Appl. Polym. Sci., Appl. Polym. Symp.*, **39**, 21 (1984).
33. P. Ehrburger and J.-B. Donnet, in *High Technology Fibers*, Part A (M. Lewin and J. Preston, eds.), Dekker, New York, 1985.
34. G. Henrici-Olivé and S. Olivé, *Adv. Polym. Sci.*, **51**, 1 (1983).
35. E. Fitzer, *Angew. Chem. Intern. Ed. English*, **19**, 375 (1980).
36. T. Ogawa, R. Cedeno, and E. T. Herrera, *Makromol. Chem.*, **180**, 785 (1979).
37. G. J. Smets, in *Chemical Reactions of Polymers* (E. M. Fettes, ed.), Wiley-Interscience, New York, 1964, Chap. 1D.
38. I. R. Gelling, *Rubber Chem. Technol.*, **58**, 86 (1985).
39. M. A. Golub, in *Polymer Chemistry of Synthetic Elastomers*, Part 2 (J. P. Kennedy and E. G. Tornqvist, eds.), Wiley-Interscience, New York, 1969, Chap. 10A.
40. D. Craig, in *Chemical Reactions of Polymers* (E. M. Fettes, ed.), Wiley-Interscience, New York, 1964, Chap. 9C.
41. A. Y. Coran, in *Science and Technology of Rubber* (F. R. Eirich, ed.), Academic Press, New York, 1978.
42. A. Y. Coran, *CHEMTECH*, **13**, 106 (1983).
43. G. Alliger and I. J. Sjothun (eds.), *Vulcanization of Elastomers*, Van Nostrand Reinhold, New York, 1964.
44. A. R. Schultz, in *Chemical Reactions of Polymers* (E. M. Fettes, ed.), Wiley-Interscience, New York, 1964, Chap. 9A.
45. M. Dole (ed.), *The Radiation Chemistry of Macromolecules*, Vols. 1 and 2, Academic Press, New York, 1973.
46. F. A. Bovey, *The Effects of Ionizing Radiation on Natural and Synthetic High Polymers*, Wiley-Interscience, New York, 1958, Chaps. 3–8.
47. A. Chapiro, *Radiation Chemistry of Polymeric Systems*, Wiley-Interscience, New York, 1962.
48. D. R. Randell (ed.), *Radiation Curing of Polymers*, CRC Press, Boca Raton, Fla., 1987.
49. G. E. Green, B. P. Stark, and S. A. Zahir, *J. Macromol. Sci., Revs. Macromol. Chem.*, **C21**, 187 (1981–82).
50. G. Oster and N. Yang, *Chem. Rev.*, **68**, 125 (1968).
51. M. J. Farrall, M. Alexis, and M. Trecarton, *Polymer*, **24**, 114 (1983).
52. A. O. Patil, D. D. Deshpande, and S. S. Talwar, *Polymer*, **22**, 434 (1981).
53. O. Zimmer and H. Meier, *J. Chem. Soc., Chem. Commun.*, 481 (1982).
54. M. Tsunooka, T. Ueda, and M. Tanaka, *J. Polym. Sci., Polym. Lett. Ed.*, **19**, 201 (1981).
55. M. P. Stevens and A. D. Jenkins, *J. Polym. Sci., Polym. Chem. Ed.*, **17**, 3675 (1979).
56. M. Kata, M. Hasegawa, and T. Ichijo, *J. Polym. Sci. B*, **8**, 263 (1970).
57. N. J. Gaylord and F. S. Ang, in *Chemical Reactions of Polymers* (E. M. Fettes, ed.), Wiley-Interscience, New York, 1964, Chap. 10B.
58. N. Grassie and J. Gilks, *J. Polym. Sci., Polym. Chem. Ed.*, **11**, 1531 (1973).
59. J. K. Stille and L. D. Gotter, *Macromolecules*, **2**, 468 (1969).
60. R. Longwirth, in *Ionic Polymers* (L. Holliday, ed.), Applied Science, London, 1975, Chap. 2.

61. R. H. Kinsey, *J. Appl. Polym. Sci., Appl. Polym. Symp.*, **11**, 77 (1969).
62. A. Noshay and J. E. McGrath, *Block Copolymers: Overview and Critical Survey*, Academic Press, New York, 1977.
63. H. A. J. Battaerd and G. W. Tregear, *Graft Copolymers*, Wiley-Interscience, New York, 1967, Chap. 3.
64. R. J. Ceresa (ed.), *Block and Graft Copolymerization*, Wiley-Interscience, New York, 1973.
65. G. E. Ham (ed.), *Copolymerization*, Wiley-Interscience, New York, 1964: (a) A. S. Hoffman and R. Bacskai, Chap. 6; (b) G. Bier and G. Lehman, Chap. 4B.
66. R. Jerome, R. Fayt, and T. Ouhadi, *Progr. Polym. Sci.*, **10**, 87 (1984).
67. A. D. Jenkins, *J. Appl. Polym. Sci., Appl. Polym. Symp.*, **36**, 185 (1981).
68. H. C. Hass, P. M. Kamath, and N. W. Schuler, *J. Polym. Sci.*, **24**, 85 (1957).
69. H. C. Hass and P. M. Kamath, *J. Polym. Sci.*, **24**, 143 (1957).
70. N. G. Gaylord and A. Takahashi, *J. Polym. Sci., B*, **8**, 361 (1970).
71. G. Greber, J. Tölle, and W. Burchard, *Makromol. Chem.*, **71**, 47 (1964).
72. G. Greber and J. Tölle, *Makromol. Chem.*, **53**, 208 (1962).
73. G. Greber and G. Egle, *Makromol. Chem.*, **53**, 206 (1962); **64**, 68 (1963).
74. H. Mark, *Textile Res. J.*, **23**, 294 (1954).
75. O. Cicchetti, *Adv. Polym. Sci.*, **7**, 70 (1970).
76. L. Reich and S. S. Stivala, *Elements of Polymer Degradation*, McGraw-Hill, New York, 1971.
77. S. L. Madorsky, *Thermal Degradation of Organic Polymers*, Wiley-Interscience, New York, 1964.
78. A. L. Bhuiyan, *Adv. Polym. Sci.*, **47**, 43 (1982).
79. J. H. Flynn and R. E. Florin, in *Pyrolysis and GC in Polymer Analysis* (S. A. Liebman and E. J. Levy, eds.), Dekker, New York, 1985, Chap. 4.
80. J. H. Edwards, W. J. Feast, and D. C. Bott, *Polymer*, **25**, 395 (1984).
81. W. J. Feast and J. N. Winter, *J. Chem. Soc., Chem. Commun.*, 202 (1985).
82. D. White and D. C. Bott, *Polym. Commun.*, **25**, 98 (1984).
83. S. A. Jenekhe, in *Chemical Reactions on Polymers* (J. L. Benham and J. F. Kinstle, eds.), ACS Symp. Ser. 364, American Chemical Society, Washington, D.C., 1988, Chap. 32.
84. L. A. Wall, in *Analytical Chemistry of Polymers* (G. M. Kline, ed.), Wiley-Interscience, New York, 1964, Part 2, Chap. 5.
85. M. P. Stevens, *Characterization and Analysis of Polymers by Gas Chromatography*, Dekker, New York, 1969, Chap. 4.
86. F. A. Lehmann and G. M. Brauer, *Anal. Chem.*, **33**, 673 (1961).
87. M. A. Golub and M. L. Rosenberg, *J. Polym. Sci., Polym. Chem. Ed.*, **18**, 2543 (1980).
88. S. W. Shalaby, *Macromol. Rev.*, **14**, 419 (1979).

Review exercises

1. Write a concise definition of the following terms: (a) depropagation; (b) ionomer; (c) nonchain scission; (d) random chain scission; (e) vulcanization.

2. Write equations for the expected reaction of polystyrene with (a) HNO_3/H_2SO_4; (b) $Br_2/FeBr_3$; (c) H_2SO_4/SO_3; (d) N-bromosuccinimide (NBS); (e) $CH_3COCl/SnCl_4$.

3. Write equations for the expected reaction of 1,4-polyisoprene with the following reagents. (You may wish to consult your organic chemistry text.) For those reactions that do not result in degradation, indicate how the polymer's properties might be altered. (a) Br_2; (b) F_2; (c) H_2O, H^+; (d) H_2, Ni; (e) $CHCl_3$, base; (f) N-bromosuccinimide; (g) *m*-Chloroperbenzoic acid; (h) $KMnO_4$, base; (i) $KMnO_4$, H^+; (j) 1. $CH_3OCH_2OCH_2CH_2OH$, $Hg(OAc)_2$; 2. NaBH. (The product of this two-step reaction sequence on 1,3-butadiene has been suggested as a polyelectrolyte by S. E. Bell and D. J. Bannister, *J. Polym. Sci. C., Polym. Lett.*, **24**, 165, 1985.)

4. Commercial styrene–diene AB and ABA block copolymers are often hydrogenated. What advantage does this procedure offer?

5. When colorless poly(1,3-butadiene) is treated with potassium *t*-butoxide, the polymer turns blue-black. Speculate on what chemical change might have occurred. Propose a mechanism for the change. (A. J. Dias and T. J. McCarthy, *Macromolecules*, **18**, 869, 1985)

6. Suggest a way to synthesize (a) head-to-head polypropylene (O. Vogl, *Polym. Preprints*, **20**(1), 154, 1979); (b) head-to-head poly(vinyl alcohol); (c) the "alternating copolymer" of ethylene and vinylidene chloride (1,1-dichloroethene).

7. Using designations of the type used in exercise 6, how might you describe (a) the polymer prepared by oxidative cleavage of 1,4-poly(1,3-cyclohexadiene) (see reaction 8.20); (b) the polymer formed by hydration (H^+, H_2O) of 1,4-polyisoprene; (c) the product of hydroboration of 1,4-polyisoprene (reaction 9.14)?

8. Show how polyethylene modified according to reaction (9.6) undergoes photodegradation. Do the same for poly(methyl vinyl ketone).

9. Cyclization of 1,4-polyisoprene under the influence of acid is shown in reaction (9.21). Under acidic conditions, 3,4-polyisoprene also undergoes cyclization. Give the likely structure of the product and show how it is formed.

10. In reaction 9.38, explain why linear rather than crosslinked polymer is formed.

11. The chloro group of chloromethylated polystyrene (see reaction 9.3) undergoes facile displacement with SH^-. Show how this could be used to advantage to prepare polystyrene-*graft*-poly(methyl methacrylate).

12. Show how chloromethylated polystyrene could be converted to polystyrene-*graft*-polyisobutylene.

13. What advantages does the macromonomer route (Chapter 6) have over conventional methods of preparing graft copolymers?

14. Ease of oxidation of vinyl polymers is in the order polybutadiene > polypropylene > polyethylene > polyisobutylene. Explain why.

15. List at least two compounds that can be used to formulate more ozone-resistant elastomers, and write the structural units for the compounds as they are incorporated into the polymer.

16. Give the products of reductive ozonolysis of (a) 1,4-poly(1,3-butadiene); (b) 1,2-poly(1,3-butadiene); (c) poly(isobutylene-*co*-cyclopentadiene).

17. How might one conveniently determine whether thermal depropagation of a vinyl polymer is initiated at the chain ends or at random sites along the backbone?

18. On exposure to uv light, natural rubber (1,4-polyisoprene) undergoes extensive rearrangement leading to (among other structures) some 1,2- and 3,4-polyisoprene linkages. Show how these structures are likely to arise.

19. An ionic polymer behaves as a thermoplastic elastomer even if the counterion is monovalent. Speculate on the reason. (V. K. Daltye and R. L. Taylor, *Macromolecules*, **17**, 414, 1984)

Part III
NONVINYL POLYMERS

10

Step-reaction and ring-opening polymerization

10.1 Introduction

The preceding four chapters dealt with polymers having carbon–carbon bonds in the backbone prepared from alkene and diene monomers. With the exception of ring-opening metathesis, such polymers are almost always prepared by chain-reaction addition polymerization. The remainder of this book is devoted almost exclusively to polymers having functional groups in the backbone that, in most instances, are prepared by step-reaction or ring-opening polymerization. The purpose of this chapter is to explore briefly the characteristics of these two polymerization processes. More detailed mechanistic treatments of individual polymer types are provided in the appropriate chapters.

10.2 Step-reaction polymerization — kinetics

At this point it behooves the student to review the material in Chapter 1 (Sections 1.4 and 1.5) that compares step-reaction and chain-reaction polymerizations. Among the distinctive features[1-5] of the former are:

1. Linear polymers are synthesized either from difunctional monomers of the AB type (where A and B represent coreactive functional groups) or from a combination of AA and BB difunctional monomers.
2. Network polymers are formed from monomers having a functionality greater than two..
3. Polymers retain their functionality as end groups at the completion of polymerization.
4. A single reaction (or reaction sequence) is responsible for all steps contributing to polymer formation (in contrast to initiation, propagation, and termination in chain-reaction polymerization).
5. Molecular weight increases slowly even at high levels of conversion. This precept is embodied in the Carothers equation relating $\overline{DP}$ to monomer

conversion (p):

$$\overline{DP} = \frac{1}{1-p}$$

6. High yield reactions and an exact stoichiometric balance are necessary to obtain high-molecular-weight linear polymer. While an exact stoichiometric balance is readily achieved by using highly purified AB-type monomer, this achievement is much more difficult when using AA and BB monomers. And in both cases, side reactions (e.g., dehydration of alcohol in polyesterification) will upset the balance.

From the practical standpoint, it should be realized that the very high molecular weights one can obtain routinely by chain-reaction polymerization are very difficult to achieve by step polymerization of difunctional monomers. Furthermore, many step polymerization reactions are equilibrium condensation processes; it is necessary in these cases to provide an efficient means of removing by-products to drive the reaction to high conversion. A further complication arises from the tendency of certain monomers to form cyclic compounds, particularly where five- or six-membered rings are involved. Thus ethylene glycol, in the presence of acid, forms 1,4-dioxane (10.1) in preference to polyether (10.2), and 4-aminobutanoic acid forms cyclic amide (lactam) (10.3) in addition to polyamide (10.4). More is said about ring stability later in the chapter.

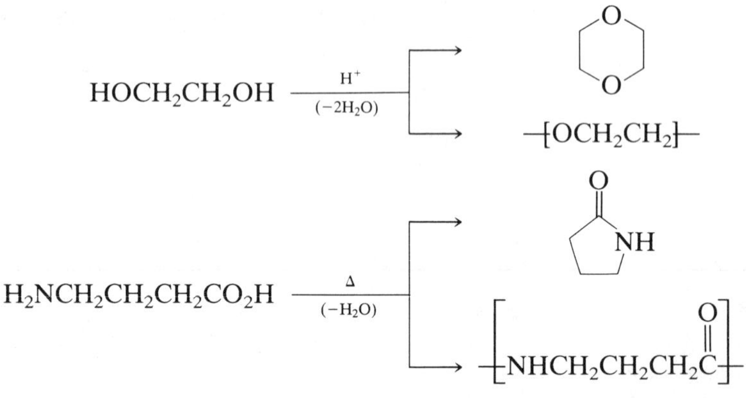

$$HOCH_2CH_2OH \xrightarrow[(-2H_2O)]{H^+}$$

(10.1)

$$-\!\!\left[OCH_2CH_2\right]\!\!-$$ (10.2)

$$H_2NCH_2CH_2CH_2CO_2H \xrightarrow[(-H_2O)]{\Delta}$$

(10.3)

$$-\!\!\left[NHCH_2CH_2CH_2\overset{\displaystyle O}{\overset{\|}{C}}\right]\!\!-$$ (10.4)

Table 10.1 lists most of the commercially important polymers prepared by step-reaction polymerization in the order in which they appear in subsequent chapters. A wealth of step-reaction polymers have been investigated besides those listed. From the standpoint of mechanism and kinetics, polyesters and polyamides have been most thoroughly studied. Such studies allow us to make certain simplifying assumptions in order to relate $\overline{DP}$ to polymerization rate in a manageable way. One of the most important is that the rate constant is independent of chain length. Indeed, this has been shown to be the case for

Table 10.1. Commercially important polymers prepared by step-reaction polymerization

Polymer type	Repeating functional unit	Chapter
Polyether	—Ar—O—	11
Polyether [poly(phenylene oxide)]	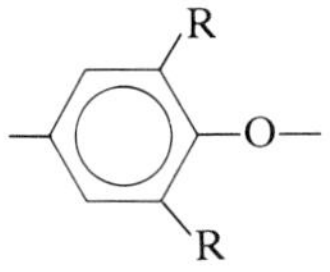 (aromatic ring with R at top and R at bottom, —O—)	11
Polyether (epoxy)[a]	$-CH_2CHCH_2OAr-$ with OH	11
Polysulfide	—ArS—	11
Poly(alkylene polysulfide)	$-RS_x-$	11
Polysulfone	$-ArSO_2-$	11
Polyester	$-R\overset{O}{\overset{\|}{C}}O-$	12
Polycarbonate	$-RO\overset{O}{\overset{\|}{C}}O-$	12
Polyamide	$-R\overset{O}{\overset{\|}{C}}NH-$	13
Polyurea	$-RNH\overset{O}{\overset{\|}{C}}NH-$	13
Polyurethane	$-RO\overset{O}{\overset{\|}{C}}NH-$	13
Polyhydrazide	$-R\overset{O}{\overset{\|}{C}}NHNH-$	13
Polyimide	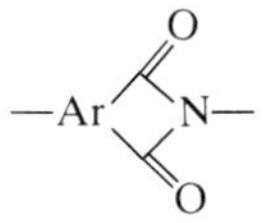 —Ar (imide ring with two C=O, N)—	13
Polybenzimidazole	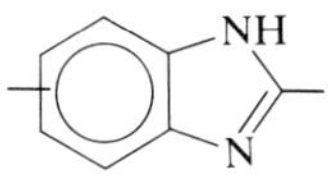 (benzimidazole ring, NH, N)	13

(*continues*)

Table 10.1 (*Continued*)

Polymer type	Repeating functional unit	Chapter
Phenol–formaldehyde[b]	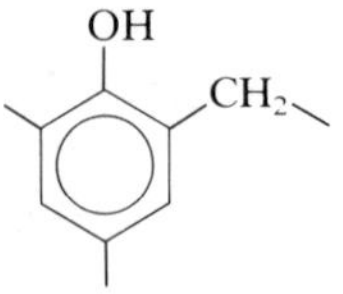 OH / benzene ring with CH₂ substituent	14
Urea–formaldehyde[c]	 $>NCNHCH_2-$ with O above C (carbonyl)	14
Melamine–formaldehyde[b]	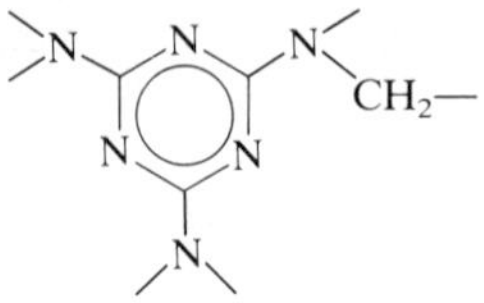 triazine ring with N substituents and CH_2-	14
Polyphenylene	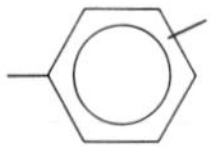 benzene ring	16
Poly(*p*-xylylene)	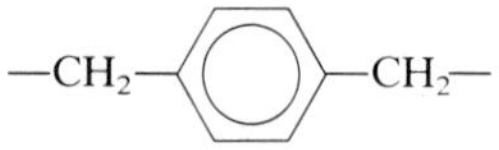 $-CH_2-$ benzene ring $-CH_2-$	16

[a] Formed from epoxide-terminated oligomer.
[b] Complex network structure.
[c] Network or linear; see Chap. 14 for discussion.

simple esterification of a homologous series of linear carboxylic acids in which the rate becomes virtually constant once the chain length reaches four carbons. Clearly the situation is more complicated where viscosity effects might hinder the rate at which carboxyl and hydroxyl groups diffuse together. Because most step polymerization reactions have relatively high activation energies, the rate of diffusion is usually much higher than the rate of the functional group reaction. Where diffusion rate might be more critical, however, it is assumed that once the groups are close enough to react, they are more favorably disposed to do so because the diffusion of groups *away* from one another is equally slowed by viscosity effects. Another simplifying assumption, which is more intuitively reasonable, is that the *mechanism* of the reaction remains constant throughout.

If the polymerization reaction is first order with respect to each functional group reactant, A and B, then the rate expression is

$$-\frac{d[A]}{dt} = k[A][B]$$

Given the fact that an exact stoichiometric balance is usually a necessary condition for achieving high molecular weight, we can equate [A] and [B], and the rate expression then becomes

$$-\frac{d[A]}{dt} = k[A]^2$$

or, by integration,

$$\frac{1}{[A]} - \frac{1}{[A_0]} = kt$$

At any particular time, t, in the polymerization process, $\overline{DP}$ is equal to the ratio of monomer molecules present initially to the total number at that time; that is,

$$\overline{DP} = \frac{[A_0]}{[A]}$$

Combining this expression with the Carothers equation and solving for [A], we have

$$[A] = [A_0](1 - p)$$

By substitution in the second-order rate expression, we obtain

$$\frac{1}{[A_0](1 - p)} - \frac{1}{[A_0]} = kt$$

Since $\overline{DP} = 1/(1 - p)$,

$$\frac{\overline{DP}}{[A_0]} - \frac{1}{[A_0]} = kt$$

or

$$\overline{DP} = [A_0]kt + 1$$

Thus if one knows the rate constant, one can calculate the time needed to reach any number average molecular weight for a given concentration of monomer.

While the foregoing relationships are valid for a great many step-reaction polymerizations, they do not apply to the very important uncatalyzed polyesterification reactions in which carboxylic acid is monomer. It may be recalled from organic chemistry that esterification is normally acid catalyzed. In preparing polyesters, however, where high temperatures are usually necessary to reduce viscosity and facilitate removal of water, an acid catalyst might well have an adverse effect, causing discoloration and even decomposition of monomer or polymer. In the absence of added acid the carboxylic acid assumes the role of catalyst, and the rate expression then becomes second order in acid, or third order overall:

$$-\frac{d[A]}{dt} = k[A]^2[B]$$

Again, assuming $[A]$ and $[B]$ are equal,

$$-\frac{d[A]}{dt} = k[A]^3$$

Integration then gives

$$\frac{1}{[A]^2} - \frac{1}{[A_0]^2} = 2kt$$

Substituting for $[A]$ as above and rearranging,

$$\frac{1}{[A_0]^2(1-p)^2} - \frac{1}{[A_0]^2} = 2kt$$

or

$$\frac{1}{(1-p)^2} = 2kt[A_0]^2 + 1$$

and

$$(\overline{DP})^2 = 2kt[A_0]^2 + 1$$

Thus, in the case of uncatalyzed polyesterification, the molecular weight increases more gradually than when acid catalyst is added.

These relationships, which have been confirmed experimentally, do not apply over the full polymerization reaction. Plots of $\overline{DP}$ versus t deviate from linearity at both low and high levels of functional group conversion. The reason cited most often for deviations in the early stages of polyesterification has to do with changes in polarity of the reaction medium as the highly polar alcohol and acid are converted to the much lower polarity ester. The more polar the medium, the more association through hydrogen bonding inhibits reactivity. In the uncatalyzed polyesterification of adipic acid and diethylene glycol (10.5), for example, 80% of the functional groups are consumed before the third-order relationship given assumes linearity. (As may be seen by application of the Carothers equation, this still represents a very low molecular weight.) Deviations at very high levels of conversion arise partly from difficulties in removing water from the highly viscous reaction medium and, in the case of acid-catalyzed reactions, from the reaction of catalyst with hydroxyl end groups.

$$HO_2C(CH_2)_4CO_2H + HOCH_2CH_2OH \longrightarrow$$

$$\left[\begin{matrix} O & O \\ \| & \| \\ C(CH_2)_4COCH_2CH_2O \end{matrix} \right] + H_2O \quad (10.5)$$

10.3 Stoichiometric imbalance

Despite the fact that an exact stoichiometric balance is necessary to achieve high molecular weights in most instances of step-reaction polymerization, in some important applications it is necessary to limit the molecular weight. Formation of epoxy resins (Chapter 11) is one. As footnoted in Table 10.1, these are polyethers that are processed to an oligomer stage and are subsequently converted to network polymer by appropriate reactions of terminal epoxide groups. Similarly, unsaturated polyesters (Chapter 12) are processed to moderately low molecular weights before being crosslinked. With polyamides for fiber applications, molecular weight must often be limited because too high a viscosity is detrimental to extrusion of filaments through the fine holes of a spinneret.

There are three ways to limit molecular weight in step-reaction polymerization. One is to quench the polymerization reaction by rapid cooling when the desired molecular weight (or viscosity) is attained. This is routinely done in the manufacture of unsaturated polyesters. Another is to use an excess of one monomer when two difunctional monomers (AA and BB) are polymerized. This is the method of choice with epoxy resins to ensure that epoxide end groups are present in the oligomer. The third method is to use small amounts of monofunctional reactant. Addition of acetic acid, for example, is one way to limit the molecular weight of nylon 66 (prepared from adipic acid and 1,6-hexanediamine) for fiber manufacture.

When a nonstoichiometric amount of functional groups is used, the relationship between $\overline{DP}$ and reaction conversion can be quantified by a modification of the Carothers equation. At this point we introduce a factor, r, representing the *stoichiometric imbalance*. If the number of A and B functional groups present at the beginning of an AA + BB polymerization is N_A^0 and N_B^0, respectively,

$$r = \frac{N_A^0}{N_B^0}$$

By convention, r is always less than unity (except, of course, when $N_A^0 = N_B^0$). As before, p is the reaction conversion, which in this case represents the fraction of A groups that have reacted. Because the reaction of each A group consumes one B group, the fraction of B reacted at conversion p is equal to pN_A^0, or prN_B^0. The number of *unreacted* groups, N_A and N_B, is then given by

$$N_A = (1 - p)N_A^0 \quad \text{and} \quad N_B = (1 - pr)N_B^0 = (1 - pr)\frac{N_A^0}{r}$$

At this point, the number of A and B end groups is equal to $N_A + N_B$ and, because there are two end groups per molecule, the number of molecular chains, N, is given by

$$N = \frac{1}{2}(N_A + N_B)$$

that is,

$$N = \frac{1}{2}\left[(1-p)N_A^0 + (1-pr)\frac{N_A^0}{r}\right]$$

which reduces to

$$N = \frac{N_A^0}{2}\left(1 + \frac{1}{r} - 2p\right)$$

We know that one repeating unit is formed every time an A and a B functional group react; therefore, the total number of repeating units, N_r, is given by

$$N_r = \frac{1}{2}(N_A^0 + N_B^0)$$

Since $r = N_A^0/N_B^0$,

$$N_r = \frac{1}{2}\left(N_A^0 + \frac{N_A^0}{r}\right) = \frac{N_A^0}{2}\left(\frac{r+1}{r}\right)$$

The average degree of polymerization is equal to the number of monomeric units divided by the number of chains; that is

$$\overline{DP} = \frac{N_r}{N} = \frac{(N_A^0/2)[(r+1)/r]}{N_A^0/2[1+(1/r)-2p]}$$

which reduces to

$$\overline{DP} = \frac{1+r}{r+1-2rp}$$

Given the stoichiometric imbalance, therefore, we can calculate the extent of reaction necessary to achieve a given degree of polymerization. If $r = 1$, the relationship reduces to

$$\overline{DP} = \frac{1}{1-p}$$

which is, of course, the Carothers equation. When monomer A is completely used up in the polymerization (that is, when $p = 1$), the equation becomes

$$\overline{DP} = \frac{1+r}{1-r}$$

Now, what about the situation where monofunctional reagent is added to control molecular weight? According to our definition of r, these must necessarily be B groups. The imbalance factor is then redefined as

$$r' = \frac{N_A^0}{N_B^0 + 2N_B^{0'}}$$

where $N_B^{0'}$ is the number of monofunctional B groups. The factor 2 takes into account the fact that each monofunctional B' molecule is equally as effective as one excess BB monomer in limiting the molecular weight. The factor r' is substituted for r in the previous equations.

Up to this point we have only considered reactions involving AA and BB. For the AB type, $r = 1$ and molecular weight is controlled by addition of monofunctional reagent. The same relationships apply, except that in the definition of r', N_A^0 is necessarily always equal to N_B^0.

10.4 Molecular weight distribution

Using statistical methods derived by Flory,[1,2,6] it is possible to relate the molecular weight distribution in step polymerization to the reaction conversion. In any polymerization involving difunctional monomer (AB or AA + BB), each step reaction links two molecules together; thus the number of monomer units in the polymer is always one more than the number of functional group reactions. If a polymer contains x monomer units ($\overline{DP} = x$), the number of A or B groups that have reacted is $x - 1$. The remaining unreacted A or B groups are at the ends of the chain. It is assumed that each reaction occurs in random fashion throughout the polymerization medium. The problem is to determine the probability of finding a chain containing x monomer units and a single unreacted A or B group at time t.

The probability that $x - 1$ A or B groups has reacted is p^{x-1} where p is the reaction conversion, defined previously as

$$p = \frac{N_0 - N}{N_0}$$

The probability of finding an *unreacted* group is $1 - p$. The probability of finding a molecule containing x units and an unreacted A or B group is then $p^{x-1}(1-p)$. If the total number of molecules present at time t is N, then the fraction that contains x units, N_x, is given by

$$N_x = Np^{x-1}(1-p)$$

Knowing that $N/N_0 = 1 - p$ (Carothers equation), we can rewrite the expression for N_x as

$$N_x = N_0(1-p)^2 p^{x-1}$$

where N_0 is the number of monomer units present initially. This relationship defines the number or mole fraction distribution at time t. Plots of N_x versus x at three different reaction conversions are shown in Figure 10.1.[2] Interestingly, even at 99% conversion, monomer still represents the most abundant species present. This is misleading, however, because monomer contributes a small percentage to the overall mass. A more reasonable picture is obtained by expressing the distribution in terms of the weight fraction. The weight

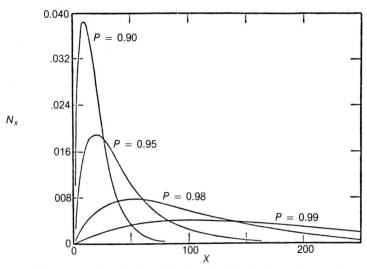

Figure 10.1. Mole fraction distribution of chain molecules in linear step-reaction polymers. [From Flory,[2] copyright 1946. Reprinted by permission of the American Chemical Society.]

fraction, w_x, of molecules containing x units is equal to the ratio of the mass of those molecules divided by the total mass; that is

$$w_x = \frac{xN_xM_0}{N_0M_0} = \frac{xN_x}{N_0}$$

where M_0 is the mass of the repeating unit. Substituting in the previous expression for N_x, we obtain

$$w_x = x(1-p)^2 p^{x-1}$$

which is the weight fraction distribution. A plot of w_x versus x at four levels of conversion are shown in Figure 10.2.[2] Both figures confirm what we already know from the Carothers equation, that very high conversions are necessary to obtain high degrees of polymerization.

To determine the polydispersity index $(\bar{M}_w/\bar{M}_n)$ at a given conversion, we need to define $\bar{M}_w$ and $\bar{M}_n$ in terms of p. Given that $\bar{M}_n$ is the product of $\overline{DP}$ and M_0, and $\overline{DP} = 1/(1-p)$, we can write

$$\bar{M}_n = \frac{M_0}{1-p}$$

To obtain a similar expression for $\bar{M}_w$, we take the expression

$$\bar{M}_w = \Sigma w_x M_x$$

and rewrite it for x units as

$$\bar{M}_w = \Sigma w_x x M_0$$

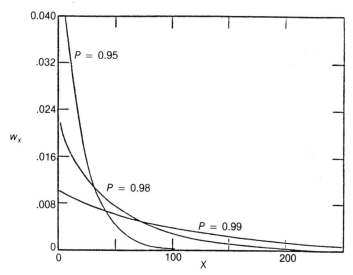

Figure 10.2. Weight fraction distribution of chain molecules in linear step-reaction polymers. [From Flory,[2] copyright 1946. Reprinted by permission of the American Chemical Society.]

Substituting the expression for w_x above, we obtain

$$\bar{M}_w = M_0(1-p)^2\Sigma x^2 p^{x-1}$$

The series $\Sigma x^2 p^{x-1}$ reduces to $(1+p)/(1-p)^3$, therefore

$$\bar{M}_w = \frac{M_0(1+p)}{1-p}$$

Dividing $\bar{M}_w$ by $\bar{M}_n$ we arrive at the polydispersity index

$$\frac{\bar{M}_w}{\bar{M}_n} = 1+p$$

which tells us that at very high levels of conversion (as p approaches 1), the most probable polydispersity approaches 2. This has been confirmed experimentally.

It should be noted that in the case of condensation polymerization, each polymer molecule contains the mass of the by-product in its end groups. This usually represents a minor contribution to the overall mass, however, and is ignored in the above equations.

10.5   Network step polymerization

If monomers containing a functionality greater than two are used in step polymerization, chain branching results. An important application of this type of polymerization is the manufacture of alkyd-type polyester resins

(Chapter 12) in which polyfunctional alcohols such as glycerol are used. Phenol–, urea–, and melamine–formaldehyde polymers (Chapter 14) are other important examples. If the reaction is carried to a high enough conversion, gelation occurs. The onset of gelation, or *gel point*, is accompanied by a sudden increase in viscosity such that the polymer undergoes an almost instantaneous change from a liquid to a gel. Bubbles no longer rise in the mixture, and stirring becomes impossible. If the polymerization is being done in solution, the gel is swollen with solvent but is otherwise insoluble. At this point the polymer is a true network polymer, and the crosslink density will continue to rise with a concomitant increase in rigidity if the reaction is allowed to continue. Gelation during polymer synthesis is normally something to be avoided because it can have disastrous consequences if it occurs in a large-scale reactor. Clearly it is desirable to be able to predict the gel point.

One way to predict the onset of gelation is to use a modified form of the Carothers equation that takes into account the *average functionality* of the monomers involved.[7] Suppose, for example, one wanted to prepare a polyester using an equimolar mixture of phthalic anhydride (**1**) (which reacts as a diacid), trimellitic acid (**2**), ethylene glycol (**3**), and glycerol (**4**).

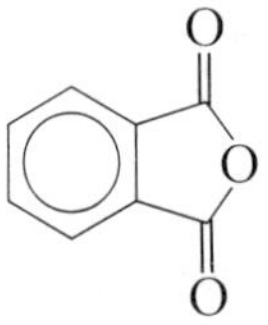

1 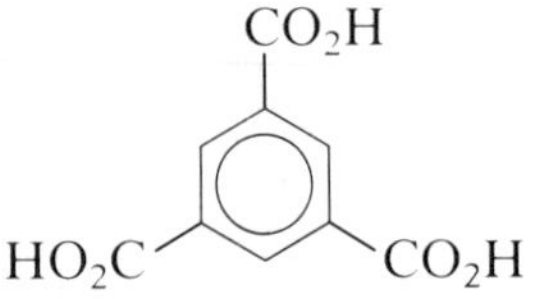   2

HOCH$_2$CH$_2$OH   HOCH$_2$CHCH$_2$OH

3   4

The average functionality, f_{av}, is $(2 + 3 + 2 + 3)/4 = 2.5$. Given that

$$p = \frac{N_0 - N}{N_0}$$

where N_0 and N are the number of monomer molecules initially and at conversion p, respectively, then the number of functional groups that have reacted is $2(N_0 - N)$. The number of functional groups initially is $N_0 f_{av}$. Thus,

$$p = \frac{2(N_0 - N)}{N_0 f_{av}}$$

Since $\overline{DP} = N_0/N$, the above expression may be rewritten as

$$p = \frac{2}{f_{av}} - \frac{2}{\overline{DP} f_{av}}$$

It is assumed that gelation occurs when $\overline{DP}$ becomes infinite, at which point the second term becomes zero. Then

$$p_c = \frac{2}{f_{av}}$$

where p_c denotes the *critical reaction conversion* at which gelation occurs. For the system given above, this relationship predicts that gelation would occur at a reaction conversion of 80%. This may be contrasted with a system consisting solely of difunctional monomers in which an 80% conversion leads to a $\overline{DP}$ of only 5.

Similarly, a mixture of 3 mol of **1** and 2 mol of **4**, which has a functionality of

$$\frac{(3 \times 2) + (2 \times 3)}{5} = 2.4$$

is predicted to gel at 83% conversion. In practice, such a mixture gels at about 77% conversion. The discrepancy arises primarily from the greater contribution of higher molecular weight fractions.

Both examples given have *equivalent* amounts of acid and alcohol functional groups. Where nonequivalent amounts are involved, the determination of f_{av} takes a different form.[8] Let's consider a mixture consisting of three monomers A, B, and C, in which A and C have the same functional groups but different functionalities (e.g., a mixture of two acids such as **1** and **2**), B contains a different functional group (e.g., an alcohol such as **3** or **4**), and the B groups are in excess. Then the average functionality is given by

$$f_{av} = \frac{2rf_A f_B f_C}{f_A f_C + r\rho f_A f_B + r(1-\rho)f_B f_C}$$

where the constants r and ρ are given by

$$r = \frac{N_A f_A + N_C f_C}{N_B f_B}$$

and

$$\rho = \frac{N_C f_C}{N_A f_A + N_C f_C}$$

The critical conversion, p_c, then refers to the extent of reaction of the A groups only.

Statistical methods have also been developed that predict gelation at a lower level of conversion than that predicted by the Carothers equation.[1,9] For this case, with f_A and f_B each equivalent to 2, and $f_C > 2$,

$$p_c = \frac{1}{[r + r\rho(f-2)]^{1/2}}$$

where f is the functionality of C. Experimental values of p_c fall between the values calculated by the statistical and nonstatistical methods. Deviations from the former have been attributed to intramolecular branching reactions leading to what are called *wasted loops*, and to differing reactivities of functional groups, for example, the lower reactivity of the secondary alcohol group of glycerol relative to that of the two primary groups.

10.6   Step-reaction copolymerization

A copolymer is defined in step polymerization as one having more than one repeating unit. Thus a polyester (**5**) prepared from terephthalic acid and ethylene glycol is a homopolymer, but a polyester (**6**) made with a $1:1:2$ mixture of terephthalic acid, isophthalic acid, and ethylene glycol is a copolymer. A variety of copolymer compositions is employed in industrial step polymerization. An example that appeared earlier in the book (Chapter 3) was a liquid crystalline copolyester prepared from terephthalic acid, *p*-hydroxybenzoic acid, and ethylene glycol.

$$\left[\ \overset{\overset{\displaystyle O}{\|}}{C}\!-\!\bigcirc\!-\!\overset{\overset{\displaystyle O}{\|}}{C}OCH_2CH_2O\ \right]$$

5

$$-\overset{\overset{\displaystyle O}{\|}}{C}\!-\!\bigcirc\!-\!\overset{\overset{\displaystyle O}{\|}}{C}OCH_2CH_2O\ \text{www}\ \overset{\overset{\displaystyle O}{\|}}{C}\!\diagdown\!\bigcirc\!\diagup\!\overset{\overset{\displaystyle O}{\|}}{C}OCH_2CH_2O-$$

6

In synthesizing copolymers such as **6**, the distribution of monomer units is random because the two dicarboxylic acids have virtually equal reactivity. It is possible in many instances to synthesize alternating step copolymers. Consider the hypothetical case of two different monomers, AA and BB, both of which react with monomer CC. A $1:1:2$ mixture of AA, BB, and CC yields a random copolymer. If, however, AA is first reacted with the two equivalents of CC and then in a separate step the product is reacted with BB, an alternating copolymer is formed. This is shown schematically in reaction (10.6). Such reactions are used to synthesize some important

$$AA + 2CC \longrightarrow CC\!-\!AA\!-\!CC \xrightarrow{\text{BB}} \{CC\!-\!AA\!-\!CC\!-\!BB\} \quad (10.6)$$

aromatic polyamides, as described in Chapter 13. The method does not *always* lead to alternating copolymer. Polyesters prepared with dibasic acids and diols may undergo transesterification in the second step and scramble the monomer distribution, but where such complications can be avoided, alternating polymers are formed without difficulty.

Because step polymers are true telechelic polymers, it is a relatively straightforward matter to prepare block copolymers by linking homopolymers together through coreactive functional groups. Typical examples are the reaction of a hydroxyl-terminated polyether and an isocyanate-terminated polyurethane (10.7), and the reaction of an acid chloride-terminated polyester with an amine-terminated polyamide (10.8). Alternatively, one could react an isocyanate-terminated polyurethane with two equivalents of hydroxyl-terminated polyester to form an ABA block copolymer (10.9).

HO-polyether-OH + OCN-polyurethane-NCO $\longrightarrow$

$$\left[\text{O-polyether-O}\overset{\overset{\text{O}}{\|}}{\text{C}}\text{NH-polyurethane-NH}\overset{\overset{\text{O}}{\|}}{\text{C}} \right] \quad (10.7)$$

$$\overset{\overset{\text{O}}{\|}}{\text{ClC}}\text{-polyester-}\overset{\overset{\text{O}}{\|}}{\text{C}}\text{Cl} + \text{H}_2\text{N-polyamide-NH}_2 \longrightarrow$$

$$\left[\overset{\overset{\text{O}}{\|}}{\text{C}}\text{-polyester-}\overset{\overset{\text{O}}{\|}}{\text{C}}\text{NH-polyamide-NH} \right] \quad (10.8)$$

2HO-polyester-OH + OCN-polyurethane-NCO $\longrightarrow$

$$\text{HO-polyester-O}\overset{\overset{\text{O}}{\|}}{\text{C}}\text{NH-polyurethane-NH}\overset{\overset{\text{O}}{\|}}{\text{C}}\text{O-polyester-OH} \quad (10.9)$$

Irreversible reactions of the types depicted here are necessary to control placement of the blocks. Numerous examples of AB diblock, $+$AB$+$ multiblock, and ABA triblock copolymers have been reported.[10,11] A number of these have commercial application.

10.7 Step polymerization techniques

A significant difference between vinyl and nonvinyl polymerization is that with the former, a large enthalpy factor contributes to the change from double to single bonding, resulting in generally low activation energies and significant reaction exotherms. Dissipation of the exotherm is often a critical factor in large-scale bulk polymerization of vinyl monomers. Nonvinyl polymerization, on the other hand, is more often than not characterized by high activation energies and low reaction exotherms. As a result, a good many step polymerizations are run at elevated temperatures.

There are basically four processing techniques used to make step-reaction polymers, two homogeneous (bulk and solution), and two heterogeneous (interfacial and phase-transfer catalyzed). Dispersion or emulsion methods have been used rarely. Of the four principal techniques, bulk and solution are by far the most important from the standpoint of commercial production.

Bulk polymerization has the obvious advantage of providing a product free of contaminants other than residual by-products or impurities arising from side reactions. The major disadvantage is that high viscosities necessitate the use of elevated temperatures for effective stirring and, in the case of condensation polymerization, removal of by-product. As a rule, bulk polymerizations are done under an inert atmosphere. To facilitate removal of by-product, large volumes of inert gas, most often nitrogen, are blown through the reaction mixture in the later stages of polymerization, a process called *sparging*.

Solvent polymerization minimizes the problems arising from high viscosity and can assist in removal of by-product by azeotropic distillation. As will be seen in later chapters, methods have been developed to remove water in situ by use of effective dehydrating agents, which often allow the polycondensation to proceed at ambient temperatures. The major disadvantage of solvent processing is the necessity of removing solvent once the reaction is finished. The problem is obviated when solvent-based coatings or solution-extruded fibers are the final product.

Interfacial polymerization[12,13] involves solutions of the two monomers in separate, immiscible solvents, one usually being water. When the two solutions are brought into contact, polymer is formed at the interface. A typical example is the Schotten–Baumann synthesis of polyamide from a diacid chloride dissolved in an organic solvent, and a diamine dissolved in aqueous base (10.10). The base is needed to neutralize the by-product HCl, which would otherwise react with the diamine to form amine hydrochloride. If the

$$Cl-\overset{\overset{\displaystyle O}{\|}}{C}-R-\overset{\overset{\displaystyle O}{\|}}{C}-Cl + H_2N-R'-NH_2 \xrightarrow[\text{H}_2\text{O. CH}_2\text{Cl}_2]{\text{NaOH}}$$

$$\left[\!\!\begin{array}{c} \overset{\overset{\displaystyle O}{\|}}{C}-R-\overset{\overset{\displaystyle O}{\|}}{C}-NH-R'-NH \end{array}\!\!\right]\!\!- \quad (10.10)$$

reaction is done in an open beaker, the polymer film at the interface may be grasped with tweezers and pulled out as a continuous "filament" of collapsed film until monomer is used up. This very effective demonstration (Figure 10.3) is aptly called the "nylon rope trick."[14] Rapid stirring to maximize the interfacial area increases the yield of polymer. (A kitchen blender is often used for interfacial laboratory preparations.) The method may also be used to synthesize polyesters from diacid chlorides and bisphenols; however, dialcohols are not sufficiently reactive, and hydrolysis of the diacid chloride takes precedence. Apart from some miscellaneous aromatic polyamides and polyesters, commercial exploitation of interfacial polymerization has been primarily limited to polycarbonates (Chapter 12).[15]

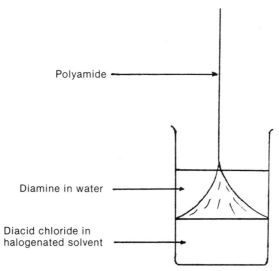

Figure 10.3. Interfacial synthesis of polyamide.

Interfacial polymerization differs significantly from bulk or solution polymerization:

1. The reaction goes rapidly at low temperatures.
2. Because the reaction is so rapid, diffusion of monomer to the interface is rate determining.
3. The polymerization takes on some of the characteristics of chain polymerization as monomer reacts with the growing chains at the interface more rapidly than it diffuses through the polymer film to initiate new chains, hence molecular weights tend to be significantly higher.
4. As a consequence of the last point, an exact stoichiometric balance is not necessary.

With so many apparent advantages, the obvious question is, why is interfacial polymerization not used more widely in industry? The answer has to do with the economics of the process. The high cost of acid chlorides and the large volumes of solvent required make the interfacial method prohibitively expensive for many polymers relative to bulk or solution processes.

Phase-transfer catalysis (PTC), although commonly employed in organic synthesis, has had more limited application in polymerization reactions.[16–18] In its most common form, the PTC method involves an aqueous phase and an organic phase, each containing one of the monomers. Hence it is also an interfacial technique. The catalyst, often a quaternary ammonium salt, functions by transporting a nucleophilic monomer from the aqueous phase to the organic phase, where its nucleophilicity is greatly enhanced because of reduced solvation effects. A case in point is the synthesis of an unusual carbon chain polymer (**9**) by reaction of α,α'-dichloro-*p*-xylene (**7**) and *t*-butyl cyanoacetate (**8**) using benzyltriethylammonium chloride catalyst (10.11).[17] In this case, the anion derived from **8** by reaction with NaOH is transported to the benzene solution as the soluble benzyltriethylammonium salt, where it reacts rapidly with **7** by nucleophilic displacement. Polymer is formed because **8** contains two active hydrogens. Other examples of PTC polymerization are given in later chapters.

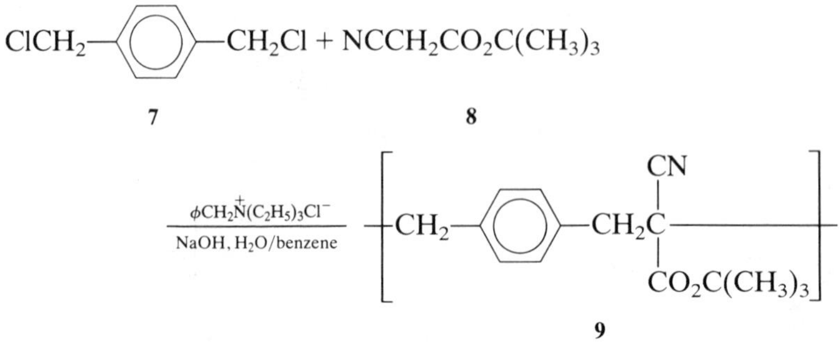

$$\text{ClCH}_2\text{---}\langle\bigcirc\rangle\text{---CH}_2\text{Cl} + \text{NCCH}_2\text{CO}_2\text{C(CH}_3)_3$$

7   **8**

$$\xrightarrow[\text{NaOH, H}_2\text{O/benzene}]{\phi\text{CH}_2\overset{+}{\text{N}}(\text{C}_2\text{H}_5)_3\text{Cl}^-} \left[\text{CH}_2\text{---}\langle\bigcirc\rangle\text{---CH}_2\overset{\displaystyle \text{CN}}{\underset{\displaystyle \text{CO}_2\text{C(CH}_3)_3}{\text{C}}}\right] \qquad (10.11)$$

9

In addition to polymer synthesis, PTC is an effective method of modifying polymers.[16,19,20] Reaction of chloromethylated polystyrene with carbazole[19] using tetrabutylammonium hydrogen sulfate (TBAH) catalyst in an N,N-dimethylformamide (DMF)–water system (10.12) is an example. (Of interest is the fact that polymers themselves are used as supports for solid-phase PTC reactions.[16,21])

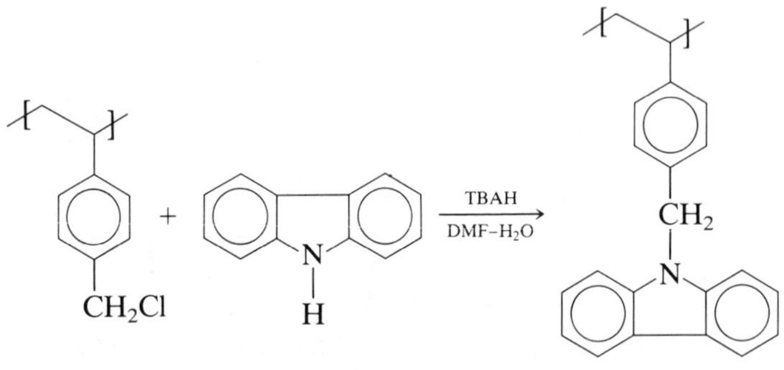

$$\xrightarrow[\text{DMF--H}_2\text{O}]{\text{TBAH}} \qquad (10.12)$$

10.8   Ring-opening polymerization

Ring-opening is the third general method of preparing linear polymers.[22–25] Examples of commercially important types are given in Table 10.2 in the order in which they appear in later chapters. Of those listed, only the metathesis-derived polymers (discussed in Chapter 8) are composed solely of carbon chains. Those that have enjoyed the longest history of commercial exploitation are polyethers prepared from three-membered ring cyclic ethers (epoxides), polyamides from cyclic amides (lactams), and polysiloxanes from

Table 10.2. Commercially important polymers prepared by ring-opening polymerization

Polymer type	Polymer repeating group	Monomer structure	Monomer type	Chapter
Polyalkene	$-[CH=CH(CH_2)_x]-$	 $(CH_2)_x \stackrel{CH}{\underset{CH}{\parallel}}$	Cycloalkene	8
Polyether	$-[CH_2O]-$	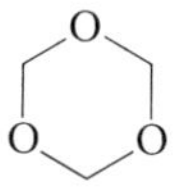 trioxane ring (O, O, O)	Trioxane	11
Polyether	$-[(CH_2)_xO]-$	 $(CH_2)_xO$	Cyclic ether[a]	11
Polyester	$-\left[(CH_2)_x\overset{O}{\overset{\parallel}{C}}O\right]-$	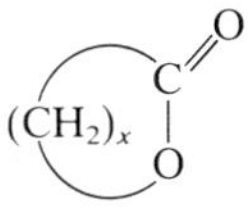 $(CH_2)_x \overset{O}{\underset{O}{C}}$	Lactone	12
Polyamide	$-\left[(CH_2)_x\overset{O}{\overset{\parallel}{C}}NH\right]-$	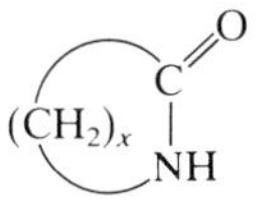 $(CH_2)_x \overset{O}{\underset{NH}{C}}$	Lactam	13
Polysiloxane	$-\left[\underset{CH_3}{\overset{CH_3}{\underset{\mid}{\overset{\mid}{Si}}}}O\right]-$	 $[Si(CH_3)_2]_x$	Cyclic siloxane	15
Polyphosphazene	$-\left[\underset{Cl}{\overset{Cl}{\underset{\mid}{\overset{\mid}{P}}}}=N\right]-$	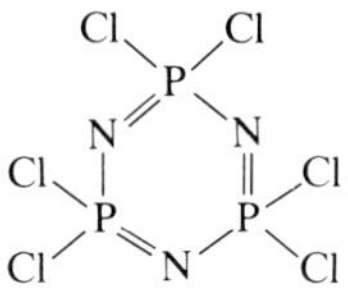 hexachlorocyclotriphosphazene ring	Hexachloro-cyclotriphos-phazene[b]	15
Polyamine	$-[CH_2CH_2NH]-$	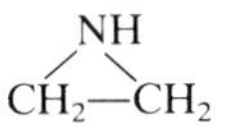 CH_2-CH_2 with NH bridge	Aziridene[c]	16

[a] Epoxide ($x = 2$); oxetane ($x = 3$).
[b] Phosphonitrilic chloride trimer.
[c] Also called alkyleneimine.

cyclic siloxanes. It should be remembered that although these reactions (other than metathesis) lead to condensation-type polymers, they are not polycondensation reactions because no by-product is formed. (Exceptions to this general rule are found in Chapters 12 and 13.)

Mechanisms of ring-opening polymerization vary according to monomer type and initiator, but in most instances they fit one of two general forms:

1. Monomer is attacked by some ionic or coordination species, designated X*, at the functional group (G) that causes ring-opening. This is followed

by attack of the ring-opened monomer or another cyclic unit, and so on (10.13).

$$\text{G} \xrightarrow{\text{X*}} \text{X} \text{\footnotesize\textasciitilde\textasciitilde\textasciitilde} \text{G*} \xrightarrow{\text{G}} \text{X} \text{\footnotesize\textasciitilde\textasciitilde\textasciitilde} \text{G} \text{\footnotesize\textasciitilde\textasciitilde\textasciitilde} \text{G*} \xrightarrow{\text{G}} \text{etc.} \quad (10.13)$$

2. Monomer is attacked by X^* to form a coordination species (most frequently a cation) that undergoes reaction with a second monomer molecule to open the ring, and so on (10.14).

$$\text{G} \xrightarrow{\text{X*}} \text{G*—X} \xrightarrow{\text{G}} \text{G*} \text{\footnotesize\textasciitilde\textasciitilde\textasciitilde} \text{G—X} \longrightarrow$$
$$\text{G*} \text{\footnotesize\textasciitilde\textasciitilde\textasciitilde} \text{G} \text{\footnotesize\textasciitilde\textasciitilde\textasciitilde} \text{G—X} \quad \text{etc.} \quad (10.14)$$

Although these mechanisms suggest a chain-growth process, this is not necessarily the case. Epoxide polymerization under anionic conditions, for example, exhibits characteristics of both step and chain growth depending on choice of initiator. Lactams may polymerize by a combination of step and chain growth with a single initiator. Thus it is impossible to generalize in terms of an overall kinetics scheme for ring-opening polymerization. As indicated above, ionic or coordination compounds are used most commonly to initiate ring-opening polymerization, although molecular compounds (water, alcohols, amines, etc.) are also used. In such cases, the initiator (XY) serves to open the ring, then polymerization proceeds by step growth (10.15). Free radical ring-opening reactions are relatively rare, although an important example leading to polyester formation is given in Chapter 12.[26,27]

$$\text{G} \xrightarrow{\text{XY}} \text{X} \text{\footnotesize\textasciitilde\textasciitilde\textasciitilde} \text{G—Y} \xrightarrow{\text{G}} \text{X} \text{\footnotesize\textasciitilde\textasciitilde\textasciitilde} \text{G} \text{\footnotesize\textasciitilde\textasciitilde\textasciitilde} \text{G—Y} \xrightarrow{\text{G}} \text{etc.} \quad (10.15)$$

As might be expected from considerations of ring strain, the most reactive monomers are usually those containing three- and four-membered rings.[22,28] With seven-membered and larger rings, transannular interactions provide the thermodynamic favorability for ring-opening. Five- and six-membered rings are virtually free of angle strain; however, torsional strain arising from conformational eclipsing of C–H bonds is the major factor responsible for the lower thermodynamic stability of cyclopentane relative to cyclohexane. Incorporation of heteroatoms into the ring influences ring stability in ways that are not always predictable. Six-membered cyclic esters (lactones) form polyester quite readily, for example, whereas six-membered lactams are reluctant to form polyamide. The reverse is the case with five-membered rings. Cyclic ethers, which exhibit ring strain similar to that encountered in cycloalkanes, exhibit the following order of reactivity in terms of ring size:

$$3 > 4 > 8 > 7 > 5 > 6$$

Tetrahydropyran, the six-membered ring, resists polymerization. Ring strain does not, of itself, guarantee that ring-opening polymerization will occur. Cycloalkanes, which have no functional groups to provide a kinetic pathway, do not polymerize apart from some highly strained cyclic or polycyclic compounds (Chapter 16). Ceiling temperatures are often quite low in ring-opening polymerizations compared with vinyl polymerization, particularly where five- or six-membered rings are involved.

Copolymers may be prepared by ring-opening polymerizations, including some with different functional groups. Large differences in reactivity, however, often result in significant block formation. Deliberately formed block copolymers represent the most interesting types. A wide variety of AB, ┼AB┼, and ABA block copolymers have been prepared by the living polymer method, including some containing both vinyl and nonvinyl blocks.[11]

Ring-opening polymerizations are generally restricted to solution or bulk processes.

References

1. P. J. Flory, *Principles of Polymer Chemistry*, Cornell University Press, Ithaca, N.Y., 1953, Chaps. 3, 8, and 9.
2. P. J. Flory, *Chem. Rev.*, **39**, 137 (1946).
3. D. H. Solomon (ed.), *Step-Growth Polymerization*, Dekker, New York, 1972.
4. R. W. Lenz, *Organic Chemistry of Synthetic High Polymers*, Wiley-Interscience, New York, 1967, Chap. 3.
5. P. E. M. Allan and C. R. Patrick, *Kinetics and Mechanisms of Polymerization Reactions*, Wiley-Interscience, New York, 1974, Chap. 5.
6. L. H. Peebles Jr., *Molecular Weight Distributions in Polymers*, Wiley-Interscience, New York, 1971.
7. W. H. Carothers, *Trans. Faraday Soc.*, **32**, 39 (1936).
8. S. H. Pinner, *J. Polym. Sci.*, **21**, 153 (1956).
9. W. H. Stockmayer, *J. Polym. Sci.*, **9**, 69 (1952); **11**, 424 (1953).
10. R. Jerome, R. Fayt, and T. Ouhadi, *Progr. Polym. Sci.*, **10**, 87 (1984).
11. A. Noshay and J. E. McGrath, *Block Copolymers: Overview and Critical Survey*, Academic Press, New York, 1977.
12. F. Millich and C. E. Carraher Jr. (eds.), *Interfacial Synthesis*, Vols. 1 and 2, Dekker, New York, 1975 and 1977.
13. P. W. Morgan, *J. Macromol. Sci.-Chem.*, **A15**, 683 (1981).
14. P. W. Morgan and S. L. Kwolek, *J. Chem. Educ.*, **36**, 182 (1959); G. C. East and S. Hassell, ibid, **60**, 69 (1983).
15. E. D. Oliver and Y.-C. Chen, in *Interfacial Synthesis*, Vol. 2 (F. Millich and C. E. Carraher Jr., eds.), Dekker, New York, 1977, Chap. 11.
16. L. J. Mathias and C. E. Carraher Jr. (eds.), *Crown Ethers and Phase Transfer Catalysis in Polymer Science*, Plenum Press, New York, 1984.
17. Y. Imai, *J. Macromol. Sci.-Chem.*, **A15**, 833 (1981).
18. L. J. Mathias, *J. Macromol. Sci.-Chem.*, **A15**, 853 (1981).
19. T. D. N'Guyen, A. Deffieux, and S. Boileau, *Polymer*, **19**, 423 (1978).

20. J. M. J. Fréchet, *J. Macromol. Sci.-Chem.*, **A15**, 877 (1981).
21. W. T. Ford and M. Tomoi, *Adv. Polym. Sci.*, **55**, 49 (1984).
22. K. C. Frisch and S. L. Reegan (eds.), *Ring-Opening Polymerization*, Dekker, New York, 1969.
23. H. R. Allcock, *Heteroatom Ring Systems and Polymers*, Academic Press, New York, 1967.
24. O. Vogl and J. Furukawa (eds.), *Polymerization of Heterocycles*, Dekker, New York, 1973.
25. J. E. McGrath (ed.), *Ring-Opening Polymerization*, ACS Symp. Ser. 286, American Chemical Society, Washington, D.C., 1985.
26. W. J. Bailey et al., *J. Macromol. Sci.-Chem.*, **A21**, 1611 (1984).
27. W. J. Bailey, in *Ring-Opening Polymerization* (J. E. McGrath, ed.), ACS Symp. Ser. 286, American Chemical Society, Washington, D.C., 1985, pp. 46–65.
28. J. E. McGrath, in *Ring-Opening Polymerization* (J. E. McGrath, ed.), ACS Symp. Ser. 286, American Chemical Society, Washington, D.C., 1985, pp. 1–22.

Review exercises

1. Considering the reaction conditions for bulk polycondensations, suggest reasons other than those given in the chapter for deviations in $\overline{DP}$ versus t plots for polyesters.

2. Plots of $1/(1-p)^2$ versus t for the uncatalyzed polyesterification of adipic acid and ethylene glycol are linear between 80 and 93% conversion of monomer. To what number average molecular weights do these limits correspond?

3. To what percent conversion should a batch of 11-aminoundecanoic acid be processed to yield polyamide having a number average molecular weight of 2.50×10^4?

4. Rate constants for polyesterification of adipid acid and 1,10-decanediol at 161°C are 0.0012 (kg/equiv)^2/min (uncatalyzed) and 0.079 (kg/equiv)/min (catalyzed) (S. D. Hamann et al., *J. Macromol. Sci.-Chem.*, **A2**, 153 1968). How long would it take to prepare a polyester of number average molecular weight 15,000 (bulk processing) under each set of conditions? (Assume equimolar amounts of each reactant and 2 equivalents per mole, and that bulk mass remains constant.) What would the polydispersity index be for the polymer?

5. If 5 mol % excess diol were to be used in the previous problem, what number average molecular weight would be obtained if the reaction were carried to the same conversion?

6. Using the Carothers relationship, calculate the percent conversion at the gel point for (a) a 3:1:3 and (b) a 1:1:2 molar ratio of phthalic anhydride, trimellitic acid, and glycerol.

7. Explain why aliphatic polyamides but not aliphatic polyesters can be prepared by interfacial polymerization.

8. Write equations illustrating the synthesis of (a) an alternating copolyamide from isophthaloyl chloride, terephthaloyl chloride, and *m*-diaminobenzene, and (b) an AB block copolymer of poly(11-undecanoamide) with the polyamide formed in (a).

9. Which reactions would be expected to yield higher molecular weight polyester: the polycondensation of 5-hydroxypentanoic acid or 6-hydroxyhexanoic acid? Explain.

10. Give structures of cyclic monomers that might be used to prepare:

(a) $\left[\text{NHCH}_2\text{CH}_2\text{CH}_2\overset{\displaystyle \overset{\text{O}}{\|}}{\text{C}}\right]$ (b) $\left[\overset{\displaystyle \overset{\text{CH}_3}{|}}{\underset{\displaystyle \underset{\text{CH}_3}{|}}{\text{C}}}\text{CH}_2\overset{\displaystyle \overset{\text{O}}{\|}}{\text{C}}\text{O}\right]$

(c) 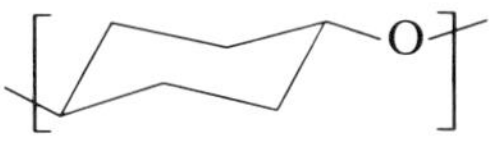 $\left[\underset{\hspace{3em}}{\bigcirc}\text{O}\right]$

11
Polyethers, polysulfides, and related polymers

11.1 Introduction

Polyethers and related sulfur-containing polymers are the most diverse of the nonvinyl polymers in the structural variety that have achieved commercial prominence (Table 11.1). They include polyacetals synthesized by chain-reaction polymerization of aldehydes (a reaction akin to chain polymerization of vinyl monomers), polyethers prepared by ring-opening and step-reaction polymerization, and the structurally related polysulfides and other sulfur-containing polymers. Many of these polymers fit the definition of engineering plastics, and their use is expanding accordingly.

This chapter is divided into three sections. Section 11.1 includes polyethers made by chain polymerization of carbonyl compounds and ring-opening polymerization of cyclic ethers. Section 11.2 is concerned with step-polymerization methods that have proven most fruitful in development of epoxy polymers and a number of aromatic polyethers for engineering applications. Certain types of polyacetals and polyketals are also covered in this section. Section 11.3 deals with sulfur-containing polymers: polysulfides (sometimes called *polythioethers*), poly(alkylene polysulfides), and polysulfones, which represent an oxidized form of polysulfides. It should be noted that there are no oxygen-containing counterparts to these last two polymers, although unstable copolymers of oxygen and vinyl monomers containing peroxide groups in the backbone have been used to synthesize block copolymers (see Chapter 9) and poly(styrene peroxide)[1] has been suggested as a solid propellant because of its autopyrolytic properties.[2]

11.2 Preparation of polyethers by chain-reaction and ring-opening polymerization

11.2.1 Polymerization of carbonyl compounds[3,4]

Aldehydes have a tendency to form cyclic trimers or tetramers, and polymerization conditions are chosen to circumvent this reaction. Formaldehyde is an exception in that while its cyclic trimer, *trioxane* (**1**), is well known,

352

Table 11.1. Commercially available polyethers and sulfur-containing polymers

Type	Structure	Typical uses
Polyacetal[a,b]	$\mathrm{+[CH_2O]+}$	Engine, appliance, and plumbing parts; electronics components; zippers; buckles
Poly(ethylene oxide)[c]	$\mathrm{+[CH_2CH_2O]+}$	Surfactants, hydraulic fluids, lubricants, water-soluble packaging film, textile size
Poly(propylene oxide)[c]	$\mathrm{+[CH_2CHO]+}$ $\quad\;\;\mathrm{CH_3}$	Polyurethane intermediates, lubricants, surfactants
Poly(hexafluoropropylene oxide)	$\mathrm{+[CF_2CFO]+}$ $\quad\;\;\mathrm{CF_3}$	Lubricating oils, greases
Poly[3,3-(dichloromethyl)trimethylene oxide]	$\mathrm{CH_2Cl}$ $\mathrm{+[CH_2CCH_2O]+}$ $\mathrm{CH_2Cl}$	Corrosion-resistant tanks and tank liners, gaskets, machine parts
Polytetrahydrofuran	$\mathrm{+[(CH_2)_4O]+}$	Thermoplastic elastomers, artificial leather
Polyetherimide[a]	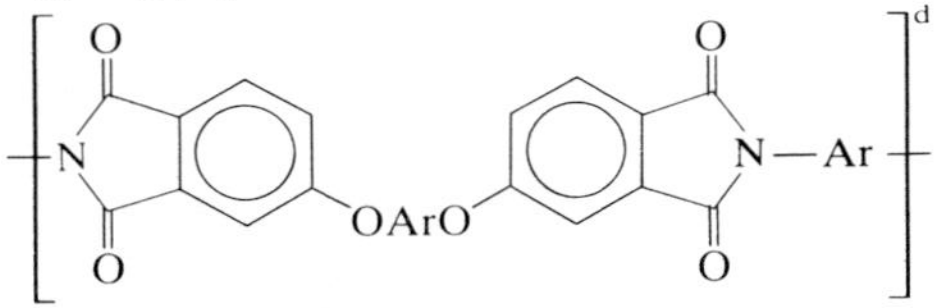	Automotive and electronics parts, composites, wire and cable insulation

(continues)

Table 11.1. (*Continued*)

Type	Structure	Typical uses
Polyetheretherketone[a] (PEEK)	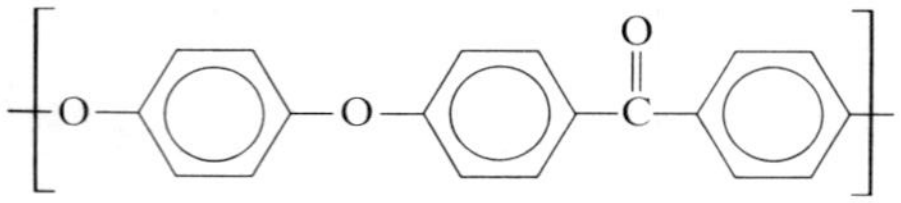	Wire and cable insulation, radiation-resistant parts for nuclear plants, machine parts
Poly(phenylene oxide)[a] (PPO)	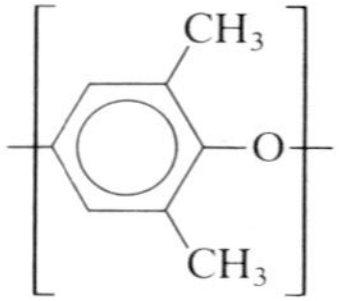	Telecommunications and computer equipment, automotive parts, appliance housing, electrical equipment
Epoxy	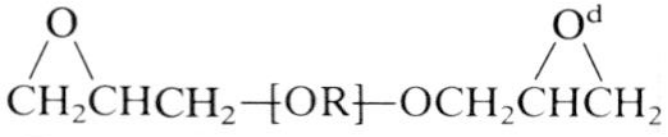 CH_2CHCH_2—$[OR]$—OCH_2CHCH_2 with epoxide group O^d	Coatings, laminated circuit boards, adhesives, composites, bridge road surfaces, highway "rumble strips"
Polysulfide[a]	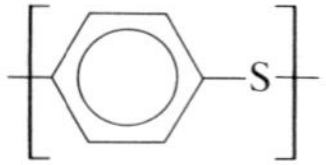 $[RS_x]^d$ (with phenylene–S structure)	Electronics, electrical, automotive, and machine parts; small appliances; chemical processing equipment
Poly(alkylene polysulfide)	$[RS_x]^d$	Sealants, gaskets, hose, solid rocket fuel binder
Polysulfone[a]	$[ArSO_2]^d$	Medical and food-processing equipment, electrical and electronics components, camera cases, piping

[a] Engineering plastics.

[b] Homopolymer and copolymer.

[c] Block copolymers of ethylene oxide and propylene oxide are also available as surfactants.

[d] See text for details of structural units.

it is not so easy to prepare, although it can be made by condensing formal-dehyde vapors at low temperatures. Acetaldehyde, on the other hand, forms cyclic trimer (*paraldehyde*, **2**) and tetramer (*metaldehyde*, **3**) under the in-fluence of acids.

Formaldehyde is by far the most important aldehyde monomer because it forms polymers having the combination of physical and mechanical properties and economy of processing to make it attractive for commercial use. *Polyox-ymethylene* (**4**), alternatively referred to as *polyformaldehyde, polyformal,* or *acetal* polymer,[5,6] is a highly crystalline (75–80%), high-strength polymer that replaces metals in machine parts, springs, zippers, and a host of other applica-tions. Commercialized in 1960, it is marketed under the du Pont trade name Delrin.[7a] Polyoxymethylene has a relatively low ceiling temperature (about 120°C) and undergoes depolymerization initiated at the hydroxyl end groups, hence the commercial product is stabilized by end-capping with acetic anhy-dride (11.1).

$$\text{HO}\text{+}\text{CH}_2\text{O}\text{+}\text{CH}_2\text{OH} \xrightarrow{\text{(CH}_3\text{CO)}_2\text{O}} \text{CH}_3\overset{\overset{\text{O}}{\|}}{\text{C}}\text{O}\text{+}\text{CH}_2\text{O}\text{+}\text{CH}_2\text{O}\overset{\overset{\text{O}}{\|}}{\text{C}}\text{CH}_3 \quad (11.1)$$

4

A number of methods have been employed to prepare poly-oxymethylene.[4a,5] One is to condense gaseous formaldehyde on a cold surface in the presence of such initiators as metal alkoxides and metal alkyls, Lewis acids, and HCl. Free radical processes do not appear to be significant in polymerization of carbonyl compounds. The commercial product is prepared under anionic or cationic conditions in a hydrocarbon solvent from which the polymer precipitates. Interestingly, formaldehyde polymerizes spontaneously in aqueous solution (11.2) by successive condensation reactions of the hy-drate, **5**. Evaporation of the solvent yields solid polymer. Known in this form as *paraformaldehyde*, the polymer serves as a convenient source of anhy-drous, gaseous formaldehyde because it depolymerizes at 180 to 200°C.

$$n\text{CH}_2(\text{OH})_2 \rightarrow \text{HO}[\text{CH}_2\text{O}]_n\text{H} + (n-1)\text{H}_2\text{O} \quad (11.2)$$

5

Cationic polymerization,[8a] illustrated in reaction (11.3) for the BF_3–H_2O initiating system, is analogous to cationic vinyl polymerization. The carbon–oxygen double bond is highly polar, and addition of the cationic species is regiospecific at the electronegative oxygen. Chain termination occurs primarily by transfer reactions such as (11.4).

$$F_3\bar{B}:\overset{+}{O}H_2 + RCHO \longrightarrow \left[HO-\overset{+}{C}H \longleftrightarrow H\overset{+}{O}=CH \right] F_3\bar{B}:OH \xrightarrow{nRCHO}$$

$$HOCH\overset{\displaystyle }{\underset{\displaystyle R}{\left[OCH \right.}} \underset{R}{\left. \right]_{n-1}} \overset{+}{OCH}_R F_3\bar{B}:OH \quad (11.3)$$

$$\text{wwm}\overset{+}{O}\underset{R}{C}H \ F_3\bar{B}:OH + H_2O \longrightarrow \text{wwm}\underset{R}{O}CHOH + F_3\bar{B}:\overset{+}{O}H_2 \quad (11.4)$$

Alkoxides and metal alkyls are the best anionic initiators for aldehyde polymerization (11.5)[4a]; however, the aldol condensation is a serious side reaction with aldehydes higher than formaldehyde that contain alpha hydrogen. Spectroscopic detection of ester end groups suggests that chain transfer occurs by hydride transfer (11.6), a reaction analogous to the crossed Cannizzaro reaction. As is the case with anionic vinyl polymerization, there is no inherent termination reaction. Living polymers that lend themselves to block copolymerization may be formed.

$$RCHO + R'M \longrightarrow R'\underset{R}{C}HO^-M^+ \xrightarrow{nRCHO} R'\underset{R}{\left[CHO \right]_{n-1}} \underset{R}{CHO}^-M^+$$

$$(11.5)$$

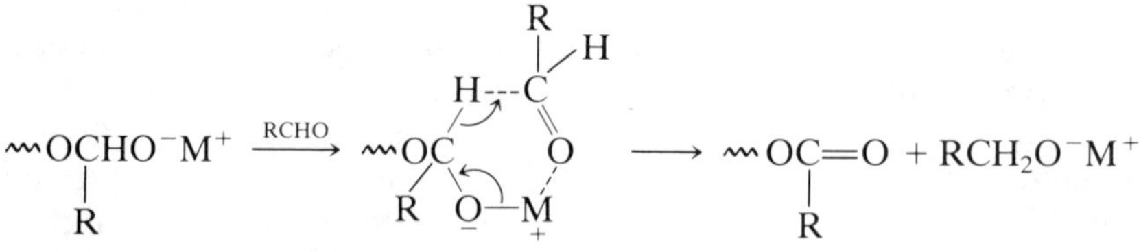

$$\text{wm}\underset{R}{O}CHO^-M^+ \xrightarrow{RCHO} \text{wm}OC \cdots \longrightarrow \text{wm}\underset{R}{O}C{=}O + RCH_2O^-M^+$$

$$(11.6)$$

Polymers of acetaldehyde and higher aldehydes have considerably lower ceiling temperatures than polyoxymethylene (usually below 0°C), although stability can be improved by end-capping. Polychloral (polytrichloroacetaldehyde, **6**), of interest because of its flame resistance and good mechanical properties, has been formed into objects by an interesting monomer casting technique that takes advantage of the low ceiling temperature of 58°C.[9] Monomer and initiator are mixed above 58°C, poured into a mold, then cooled to room temperature or below to effect polymerization. Halogen substituents, particularly fluorine,[10] facilitate anionic polymerization of aldehydes by inductive stabilization of the propagating alkoxide ion.

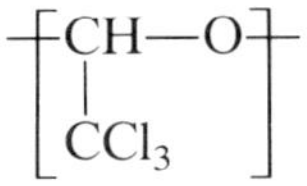

$$\left[\begin{matrix} CH-O \\ | \\ CCl_3 \end{matrix}\right]$$

6

Unsaturated aldehydes, especially acrolein (**7**), have also been studied; however, the resultant polymers, prepared under free radical or ionic conditions, are insoluble, infusible, and complex in structure. FT-IR and solid-phase ^{13}C NMR indicates[11] that under free radical conditions, acrolein polymerizes mainly through the vinyl groups to yield a polymer having intra-molecular and intermolecular acetal structures (**8**). Under alkaline conditions, a combination of Michael addition to the carbon–carbon double bond and carbonyl addition gives rise to polymer (**9**), which has significant amounts of residual vinyl groups. Polyacrolein is of interest because of its ability to bind antibodies and enzymes with little loss of biological activity.[12]

$$CH_2{=}CH{-}CH{=}O$$

7

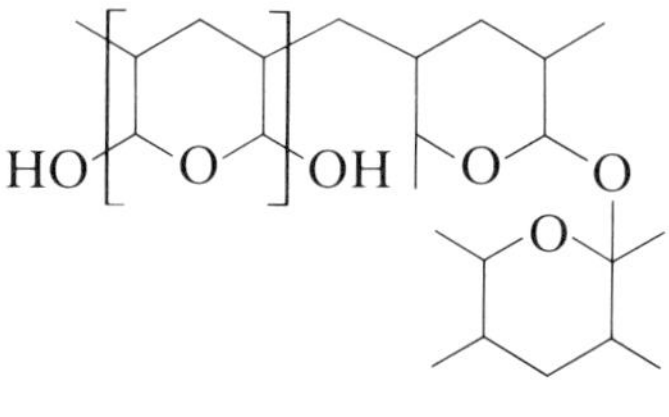

8 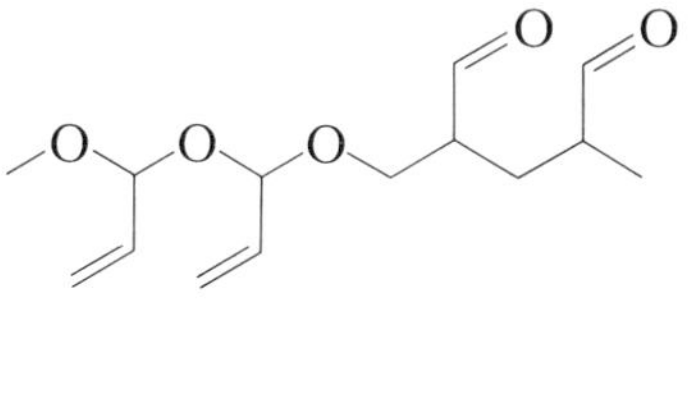   **9**

Dialdehydes also form polymers containing cyclic ether units.[8a] Gluturaldehyde, for example, polymerizes under cationic conditions (11.7) by a mechanism analogous to that of polymerization of isolated dienes.

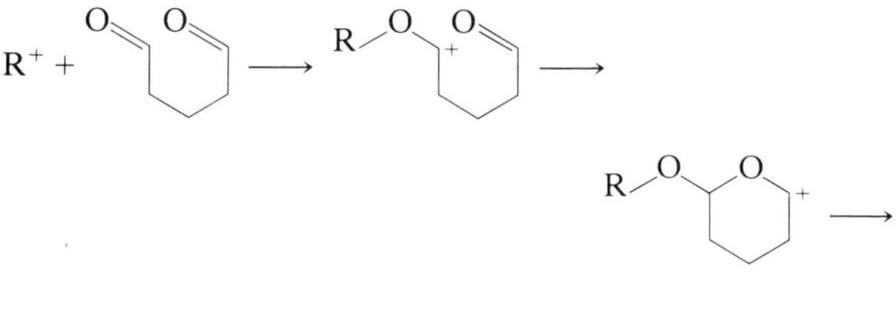

etc.   (11.7)

Ketones are polymerizable at low temperatures, but the polymers are unstable. Polyacetone (**10**) prepared at liquid nitrogen temperatures, for example, decomposes rapidly at room temperature.[4a] Copolymers of acetone with formaldehyde[13] and hexafluoroacetone with ethylene[14] have been prepared, as have block copolymers of acetone and propylene using Ziegler–Natta catalysts.[4b] Numerous other copolymers of aldehydes, vinyl monomers, and epoxides have been reported.[8a]

$$\left[\begin{array}{c} CH_3 \\ | \\ -C-O- \\ | \\ CH_3 \end{array}\right]$$

10

11.2.2 Stereochemistry of aldehyde polymerization

Polymerization of acetaldehyde and the higher aldehydes leads to polymers having chiral carbon atoms and the resultant potential for stereoregularity. Not as much work has been done in this area as with stereoregular vinyl polymerization, but it is apparent that isotactic sequences are favored by anionic initiators and by bulky monomers.[8a] Syndiotactic polyaldehydes have not been reported at the time of this writing.

With acetaldehyde, cationic initiators give amorphous atactic polymer, or, at best, very small amounts of crystalline material. Crystalline polymers are formed, however, with higher aldehydes under cationic conditions. This may not necessarily be a function of the cationic mechanism but may reflect conformational effects resulting from the size of the monomer, because some higher aldehydes such as isobutyraldehyde and *n*-heptaldehyde form crystalline polymer under high pressure in the absence of catalyst.

Anionic polymerization using metal alkyls results in isotactic polyaldehyde with the degree of stereoregularity increasing with increasing size of the aldehyde. Because the reaction is believed to occur via alkoxide addition reactions (11.4), a mechanism of stereoregular polymerization has been proposed[4a] in which stereochemistry is controlled by coordination of the metal atom of the alkoxide chain end and the carbonyl oxygen of monomer. The resultant complex (**11**) can then follow the pathway of hydride transfer leading to the well-known Meerwein–Ponndorf–Verley reduction (11.8), or it can polymerize through alkoxide addition (11.9). At low temperatures polymerization is favored, but chain transfer can occur by the hydride transfer mechanism (see reaction 11.6).

Stereoregularity is also influenced by the size of the initiator. This has been demonstrated using triethylaluminum and equimolar amounts of different alcohols as the initiating system for polymerizing acetaldehyde.[4c] These re-

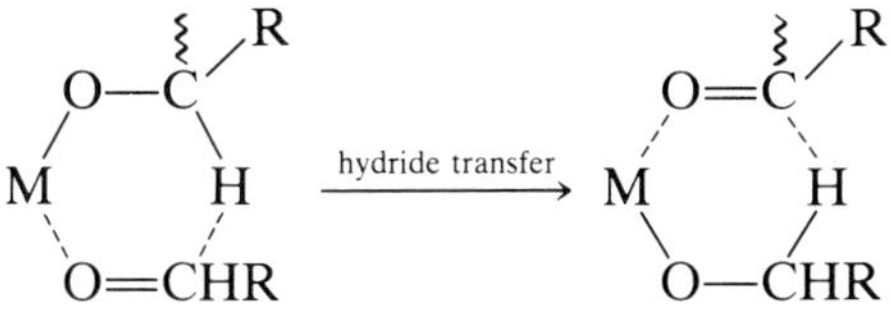

$$(11.8)$$

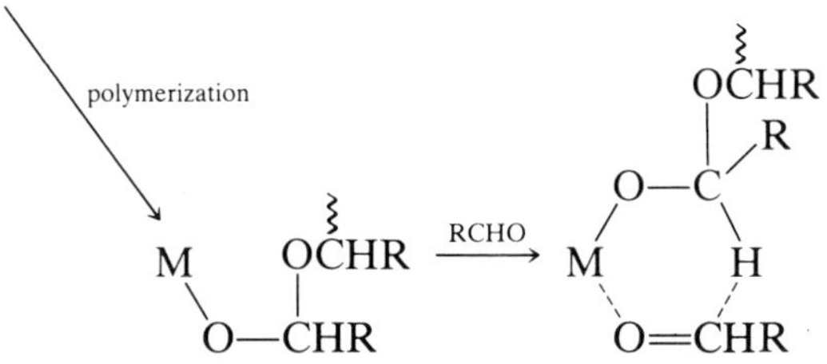

$$(11.9)$$

agents react according to reaction (11.10). As the size of the alcohol is increased, the degree of isotactic placement increases, with best results being

$$Al(C_2H_5)_3 + ROH \rightarrow (C_2H_5)_2AlOR + C_2H_6 \qquad (11.10)$$

obtained with tertiary alcohols. Results are given in Table 11.2. Similar types of coordination mechanisms have been proposed to explain stereoregular polymerization with aluminum-based initiators.[8a]

Table 11.2. Effect of alcohol on stereoregularity of polyacetaldehyde using triethylaluminum–alcohol initiator[a,b]

Alcohol	Total yield (%)	Insoluble fraction (%)
None	39	16
Methyl	44	32
Ethyl	36	14
n-Propyl	59	22
n-Butyl	45	22
n-Octyl	44	34
i-Butyl	35	29
i-Propyl	43	49
s-Butyl	62	79
Cyclohexyl	29	45
t-Butyl	12	92

[a] Data from Furukawa and Saegusa,[4c] courtesy of Wiley-Interscience.

[b] Acetaldehyde, 0.25 mol; initiator, 0.0025 mol; hexane, 20 ml; temperature, −78°C for 42 hours.

11.2.3   Polymerization of cyclic ethers

The history of cyclic ether polymerization dates back to some early experiments by Wurtz on ethylene oxide in the last century.[15,16] Earlier in this century Staudinger did considerable work on polyether synthesis using both homogeneous and heterogeneous initiator systems.[17,18] Much of the impetus for research in this area in recent years has resulted from the development of polyurethane foams, which are made from hydroxyl-terminated polyethers and diisocyanates. (Polyurethane chemistry is discussed in Chapter 13.)

Trioxane (**1**), used in commercial production of polyoxymethylene under cationic conditions, is one of the most important cyclic ethers.[19a] The probable polymerization mechanism, illustrated for the BF_3–H_2O initiating system, involves initial protonation and ring-opening to form a resonance-stabilized cation (11.11), and propagation (11.12) by successive ring-opening steps. The commercial product (Celanese trade name Celcon) is actually a random copolymer of trioxane and ethylene oxide. The —CH_2CH_2O— units arising from the latter prevent the polymer from unzipping at elevated temperatures.

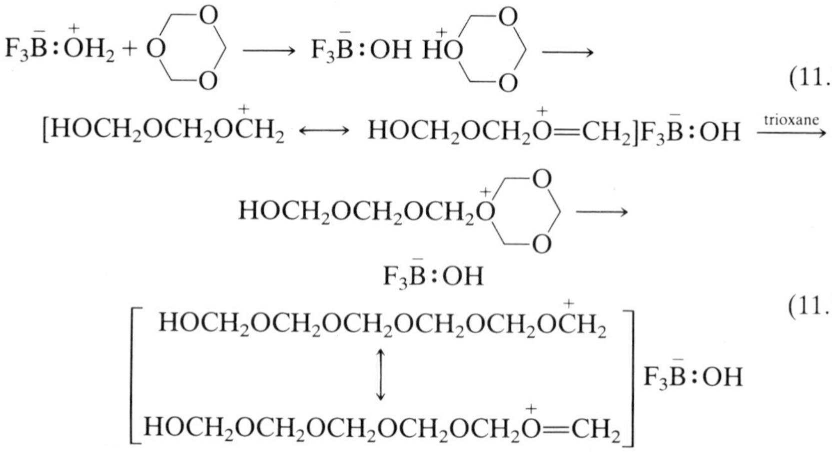

$$F_3\bar{B}:\overset{+}{O}H_2 + \text{(trioxane)} \longrightarrow F_3\bar{B}:OH \ H\overset{+}{O}\text{(ring)} \longrightarrow$$

$$(11.11)$$

$$[HOCH_2OCH_2O\overset{+}{C}H_2 \longleftrightarrow HOCH_2OCH_2\overset{+}{O}{=}CH_2]F_3\bar{B}:OH \xrightarrow{\text{trioxane}}$$

$$HOCH_2OCH_2OCH_2\overset{+}{O}\text{(ring)} \longrightarrow$$

$$F_3\bar{B}:OH$$

$$\begin{bmatrix} HOCH_2OCH_2OCH_2OCH_2OCH_2O\overset{+}{C}H_2 \\ \updownarrow \\ HOCH_2OCH_2OCH_2OCH_2OCH_2\overset{+}{O}{=}CH_2 \end{bmatrix} F_3\bar{B}:OH$$

$$(11.12)$$

Apart from trioxane, the most important cyclic ethers are ethylene oxide[20] (**12**), propylene oxide (**13**), hexafluoropropylene oxide (**14**), 3,3-di(chloromethyl)oxacyclobutane (**15**), and tetrahydrofuran (oxacyclopentane) (**16**).

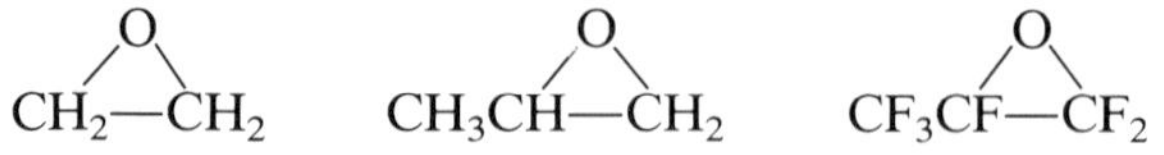

$$\underset{\textbf{12}}{CH_2{-}CH_2} \qquad \underset{\textbf{13}}{CH_3CH{-}CH_2} \qquad \underset{\textbf{14}}{CF_3CF{-}CF_2}$$

$$\underset{\textbf{15}}{O\diamond\begin{smallmatrix}CH_2Cl \\ CH_2Cl\end{smallmatrix}} \qquad \underset{\textbf{16}}{\text{(oxacyclopentane)}}$$

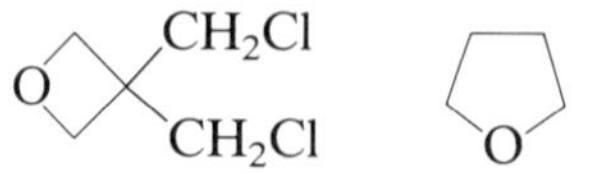

Structures of the corresponding polymers are given in Table 11.1. Although the four- and five-membered rings are occasionally called 1,3-epoxypropane and 1,4-epoxybutane, respectively, the term *epoxide* is more commonly reserved for the three-membered ring, and this is how it is used in this book. The terms *oxirane* and *oxetane* are also used to denote ethylene oxide and oxacyclobutane, respectively.

Depending on molecular weight, poly(ethylene oxide) and poly(propylene oxide) range in properties from liquids to waxy solids. A significant difference between the two polymers is that the former is water soluble and the latter is insoluble. Poly(hexafluoropropylene oxide) has the advantage of retaining flexibility at temperatures as low as −50°C and exhibiting excellent high temperature stability.[21] Poly[3,3-(dichloromethyl)trimethylene oxide] and polytetrahydrofuran are semicrystalline polymers having T_m's of 177 and 45°C, respectively.

Because of the strain inherent in the three-membered ring, epoxides are particularly amenable to ring-opening polymerization and can be converted to polyether with cationic, anionic, or coordination initiators.[3b,4d,19b,22,23a] Free radical initiation occurs with difficulty and is complicated by competing hydrogen abstraction reactions.[19b] The less reactive larger ring cyclic ethers polymerize well only under cationic conditions. The chemistry of these various processes follows. Stereospecific polymerization of epoxides is covered in the next section.

11.2.3.1 Cationic polymerization of epoxides[19b,24] Various types of protonic and Lewis acid systems initiate the polymerization of epoxides. Mechanisms involving the latter type are frequently complex and not well understood; some, in fact, are believed to react with monomer to form metal alkoxides that cause polymerization to occur via an *anionic* coordination mechanism. In the case of proton donors the mechanism is straightforward, but not very high molecular weights are obtainable. The initial step is formation of an oxonium ion (11.13). Propagation (11.14) involves ring-opening attack of monomer to form a new oxonium ion that competes with chain termination by reaction with water (11.15).

$$\underset{\text{CH}_2-\text{CH}_2}{\overset{\text{O}}{\triangle}} + \text{HA} \rightleftharpoons \underset{\text{CH}_2-\text{CH}_2}{\overset{\overset{\text{H}}{|}}{\overset{\text{O}^+\quad \text{A}^-}{\triangle}}} \qquad (11.13)$$

$$\underset{\text{CH}_2-\text{CH}_2}{\overset{\overset{\text{H}}{|}}{\overset{\text{O}^+\,\text{A}^-}{\triangle}}} + \underset{\text{CH}_2-\text{CH}_2}{\overset{\text{O}}{\triangle}} \rightleftharpoons \text{HOCH}_2\text{CH}_2\overset{+}{\text{O}}\underset{\text{CH}_2}{\overset{\text{A}^-\ \text{CH}_2}{\langle\ |}} \qquad (11.14)$$

$$H\text{---}[OCH_2CH_2]_n\text{---}\overset{+}{O}\overset{A^-}{\underset{CH_2}{\overset{CH_2}{\diagdown}}}\Big| \quad + H_2O \longrightarrow$$

$$H\text{---}[OCH_2CH_2]_n\text{---}OCH_2CH_2OH + HA \quad (11.15)$$

Propylene oxide and other substituted epoxides undergo similar reactions except that ring-opening is regiospecific at the carbon bearing the substitutent group (11.16).

$$\underset{CH_2\text{---}CHCH_3}{\overset{\overset{\displaystyle H}{\overset{|}{\overset{+}{O}}}\;\; A^-}{\diagup\diagdown}} + \underset{CH_2\text{---}CHCH_3}{\overset{O}{\diagup\diagdown}} \longrightarrow \underset{\underset{CH_3}{\overset{|}{CH}}}{HOCH_2\overset{+}{CH}\overset{A^-}{\underset{CH_2}{\overset{CHCH_3}{\diagdown}}}}\Big| \quad (11.16)$$

Higher molecular weight polymer is obtained using stannic chloride initiator, but the mechanism appears to be quite different. Kinetics studies indicate that each stannic chloride molecule initiates two polymer chains, which has led to a suggestion[4d,19a] that initiation proceeds according to reaction (11.17). Such a scheme has the disadvantage of requiring considerable charge separation in the growing chains unless chain coiling allows some ion-counterion interaction. More work is necessary to elucidate these mechanisms.

$$SnCl_4 + 4\underset{CH_2\text{---}CH_2}{\overset{O}{\diagup\diagdown}} \longrightarrow$$

$$\overset{CH_2}{\underset{CH_2}{\diagdown}}\!\!\Big|\!\!\overset{}{\diagup}O^+CH_2CH_2O\text{---}(SnCl_4)^{2-}\text{---}OCH_2CH_2\overset{+}{O}\overset{CH_2}{\underset{CH_2}{\diagdown}}\Big| \quad (11.17)$$

11.2.3.2   Cationic polymerization of other cyclic ethers[3c,4e,19c,24-26]   Oxacyclobutanes and tetrahydrofuran can be polymerized by a variety of cationic initiating systems, and in each case the initiating species is the corresponding oxonium ion. Propagation undoubtedly involves nucleophilic attack of monomer on the α-carbon with simultaneous ring-opening (11.18, 11.19), a reaction analogous to the normal S_N2 displacement of oxonium ions.

$$\square\overset{+}{\overset{\diagup}{O}}^{H} + \square O \longrightarrow \square\overset{X^-}{\overset{\pm}{O}}\diagdown\diagdown OH \quad (11.18)$$

$$\underset{\overset{+}{O}}{\overset{\overset{\displaystyle H}{\overset{|}{}}}{\bigcirc}} + \underset{O}{\bigcirc} \longrightarrow \underset{\overset{+}{O}}{\bigcirc}\diagup\diagdown\diagdown^{X^-}OH \quad (11.19)$$

As expected from considerations of ring strain, these monomers are not as reactive as epoxides, although the initiation rate can be increased by addition of an epoxide "promoter." Substituted oxetanes such as **15** polymerize readily. Substituted tetrahydrofurans, on the other hand, resist polymerization, with the exception of some bicyclic ethers such as 7-oxabicyclo[2.2.1] heptane (**17**). This monomer, which may be considered a disubstituted tetrahydrofuran, owes its reactivity (11.20) to strain associated with the bicyclic ring system.[27]

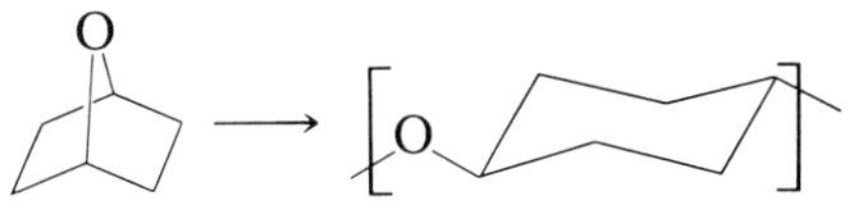

$$\text{17} \longrightarrow \qquad\qquad\qquad (11.20)$$

17

11.2.3.3 Anionic polymerization of epoxides Low-molecular-weight polyethers can be prepared with basic catalysts such as sodium or potassium hydroxides or alkoxides. Mechanistic studies indicate that initiation (11.21) involves S_N2 displacement to form alkoxide ion and that propagation occurs by nucleophilic displacement involving the new alkoxide ion (11.22). The low molecular weight is due to the fact that initiation occurs rapidly in competition with propagation. As with other anionic polymerizations, the chain end remains active and higher molecular weights can be obtained by addition of more monomer.

$$RO^- + CH_2\!\!-\!\!CH_2 \longrightarrow ROCH_2CH_2O^- \qquad (11.21)$$

$$ROCH_2CH_2O^- + CH_2\!\!-\!\!CH_2 \longrightarrow ROCH_2CH_2OCH_2CH_2O^- \quad (11.22)$$

The mechanism of polymerization is not as straightforward as that shown above. Kinetics studies suggest that protonation of the epoxide or possibly hydrogen bonding involving the alcohol (11.23) catalyzes ring-opening. Phen-

$$\text{\textasciitilde}CH_2CH_2O^- + \begin{array}{c} CH_2 \\ | \quad \rangle O \cdots H\!\!-\!\!OR \longrightarrow \\ CH_2 \end{array}$$

$$\text{\textasciitilde}CH_2CH_2OCH_2CH_2OH + {}^-OR \quad (11.23)$$

oxide ion, $C_6H_5O^-$, will not initiate ethylene oxide polymerization unless free phenol is present, which again suggests that proton donation is important. When tertiary alkoxides are used as initiators, higher molecular weight polymer is formed, presumably because initiation leads to a more reactive primary alcohol.

In the case of propylene oxide, nucleophilic substitution occurs preferentially at the less hindered carbon (11.24), and the resultant polymer, having

$$RO^- + \overset{\displaystyle O}{\overset{\diagup\diagdown}{CH_2-CHCH_3}} \longrightarrow ROCH_2\underset{\underset{\displaystyle CH_3}{|}}{CHO^-} \longrightarrow$$

$$RO\!\!-\!\!\left[\!\!\begin{array}{c} CH_2CHO \\ | \\ CH_3 \end{array}\!\!\right]\!\!- \quad (11.24)$$

pendant methyl groups, is capable of exhibiting stereoregularity. Again, only low-molecular-weight polymers are formed, and this, together with the fact that rather broad molecular weight distributions are observed, suggests that chain transfer processes occur easily. Observations of both allyl ether and propenyl ether end groups from infrared spectra of poly(propylene oxide) suggest two types of chain transfer, one (11.25) involving an E2-type elimination (probably analogous to dehydration, because catalysis by proton donors is important) and the other (11.26) an intramolecular transfer of an allylic proton from a chain end formed in reaction (11.25).

$$RO^- + \overset{\displaystyle O}{\overset{\diagup\diagdown}{CH_3CH-CH_2}} \longrightarrow ROH + CH_2{=}CH\overset{\displaystyle O^-}{\overset{|}{CH_2}} \longrightarrow$$

$$CH_2{=}CHCH_2O\!\!-\!\!\left[\!\!\begin{array}{c} CH_2CHO \\ | \\ CH_3 \end{array}\!\!\right]_n\!\!\!\!-\!\!CH_2\underset{\underset{\displaystyle CH_3}{|}}{CHO^-} \quad (11.25)$$

$$CH_2{=}CHCH_2 \quad \overset{\displaystyle \cdot\cdot}{O}CHCH_3 \longrightarrow$$
$$\diagdown O\text{\small www}CH_2$$

$$\left[\!CH_2{=}CH\overset{\displaystyle \cdot\cdot}{C}H \quad HOCHCH_3 \longleftrightarrow \; :CH_2{-}CH{=}CH \quad HOCHCH_3\!\right] \longrightarrow$$
$$\diagdown O\text{\small www}CH_2 \qquad\qquad\qquad \diagdown O\text{\small www}CH_2$$

$$CH_3{-}CH{=}CH \quad \overset{\displaystyle -}{O}CHCH_3 \quad (11.26)$$
$$\diagdown O\text{\small www}CH_2$$

11.2.3.4 Polymerization of epoxides with coordination catalysts Highest molecular weights have been achieved with coordination catalysts, and the different catalyst systems that have been investigated are far too numerous to list here. They can, however, be broken down into three general types: (1) alkaline earth (group II) compounds, (2) mixtures of transition metal compounds with compounds of neighboring metals; and (3) metal–porphyrin complexes. Examples of the first type[4d,8b] include alkaline earth oxides, carbonates, sulfates, alkoxides, amides, carboxylates, and chelates, with the most reactive being amides or amide–alkoxides such as **18** and **19**. Chelates of β-keto esters, β-diketones, o-hydroxyphenyl ketones, and β-keto amides have the advantage of being soluble in monomer. The mechanism of

$$Ca(NH_2)_2 \qquad\qquad Ca(NH_2)OC_2H_5$$

18   **19**

polymerization by these initators is not known, but it may involve propagation by coordination of monomer with a complex alkoxide ion-pair chain end followed by anionic attack of the alkoxide ion on the ring (11.27).

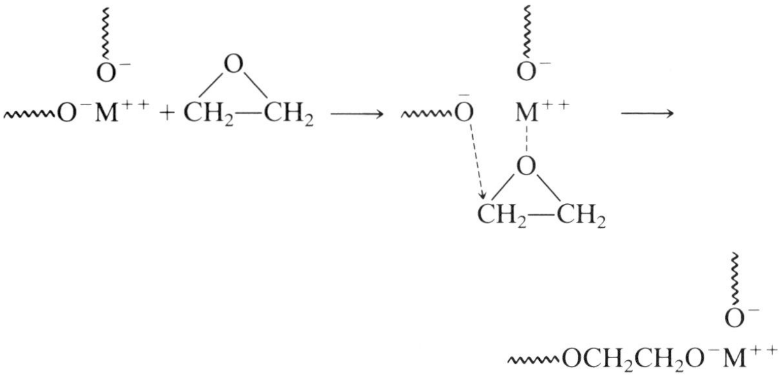

$$\text{wwwOCH}_2\text{CH}_2\text{O}^-\text{M}^{++} \qquad (11.27)$$

Examples of the second type[4d,8b] include typical Ziegler–Natta systems; aluminum, magnesium, zinc, ferric alkoxides; aluminum and zinc alkyls; and ferric chloride–propylene oxide complex. Polymerization with metal alkyls presumably proceeds by a mechanism similar to that in reaction (11.27). The ferric chloride–propylene oxide complex, an active catalyst that gives high-molecular-weight crystalline polymer with propylene oxide, is believed to have the structure **20**, where $m + n = 4$ or 5.

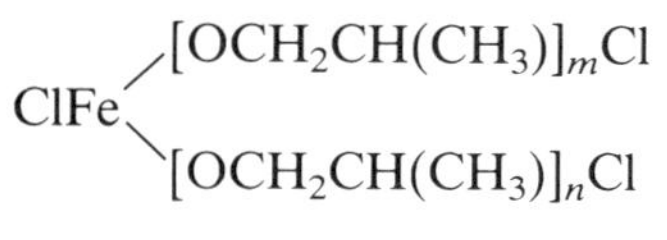

$$\text{ClFe}\begin{array}{l}\diagup[\text{OCH}_2\text{CH}(\text{CH}_3)]_m\text{Cl}\\[4pt]\diagdown[\text{OCH}_2\text{CH}(\text{CH}_3)]_n\text{Cl}\end{array}$$

20

The metal alkoxides and metal alkyls are generally much more reactive in the presence of water or alcohols. The function of these coreactants is not simply that of proton donor as it is in normal cationic polymerization. Instead they react to modify the organometallic compound. For the diethylzinc–water system, for example, best results are obtained at catalyst water ratios of $1:1$. It is believed that the active species is probably **21**; however, the mechanism of polymerization is by no means clear.[8b] A possible pathway for propylene oxide polymerization involves initial coordination (11.28) with propagation proceeding by intramolecular rearrangement (11.29).

$$C_2H_5ZnOZnC_2H_5$$
21

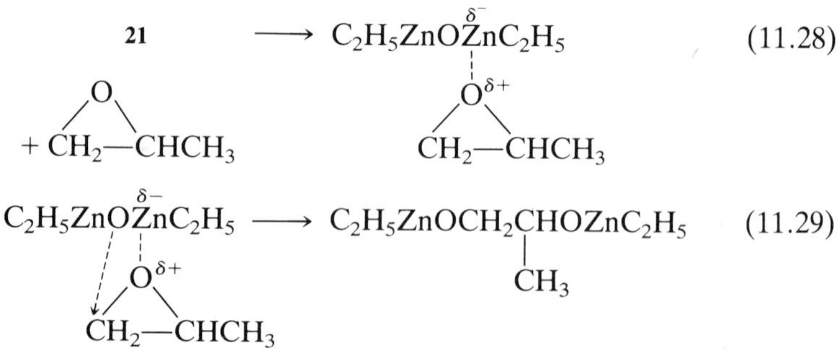

$$\mathbf{21} \longrightarrow C_2H_5ZnO\overset{\delta-}{Z}nC_2H_5 \tag{11.28}$$

$$+ \underset{CH_2-CHCH_3}{\overset{O}{\triangle}}$$

$$\underset{CH_2-CHCH_3}{\overset{O^{\delta+}}{\triangle}}$$

$$C_2H_5ZnO\overset{\delta-}{Z}nC_2H_5 \longrightarrow C_2H_5ZnOCH_2\underset{CH_3}{\overset{|}{C}HOZnC_2H_5} \tag{11.29}$$

$$\underset{CH_2-CHCH_3}{\overset{O^{\delta+}}{\triangle}}$$

The trimethylaluminum–water system is most active at catalyst/water ratios of about 1:1.5. A structure (**22**) similar to that for the diethylzinc–water system has been proposed as the reactive species, and analogous

$$(CH_3)_2AlOAl(CH_3)_2$$
22

mechanisms are assumed for propylene oxide or ethylene oxide polymerization. Both the zinc and the aluminum catalysts cause propylene oxide to form very-high-molecular-weight crystalline polymer.

Metalloporphyrin initiators, such as tetraphenylporphinatoaluminum chloride (**23**), were developed more recently.[28,29] They polymerize epoxides to very-low-polydispersity ($\bar{M}_w/\bar{M}_n = 1.1$) living polymer, with the propagat-

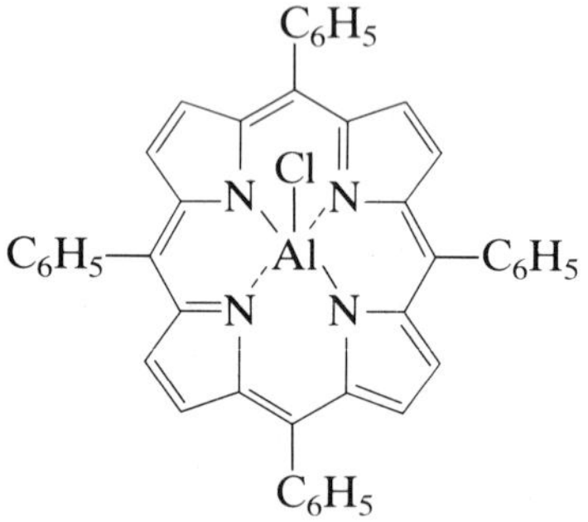

23

ing chain being the complex aluminum alkoxide, as shown for propylene oxide in reaction (11.30). The metalloporphyrin end group is not fixed, but undergoes rapid exchange with other polymer molecules. Particularly in-

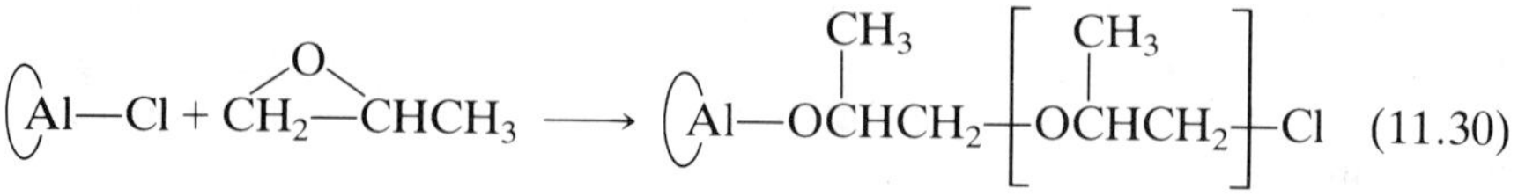

$$\overset{\frown}{Al}-Cl + \underset{CH_2-CHCH_3}{\overset{O}{\triangle}} \longrightarrow \overset{\frown}{Al}-O\underset{\overset{|}{CH_3}}{CH}CH_2 \left[O\underset{\overset{|}{CH_3}}{CH}CH_2 \right] Cl \tag{11.30}$$

teresting is the fact that even in the presence of chain transfer agents such as alcohols, the low polydispersity is retained even as the molecular weight is reduced. This results from the reversible exchange of alkoxide (11.31), which

$$\left(Al\!-\!O\text{\scriptsize www} + ROH \rightleftharpoons \left(Al\!-\!OR + HO\text{\scriptsize www} \right.\right. \tag{11.31}$$

is much faster than propagation. Thus the forward reaction (transfer) kills the chain, but the reverse reaction revives it, a phenomenon that led to the resultant polymer being characterized as *immortal*! The exchange reaction works equally well with such alcohols as ethylene glycol or hydroxyethyl methacrylate to yield polymers having well-defined molecular weights containing reactive end groups.

11.2.4 Stereochemistry of epoxide polymerization

Simple cationic or anionic polymerization of propylene oxide does not lead to stereoregular polyether, although a crystalline, optically active polymer has been obtained from levorotatory propylene oxide using potassium hydroxide as initiator.[3b,4d] In this case the stereoregularity arises because the S_N2 displacement at the less hindered carbon (11.32) does not disturb the chiral center (designated with an asterisk). Potassium hydroxide itself is not capable of inducing stereoregularity, as is shown by the fact that *d,l*-propylene oxide forms amorphous, inactive polymer under the same conditions. Infrared studies of polyethers prepared from *cis*- and *trans*-1,2-dideuteroethylene oxide have shown quite conclusively that inversion always occurs at the carbon undergoing nucleophilic attack.

$$HO^- + CH_2\!\!-\!\!CHCH_3 \longrightarrow HOCH_2\overset{*}{C}HO^- \tag{11.32}$$

Only with coordination catalysts have stereoregular high-molecular-weight polymers been obtained,[4d,8b] and the mechanism of their formation has been the subject of much speculation. Presumably, the stereoregular placement is influenced by the nature of the active sites on the catalyst surface, as it is with Ziegler–Natta polymerization of vinyl monomers (Chapter 8).

Another epoxide that has been considered for market development is 1,4-dichloro-2,3-epoxybutane (**24**). Both *cis* and *trans* isomers of **24** undergo stereospecific cationic polymerization to yield semicrystalline polymer, the former *racemic*-diisotactic (all R or all S), and the latter *meso*-diisotactic

$$ClCH_2CH\!\!-\!\!CHCH_2Cl$$
$$O$$

24

(RS-RS sequences). It has been suggested that coordination between the propagating oxonium ion and the chlorine of the β-chloromethyl group of the last chain unit plays a role in steric control.[30]

11.3 Preparation of polyethers by step-reaction polymerization

11.3.1 Synthesis of polyethers from glycols and bisphenols

In the preceding chapter (Section 10.2), we learned that self-condensation reactions of glycols to form polyethers are complicated by the tendency to form cyclic ethers. Dehydration to form alkene is another limiting factor. Thus, while it is possible to form very-low-molecular-weight polymers by reactions such as that shown for 1,3-propanediol (11.33) using dehydrating reagents, and given the fact that such polymers have found occasional use as lubricants, the reaction is of very limited utility.[3d]

$$n\text{HOCH}_2\text{CH}_2\text{CH}_2\text{OH} \xrightarrow{(-\text{H}_2\text{O})} \text{H}-\!\!\left[\text{OCH}_2\text{CH}_2\text{CH}_2\right]_{\!n}\!\!-\text{OH} \quad (11.33)$$

Condensation of benzylic glycols such as **25** occurs more readily because the reaction (11.34) involves resonance-stabilized carbocation intermediates. Even so, the polymer (**26**) obtained is of relatively low molecular weight, having a $\overline{\text{DP}}$ of about 50. A related reaction, the exchange of glycol and acetal groups to form polyacetals, is discussed in the next section.

$$(11.34)$$

Most success is preparing polyethers from dihydroxy compounds has been achieved by nucleophilic displacement reactions analogous to the classical Williamson synthesis. An example is the reaction of a bisphenol with methylene chloride under basic conditions, which leads to high-molecular-weight aromatic polyformal[31,32] (11.35).

$$\text{HO}-\text{Ar}-\text{OH} + \text{CH}_2\text{Cl}_2 \xrightarrow{\text{base}} -\!\!\left[\text{OArOCH}_2\right]\!\!- \quad (11.35)$$

Nucleophilic aromatic substitution occurs when the leaving group is activated by electron-withdrawing substituents. Examples of three commercial engineering plastics (**27–29**) prepared this way are given in reactions (11.36), (11.37), and (11.38). The first (**27**), involving displacement of chloro groups activated by SO_2,[33] is referred to as a *polysulfone* even though ether linkages are formed in the polymerization reaction. The second (**28**) is known as *polyetherimide*[34,35] (General Electric trade name Ultem) because it combines the two functionalities. In this case displacement of the nitro groups is promoted by the electron-withdrawing imide. The third reaction (11.38), involving displacement of fluoro groups activated by ketone, gives a polymer (**29**) called a *polyetheretherketone* (designated PEEK).[36,37] Polymers **27** and **28** are amorphous with T_g's of about 220 and 215°C, respectively, whereas **29** is semicrystalline with a T_g of about 144°C and a T_m of about 334°C.

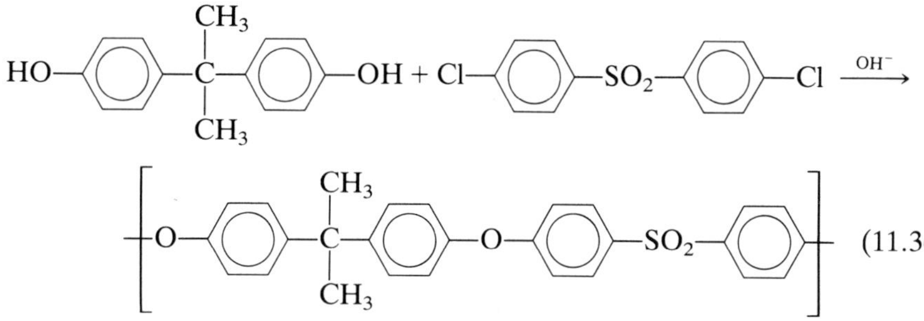

(11.36)

27

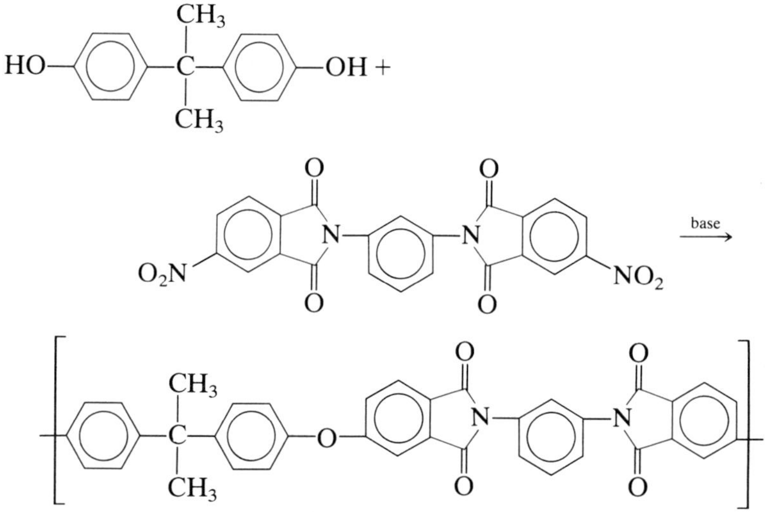

(11.37)

28

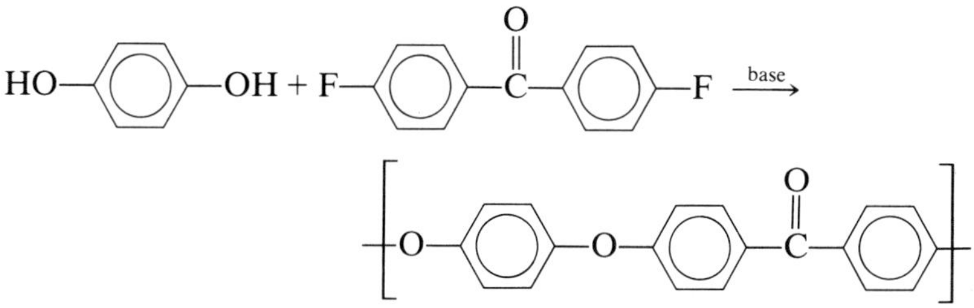

(11.38)

29

Phase-transfer methods have also been applied to polyether synthesis.[38] An example is the reaction (11.39) of 4,4'-dihydroxy-α-methylstilbene (**30**) with α,ω-dibromoalkanes catalyzed by tetrabutylammonium hydrogen sulfate (TBAH) to yield thermotropic liquid crystalline polymer (**31**) in which the methylene groups serve as flexible spacers between the mesogenic stilbene groups.[39]

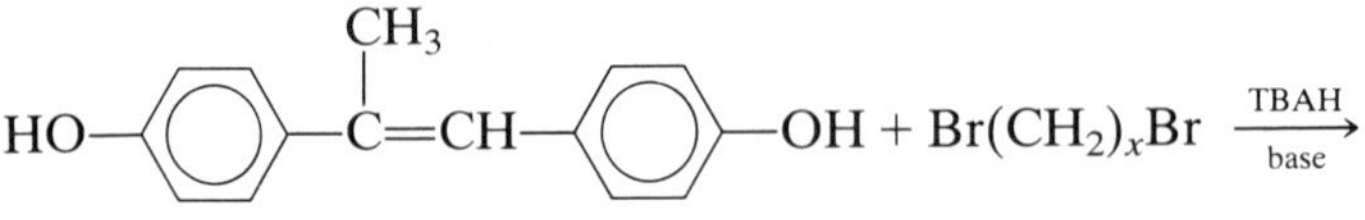

$$HO-\!\!\bigcirc\!\!-\overset{\overset{\displaystyle CH_3}{|}}{C}\!=\!CH-\!\!\bigcirc\!\!-OH + Br(CH_2)_xBr \xrightarrow[\text{base}]{\text{TBAH}}$$

30

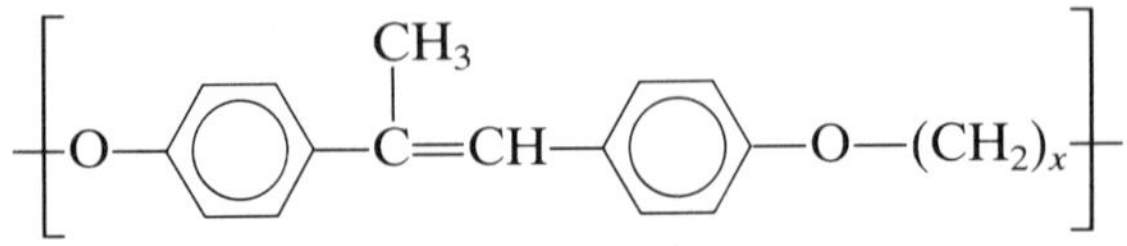

$$\left[-O-\!\!\bigcirc\!\!-\overset{\overset{\displaystyle CH_3}{|}}{C}\!=\!CH-\!\!\bigcirc\!\!-O-(CH_2)_x-\right] \quad (11.39)$$

31

11.3.2 Polyacetals and polyketals

Apart from chain-reaction polymerization of aldehydes and methylene chloride displacement reactions described earlier, there are three general methods for synthesizing polymers containing acetal or ketal groups[3d]: (1) reaction of diols with carbonyl compounds, an extension of the classical acetal (or ketal) synthesis, (2) acetal (or ketal) exchange, and (3) addition of diols to dialkenes.

Formation of acetals and ketals by the acid-catalyzed reaction of excess alcohol with aldehydes and ketones, respectively, is commonly employed in organic chemistry to protect the carbonyl function. By reacting equimolar amounts of aldehyde and diol and removing by-product water by azeotropic distillation, one forms polyacetals (11.40). A serious limitation is the tendency for cyclic acetals also to be formed; nevertheless, high-melting polyformals (**32** and **33**) have been prepared by heating the corresponding cyclic diols with paraformaldehyde in the presence of acid catalysts.[40]

$$RCHO + HO-R'-OH \xrightarrow{H^+} \left[\begin{array}{c} CH-O-R'-O \\ | \\ R \end{array}\right] + H_2O \quad (11.40)$$

$$\left[-O-\underset{\underset{\displaystyle CH_3CH_3}{\diagup\diagdown}}{\overset{\overset{\displaystyle CH_3CH_3}{\diagdown\diagup}}{}}-OCH_2-\right] \qquad \left[-O-\!\!\bigcirc\!\!-OCH_2-\right]$$

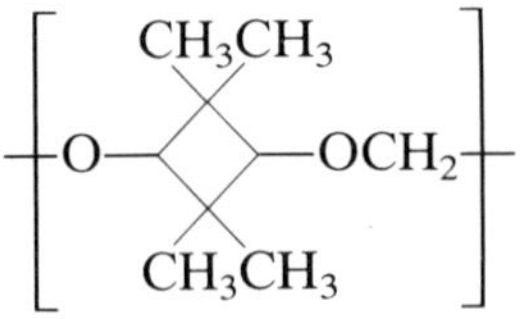

32 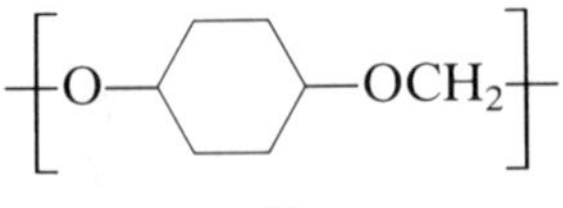 **33**

The tendency to form cyclic acetals can be used to advantage by reacting tetrols such as pentaerythritol (**34**) with dialdehydes to form cyclic polyacetals (**35**) (11.41). The polymer prepared from glutaraldehyde ($R = -CH_2CH_2CH_2-$), for example, has been investigated for fiber and wire coating applications. Particularly interesting is the reaction of **34** with 1,4-cyclohexanedione (**36**) to form *spiro* polyketal (**37**) (11.42).[41] Spiro polymers (i.e., polymers having recurring ring units joined through a common carbon atom) exhibit the same type of backbone rigidity and thermal stability as ladder polymers.

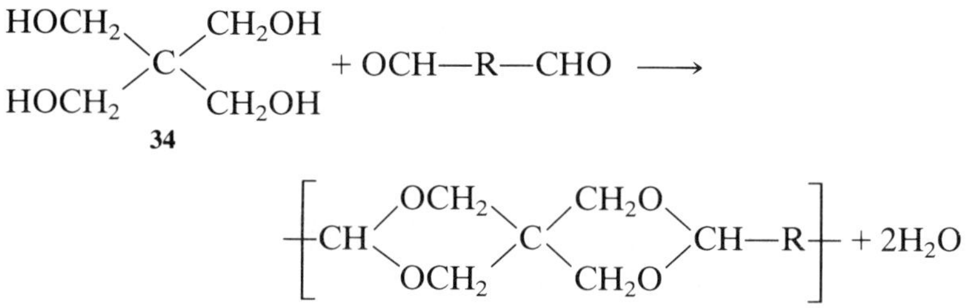

$$
\begin{array}{c}
\text{HOCH}_2 \\
 \\
\text{HOCH}_2
\end{array}
\!\!C\!\!
\begin{array}{c}
\text{CH}_2\text{OH} \\
 \\
\text{CH}_2\text{OH}
\end{array}
+ \text{OCH}-\text{R}-\text{CHO} \longrightarrow
$$

34

$$
\left[-\text{CH} \underset{\text{OCH}_2}{\overset{\text{OCH}_2}{<\quad>}} C \underset{\text{CH}_2\text{O}}{\overset{\text{CH}_2\text{O}}{<\quad>}} \text{CH}-\text{R} - \right] + 2\text{H}_2\text{O} \quad (11.41)
$$

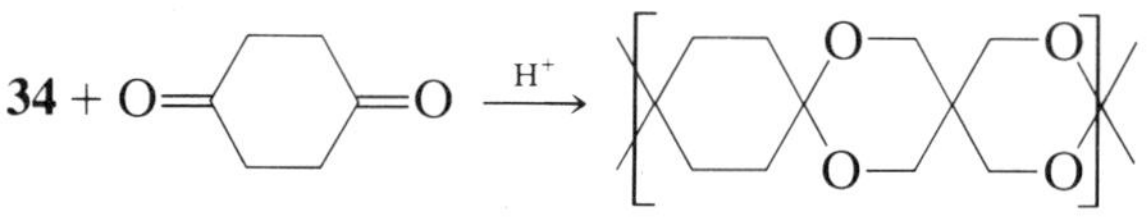

35

$$
\textbf{34} + \text{O}\!=\!\!\bigcirc\!\!=\!\!\text{O} \xrightarrow{\text{H}^+} \left[\text{spiro polyketal structure} \right] + 2\text{H}_2\text{O} \quad (11.42)
$$

36   **37**

Acid-catalyzed acetal or ketal exchange reactions, which are mechanistically similar to the self-condensation of glycol discussed in the preceding section, give rise to the same types of polyacetals and polyketals shown above, except that alcohol is the by-product. Polymer **35**, for example, may be prepared by reacting tetrol **34** with the acetal of the dialdehyde (11.43).

$$
\textbf{34} + (\text{R}'\text{O})_2\text{CH}-\text{R}-\text{CH}(\text{OR}')_2 \xrightarrow{\text{H}^+} \textbf{35} + 4\text{R}'\text{OH} \quad (11.43)
$$

Linear polyformals can also be prepared by an entirely different route, the acid-catalyzed stepwise addition polymerization of glycols to divinyl ethers[3d] (11.44). Polyacetals of this type have sparked interest as bioerodable polymers for controlled release of drugs.[42]

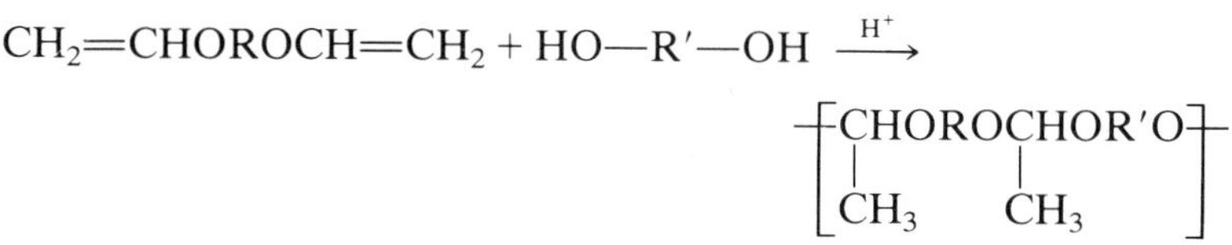

$$
\text{CH}_2\!=\!\text{CHOROCH}\!=\!\text{CH}_2 + \text{HO}-\text{R}'-\text{OH} \xrightarrow{\text{H}^+}
$$

$$
\left[\begin{array}{c} -\text{CHOROCHOR}'\text{O}- \\ || \\ \text{CH}_3 \text{CH}_3 \end{array} \right] \quad (11.44)
$$

11.3.3 Poly(phenylene oxide)s

Aromatic polyethers have received considerable attention in recent years because of the current interest in thermally stable polymers. The most successful method of preparation involves oxidative coupling of 2,6-disubstituted phenols.[33,43] If the substituents are methyl groups (reaction 11.45), the polymerization can be done at room temperature by bubbling oxygen through

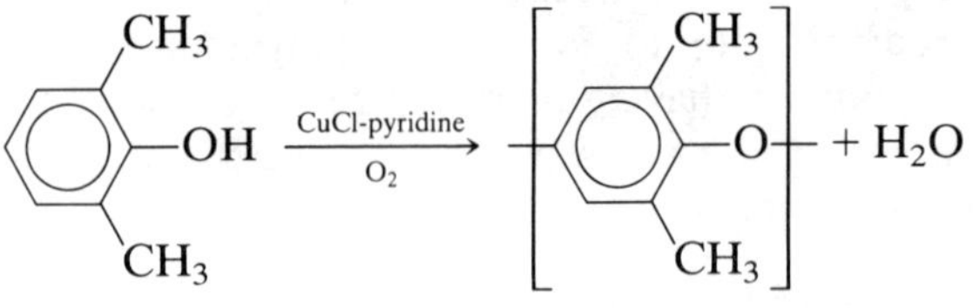

(11.45)

38

a solution of the phenol in the presence of a cuprous salt–amine catalyst. Pyridine is commonly used as amine and solvent and cuprous chloride as the copper salt. Somewhat higher temperatures are required for halogen-substituted phenols. The polymer (**38**) obtained from 2,6-dimethylphenol, known commercially as poly(phenylene oxide) (PPO), has good thermal and hydrolytic stability. It may be recalled from Chapter 3 that **38** is used primarily as an engineering plastic in a homogeneous blend with polystyrene. Co-polymers with 2,3,6-trimethylphenol, which exhibit a higher T_g and greater strength than the homopolymer, are also used commercially as polyblends. If the substituent alkyl groups are sufficiently bulky, carbon–carbon rather than carbon–oxygen coupling occurs; 2,6-di-*t*-butylphenol, for example, gives the diphenoquinone (**39**) rather than polymer.

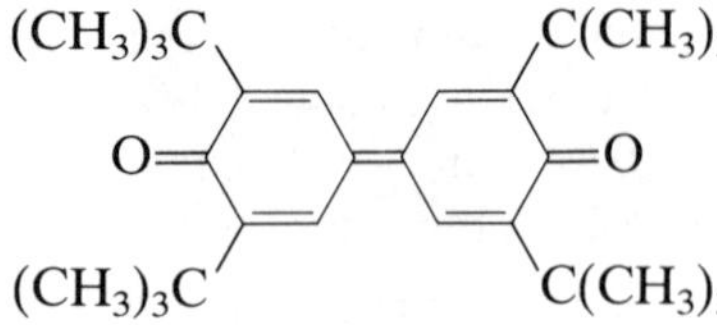

39

The mechanism of this reaction has been investigated in some detail. It is believed that the active catalyst is a basic cupric salt (resulting from oxidation of cuprous chloride) complexed with two amine groups (**40**). Initiation might then involve reaction between catalyst and phenol according to reaction

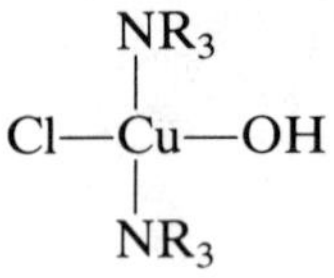

40

(11.46). Electron transfer to copper would generate phenoxy radicals that might be expected to lead to chain-growth polymerization by successive

$$-\overset{|}{\underset{|}{Cu}}-OH + ArOH \longrightarrow -\overset{|}{\underset{|}{Cu}}-OAr + H_2O \qquad (11.46)$$

radical additions at the *para* position; however, the reaction has the characteristics of a step-growth rather than a chain-growth process, as evidenced by the fact that high molecular weight is achieved only at high degrees of polymerization. It is now believed that radical coupling to give quinol ethers occurs (11.47) followed by dissociation to give the original aryloxy radicals or two new radicals (11.48) which, in turn, could couple. Chain buildup thus occurs in a stepwise fashion, and a broad molecular weight distribution results from hydrogen transfer reactions.

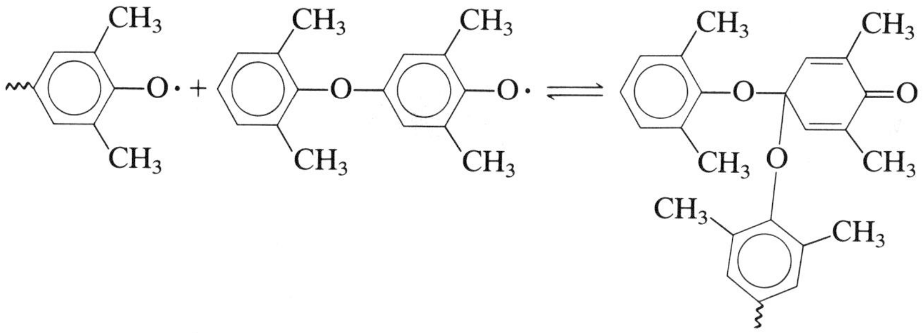

$$(11.47)$$

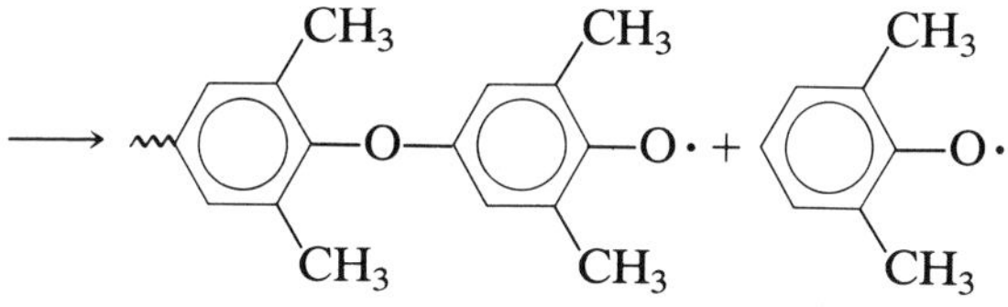

$$(11.48)$$

Block copolymers have been prepared from 2,6-diphenylphenol and 2,6-dimethylphenol by oxidative coupling.[44] The diphenyl derivative is polymerized first; then the dimethyl derivative is added to the reaction mixture. If the order is reversed, or if a mixture is used, a random copolymer results.

Reactions (11.49) through (11.52) illustrate additional methods of synthesizing poly(phenylene oxide)s,[33] none of which has been developed commercially.

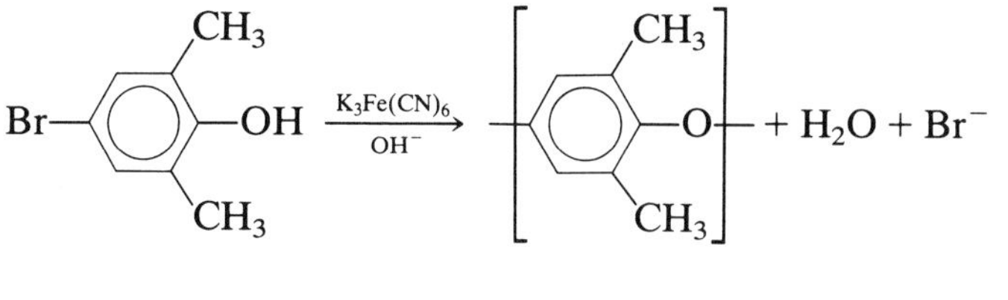

$$(11.49)$$

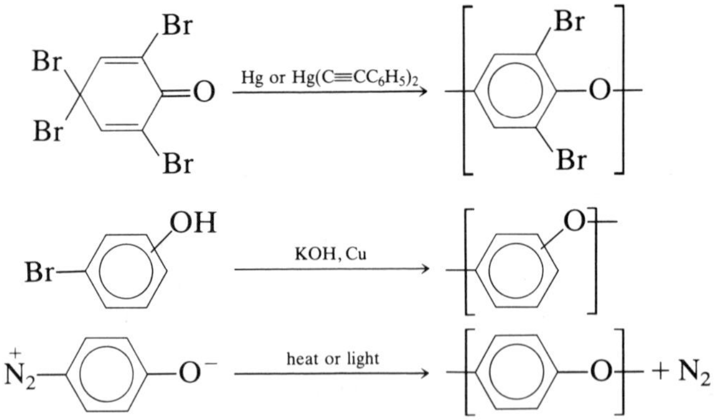

$$(11.50)$$

$$(11.51)$$

$$(11.52)$$

11.3.4 Epoxy resins

From the commercial standpoint, epoxy polymers or resins are among the most important of the nonvinyl polymers; their commercial significance is, in fact, exemplified by the publication of several books devoted entirely to the topic.[45-50] They represent a special type of polyether prepared by a step-polymerization reaction between an epoxide and a dihydroxy compound, usually a bisphenol, in the presence of a base. It might appear that a di-epoxide would be necessary to form polymer; in practice, however, epichlorohydrin (**41**) is most commonly used because it reacts in the manner of a diepoxide.

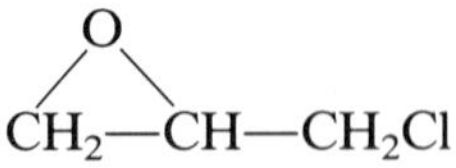

41

The polymerization sequence involves formation of alkoxide ion (11.53), nucleophilic addition of alkoxide to the least hindered carbon of the epoxide ring (11.54), then ring closure by internal displacement of chloride ion (11.55). This regenerates the epoxide ring, allowing further reaction with the hydroxy compound to eventually give a product having pendant hydroxy groups (11.56). With use of excess epichlorohydrin, a polyether having *glycidyl ether* end groups is formed (11.57).

$$HO—R—OH + OH^- \rightleftharpoons HO—R—O^- + H_2O \quad (11.53)$$

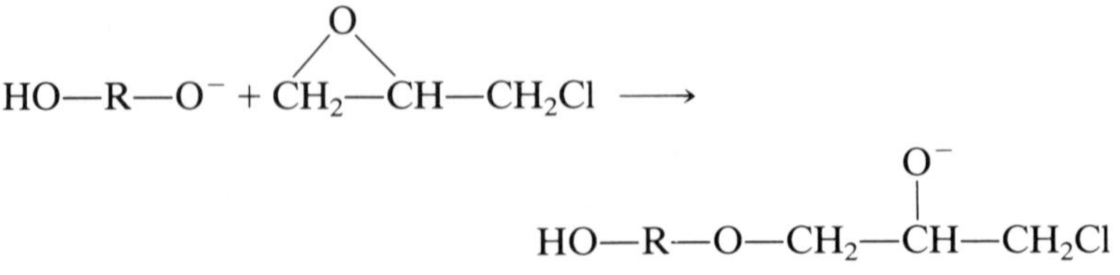

$$(11.54)$$

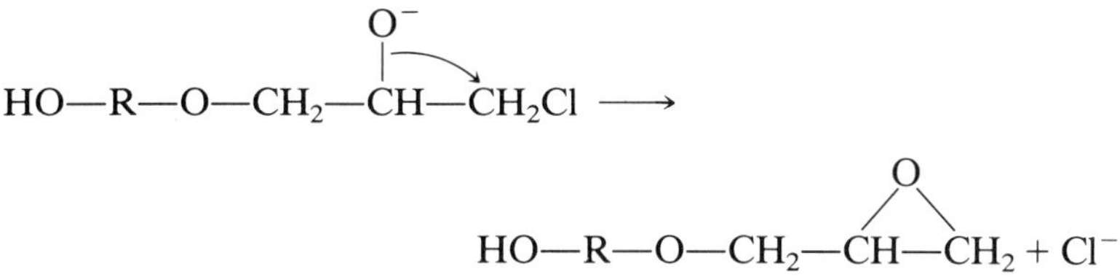

$$HO-R-O-CH_2-\overset{\overset{\displaystyle O^-}{|}}{CH}-CH_2Cl \longrightarrow$$

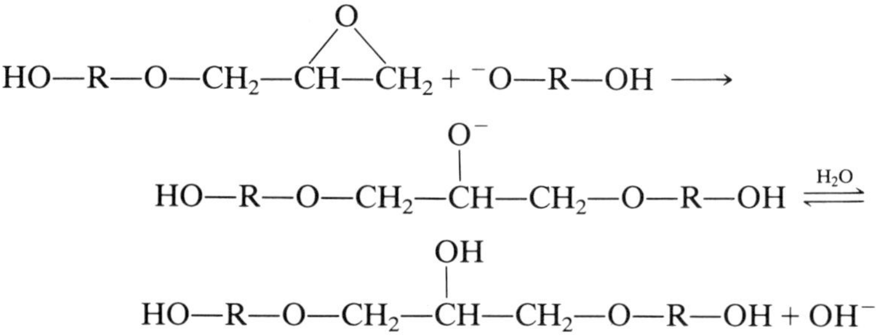

$$HO-R-O-CH_2-\overset{O}{\overset{\triangle}{CH-CH_2}} + Cl^- \quad (11.55)$$

$$HO-R-O-CH_2-\overset{O}{\overset{\triangle}{CH-CH_2}} + {}^-O-R-OH \longrightarrow$$

$$HO-R-O-CH_2-\overset{\overset{\displaystyle O^-}{|}}{CH}-CH_2-O-R-OH \overset{H_2O}{\rightleftharpoons}$$

$$HO-R-O-CH_2-\overset{\overset{\displaystyle OH}{|}}{CH}-CH_2-O-R-OH + OH^- \quad (11.56)$$

$$\xrightarrow[\underset{(excess)}{ClCH_2CH-CH_2}]{\overset{O}{\triangle}} \overset{O}{\overset{\triangle}{CH_2-CHCH_2}}\left[OROCH_2\overset{\overset{\displaystyle OH}{|}}{CHCH_2}\right]_n OROCH_2\overset{O}{\overset{\triangle}{CH-CH_2}}$$

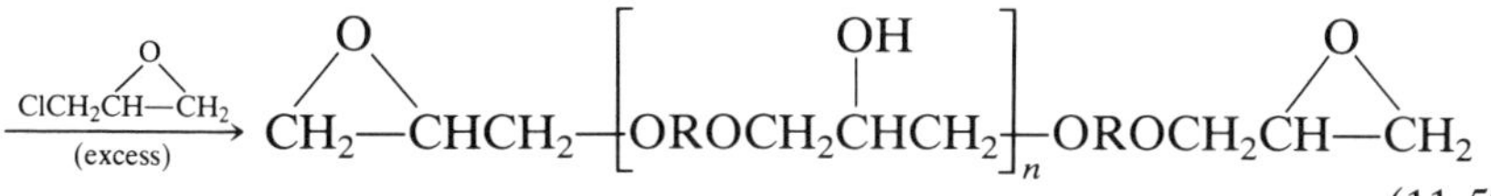

$$(11.57)$$

Initially the product formed is a low-molecular-weight *prepolymer*. It is capable of undergoing further polymerization through either the terminal epoxide or the pendant hydroxyl groups. The molecular weight of the prepolymer is varied by suitable adjustment of the epichlorohydrin excess. Depending on the number of repeating units and the structure of the dihydroxy compound, the product varies from a viscous liquid to a solid. Optimum properties of the prepolymer depend on the application, but generally, very high molecular weights are not desirable because of handling difficulties.

The commonest dihydroxy compound in commercial use is bisphenol A, giving prepolymer structure **42**. Other bisphenols such as hydroquinone (**43**) and resorcinol (**44**) have also been used, as have other aldehyde– or ketone–phenol condensation products similar to bisphenol A. In addition to the simple bisphenol (**45**), low-molecular-weight novolacs (**46**), formed by polycondensation of phenol and formaldehyde, have also been studied. (Novolac chemistry is discussed in Chapter 14.)

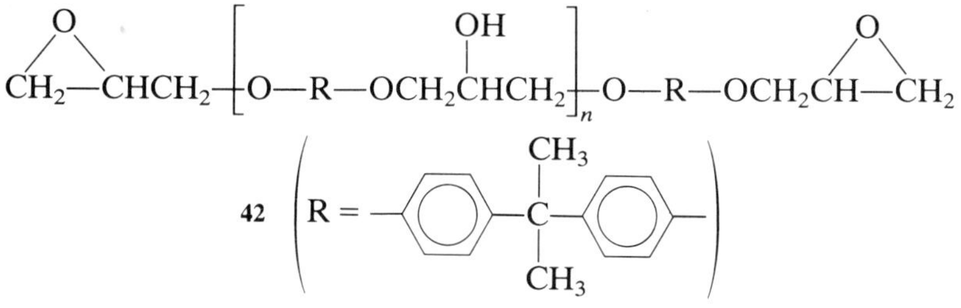

$$\overset{O}{\overset{\triangle}{CH_2-CHCH_2}}\left[O-R-OCH_2\overset{\overset{\displaystyle OH}{|}}{CHCH_2}\right]_n O-R-OCH_2\overset{O}{\overset{\triangle}{CH-CH_2}}$$

$$\mathbf{42}\quad \left(R = -\!\!\left\langle \bigcirc \right\rangle\!\!-\overset{\overset{\displaystyle CH_3}{|}}{\underset{\underset{\displaystyle CH_3}{|}}{C}}-\!\!\left\langle \bigcirc \right\rangle\!\!- \right)$$

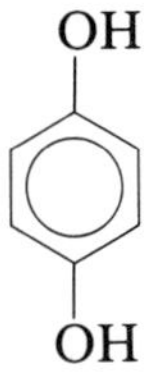

43 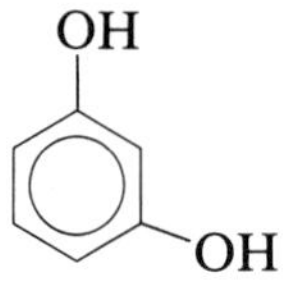   **44** 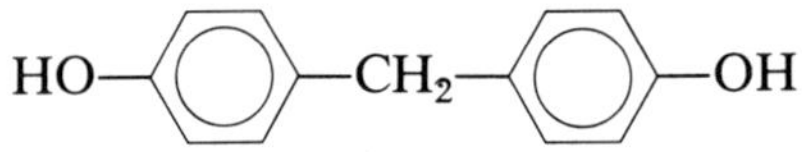   **45**

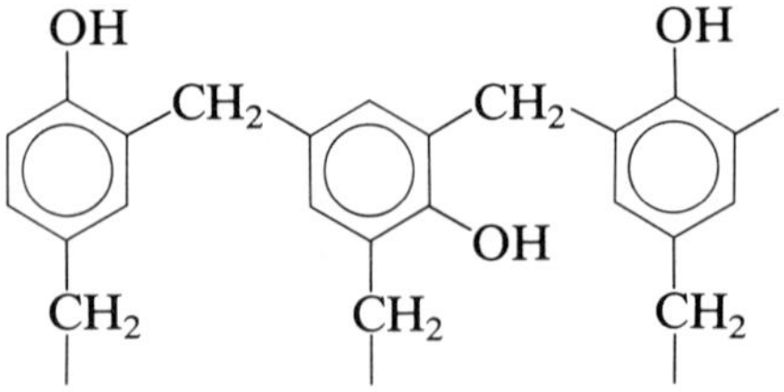

46

While the glycidyl ethers represent the most common type of epoxy prepolymer, other epoxide structures are used commercially, amount them multifunctional glycidyl amines such as **47** and **48**, cycloaliphatic compounds such as vinylcylohexene dioxide (**49**) and dicyclopentadiene dioxide (**50**), epoxidized diene polymers, and epoxidized oils.

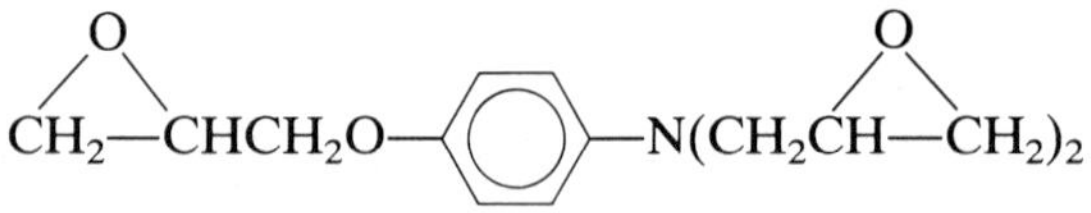

47

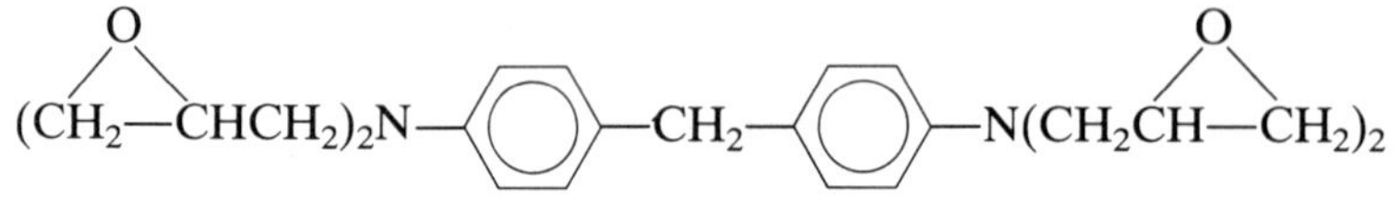

48

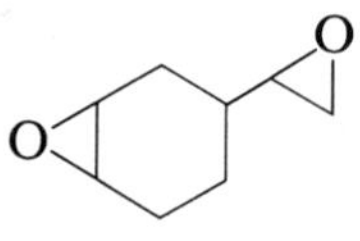

49 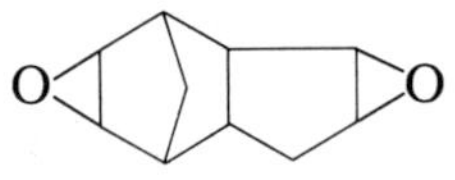   **50**

Further polymerization of prepolymer to form crosslinked high polymer (*curing*) can be accomplished several ways, although amines and carboxylic acid anhydrides are most commonly used. Amines react by nucleophilic addition to the epoxide ring (11.58). Diamines and polyamines are also used as curing agents and, in some applications, the epoxy resin is blended with other polymers such as melamine– or urea–formaldehyde resins (see Chapter 4), which cause crosslinking by essentially the same reaction to give network copolymer. Tertiary amines may also be used, but the mechanism here involves ring-opening polymerization of the epoxy groups (Section 11.2.3).

Reaction with amines is exothermic and is known to be catalyzed by proton donors.

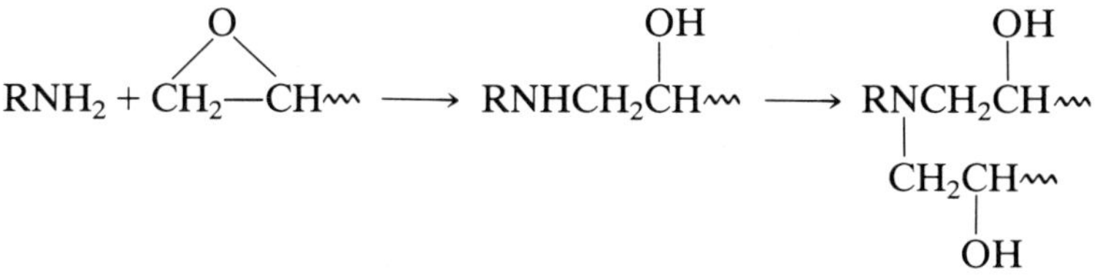

$$RNH_2 + CH_2\!-\!CH\text{\large\char"007E} \longrightarrow RNHCH_2\overset{\displaystyle OH}{\underset{\displaystyle |}{CH}}\text{\large\char"007E} \longrightarrow RN\overset{\displaystyle OH}{\underset{\displaystyle |}{CH_2CH}}\text{\large\char"007E}\quad etc.$$

$$\begin{array}{c} | \\ CH_2CH\text{\large\char"007E} \\ | \\ OH \end{array}\qquad (11.58)$$

Partially cured epoxy resins that are stable in storage but cure further on heating can be prepared[51] by adding hydrazides of structure **51**. Resins of this type are called *B-stage* resins. (A-stage refers to uncured resins, B-stage to partially cured, and C-stage to fully cured.)

$$RNHCH_2CH_2\overset{\displaystyle O}{\overset{\displaystyle \|}{C}}NHNH_2$$

51

Anhydrides—for example, phthalic anhydride (**52**)—react with pendant hydroxyl groups to give ester acids (11.59). The acid group then reacts with epoxide groups (11.60) or other hydroxyls (11.61) to form more ester groups. Ether groups also form by the acid-catalyzed reaction between hydroxyl and epoxide groups. Obviously crosslinking of epoxy resins is a very complex process, and it is impossible to give a definitive structure for the final product.

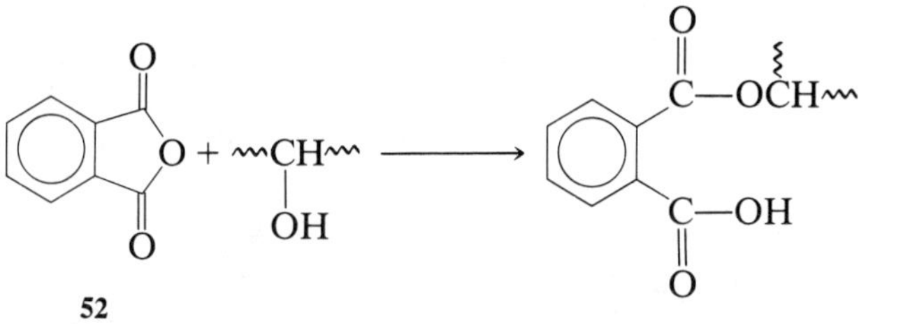

(11.59)

52

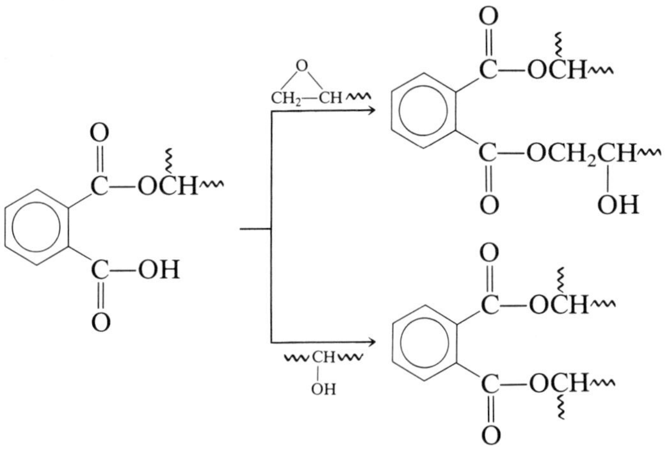

(11.60)

(11.61)

Curing with anhydrides is usually catalyzed by amine. Some of the more common anhydrides in commercial use include phthalic (**52**), maleic (**53**), and chlorendic (**54**), and the liquid anhydrides, dodecenylsuccinic anhydride (**55**) and methylbicyclo[2.2.1]hept-5-ene-2,3-dicarboxylic acid anhydride (**56**) (nadic methyl anhydride, the Diels–Alder adduct of mixed methylcyclopentadienes and maleic anhydride). Liquid Diels–Alder adducts derived from maleic anhydride and terpenes have also been reported.[52] It appears that Diels–Alder adducts improve the high-temperature stability of epoxy resins because the retrograde Diels–Alder reaction generates alkenes in the polymeric network, which crosslink further and cause more char formation.[53]

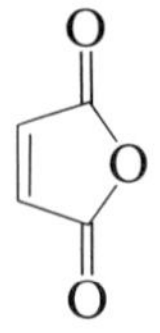

53 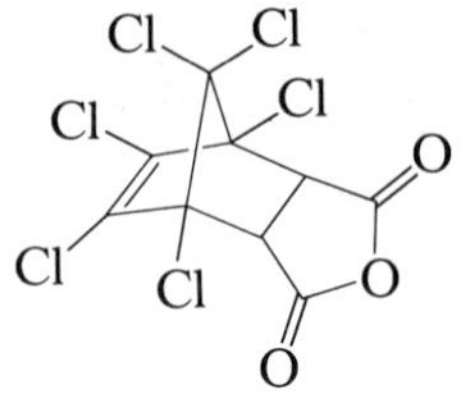   54

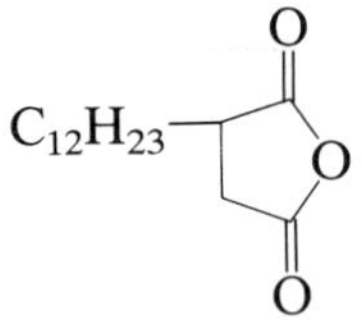

55 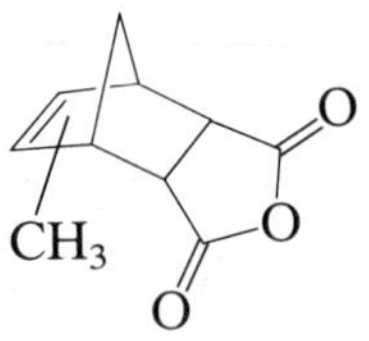   56

Dianhydrides are also employed as curing agents. They have two distinct advantages over simple anhydrides: (1) they cause an increase in cure rate, and (2) they form a more densely crosslinked product. The increase in cure rate is desirable because the reactions of the initially formed ester acid described above are much slower than those of the anhydride. The most commonly used dianhydride is pyromellitic dianhydride (**57**); however, **57** has the disadvantage of being difficult to dissolve in the resin before gelation occurs. Better solubility, with no sacrifice in mechanical properties, has been claimed for the photoaddition products (**58**) of alkylbenzenes (notably ethylbenzene and cumene) and maleic anhydride.[54]

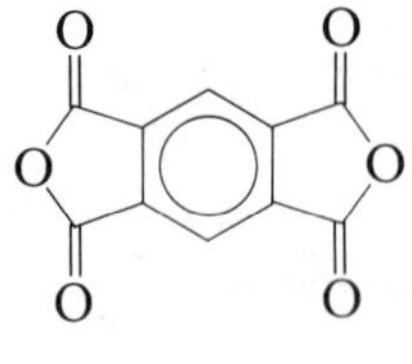

57 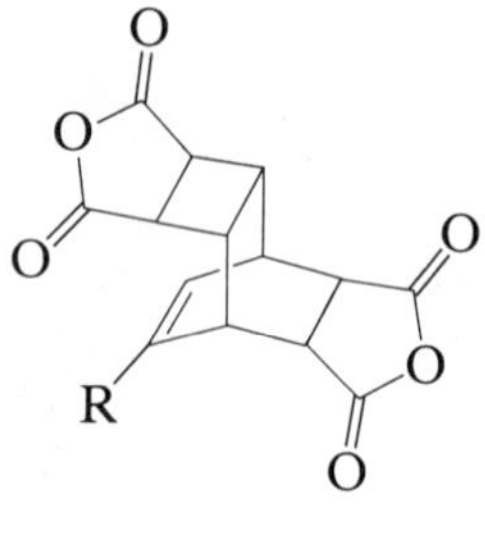   58

Because of the reactive end groups, epoxy resins can be grafted onto other polymers. Thermosetting acrylic polymers have been prepared, for example, by reaction of an epoxy resin with an acrylic to form a graft copolymer (11.62) capable of being crosslinked. For certain coatings applications, epoxies are also reacted in analogous fashion with drying oil fatty acids to take advantage of the free radical crosslinking proclivity of the fatty acid. (Drying oil polymerization is discussed in the next chapter.)

$$\text{wCH}_2\overset{\overset{\displaystyle CH_3}{|}}{\underset{\underset{\displaystyle CO_2H}{|}}{C}}\text{w} \;+\; \overset{\displaystyle O}{\overset{\displaystyle /\backslash}{CH_2\text{—}CH}}\text{w} \longrightarrow \text{wCH}_2\overset{\overset{\displaystyle CH_3}{|}}{\underset{\underset{\displaystyle \underset{O}{\overset{\displaystyle \|}{C}}OCH_2\underset{OH}{\overset{|}{CH}}\text{w}}{|}}{C}}\text{w} \qquad (11.62)$$

11.4 Polysulfides, poly(alkylene polysulfide)s, and polysulfones

11.4.1 Polysulfides

Polysulfides, or poly(alkylene sulfide)s as they are sometimes called, are the sulfur analogs of polyethers and, as might be expected, they are prepared by similar reactions. Polymerization of thiocarbonyl compounds has not been extensively investigated with the exception of thiocarbonyl fluoride (**59**) and related compounds.[55] This monomer is extremely reactive and undergoes polymerization (11.61) at −78°C under anionic or free radical conditions to give a high-molecular-weight (300,000 to 400,000), tough, elastomeric product that resists hydrolysis by acid or base, although it is degraded rapidly by amines. Copolymerization of **59** with vinyl monomers can also be brought about with free radical initiators. Related monomers such as $CF_3CF{=}S$,

$$CF_2{=}S \longrightarrow {-}[CF_2{-}S]{-} \qquad (11.63)$$
$$\mathbf{59}$$

$ClCF_2CF{=}S$, and $ClCHFCF{=}S$ also undergo anionic polymerization, but the products are not as useful. Hexafluorothioacetone polymerizes at −110°C, but the polymer decomposes at room temperature.

With the exception of hexafluorothioacetone, these compounds are not, of course, the sulfur analogs of aldehydes or ketones but are related in structure to the acyl halides. Their polymerization through the carbon–sulfur double bond is, however, more typical of aldehyde or ketone polymerization. Little has been done in the area of polymerization of ordinary thioaldehydes and thioketones,[56a] although some (for example, the unstable polythioacetophenone[57]) have been reported.

Episulfides, the sulfur analogs of epoxides, undergo ring-opening polymerization to form polysulfide (11.64),[19c,23b] although their reactivity is generally higher than that of epoxides. Cationic, anionic, and coordination

$$\underset{\text{CH}_2\text{---CH}_2}{\overset{\text{S}}{\triangle}} \longrightarrow -\!\!\left[\text{CH}_2\text{CH}_2\text{S}\right]\!\!- \tag{11.64}$$

initiators may be used. Although mechanisms parallel those of epoxides, for the most part, anionic initiation with alkyllithium compounds appears to follow a different pathway.[58] With propylene sulfide (**60**), for example, stoichiometric formation of propylene occurs in a first step (11.65), followed by initiation of polymerization with the resultant lithium thiolate (11.66).

$$\text{RLi} + \underset{\mathbf{60}}{\overset{\text{S}}{\overset{\triangle}{\text{CH}_2\text{---CHCH}_3}}} \longrightarrow \text{CH}_2\!\!=\!\!\text{CHCH}_3 + \text{RSLi} \tag{11.65}$$

$$\text{RSLi} + \mathbf{60} \longrightarrow \text{RS}\!\!\left[\!\!\begin{array}{c}\text{CH}_3\\|\\\text{CH}_2\text{CHS}\end{array}\!\!\right]\!\!\text{Li} \tag{11.66}$$

Step-reaction polymerization has been more widely used for polysulfide synthesis. The reaction of dithiols with dihalides under basic conditions (11.67), like the analogous Williamson synthesis of polyethers, leads to low-molecular-weight polymers in most instances, but the reaction proceeds more readily because of the greater nucleophilicity of sulfide ion relative to alkoxide. Polymerizations of this type have been carried out successfully under phase-transfer conditions.[38,39]

$$\text{NaS(CH}_2)_6\text{SNa} + \text{Br(CH}_2)_6\text{Br} \rightarrow -\!\!\left[\text{S(CH}_2)_6\text{S(CH}_2)_6\right]\!\!- + 2\text{NaBr} \tag{11.67}$$

Polymers of like structure are obtained by free radical or ionic addition of dithiols to dialkenes,[56a] reactions that proceed far more readily than addition of glycol to dialkenes. Free radical addition, initiated by peroxides, follows the anti-Markovnikov pathway (11.68) to yield the same polymer as that formed in reaction (11.67), whereas ionic addition, catalyzed by acid or base, yields the Markovnikov product (11.69). Reaction (11.69) is unusual in that

$$\text{HS(CH}_2)_6\text{SH}$$
$$+$$
$$\text{CH}_2\!\!=\!\!\text{CHCH}_2\text{CH}_2\text{CH}\!\!=\!\!\text{CH}_2$$

$$\xrightarrow{\text{peroxide}} -\!\!\left[\text{S(CH}_2)_6\text{S(CH}_2)_6\right]\!\!- \tag{11.68}$$

$$\xrightarrow{\text{H}^+ \text{ or OH}^-} \left[\!\!\begin{array}{c}\text{S(CH}_2)_6\text{SCHCH}_2\text{CH}_2\text{CH}\\ \quad\quad\quad\quad |\quad\quad\quad\quad\quad\quad |\\ \quad\quad\quad\quad \text{CH}_3\quad\quad\quad\quad\quad \text{CH}_3\end{array}\!\!\right] \tag{11.69}$$

it involves a free radical chain reaction combined with step-reaction polymerization (see review exercise 16). Polysulfides having molecular weights of 20,000 to 60,000 have been prepared by these methods. If the carbon–carbon double bonds are electron deficient, uncatalyzed nucleophilic addition occurs. Thus, polyimidesulfides have been synthesized by Michael addition of dithiols to bismaleimides (11.70),[60-63] and polyestersulfides have been made by the amine-catalyzed reaction of acrylic esters with hydrogen sulfide (11.71).[64] High-molecular-weight polyenonesulfides are formed by the analogous addition of disulfides to dipropynones (11.72).[65,66]

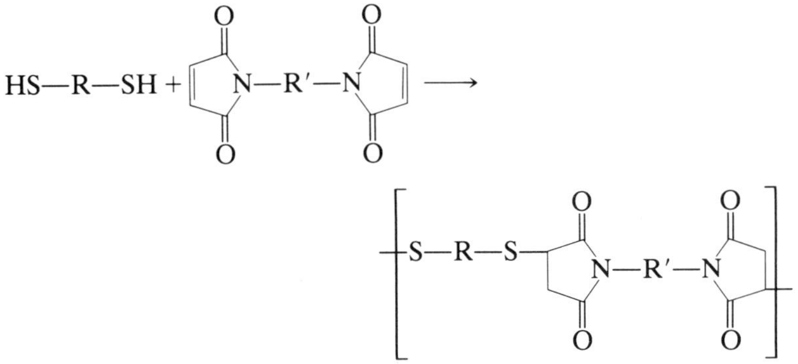

$$HS-R-SH + \text{(bismaleimide)} \longrightarrow \qquad (11.70)$$

$$CH_2=CHCOCH_2-\text{(ring)}-CH_2OCCH=CH_2 + H_2S \xrightarrow{R_3N} \qquad (11.71)$$

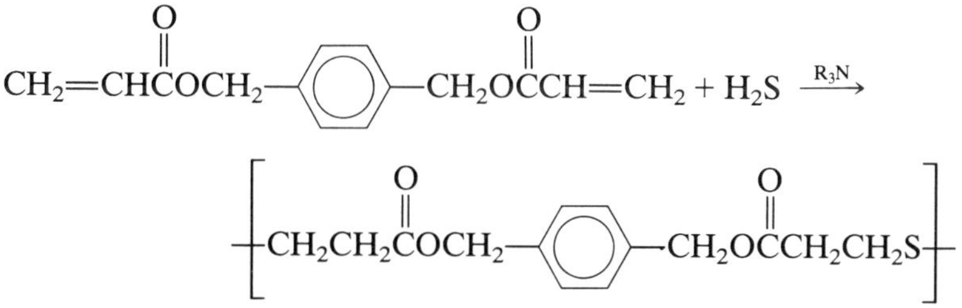

$$Ar-C\equiv C-C-\text{(ring)}-C-C\equiv C-Ar + HS-Ar'-SH \xrightarrow{amine} \qquad (11.72)$$

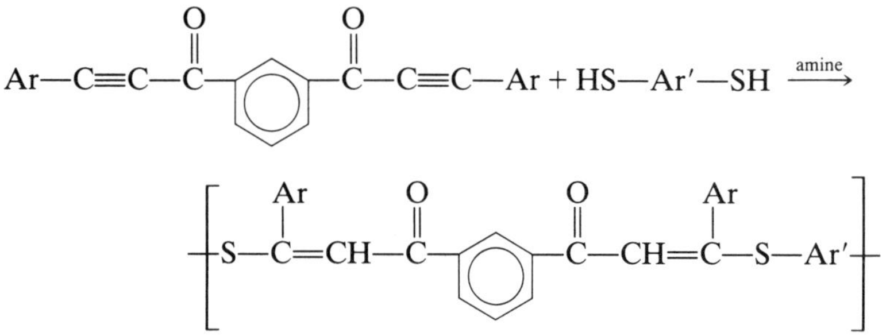

Thiol-terminated, low-molecular-weight polysulfides useful as adhesives and caulking compounds have been prepared by the addition of dithiols to 1-alkynes.[67] The reaction is similar to that of addition to dienes, except that a very reactive vinyl sulfide is formed initially (11.73), which reacts rapidly with dithiol (11.74). Further reaction (11.75) leads to polymer. Ionizing radiation from a cobalt-60 source is the preferred mode of initiation.

$$HS-R-SH + HC\equiv CR' \longrightarrow HS-R-S-CH=CHR' \quad (11.73)$$

$$\xrightarrow[\text{fast}]{HS-R-SH} HS-R-S-CH_2-\underset{\underset{R'}{|}}{CH}-S-R-SH \quad (11.74)$$

$$\xrightarrow[\text{(2) } HS-R-SH]{\text{(1) } HC\equiv CR'} HS\left[R-S-CH_2-\underset{\underset{R'}{|}}{CH}-S\right]R-SH \quad (11.75)$$

Reaction of aldehydes and ketones with dithiols (11.76) leads to low-molecular-weight polythioacetals or polythioketals (compare Section 11.3.2), but, as in the case of the corresponding polyethers, ring formation is a problem.[56a] The reaction of dithiols has two main advantages over that of diols: (1) the resultant polymers are much more resistant to hydrolysis and (2) the reaction works well with ketones as well as with aldehydes. Thiol–acetal exchange reactions (11.77) may also be used. As was the case with polyketals, best results are obtained with cyclic polythioketals (11.78).

$$R-CHO + HS-R'-SH \longrightarrow \left[\underset{\underset{R}{|}}{CH}-S-R'-S\right] + H_2O \quad (11.76)$$

$$RCH\overset{OR'}{\underset{OR'}{\diagdown}} + HS-R''-SH \longrightarrow \left[\underset{\underset{R}{|}}{CH}-S-R''-S\right] + 2R'OH \quad (11.77)$$

$$O=\!\!\!\!\bigcirc\!\!\!\!=O + \overset{HSCH_2}{\underset{HSCH_2}{\diagdown}}C\overset{CH_2SH}{\underset{CH_2SH}{\diagup}} \longrightarrow$$

$$\left[\bigcirc\!\!\overset{S}{\underset{S}{\diagdown\!\!\diagup}}\!\!\bigcirc\!\!\overset{S}{\underset{S}{\diagdown\!\!\diagup}}\!\!\bigcirc\right] + 2H_2O \quad (11.78)$$

Of the numerous polysulfides that have been synthesized, the one that has achieved the greatest commercial success is the important engineering plastic poly(phenylene sulfide)[56b,68] (61) (Phillips Petroleum trade name Ryton). The polymer can be prepared by the reaction of *p*-dichlorobenzene with sodium sulfide (11.79) or by self-condensation of *p*-chlorothiophenol or *p*-bromothiophenol in the presence of base (11.80), the former being used in the commercial process. Polymer 61 is highly crystalline with a T_m of about

$$Cl-\!\!\bigcirc\!\!-Cl \xrightarrow{Na_2S} \quad (11.79)$$

$$\left.\begin{array}{c}\\\\\end{array}\right\} \longrightarrow \left[\!\!\bigcirc\!\!-S\right]$$

$$Br-\!\!\bigcirc\!\!-S^-Na^+ \xrightarrow{NaOH} \quad (11.80)$$

61

288°C and, besides its engineering properties, is of interest because it can be made highly conducting by addition of dopant.[69,70] (see Chapter 4, Section 4.8).

11.4.2 Poly(alkylene polysulfide)s

Related to the polysulfides are the poly(alkylene polysulfide)s prepared from a dihalide and a sodium polysulfide.[56c] The most common dihalides are 1,2-dichloroethane and bis(2-chloroethyl)formal (62).

Typically the sodium polysulfide is prepared from sodium sulfide and sulfur. The polysulfide is then reacted with dihalide in water to give a dispersion of polymer (11.81). In these polymers, x usually averages between 2 and 4.

$$ClCH_2CH_2OCH_2CH_2Cl + Na_2S_x \rightarrow \underset{62}{} -\!\!\left[CH_2CH_2OCH_2CH_2S_x\right]\!\!- + 2NaCl$$

$$(11.81)$$

Primary halides give the best yield of polymer, which is to be expected because the mechanism involves nucleophilic displacement of halide by polysulfide anion. Secondary and tertiary halides, particularly the latter, tend to undergo elimination rather than substitution.

Poly(alkylene polysulfide)s (sometimes referred to as Thiokol rubbers) are useful elastomers. Their properties can be modified by varying the number of carbon atoms in the repeating unit or the number of sulfur atoms. Increasing either will increase the elastomeric quality of the polymer.

Crosslinkable polymers can be prepared by adding small amounts of polyhalides such as trichloropropane or by introducing into the polymer backbone some other functional group like hydroxyl or alkene. Typical monomers for this purpose include glycerol dichloroacetate (63) and 1,4-dichloro-2-butene (64).

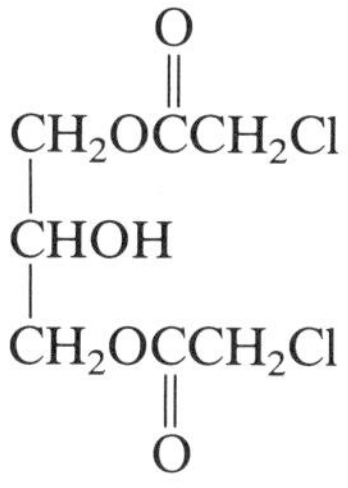

$$\begin{array}{c} O \\ \parallel \\ CH_2OCCH_2Cl \\ | \\ CHOH \\ | \\ CH_2OCCH_2Cl \\ \parallel \\ O \end{array} \qquad ClCH_2CH{=}CHCH_2Cl$$

63   64

The most important method of manufacturing curable elastomers involves first forming the crosslinked polymer using a polyhalide monomer, then degrading the product to a thiol-terminated fusible polymer. This is accomplished by reduction with sodium hydrosulfide in the presence of sodium sulfite (11.82). The average molecular weight of the degraded polymer depends on the amount of sodium hydrosulfide used.

$$\mathrm{-\!\!\!+RSSR\!\!\!+\!} + NaSH + Na_2SO_3 \rightarrow \mathrm{-\!\!\!+RSH + NaSR\!\!\!+\!} + Na_2S_2O_3 \quad (11.82)$$

Numerous oxidizing agents are used to cure the thiol-terminated polymer, including lead or zinc peroxides, cumene hydroperoxide, or even air in thin-film applications. Nonoxidative curing may be accomplished with zinc oxide or other suitable oxides that form stable mercaptide salts, or with aldehydes such as furfural that cure by thioacetal formation. Curing can also be effected with diisocyanates that react to form the sulfur analogs of urethanes (Chapter 13). The polymers can also be cured by blending with other types of polymers including polyesters, epoxy resins, and phenolic resins.[56d]

Poly(alkylene polysulfide)s are among the oldest synthetic polymers, having been in commercial use since 1930. They lack the strength of most other synthetic elastomers and tend to have unpleasant odors because of the presence of low-molecular-weight mercaptans and disulfides. Nevertheless, they are exceptionally resistant to oils and solvents and to weathering, and hence are widely used in gaskets, gasoline hoses, and sealants. Large volumes are now being used as binders for solid rocket propellants.

Ring-opening reactions of cyclic disulfides have been used to prepare polydisulfides, but such processes have not been exploited commercially.[19c]

11.4.3   *Polysulfones*

Polysulfones[56e] are derivatives of polysulfides and can be prepared from them by oxidation—for example, with hydrogen peroxide (11.83). Introduction of the sulfone group normally results in a much higher polymer melt temperature relative to the polysulfide.

$$\mathrm{-\!\!\!+RS\!\!\!+\!} \xrightarrow{\ H_2O_2\ } \begin{bmatrix} O \\ \| \\ \mathrm{-\!\!\!+RS\!\!\!+\!} \\ \| \\ O \end{bmatrix} \quad (11.83)$$

In recent years attention has centered on aromatic polysulfones as thermally stable engineering plastics.[7c,71,72] One method of preparing them by a nucleophilic displacement reaction was described earlier (reaction 11.36). Another commercial route to polysulfones involves a Friedel–Crafts-type reaction using sulfonyl chlorides (11.84). The product in this case (65) is referred to in the plastics industry as a *polyethersulfone*, whereas that described earlier (27) is known simply as *polysulfone* (regardless of the fact that ether formation is the focal point of its synthesis). If diphenyl ether is replaced with biphenyl in reaction (11.84), the product (66) is called a *polyarylsulfone*. Polysulfones 27, 65, and 66, are manufactured by Union Carbide, Imperial Chemical Industries, (ICI), and Minnesota Mining and Manufacturing (3M), respectively.

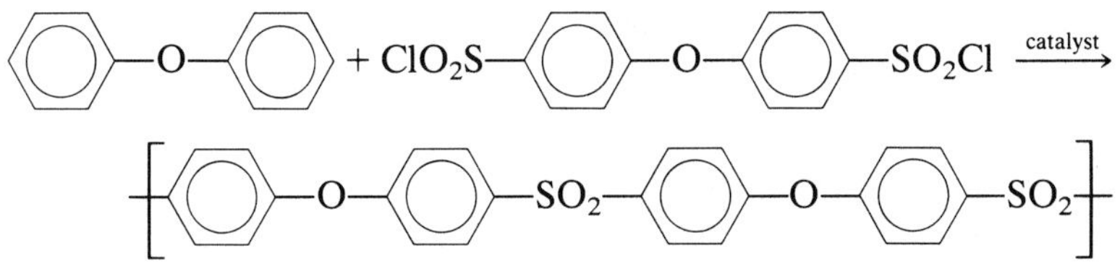

$$(11.84)$$

65

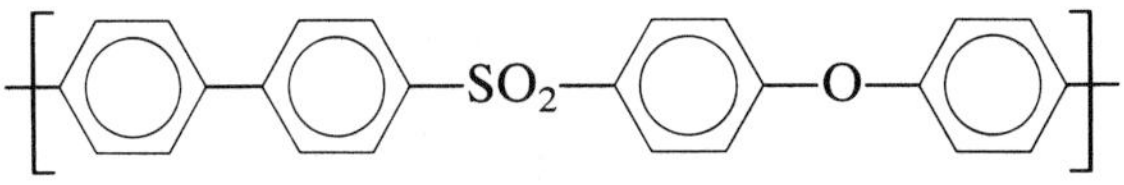

66

Two additional routes to polysulfones are the alternating copolymerization of sulfur dioxide[73] with vinyl monomers (11.85) and the reaction of aromatic disulfinates with aliphatic dihalides (11.86).[56a] The polysulfones formed by reaction (11.85) exhibit relatively low ceiling temperatures ($\sim 100°C$ when R = H, $\sim 0°C$ when R = CH_3). Depolymerization occurs on heating or on exposure to ionizing radiation. As a consequence, such polysulfones are useful in electron beam resist applications.

$$CH_2 = CR_2 + SO_2 \xrightarrow{\text{free radical initiator}} -[CH_2CR_2SO_2]- \qquad (11.85)$$

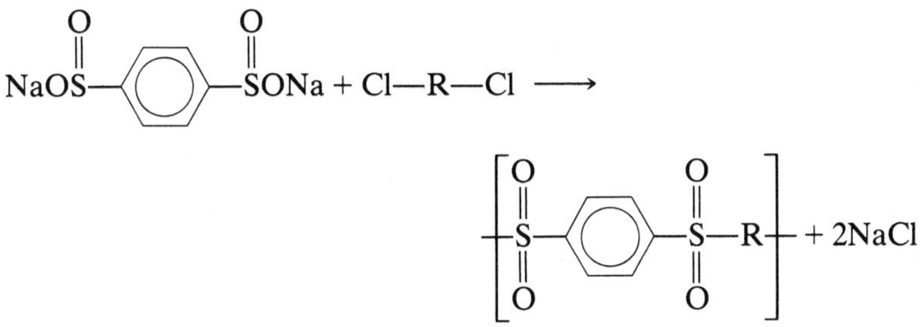

$$+ 2NaCl \qquad (11.86)$$

References

1. A. A. Miller and F. R. Mayo, *J. Am. Chem. Soc.*, **78**, 1017 (1956).
2. K. Kishore and T. Mukundan, *Nature*, **324**, 130 (1986).
3. N. G. Gaylord (ed.), *Polyethers*, Part 1, Wiley-Interscience, New York, 1963: (a) J. C. Bevington, Chap. 2; (b) L. E. St. Pierre, Chap. 3; (c) A. C. Farthing, Chap. 5; (d) N. G. Gaylord, Chap. 7.
4. J. Furukawa and T. Saegusa, *Polymerization of Aldehydes and Oxides*, Wiley-Interscience, New York, 1963: (a) Chap. 2; (b) pp. 413ff; (c) pp. 470ff; (d) Chap. 3; (e) Chaps. 4 and 5.
5. R. B. Akin, *Acetal Resins*, Reinhold, New York, 1962.
6. S. J. Barker and M. B. Bruce, *Polyacetals*, Elsevier, New York, 1970.

386 *Nonvinyl Polymers*

7. R. B. Seymour and G. S. Kirshenbaum (eds.), *High Performance Polymers: Their Origin and Development*, Elsevier, New York, 1986: (a) K. J. Persak and R. A. Fleming, pp. 105–114; (b) H. W. Hill Jr., pp. 135–148; (c) R. A. Clendinning, A. G. Farnham, and R. N. Johnson, pp. 149–158.
8. A. D. Ketley (ed.), *The Stereochemistry of Macromolecules*, Vol. 2, Dekker, New York, 1967: (a) G. F. Pregalia and M. Binaghi, Chap. 3; (b) T. Tsuruta, Chap. 4.
9. O. Vogl, H. C. Miller, and W. H. Sharkey, *Macromolecules*, **5**, 658 (1972).
10. K. Neeld and O. Vogl, *Macromol. Rev.*, **16**, 1 (1981).
11. M. Chang, A. Rembaum, and C. J. McDonald, *J. Polym. Chem., Polym. Lett. Ed.*, **22**, 279 (1984).
12. E. Brown, A. Racois, M. Grimaud, J. C. Lecog, R. Tixien, and M. Corgier, *Makromol Chem.*, **179**, 2887 (1978).
13. H. O. Colomb and F. E. Bailey Jr., *J. Poly. Sci., Polym. Lett. Ed.*, **16**, 507 (1978).
14. G. E. Gerhardt, E. T. Dumitru, and R. J. Lagow, *J. Polym. Sci., Polym. Chem. Ed.*, **18**, 157 (1980).
15. A. Wurtz, *Ann. Chim. Phys.*, **69**, 330, 334 (1863).
16. A. Wurtz, *Chem. Ber.*, **10**, 90 (1879).
17. H. Staudinger and O. Schweitzer, *Chem. Ber.*, **62**, 2395 (1929).
18. H. Staudinger and H. Lehmann, *Liebigs Ann. Chem.*, **505**, 41 (1933).
19. K. C. Frisch and S. L. Reegen (eds.), *Ring-Opening Polymerization*, Dekker, New York, 1969: (a) Y. Ishii and S. Sakai, Chap. 1; (b) P. Dreyfuss and M. P. Dreyfuss, Chap. 2; (c) P. Sigwalt, Chap. 4.
20. F. E. Bailey and J. V. Koleske, *Poly(ethylene oxide)*, Academic Press, New York, 1976.
21. J. T. Hill, *J. Macromol. Sci.-Chem.*, **8**, 499 (1974).
22. T. Saegusa, in *Polymerization of Heterocyclics* (O. Vogl and J. Furukawa, eds.), Dekker, New York, 1973, pp. 5–34.
23. K. C. Frisch (ed.), *Cyclic Monomers*, Wiley-Interscience, New York, 1972: (a) G. Pruckmayr, Chap. 1; (b) F. Lautenschlaeger and R. T. Woodhaus, Chap. 5.
24. S. Penczek, P. Kubisa, and K. Matyjaszewski, *Adv. Polym. Sci.*, **37**, 1 (1980).
25. P. Dreyfuss, *Poly(tetrahydrofuran)*, Gordon and Breach, New York, 1982.
26. P. Dreyfuss and M. P. Dreyfuss, *Adv. Polym. Sci.*, **4**, 528 (1967).
27. Y. Yokoyama and H. K. Hall Jr., *Adv. Polym. Sci.*, **42**, 107 (1982).
28. T. Aida, Y. Maekawa, S. Asano, and S. Inoue, *Macromolecules*, **21**, 1195 (1988).
29. S. Inoue, *J. Macromol. Sci-Chem.*, **A25**, 571 (1988).
30. E. J. Vandenberg, *J. Polym. Sci., Polym. Chem. Ed.*, **13**, 2221 (1975).
31. A. S. Hay et al. *J. Polym. Sci., Polym. Lett. Ed.*, **21**, 449 (1983).
32. T. J. Shea and G. M. Loucks, *Polymer Preprints*, **28**(2), 354 (1987).
33. A. S. Hay, *Adv. Polym. Sci.*, **4**, 496 (1967).
34. R. O. Johnson and H. S. Burlhis, *J. Polym. Sci., Polym. Symp.*, **70**, 129 (1983).
35. D. M. White et al., *J. Polym. Sci., Polym. Chem. Ed.*, **19**, 1635 (1981).
36. D. J. Blundell and B. N. Osborn, *Polymer*, **24**, 953 (1983).
37. T. E. Attwood, P. C. Dawson, J. L. Freeman, L. R. J. Hoy, J. B. Rose, and P. A. Staniland, *Polymer*, **22**, 1096 (1981).
38. Y. Imai, *J. Macromol. Sci.-Chem.*, **A15**, 833 (1981).
39. V. Percec, T. D. Shaffer, and H. Nava, *J. Polym. Sci., Polym. Lett. Ed.*, **22**, 637 (1984).

40. W. J. Jackson and J. R. Caldwell, in *Polymerization and Polycondensation Processes* (R. F. Gould, ed.), Adv. Chem. Ser. 34, American Chemical Society, Washington, D.C., 1962, pp. 200ff.
41. W. J. Bailey and A. A. Volpe, *J. Polym. Sci. A-1*, **8**, 2109 (1970).
42. J. Heller, D. W. H. Penhale, and R. F. Helwing, *J. Polym. Sci., Polym. Lett. Ed.*, **18**, 293 (1980).
43. A. S. Hay, *Polym. Eng. Sci.*, **16**, 1 (1976).
44. J. G. Bennett Jr., and G. D. Cooper, *Macromolecules*, **3**, 101 (1970).
45. W. G. Potter, *Epoxide Resins*, Springer-Verlag, New York, 1970.
46. I. Skeist, *Epoxy Resins*, Reinhold, New York, 1958.
47. C. A. May (ed.), *Epoxy Resins: Chemistry and Technology*, 2nd ed., Dekker, New York, 1988.
48. P. F. Bruins, *Epoxy Resin Technology*, Wiley-Interscience, New York, 1968.
49. B. Sedláček and J. Kahovek (eds.), *Crosslinked Epoxies*, de Gruyler, New York, 1986.
50. R. S. Bauer (ed.), *Epoxy Resin Chemistry*, 2 vols., ACS Symp. Ser. 114 and 221, American Chemical Society, Washington, D.C., 1979 and 1983.
51. D. Aelony, *J. Appl. Polym. Sci.*, **13**, 227 (1969).
52. T. Matynia, *J. Appl. Polym. Sci.*, **25**, 1 (1980).
53. G. J. Fleming, *J. Appl. Polym. Sci.*, **13**, 989 (1969).
54. J. S. Bradshaw and M. P. Stevens, *J. Appl. Polym. Sci.*, **10**, 1809 (1966).
55. W. H. Sharkey, in *Polyaldehydes* (O. Vogl, ed.), Dekker, New York, 1967, pp. 91ff.
56. N. G. Gaylord (ed.), *Polyethers*, Part 3, Wiley-Interscience, New York, 1962: (a) F. O. Davis and E. M. Fettes, Chap. 12; (b) R. W. Lenz, pp. 30–37; (c) M. B. Berenbaum, Chap. 12; (d) J. R. Panek, Chap. 14; (e) E. M. Fettes and F. O. Davis, Chap. 15.
57. T. Kunitake, M. Kasumatsu, and C. Aso, *J. Polym. Sci. A-1*, **9**, 3675 (1971).
58. M. Morton and R. F. Kammereck, *J. Am. Chem. Soc.*, **92**, 3217 (1970).
59. V. Percec, H. Nava, and B. C. Auman, *Polym. J.*, **16**, 681 (1984).
60. J. E. White and M. D. Scaia, *Polymer*, **25**, 850 (1984).
61. V. A. Sergeyev, V. I. Nedel'kin, and S. S. Arustamyan, *Polymer*, **26**, 365 (1985).
62. J. T. Crivello, *J. Polym. Sci., Polym. Chem. Ed.*, **14**, 159 (1976).
63. N. D. Ghatge and R. A. N. Murphy, *Polymer*, **22**, 1250 (1981).
64. J. G. Erickson, *J. Polym. Sci. A-1*, **4**, 519 (1966).
65. R. G. Bass, E. Cooper, and J. W. Connell, *Polym. Preprints*, **21**(1), 313 (1986).
66. F. W. Harris and M. W. Beltz, *Polym. Preprints*, **27**(1), 114 (1986).
67. A. A. Oswald, W. H. Mueller, K. Griesbaum, and D. N. Hall, *Polym. Preprints*, **9**(1), 657 (1968).
68. H. W. Hill, Jr., in *High Performance Polymers: Their Origin and Development* (R. B. Seymour and G. S. Kirshenbaum, eds.), Elsevier New York, 1986, pp. 135–148.
69. K. F. Schoch Jr., J. F. Chance, and K. E. Pfeiffer, *Macromolecules*, **18**, 2389 (1985).
70. T. C. Clarke, K. K. Kanazawa, V. Y. Lee, J. F. Raybolt, J. R. Reyonds, and G. B. Street, *J. Polym. Sci., Polym. Phys. Ed.*, **20**, 117 (1982).
71. K. J. Ivin and J. B. Rose, *Adv. Macromol. Chem.*, **1**, 336 (1968).
72. C. Arnold, *J. Polym. Sci., Macromol. Rev.*, **14**, 265 (1979).

73. A. H. Fawcett, in *Encyclopedia of Polymer Science and Engineering*, 2nd ed., Vol. 10 (H. F. Mark, N. M. Bikales, C. G. Overberger, G. Menges, and J. I. Kroschwitz, eds.), Wiley-Interscience, New York, 1988, pp. 408–432.

Review exercises

1. The following polyethersulfide is prepared by reacting diphenyl ether with SCl_2 in chloroform. Give the IUPAC name of the polymer. (T. Fujisawa and M. Kakatuni, *J. Polym. Sci., B*, **8**, 19, 1970)

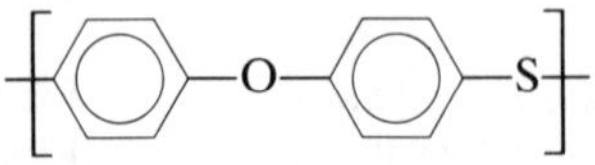

2. Assuming that the polymerization reactions are feasible, write repeating units for the polymers prepared from the following monomers:

(a) $(CH_3)_2CHCHO$ (b) 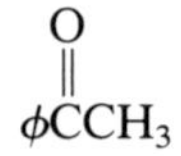 ϕCCH_3 (with =O)   (c) 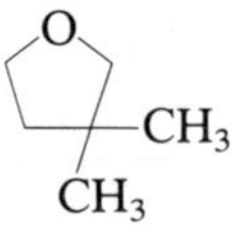   (d)

3. Predict the structure of the soluble polymer prepared by polymerization of *o*-phthaladehyde. (C. Aso and S. Tagami, *Macromolecules*, **2**, 414, 1969)

4. Write the repeating units of the polyethers prepared by the Williamson synthesis
   (a) from bisphenol A and

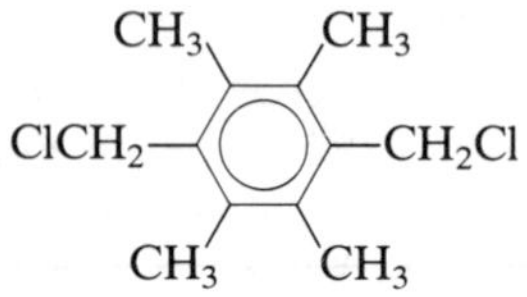

   (b) from

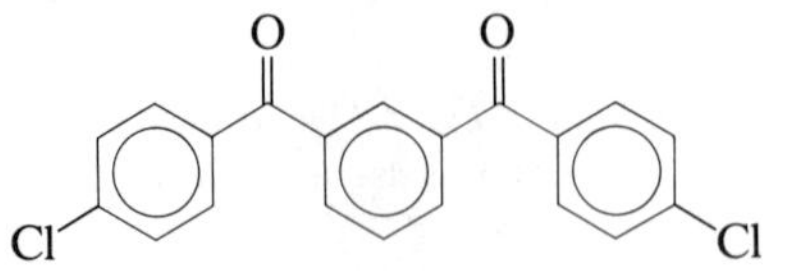

and

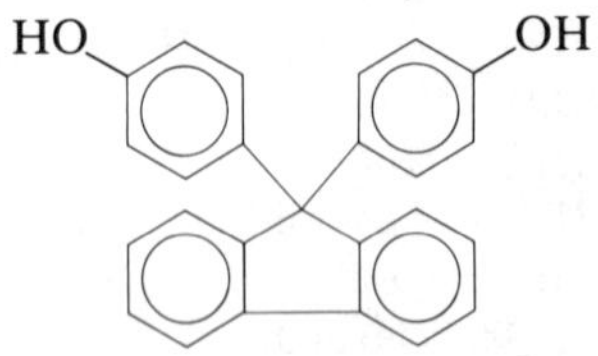

   In part (b), what advantage might be gained by using this particular bisphenol? (You may wish to refer to Chapter 4.) Suggest a synthesis of the dichloride. [(a) ref. 6, p. 168; (b) P. M. Hergenrother and B. J. Jensen, *Polym. Preprints*, **26**(2), 174, 1985]

5. Synthesis of polyether by reaction (11.34) does not work well if the methyl groups are missing from the ring. Explain.

6. Write a reaction showing the synthesis of spiro polymer **37** by a ketal exchange reaction.

7. Acetals and ketals are usually stable to base but undergo rapid hydrolysis in the presence of dilute acid. Polyformal **32**, however, is quite stable to acid. A polymer film began to disintegrate in 10% HCl only after 4 hours at 100°C. Rationalize the polymer's good acid resistance. (ref. 40)

8. It was mentioned in this chapter that aldol condensation is a side reaction in the polymerization of aldehydes containing alpha hydrogen. Low-molecular-weight polymers have actually been prepared by the aldol condensation reaction of cyclohexanone for use in coatings and inks to improve gloss and adhesion. Show with equations how such a polymer might be formed from cyclohexanone.

9. A polymer related in structure to PEEK (see reaction 11.38) is PAEK (polyaryletherketone), also called PEK (polyetherketone).

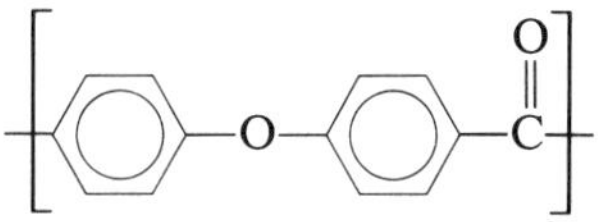

   (a) Suggest two possible reactions leading to PAEK.
   (b) Write IUPAC names for PEEK and PAEK.

10. A commercially available polyethersulfone not described in this chapter has the structure

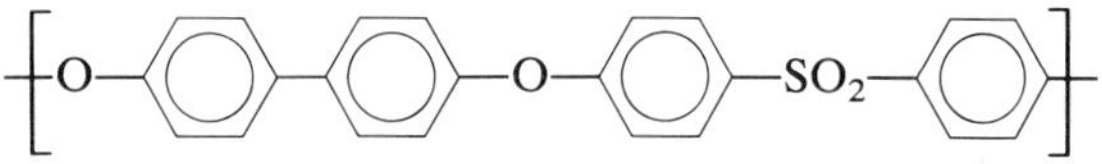

   Which two compounds are most likely used in its synthesis?

11. In the Friedel–Crafts-type synthesis of polysulfones (reaction 11.84), crosslinking is not a problem, even though the ether group is *ortho, para* directing. Explain why. Suggest a single monomer that would yield the same polysulfone as that shown in the reaction.

12. A polysulfone exhibiting no olefinic absorption in its IR spectrum results from the alternating free radical copolymerization of SO_2 with such diallyl monomers as

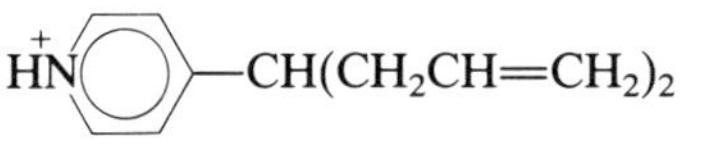

   Propose possible structures for the polymers (S. Harada et al., *Makromol. Chem.*, **90**, 177, 1966; **107**, 64, 78, 1967; **176**, 1289, 1975)

13. Phase-transfer catalysis has been used to synthesize the polymer

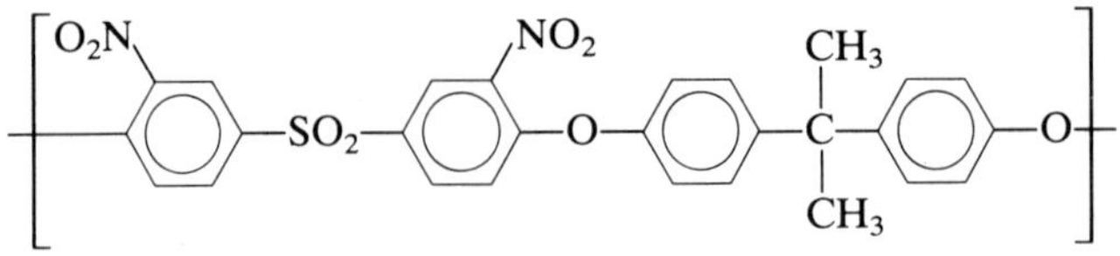

A methylene chloride–water system was used.

(a) Give structures for the most likely monomers used.

(b) Given the reaction conditions, what side reaction might have been expected? (Show the product.) Why is the side reaction not favored? (Y. Imai et al., *J. Polym. Sci., Polym. Lett. Ed.*, **17**, 85, 1979)

14. Percec and coworkers (ref. 59) have prepared sequential (-AB-) block copolymers under phase-transfer conditions by reacting bromine-terminated polysulfides with phenol-terminated polyethers. Choosing any combination of monomers, write reactions illustrating such a synthesis.

15. Give the structure of the polyetherimide prepared from

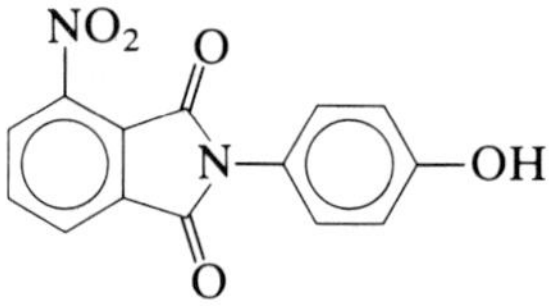

This synthesis is less satisfactory than the one shown in reaction (11.37). What factors might be responsible? (S. Maiti and B. K. Mandel, *Makromol. Chem., Rapid Commun.*, **6**, 841, 1985)

16. Write plausible mechanisms for the reaction of 1,6-hexanedithiol with 1,5-hexadiene under acidic, basic, and free radical conditions. (See reactions 11.68 and 11.69.)

17. The following polymer (a polyestersulfideamine) was prepared without added initiator or catalyst. (The amine function was sufficiently activating.)

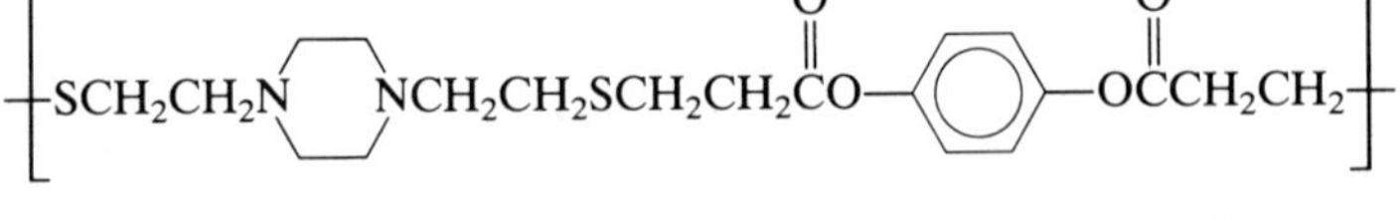

In light of your answer to exercise 16, which monomers were most likely used? (P. Ferruti and E. Ranucci, *J. Polym. Sci. C: Polym. Lett.*, **26**, 357, 1988)

18. Block copolymers of poly(alkylene disulfide)s and vinyl polymers have been prepared by reaction of the former with vinyl monomers under free radical conditions. Propose a mechanism. (M. L. Hallensleben, *Eur. Polym. J.*, **13**, 437, 1977)

19. Ketals have been incorporated into vinyl polymers as side chains using the monomer

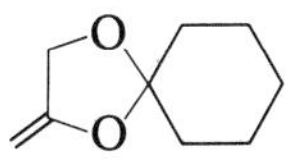

(a) What conditions (free radical, cationic, etc.) would you expect to be most favorable for polymerization? (b) Give the repeat unit of the polymer. (c) Starting materials for synthesis of the monomer are cyclohexanone and 3-chloro-1,2-propanediol. Outline the synthesis. (d) What commercially important polymer contains pendant acetal groups? (H. Orth, *Angew. Chem.*, **64**, 544, 1952)

12
Polyesters

12.1 Introduction

Polyesters[1-3] are among the more versatile synthetic polymers in that they find wide commercial use as fibers, plastics, and coatings. The common methods of synthesizing simple esters are used to make polyesters. These include direct esterification (12.1), transesterification (12.2), and the reaction of alcohols with acyl chlorides (12.3) or anhydrides (12.4).

$$RCO_2H + R'OH \rightleftharpoons RCO_2R' + H_2O \tag{12.1}$$

$$RCO_2R'' + R'OH \rightleftharpoons RCO_2R' + R''OH \tag{12.2}$$

$$RCOCl + R'OH \rightarrow RCO_2R' + HCl \tag{12.3}$$

$$(RCO)_2O + R'OH \rightarrow RCO_2R' + RCO_2H \tag{12.4}$$

These reactions are mechanistically well understood. Each involves nucleophilic addition to the carbonyl group, addition that is facilitated by the polar nature of the carbon–oxygen double bond, the ability of the carbonyl oxygen atom to assume a formal negative charge, and the planar configuration of the trigonal carbon that minimizes steric interference. A general mechanism may be written (reaction 12.5: Y = OH, OR″, Cl, or OCOR) showing each step as reversible, although it should be recognized that in the case of acid chlorides and anhydrides the overall process is, for practical purposes, irreversible.

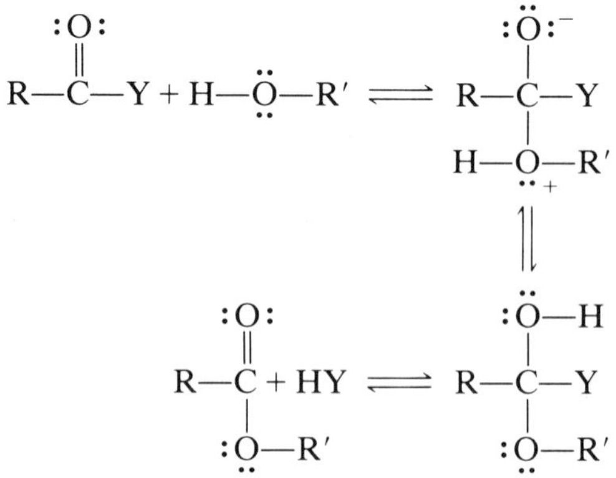

$$
\begin{array}{c}
\overset{:O:}{\underset{\parallel}{R-C-Y}} + H-\overset{\cdot\cdot}{\underset{\cdot\cdot}{O}}-R' \rightleftharpoons
\overset{:\overset{\cdot\cdot}{O}:^-}{\underset{\mid}{R-C-Y}} \\[2mm]
\underset{\overset{\mid}{H-\underset{\cdot\cdot\;+}{O}-R'}}{} \\[2mm]
\Updownarrow \\[2mm]
\overset{:O:}{\underset{\parallel}{R-C}} + HY \rightleftharpoons
\overset{:\overset{\cdot\cdot}{O}-H}{\underset{\mid}{R-C-Y}} \\
\underset{:\overset{\cdot\cdot}{O}-R'}{\mid} \qquad\qquad \underset{:\overset{\cdot\cdot}{O}-R'}{\mid}
\end{array}
\tag{12.5}
$$

392

Less traditional methods are also used in polyester manufacturing, among them *acidolysis* (12.6) (a variation of transesterification), the reaction of carboxylic acids with epoxides (12.7), nucleophilic displacement (12.8), and ring-opening reactions of cyclic esters. Application of these reactions to the

$$RCO_2H + R'CO_2R'' \rightleftharpoons RCO_2R'' + R'1CO_2H \qquad (12.6)$$

$$\overset{\displaystyle O}{\overset{\displaystyle /\,\backslash}{RCO_2H + CH_2\!\!-\!\!CH_2}} \rightarrow RCO_2CH_2CH_2OH \qquad (12.7)$$

$$RCO_2^- + R'Br \rightarrow RCO_2R' + Br^- \qquad (12.8)$$

synthesis of linear polyesters is covered in Section 12.2 of this chapter. Section 12.3 is concerned with polyesters that are crosslinked by use of polyfunctional monomers, or by incorporation of reactive double bonds into the polymer backbone.

Commercially important linear polyesters are shown in Table 12.1. No attempt is made to list applications of these polymers, because they are so diverse. Suffice it to say that the aromatic polyesters (including polycarbonate) are used primarily as engineering plastics with the exception of poly(1,4-dihydroxymethylcyclohexyl terephthalate) and poly(ethylene terephthalate), the former being used as a textile fiber and the latter as both fiber and plastic.* Indeed, poly(ethylene terephthalate) may be considered the workhorse of the polyester industry, having enjoyed widespread commercial use since the 1940s. Other aromatic polyesters and the aliphatic ones are of more recent vintage. The latter are of principal interest because of their degradability. As indicated in the footnotes of Table 12.1, a variety of copolyesters are also available. Not included in Table 12.1 are hydroxy-terminated aliphatic polyesters that are used in the manufacture of polyurethanes (Chapter 13).

12.2 Linear polyesters

12.2.1 Preparation of polyesters by polycondensation reactions

In principle any dibasic acid will condense with any glycol or any hydroxy acid will self-condense to form linear polyester.[4] In practice very few have achieved commercial importance, the major limitations being availability of monomer and chemical and physical properties of the finished product.

The most widely used linear polyester is poly(ethylene terephthalate) (**1**) prepared from terephthalic acid and ethylene glycol (12.9). The polymer can be prepared in bulk or in solution using an excess of ethylene glycol to increase esterification rate. This leads initially to low-molecular-weight

---

* Well-known trade names for the two polymeric fibers are Kodel (Eastman Kodak) and Dacron (du Pont), respectively. The latter polymer is also marketed in film form as Mylar (du Pont).

Table 12.1. Commercially available linear polyesters

Common name	Principal structure
Poly(ethylene terephthalate) (PET)	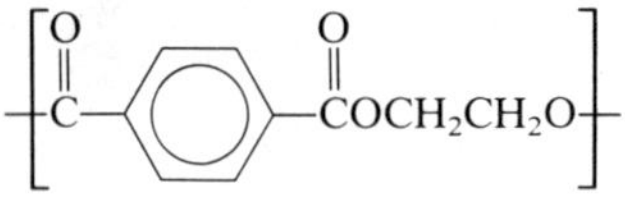 $\left[\!\!-\overset{O}{\overset{\|}{C}}-C_6H_4-\overset{O}{\overset{\|}{C}}OCH_2CH_2O-\!\!\right]$
Poly(butylene terephthalate)[a] (PBT)	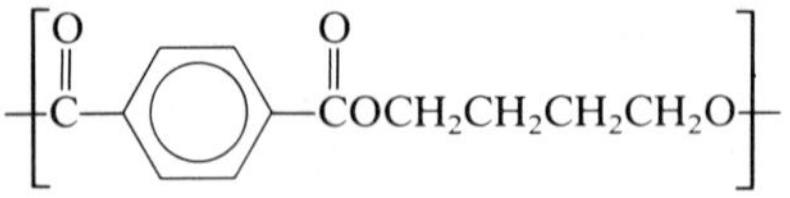 $\left[\!\!-\overset{O}{\overset{\|}{C}}-C_6H_4-\overset{O}{\overset{\|}{C}}OCH_2CH_2CH_2CH_2O-\!\!\right]$
Poly(1,4-dihydroxymethylcyclohexyl terephthalate)	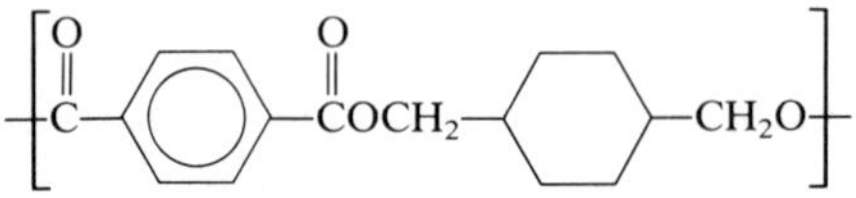 $\left[\!\!-\overset{O}{\overset{\|}{C}}-C_6H_4-\overset{O}{\overset{\|}{C}}OCH_2-C_6H_{10}-CH_2O-\!\!\right]$
Poly(4-hydroxybenzoate)[b,c]	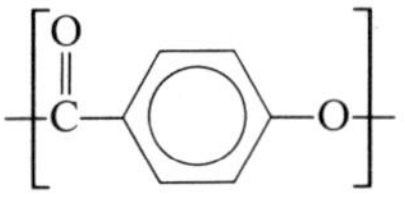 $\left[\!\!-\overset{O}{\overset{\|}{C}}-C_6H_4-O-\!\!\right]$
Poly(bisphenol A terephthalate/isophthalate)[d]	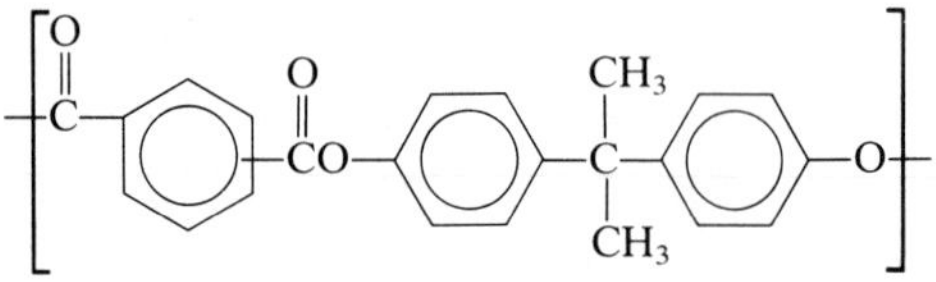 $\left[\!\!-\overset{O}{\overset{\|}{C}}-C_6H_4-\overset{O}{\overset{\|}{C}}O-C_6H_4-\overset{CH_3}{\underset{CH_3}{C}}-C_6H_4-O-\!\!\right]$
Polycarbonate[a]	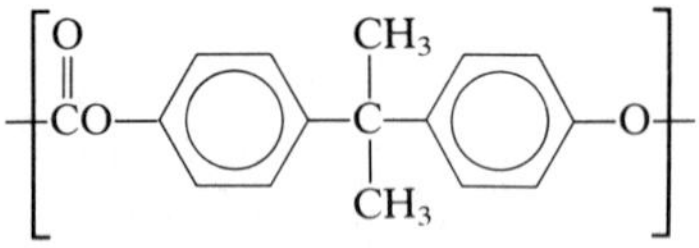 $\left[\!\!-\overset{O}{\overset{\|}{C}}O-C_6H_4-\overset{CH_3}{\underset{CH_3}{C}}-C_6H_4-O-\!\!\right]$
Polycaprolactone[e]	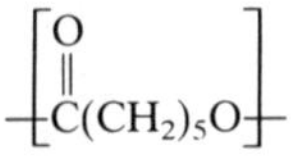 $\left[\!\!-\overset{O}{\overset{\|}{C}}(CH_2)_5O-\!\!\right]$
Poly(2-hydroxybutyric acid)[f](PHB)	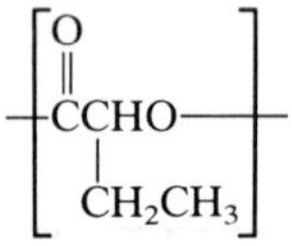 $\left[\!\!-\overset{O}{\overset{\|}{C}}\underset{CH_2CH_3}{C}HO-\!\!\right]$
Poly(glycolic acid)[e]	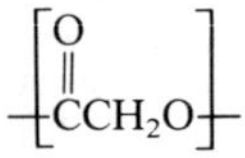 $\left[\!\!-\overset{O}{\overset{\|}{C}}CH_2O-\!\!\right]$
Poly(lactic acid)[e]	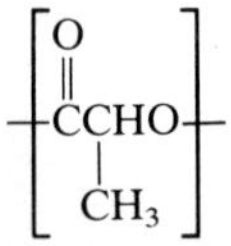 $\left[\!\!-\overset{O}{\overset{\|}{C}}\underset{CH_3}{C}HO-\!\!\right]$

[a] A heterogeneous polyblend of PBT and polycarbonate (50:50 wt %) is commercially available.

[b] Also copolymers with bisphenol A, terephthalic acid, and isophthalic acid.

[c] A thermotropic liquid crystalline copolymer with terephthalic acid and 4,4'-dihydroxybiphenyl is also available (see Chap. 3, Sect. 3.7)

[d] Homopolymers and copolymers are available.

[e] Aliphatic copolyesters are also available.

[f] Bacterial metabolism product of carbohydrates.

hydroxyl-terminated polyester, which is then transesterified with removal of excess glycol to attain high molecular weight.

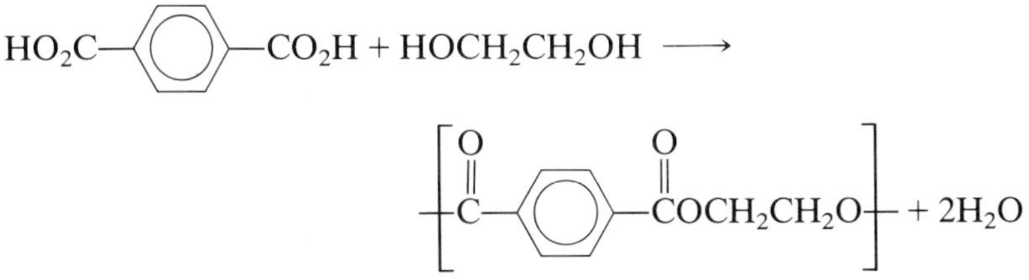

$$HO_2C-\langle\bigcirc\rangle-CO_2H + HOCH_2CH_2OH \longrightarrow$$

$$\left[\begin{matrix} O \\ \| \\ C-\langle\bigcirc\rangle-COCH_2CH_2O \end{matrix} \right] + 2H_2O \quad (12.9)$$

1

Terephthalic acid is high melting (sublimes at 300°C) and quite insoluble in most common solvents. For these reasons, the transesterification (*alcoholysis*) reaction of dimethyl terephthalate with ethylene glycol has been used more frequently to make poly(ethylene terephthalate), although use of the cheaper terephthalic acid is increasing. The alcoholysis reaction involves an initial ester interchange to form the diester **2** and to remove methanol (12.10), followed by a second interchange to form polymer (12.11).

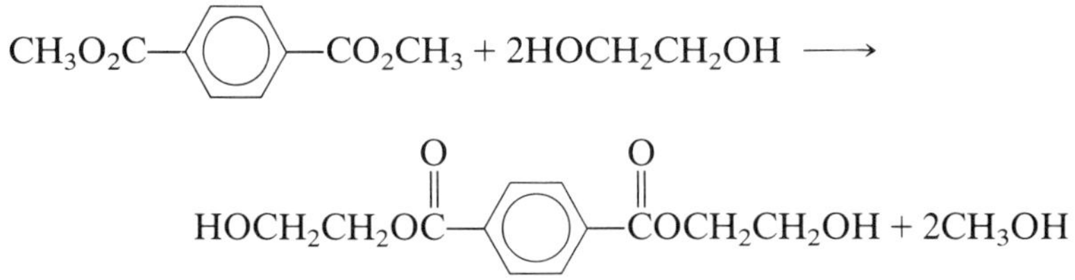

$$CH_3O_2C-\langle\bigcirc\rangle-CO_2CH_3 + 2HOCH_2CH_2OH \longrightarrow$$

$$HOCH_2CH_2O\overset{O}{\underset{\|}{C}}-\langle\bigcirc\rangle-\overset{O}{\underset{\|}{C}}OCH_2CH_2OH + 2CH_3OH \quad (12.10)$$

2

$$\longrightarrow \mathbf{1} + HOCH_2CH_2OH \quad (12.11)$$

More recently a process based on ethylene oxide instead of ethylene glycol has been commercialized in Japan.[5] Ethylene oxide reacts rapidly with terephthalic acid to form intermediate **2**, which then undergoes alcoholysis (12.11) to form polyester.

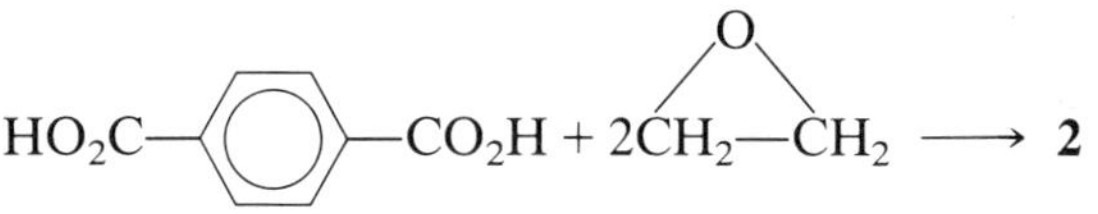

$$HO_2C-\langle\bigcirc\rangle-CO_2H + 2CH_2-CH_2 \longrightarrow \mathbf{2} \quad (12.12)$$

Polyesters prepared from phenolic monomers are commonly called *polyarylates*, a term coined by Russian workers who have been particularly active in this field.[6] Polyarylates are difficult to prepare by direct polycondensation with diacids because the high temperatures necessary for achieving practical molecular weights frequently lead to decarboxylation or carboxyl-catalyzed ether formation. Consequently, they are usually made by transesterification methods[7] and, less commonly (because of high cost), by interfacial or solution polymerization of the acid chlorides.[8a] The commercially important poly(4-hydroxybenzoate) (**3**), for example, is made by alcoholysis of the phenyl ester (12.13) or acidolysis of the acetyl derivative (12.14). The synthesis of poly(bisphenol A isophthalate) (**4**) (12.15) illustrates an interfacial synthesis of a polyarylate.

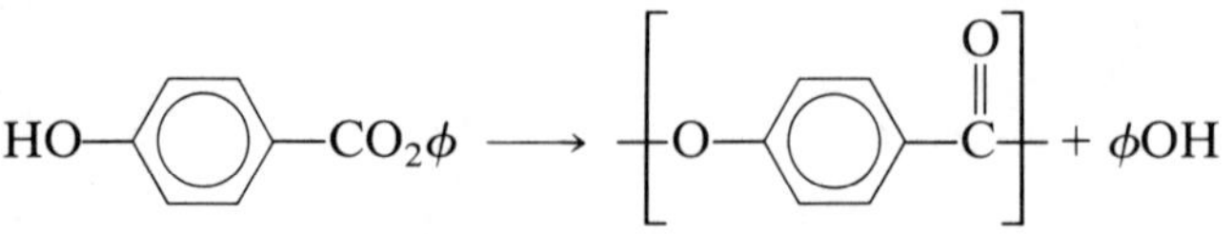

$$HO-\underset{}{\bigcirc}-CO_2\phi \longrightarrow \left[O-\underset{}{\bigcirc}-\overset{O}{\overset{\|}{C}} \right] + \phi OH \quad (12.13)$$

3

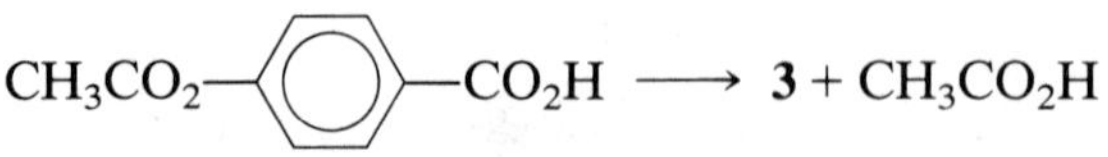

$$CH_3CO_2-\underset{}{\bigcirc}-CO_2H \longrightarrow 3 + CH_3CO_2H \quad (12.14)$$

$$\underset{}{\overset{O}{\overset{\|}{ClC}}}\underset{}{\bigcirc}\underset{}{\overset{O}{\overset{\|}{CCl}}} + Na^+ \; ^-O-\underset{}{\bigcirc}-\overset{CH_3}{\underset{CH_3}{\overset{|}{\underset{|}{C}}}}-\underset{}{\bigcirc}-O^-Na^+ \xrightarrow{H_2O/CH_2Cl_2}$$

$$\left[\overset{O}{\overset{\|}{C}}\underset{}{\bigcirc}\overset{O}{\overset{\|}{CO}}-\underset{}{\bigcirc}-\overset{CH_3}{\underset{CH_3}{\overset{|}{\underset{|}{C}}}}-\underset{}{\bigcirc}-O \right] + 2\,NaCl \quad (12.15)$$

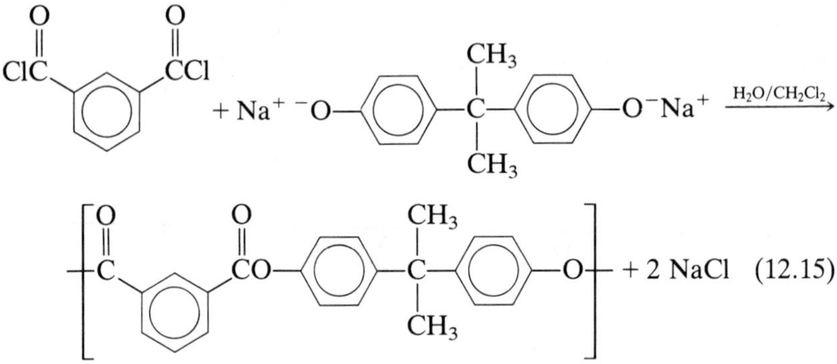

4

A number of attempts have been made to prepare polyesters directly from dicarboxylic acids and dihydroxy compounds at low to moderate temperatues using activating agents. Poly(ethylene terephthalate), for example, may be synthesized at room temperature from terephthalic acid and ethylene glycol in the presence of picryl chloride[9] (**5**), whereas polyarylates are formed directly at about 120°C using diphenyl chlorophosphate[10] (**6**) or *p*-toluenesulfonyl chloride[11] (**7**). Pyridine is used as a solvent or cosolvent in each case to remove by-product HCl or to assist in the activation process. These reactions contrast with direct polycondensation or transesterification reactions, which are normally run at temperatures in excess of 200°C. The role of the activating agent is to convert the acid into a highly reactive ester intermediate.

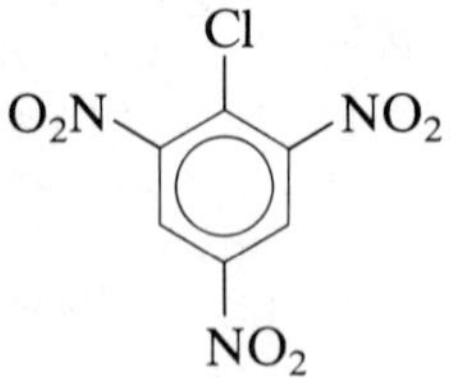

$$O_2N\underset{NO_2}{\overset{Cl}{\bigcirc}}NO_2 \qquad \underset{}{\overset{O}{\overset{\|}{Cl-P(O\phi)_2}}} \qquad \underset{CH_3}{\overset{SO_2Cl}{\bigcirc}}$$

5 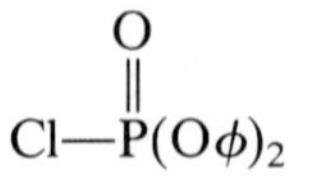   **6** 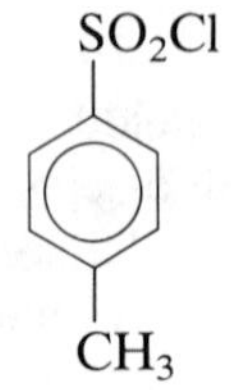   **7**

Polyester synthesis by nucleophilic displacement reactions are much less common because of the low nucleophilicity of carboxylate ion. One such reaction, the synthesis of the biodegradable poly(glycolic acid) (**8**), has been accomplished by reacting bromoacetic acid with an equimolar amount of triethylamine in nitromethane or ether at ambient temperatures (12.16).[12] Phase-transfer catalysis facilitates this type of reaction. Phase-transfer methods have also been used to make sulfonate (**9**) and phosphonate (**10**) polyesters[13] from the corresponding sulfonyl chlorides (12.17) and phenyl-phosphonic dichloride (12.18).

$$BrCH_2CO_2H + (C_2H_5)_3N \longrightarrow \left[OCH_2\overset{\overset{O}{\|}}{C} \right] + (C_2H_5)_3\overset{+}{N}H\ Br^- \quad (12.16)$$

$$\textbf{8}$$

$$HO—Ar—OH + ClSO_2—Ar'—SO_2Cl \longrightarrow$$

$$\left[O—Ar—OSO_2—Ar'—SO_2 \right] \quad (12.17)$$

$$\textbf{9}$$

$$HO—Ar—OH + Cl—\overset{\overset{O}{\|}}{\underset{\underset{\phi}{|}}{P}}—Cl \longrightarrow \left[O—Ar—O—\overset{\overset{O}{\|}}{\underset{\underset{\phi}{|}}{P}} \right] \quad (12.18)$$

$$\textbf{10}$$

Anhydrides may also be used as monomers for polyesters. Succinic anhydride and ethylene glycol react initially to form hydroxy acid, which undergoes subsequent esterification to form poly(ethylene succinate) (12.19). Such reactions are used most commonly in the preparation of polyester resins

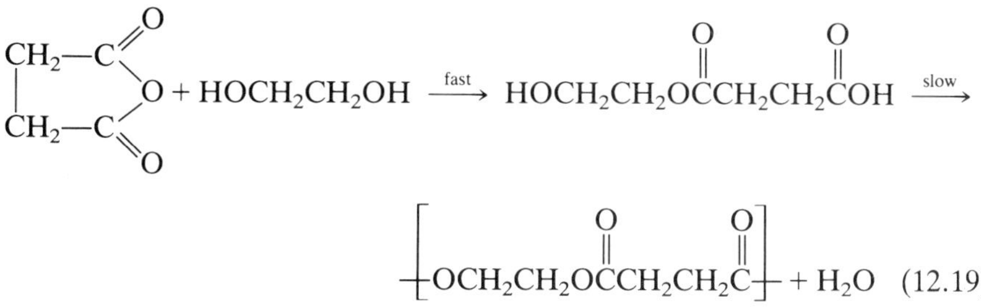

$$\begin{matrix} CH_2—C \\ | \quad\quad\quad \\ CH_2—C \end{matrix} \overset{O}{\underset{O}{}} O + HOCH_2CH_2OH \xrightarrow{\text{fast}} HOCH_2CH_2O\overset{\overset{O}{\|}}{C}CH_2CH_2\overset{\overset{O}{\|}}{C}OH \xrightarrow{\text{slow}}$$

$$\left[OCH_2CH_2O\overset{\overset{O}{\|}}{C}CH_2CH_2\overset{\overset{O}{\|}}{C} \right] + H_2O \quad (12.19)$$

(Section 12.3), where high degrees of polymerization are not so important. Under very mild conditions, dianhydrides form base-soluble polyesters (12.20), which lend themselves to crosslinking or grafting through the free carboxyl groups. A wide variety of block copolymers of polyesters have been reported.[14a] The reaction of hydroxyl-terminated polyester with an epoxide-terminated polyether (12.21) to form an $[AB]$ multiblock copolymer is an example.

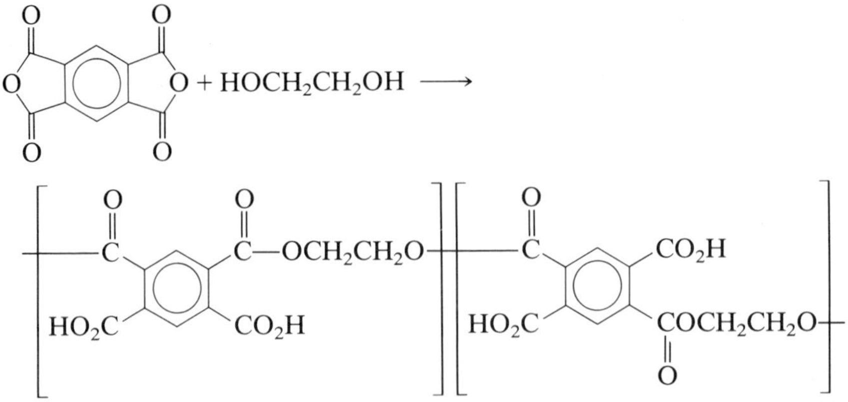

$$(12.20)$$

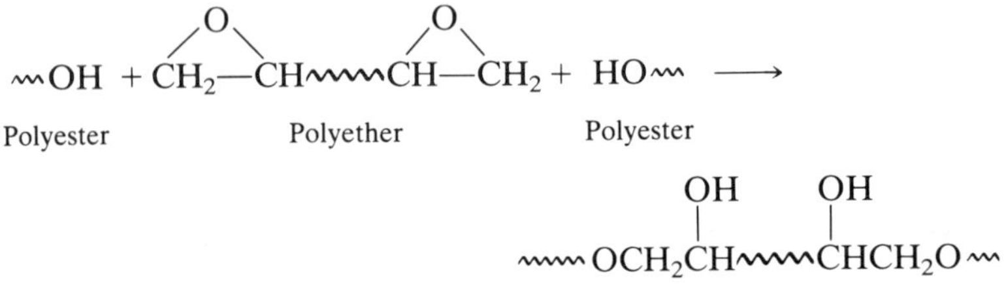

| Polyester | Polyether | Polyester |

$$(12.21)$$

Block copolymer

The effect of structure on polyester properties is evident in the melting points presented in Table 12.2.[15,16] The first three entries illustrate why aromatic units in the backbone are usually necessary to give the polymers

Table 12.2. Representative melting points of linear polyesters[a]

Structure	Approximate melting temperature (T_m) (°C)
	50
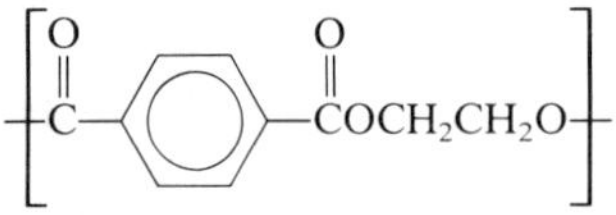	265
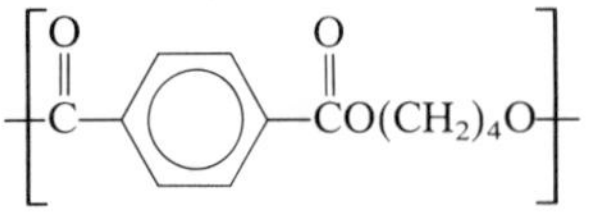	232–267
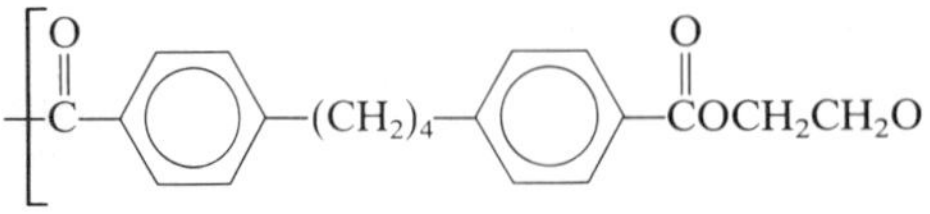	170

Table 12.2. (*Continued*)

Structure	Approximate melting temperature (T_m) (°C)

	T_m (°C)
$\left[\!\!\begin{array}{c}\mathrm{O}\\\|\\\mathrm{C}\end{array}\!\!-\!\!\bigcirc\!\!-\mathrm{O(CH_2)_2O}\!\!-\!\!\bigcirc\!\!-\!\!\begin{array}{c}\mathrm{O}\\\|\\\mathrm{C}\end{array}\!\mathrm{OCH_2CH_2O}\right]$	240
$\left[\!\!\begin{array}{c}\mathrm{O}\\\|\\\mathrm{C}\end{array}\!\!-\!\!\bigcirc\!\!-\mathrm{NH(CH_2)_2NH}\!\!-\!\!\bigcirc\!\!-\!\!\begin{array}{c}\mathrm{O}\\\|\\\mathrm{C}\end{array}\!\mathrm{OCH_2CH_2O}\right]$	273
$\left[\!\!\begin{array}{c}\mathrm{O}\\\|\\\mathrm{C}\end{array}\!\!-\!\!\bigcirc\!\!-\mathrm{CH{=}CH}\!\!-\!\!\bigcirc\!\!-\!\!\begin{array}{c}\mathrm{O}\\\|\\\mathrm{C}\end{array}\!\mathrm{OCH_2CH_2O}\right]$	420
$\left[\!\!\begin{array}{c}\mathrm{O}\\\|\\\mathrm{C}\end{array}\!\!-\!\!\bigcirc\!\!-\!\!\begin{array}{c}\mathrm{O}\\\|\\\mathrm{C}\end{array}\!\!-\mathrm{O}\!\!-\!\!\bigcirc\!\!-\mathrm{O}\right]$	209
$\left[\!\!\begin{array}{c}\mathrm{O}\\\|\\\mathrm{C}\end{array}\!\!-\!\!\bigcirc\!\!-\!\!\begin{array}{c}\mathrm{O}\\\|\\\mathrm{C}\end{array}\!\!-\mathrm{O}\!\!-\!\!\bigcirc\!\!-\mathrm{O}\right]$ with $\mathrm{CH_2(CH_2)_3CH_3}$ substituent	199
$\left[\!\!\begin{array}{c}\mathrm{O}\\\|\\\mathrm{C}\end{array}\!\!-\!\!\bigcirc\!\!-\!\!\begin{array}{c}\mathrm{O}\\\|\\\mathrm{C}\end{array}\!\!-\mathrm{O}\!\!-\!\!\bigcirc\!\!-\mathrm{SO_2}\!\!-\!\!\bigcirc\!\!-\mathrm{O}\right]$	330
$\left[\!\!\begin{array}{c}\mathrm{O}\\\|\\\mathrm{C}\end{array}\!\!-\!\!\bigcirc\!\!-\!\!\begin{array}{c}\mathrm{O}\\\|\\\mathrm{C}\end{array}\!\!-\mathrm{O}\!\!-\!\!\bigcirc\!\!-\mathrm{SO_2}\!\!-\!\!\bigcirc\!\!-\mathrm{O}\right]$ with $\mathrm{CH_2(CH_2)_3CH_3}$ substituent	246
$\left[\!\!\begin{array}{c}\mathrm{O}\\\|\\\mathrm{C}\end{array}\!\!-\!\!\bigcirc\!\!-\mathrm{O}\right]$	610
$\left[\!\!\begin{array}{c}\mathrm{O}\\\|\\\mathrm{C}\end{array}\!\!-\!\!\bigcirc\!\!-\!\!\begin{array}{c}\mathrm{O}\\\|\\\mathrm{C}\end{array}\!\!-\mathrm{O}\!\!-\!\!\bigcirc\!\!-\mathrm{O}\right]$	600

[a] Data from Wilfong[15]; Ehlers, Evers, and Fisch[16]; and Jackson[17]; and from *Modern Plastics Encyclopedia*, 1984–85, McGraw-Hill, New York, 1984, pp. 466–467.

other than elastomeric properties. The next four indicate how different functionalities influence thermal properties. The particularly high melting point resulting from incorporation of a double bond between the *p*-phenylene units reflects the significance of restricting rotation. The next four polyesters show the effect of including a "plasticizing" five-carbon chain in the backbone.

The remaining two polyesters in Table 12.2 represent extreme examples of chain stiffening. Because the melting points are too high for most molding applications, high-temperature powder sintering techniques are used to fabricate the polymers. Attempts to improve the solubility and lower the melting point include copolymerization with other polyarylates[17] and introduction of trifluoromethyl substituents.[18] It may be recalled from Chapter 3 (Section 3.7) that very-high-strength thermotropic liquid crystalline polyesters containing 4-hydroxybenzoate units have been developed. Hence intense interest in this aspect of polyester morphology is continuing.[19]

12.2.2 Polycarbonates

Polycarbonates[8b,20,21] are polyesters of carbonic acid. Although carbonic acid itself is not a stable compound, its derivatives (phosgene, urea, carbonates) are commercially available. The reaction of gaseous phosgene with bisphenols, notably bisphenol A (12.22), was developed into commercial production independently by Farbenfabriken Bayer in Germany and General Electric in the United States. The product (**11**) is marketed under such trade names as Lexan (General Electric) and Merlon (Mobay).

$$(12.22)$$

11

While any dihydroxy compound can, in principle, form a polycarbonate with phosgene, only bisphenol A has gained commercial importance, although other bisphenols are sometimes incorporated as property modifiers. The polymer is partially crystalline, is transparent, and has unusually high impact strength (see Chapter 4, Table 4.1), which makes it ideal for laboratory safety shields, bullet-proof windows, safety helmets, and the like. Polycarbonate is also thermally stable (it can be molded at temperatures as high as 550 to 600°C) and self-extinguishing. Other uses include gears, bushings, automotive parts, tableware, food containers, machine housings, medical appliances, and telephone and electronics parts.[22] Several polyblends of polycarbonate have been developed as engineering plastics, the most important being those with poly(butylene terephthalate) and acrylonitrile–butadiene–styrene (ABS) copolymer.[22,23]

Table 12.3 compares the glass transition temperatures of polycarbonates of varying structure. Interestingly, replacing the methyl groups of bisphenol A polycarbonate has only a marginal effect on the T_g except where a five- or six-membered ring is incorporated. The greatest effect occurs when all ring positions *ortho* to the carbonate group are substituted, and when the isopropylidene group is replaced by sulfonyl, both of which severely restrict rotational freedom.

Although interfacial polymerization is the preferred method of synthesis, transesterifcation (12.23) is sometimes used, with diphenyl carbonate being the ester of choice. Transesterifcation generally results in lower molecular weight polymer because of difficulties in removing phenol from the viscous polymer mass. Furthermore, the polymerization temperature must be carefully controlled to avoid rearrangement (12.24) that could result in crosslinking or chain branching.

Table 12.3. Glass transition temperatures (T_g's) of representative polycarbonates[a]

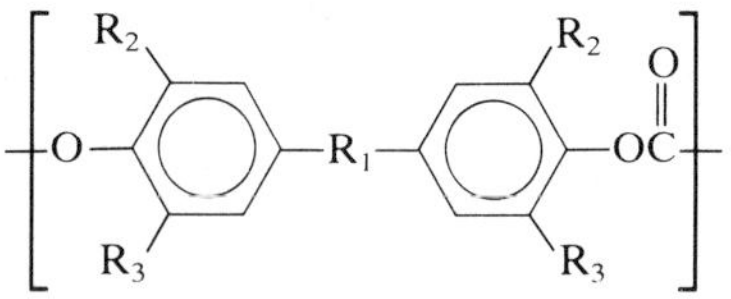

R_1	R_2	R_3	$T_g(°C)$
CH_2	H	H	147
$C(CH_3)_2$	H	H	149
$C(CH_3)_2$	CH_3	H	125
$C(CH_3)_2$	CH_3	CH_3	210
SO_2	CH_3	CH_3	260
$C(CH_2CH_2CH_3)_2$	H	H	148
$C(C_6H_5)_2$	H	H	121
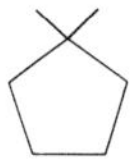	H	H	167
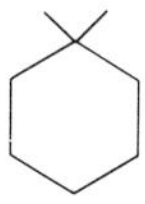	H	H	171

[a] Data from Temin,[8b] and Schnell.[21]

$$\text{HO}-\langle\bigcirc\rangle-\underset{\underset{\text{CH}_3}{|}}{\overset{\overset{\text{CH}_3}{|}}{\text{C}}}-\langle\bigcirc\rangle-\text{OH} + \langle\bigcirc\rangle-\text{O}\overset{\overset{\text{O}}{\|}}{\text{C}}\text{O}-\langle\bigcirc\rangle \longrightarrow$$

$$\left[\text{O}-\langle\bigcirc\rangle-\underset{\underset{\text{CH}_3}{|}}{\overset{\overset{\text{CH}_3}{|}}{\text{C}}}-\langle\bigcirc\rangle-\text{O}\overset{\overset{\text{O}}{\|}}{\text{C}}\right] + 2\,\langle\bigcirc\rangle-\text{OH} \quad (12.23)$$

$$\left[\text{O}-\langle\bigcirc\rangle-\underset{\underset{\text{CH}_3}{|}}{\overset{\overset{\text{CH}_3}{|}}{\text{C}}}-\langle\bigcirc\rangle-\text{O}\overset{\overset{\text{O}}{\|}}{\text{C}}\text{O}-\langle\bigcirc\rangle-\underset{\underset{\text{CH}_3}{|}}{\overset{\overset{\text{CH}_3}{|}}{\text{C}}}-\langle\bigcirc\rangle-\text{O}\overset{\overset{\text{O}}{\|}}{\text{C}}\right] \xrightarrow{\text{heat}}$$

$$\left[\text{O}-\langle\bigcirc\rangle-\underset{\underset{\text{CH}_3}{|}}{\overset{\overset{\text{CH}_3}{|}}{\text{C}}}-\langle\overset{}{\underset{\text{CO}_2\text{H}}{\bigcirc}}\rangle-\text{O}-\langle\bigcirc\rangle-\underset{\underset{\text{CH}_3}{|}}{\overset{\overset{\text{CH}_3}{|}}{\text{C}}}-\langle\bigcirc\rangle-\text{O}\overset{\overset{\text{O}}{\|}}{\text{C}}\right] \quad (12.24)$$

Polycarbonates have also been synthesized from dihydroxy compounds and bischloroformates[21] (12.25) and by alternating copolymerization of epoxides with carbon dioxide (12.26).[24,25] A considerable amount of developmental work has been devoted to the latter,[26] not surprising considering the obvious

$$\text{Cl}-\overset{\overset{\text{O}}{\|}}{\text{C}}-\text{O}-\text{R}-\text{O}-\overset{\overset{\text{O}}{\|}}{\text{C}}-\text{Cl} + \text{HO}-\text{R}'-\text{OH} \longrightarrow$$

$$\left[\text{O}-\text{R}'-\text{O}\overset{\overset{\text{O}}{\|}}{\text{C}}\text{O}-\text{R}-\text{O}\overset{\overset{\text{O}}{\|}}{\text{C}}\right] + 2\text{HCl} \quad (12.25)$$

$$\underset{\diagdown\!\diagup}{\overset{}{\underset{\text{O}}{\text{CH}_2-\text{CHR}}}} + \text{CO}_2 \xrightarrow{\text{Zn(C}_2\text{H}_5)_2\cdot\text{H}_2\text{O}} \left[\text{CH}_2\overset{\overset{\text{R}}{|}}{\text{CH}}\text{O}\overset{\overset{\text{O}}{\|}}{\text{C}}\text{O}\right] \quad (12.26)$$

economic advantage of low monomer cost. Carbon dioxide may also be used in the form of carbonate salt, as shown in reaction (12.27).[27] In this instance, the cyclic ether 18-crown-6* is used to bind the potassium ions, thus increas-

---

*The catalyst, 18-crown-6, is an 18-membered ring macrocyclic polyether containing six oxygen atoms. Such compounds are unusual in that they complex alkali metal cations strongly. They were designated *crown ethers* because in the complexed form, their shape bears some resemblance to a crown.

ing the nucleophilicity of carbonate ion. Additionally, polythiocarbonates may be prepared using carbon disulfide as a comonomer.[28]

$$BrCH_2\text{—}\langle\bigcirc\rangle\text{—}CH_2Br + K_2CO_3 \xrightarrow{\text{18-crown-6}}$$

$$\left[\!\!\left[CH_2\text{—}\langle\bigcirc\rangle\text{—}CH_2O\overset{\displaystyle O}{\underset{\displaystyle \|}{C}}O\right]\!\!\right] \quad (12.27)$$

12.2.3 *Preparation of polyesters by ring-opening polymerization*

Ring-opening polymerization of cyclic esters (lactones) is brought about by cationic, anionic, and complex coordination initiators.[3,29–31] The mechanism of cationic polymerization appears in most instances to involve an intermediate acylium ion with each propagation step proceeding with acyl–oxygen bond cleavage (12.28, $Z^+ = $ initiator). An exception is β,β-dimethyl-β-propiolactone (**12**), which undergoes alkyl–oxygen cleavage to give the preferred tertiary carbocation intermediate (12.29).

$$(CH_2)_x\overset{\displaystyle C\!=\!O}{\underset{\displaystyle O}{|}} \xrightarrow{Z^+} ZO(CH_2)_x\overset{\displaystyle O}{\underset{\displaystyle \|}{C}} + \longrightarrow$$

$$ZO(CH_2)_x\overset{\displaystyle O}{\underset{\displaystyle \|}{C}}\!\!\left[\!\!O(CH_2)_x\overset{\displaystyle O}{\underset{\displaystyle \|}{C}}\!\!\right]\!\!O(CH_2)_x\overset{\displaystyle O}{\underset{\displaystyle \|}{C}} + \quad (12.28)$$

$$\underset{\mathbf{12}}{\overset{\displaystyle CH_3\ CH_3}{\underset{\displaystyle O}{\underset{\displaystyle \|}{C}}}\!\!\langle\!\!\begin{array}{c}C\\CH_2\end{array}\!\!\rangle} \xrightarrow{Z^+} ZO\!\!\langle\!\!\begin{array}{c}\overset{CH_3\ CH_3}{C}\\ \overset{+}{C}\!\!=\!\!O\end{array}\!\!CH_2\rangle \longrightarrow ZOCCH_2\overset{\displaystyle CH_3}{\underset{\displaystyle CH_3}{C}} + \longrightarrow$$

$$ZOCCH_2\overset{\displaystyle CH_3}{\underset{\substack{\| \\ O}}{C}}\!\!\left[\!\!OCCH_2\overset{\displaystyle CH_3}{\underset{\substack{\| \\ O}}{C}}\!\!\right]\!\!OCCH_2\overset{\displaystyle CH_3}{\underset{\substack{\| \\ O}}{C}} + \quad (12.29)$$

Two mechanisms have been suggested for anionic polymerization of lactones. Nucleophilic addition to the carbonyl group followed by acyl–oxygen cleavage (analogous to ordinary ester saponification) leads to propagation through alkoxide ion chain ends (12.30, $B^- = $ initiator). As in the case of epoxide polymerization, living polymers may be formed. There is, however, a tendency for the growing chain to depropagate via a back-biting mechanism to form cyclic oligomers. The resultant chain-ring equilibrium, which may severely limit the molecular weight, is less prevalent when Li^+ is the counter-ion, and high-molecular-weight polyesters have been prepared with alkyl-lithium and lithium alkoxide initiators.[32a]

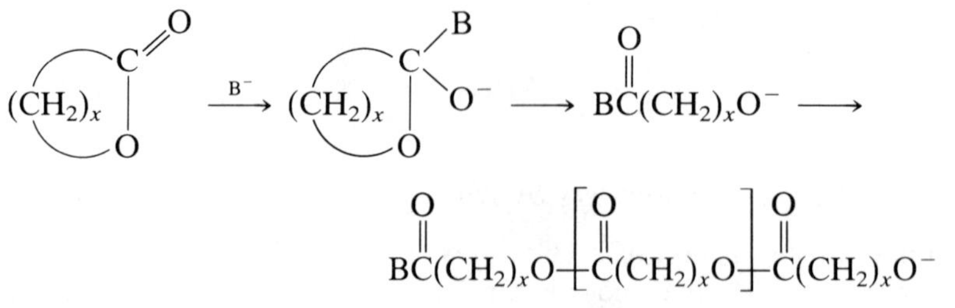

$$BC(CH_2)_xO \xrightarrow{\quad} BC(CH_2)_xO \left[C(CH_2)_xO \right] C(CH_2)_xO^- \quad (12.30)$$

The other anionic mechanism involves nucleophilic attack on the carbon adjacent to the ring oxygen, giving alkyl–oxygen cleavage and propagating carboxylate ion chain ends. Such a mechanism (12.31) applies to the more highly strained β-propiolactone (**13**).* This reaction is subject to chain transfer reactions resulting in unsaturated chain ends (12.32) and low molecular weights.

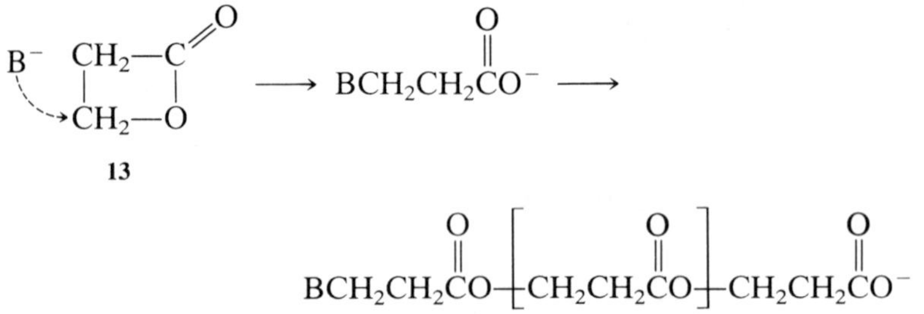

13

$$BCH_2CH_2CO \left[CH_2CH_2CO \right] CH_2CH_2CO^- \quad (12.31)$$

$$\text{mCO}^- \cdots H-CH-C \longrightarrow \text{mCOH} + CH-C-O^- \longrightarrow$$

$$CH_2=CHCO \left[CH_2CH_2CO \right] CH_2CH_2CO^- \quad (12.32)$$

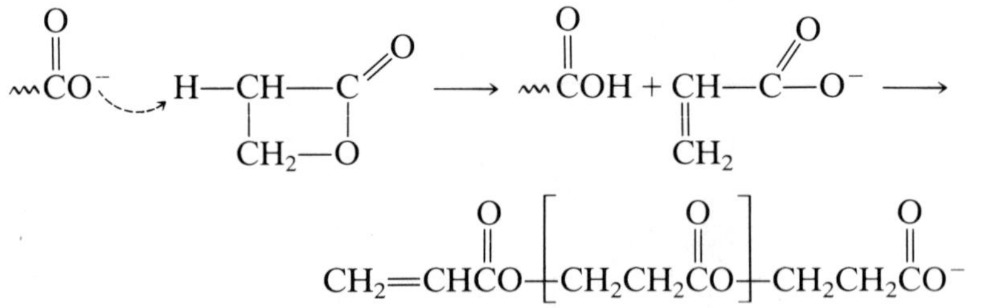

*Supporting evidence is derived from studies of the base-catalyzed hydrolysis of β-lactones in ^{18}O-enriched water.

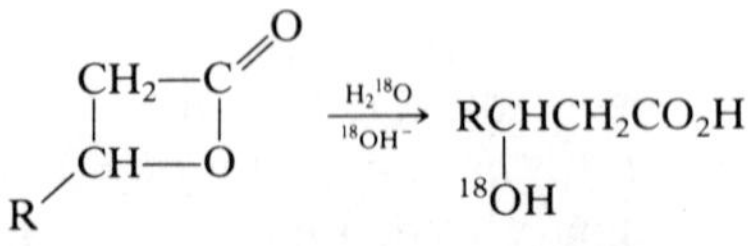

$$\xrightarrow{H_2^{18}O,\ ^{18}OH^-} RCHCH_2CO_2H$$
$$\qquad\qquad\qquad ^{18}OH$$

Nucleophilic attack at the carbonyl carbon would result in the ^{18}O in the carboxylic acid group.

Polymerization of β-lactams may also be initiated with metalloporphyrin initiators of the type encountered in the previous chapter.[32b,33] Ring-opening takes place at the alkyl–oxygen bond (12.33) and—as was the case with epoxides—the result is an "immortal" polymer suitable for forming block copolymers. Another example of complex coordination catalysis is the ring-opening polymerization of ε-caprolactone (**14**) using the dibutylzinc–triiso-butylaluminum system (12.34).[34]

$$\left(\underset{Al}{\bigcirc}\right) + 13 \longrightarrow \left(\underset{Al}{\bigcirc}\right)\!\!-\!\!\left[OCCH_2CH_2\right]\!\!-\!\!Cl \qquad (12.33)$$

$$\underset{14}{\left(\bigcirc\!\!\!=\!\!O\right)} \xrightarrow{(n\text{-}C_4H_9)_2Zn-(i\text{-}C_4H_9)_3Al} \underset{15}{\left[(CH_2)_5\overset{O}{\overset{\|}{C}}\!\!-\!\!O\right]} \qquad (12.34)$$

With the exception of the five-membered γ-butyrolactone, most lactones appear to form polyester, although at the present time only polycaprolactone (**15**) has achieved any commercial significance, especially in the medical field[35] as biodegradable surgical sutures and postoperative support pins and splints, or as a delivery medium for controlled release of drugs. Two other polyesters that find similar application[36] are poly(glycolic acid) (**8**) and poly-(lactic acid) (**18**), which are prepared by the Lewis acid-catalyzed ring-opening polymerization of glycolide (**16**) (12.35) and lactide (**17**), (12.36) respectively.[37] Glycolide and lactide are the cyclic dimers of the corresponding hydroxy acids, glycolic acid (hydroxyethanoic acid) and lactic acid (2-hydroxypropanoic acid), and are prepared by pyrolysis of the alkali metal salts of the corresponding halo acids. Polymers **8** and **18** are sometimes called polyglycolide and polylactide, respectively.

$$\underset{16}{O\!=\!\!\left\langle\!\!\begin{array}{c} O\!-\! \\ \\ \!-\!O \end{array}\!\!\right\rangle\!\!=\!\!O} \xrightarrow{SbF_3} \underset{8}{\left[CH_2\overset{O}{\overset{\|}{C}}O\right]} \qquad (12.35)$$

$$\underset{17}{O\!=\!\!\left\langle\!\!\begin{array}{c} O\!-\!CH_3 \\ \\ CH_3\!-\!O \end{array}\!\!\right\rangle\!\!=\!\!O} \xrightarrow{ZnCl_2} \underset{18}{\left[\underset{CH_3}{\overset{O}{\overset{\|}{CHCO}}}\right]} \qquad (12.36)$$

A variety of copolymers of lactones with other monomer types have been prepared, including vinyl monomers, epoxides, and formaldehyde. In the case of vinyl monomers, there is little tendency for cross-propagation, and largely block copolymers are formed. Indeed, a number of lactone-based AB block copolymers have been investigated for adhesive, film, and fiber applications.[14b] Some increase in cross-propagation occurs with styrene when high concentrations of stannic chloride are used as initiator or when solvent polarity is high, suggesting a complex between initiator and lactone may be initiating styrene polymerization.

A more recent method of preparing polyesters is by the free radical ring-opening polymerization (12.37) of 2-methylene-1,3-dioxolane (**19**) (a ketene acetal) to form polymer **20**.[32c,38] This represents one of the few examples of a ring-opening polymerization under free radical conditions. Other ketene acetals react similarly; for example, polycaprolactone (**15**) of higher molecular weight than is formed by ring-opening polymerization of caprolactone (**14**) is obtained from **21** (12.38). The proposed mechanism of polymerization involves initial reaction of **19** with radical RO· to form **22** followed by ring-opening (12.39) to give **23** which, in turn, initiates ring-opening (12.40). Ordinary vinyl polymerization (nonring-opening) is a competing reaction in the case of **19**, especially at lower temperatures (about 50:50 at 60°C), but this is much less of a problem with the more sterically hindered **21**.

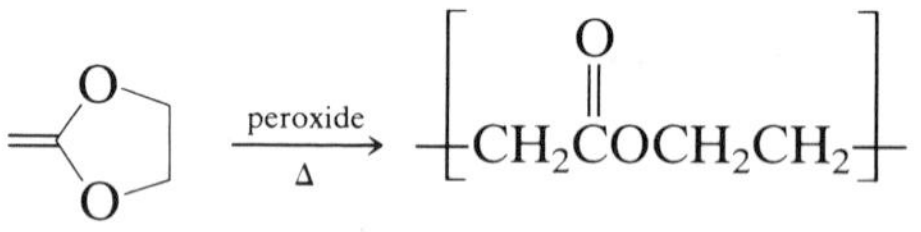

$$\text{19} \xrightarrow[\Delta]{\text{peroxide}} \left[\text{CH}_2\overset{\displaystyle O}{\overset{\displaystyle \|}{\text{C}}}\text{OCH}_2\text{CH}_2 \right] \qquad (12.37)$$

19   20

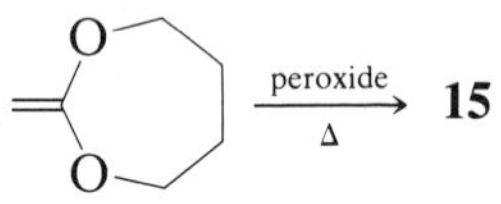

$$\text{21} \xrightarrow[\Delta]{\text{peroxide}} \text{15} \qquad (12.38)$$

21

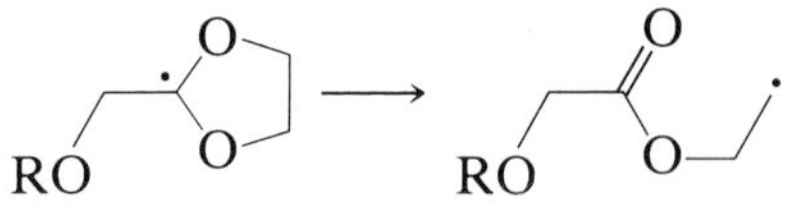

$$\text{22} \longrightarrow \text{23} \qquad (12.39)$$

22   23

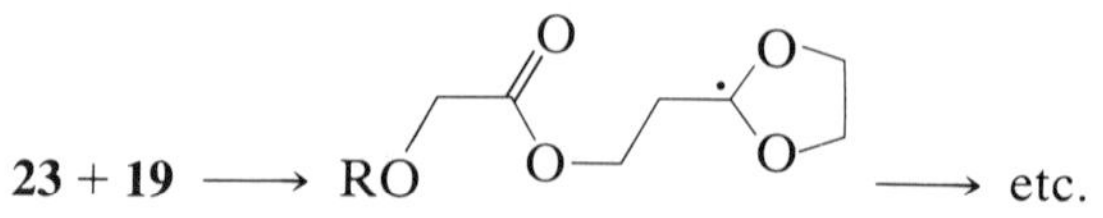

$$\text{23} + \text{19} \longrightarrow \text{RO} \cdots \longrightarrow \text{etc.} \qquad (12.40)$$

Ketene acetal monomers copolymerize readily with vinyl monomers. Introduction of polyester **20** units into polyethylene yields a copolymer (**24**) that is biodegradable by initial breakdown to form fatty acid-type residues.

$$-\!\!\left[CH_2CH_2\right]_{\!x}\!\!CH_2\overset{\displaystyle O}{\overset{\displaystyle \|}{C}}OCH_2CH_2\!\!\left[CH_2CH_2\right]_{\!y}\!\!-$$

24

12.2.4 *Miscellaneous methods of preparing linear polyesters*

Table 12.4 lists several miscellaneous methods reported to yield polyesters.[4,39–47] Most are of academic interest, but all serve to demonstrate the remarkable diversity of polymer chemistry as related to a single polymer type. The first two are chain polymerization reactions, one (12.41) employing crown ether to increase the nucleophilicity of carboxylate ion for Michael addition, the other (12.42) employing free radical initiators to promote alternating copolymerization of cyclohexene and formic acid.

The next three reactions are step-growth addition polymerizations. Carbon suboxide not only undergoes addition as shown (12.43), but under the influence of heat or light it forms a network homopolymer containing α-pyrone units[48] (**25**). Reaction (12.44), an application of the Tischenko reaction, includes an intermolecular hydride transfer step. Addition of dithioacids to dienes (12.45) is analogous to dithiol addition to form polysulfide (see Chapter 11).

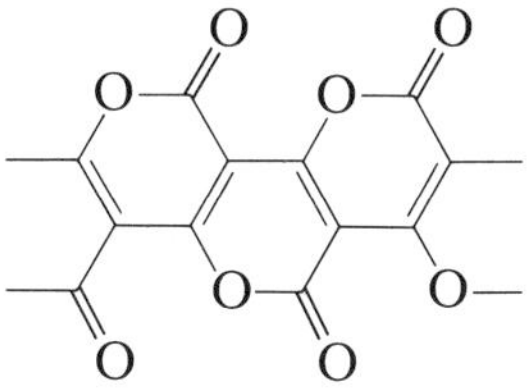

25

Reactions (12.46), (12.47), and (12.48) are condensation processes. Addition of dihydroxy compounds to dinitriles (12.46) involves an intermediate imino derivative that undergoes hydrolysis to polyester. Diimidazole monomers (12.47) react much like diacid chlorides. Reaction (12.48) represents one of the few examples of ring-opening condensation polymerization. The last reaction (12.49) is unusual in that it represents an industrial-scale fermentation process (Table 12.1). The product, poly(2-hydroxybutyrate), is biodegradable and exhibits piezoelectric properties (i.e., it generates a slight voltage under pressure). The polymer promises to be a worthy competitor to the other commercially available biodegradable polyesters.

Table 12.4. Miscellaneous methods of forming polyesters

Reaction number	Reaction	Reference number
(12.41)	$CH_2{=}CHCO_2H \xrightarrow[\text{crown ether}]{\text{KOAc}} \left[CH_2CH_2\overset{\displaystyle O}{\overset{\|}{C}}O \right]$	39
(12.42)	$HCO_2H + \text{(cyclohexene)} \xrightarrow[\text{ROOH}]{I_2} \text{(cyclohexene with } \overset{O}{\overset{\|}{C}}O\text{)}$	40
(12.43)	$HO{-}R{-}OH + O{=}C{=}C{=}C{=}O \longrightarrow \left[OROC\overset{\displaystyle O}{\overset{\|}{}}CH_2\overset{\displaystyle O}{\overset{\|}{C}} \right]$	41
(12.44)[a]	$OCH{-}R{-}CHO \xrightarrow{\text{base}} \left[OCH_2R\overset{\displaystyle O}{\overset{\|}{C}} \right]$	4, 41–43
(12.45)	$HS\overset{O}{\overset{\|}{C}}R\overset{O}{\overset{\|}{C}}SH + CH_2{=}CH{-}R{-}CH{=}CH_2 \xrightarrow{\text{ROOR}} \left[\overset{O}{\overset{\|}{C}}R\overset{O}{\overset{\|}{C}}SCH_2CH_2RCH_2CH_2S \right]$	44

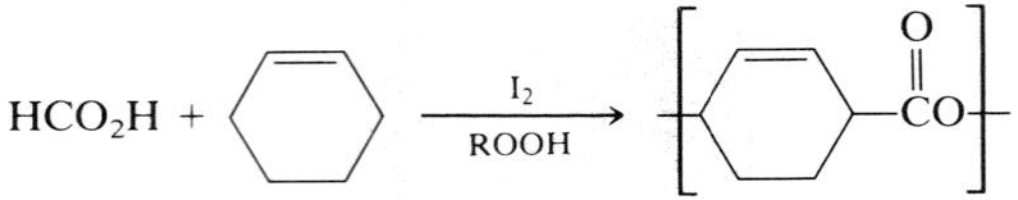

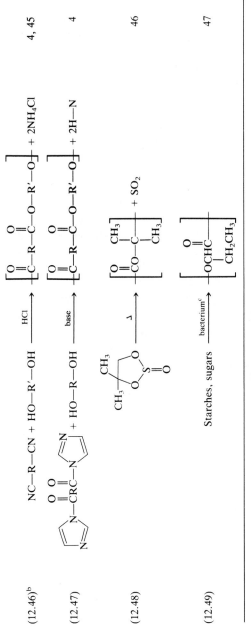

$$(12.46)^b \quad NC-R-CN + HO-R'-OH \xrightarrow{HCl} \left[\overset{O}{\underset{\|}{C}}-R-\overset{O}{\underset{\|}{C}}-O-R'-O \right] + 2NH_4Cl \qquad 4, 45$$

$$(12.47) \quad \text{(imidazolide)} + HO-R-OH \xrightarrow{base} \left[\overset{O}{\underset{\|}{C}}-R-\overset{O}{\underset{\|}{C}}-O-R'-O \right] + 2H-N \qquad 4$$

$$(12.48) \quad \text{(cyclic sultine)} \xrightarrow{1} \left[\overset{O}{\underset{\|}{CO}}-\overset{CH_3}{\underset{CH_3}{C}} \right] + SO_2 \qquad 46$$

$$(12.49) \quad \text{Starches, sugars} \xrightarrow{bacterium^c} \left[O\overset{O}{\underset{\|}{CHC}} \atop CH_2CH_3 \right] \qquad 47$$

[a] R = aryl or alkyl containing no α-hydrogen. Other repeating units are formed.

[b] Low-molecular-weight polyester is formed.

[c] *Alcaligenes eutrophus*.

12.3   Crosslinked polyesters

There are two types of crosslinked or network polyesters:

1. Those prepared from polyfunctional monomers so that crosslinking occurs during the polyesterification reaction. (These are referred to below as *saturated crosslinked polyesters*.)
2. Those crosslinked by a separate addition polymerization reaction through reactive double bonds incorporated into the polyester backbone.

The latter type are commonly known as *unsaturated polyesters*. Both are referred to as *alkyds* or *polyester resins*, and both are widely used in a variety of commercial applications. Their chemistry will be discussed in some detail.

12.3.1   Saturated polyester resins

Some of the common polyfunctional alcohols and acids or acid derivatives used to make saturated alkyd resins[49] include glycerol, pentaerythritol, sorbitol, citric acid, trimellitic acid, and pyromellitic dianhydride. Phthalic anhydride is by far the most common dibasic acid derivative used.

The polyester prepared from phthalic anhydride and glycerol is called *glyptal*, and its network structure typifies this type of resin (reaction 12.50). Structures are complex and highly crosslinked, the degree of crosslinking depending on the reactants employed and the extent of reaction. Normally, polymerization is carried to the point of forming a viscous liquid that can be hardened later to a more fully crosslinked product, for example in a mold. In production the process is monitored by periodically checking the viscosity and acid number. (See Chapter 2, Section 2.3.)

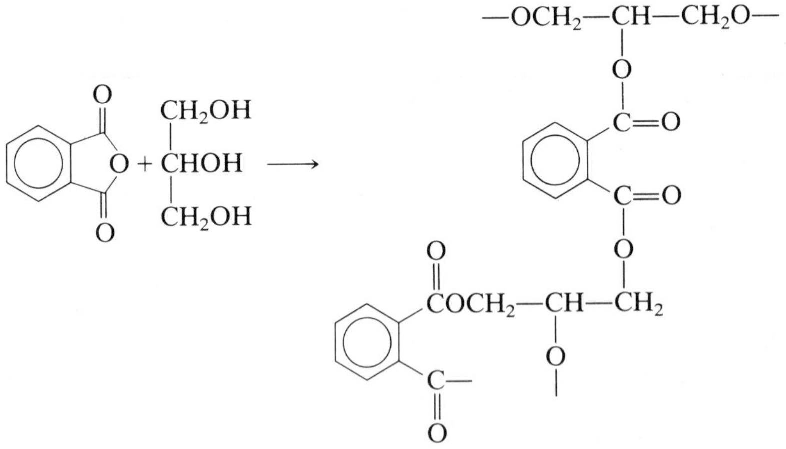

$$(12.50)$$

Obviously one can vary the structure at will by changing the type and concentration of reactant, or by adding some diol or monofunctional acid or alcohol. These types of alkyds have little use other than as adhesives or as additives in certain types of coatings, such as nitrocellulose. They are, however, extremely important when modified with natural or synthetic oils (triglycerides). Such products are called *oil-modified alkyds*.

Prior to latex emulsion paints, oil-modified alkyds were the most important type of coating. Oils are incorporated into polyesters resins by two general methods:

1. Polyol and oil are transesterified (12.51) so that mixing occurs; the mixed product is then reacted with dibasic acid or anhydride in a second step (12.52) to give the final complex crosslinked polymer.
2. Dibasic acid or anhydride, polyol, and oil fatty acids (called "split oils") are reacted together in a single reaction.

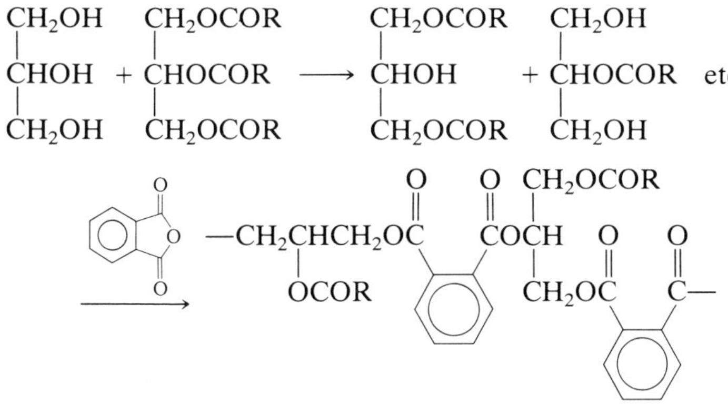

$$
\begin{array}{ccccc}
CH_2OH & CH_2OCOR & & CH_2OCOR & CH_2OH \\
| & | & & | & | \\
CHOH & + \ CHOCOR & \longrightarrow & CHOH & + \ CHOCOR \quad \text{etc.} \quad (12.51) \\
| & | & & | & | \\
CH_2OH & CH_2OCOR & & CH_2OCOR & CH_2OH
\end{array}
$$

$$\qquad\qquad\qquad (12.52)$$

The net effect of either process is to incorporate the long-chain oil structure into the polyester. If the oil itself is reacted with polyol and dibasic acid in one step, a two-phase product results. After initial esterification, the viscosity is increased by heating to a high temperature. This process, known as *bodying*, results from a combination of further esterification and polymerization through the double bonds of the oil.

The type of oil used may contain varying amounts of unsaturation depending on the application. Examples of the more unsaturated oils (*drying oils*) are linseed oil, which contains principally linoleic (**26**) and linolenic (**27**) acids, and tung oil, a major constituent of which is elaeostearic acid (**28**). Dehydrated castor oil, a synthetic oil, also contains linoleic acid. It results from dehydration of ricinoleic acid (**29**) in castor oil. Olive oil and castor oil are examples of nondrying oils.

$$CH_3(CH_2)_4CH{=}CHCH_2CH{=}CH(CH_2)_7CO_2H$$

26

$$CH_3(CH_2CH=CH)_3(CH_2)_7CO_2H$$
27

$$CH_3(CH_2)_3CH=CH-CH=CH-CH=CH(CH_2)_7CO_2H$$
28

$$CH_3(CH_2)_5\underset{\underset{OH}{|}}{C}HCH_2CH=CH(CH_2)_7CH_2H$$
29

The oil-modified alkyds, dissolved in suitable solvents (the choice depending usually on the amount of oil used), together with appropriate additives such as pigments and metal naphthenates (driers), constitute *alkyd paints*. The function of the oil is to cause *drying* by a free radical crosslinking reaction involving allylic radicals. (Oils "dry" by themselves when spread as a film.) The reaction depends on the presence of atmospheric oxygen and is speeded up by the presence of the driers, often cobalt or manganese naphthenates.*

The rapidity of drying depends on the amount of unsaturation and the amount of oil incorporated into the alkyd (referred to as *oil length*). Oil lengths are commonly designated as follows:

> Short-oil alkyds, 30 to 50% oil
> Medium-oil alkyds, 50 to 65% oil
> Long-oil alkyds, 65 to 80% oil

The short-oil alkyds usually have to be baked (baking enamels) to effect drying in a reasonable time, although the coating is generally more durable. Alkyds modified with nondrying oils will always remain tacky; they are usually used for blending with other types of resins.

Oil-modified alkyds are sometimes modified further by addition of styrene. The styrene is copolymerized with the oil portion of the alkyd to give a tougher, more durable film by reactions similar to those discussed in the next section. The products are referred to as *styrenated alkyds*.

Although the above-mentioned alkyds have a long history of commercial success, other attempts have been made to synthesize useful, crosslinked polyesters. One of the more notable ones has been with poly(ester acetals)[50] such as 30. Polyesters of this type can be crosslinked by acid-catalyzed ring-opening acetal interchange reactions.

---

* Napthenic acids are petroleum-derived complex mixtures consisting primarily of fused cyclopentane rings with aliphatic side chains and carboxylic acid groups. Metal naphthenates are viscous liquids.

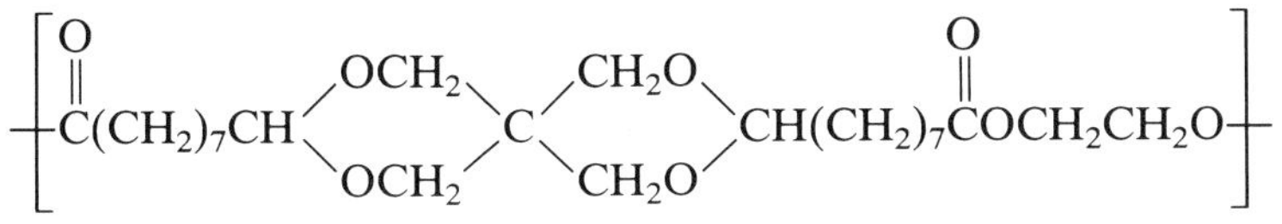

$$\left[\begin{array}{c} O \\ \parallel \\ -C(CH_2)_7CH \end{array}\begin{array}{c} OCH_2 \\ \diagup \\ \diagdown \\ OCH_2 \end{array}C\begin{array}{c} CH_2O \\ \diagup \\ \diagdown \\ CH_2O \end{array}CH(CH_2)_7\begin{array}{c} O \\ \parallel \\ COCH_2CH_2O \end{array}\right]$$

30

12.3.2 Unsaturated polyesters

Unsaturated polyesters[51,52] are among the most common polymers used in conjunction with glass fiber reinforcing. They are prepared from difunctional monomers, one of which contains a double bond that is capable of undergoing addition polymerization in a subsequent crosslinking reaction. Typically, the linear unsaturated polyester is processed to a relatively low molecular weight; then it is dissolved in a monomer such as styrene to form a viscous solution. Crosslinking, usually initiated with free radical initiators, is thus a vinyl copolymerization between the polyester and the solvent monomer. Styrene is by far the most commonly used solvent monomer, although one can use such others as vinyl acetate or methyl methacrylate or, for improved flame-resistance, halogenated monomers such as *ortho-* and *para-*bromostyrene.[53]

The only ingredients of commercial significance for introducing unsaturation into the polymer backbone are maleic anhydride and fumaric acid, particularly the former because of its lower cost.[54] If only unsaturated acid and glycol are used, the final product is too highly crosslinked and brittle to be useful. For this reason, copolyesters are normally prepared containing both unsaturated acid and a noncrosslinkable acid. Reaction (12.53) shows a

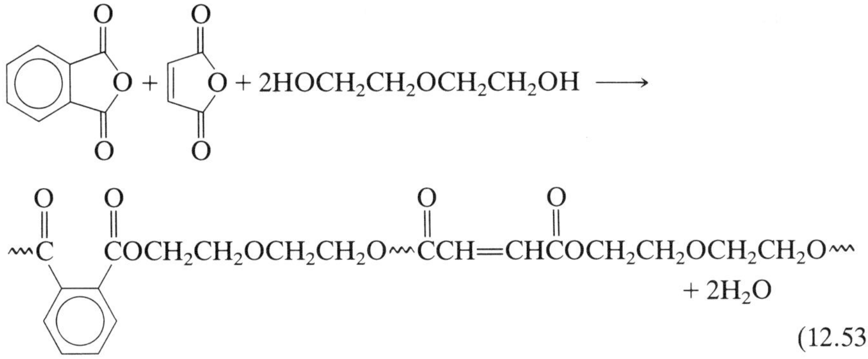

$$\text{phthalic anhydride} + \text{maleic anhydride} + 2HOCH_2CH_2OCH_2CH_2OH \longrightarrow$$

$$\begin{array}{cc} O & O \\ \parallel & \parallel \\ \sim\!\!C & COCH_2CH_2OCH_2CH_2O\!\!\sim \end{array}\begin{array}{cc} O & O \\ \parallel & \parallel \\ CCH\!\!=\!\!CHCOCH_2CH_2OCH_2CH_2O\!\!\sim \end{array}$$
$$+ 2H_2O$$

$$(12.53)$$

typical unsaturated polyester synthesis from maleic and phthalic anhydrides (in a 1:1 molar ratio) and diethylene glycol. The phthalic and maleic units would be randomly distributed. Crosslinking with styrene is depicted in reaction (12.54).

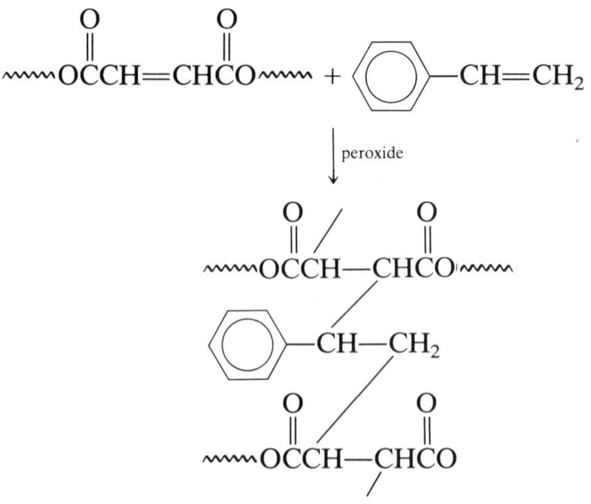

$$\text{wwwOCCH=CHCOwww} + \underset{\text{}}{\bigcirc}-\text{CH=CH}_2$$

$$\downarrow \text{peroxide}$$

(12.54)

Properties of unsaturated polyesters can be readily varied by using different types of diacids and glycols or different reactant ratios. Solvent monomer may also be varied in type as well as concentration.

Some of the dibasic acids or anhydrides commonly used in addition to those mentioned above are isophthalic acid, adipic acid, and succinic anhydride. Work has also been done on polyesters prepared from 2,2-diphenylpropane dioxyacetic acid (**31**), a diacid prepared from bisphenol A and chloroacetic acid.[55] Common glycols include ethylene, propylene, diethylene, and dipropylene glycols. To obtain maximum hardness and flexural strength, "rigid" monomers such as phthalic anhydride, isophthalic acid, and propylene glycol are normally used. Greater flexibility and higher impact strengths are obtained from the longer chain monomers. Flame resistance is imparted by halogenated monomers such as chlorendic anhydride and tetrabromophthalic anhydride. (Refer to the discussion in Section 4.5, Chapter 4.) Polyesters with good chemical resistance are prepared with glycols that affords steric hindrance to the ester linkage. (See Section 4.6, Chapter 4.)

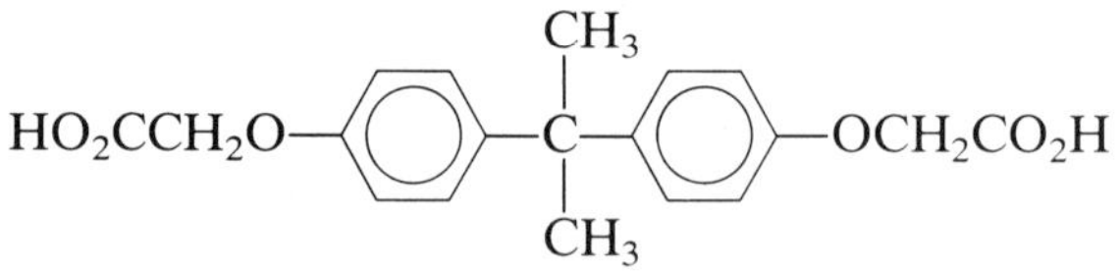

$$\text{HO}_2\text{CCH}_2\text{O}-\underset{\text{}}{\bigcirc}-\underset{\underset{\text{CH}_3}{|}}{\overset{\overset{\text{CH}_3}{|}}{\text{C}}}-\underset{\text{}}{\bigcirc}-\text{OCH}_2\text{CO}_2\text{H}$$

31

Most unsaturated polyesters are not suitable as coatings because atmospheric oxygen inhibits the crosslinking reaction. Two approaches have been used to circumvent this difficulty. One is to mix wax into the polyester–styrene solution. Not being completely soluble, the wax rises to the surface and excludes oxygen and, incidentally, reduces loss of styrene by evaporation. Wax has the disadvantage of imparting a dull finish to the film.

The other approach is to incorporate into the polyester backbone monomers such as tetrahydrophthalic anhydride (**32**) and dodecenylsuccinic anhydride (**33**), which contain isolated double bonds. These allow crosslinking by the same oxygen-induced mechanism that causes crosslinking of oil-modified alkyds.

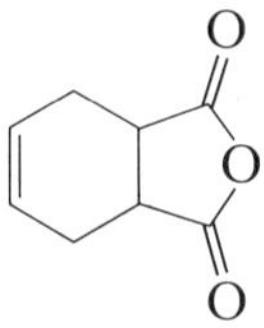

$CH_3(CH_2)_8CH{=}CHCH_2$—

32 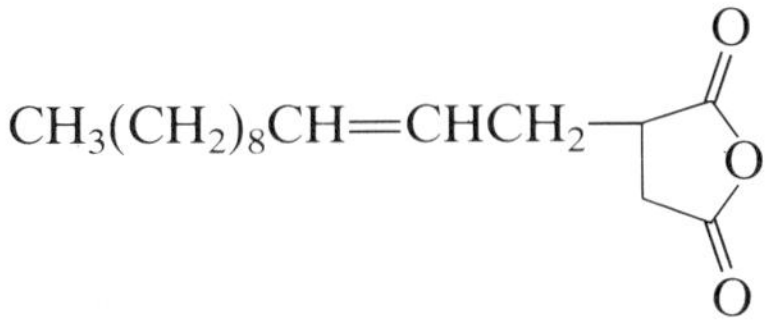   **33**

A number of photochemical methods have been developed to cure unsaturated polyester coatings or transparent articles, among them the use of disulfides[56] or dyes[57] to initiate photocrosslinking, or the incorporation of photosensitive groups such as chalconyl or cinnamyl into the polymer[58] (see Table 9.1, Chapter 9). Photochemical curing not only allows the formulation of *one-pot* systems, but B-stage (partially cured) resins can be formed simply by removing the light source once the appropriate viscosity is reached.

References

1. Bjorksten Research Laboratories, *Polyesters and Their Applications*, Reinhold, New York, 1956.
2. V. V. Korshak and S. V. Vinogradova, *Polyesters*, Pergamon Press, New York, 1965.
3. I. Goodman and J. A. Rhys, *Polyesters*, Vol. 1, *Saturated Polymers*, Iliffe Books, London, 1965.
4. P. W. Morgan, *Condensation Polymers: By Interfacial and Solution Methods*, Wiley-Interscience, New York, 1965, Chap. 8.
5. Anon., *Chem. Eng. News*, **48**(49), 42 (1970).
6. V. V. Korshak, *Heat-Resistant Polymers*, Halsted Press, New York, 1972.
7. J. P. Critchley, G. J. Knight, and W. W. Wright, *Heat-Resistant Polymers*, Plenum Press, New York, 1983, Chap. 4.
8. F. Millich and C. E. Carraher Jr., *Interfacial Synthesis*, Vol. 2, Dekker, New York, 1977: (a) S. C. Temin, Chap. 12; (b) H. Vernaleken, Chap. 13.
9. H. Tanaka, Y. Iwanaga, G. Wu, K. Sanui, and N. Ogata, *Polym. J.*, **14**, 648 (1982).
10. F. Higashi, A. Hoshio, Y. Yamada, and M. Ozawa, *J. Polym. Chem., Polym. Lett. Ed.*, **23**, 69 (1985).
11. F. Higashi, N. Akiyama, I. Takahashi, and T. Koyama, *J. Polym. Sci., Poly. Chem. Ed.*, **22**, 1653 (1984).
12. A. G. Pinkus and R. Subramanyam, *J. Polym. Sci., Polym. Chem. Ed.*, **22**, 1131 (1984).

13. Y. Imai and M. Ueda, in *Crown Ethers and Phase Transfer Catalysis in Polymer Science* (L. J. Mathias and C. E. Carraher Jr., eds.), Plenum Press, New York, 1984, pp. 121–137.
14. A. Noshay and J. E. McGrath, *Block Copolymers: Overview and Critical Survey*, Academic Press, New York, 1977: (a) Chap. 7; (b) Chap. 5.
15. R. E. Wilfong, *J. Polym. Sci.*, **54**, 385 (1961).
16. G. F. L. Ehlers, R. C. Evers, and K. R. Fisch, *J. Polym. Sci., A-1*, **7**, 3413 (1969).
17. W. J. Jackson Jr., *Br. Polym. J.*, **12**, 154 (1980).
18. R. Sinta, R. A. Minns, R. A. Gaudiana, and H. G. Rogers, *J. Polym. Sci. C, Polym. Lett.*, **25**, 11 (1987).
19. W. J. Jackson Jr., *J. Appl. Polym. Sci., Appl. Polym. Symp.*, **41**, 25 (1985).
20. W. F. Christopher and D. W. Fox, *Polycarbonates*, Reinhold, New York, 1962.
21. H. Schnell, *Chemistry and Physics of Polycarbonates*, Wiley-Interscience, New York, 1964.
22. S. K. Sikdar, *CHEMTECH*, **17**, 112 (1987).
23. B. D. Hanrahan, S. R. Angeli, and J. Runt, *Polym. Bull.*, **14**, 399 (1985).
24. M. Kobayashi, S. Inoue, and T. Tsuruta, *J. Polym. Sci., Polym. Chem. Ed.*, **11**, 2383 (1973).
25. R. Rokicki and W. Kuran, *J. Macromol. Sci., Rev. Macromol. Chem.*, **C21**, 135 (1981).
26. S. Inoue, *CHEMTECH*, **6**, 588 (1976).
27. K. Soga, S. Hosoda, and S. Ikeda, *J. Polym. Sci., Polym. Chem. Ed.*, **17**, 517 (1979).
28. G. Rokicki, W. Kuran, and J. Kieliewicz, *Makromol. Chem.*, **180**, 2779 (1979).
29. G. L. Brode and J. V. Koleske, in *Polymerizations of Heterocyclics*, (O. Vogl and J. Furukawa, eds.), Dekker, New York, 1973, pp. 117–152.
30. J. O. Brash and D. J. Lyman, in *Cyclic Monomers*, (K. C. Frisch, ed.), Wiley-Interscience, New York, 1972, Chap. 2.
31. R. D. Lundberg and E. F. Cox, in *Ring-Opening Polymerization*, (K. C. Frisch and S. L. Reegen, eds.), Dekker, New York, 1969, Chap. 6.
32. J. E. McGrath (ed.), *Ring-Opening Polymerization*, ACS Symp. Ser. 286, American Chemical Society, Washington, D.C., 1985: (a) M. Morton and M. Wu, pp. 175–182; (b) S. Inoue and T. Aida, pp. 137–146; (c) W. J. Bailey, pp. 47–65.
33. S. Inoue, *Makromol. Chem., Macromol. Symp.*, **3**, 295 (1986).
34. R. D. Lundberg, J. V. Koleske, and K. B. Wischmann, *J. Polym. Sci A-1*, **7**, 2915 (1969).
35. H. J. Sanders, *Chem. Eng. News*, **63**(13), 30 (1985).
36. K. S. Devi and P. Vasudevan, *J. Macromol. Sci., Rev. Macromol. Chem. Phys.*, **C25**, 315 (1985).
37. K. Chujo, H. Kobayashi, J. Suzuki, S. Tokuhara, and M. Tanabi, *Makromol. Chem.*, **100**, 262 (1967).
38. W. J. Bailey et al., *J. Macromol. Sci.-Chem.*, **A25**, 781 (1988).
39. B. Yamada, Y. Yasuda, T. Matushita, and T. Ostsu, *J. Polym. Sci., Polym. Lett. Ed.*, **14**, 277 (1976).
40. C. G. Gebelein, *J. Polym. Sci. A-1*, **10**, 1763 (1972).
41. V. V. Korshak, S. V. Rogozhin, and V. I. Volkov, *Vysokomolekul. Soedin.*, **1**, 804 (1959) (*Chem. Abstr.*, **54**, 12643e, 1960).
42. W. Sweeny, *J. Appl. Polym. Sci.*, **7**, 1983 (1963).

43. K. Harashi, *Macromolecules*, **3**, 5 (1970).

44. C. S. Marvel and E. A. Kraiman, *J. Org. Chem.*, **18**, 707, 1664 (1953).

45. E. N. Zil'berman, A. E. Kulikova, and N. M. Teplyakov, *J. Polym. Sci.*, **56**, 417 (1962).

46. W. R. Sorenson and T. W. Campbell, *Preparative Methods of Polymer Chemistry*, 2nd ed., Wiley-Interscience, New York, 1968, pp. 359ff.

47. Anon., *Chem. Eng. News*, **60**(47), 6 (1982).

48. N.-L. Yang, A. Snow, and H. Haubenstock, *J. Polym. Sci., Polym. Chem. Ed.*, **16** 1909 (1978).

49. C. R. Martens, *Alkyd Resins*, Reinhold, New York, 1961.

50. R. W. Lenz, *Macromolecules*, **2**, 129 (1969).

51. H. V. Boenig, *Unsaturated Polyesters: Structure and Properties*, Elsevier, Amsterdam, 1964.

52. B. Parkyn, F. Lamb, and B. V. Clitton, *Polyesters*, Vol. 2, *Unsaturated Polyesters and Polyester Plasticizers*, Elsevier, New York, 1967.

53. D. Alsheh and G. Marom, *J. Appl. Polym. Sci.*, **22**, 3177 (1978).

54. B. C. Trivedi and B. M. Culbertson, *Maleic Anhydride*, Plenum Press, New York, 1982, Chap. 12.

55. R. S. Lenk, *J. Appl. Polym. Sci.*, **15**, 2211 (1971).

56. Neth. Pat. Appl. 6,517,086 (to Farbenfabriken Bayer, A. G.), 1966 (*Chem. Abstr.*, **66**, 11427h, 1967).

57. I. J. Alexander and R. J. Scott, *Br. Polym. J.*, **15**, 30 (1983).

58. D. S. Sadafule and S. P. Panda, *J. Appl. Polym. Sci.*, **24**, 511 (1979).

Review exercises

1. Write IUPAC names for the polyesters in Table 12.1.

2. Low-molecular-weight polyesters used in the manufacture of polyurethanes (Chapter 13) are poly(ethylene adipate), poly(butylene adipate), and poly-(diethylene glycol adipate). Write the repeating units of the polymers.

3. Write equations for three ways to synthesize poly(hydroquinone terephthalate).

4. The nucleic acids DNA and RNA (see Chapter 17) are polyesters of phosphoric acid (i.e., polyphosphates). Write a reaction for a laboratory preparation of a polyphosphate.

5. Although γ-butyrolactone resists polymerization, the following five-membered ring lactones polymerize readily. Explain. (H. K. Hall Jr., et al., *Macromolecules*, **4**, 142, 1971; *J. Am. Chem. Soc.*, **80**, 6412, 1958)

6. The following lactones undergo base-catalyzed polymerization to yield high molecular weight polyester. Write the repeating units, including end groups, when

[18]OH$^-$ is used as initiator. Why is chain transfer not a problem with these monomers? (R. P. J. G. Thiebaut et al., *Ind. Plastiques Mod.* (Paris), **14**(2), 13, 1962; *Chem. Abstr.*, **57**, 6118e, 1962; H. K. Hall Jr., et al., *Macromolecules* **2**, 475, 1969)

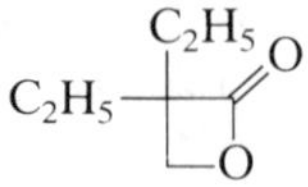

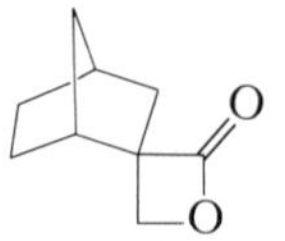

7. Monomer 19 (reaction 12.37) undergoes some vinyl polymerization in competition with free radical ring-opening polymerization (W. J. Bailey et al., *J. Macromol. Sci.-Chem.*, **A21**, 1611 1984). (a) Give the structure of the expected vinyl polymer. (b) If one of the ring hydrogens of **19** is replaced with a phenyl, the monomer undergoes only ring-opening polymerization. Give a plausible explanation. (c) Under what other conditions might you expect polymerization to occur? Explain. (d) In principle, the polyester formed in reaction (12.37) could be prepared by step-reaction polymerization of 4-hydroxybutanoic acid. Can you foresee any difficulties with this approach? Explain.

8. Give the structure of the polymer expected from the anionic polymerization of

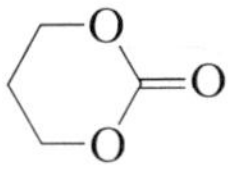

Write a mechanism for the polymerization. (ref. 21)

9. Metalloporphyrin initiators bring about the copolymerization of epoxides with carbon dioxide to form polycarbonates (see reaction 12.26). Additionally they initiate alternating copolymerization of epoxides with cyclic anhydrides (refs. 32a, 33). What type of polymer would be formed in the latter case? Explain.

10. Reaction (12.44) (Table 12.4) is an application of the Tischenko reaction in which the basic catalyst promotes intermolecular hydride transfer. Show mechanistically how the ester linkage is formed using terephthalaldehyde as monomer. What other polyester repeating unit might you expect to be present besides the one shown? Why wouldn't the reaction work equally as well for an aliphatic dialdehyde such as succinaldehyde?

11. Write the repeating units expected from the following polymerization reactions:

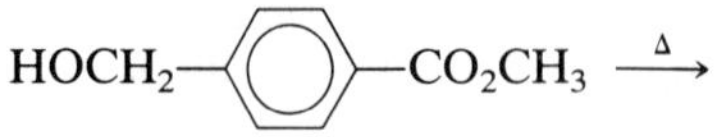

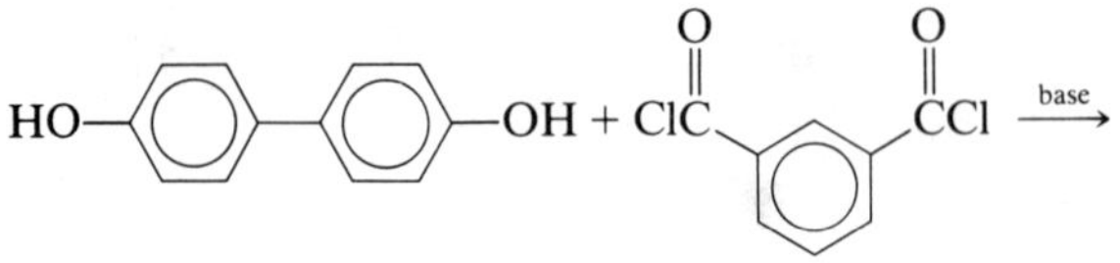

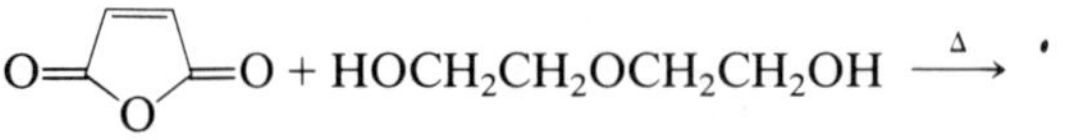

(d)

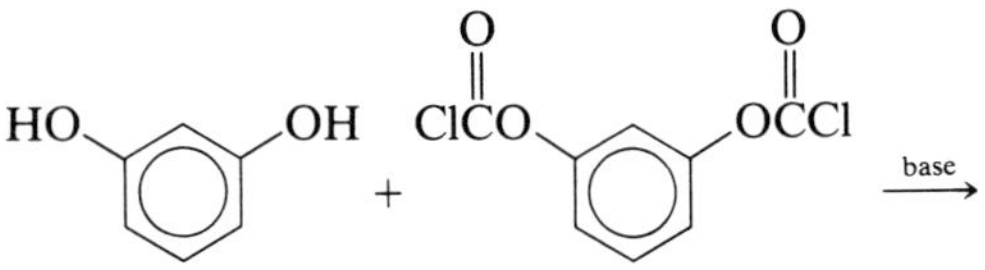

(e)

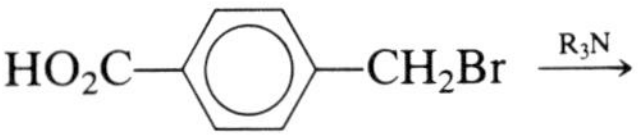

(f) 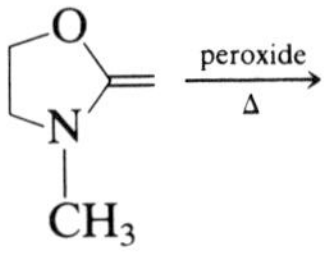
(ref. 32c)

(g) 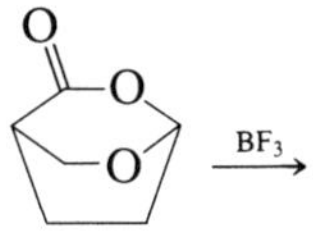
(M. Okada et al., *Macromolecules*, **19**, 503, 1986)

(h) 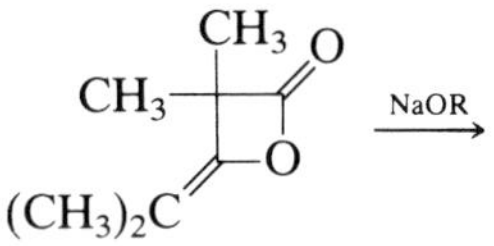
(R. H. Hasek et al., *J. Org. Chem.*, **27**, 60, 1962)

(i) 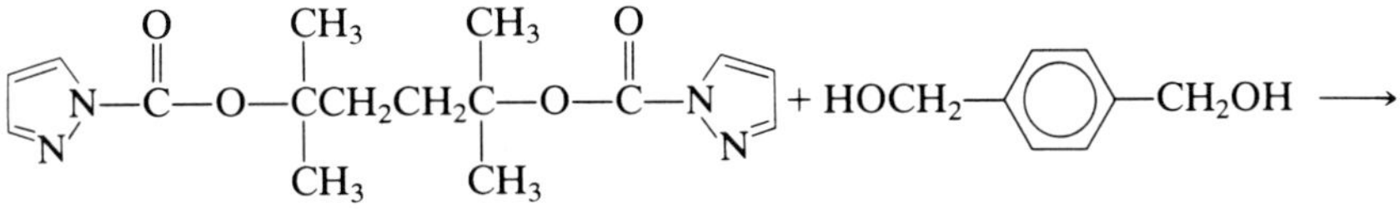
(F. M. Houlihan et al., *Macromolecules*, **19**, 13, 1986)

(j) 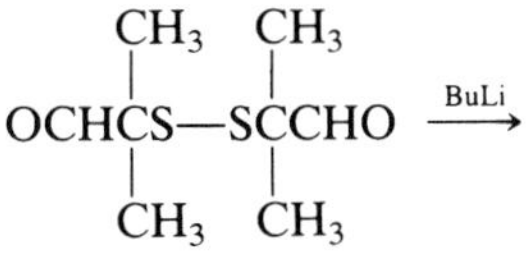
(ref. 43)

(k) 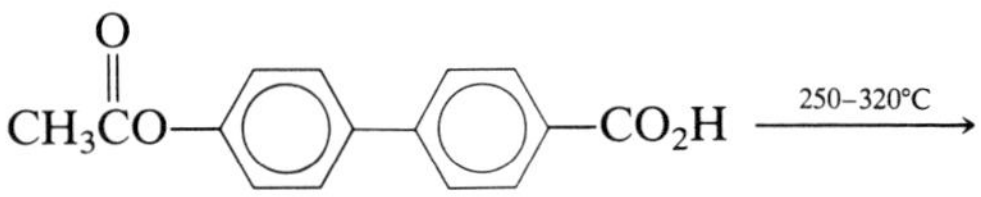
(G. Schwarz and H. R. Kricheldorf, *Makromol. Chem., Rapid Commun.*, **9**, 717, 1988)

(l) 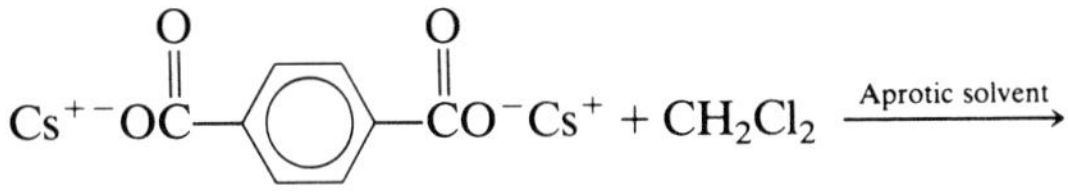
(A. L. Cimecioglu et al., *J. Polym. Sci. A: Polym. Chem.*, **26**, 2129, 1988)

(m)

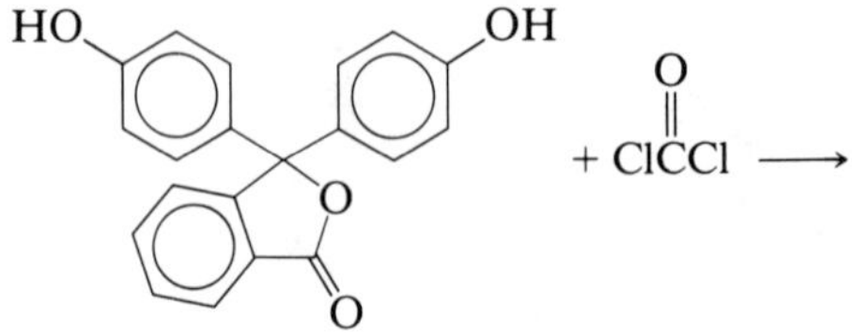

(J. M. Mikroyannidis, *Eur. Polym. J.*, **22**, 185, 1986)

12. Referring to exercise 11(g), what stereochemistry would be expected in the polymer if ring opening involved (a) acyl–oxygen cleavage, and (b) alkyl–oxygen cleavage of the ester group?

13. The *t*-butyloxycarbonyl (BOC) group is commonly used to protect amines (e.g., in proteins) and alcohols. The BOC group may be removed by heating:

$$RX\overset{\overset{\textstyle O}{\|}}{C}OC(CH_3)_3 \xrightarrow{\Delta} RXH + CO_2 + CH_2{=}C(CH_3)_2$$

(X = NH or O)

(a) Predict the outcome of heating the polymer prepared in exercise 11(i).
(b) Does this exercise suggest a possible alternate synthesis of poly(vinyl alcohol)? (See Chapter 9.) Explain. (Y. Yiang et al., *Polym. Bull.*, **17**, 1, 1987; H. W. Gibson and P. R. Kurek, *Polym. Commun.*, **28**, 97, 1987)

14. Write plausible mechanisms for reactions (12.41) and (12.45) (Table 12.4). Suggest alternative routes to the two polymers.

15. A practical route to unsaturated polyesters has been developed that makes use of scrap poly(ethylene terephthalate) bottles. The method involves digestion of the bottles with propylene glycol followed by reaction with maleic anhydride. Write reactions for the process.

13
Polyamides and related polymers

13.1 Introduction

Historically and commercially, polyamides [1-4] occupy an important place in the world of polymers. The early work of Carothers at du Pont dealt with both polyesters and polyamides, and emphasis on the latter led to the first commercial development of synthetic fibers, known commonly as nylons. Nylon fibers characteristically have good tensile strength and elasticity and are widely used in the manufacture of clothing, tire cord, carpets, and rope. In plastic form polyamides are used in making such items as gears, piping, wire insulation, zippers, and brush bristles.

Table 13.1 lists the numerous polyamides for fibers or plastics applications that have been made available commercially. As impressive as the list is, however, it should be stressed that a large percentage of the total production (about 80%) comprises just two polymers, nylons 6 and 66, the remainder being reserved for more specialized applications. (Refer to Chapter 1 for the nylon numbering system.) Not all those listed enjoy worldwide production; nylons 7 and 9, for example, have been produced only in the Soviet Union, and nylon 8 has never gained a sound commercial foothold.

According to the data given in Chapter 1, nylons represent roughly 25% of total fibers production and about 40% of all engineering plastics. Of all the nylon polymers produced, between 80 and 90% are used as fibers.

Ordinary polyamides, having the functional group

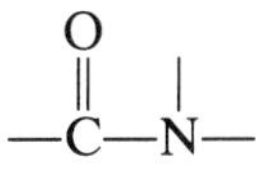

$$\begin{matrix} & O & \\ & \| & | \\ -&C&-N- \end{matrix}$$

in the polymer backbone are covered in Section 13.2 of this chapter. Also included are polyureas, which, like polycarbonates, are derived from the simplest dibasic acid, carbonic acid. Natural polyamides (proteins) are discussed in Chapter 17.

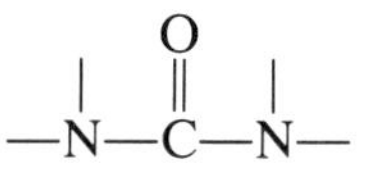

$$\begin{matrix} & & O & & \\ | & & \| & & | \\ -&N&-C&-N&- \end{matrix} \quad \text{Urea}$$

421

Table 13.1. Polyamides developed for commercial use[a]

Structure	Generic and/or common name[b]	Type[c]
$[NH(CH_2)_5C{=}O]$	Nylon 6 (polycaprolactam)	F, P
$[NH(CH_2)_6C{=}O]$	Nylon 7 [poly(7-heptanoamide)]	F, P
$[NH(CH_2)_7C{=}O]$	Nylon 8 (polycapryllactam)	F, P
$[NH(CH_2)_8C{=}O]$	Nylon 9 [poly(9-nonanoamide)]	F
$[NH(CH_2)_{10}C{=}O]$	Nylon 11 [poly(11-undecanoamide)]	P
$[NH(CH_2)_{11}C{=}O]$	Nylon 12 (polylauryllactam)	P
$[NH(CH_2)_4NHC(CH_2)_4C]$ (with two $=O$)	Nylon 46 [poly(tetramethylene adipamide)]	F, P
$[NH(CH_2)_6NHC(CH_2)_4C]$ (with two $=O$)	Nylon 66 [poly(hexamethylene adipamide)]	F, P
$[NH(CH_2)_6NHC(CH_2)_7C]$ (with two $=O$)	Nylon 69 [poly(hexamethylene azelamide)]	P
$[NH(CH_2)_6NHC(CH_2)_8C]$ (with two $=O$)	Nylon 610 [poly(hexamethylene sebacamide)]	P

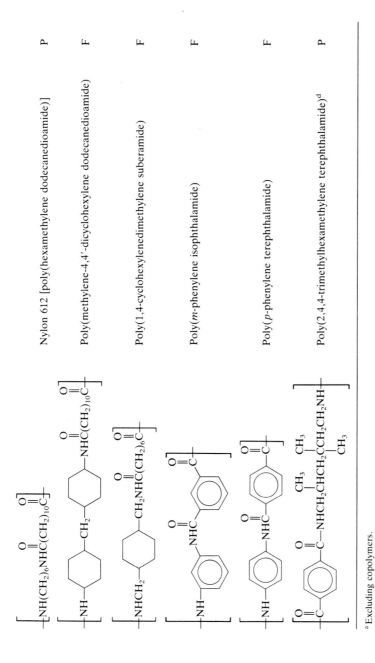

Nylon 612 [poly(hexamethylene dodecanedioamide)] P

Poly(methylene-4,4'-dicyclohexylene dodecanedioamide) F

Poly(1,4-cyclohexylenedimethylene suberamide) F

Poly(m-phenylene isophthalamide) F

Poly(p-phenylene terephthalamide) F

Poly(2,4,4-trimethylhexamethylene terephthalamide)[d] P

[a] Excluding copolymers.

[b] Nylons prepared from lactams are named accordingly; others are prepared by polycondensation.

[c] F = fiber, P = plastic.

[d] The 2,2,4-trimethyl isomer is also used.

A number of polymers related in structure to polyamides have also been developed, including polyhydrazides, polyurethanes, and polyimides. The characteristic functional groups of these polymers, all of which are covered in Section 13.3, are as follows:

$$
\begin{array}{c} \text{O} \\ \| \\ \text{—C—NH—NH—} \end{array} \qquad \text{Hydrazide}
$$

$$
\begin{array}{c} \text{O} \\ \| \quad | \\ \text{—O—C—N—} \end{array} \qquad \text{Urethane}
$$

$$
\begin{array}{c} \text{O} \quad\ \text{O} \\ \| \quad | \quad \| \\ \text{—C—N—C—} \end{array} \qquad \text{Imide}
$$

All three have achieved commercial importance.

As was mentioned earlier, there has been considerable interest in recent years in aromatic polymers for applications requiring good thermal stability. As a result, a number of interesting heterocyclic polymers have been developed. Several of these, which are obtainable from polyamides or from polymers of related structure, are discussed at the end of this chapter.

13.2   Polyamides

13.2.1   *Preparation of polyamides by polycondensation reactions*

There are three common condensation reactions employed in synthesis of amides: the reaction of acid chlorides with amines (13.1), dehydration of amine salts of carboxylic acids (13.2), and, less commonly, aminolysis of esters (13.3). All three methods have been extended to polyamide synthesis.

$$
\begin{array}{c} \text{O} \\ \| \\ \text{R—C—Cl} + \text{R}_2'\text{NH} \end{array} \longrightarrow \begin{array}{c} \text{O} \\ \| \\ \text{R—C—NR}_2' \end{array} + \text{HCl} \qquad (13.1)
$$

$$
\begin{array}{c} \text{O} \\ \| \\ \text{R—C—O}^-\overset{+}{\text{N}}\text{H}_2\text{R}_2' \end{array} \overset{\text{heat}}{\longrightarrow} \begin{array}{c} \text{O} \\ \| \\ \text{R—C—NR}_2' \end{array} + \text{H}_2\text{O} \qquad (13.2)
$$

$$
\begin{array}{c} \text{O} \\ \| \\ \text{R—C—OR''} + \text{R}_2'\text{NH} \end{array} \longrightarrow \begin{array}{c} \text{O} \\ \| \\ \text{R—C—NR}_2' \end{array} + \text{R''OH} \qquad (13.3)
$$

Direct amidation between dibasic acid and diamine requires the usual stoichiometric balance, but since the reaction proceeds via the intermediate ammonium salt, purification of the salt guarantees this balance. Nylon 66

(**1**, $x = 4$, $y = 6$) originally developed by Carothers, is manufactured this way (13.4), as is the more recently developed nylon 46[5] (**1**, $x = 4$, $y = 4$). Polymerization temperatures in the range of 260 to 280°C are used to obtain

$$HO_2C(CH_2)_xCO_2H + H_2N(CH_2)_yNH_2 \longrightarrow$$

$$[^-O_2C(CH_2)_xCO_2^-][H_3\overset{+}{N}(CH_2)_y\overset{+}{N}H_3] \xrightarrow{\text{heat}}$$

$$\left[\begin{array}{c} O \quad\quad O \\ \parallel \quad\quad \parallel \\ -C(CH_2)_x CNH(CH_2)_y NH- \end{array}\right] + 2H_2O \quad (13.4)$$

1

practical molecular weights. This method is also applied to amino acids as long as the two functional groups are separated by at least six carbon atoms to prevent cyclization to lactam. An example is the formation of nylon 11 (**2**) from 11-aminoundecanoic acid (13.5).* Nylons 7 and 9 are made similarly.

$$H_2N(CH_2)_{10}CO_2H \xrightarrow{\text{heat}} \left[\begin{array}{c} O \\ \parallel \\ -NH(CH_2)_{10}C- \end{array}\right] + H_2O \quad (13.5)$$

2

Low-temperature solution methods have been developed to prepare polyamides from diacids and diamines or from amino acids, among them the important *phosphorylation* reaction[6] (13.6). It has been proposed that an intermediate phosphorus complex (**3**) is formed, which undergoes rapid displacement by amine. The more basic aliphatic diamines are not suitable for this reaction because they displace pyridine in the complex. Another example[7] is the room-temperature synthesis of poly(*p*-benzamide) (**4**) using picryl chloride in pyridine (13.7).

---

* Nylon 11 is produced by the French company Aquitaine-Organico. The monomer, 11-aminoundecanoic acid, is prepared by an interesting process that begins with castor oil. Ricinoleic acid, obtained by hydrolysis of the oil, is thermally degraded to undecylenic acid; the latter is converted to 11-bromoundecanoic acid by free radical hydrobromination, and this, in turn, is converted to the corresponding amino acid with ammonia.

$$CH_3(CH_2)_5\overset{\overset{\displaystyle OH}{|}}{C}HCH=CH(CH_2)_7CO_2H \xrightarrow{\text{heat}} CH_2=CH(CH_2)_8CO_2H + C_6H_{13}CHO$$

$$\downarrow \text{HBr, peroxide}$$

$$H_2N(CH_2)_{10}CO_2H \xleftarrow{NH_3} BrCH_2CH_2(CH_2)_8CO_2H$$

The polymer is sold under the trade name Rilsan 11.

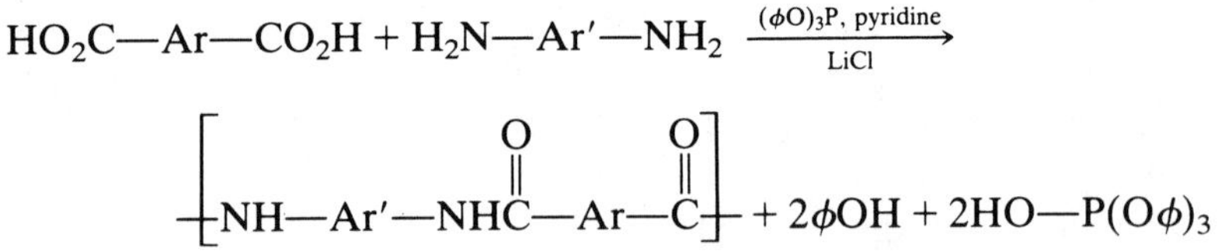

$$HO_2C—Ar—CO_2H + H_2N—Ar'—NH_2 \xrightarrow[LiCl]{(\phi O)_3P,\ pyridine}$$

$$\left[NH—Ar'—NH\overset{O}{\underset{\|}{C}}—Ar—\overset{O}{\underset{\|}{C}}\right] + 2\phi OH + 2HO—P(O\phi)_3 \quad (13.6)$$

$$(\phi O)_2PH—\overset{O}{\underset{\|}{C}}—Ar—\overset{O}{\underset{\|}{C}}—PH(O\phi)_2$$

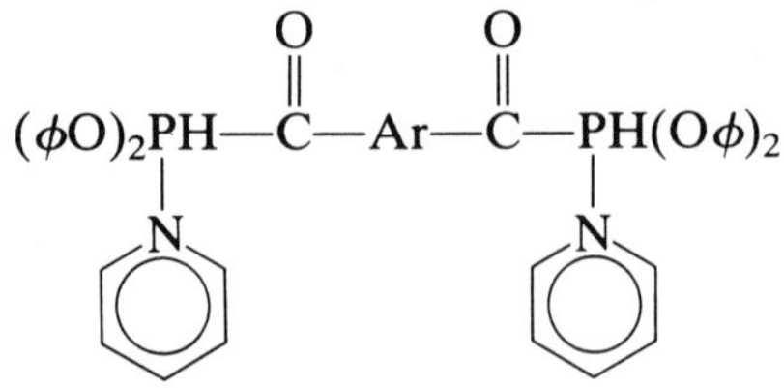

3

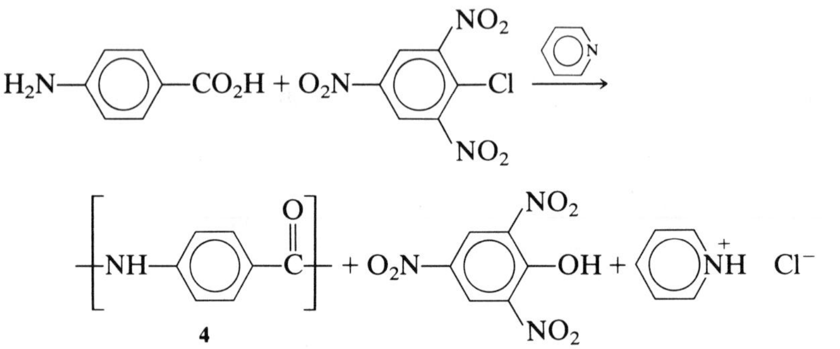

$$\left[NH—\overset{O}{\underset{\|}{C}}\right] + O_2N—C_6H_2(NO_2)_2—OH + \text{pyridinium } Cl^- \quad (13.7)$$

4

Although the acid chloride route (13.1) is most versatile for preparing polyamides,[8a,9a] cost considerations have limited its use primarily to aromatic polyamides,[10,11a,12,13a] or *aramids*, as they are now called. Both solution and interfacial techniques are used. A case in point is the synthesis (13.8) of poly(*m*-phenylene isophthalamide) (**5**) (marketed under the du Pont trade name Nomex). It may be recalled from Chapter 3 (Section 3.7) that the *para,para* isomer of **5** is spun into extremely high-modulus fibers from a lyotropic liquid crystalline solution in sulfuric acid (du Pont's Kevlar).

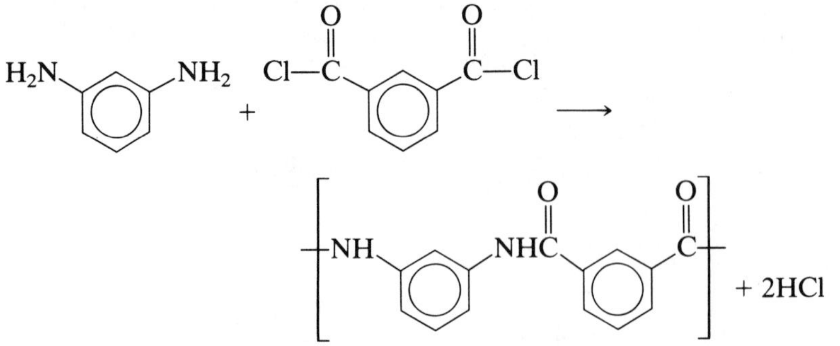

$$\left[NH—C_6H_4—NH\overset{O}{\underset{\|}{C}}—C_6H_4—\overset{O}{\underset{\|}{C}}\right] + 2HCl \quad (13.8)$$

5

Polysulfonamides and polyphosphonamides are also synthesized from the corresponding acid chlorides,[8b] as shown for poly(hexamethylene-*m*-benzenedisulfonamide) (**6**) (13.9) and poly(hexamethylene phosphonamide) (**7**) (13.10). In general, the polyphosphonamide synthesis is less successful because of differences in reactivity between the two chlorine atoms and because of a tendency to form cyclic structures.

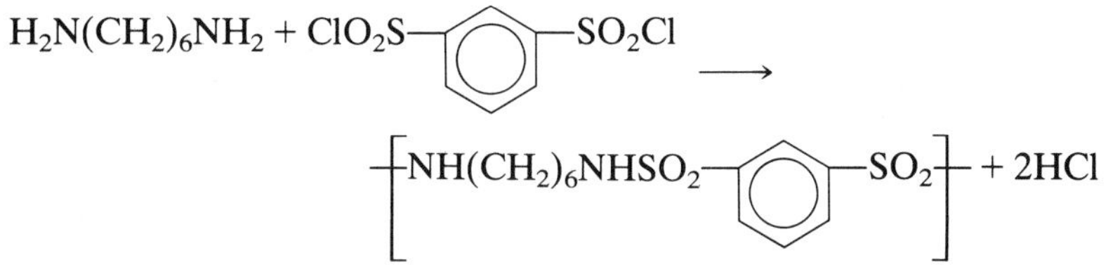

$$H_2N(CH_2)_6NH_2 + ClO_2S\!-\!\!\bigcirc\!\!-\!SO_2Cl \longrightarrow$$

$$\left[NH(CH_2)_6NHSO_2\!-\!\!\bigcirc\!\!-\!SO_2\right] + 2HCl \quad (13.9)$$

6

$$H_2N(CH_2)_6NH_2 + Cl\!-\!\overset{\overset{O}{\|}}{\underset{\bigcirc}{P}}\!-\!Cl \longrightarrow \left[\overset{\overset{O}{\|}}{\underset{\bigcirc}{P}}\!-\!NH(CH_2)_6NH\right] + 2HCl$$

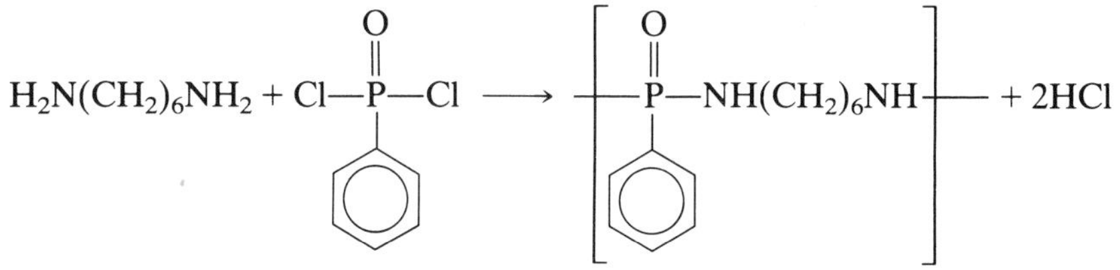

7 (13.10)

Aminolysis of diesters is mainly useful for preparing polyamides from thermally unstable diacids, such as malonic and oxalic acids (13.11).[8a,14] If the esters are sufficiently reactive (for example, the 2,4-dinitrophenyl ester), polymerization occurs at room temperature.[15]

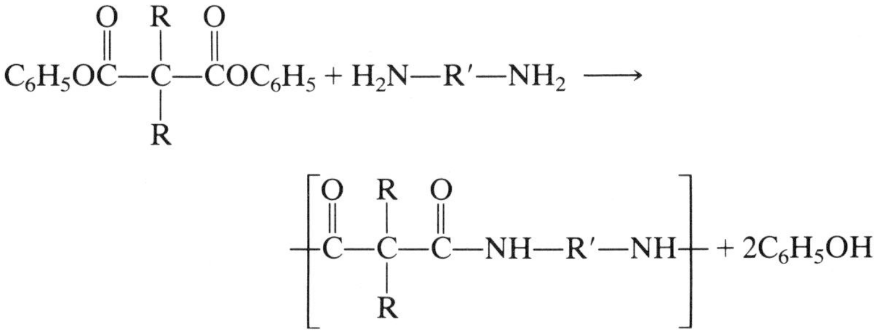

$$C_6H_5O\overset{\overset{O}{\|}}{C}\!-\!\overset{\overset{R}{|}}{\underset{R}{C}}\!-\!\overset{\overset{O}{\|}}{C}OC_6H_5 + H_2N\!-\!R'\!-\!NH_2 \longrightarrow$$

$$\left[\overset{\overset{O}{\|}}{C}\!-\!\overset{\overset{R}{|}}{\underset{R}{C}}\!-\!\overset{\overset{O}{\|}}{C}\!-\!NH\!-\!R'\!-\!NH\right] + 2C_6H_5OH \quad (13.11)$$

13.2.2 Polymerization of lactams

Ring-opening polymerization of lactams[16,17a] is a commercially important process, more so than the analogous polymerization of lactones to give polyesters. Most important is the polymerization of caprolactam (**8**) (13.12) to give nylon 6 (**9**). The reaction can be brought about with cationic or anionic initiators or with water.

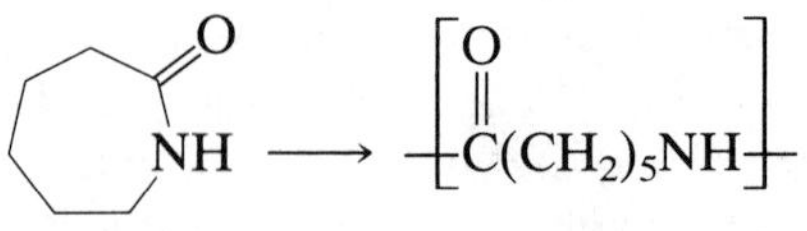

$$\text{8} \qquad \longrightarrow \qquad \left[\text{C(CH}_2)_5\text{NH}\right] \qquad (13.12)$$

8   **9**

Cationic polymerization[18,19] with acid, HA, is believed to involve initial protonation to give both O-protonated and N-protonated amide (13.13), mainly the former because of resonance stabilization. Nucleophilic attack by monomer on the more reactive N-protonated acid (13.14), followed by successive ring-opening acylation of the primary amine function, results in formation of polyamide (13.15).

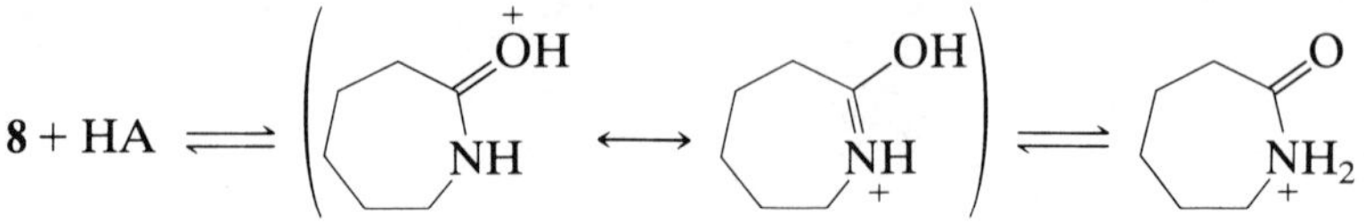

$$\text{8} + \text{HA} \rightleftharpoons \left(\begin{array}{cc} & \end{array} \right) \rightleftharpoons \qquad (13.13)$$

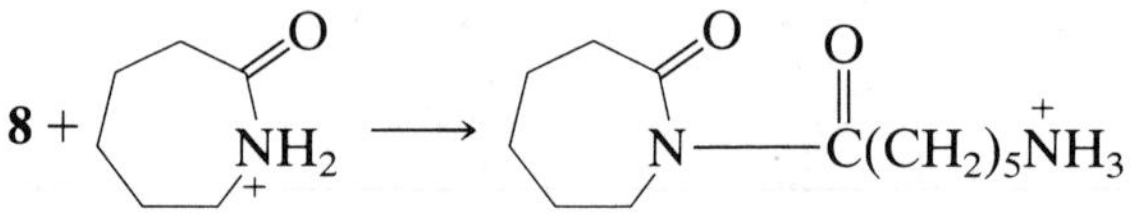

$$\text{8} + \qquad \longrightarrow \qquad (13.14)$$

$$ + n\ \text{8} \longrightarrow $$

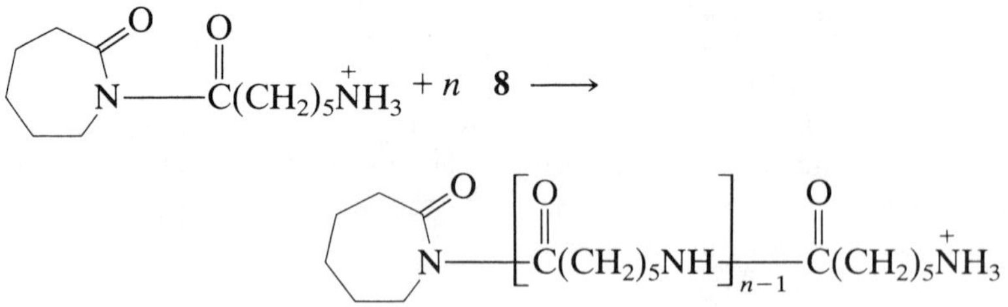

$$ (13.15) $$

Generally, high molecular weights are not obtainable under cationic conditions, possibly because of intramolecular cyclization reactions involving terminal amino groups (for example, 13.16).

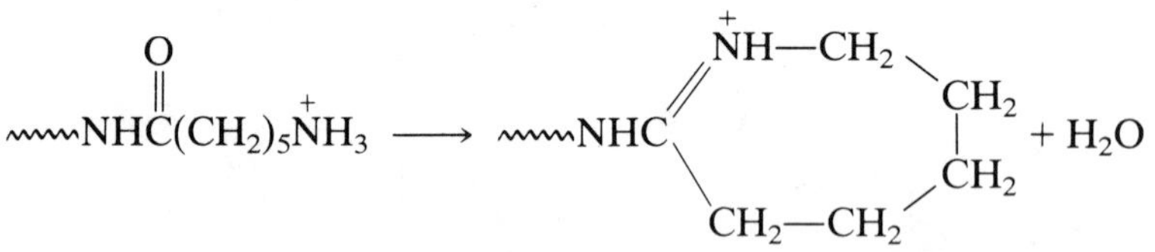

$$ \text{wwwNHC(CH}_2)_5\overset{+}{\text{NH}}_3 \longrightarrow \text{wwwNHC} \qquad + \text{H}_2\text{O} \quad (13.16) $$

Strong bases such as sodium hydride, sodium amide, or alkali metals are effective anionic initiators. The mechanism, using base B:⁻, involves initial formation of resonance-stabilized anion (13.17) which reacts with monomer in a slow step (13.18), slow because the new anion is not stabilized by resonance. Once formed, however, it undergoes rapid proton transfer (13.19) and propagation (13.20) by successive formation of resonance-stabilized anions.

$$8 + B:^- \; \rightleftharpoons \; \left(\underset{\substack{\\ N:_-}}{\overset{\substack{O}}{\bigcirc}} \longleftrightarrow \underset{\substack{\\ N}}{\overset{\substack{O^-}}{\bigcirc}} \right) + BH \qquad (13.17)$$

$$\underset{N:^-}{\overset{O}{\bigcirc}} + 8 \; \overset{slow}{\longrightarrow} \; \underset{N}{\overset{O}{\bigcirc}}\!\!-\!\!\overset{O}{\overset{\|}{C}}(CH_2)_5\ddot{N}H \qquad (13.18)$$

$$\overset{8}{\underset{fast}{\rightleftharpoons}} \; \underset{N}{\overset{O}{\bigcirc}}\!\!-\!\!\overset{O}{\overset{\|}{C}}(CH_2)_5NH_2 + \underset{N:^-}{\overset{O}{\bigcirc}} \qquad (13.19)$$

$$\overset{fast}{\longrightarrow} \; \underset{N}{\overset{O}{\bigcirc}}\!\!-\!\!\overset{O}{\overset{\|}{C}}(CH_2)_5\ddot{N}\!\!-\!\!\overset{O}{\overset{\|}{C}}(CH_2)_5NH_2 \qquad (13.20)$$

Because the second step is slow, this type of ring-opening polymerization is characterized by an induction period. The induction period can, however, be eliminated by addition of coinitiators (called *activators*) such as acyl halides or anhydrides that ensure formation of a stabilized anion in the second step. This is illustrated in reactions (13.21) and (13.22) using an acid chloride.

$$\underset{N:^-}{\overset{O}{\bigcirc}} + R\!\!-\!\!\overset{O}{\overset{\|}{C}}\!\!-\!\!Cl \longrightarrow \underset{N}{\overset{O}{\bigcirc}}\!\!-\!\!\overset{O}{\overset{\|}{C}}\!\!-\!\!R + Cl^- \qquad (13.21)$$

$$\underset{N}{\overset{O}{\bigcirc}}\!\!-\!\!\overset{O}{\overset{\|}{C}}\!\!-\!\!R + \underset{N:^-}{\overset{O}{\bigcirc}} \longrightarrow \underset{N}{\overset{O}{\bigcirc}}\!\!-\!\!\overset{O}{\overset{\|}{C}}(CH_2)_5\ddot{N}\overset{O}{\overset{\|}{C}}R \qquad (13.22)$$

One of the difficulties associated with anionic lactam polymerization is that proton transfer can occur with the polymer chain, particularly at low monomer concentrations. This results in chain branching and a polydisperse polymer.

Water is also used to initiate polymerization, a process referred to as *hydrolytic polymerization*. It is believed that initially hydrolysis occurs to form amino acid (13.23) which reacts with lactam by ring cleavage (13.24). Although some polymerization undoubtedly occurs through the free amino

$$8 + H_2O \longrightarrow HO_2C(CH_2)_5NH_2 \qquad (13.23)$$

$$HO_2C(CH_2)_5NH_2 + \underset{NH}{\overset{O}{\bigcirc}} \longrightarrow HO_2C(CH_2)_5NH\overset{O}{\overset{\|}{C}}(CH_2)_5NH_2$$

$$(13.24)$$

and carboxyl groups of the intermediate amino acid, the primary propagation process is ring-opening.

Besides nylon 6 from anionic or hydrolytic polymerization of caprolactam,* nylon 8 and nylon 12 are made by ring-opening polymerization of capryllactam (**10**) and lauryllactam (**11**).† The five-membered butyrolactam (**12**) (commonly called 2-pyrrolidone) polymerizes slowly under the usual anionic conditions to yield nylon 4; however, nylon 4 having $\bar{M}_w$ greater than 1 million is formed rapidly under conditions of phase-transfer catalysis.[20] Valerolactam (**13**) (also called α-piperidone) forms relatively low-molecular-weight nylon 5 slowly under basic conditions. Both nylons 4 and 5 revert to monomer at elevated temperatures.

* Caprolactam is extensively used because its polymer has excellent fiber properties and because the monomer is made cheaply from phenol by the following reaction sequence:

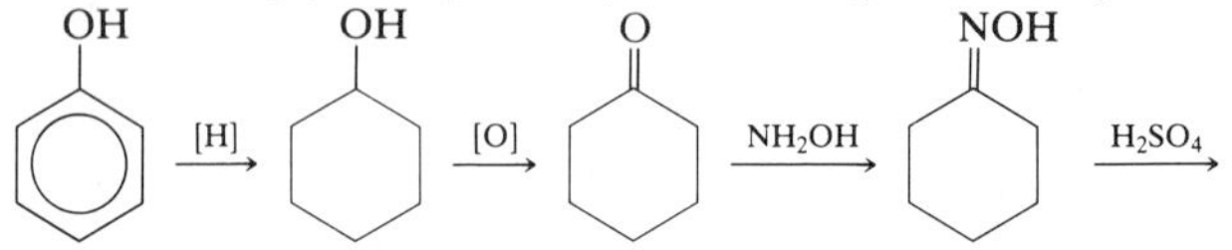

8

The last step is an example of a Beckmann rearrangement. Other methods involve converting cyclohexanone to caprolactone using peracetic acid, followed by reaction of the cyclic ester with ammonia to give lactam; forming cyclohexanone oxime directly from cyclohexane by a light-catalyzed process; and forming caprolactam by an acetyl exchange reaction between cyclohexanone oxime and acetyl caprolactam. The resultant acetylcyclohexanone oxime is then converted to acetyl caprolactam for recycling:

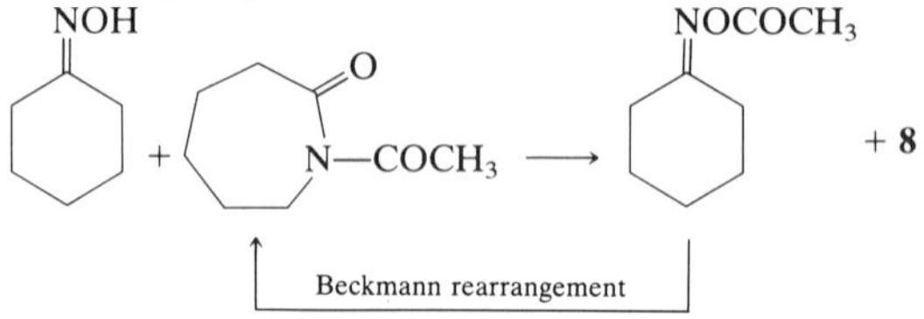

† Both capryllactam and lauryllactam are derived from butadiene according to the following schemes:

$$2CH_2{=}CH{-}CH{=}CH_2 \xrightarrow{\text{catalyst}}$$

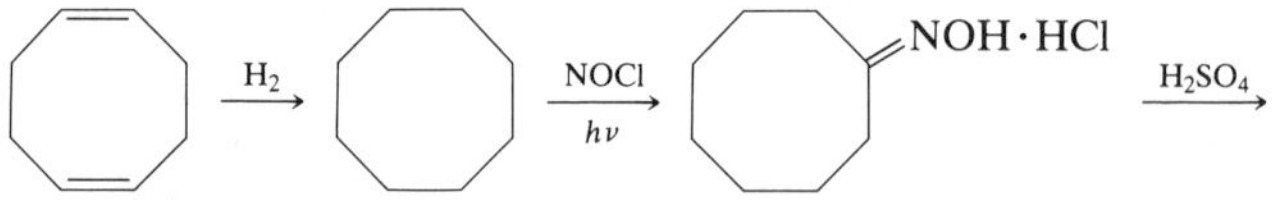

10

$$3CH_2{=}CH{-}CH{=}CH_2 \xrightarrow{\text{catalyst}}$$

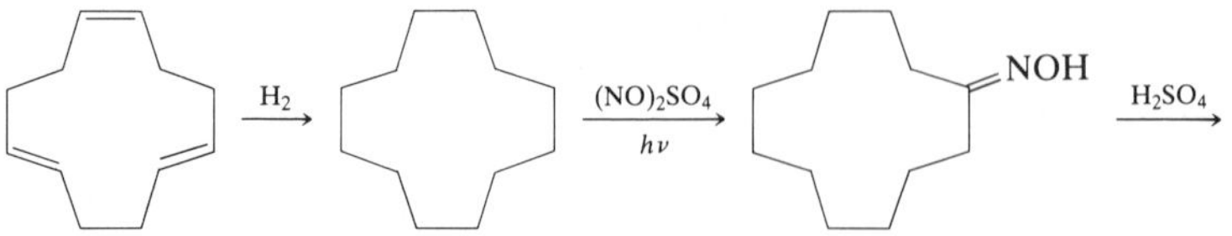

11

Lauryllactam is produced by the French company Aquitaine-Organico, which markets nylon 12 under the trade name Rilsan 12. The Japanese firm, Mitsubishi Petrochemical Company, produces nylon 12 by a similar route.

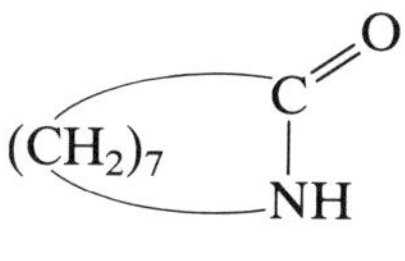

$(CH_2)_7$ — C=O —NH

10

$(CH_2)_{11}$ — C=O —NH

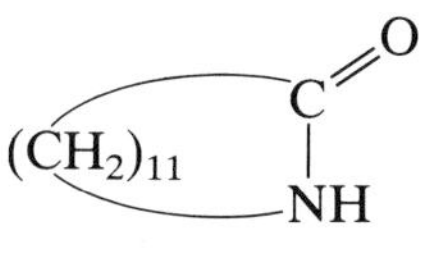

11

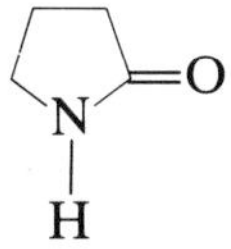

12

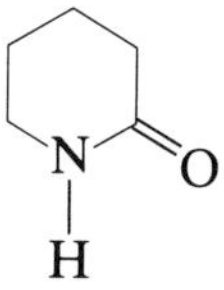

13

Substituent groups on lactam rings, particularly if they are on the nitrogen atom, tend to reduce the activity of the monomers. In the case of β-lactams, however, substituents are helpful in that the parent monomer is very difficult to synthesize without them.[21a] As an example, α,α-diethyl-β-propiolactam (**14**) may be prepared in high yield from the corresponding amino ester (13.25). The disubstituted nylon 3 (**15**) is prepared from **14** under anionic conditions (13.26). Cycloaddition of chlorosulfonylisocyanate (**16**) to alkenes (13.27) is an alternative route to β-lactams,[22] and is probably the best general method.

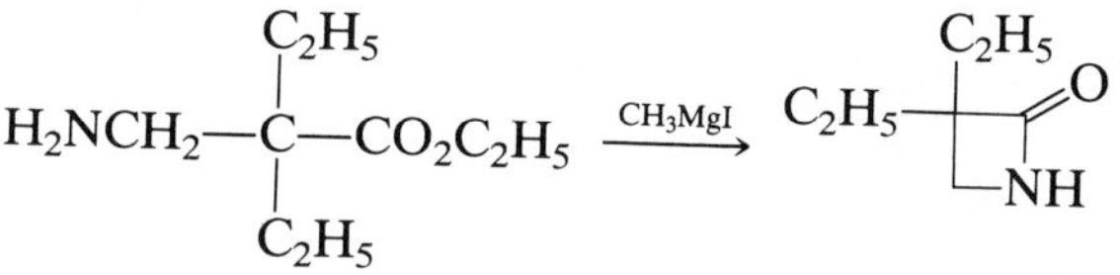

$$H_2NCH_2 - \underset{\underset{C_2H_5}{|}}{\overset{\overset{C_2H_5}{|}}{C}} - CO_2C_2H_5 \xrightarrow{CH_3MgI} \quad (13.25)$$

14

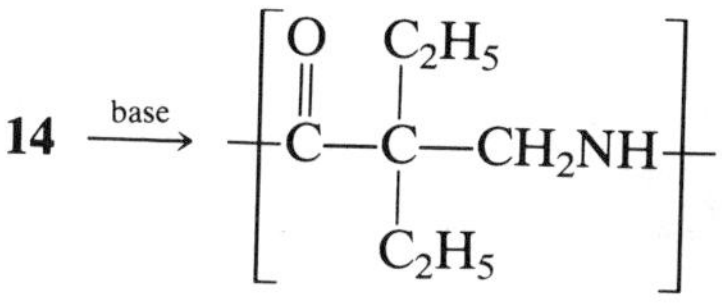

$$\mathbf{14} \xrightarrow{\text{base}} \left[\overset{O}{\overset{||}{C}} - \underset{\underset{C_2H_5}{|}}{\overset{\overset{C_2H_5}{|}}{C}} - CH_2NH \right] \quad (13.26)$$

15

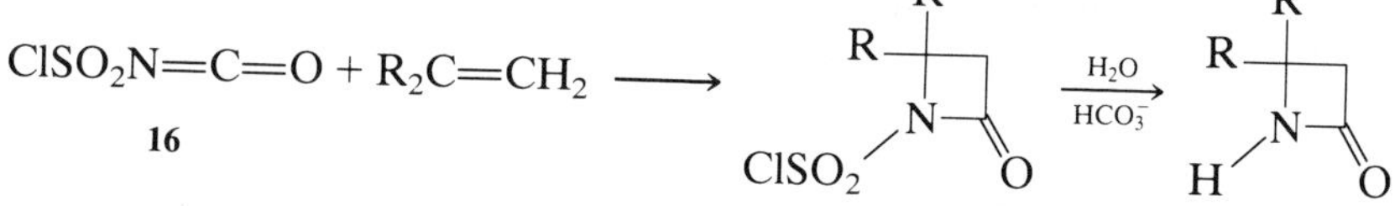

$$ClSO_2N{=}C{=}O + R_2C{=}CH_2 \longrightarrow \quad \xrightarrow[HCO_3^-]{H_2O} \quad (13.27)$$

16

Some other ring-opening reactions related to lactam polymerization include the base-catalyzed reaction of N-carboxy-α-amino acid anhydrides[17b,23–25] (**17**) (prepared from α-amino acid and phosgene) to give the corresponding nylon 2 (**18**) (13.28), and the preparation of a polysulfonamide **20** from propiosultam (**19**) (13.29).[21b] The former reaction (13.28) is of particular interest to protein chemists because the products, when prepared from naturally occurring α-amino acids, provide useful models for protein studies.

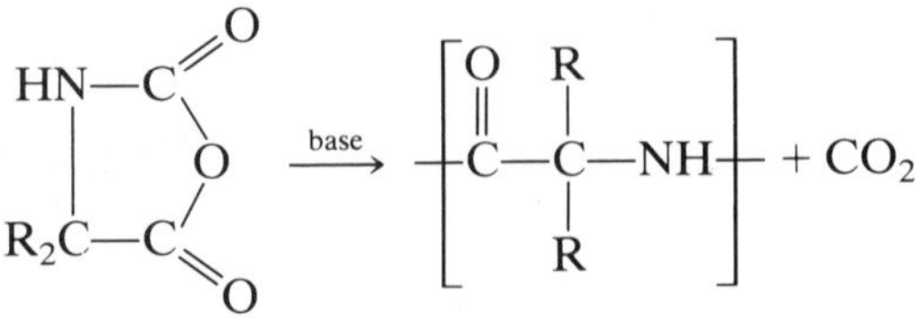

$$
\underset{\textbf{17}}{\left. \begin{array}{c} HN-C \\ | \\ R_2C-C \end{array} \right.} \quad \xrightarrow{\text{base}} \quad \underset{\textbf{18}}{\left[\begin{array}{cc} O & R \\ || & | \\ C-C-NH \\ & | \\ & R \end{array} \right]} + CO_2 \qquad (13.28)
$$

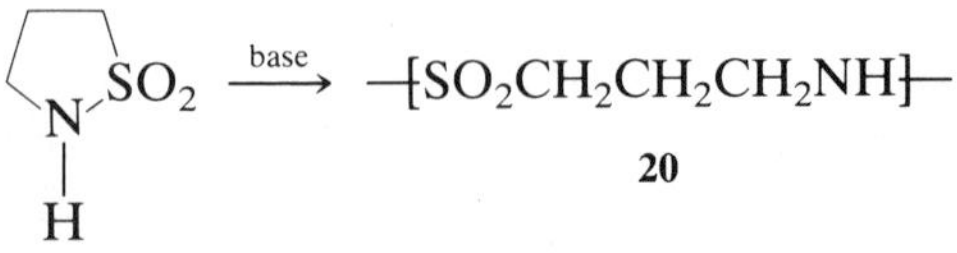

$$
\underset{\textbf{19}}{\left. \begin{array}{c} \\ N \\ | \\ H \end{array} SO_2 \right.} \xrightarrow{\text{base}} \underset{\textbf{20}}{-[SO_2CH_2CH_2CH_2NH]-} \qquad (13.29)
$$

13.2.3   Miscellaneous methods of preparing polyamides

Besides the polycondensation and ring-opening methods described above, a variety of other routes to polyamides have been described, several of which are given in Table 13.2. One of the most useful is the reaction of diisocyanates with dicarboxylic acids[26] (13.30); it is the only one of those listed that enjoys widespread commercial use. The carboxylic acid adds across the carbon–nitrogen double bond to give initially an unstable polymeric anhydride (**21**) which undergoes decarboxylation to form polyamide. If diisothiocyanate,

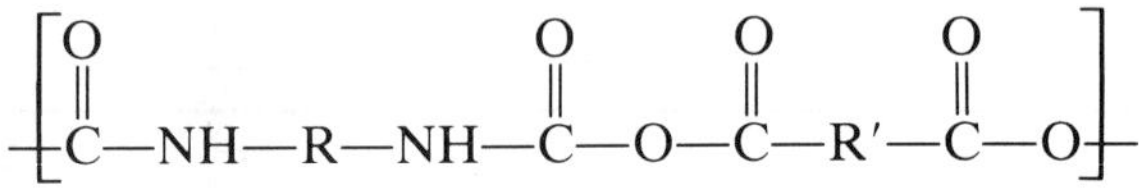

$$
\underset{\textbf{21}}{\left[\begin{array}{c} O \\ || \\ C-NH-R-NH-C-O-C-R'-C-O \\ \end{array} \right]}
$$

SCN—R—NCS, is used instead of diisocyanate, carbon oxysulfide is evolved; hence loss of carbon dioxide (or carbon oxysulfide) must involve the carbon attached to nitrogen. Use is made of this reaction in generating foams by treating carboxyl-terminated low-molecular-weight polyesters with diisocyanates such as 2,4-toluenediisocyanate (**22**) (13.39). The evolved carbon dioxide is trapped as bubbles in the viscous medium, and thereby produces the foam. Once set, such foams are flexible and find use in cushions and other upholstery. Sometimes they are referred to erroneously as "urethane" foams. More is said on this subject later in the chapter.

Table 13.2. Miscellaneous methods of synthesizing polyamides

Reaction	Number

$OCN-R-NCO + HO_2C-R'-CO_2H \longrightarrow$

$$\left[NH-R-NH-\overset{\overset{\displaystyle O}{\|}}{C}-R'-NH \right] + 2CO_2 \qquad (13.30)$$

$$OCN-R-NCO + CH_2{=}C(NR'_2)_2 \longrightarrow \left[\overset{\overset{\displaystyle O}{\|}}{C}NH-R-\overset{\overset{\displaystyle O}{\|}}{C}-\underset{\underset{\displaystyle C(NR'_2)_2}{\|}}{C} \right] \qquad (13.31)$$

$$NC-R-CN + HO-R'-OH^a \longrightarrow \left[NH-\overset{\overset{\displaystyle O}{\|}}{C}-R-\overset{\overset{\displaystyle O}{\|}}{C}-NH-R' \right] \qquad (13.32)$$

$$NC-R-CN + CH_2O^b \longrightarrow \left[NH-\overset{\overset{\displaystyle O}{\|}}{C}-R-\overset{\overset{\displaystyle O}{\|}}{C}-NH-CH_2 \right] \qquad (13.33)$$

$$RN{=}C{=}O \xrightarrow{\text{base}^c} \left[\underset{\underset{\displaystyle N}{|}}{\overset{\overset{\displaystyle R}{|}}{}}-\overset{\overset{\displaystyle O}{\|}}{C} \right] \qquad (13.34)$$

$$\overset{\overset{\displaystyle O}{\|}}{CH_2{=}CHCNH_2} \xrightarrow{\text{base}} \left[CH_2CH_2\overset{\overset{\displaystyle O}{\|}}{C}NH \right] \qquad (13.35)$$

$$H_2N-R-NH_2 + O{=}C{=}C{=}C{=}O \longrightarrow \left[NH-R-NH\overset{\overset{\displaystyle O}{\|}}{C}-CH_2-\overset{\overset{\displaystyle O}{\|}}{C} \right] \qquad (13.36)$$

$$H_2N-R-NH_2 + Ar\overset{\overset{\displaystyle O}{\|}}{C}\overset{\overset{\displaystyle O}{\|}}{C}-R'-\overset{\overset{\displaystyle O}{\|}}{C}\overset{\overset{\displaystyle O}{\|}}{C}Ar^d \longrightarrow$$

$$\left[\overset{\overset{\displaystyle O}{\|}}{C}-R-\overset{\overset{\displaystyle O}{\|}}{C}-NH-R'-NH \right] + 2ArCO_2H \qquad (13.37)$$

$$Br-Ar-Br + H_2N-Ar'-NH_2 + 2CO \xrightarrow[R_3N]{\text{Pt catalyst}}$$

$$\left[\overset{\overset{\displaystyle O}{\|}}{C}-Ar-\overset{\overset{\displaystyle O}{\|}}{C}-NH-Ar'-NH \right] \qquad (13.38)$$

[a] Both OH groups are tertiary.

[b] Trioxane is used as the source of CH_2O.

[c] Typical bases: RLi, ROLi, RNa, NaCN. Crystalline polymer is obtained at temperatures of -100 to $-80°C$.

[d] Ar = 2,4,6-trimethylphenyl.

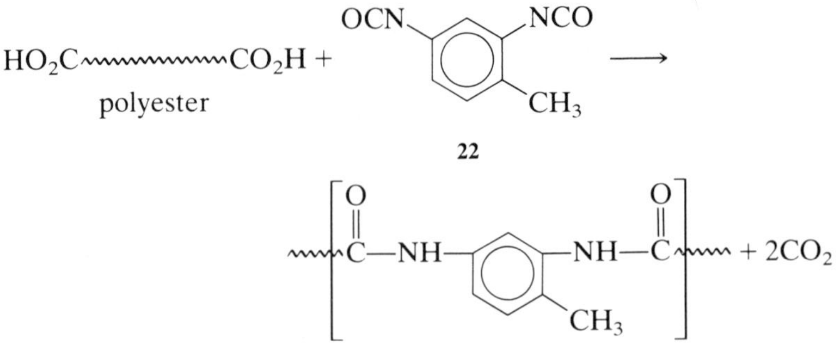

$$HO_2C\text{\Large\char`\~}CO_2H + \quad \underset{\text{\Large 22}}{\overset{\displaystyle OCN\underset{\qquad\qquad CH_3}{\underset{\displaystyle\bigcirc}{}}NCO}{}} \quad \longrightarrow$$

$$\left[\!\!\begin{array}{c} \underset{\displaystyle C}{\overset{\displaystyle O}{\|}}\!-\!NH\!-\!\underset{\displaystyle CH_3}{\bigcirc}\!-\!NH\!-\!\underset{\displaystyle C}{\overset{\displaystyle O}{\|}} \end{array}\!\!\right] + 2CO_2 \quad (13.39)$$

The second reaction (13.31), a more recent commercial development, involves a nucleophilic addition to the isocyanate to yield a product, **23**, which is capable of further addition by the same mechanism.[27] The commercial process involves the simultaneous reaction with a hydroxy-terminated prepolymer to form a urethane-amide copolymer suitable for RIM applications. The nucleophilic monomer (a *ketene aminal*) behaves like an enamine, but is highly reactive by virtue of having two nitrogen atoms to increase the nucleophilicity of the double bond.

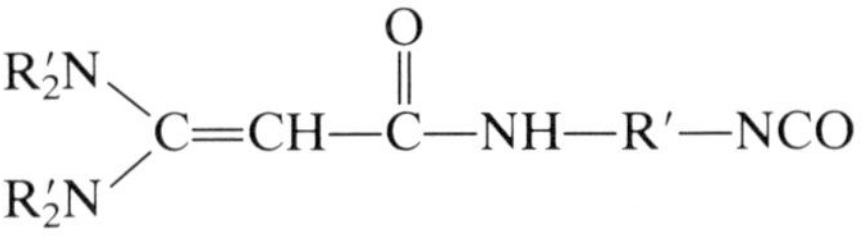

$$\underset{R_2'N}{\overset{R_2'N}{\diagdown\!\!\!\diagup}} C\!=\!CH\!-\!\overset{\displaystyle\overset{O}{\|}}{C}\!-\!NH\!-\!R'\!-\!NCO$$

$$\text{\Large 23}$$

The acid-catalyzed polyaddition of tertiary diols to dinitriles (13.32) (an extension of what is known as the Ritter reaction)[8a] provides a route to polyamides having substituents on the carbon adjacent to the amide nitrogen. The mechanism involves addition of an alcohol-derived carbocation to the carbon–nitrogen triple bond, hence any species capable of generating a carbocation (alcohol, alkene, aldehyde, alkyl halide) can in principle undergo this reaction. The following reaction (13.33) is a case in point. Reaction (13.33) leads to relatively high-molecular-weight polyamide of the inherently unstable diaminomethane.[8a]

Nylon 1 derivatives are obtained by the base-catalyzed homopolymerization of isocyanates (13.34).[28,29] Very-high-molecular-weight crystalline polymers are obtainable by this reaction; however, the polymers tend to depolymerize at elevated temperatures.

Nylon 3 is obtained by the base-catalyzed step-growth addition polymerization of polyacrylamide (13.35).[30,31] This Michael addition reaction is analogous to the formation of polyester from acrylic acid, as described in Chapter 12 (Table 12.4). Addition of diamines to carbon suboxide to form polymalonamides (13.36)[8a] also has its counterpart in polyester chemistry (Table 12.4). The polycondensation of diamine with dianhydride (13.37), is related to the acid chloride synthesis, and is likewise adaptable to both solution and interfacial techniques.[32] The last reaction in Table 13.2 (13.38) represents a novel and potentially attractive carbonylation route to aramids.[33]

13.2.4 Properties of polyamides

Like polyesters, polamides lend themselves to a wide variety of structural variations. They melt at much higher temperatures than polyesters, however, because of intermolecular hydrogen bonding. Nylon 66, for example, melts at about 265°C, whereas the corresponding poly(hexamethylene adipate) melts below 60°C. Table 13.3 lists the approximate melting points of several linear aliphatic polyamides, the first set representing those having one amide group per repeating unit, and the second set, two. Both series exhibit an odd-even relationship typical of homologous series. Polyamides having an even number of carbon atoms between amide groups tend to be higher melting, a phenomenon reflecting the better packing ability within the crystalline lattice. Similar behavior is seen with the partially aromatic poly(alkylene terephthalamides) in Table 13.4. Fully aromatic polyamides[10,11a,34a] containing *para-* or *meta*-phenylene linkages do not melt below 400°C, whereas those having *ortho* arrangements are considerably lower melting. The *ortho*-substituted polymers are also significantly less thermally stable, exhibiting up to 90% weight loss between 200 and 300°C, compared with the *m-* and *p*-substituted polymers, which are stable up to 400°C.[35]

A major problem with wholly aromatic polyamides having high softening temperatures (particularly those prepared with *p*-phenylene units) is their low solubility and resultant difficulty in fabrication. One approach to solving this

Table 13.3. Approximate melting points of aliphatic polyamides[a]

Nylon	Melting point (°C)	Nylon	Melting point (°C)
3	>320	46	295
4	265	56	230
5	250–270	66	265
6	233	76	220
7	233	68	240
8	200	69	205
9	188	410	236
10	188	510	186
11	190	610	225
12	179	612	212

[a] Data from J. Zimmerman, in *Encyclopedia of Polymer Science and Engineering*, Vol. 2 (H. F. Mark, N. M. Bikales, C. G. Overberger, G. Menges, and J. I. Kroschwitz, eds.), Wiley-Interscience, New York, 1988, pp. 515–581.

Table 13.4. Approximate melting temperatures of terephthalamides[a]

Diamine	Melting temperature (°C)
Ethylene	455
Trimethylene	399
Tetramethylene	436
Pentamethylene	353
Hexamethylene	371
Heptamethylene	341

[a] Data from Frazer.[34a]

has been the development of highly regular copolyamides, referred to as *ordered copolyamides*,[36] which are synthesized by treating diamine monomers containing one type of repeating unit with diacid chlorides of another. Reaction (13.40), for example, illustrates the formation of an ordered copolyamide containing *m*-benzamide and isophthalamide units. The corresponding random copolymer has a melting point (about 300°C) about 100°C lower than that of the ordered copolymer. Compared with polyamide **5** or its *para* analog, the ordered copolyamides are generally more soluble and less crystalline. Films and fibers of ordered copolyamides can be oriented, however, to induce greater crystallinity.

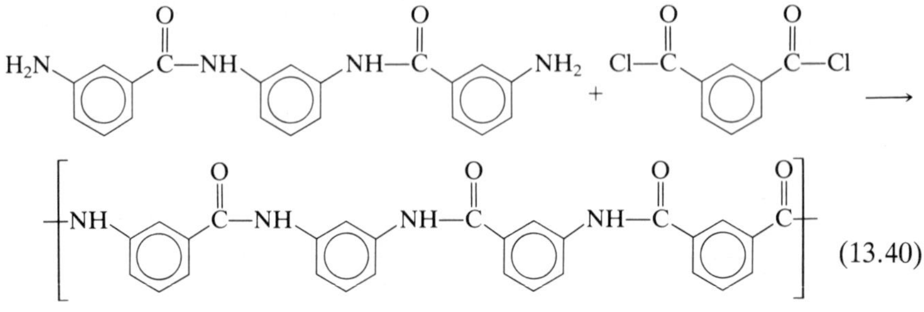

(13.40)

Substituent groups on the polymer backbone reduce the melting point of polyamides significantly. Nylon 66 containing α-methyl groups on the carboxyl portion of the polymer melts at about 166°C, compared with 265°C for the unsubstituted polymer. Also, substituting about 50% of the hydrogen atoms attached to nitrogen with isobutyl groups reduces the melting point of nylon 610 more than 70°C. If the substituent groups are sufficiently bulky, the polymers are amorphous.

Polyamides having substituent groups on carbon are usually prepared from the appropriate monomers. A case in point is polyamide **24**, which is synthesized by the aminolysis reaction of dimethyl terephthalate with 2,2,4- and

$$\left[\begin{array}{c} O \\ \parallel \\ {-}C{-} \end{array} \bigcirc \begin{array}{ccc} & CH_3 & CH_3 \\ O & | & | \\ \parallel & & \\ {-}CNHCH_2CHCH_2CCH_2CH_2NH{-} \\ & | \\ & CH_3 \end{array} \right]$$

24

2,4,4-trimethyl-1,3-hexanediamine. (Only the latter diamine unit is shown.) The amorphous polymer is marketed by the German firm Dynamit Nobel (trade name Trogamid T). Alternatively, substituent groups can be introduced at the nitrogen atoms by such reactions as methoxymethylation (13.41), alkylation (13.42), and polyetherification (13.43). Methoxymethylation of 36% of the nitrogens reduces the melting point of nylon 66 by about 115°C and increases its solubility. Alkylation of amorphous **24** with long-chain alkyl groups leads to a comb structure with crystallinity in the side chains.[37] Considerable work has been done in these areas, as well as with radiation grafting, to improve flexibility or solubility and, in the case of fibers, elasticity and wash-and-wear properties.[38,39] Random copolymerization is another strategy employed commercially for reducing crystallinity.

$$\begin{array}{ccc} & \xrightarrow[CH_3OH]{CH_2O,\ OH^-} & \begin{array}{c} O \\ \parallel \\ \sim\sim N{-}C\sim\sim \\ | \\ CH_2OCH_3 \end{array} & (13.41) \end{array}$$

$$\begin{array}{c} O \\ \parallel \\ \sim\sim NHC\sim\sim \end{array} \quad \xrightarrow[2.\ RBr]{1.\ NaH} \quad \begin{array}{c} O \\ \parallel \\ \sim\sim N{-}C\sim\sim \\ | \\ R \end{array} \quad (13.42)$$

$$\xrightarrow{CH_2{-}CH_2} \quad \begin{array}{c} O \\ \parallel \\ \sim\sim N{-}C\sim\sim \\ | \\ [CH_2CH_2O]{-}H \end{array} \quad (13.43)$$

Another important characteristic of polyamides is a propensity for absorbing moisture from their immediate environment, which generally has a plasticizing effect.[40] Because properties may thus vary with climate, moisture absorption must be taken into account in commercial applications. This is of less concern with the more hydrophobic polyamides such as nylons 11 and 12.

13.2.5 Polyureas

Polyureas[8a] are often treated as a separate class of polymers; they are, however, polyamides of carbonic acid, just as polycarbonates are polyesters of carbonic acid. Like the more conventional polyamides, they are tough, high melting, and suitable for fiber applications. They are usually higher melting than polyamides of like structure.

A number of methods have been employed to prepare polyureas, some of which bear analogy with polycarbonate syntheses, with diamines being used instead of diols. The reaction of diamines with phosgene (13.44) can be accomplished by bubbling the latter through an aqueous solution of diamine

$$H_2N-R-NH_2 + Cl-\overset{\overset{\displaystyle O}{\|}}{C}-Cl \longrightarrow \left[NH-R-NH-\overset{\overset{\displaystyle O}{\|}}{C}\right] + 2HCl \quad (13.44)$$

or by the interfacial technique, which is generally simpler and results in higher molecular weights. Polyhexamethyleneurea, prepared by interfacial polymerization, melts at about 295°C which is significantly higher than nylon 6 (215°C) and nylon 66 (265°C), which have comparable spacing between functional groups. A variation of this procedure is to treat secondary diamines with excess phosgene and to react the resultant biscarbamyl chloride with diamine. This is illustrated (13.45) for piperazine biscarbamyl chloride (25). Polyureas can also be prepared by reacting diamines with carbonate esters (13.46) or, better, with bisurethanes (13.47).

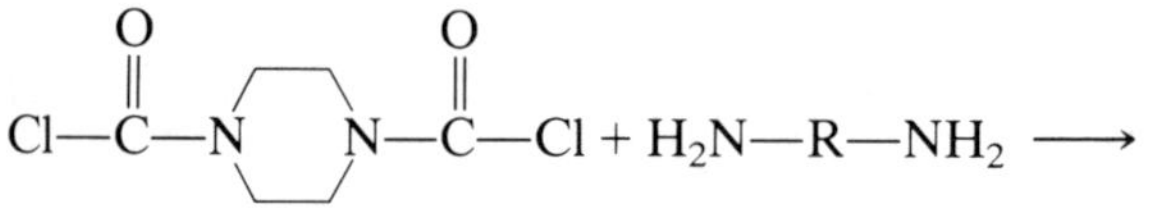

$$Cl-\overset{\overset{\displaystyle O}{\|}}{C}-N\bigcirc N-\overset{\overset{\displaystyle O}{\|}}{C}-Cl + H_2N-R-NH_2 \longrightarrow$$

25

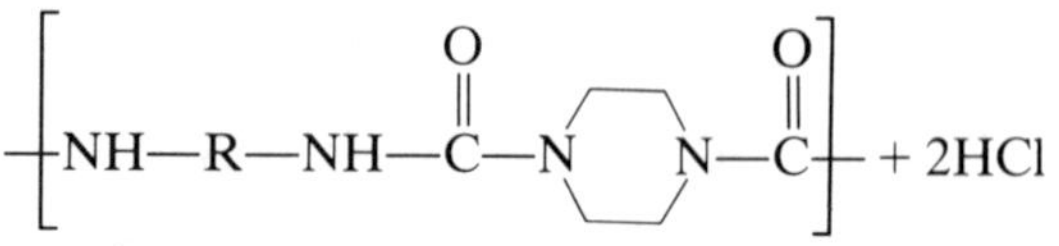

$$\left[NH-R-NH-\overset{\overset{\displaystyle O}{\|}}{C}-N\bigcirc N-\overset{\overset{\displaystyle O}{\|}}{C}\right] + 2HCl \quad (13.45)$$

$$H_2N-R-NH_2 + C_6H_5O\overset{\overset{\displaystyle O}{\|}}{C}OC_6H_5 \longrightarrow$$

$$\left[NH-R-NH-\overset{\overset{\displaystyle O}{\|}}{C}\right] + 2C_6H_5OH \quad (13.46)$$

$$H_2N-R-NH_2 + C_2H_5O\overset{\overset{\displaystyle O}{\|}}{C}NH-R'-NH\overset{\overset{\displaystyle O}{\|}}{C}OC_2H_5 \longrightarrow$$

$$\left[NH-R-NH\overset{\overset{\displaystyle O}{\|}}{C}NH-R'-NH\overset{\overset{\displaystyle O}{\|}}{C}\right] + 2C_2H_5OH \quad (13.47)$$

The best method of synthesizing polyureas is to react diamines with diisocyanates. This is an exothermic reaction that can be done in solution or by the interfacial technique. The reaction of 2,4-toluenediisocyanate (**22**) with 4,4′-diaminodiphenyl ether (**26**) (13.48), for example, gives a polyurea **27** melting at about 320°C. Because it is a step-growth addition reaction (addition of amine across the carbon–nitrogen double bond), there is no by-product, which simplifies the polymerization process.

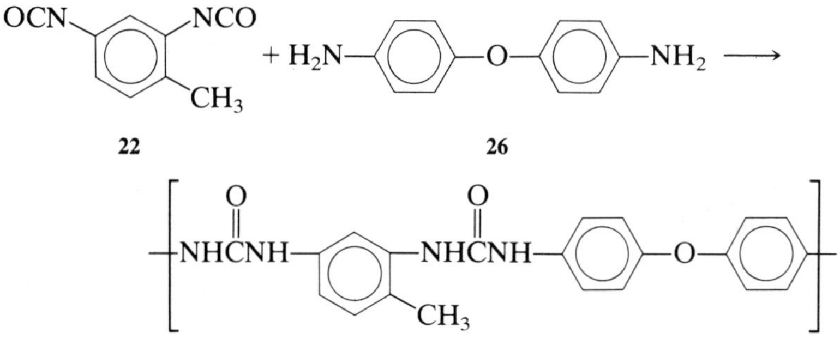

22 26 27

(13.48)

Miscellaneous methods of synthesizing polyureas include the dimethyl sulfoxide-promoted reaction of diisocyanates with monocarboxylic acids[41] (13.49) and the copolymerization of carbon dioxide with diamines[42] (13.50). In the absence of more than equimolar amounts of diphenylphosphite, the latter reaction requires the use of high temperatures and pressures.

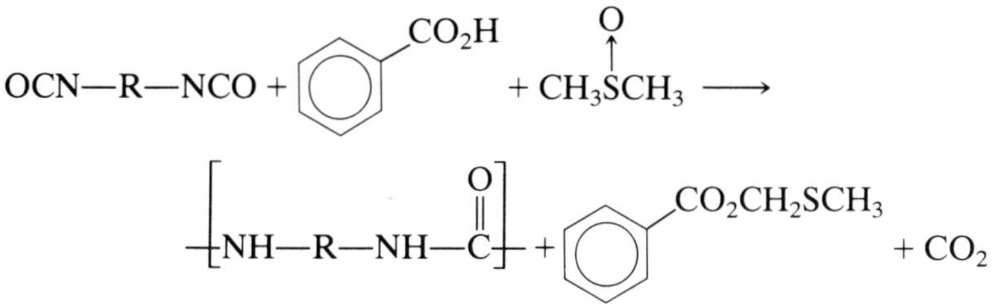

(13.49)

$$H_2N{-}R{-}NH_2 + CO_2 \longrightarrow H_3\overset{+}{N}{-}R{-}NH{-}\overset{O}{\overset{\|}{C}}{-}O^- \xrightarrow[\text{pyridine, 40°C}]{HP(O)(O\phi)_2} {+}NH{-}R{-}NH{-}\overset{O}{\overset{\|}{C}}{+} + H_2O \quad (13.50)$$

Despite their generally good properties, economic factors have restricted commercial development of polyureas. Applications include encapsulation of pharmaceuticals, inks, dyes, and so on, and the modification of wool fibers by interfacial grafting techniques.[9b]

Another type of urea polymer, prepared by condensing urea with formaldehyde, is covered in Chapter 14.

13.3   Polymers related to polyamides

13.3.1   Polyurethanes

It is appropriate to consider polyurethanes after polyureas because both are derived from carbonic acid. Polyurethanes, also called *polycarbamates* (from carbamic acid, R_2NHCO_2H), are ester-amide derivatives of carbonic acids. They are used in a wide variety of applications, including fibers (particularly the elastic type), adhesives, coatings, elastomers, and flexible and rigid foams. A number of books have been written on the subject.[43–46] There are two principal methods of forming polyurethanes: the reaction of bischlorofor-mates with diamines and, more important from the industrial perspective, the reaction of diisocyanates with dihydroxy compounds. Bischloroformates, pre-pared by reacting diols or bisphenols with excess phosgene (13.51), are less reactive than acid chlorides; nevertheless, they react with diamines at low temperatures under interfacial polymerization conditions (13.52).[8c,9c] The polyurethane formed in reaction (13.52) melts at about 180°C, compared with 295°C for the polyamide (nylon 46, Table 13.3) of comparable structure.

$$2Cl\overset{\overset{\displaystyle O}{\|}}{-C}-Cl + HO-R-OH \longrightarrow Cl\overset{\overset{\displaystyle O}{\|}}{-C}-O-R-O\overset{\overset{\displaystyle O}{\|}}{-C}-Cl + 2HCl$$

$$(13.51)$$

$$Cl\overset{\overset{\displaystyle O}{\|}}{-C}-O-(CH_2)_2-O\overset{\overset{\displaystyle O}{\|}}{-C}-Cl + H_2N-(CH_2)_6-NH_2 \longrightarrow$$

$$\left[\overset{\overset{\displaystyle O}{\|}}{C}-O-(CH_2)_2-O\overset{\overset{\displaystyle O}{\|}}{-C}-NH-(CH_2)_6-NH\right] + 2HCl \quad (13.52)$$

The addition of dihydroxy compounds to diisocyanates to form polyurethanes is similar in principle to the previously described polyurea synthesis (13.48). Among the earliest commercial products (developed in Germany as an alternative to the patent-protected nylon fibers) was polyurethane **28** (Farbenfabriken Bayer's Perlon U), prepared from 1,6-hexanediisocyanate and 1,4-butanediol (13.53). The reaction is catalyzed by amines and certain metal salts, but catalysts are not required for forming high-molecular-weight polymer. Although **28** is no longer as commercially viable as the nylons, the diisocyanate route is now used to manufacture polyurethane fibers, plastics, elastomers, and coatings. Because the reaction is so rapid, it is ideally suited to RIM technology (Chapter 4).

$$OCN(CH_2)_6NCO + HO(CH_2)_4OH \longrightarrow \left[\overset{\overset{\displaystyle O}{\|}}{C}NH(CH_2)_6NH\overset{\overset{\displaystyle O}{\|}}{C}O(CH_2)_4O\right]$$

28

$$(13.53)$$

Linear polyurethanes are normally prepared in solution because the polymer tends to dissociate into alcohol and isocyanate or to decompose into amines, olefins, and carbon dioxide at the high temperatures needed for melt polymerization. This is particularly true for those prepared with aromatic diisocyanates.

Much of the polyurethane production involves the use of low-molecular-weight hydroxyl-terminated polyesters or polyethers as the dihydroxy "monomer." Flexible copolymers of this type are not only useful as fibers but can also be converted into crosslinked elastomers by further reaction with excess diisocyanate, an addition reaction involving the nitrogen of the urethane linkage (13.54). The resultant group is an *allophonate*.

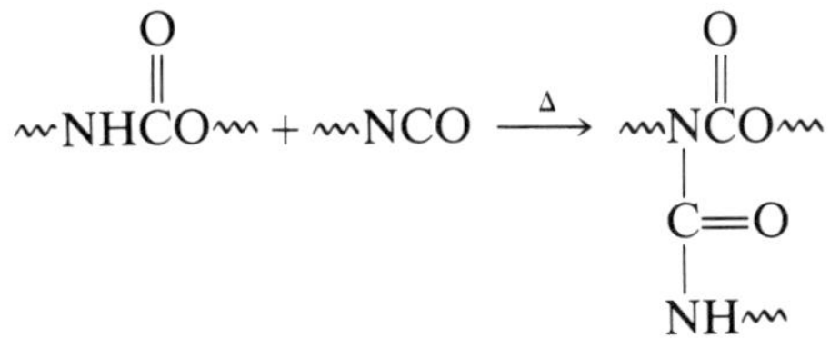

$$\text{NHCO} + \text{NCO} \xrightarrow{\Delta} \underset{\underset{\underset{NH}{|}}{\overset{|}{C=O}}}{NCO} \tag{13.54}$$

Crosslinking may also be effected by preparing the polyester portion of the polymer with a polyol such as glycerol so that the resultant pendant hydroxyl groups along the backbone can react with diisocyanate to give urethane crosslinks (13.55). The very important area of urethane foam technology, which is based on reactions of this type, is discussed in the next section.

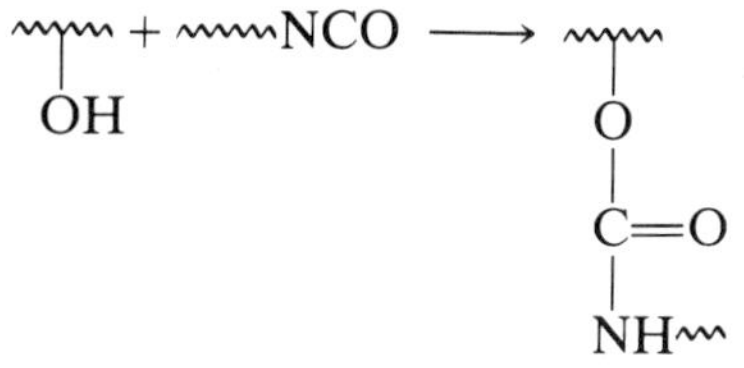

$$\underset{\underset{OH}{|}}{} + \text{NCO} \longrightarrow \underset{\underset{\underset{NH}{|}}{\overset{|}{C=O}}}{\underset{O}{|}} \tag{13.55}$$

Related copolymers are prepared by treating hydroxyl-terminated polyester or polyether with excess diisocyanate to give an isocyanate-terminated product. Curing can then be effected in several ways. As a surface coating, the isocyanate groups react with atmospheric water to give chain extension through urea linkages (13.56). The reaction involves formation of the unstable carbamic acid, which decarboxylates. Reaction between the resultant amine and unreacted isocyanate yields urea. Crosslinking may occur by

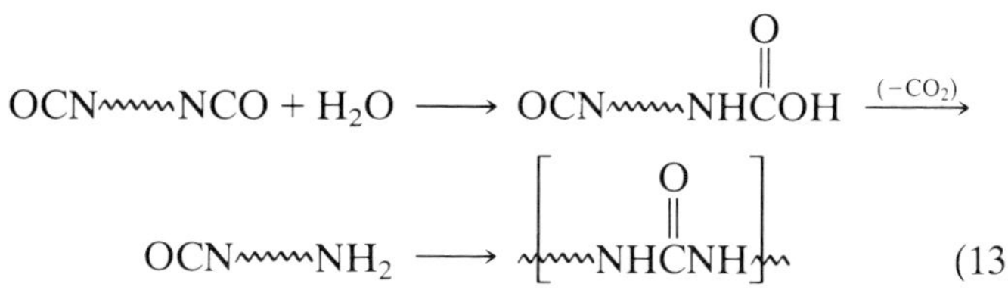

$$\text{OCN}\sim\sim\text{NCO} + H_2O \longrightarrow \text{OCN}\sim\sim\text{NHCOH} \xrightarrow{(-CO_2)}$$

$$\text{OCN}\sim\sim\text{NH}_2 \longrightarrow \left[\sim\sim\text{NHCNH} \sim \right] \tag{13.56}$$

reaction of urea groups with unreacted isocyanate (13.57) to form *biuret* crosslinks. Coatings of this type are normally formulated with polyesters prepared with polyfunctional alcohols to ensure that crosslinking will occur according to reaction (13.55). Because no additives are needed to effect cure, such coating formulations are often referred to as "one-component" or "one-pot" systems. Isocyanate-terminated polymers can be polymerized further by reactions with diols or diamines to form additional urethane or urea groups, respectively.

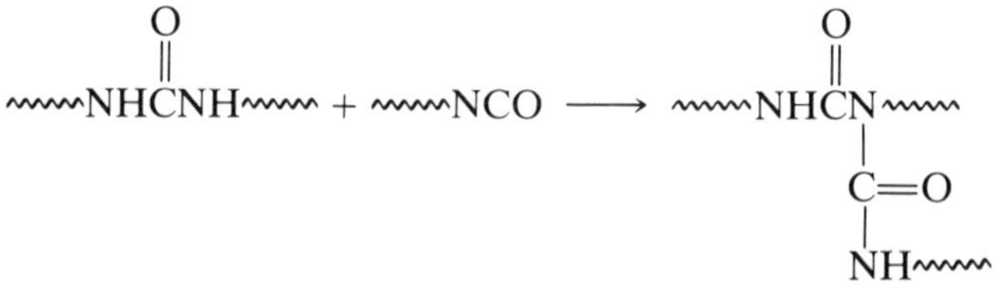

$$
\underset{\substack{\text{wwwNHCNHwww}}}{\overset{\overset{\displaystyle O}{\parallel}}{}} + \text{wwwNCO} \longrightarrow \underset{\substack{\text{wwwNHCNwww} \\ | \\ C=O \\ | \\ NHwww}}{\overset{\overset{\displaystyle O}{\parallel}}{}} \qquad (13.57)
$$

The most important applications of polyurethanes for fibers involves elastomeric $-\!\!+\!AB\!\!+\!-$ block copolymers consisting of alternating "soft" and "hard" segments.[13b,47] (Du Pont's Spandex is an example.) Typically, an aromatic diisocyanate is reacted in excess with a hydroxy-terminated polyether or polyester having an average molecular weight of 2000 to 3000 to yield an isocyanate-terminated polymer that is, in turn, reacted with a diamine chain extender. The sequence of steps is shown in Scheme 13.1. Note that chain extension occurs via urea rather than urethane linkages. Common polyethers and polyesters are poly(propylene glycol) (**29**) and poly-(diethyleneglycol adipate) (**30**), respectively. Toluenediisocyanate (**22**) (TDI) and methylene-4,4'-diphenyldiisocyanate (MDI) (**31**) are typical diisocyanates. Diamines such as ethylenediamine are used as extenders.

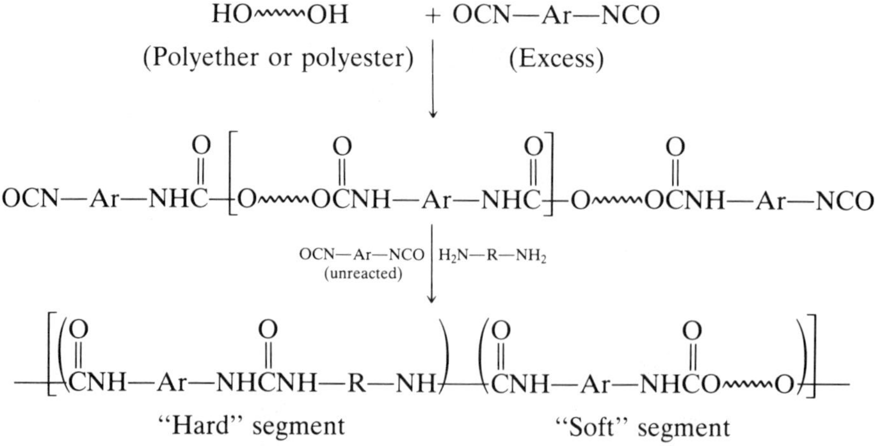

Scheme 13.1. Synthesis of an elastomeric polyurethane for fiber applications.

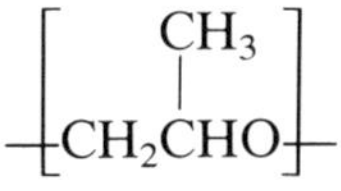

$$\left[\begin{array}{c} CH_3 \\ | \\ CH_2CHO \end{array}\right]_{29} \qquad \left[\overset{O}{\overset{||}{C}}(CH_2)_4\overset{O}{\overset{||}{C}}OCH_2CH_2OCH_2CH_2O\right]_{30}$$

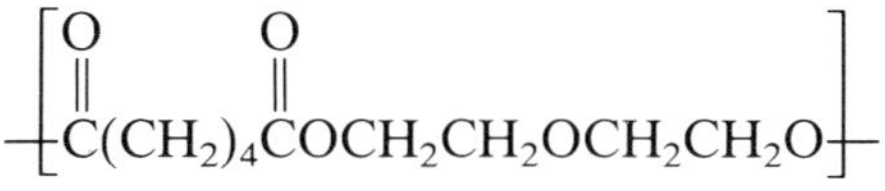

$$OCN-\!\!\bigcirc\!\!-CH_2-\!\!\bigcirc\!\!-NCO$$

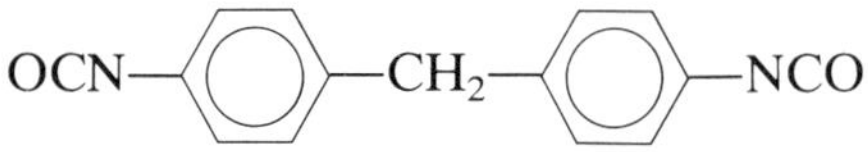

31

Elastomeric fibers exhibit morphological characteristics similar to those of thermoplastic elastomers. The hard segments associate into crystalline microdomains while the predominant soft segments, which constitute the continuous phase, remain amorphous and randomly coiled. Upon stretching, the soft segments become elongated and crystalline, but when the tension is removed they rapidly revert to the amorphous state with a force commensurate with the entropy change of the system.

A variety of conventional elastomers and thermoplastic elastomers are also marketed. The very important area of polyurethane foams is discussed in the next section.

Miscellaneous methods of preparing polyurethanes include the reaction of diisocyanates with diepoxides (13.58) to yield *poly(2-oxazolidone)s*[48,49] (**32**), which contain cyclic urethane groups; thermolysis of *p*-hydroxybenzoylazide (13.59) to form the polyurethane (**33**) derived from the intermediate *p*-hydroxyphenylisocyanate[50]; and copolymerization of aziridenes with carbon dioxide (13.60) in the presence of organozinc compounds[51] to yield polyurethane **34**.

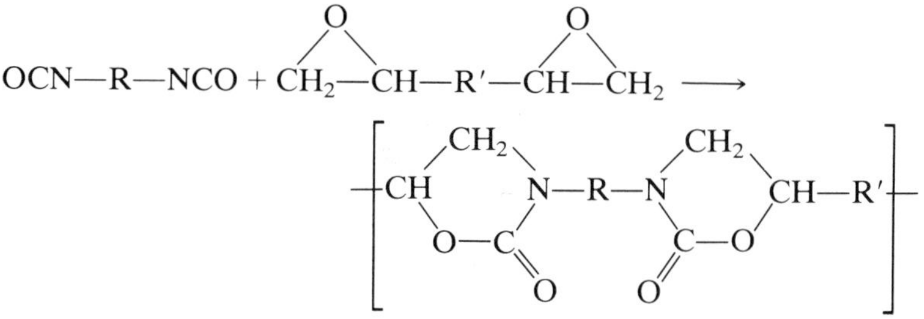

$$OCN-R-NCO + \overset{O}{\overset{\triangle}{CH_2-CH}}-R'-\overset{O}{\overset{\triangle}{CH-CH_2}} \longrightarrow$$

$$\left[CH\underset{O-C}{\overset{CH_2}{\diagup}}N-R-N\underset{C-O}{\overset{CH_2}{\diagup}}CH-R'\right] \quad (13.58)$$

32

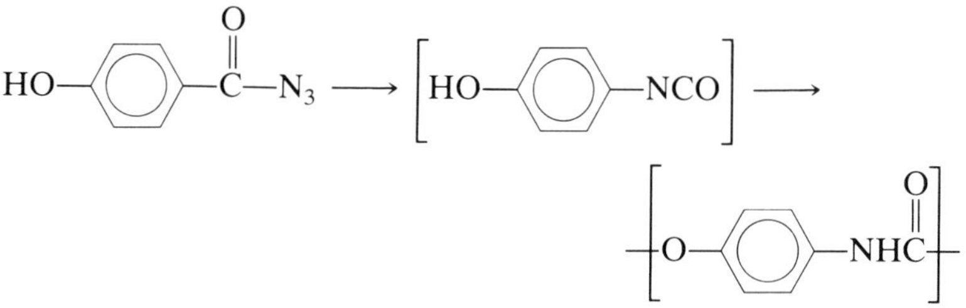

$$HO-\!\!\bigcirc\!\!-\overset{O}{\overset{||}{C}}-N_3 \longrightarrow \left[HO-\!\!\bigcirc\!\!-NCO\right] \longrightarrow$$

$$\left[O-\!\!\bigcirc\!\!-N\overset{O}{\overset{||}{HC}}\right] \quad (13.59)$$

33

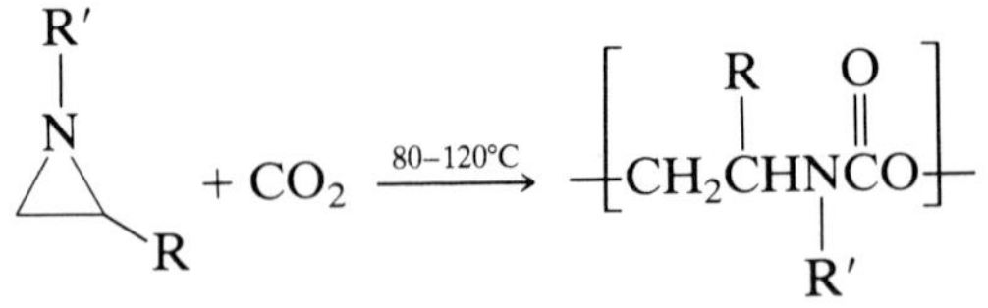

$$\underset{R}{\overset{R'}{\underset{|}{\triangle}}} + CO_2 \xrightarrow{80\text{--}120°C} \left[CH_2\underset{\underset{R'}{|}}{\overset{R}{\underset{|}{C}}}H N\overset{O}{\overset{||}{C}}O \right] \tag{13.60}$$

34

13.3.2   *Polyurethane foams*

Because much of the polyurethane production is used in making rigid and flexible foams, it would be instructive here to elaborate on foam technology. Polymeric foams[52-54] are made in a variety of ways depending on the type of polymer used and its application. For polymers such as polystyrene, *blowing agents* are used to generate foams. There are two types: physical and chemical. Physical blowing agents are gases (air, nitrogen, or carbon dioxide) dissolved in the polymer under pressure, or low-boiling liquids, such as chlorofluorocarbons or more environmentally acceptable compounds, which vaporize on heating. Chemical blowing agents decompose on heating to give off gas. Examples are *p,p'*-oxybis(benzenesulfonyl hydrazide) (**35**) and *p*-toluenesulfonyl semicarbazide (**36**), which evolve nitrogen at about 160 and 235°C, respectively. The bicyclic compound 3,7-dinitroso-1,3,5,7-tetraazobicyclo[3.3.1]nonane (**37**) is widely used as a blowing agent in foam rubber manufacture.

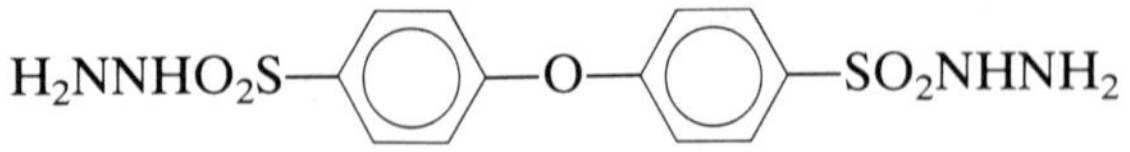

$$H_2NNHO_2S-\!\!\!\bigcirc\!\!\!-O-\!\!\!\bigcirc\!\!\!-SO_2NHNH_2$$

35

$$CH_3-\!\!\!\bigcirc\!\!\!-SO_2NH\overset{O}{\overset{||}{C}}NHNH_2$$

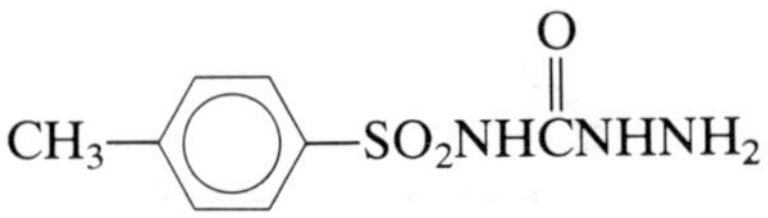

36

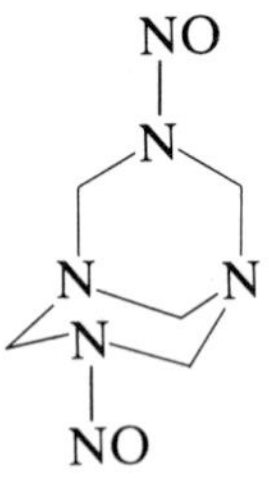

37

Polyurethanes differ in that by-product carbon dioxide is a key ingredient in the foaming process. In one method, low-molecular-weight isocyanate-terminated prepolymers (prepared as described above from diisocyanates and polyesters or polyethers) are foamed by addition of water, which causes an increase in molecular weight by formation of urea groups with simultaneous loss of carbon dioxide (see reaction 13.56). As the evolved gas causes the polymer to foam, the polymerization reaction increases the viscosity and sets the foam before it collapses. A chemical blowing agent, which is boiled by the reaction exotherm, is usually added to augment the foaming action of the carbon dioxide.

Flexible foams are usually prepared from dihydroxy polyesters or poly-ethers, rigid foams from polyhydroxy prepolymers. Rigid foams are some-times prepared without water by reacting a hydroxyl-terminated prepolymer with diisocyanate in the presence of a blowing agent. In this case, molecular weight buildup is via urethane linkages. Such foams are particularly effective insulators because the blowing agents, trapped in the foam cells, have very low thermal conductivity.

The reaction of diisocyanates with carboxyl-terminated polyesters (reaction 13.39), as discussed earlier, is also used to make flexible foams, in this case by amide formation. It is not uncommon to see such reactions included under urethane foams even though no urethane chemistry is involved. In other words, all foam manufacture that uses diisocyanates tends to be grouped under polyurethanes.

Most commonly used diisocyanates are TDI (**22**) (actually a mixture of 2,4- and 2,6-isomers) and MDI, (**31**), particularly the former. Metal salts such as stannous 2-ethylhexanoate and tertiary amines, particularly diazabi-cyclo[2.2.2]octane (**38**) (commonly called DABCO), are normally used to catalyze urethane formation. DABCO is a more effective catalyst than most other tertiary amines, possibly because the rigid cyclic system minimizes steric interference.

38

Flexible foams are used as insulators, including textile laminates for cold-weather clothing; automobile crash panels; upholstery; bedding; carpet underlays; synthetic sponges; and other miscellaneous uses. Rigid foams are most commonly used in insulated construction panels, for packaging delicate items, for lightweight furniture, and for marine flotation equipment. The use of these materials in construction has prompted attempts at preparing nonflammable polyurethanes. Phosphorus-containing polyurethanes of type **39**, for example, are self-extinguishing.[55]

A different strategy for reducing flammability of foams is to convert terminal and pendant isocyanate groups to *polyisocyanurates*.[56a] The

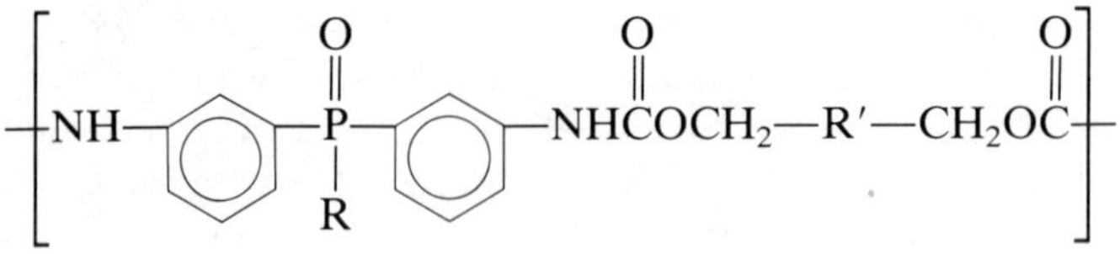

$$\left[NH-\!\!\bigcirc\!\!-\overset{\overset{\displaystyle O}{\|}}{\underset{\underset{\displaystyle R}{}}{P}}-\!\!\bigcirc\!\!-NH\overset{\overset{\displaystyle O}{\|}}{C}OCH_2-R'-CH_2O\overset{\overset{\displaystyle O}{\|}}{C}\right]$$

39

trimerization reaction (13.61) is facilitated by alkali metal alkoxide, phenoxide, and carboxylate catalysts. The isocyanurate moiety (**40**) serves the dual purpose of reducing flammability and increasing thermal stability.

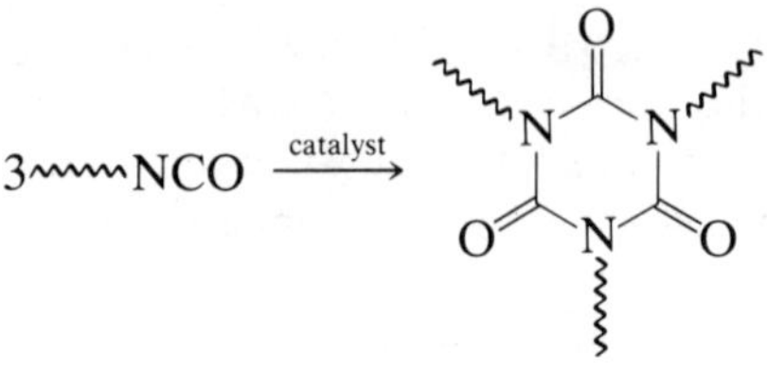

$$3\text{\footnotesize{wwww}}NCO \xrightarrow{\text{catalyst}} \quad (13.61)$$

40

13.3.3 Polyhydrazides and related polymers

There are three general methods of preparing polyhydrazides[8d]: the reaction of a diacid chloride with hydrazine or hydrazine hydrate (13.62), with a dihydrazine such as N,N'-diaminopiperazine (**41**) (13.63), and with a dihydrazide (13.64). Of the three, the dihydrazide route (13.64) is preferred because

$$H_2N-NH_2 + Cl-\overset{\overset{\displaystyle O}{\|}}{C}-R-\overset{\overset{\displaystyle O}{\|}}{C}-Cl \longrightarrow$$

$$\left[NH-NH-\overset{\overset{\displaystyle O}{\|}}{C}-R-\overset{\overset{\displaystyle O}{\|}}{C}\right] + 2HCl \quad (13.62)$$

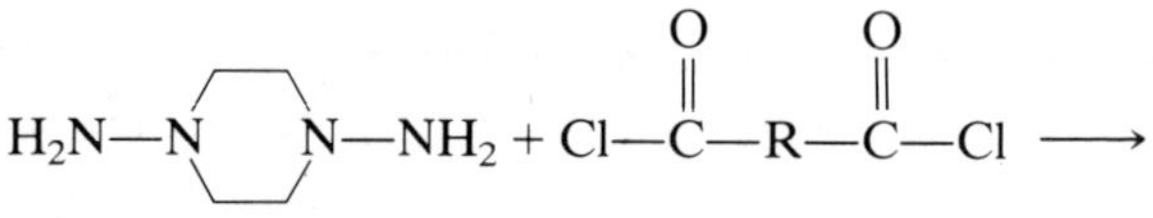

$$H_2N-N\underset{}{\frown}N-NH_2 + Cl-\overset{\overset{\displaystyle O}{\|}}{C}-R-\overset{\overset{\displaystyle O}{\|}}{C}-Cl \longrightarrow$$

41

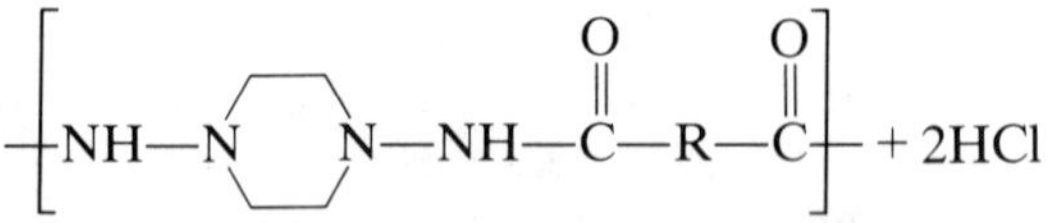

$$\left[NH-N\underset{}{\frown}N-NH-\overset{\overset{\displaystyle O}{\|}}{C}-R-\overset{\overset{\displaystyle O}{\|}}{C}\right] + 2HCl \quad (13.63)$$

$$H_2N-NH-\overset{\overset{\displaystyle O}{\|}}{C}-R'-\overset{\overset{\displaystyle O}{\|}}{C}-NH-NH_2 + Cl-\overset{\overset{\displaystyle O}{\|}}{C}-R-\overset{\overset{\displaystyle O}{\|}}{C}-Cl \longrightarrow$$

$$\left[NH-NH-\overset{\overset{\displaystyle O}{\|}}{C}-R'-\overset{\overset{\displaystyle O}{\|}}{C}-NH-NH-\overset{\overset{\displaystyle O}{\|}}{C}-R-\overset{\overset{\displaystyle O}{\|}}{C}\right] + 2HCl \quad (13.64)$$

the monomers are easily prepared from diesters and hydrazine and are easily purified. Hydrazine is readily available, but it is so hygroscopic that it is difficult to achieve an exact stoichiometric balance. Furthermore, one amino group of hydrazine is sufficiently less reactive than the other that condensations of this type (13.62) must be run in solution and cannot be done by interfacial polymerization.

In recent years there has been considerable interest in aromatic polyhydrazides, as high-modulus, thermally stable fibers[34a] and because they may exhibit liquid crystalline properties.[57] Tire cord represents a potential market for materials of this type. Polyamidehydrazides such as **42** are also of interest.[36] Later in the chapter we shall see how polyhydrazides may be converted to heterocyclic polymers.

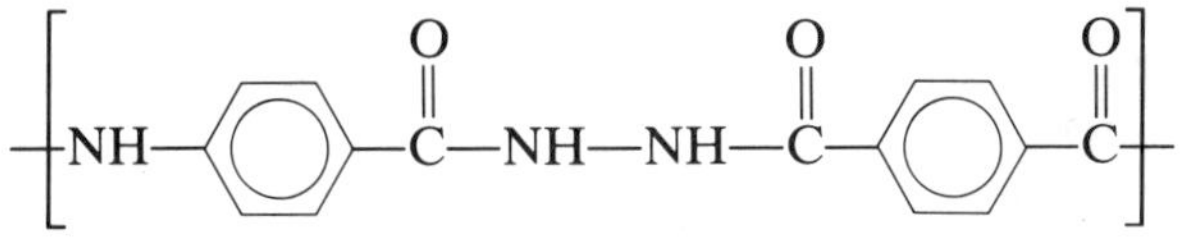

$$\left[NH - \bigotimes - \overset{\overset{\displaystyle O}{\|}}{C} - NH - NH - \overset{\overset{\displaystyle O}{\|}}{C} - \bigotimes - \overset{\overset{\displaystyle O}{\|}}{C} \right]$$

42

If diisocyanate, OCN–R–NCO, is substituted for diacid chloride in the above reactions, polymers resembling polyhydrazides are formed.[8d] Thus (13.62) would yield a *polybiurea* (also called *polyureylene*, **43**), (13.63) a *polysemicarbazide* (**44**), and (13.64) a *polyacylsemicarbazide* (**45**).

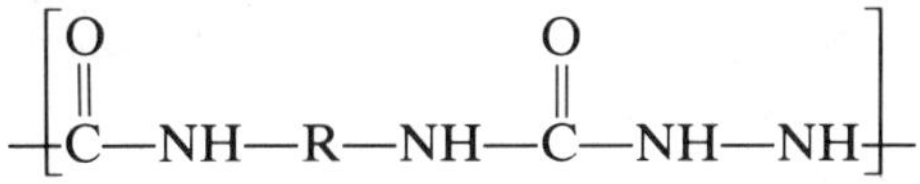

$$\left[\overset{\overset{\displaystyle O}{\|}}{C} - NH - R - NH - \overset{\overset{\displaystyle O}{\|}}{C} - NH - NH \right]$$

43

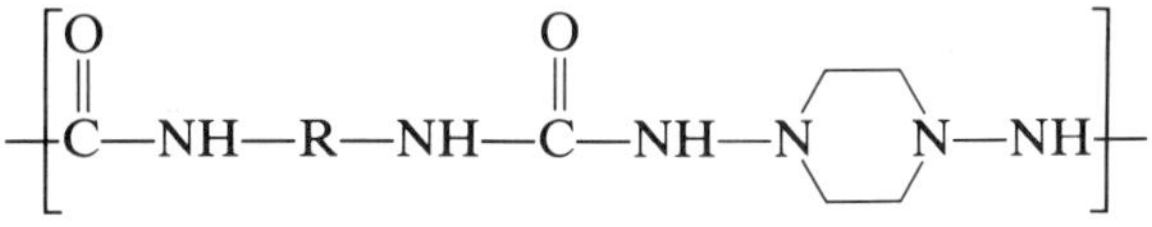

$$\left[\overset{\overset{\displaystyle O}{\|}}{C} - NH - R - NH - \overset{\overset{\displaystyle O}{\|}}{C} - NH - N \bigcirc N - NH \right]$$

44

$$\left[\overset{\overset{\displaystyle O}{\|}}{C} - NH - R - NH - \overset{\overset{\displaystyle O}{\|}}{C} - NH - NH - \overset{\overset{\displaystyle O}{\|}}{C} - R' - \overset{\overset{\displaystyle O}{\|}}{C} - NH - NH \right]$$

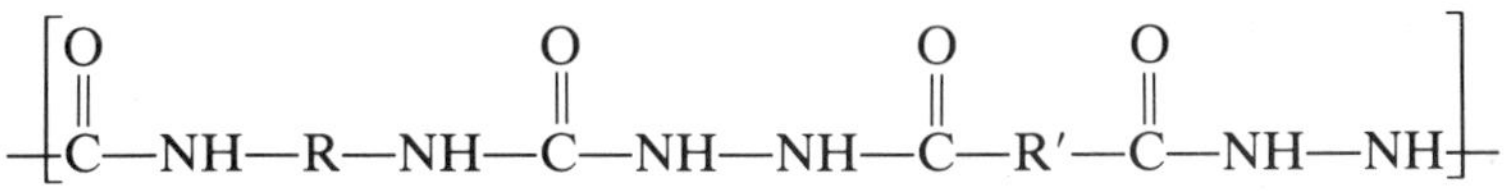

45

13.3.4 Polyimides

Although imides may have an open-chain structure, it is the cyclic imides that are of interest in polymer chemistry. Such imides are normally prepared from cyclic anhydrides and ammonia or amines via the intermediate salt (13.65).

Polyimides have experienced extremely rapid development in recent years,[58–61a] the major emphasis being on engineering applications. High-strength composites, thermally stable films, molding compounds, and adhesives are numbered among the products. Oxidative and hydrolytic stability, in addition to thermal stability, characterizes polyimides.

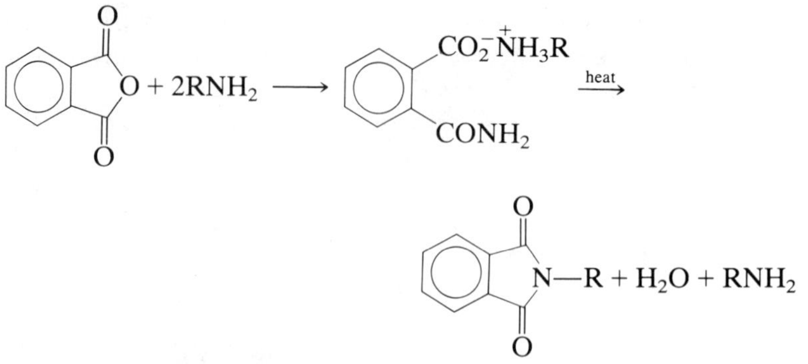

$$\text{(phthalic anhydride)} + 2RNH_2 \longrightarrow \text{(aromatic ring)} \begin{array}{l} CO_2^- \overset{+}{N}H_3R \\ CONH_2 \end{array} \xrightarrow{\text{heat}}$$

$$\text{(phthalimide ring)} N\text{—}R + H_2O + RNH_2 \quad (13.65)$$

There are basically three methods of preparing polyimides:

1. The imide functional group or its precursor is formed during the polymerization reaction.
2. A low-molecular-weight bisimide or an oligomeric bisimide serves as monomer in a polymerization reaction.
3. An imide-containing oligomer containing polymerizable end groups is converted to network polymer, usually by heating.

Type 1 polyimides were the first to be developed.[34b,62,63] The reaction, illustrated with pyromellitic dianhydride (**46**), involves initial formation of polyamide (13.66) followed by ring closure to form polyimide (13.67). The formation of a stable five-membered ring is the driving force for forming linear rather than crosslinked polymer.

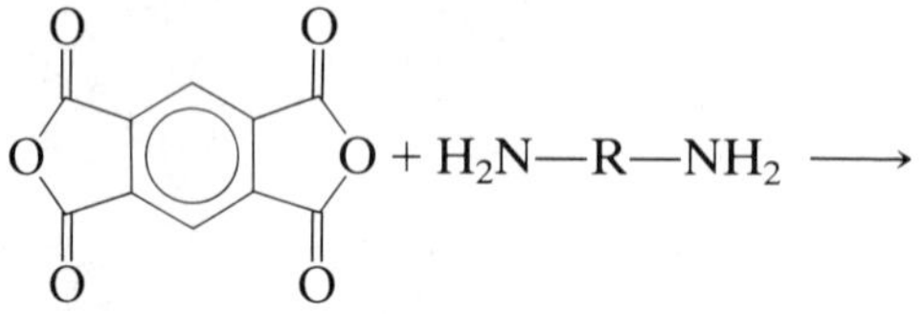

$$\text{(pyromellitic dianhydride)} + H_2N\text{—}R\text{—}NH_2 \longrightarrow$$

46

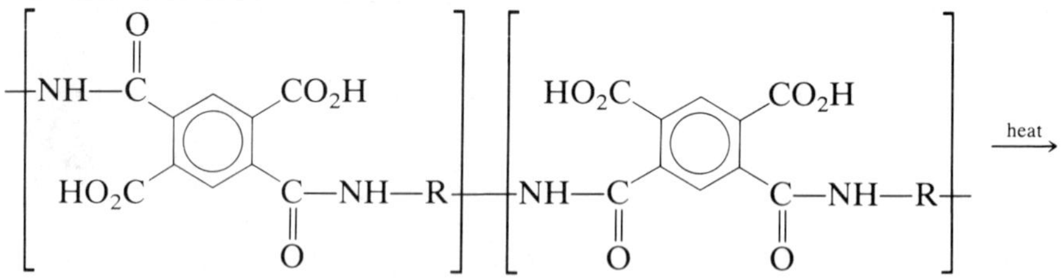

$$\left[\begin{array}{l} \text{—NH—C} \\ \quad\quad HO_2C \end{array} \begin{array}{l} CO_2H \\ C\text{—NH—R—} \end{array} \right] \left[\begin{array}{l} HO_2C \\ \text{—NH—C} \end{array} \begin{array}{l} CO_2H \\ C\text{—NH—R—} \end{array} \right] \xrightarrow{\text{heat}}$$

$$(13.66)$$

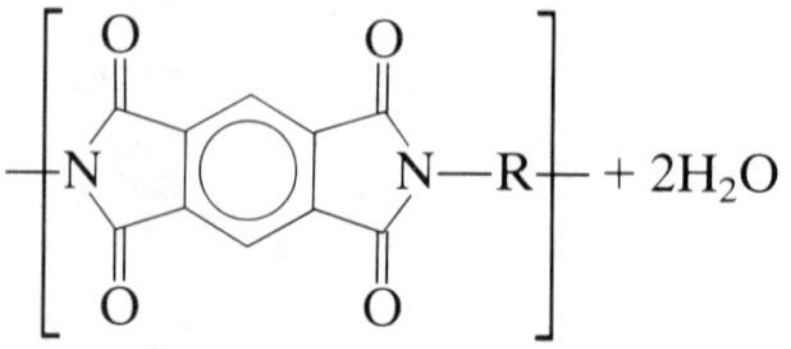

$$\left[\begin{array}{l} \text{—N} \quad\quad N\text{—R—} \end{array} \right] + 2H_2O \quad (13.67)$$

If R is aliphatic in the above equations, the two steps may be run in one operation. If R is aromatic, however, the final product is infusible and insoluble, and the second step must be carried out by a solid-state cyclization reaction—for example, with a cast film of intermediate polyamide or an impregnated fiber for manufacturing composites. Formation of by-product water in the cyclization step is of concern because it may cause voids that structurally weaken the finished product. Careful processing is necessary to obviate this problem.

Among the important polyimides prepared by the dianhydride-diamine route is **47** (du Pont's Kapton). The polymer melts in excess of 600°C and exhibits little weight loss up to 500°C in an inert atmosphere. Extensive

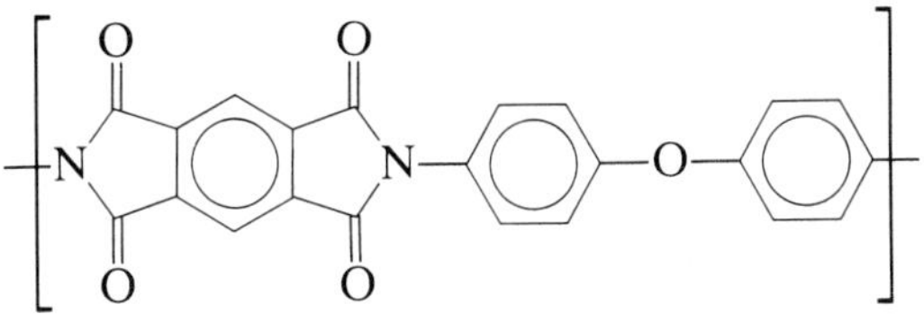

47

structural studies have been conducted on polyimides[62,64,65] and, not unexpectedly, it has been found that increasing the level of *meta* substitution in the aromatic diamine causes a reduction in both viscosity and glass transition temperature. Interestingly, polyimide **48**, which contains no hydrogen, exhibits greater thermal stability than the analogous polymer prepared from **46**, which suggests that carbon–hydrogen bonds are the sites for oxidative decomposition.[66]

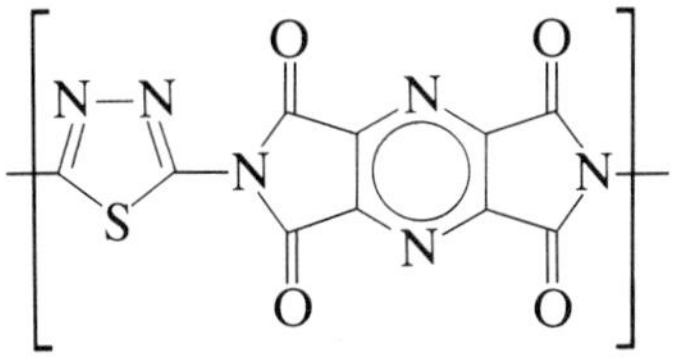

48

An alternative one-step route to polyimides is the reaction of dianhydrides with diisocyanates[67–69] (13.68). With carbon dioxide as a by-product, the reaction has obvious implications for manufacture of foams. Polymerization takes place at elevated temperatures or at moderate temperatures (50–60°C) in polar aprotic solvents if water or tetracarboxylic acid is present. Amines catalyze the reaction. The mechanism is complex and probably involves initial

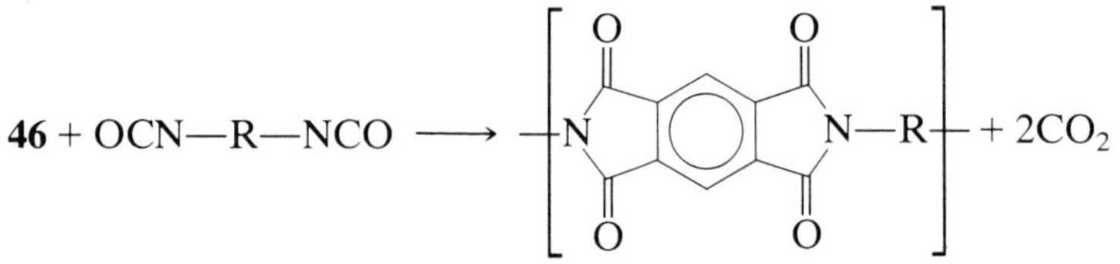

$$\mathbf{46} + \text{OCN—R—NCO} \longrightarrow \text{[imide structure]} + 2CO_2 \quad (13.68)$$

hydrolysis of isocyanate or anhydride (or both), and perhaps the intercession of an unstable seven-membered ring intermediate (**49**).

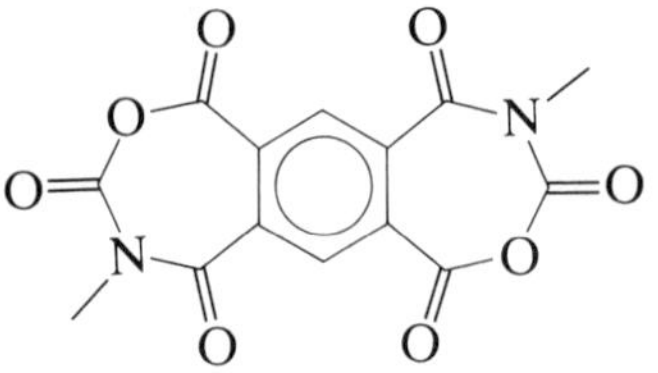

49

Although a large number of dianhydrides have been applied to polyimide synthesis, the only one in widespread use besides **46** is benzophenone-3,3',4, 4'-tetracarboxylic acid dianhydride (**50**). The fluorinated anhydride **51** finds occasional use in polyimides where high moisture resistance is a necessity (see Chapter 4, Section 4.6). One formulation that makes use of dianhydride **50** is a *polyimidesulfone* (**52**) developed by the National Aeronautics and Space Administration and marketed as an engineering plastic.

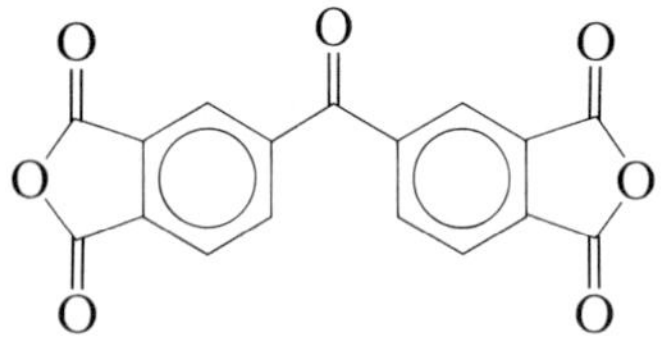

50 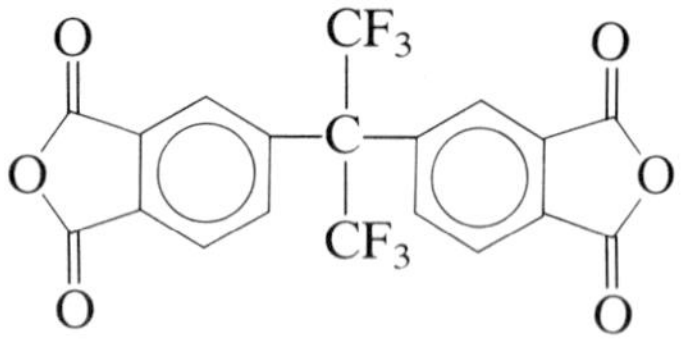   **51**

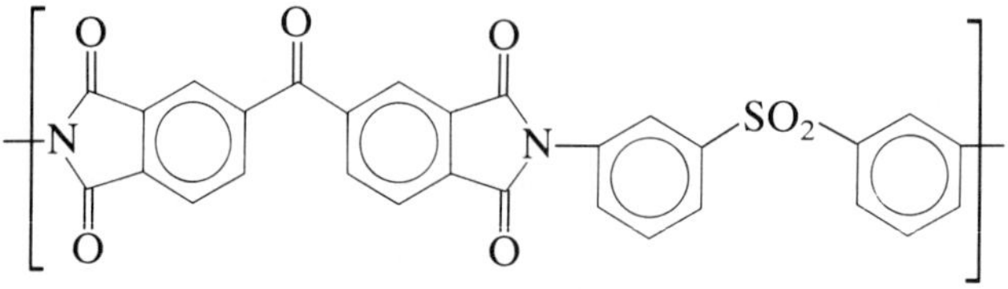

52

Two varieties of type-2 polyimides have achieved commercial acceptance,[70] one based on bismaleimides (**53**) and the other on bisnadimides*(**54**). The bismaleimide type, which is synthesized from maleic anhydride and a monomeric or oligomeric diamine, may be cured simply by heating to 200 to 400°C (13.69) to yield a highly crosslinked polymer. This represents one of the

---

Nadimide is a generic term for the 5-norbornene-2,3-dicarboximide system. It is synthesized from the corresponding nadic anhydride or acid-ester. Nadic anhydride is the Diels–Alder adduct of maleic anhydride and cyclopentadiene.

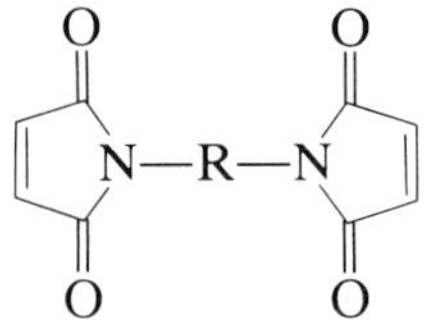

53 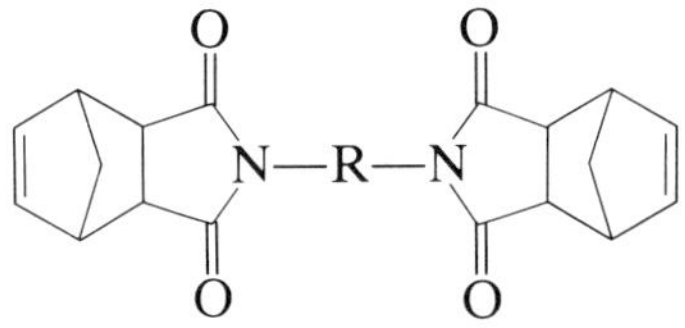   54

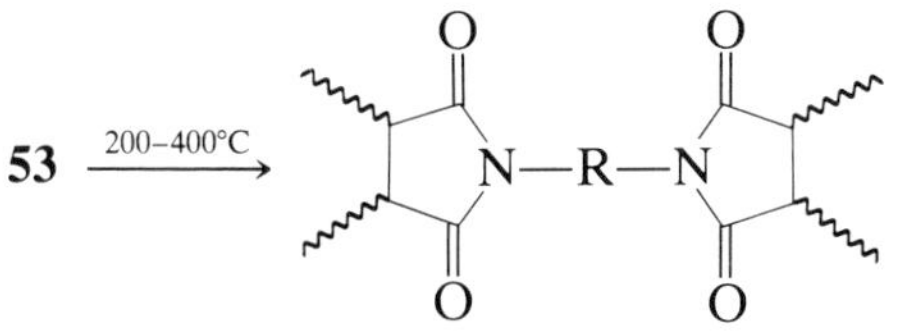

$$53 \xrightarrow{200-400°C} \qquad\qquad\qquad\qquad\qquad\qquad\qquad (13.69)$$

few instances of a vinyl monomer (maleimide) undergoing facile thermal free radical polymerization. Commercial products also make use of the Michael addition of primary or secondary diamines to effect chain extension (13.70) prior to thermal curing.[70–72] Chain extension reduces the crosslink density and brittleness of the finished product. An example is Kerimid 601 (Rhone-Poulenc trade name), which employs 4,4'-diaminodiphenylmethane (55) as the feed-

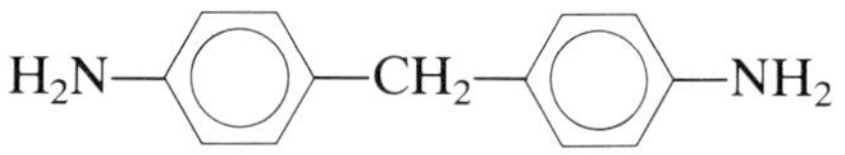

$$H_2N-\bigcirc-CH_2-\bigcirc-NH_2$$

55

stock for both the bismaleimide and diamine according to reaction (13.70). The Michael addition reaction is also used to prepare linear *polyaspartimides** (56) from equimolar quantities of bismaleimide and secondary diamines[73,74] (13.71). In Chapter 16 we shall see how bismaleimide monomers may be used for preparing polyimides by cycloaddition reactions.

$$2\ 53 + H_2N-R'-NH_2 \longrightarrow$$

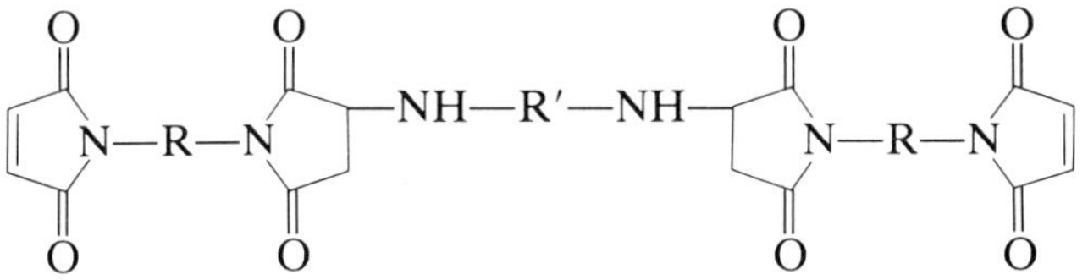

$$(13.70)$$

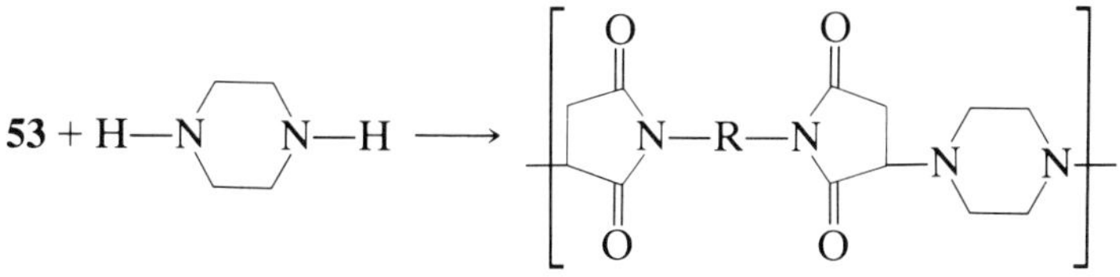

$$(13.71)$$

56

* The term *polyaspartimide* is derived from the naturally occurring amino acid aspartic acid, $HO_2CCH_2CH(NH_2)CO_2H$.

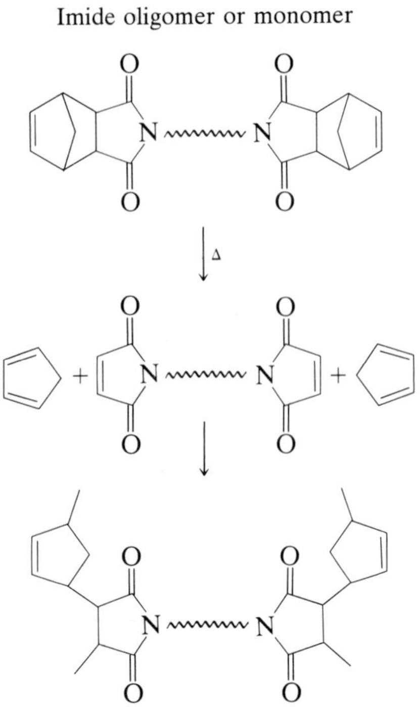

Imide oligomer or monomer

Scheme 13.2. Polymerization of a PMR-type nadimide end-capped prepolymer.

For the bisnadimide type (**54**), an amine-terminated imide prepolymer or a simple diamine is end-capped with nadimide groups by reaction with nadic anhydride or ester-acid.[75-77] Thermal curing at 270 to 310°C involves an initial retrograde Diels–Alder reaction followed by copolymerization of the resultant cyclopentadiene and maleimide end groups, as illustrated in Scheme 13.2. Curing is more complex than that shown, because homopolymerization of maleimide or copolymerization of unreacted nadimide groups may also occur.[78] The curing cycle must be controlled carefully to preclude voids arising from the highly volatile cyclopentadiene. Polymers of this type have been designated PMR (for *polymerization of monomeric reactants*).[75]

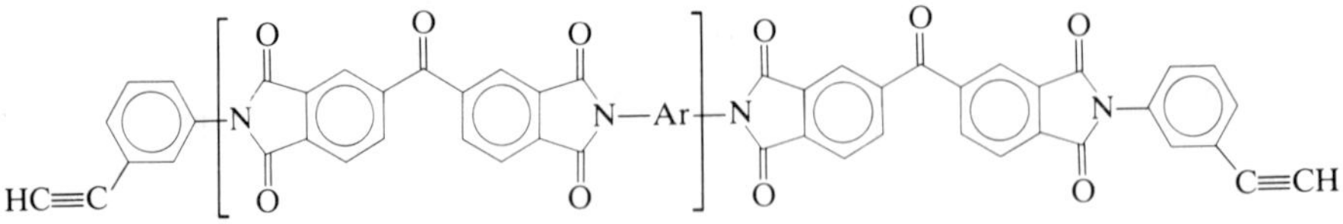

57

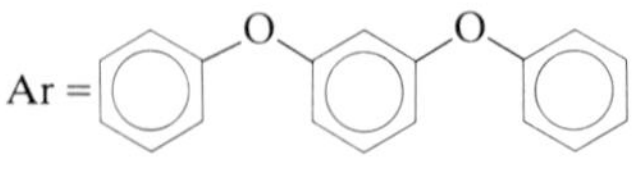

Ar =

Type 3 polyimides[79] were first encountered in Chapter 4 as one strategy for improving the processability of thermally stable polymers. An example of a commercial product (Gulf Oil Chemicals) is the ethynyl-capped polyimide oligomer **57**, which is useful as a high-temperature adhesive.

13.3.5 Heterocyclic polymers derived from polyamides and related polymers

The chemistry of heterocyclic compounds is usually treated as a special topic in the chemistry curriculum and, as such, is frequently not familiar to a great number of students. Yet the application of heterocyclic chemistry to the field of high polymers represents the most successful approach to developing thermally stable materials.[11b,34b,61b]

Table 13.5 lists a variety of heterocyclic polymers that might be considered to be derived from polymers described earlier in this chapter. In some instances they are prepared by modification of a precursor polymer; in others, a precursor polymer represents a short-lived intermediate in the synthesis. This is not an exhaustive list, but it serves to illustrate the great diversity of structures available by heterocyclic synthesis. More complete listings are available in the polymer literature.[61c] We will limit our discussion to those that have attracted commercial interest or that have been developed for commercial use.

13.3.5.1 Polybenzimidazoles[80]   These were among the first and most successful of the wholly aromatic polymers. They are normally prepared from esters of dibasic acids (particularly phenyl esters) and tetramines (reaction 13.72) or from diamino esters (13.73). The former reaction is used to prepare the commercially important polybenzimidazole **58** in which R is *m*-phenylene[13c,81] (see structure **1**, Chapter 4). Esters rather than acids are used because the latter tend to decarboxylate at the high temperatures necessary for benzimidazole formation. Polybenzimidazoles can, however, be prepared

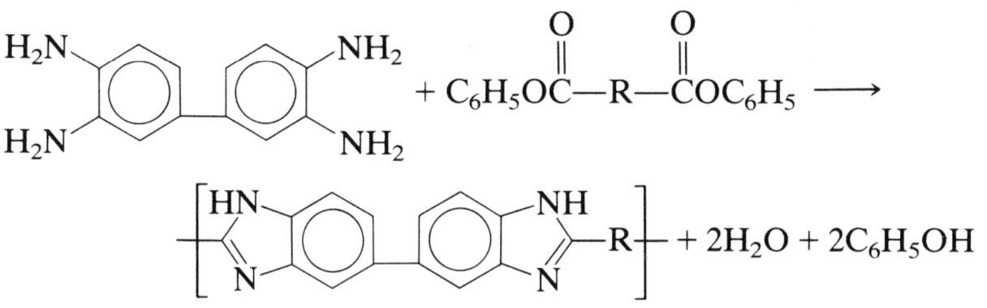

$$+ \; 2H_2O + 2C_6H_5OH \quad (13.72)$$

58

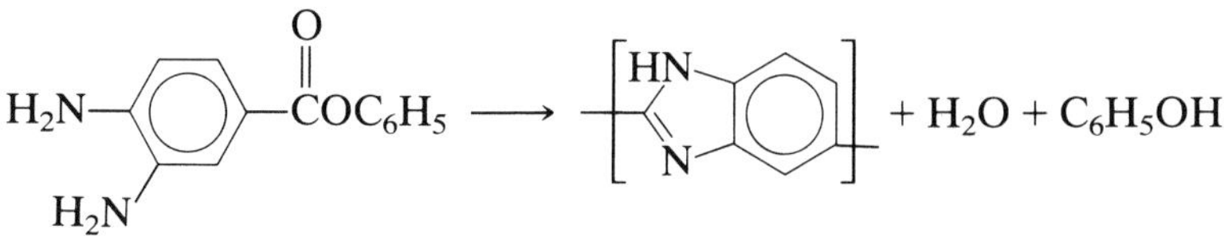

$$+ \; H_2O + C_6H_5OH \quad (13.73)$$

Table 13.5. Heterocyclic polymers derived from polyamides and related polymers

Ring system	Polymer type	Functional groups required	Intermediate stage
(benzimidazole ring structure)	Polybenzimidazole	Phenyl ester + diamine	Polyimine
(benzoxazole ring structure)	Polybenzoxazole	Phenyl ester + o-aminophenol	Polyamide
(benzothiazole ring structure)	Polybenzothiazole	Phenyl ester + o-aminothiophenol	Polyamide
(imidazopyrrolone ring structure)	Polyimidazopyrrolone	Anhydride + diamine	Polyamide
(1,3,4-oxadiazole ring structure)	Poly(1,3,4-oxadiazole)	Hydrazide	Polyhydrazide
(1,2,4-triazole ring structure)	Poly(1,2,4-triazole)	Hydrazide + amine	Polyhydrazide
(1,3,4-thiadiazole ring structure)	Poly(1,3,4-thiadiazole)	Thiohydrazide	Polythiohydrazide

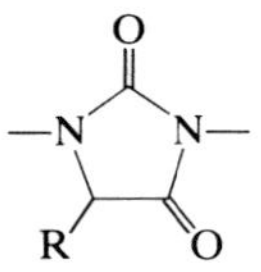 Polyhydantoin	Isocyanate + α-amino acid	Polyurea
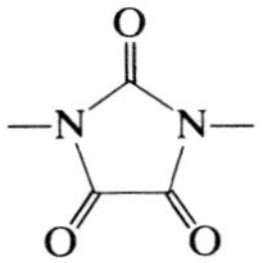 Poly(parabanic acid)	Isocyanate + HCN	Polyurea
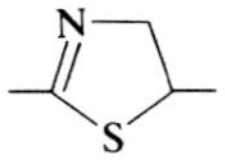 Polythiazoline	Thiourea + alcohol	Polythiourea
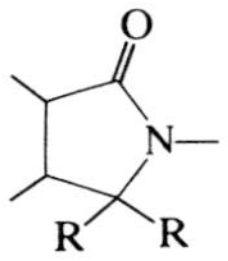 Polyimidine	Lactone + amine	Polyamide
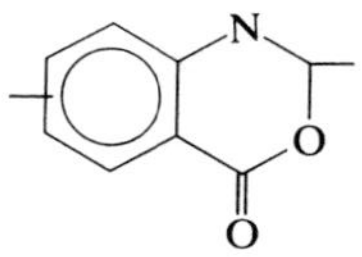 Polybenzoxazinone	Acid chloride + o-aminobenzoic acid	Polyamide

(continues)

Table 13.5. (Continued)

Ring system	Polymer type	Functional groups required	Intermediate stage
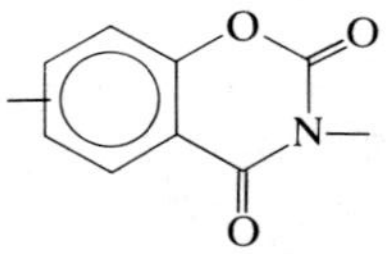	Polybenzoxazinedione	Amine + o-hydroxy benzoate ester	Polyamide
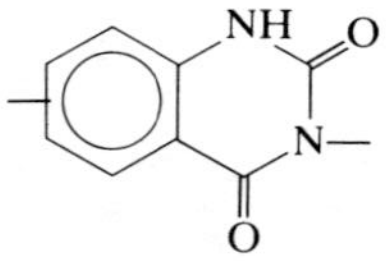	Polyquinazolinedione	Isocyanate + o-aminobenzoic acid	Polyurea
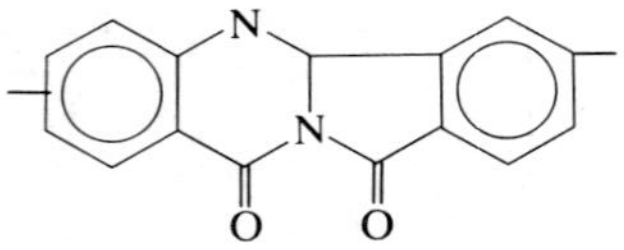	Polyisoindoloquinazolinedione	Anhydride + o-amino benzamide	Polyamide
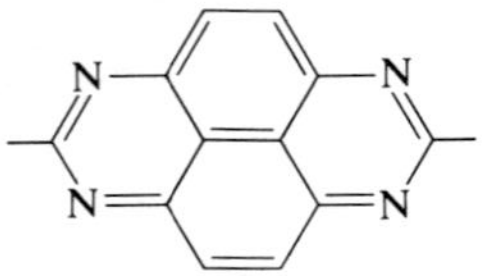	Polytetraazopyrene	Tetraamine + diphenyl ester	Polyamide

from diacids at moderate temperatures (about 140°C) when phosphorus pen-
toxide in methanesulfonic acid is used as a condensing agent.[82]

The more conventional synthetic procedure is to heat diester and tetra-
amine at temperatures below 300°C to form a solid prepolymer, with the
final polymerization being done in the solid state at temperatures up to 400°C.
It is tempting to write a mechanism involving nucleophilic attack of amine on
the ester carbonyl followed by loss of phenol to give polyamide intermediate.
Subsequent attack by the second amino group at the amide carbonyl followed
by loss of water would then give imidazole. Interestingly, however, it has
been shown[83] that water is evolved *before* phenol, suggesting a more plausible
mechanism via an intermediate Mannich base (13.74). Addition of the second
amino group to the carbon–nitrogen double bond followed by rate-
determining loss of phenol gives imidazole (13.75).

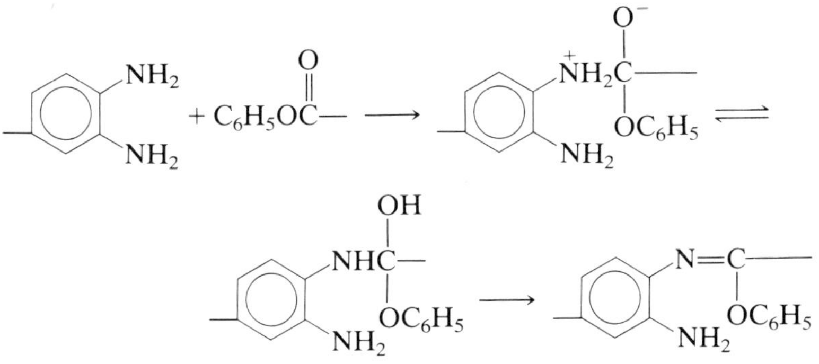

(13.74)

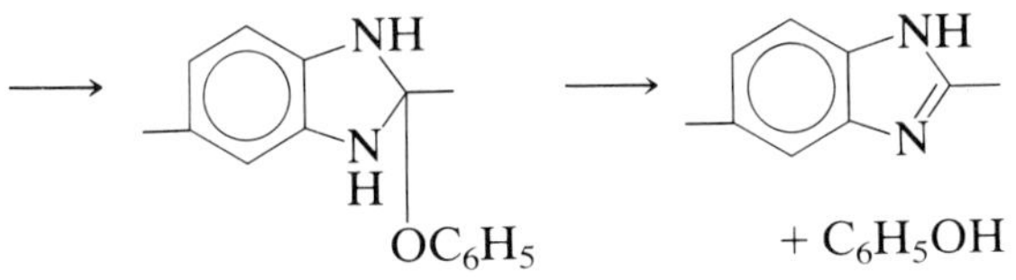

(13.75)

Interestingly, polybenzimidazoles can also be made by the reaction of
tetramines with dialdehydes[84] or dialdehyde bisulfite addition compounds.[85]
This leads to an intermediate polyimine (**59**) that undergoes cyclization on
reaction with oxygen (13.76).

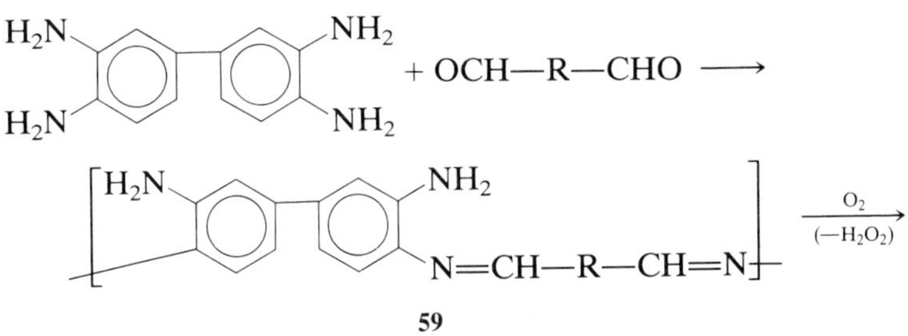

(13.76)

59

Polybenzimidazoles are highly colored (yellow to black) and exhibit good flame resistance. The limiting oxygen index (LOI) of **58** (R = *m*-phenylene) is 28, compared with values of 17 or 18 for aramids such as **5**. The polymers are porous because of loss of the more bulky phenol in the later stages of polymerization, and they retain moisture strongly, even above 100°C. Nevertheless, they exhibit very high hydrolytic stability. At elevated temperatures in the absence of air, polybenzimidazoles decompose to yield phenazine- and carbazole-type chars (**60** and **61**, respectively), but retain up to 80% of their mass up to 800°C. In air they undergo catastrophic weight loss between 450 and 500°C.

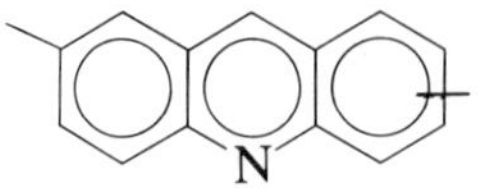

60 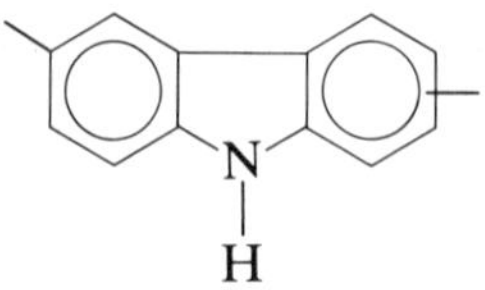   **61**

Related in structure to the polybenzimidazoles are the *polybenzoxazoles* and *polybenzothiazoles* (Table 13.5), which are prepared by analogous reactions using bis(*o*-aminophenol)s and bis(*o*-aminothiophenol)s in place of tetraamines. Both polymers are comparable in thermooxidative stability to the polybenzimidazoles.[11c]

If dianhydrides are reacted with tetraamines, ladder *polyimidazopyrrolones* (also called *ladder pyrrones*) are formed that combine the structural features of imidazole and pyrrolone rings.[86–88] Thermoplastic polyamide, which can be cast or molded prior to cyclization, is formed at room temperature (13.77). Subsequent heating results in a two-reaction sequence (13.78, 13.79) to yield blood-red thermosetting polymer (**62**). The center imidazopyrrolone structure of **62** would be inverted if the other possible polyamide structure were to form in reaction (13.77). Polyimidazopyrrolones exhibit excellent thermal stability and radiation resistance.

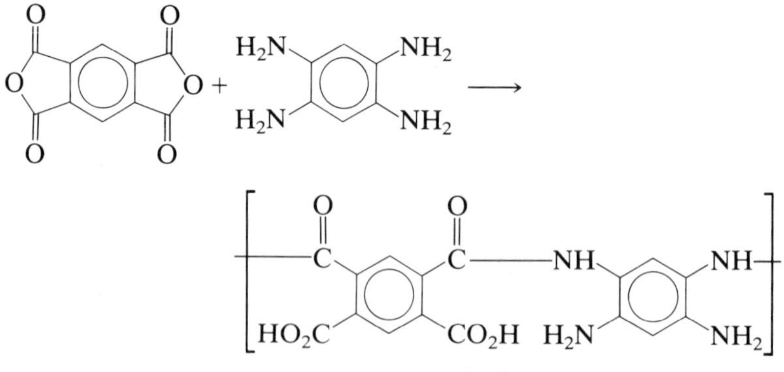

(13.77)

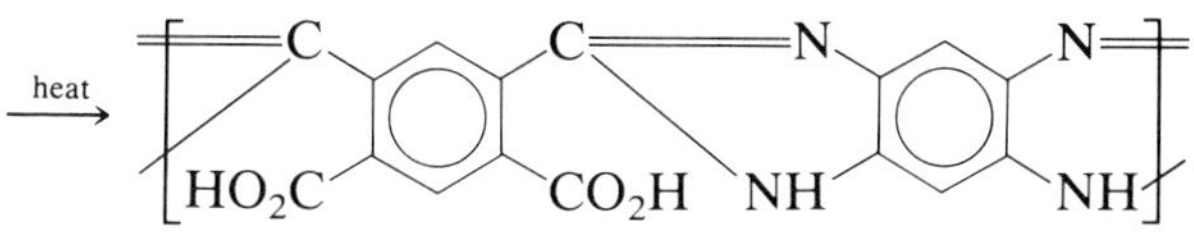

$$\xrightarrow{\text{heat}} \left[\begin{array}{c} \text{C} = \text{N} - \text{N} = \\ \text{HO}_2\text{C} \quad \text{CO}_2\text{H} \quad \text{NH} \quad \text{NH} \end{array}\right] + \text{H}_2\text{O} \quad (13.78)$$

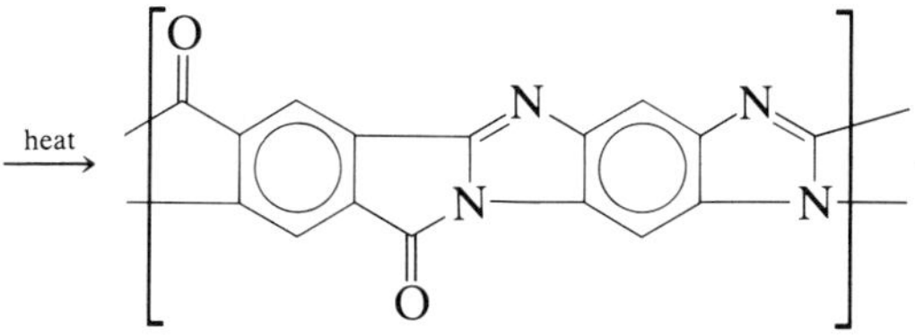

$$\xrightarrow{\text{heat}} \left[\begin{array}{c} \text{O} \\ \text{N} = \text{N} \\ \text{N} \quad \text{N} \\ \text{O} \end{array}\right] + \text{H}_2\text{O} \quad (13.79)$$

62

13.3.5.2 Poly(1,3,4-oxadiazole)s, poly(1,2,4-triazole)s, and poly(1,3,4-thiadiazole)s[34b,89,90] Polyhydrazides can be converted to poly(1,3,4-oxidiazole)s (**64**) by thermal dehydration, shown in reaction (13.80) for polyhydrazide **63**. If **63** is heated with aniline in polyphosphoric acid (PPA) (13.81), the phenyl-substituted poly(1,2,4-triazole) (**65**) is formed:

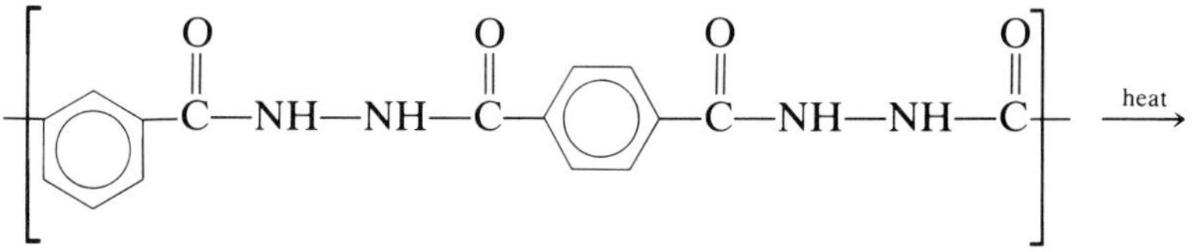

$$\left[\begin{array}{c} \text{O} \quad \text{O} \quad \text{O} \quad \text{O} \\ \text{C} - \text{NH} - \text{NH} - \text{C} - \text{C} - \text{NH} - \text{NH} - \text{C} \end{array}\right] \xrightarrow{\text{heat}}$$

63

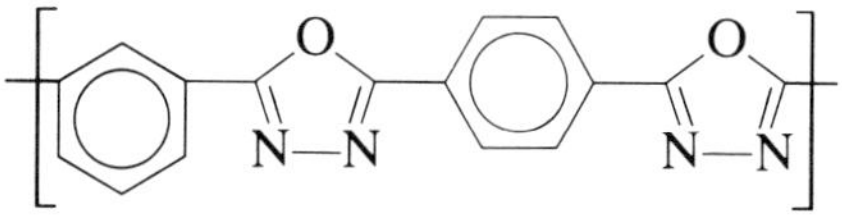

$$\left[\begin{array}{c} \text{O} \quad\quad \text{O} \\ \text{N} - \text{N} \quad \text{N} - \text{N} \end{array}\right] + 2\text{H}_2\text{O} \quad\quad (13.80)$$

64

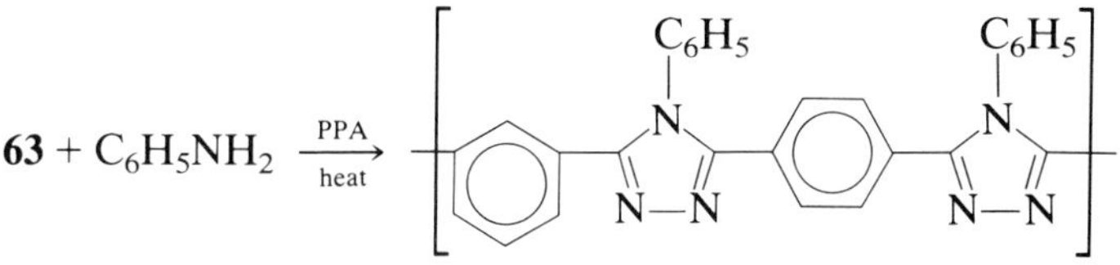

$$\mathbf{63} + \text{C}_6\text{H}_5\text{NH}_2 \xrightarrow[\text{heat}]{\text{PPA}} \left[\begin{array}{c} \text{C}_6\text{H}_5 \quad\quad \text{C}_6\text{H}_5 \\ \text{N} \quad\quad \text{N} \\ \text{N} - \text{N} \quad \text{N} - \text{N} \end{array}\right] \quad (13.81)$$

65

Poly(1,3,4-thiadiazole)s (**66**) are prepared by dehydrosulfurization (13.82) or dehydration (13.83) of dithiahydrazides or monothiahydrazides, respectively.

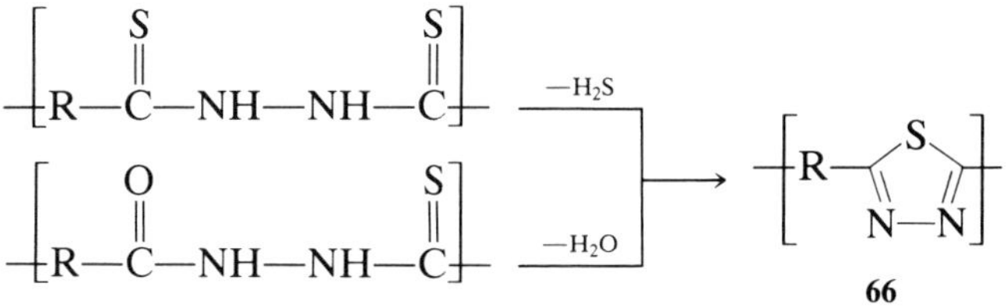

$$\left[\begin{array}{c} \text{S} \quad\quad \text{S} \\ \text{R} - \text{C} - \text{NH} - \text{NH} - \text{C} \end{array}\right] \xrightarrow{-\text{H}_2\text{S}} \quad\quad (13.82)$$

$$\left[\begin{array}{c} \text{O} \quad\quad \text{S} \\ \text{R} - \text{C} - \text{NH} - \text{NH} - \text{C} \end{array}\right] \xrightarrow{-\text{H}_2\text{O}} \left[\begin{array}{c} \text{S} \\ \text{R} \\ \text{N} - \text{N} \end{array}\right] \quad (13.83)$$

66

The three polymer types exhibit good hydrolytic and thermal stability, but, as is often the case with heterocyclic polymers, poor light stability. Poly(1,3,4-oxadiazole)s are of interest in high-modulus fiber formulations.[36]

Polymers of analogous structure can be prepared from dihydrazidine (67) monomers[91] by first reacting with diacid chloride (13.84) to form poly-(N-acylamidrazone) (68). Heating converts 68 to poly(1,3,4-oxadiazole) and poly(1,2,4-triazole) by loss of ammonia (13.85) and water (13.86), respectively:

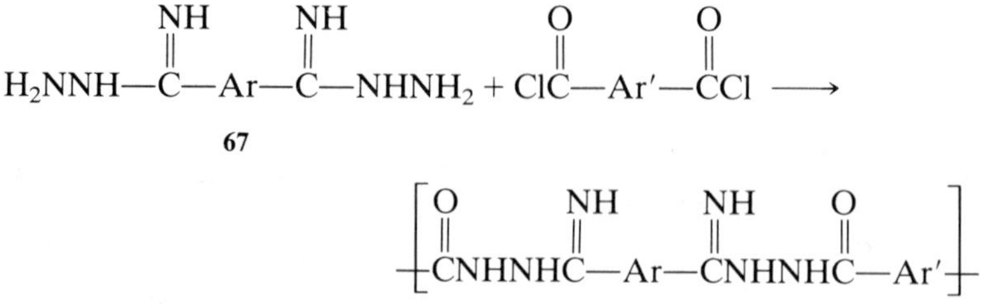

$$\underset{\text{67}}{H_2NNH-\overset{\overset{\displaystyle NH}{\|}}{C}-Ar-\overset{\overset{\displaystyle NH}{\|}}{C}-NHNH_2} + ClC-Ar'-CCl \longrightarrow$$

$$\underset{\text{68}}{\left[\overset{\overset{\displaystyle O}{\|}}{C}NHNH\overset{\overset{\displaystyle NH}{\|}}{C}-Ar-\overset{\overset{\displaystyle NH}{\|}}{C}NHNH\overset{\overset{\displaystyle O}{\|}}{C}-Ar'\right]} \quad (13.84)$$

$$\text{68} \quad \xrightarrow{-NH_3} \quad \left[\underset{N-N}{\overset{O}{\diagup\diagdown}}-Ar-\underset{N-N}{\overset{O}{\diagup\diagdown}}-Ar'\right] \quad (13.85)$$

$$\xrightarrow{-H_2O} \quad \left[\underset{N-N}{\overset{NH}{\diagup\diagdown}}-Ar-\underset{N-N}{\overset{NH}{\diagup\diagdown}}-Ar'\right] \quad (13.86)$$

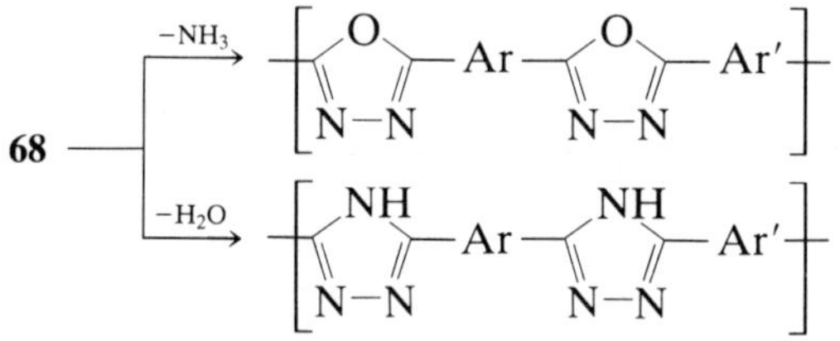

Poly(1,3,4-oxadiazole)s and poly(1,2,4-triazole)s can be made directly by treating a dihydrazide with a strong dehydrating agent such as sulfuric acid (13.87), or with excess hydrazine (13.88), respectively:

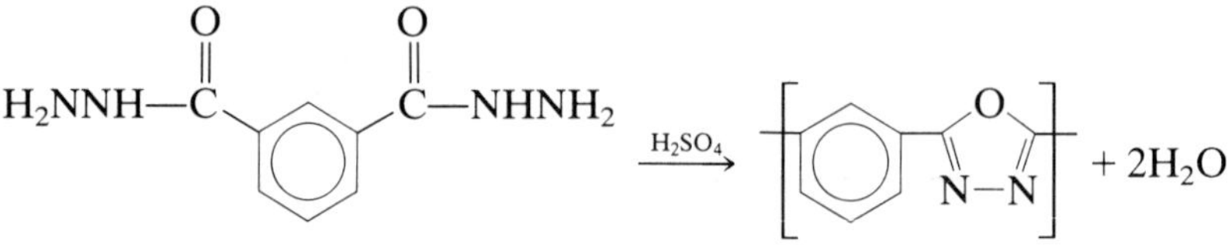

$$H_2NNH-\overset{\overset{\displaystyle O}{\|}}{C}\diagdown\!\!\!\diagup\overset{\overset{\displaystyle O}{\|}}{C}-NHNH_2 \quad \xrightarrow{H_2SO_4} \quad \left[\underset{N-N}{\overset{O}{\diagup\diagdown}}\right] + 2H_2O$$

$$(13.87)$$

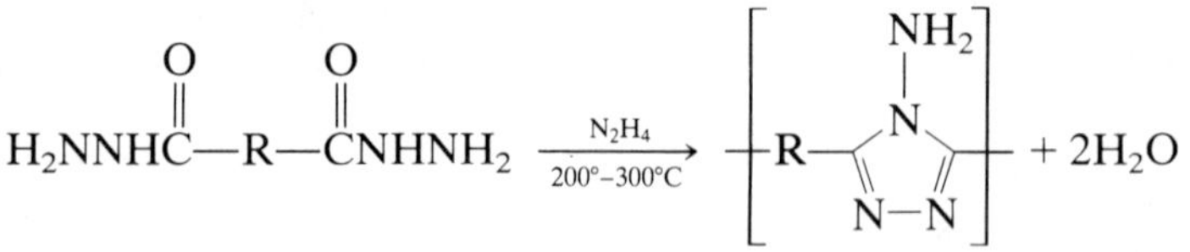

$$H_2NNHC-R-CNHNH_2 \quad \xrightarrow[200°-300°C]{N_2H_4} \quad \left[R-\underset{N-N}{\overset{\overset{\displaystyle NH_2}{|}}{\overset{N}{\diagup\diagdown}}}\right] + 2H_2O \quad (13.88)$$

13.3.5.3 Polyhydantoins[92] Several routes are available for the synthesis of polyhydantoins, among them the reaction of diisocyanates with bis(α-amino acid)s (13.89) to form polymer 69. Initially a polyurea is formed by addition of the amino group to isocyanate, which is then followed by cyclization with loss of water. Characteristically, polyhydantoins exhibit high glass transition temperatures and good chemical resistance. They have been used as wire coatings and insulating films for electric motors.[56b]

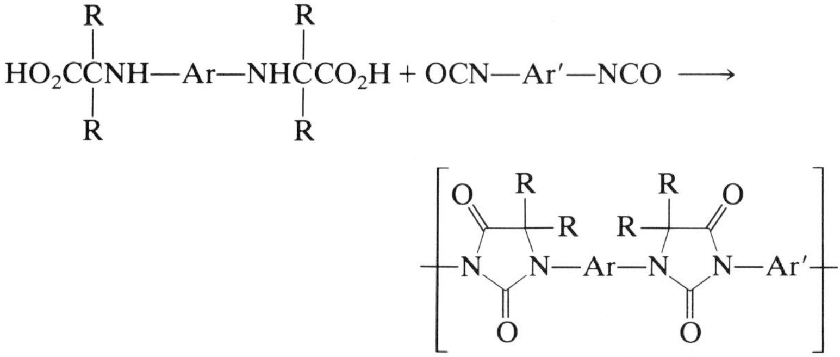

$$\underset{\underset{R}{|}}{\overset{\overset{R}{|}}{HO_2CCNH}}\!-\!Ar\!-\!\underset{\underset{R}{|}}{\overset{\overset{R}{|}}{NHCCO_2H}} + OCN\!-\!Ar'\!-\!NCO \longrightarrow$$

(13.89)

69

13.3.5.4 Poly(parabanic acid)s[61b,93]   Also called *poly(1,3-imidazolidine-2,4,5-trione)s*, these polymers are manufactured by the reaction of diisocyanates with HCN (13.90) to yield an intermediate poly(5-imino-1,3-imidazolidinedione) (**70**), which is hydrolyzed (13.91) in a separate step to the poly(parabanic acid) (**71**). The polymers may be thought of as dicarbonyl derivatives of polyureas; indeed, the parabanic acid unit can be prepared by the reaction of polyureas with oxalyl chloride[94] (13.92).

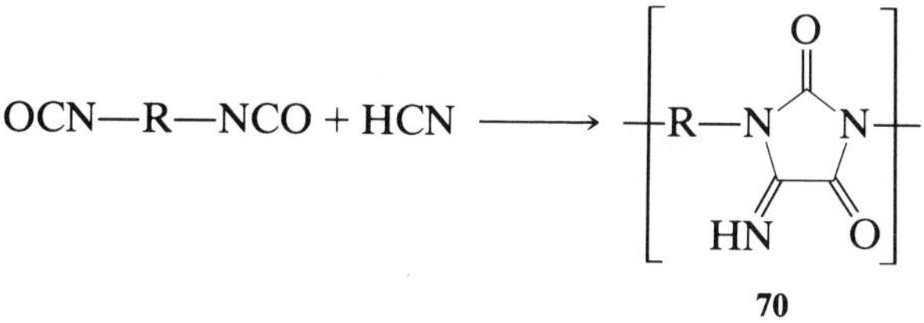

$$OCN\!-\!R\!-\!NCO + HCN \longrightarrow \quad$$ (13.90)

70

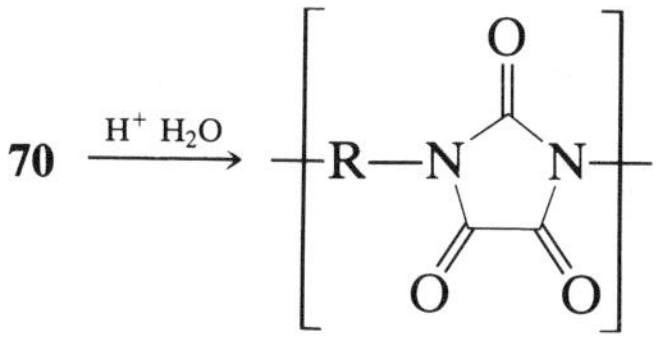

$$\mathbf{70} \xrightarrow{H^+ \ H_2O}$$ (13.91)

71

$+ \ Cl\!-\!\overset{\overset{O}{\|}}{C}\!-\!\overset{\overset{O}{\|}}{C}\!-\!Cl \xrightarrow{\Delta} \mathbf{71}$ (13.92)

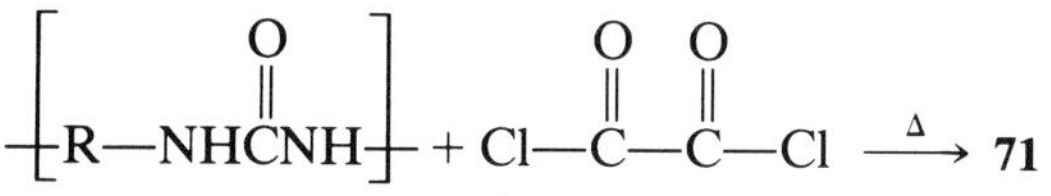

Poly(parabanic acid)s were originally considered lower cost substitutes for polyimides.

References

1. R. M. Moncrieff, *Man-Made Fibers*, Butterworth, London, 1975, Chaps. 19–22.
2. M. I. Kohan (ed.), *Nylon Plastics*, Wiley-Interscience, New York, 1973.
3. D. B. Jacobs and J. Zimmerman, in *Polymerization Processes* (C. E. Schildknecht, ed.), Wiley-Interscience, New York, 1977, Chap. 12.
4. H. K. Livingston, M. S. Sioshansi, and M. D. Glick, *J. Macromol. Sci.-Revs., Macromol. Chem.*, **C6**, 29 (1971).
5. E. Roerdink and J. M. M. Warnier, *Polymer*, **26**, 1582 (1985).
6. J. Preston and W. L. Hofferbert, *J. Polym. Sci., Poly. Symp. Ed.*, **65**, 13 (1978).
7. H. Tanaka, G. Wu, Y. Iwanage, K. Sanui, and N. Ogata, *Polym. J.*, **14**, 635 (1982).
8. P. W. Morgan, *Condensation Polymers: By Interfacial and Solution Methods*, Wiley-Interscience, New York, 1965: (a) Chap. 5; (b) Chap. 7; (c) Chap. 6; (d) Chap. 9.
9. F. Millich and C. E. Carraher (eds.), *Interfacial Synthesis*, Vol. 2, Dekker, New York, 1977: (a) V. Z. Nikonov and V. M. Savinov, Chap. 15; (b) K. C. Steuben and A. E. Barnabeo, Chap. 18; (c) T. Tanaka and T. Yokoyama, Chap. 17.
10. V. V. Korshak, *Heat-Resistant Polymers*, Halsted Press, New York, 1972.
11. J. P. Critchley, G. J. Knight, and W. W. Wright, *Heat-Resistant Polymers*, Plenum Press, New York, 1983: (a) pp. 144–150; (b) Chap. 5; (c) Chap. 9.
12. R. S. Lenk, *Macromol. Rev.*, **13**, 355 (1978).
13. M. Lewin and J. Preston (eds.), *High Technology Fibers*, Part A, Dekker, New York, 1985: (a) M. Jaffe and R. S. Jones, Chap. 9; (b) M. Couper, Chap. 2; (c) A. B. Conciatori, A. Bukley, and D. E. Stuetz, Chap. 6.
14. H. J. Chang and O. Vogl, *J. Polym. Sci., Polym. Chem. Ed.*, **15**, 311, 1043 (1977).
15. C. G. Overberger and J. Šebenda, *J. Polym. Sci. A-1*, **7**, 2875 (1969).
16. F. Millich and K. V. Seshadri, in *Cyclic Monomers* (K. C. Frisch, ed.), Wiley-Interscience, New York, 1972, Chap. 3.
17. K. C. Frisch and S. L. Reegen (eds.), *Ring-Opening Polymerization*, Dekker, New York, 1969: (a) H. K. Reimschuessel, Chap. 7; (b) Y. Shalitin, Chap. 10.
18. J. Šebenda, *J. Macromol. Sci.-Chem.*, **A6**, 1145 (1972).
19. M. Rothe and G. Bertalan, in *Ring-Opening Polymerization* (T. Saegusa and E. Goethals, eds.), ACS Symp. Ser. 59, American Chemical Society, Washington, D.C., 1977.
20. R. Bacskai, in *Crown Ethers and Phase Transfer Catalysts in Polymer Science* (L. J. Mathias and C. E. Carraher Jr., eds.), Plenum Press, New York, 1984, pp. 183–199.
21. W. R. Sorenson and T. W. Campbell, *Preparative Methods of Polymer Chemistry*, 2nd ed., Wiley-Interscience, New York, 1968: (a) pp. 377ff; (b) 349ff.
22. N. S. Isaacs, *Chem. Soc. Rev.*, **5**, 181 (1976).
23. M. Swarc, *Adv. Polym. Sci.*, **4**, 1 (1965).
24. H. R. Kricheldorf, *Alpha-Amino Acid-N- Carboxy Anhydride and Related Heterocycles*, Springer-Verlag, New York, 1987.
25. H. J. Harwood, in *Ring-Opening Polymerization* (J. E. McGrath, ed.), ACS Symp. Ser. 286, American Chemical Society, Washington, D.C., 1985, pp. 67–85.

26. J. H. Saunders and K. C. Frisch, *Polyurethanes: Chemistry and Technology*, Part 1, Wiley-Interscience, New York, 1962, p. 79.
27. D. F. Regelman, L. M. Alberino, and R. J. Lockwood, in *Reaction Injection Molding* (J. E. Kresta, ed.), ACS Symp. Ser. 270, American Chemical Society, Washington, D.C., 1985, pp. 125–134.
28. V. E. Shashoua, W. Sweeny, and R. F. Tietz, *J. Am. Chem. Soc.*, **82**, 866 (1960).
29. G. Natta, D. Pietro, and M. Cambini, *Makromol. Chem.*, **56**, 200 (1962).
30. T. Otsu, B. Yamada, M. Itahashi, and T. Mori, *J. Polym. Sci., Polym. Chem. Ed.*, **14**, 1347 (1976).
31. L. W. Bush and D. S. Breslow, *Macromolecules*, **1**, 189 (1968).
32. M. Ueda, O. Hara, A. Sato, and Y. Imai, *J. Polym. Sci., Polym. Chem. Ed.*, **17**, 769 (1979).
33. M. Yoneyama, M. Kakimoto, and Y. Imai, *Macromolecules*, **21**, 1908 (1988).
34. A. H. Frazer, *High Temperature Resistant Polymers*, Wiley-Interscience, New York, 1968: (a) Chap. 3; (b) Chap. 4.
35. R. A. Dine-Hart, B. J. C. Moore, and W. W. Wright, *J. Polym. Sci. B*, **2**, 369 (1964).
36. W. B. Black and J. Preston (eds.), *High-Modulus Wholly Aromatic Fibers*, Dekker, New York, 1973.
37. B. Espenschied and R. C. Schulz, *Makromol. Chem., Rapid Commun.*, **4**, 633 (1983).
38. A. K. Mukherjee and H. R. Goel, *J. Macromol. Sci., Rev. Macromol. Chem.*, **C25**, 99 (1985).
39. N. A. Platé and V. P. Shibaev, *J. Polym. Sci. Macromol. Rev.*, **8**, 117 (1974).
40. T. S. Ellis, *J. Appl. Polym. Sci.*, **36**, 451 (1988).
41. W. R. Sorenson, *J. Org. Chem.*, **24**, 978 (1959).
42. N. Yamazaki, T. Iguchi, and F. Higashi, *J. Polym. Sci., Polym. Chem. Ed.*, **13**, 785 (1975).
43. J. H. Saunders and K. C. Frisch, *Polyurethanes: Chemistry and Technology*, Wiley-Interscience, New York, Part 1, 1962; Part 2, 1964.
44. L. N. Phillips and D. B. Parker, *Polyurethanes—Chemistry, Technology, and Properties*, Gordon and Breach, New York, 1964.
45. K. N. Edwards (ed.), *Urethane Chemistry and Applications*, ACS Symp. Ser. 172, American Chemical Society, Washington, D.C., 1981.
46. E. N. Doyle, *The Development and Use of Polyurethane Products*, McGraw-Hill, New York, 1969.
47. D. C. Allport and A. A. Mohajer, in *Block Copolymers* (D. C. Allport and W. H. Janes, eds.), Halsted Press, New York, 1973, Chap. 5.
48. R. R. Dileone, *J. Polym. Sci. A-1*, **8**, 609 (1970).
49. J. E. Herweh and W. Y. Whitmore, *J. Polym. Sci. A-1*, **8**, 2759 (1970).
50. J. F. Kinstle and L. E. Sepulveda, *J. Polym. Sci., Polym. Lett. Ed.*, **15**, 467 (1977).
51. A. Rokicki and W. Kuran, *J. Macromol. Sci., Rev. Macromol. Chem.*, **C21**, 135 (1981).
52. T. T. Healey (ed.), *Polyurethane Foams*, Gordon and Breach, New York, 1968.
53. K. C. Frisch and J. H. Saunders, *Plastic Foams*, Dekker, New York, Part 1, 1972; Part 2, 1973.
54. J. J. Bikerman, *Foams*, Springer-Verlag, New York, 1973.
55. E. Dyer and R. A. Dunbar, *J. Polym. Sci. A-1*, **8**, 629 (1970).

464   *Nonvinyl Polymers*

56. J. A. Brydson, *Plastics Materials*, 3rd ed., Butterworth, London, 1975; (a) Chap. 27; (b) pp. 483–484.
57. P. W. Morgan, *J. Polym. Sci., Polym. Symp.*, **65**, 1 (1978).
58. K. L. Mittal (ed.), *Polyimides*, Vols. 1 and 2, Plenum-Press, New York, 1984.
59. M. I. Beesonov, M. M. Koton, V. V. Kudryavtsev, and L. A. Laius, *Polyimides—Thermally Stable Polymers*, Plenum Press, New York, 1987.
60. M. W. Ranney, *Polyimide Manufacture*, Noyes Data Corp., Park Ridge, N.J., 1971.
61. P. E. Cassidy, *Thermally Stable Polymers*, Dekker, New York, 1980: (a) Chap. 5; (b) Chaps. 6–9; (c) Appendix A.
62. C. E. Sroog, *Macromol. Rev.*, **11**, 161 (1976).
63. R. J. Cotter and M. Matzner, *Ring-Forming Polymerizations*, Part B2, Academic Press, New York, 1972; Chap. 1.
64. I. K. Varma, R. N. Goel, and D. S. Varma, *J. Polym. Sci., Polym. Chem. Ed.*, **17**, 703 (1979).
65. V. L. Bell, K. E. M. Hett, and G. M. Stokes, *J. Appl. Polym. Sci.*, **26**, 3805 (1981).
66. S. S. Hirsch, *J. Polym. Sci. A-1*, **7**, 15 (1969).
67. P. S. Carleton, W. J. Farrissey Jr., and J. S. Rose, *J. Appl. Polym. Sci.*, **16**, 2983 (1972).
68. W. M. Alvino and L. E. Edelman, *J. Appl. Polym. Sci.*, **19**, 2961 (1975).
69. N. D. Ghatge and U. P. Mulik, *J. Polym. Sci., Polym. Chem. Ed.*, **18**, 1905 (1980).
70. H. D. Stenzenberger, M. Herzog, W. Romer, R. Scheiblich, and N. J. Reeves, *Br. Polym. J.*, **15**, 2 (1983).
71. B. C. Trivedi and B. M. Culbertson, *Maleic Anhydride*, Plenum Press, New York, 1982, pp. 511ff.
72. I. K. Varma and S. Sharma, *Eur. Polym. J.*, **20**, 1101 (1984).
73. J. E. White, M. D. Scaia, and D. A. Snider, *J. Appl. Polym. Sci.*, **29**, 891 (1984).
74. V. L. Bell and P. R. Young, *J. Polym. Sci., Polym. Chem. Ed.*, **24**, 2647 (1986).
75. T. T. Serafini, P. Delvigs, and G. R. Lightsey, *J. Appl. Polym. Sci.*, **16**, 905 (1972).
76. D. A. Scola and M. P. Stevens, *J. Appl. Polym. Sci.*, **26**, 231 (1981).
77. S. E. Delos, R. K. Schellenberg, J. E. Smedley, and D. E. Kranbuel, *J. Appl. Polym. Sci.*, **27**, 4295 (1982).
78. N. G. Gaylord and M. Martan, in *Cyclopolymerization and Polymers with Ring-Chain Structures* (G. Butler and J. E. Kresta, eds.), ACS Symp. Ser 195, American Chemical Society, Washington, D.C., 1982, pp. 97–105.
79. H. Stenzenberger, *Br. Polym. J.*, **20**, 389 (1988).
80. E. W. Neuse, *Adv. Polym. Sci.*, **47**, 1 (1982).
81. G. M. Moelter, R. F. Tetrault, and N. Heffland, *Polym. News*, **9**, 134 (1983).
82. M. Ueda, M. Sato, and A. Mochizuki, *Macromolecules*, **18**, 2723 (1985).
83. W. Wrasidlio and H. H. Levine, *J. Polym. Sci. A*, **2**, 4795 (1964).
84. E. W. Neuse and M.S. Loonat, *Macromolecules*, **16**, 128 (1983).
85. J. Higgins and C. S. Marvel, *J. Polym. Sci. A-1*, **8**, 171 (1970).
86. C. S. Marvel, in *High-Temperature Polymers* (C. L. Segal, ed.), Dekker, New York, 1967, pp. 7ff.
87. B. Nartissov, *J. Macromol. Sci., Rev. Macromol. Chem.*, **C11**, 143 (1974).

88. V. V. Korshak and A. L. Rusanov, *J. Macromol. Sci., Rev. Macromol. Chem.*, **C21**, 275 (1982).
89. R. J. Cotter and M. Matzner, *Ring-Forming Polymerizations*, Part B1, Academic Press, New York, 1972, Chap. 2.
90. P. E. Cassidy and N. C. Fawcett, *J. Macromol. Sci., Rev. Macromol. Chem.*, **C17**, 209 (1979).
91. P. M. Hergenrother, in *Step-Growth Polymerizations* (D. H. Solomon, ed.), Dekker, New York, 1972, Chap. 4.
92. R. Merten, *Angew. Chem. Int. Ed. English*, **10**, 294 (1971).
93. T. L. Patton, *Polym. Preprints*, **12**(1), 162 (1971).
94. G. Caraculacu, E. Scortanu, and A. A. Caraculacu, *Eur. Polym. J.*, **19**, 143 (1983).

Review exercises

1. Write IUPAC names for as many of the polyamides in Table 13.1 as you can.

2. Ignoring steric effects, which would you expect to have the greater effect on a polyamide's melting point, substitution on nitrogen or substitution on carbon? Explain.

3. Give plausible explanations for the following facts:
   (a) Most commercially important polyamides are aliphatic, whereas commercial polyesters are mostly aromatic.
   (b) Polycondensation of α-amino acids is not a practical way to obtain nylons 4 and 5.
   (c) Polyurethanes used for organ implantation are made with polyether glycols, not the less expensive polyester glycols.
   (d) The percent crystallinity and T_m of nylon 46 are significantly higher than those of nylon 66, more so than one might expect on the basis of chain lengths alone.
   (e) Polyureas are generally higher melting than polyamides of like structure.
   (f) When primary diamines are used to make polyaspartimides (reaction 13.71), gelation occurs. (J. E. White et al., *Polym. Preprints*, **26**(1), 132, 1985)

4. Write a mechanism for the anionic polymerization of 2-pyrrolidone (**12**). Explain the function of the phase-transfer catalyst in increasing the rate. (ref. 20)

5. Polysulfonamide **6** (reaction 13.9) is soluble in base. Explain why. In basic solution the polymer can be methylated with dimethyl sulfate. Write the structure for the methylated polymer. How might the same polymer be formed directly?

6. Nylon 1313 has been investigated by the U.S. Department of Agriculture as a potential engineering plastic. Starting point for the synthesis is erucic acid, $CH_3(CH_2)_7CH=CH(CH_2)_{11}CO_2H$, obtainable from the seeds of crambe, a potential cattle feed. Beginning with erucic acid as the only source of carbon compounds, suggest a route to nylon 1313. (H. J. Nieslag, *Ind. Eng. Chem. Prod. Res. Dev.*, **16**, 101, 1977)

7. Write a plausible mechanism for the synthesis of nylon 3 from acrylamide (reaction 13.35, Table 13.2). Write reactions for two other ways that could, in principle, result in the same polymer.

8. What other reaction could, in principle, occur in reaction (13.37) (Table 13.2)? Why doesn't it? (Note the struture of the Ar groups.)

9. In reaction (13.49), what is a likely precursor to the polyurea? What is the function of the benzoic acid in the reaction? (Hint: Consider reaction 13.30.) What role must the dimethyl sulfoxide be playing? Can you suggest an overall mechanism?

10. Write a reaction illustrating the synthesis of the polyamidehydrazide **42**. (ref. 36)

11. Two commercial engineering plastics are the *polyimidesulfone* **52** and the *polyamideimide* of structure

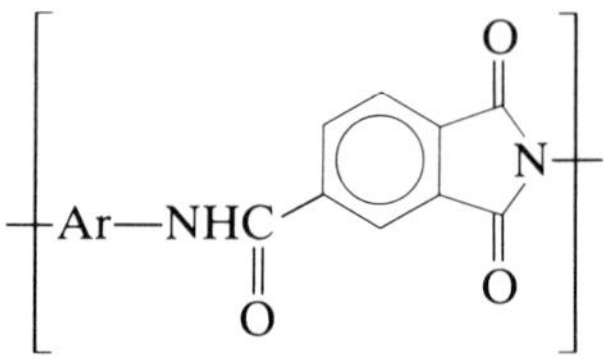

Write reactions illustrating their synthesis.

12. The commercially important polyetherimide described in Chapter 11 (reaction 11.37) is made by a displacement reaction. Suggest an alternative synthesis. (T. Takekoshi et al., *Polym. Preprints*, **24**(2), 312, 1983)

13. In Chapter 11 (reaction 11.70), the synthesis of polyimidesulfides by Michael addition was described. Suggest an alternative route that does not entail prior synthesis of bismaleimides. (V. A. Sergeyev et al., *Polym. Commun.*, **26**, 365, 1985)

14. When polyurethane **33** (reaction 13.59) was heated above 200°C, considerable weight loss occurred. A solid that condensed on the cooler surfaces of the thermolysis apparatus was found to be identical to **33**, indicating that the polymer had apparently "sublimed." What is the most likely explanation? (ref. 50)

15. For the polyurethane formed in reaction (13.52), would an alternative synthesis involving the reaction of a biscarbamyl chloride [see reaction (13.45)] with a diol or bisphenol be feasible? Explain.

16. Predict the repeating unit of the polymers formed in each of the following reactions:

    (a)

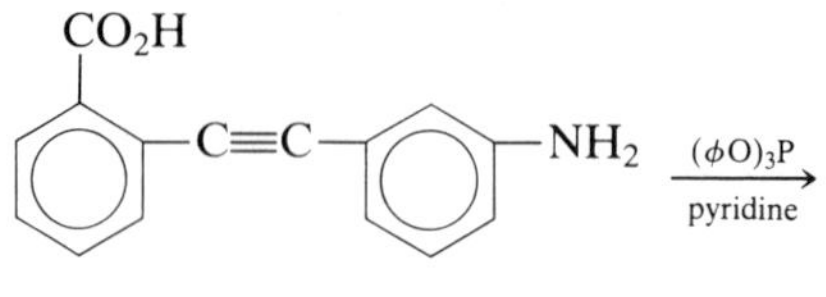

(R. C. Evers, E. J. Moore, and T. Abraham, *J. Polym. Sci. A: Polym. Chem.*, **26**, 3213, 1988)

(b)

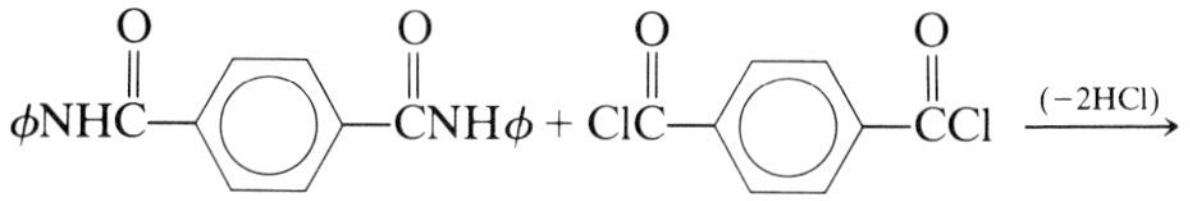

$$\phi NH\overset{O}{\underset{\|}{C}}-\!\!\!\bigcirc\!\!\!-\overset{O}{\underset{\|}{C}}NH\phi + Cl\overset{O}{\underset{\|}{C}}-\!\!\!\bigcirc\!\!\!-\overset{O}{\underset{\|}{C}}Cl \xrightarrow{(-2HCl)}$$

(V. V. Korshak et al., *Izv. Akad. Nauk SSSR. Sci. Khim.*, **6**, 1402, 1969; *Chem. Abstr.*, **71**, 81784, 1969)

(c)

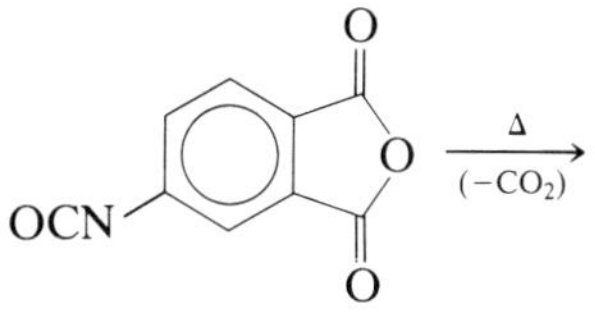

$$OCN-\text{[phthalic anhydride ring]} \xrightarrow[(-CO_2)]{\Delta}$$

(H. Ulrich and R. Richter, *J. Org. Chem.*, **38**, 2557, 1973)

(d)

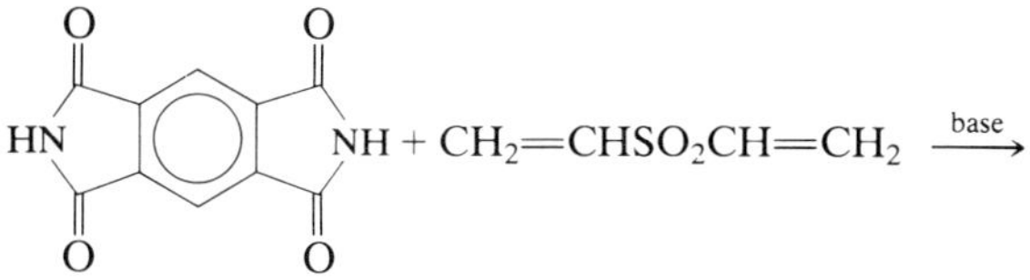

$$HN\text{[pyromellitic diimide]}NH + CH_2{=}CHSO_2CH{=}CH_2 \xrightarrow{base}$$

(M. Russo and L. Mortillaro, *J. Polym. Sci. A-1*, **7**, 3337, 1969)

(e)

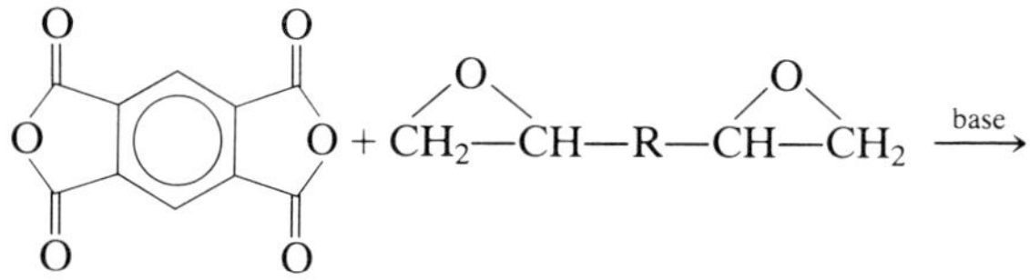

$$\text{[pyromellitic dianhydride]} + \overset{O}{CH_2{-}CH}{-}R{-}\overset{O}{CH{-}CH_2} \xrightarrow{base}$$

(Y. Iwakura and F. Hayano, *J. Polym. Sci. A-1*, **7**, 598, 1969)

(f)

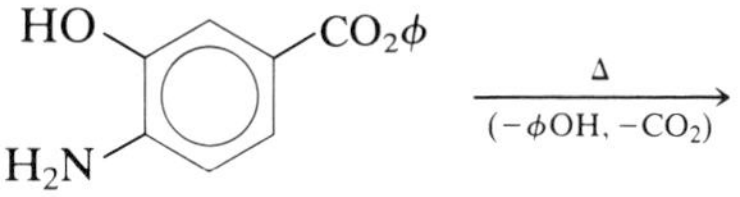

$$\underset{H_2N}{\overset{HO}{\diagdown}}\text{[benzene ring]}CO_2\phi \xrightarrow[(-\phi OH, -CO_2)]{\Delta}$$

(ref. 53b)

(g)

$$\text{[benzophenone tetracarboxylic dianhydride]}$$

$$+ H_2N{-}N{=}CH-\!\!\!\bigcirc\!\!\!-CH{=}N{-}NH_2 \longrightarrow \xrightarrow[\text{pyridine}]{Ac_2O}$$

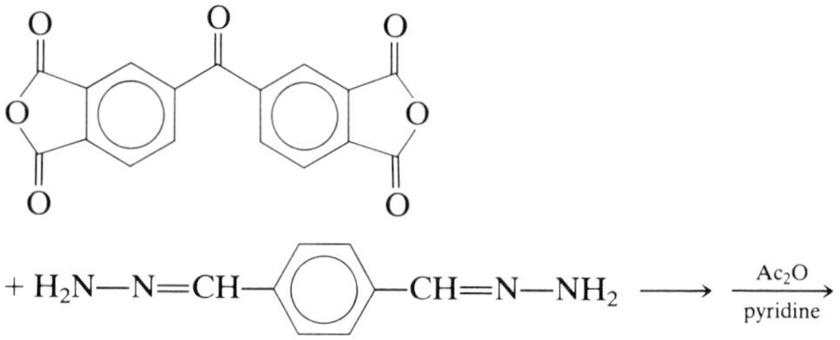

(D. B. Sobanski et al., *J. Polym. Sci., Polym. Chem. Ed.*, **A23**, 189, 1986)

(h)

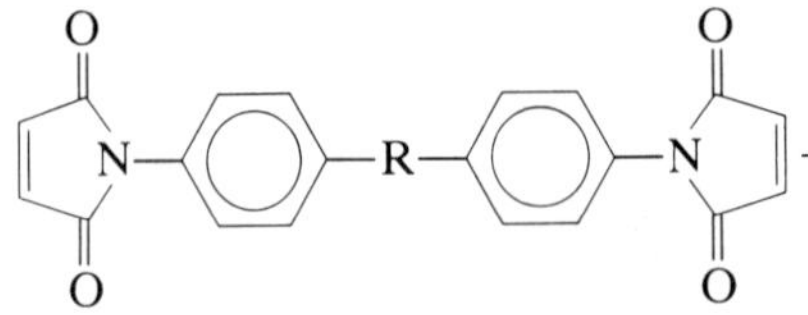

$$\text{(bismaleimide structure)} + CH_3NH(CH_2)_6NHCH_3 \longrightarrow$$

(J. E. White et al., *Polym. Preprints*, **26**(1), 132, 1985)

(i)

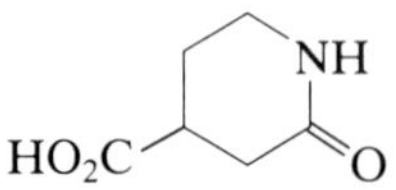

$$\xrightarrow[(-H_2O)]{\Delta}$$

(H. K. Reimschuessel et al., *Macromolecules*, **2**, 567, 1967)

(j)

$$\underset{|}{\overset{NCO}{RCHCH_2NCO}} \xrightarrow{\text{base}} \text{(linear polymer)}$$

(ref. 28)

(k) $H_2N-R-NH_2 + CS_2 \xrightarrow{(-H_2S)}$

(ref. 33 and G. J. M. van der Kerk et al., *Rec. Trav. Chim.*, **74**, 1301, 1955)

(l)

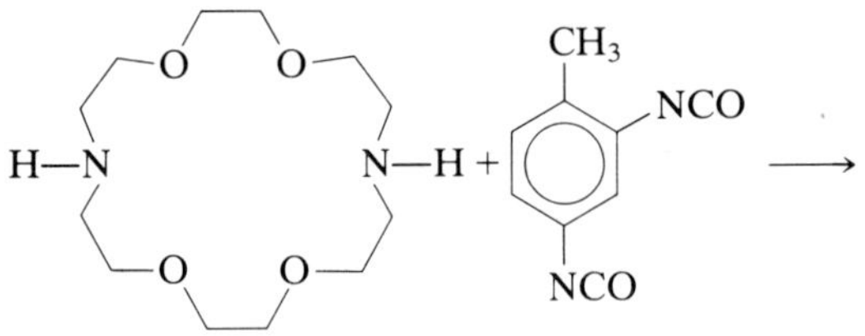

$$\text{H}-\text{N} \quad \text{(crown ether diamine)} \quad \text{N}-\text{H} + \text{(toluene diisocyanate)} \longrightarrow$$

(L. J. Mathias and K. Al-Jumah, *J. Polym. Sci., Polym. Chem. Ed.*, **18**, 2911, 1980)

(m)

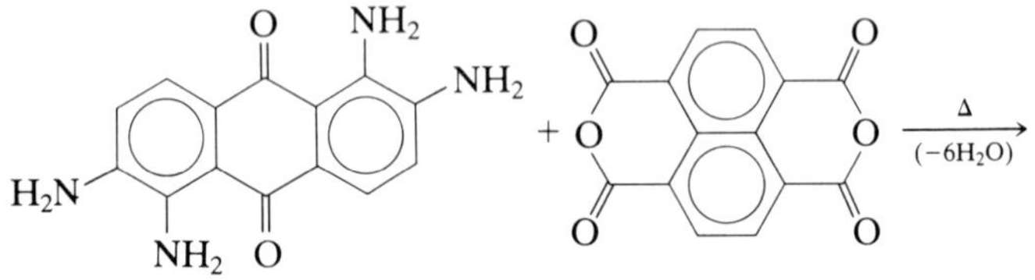

$$\text{(tetraaminoanthraquinone)} + \text{(naphthalene tetracarboxylic dianhydride)} \xrightarrow[(-6H_2O)]{\Delta}$$

(J. Szitza et al., *J. Polym. Sci. A-1*, **9**, 691, 1971)

(n) 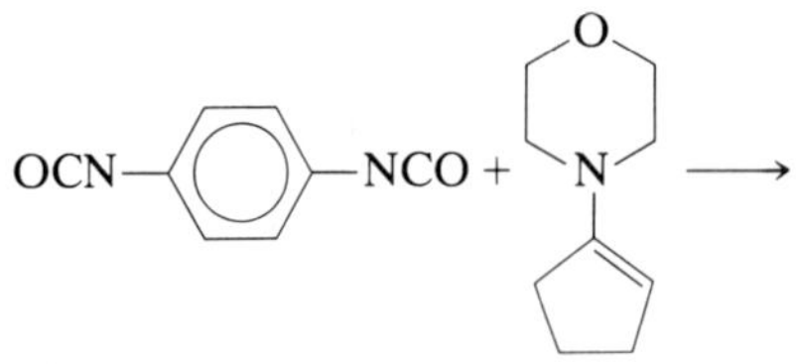 $OCN-\text{(phenylene)}-NCO + \text{(enamine)} \longrightarrow$

(ref. 27)

17. A rather unusual star-type polyamide has been prepared by a reaction sequence involving initial Michael addition of ammonia to acrylic acid (1:3 ratio) to obtain $N(CH_2CH_2CO_2H)_3$, followed by amidation with excess 1,3-diaminopropane. Repetition of these steps (Michael addition and amidation) yields the polymer that

the authors (D. A. Tomalia et al., *Polym. J.*, **17**, 117, 1985) refer to as a *starburst* polymer. Show the steps involved and decide whether the terminology is appropriate.

18. One approach to improving the tractability of aromatic polyamides has been to synthesize polymers of structure

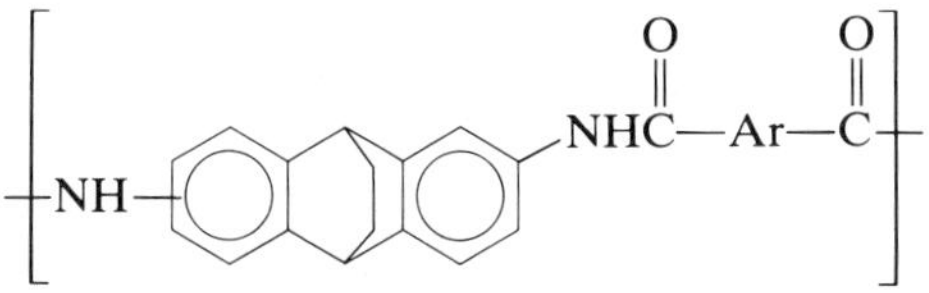

which, upon prolonged heating at 200 to 300°C under nitrogen, expel ethylene by the retrograde Diels–Alder reaction to form the anthracene system. (A. H. Frazer et al., *J. Polym. Sci., Polym. Chem. Ed.*, **23**, 2779, 2791, 1985)
   (a) Explain the rationale for this approach. (You may wish to review the discussion of thermal stability in Chapter 4.)
   (b) On the basis of what you have learned in this chapter, suggest a potential problem that might arise in processing (other than the usual problems of availability of monomers).

19. When aromatic amides substituted in the ortho position with alkynyl groups are heated, they undergo a cyclization reaction:

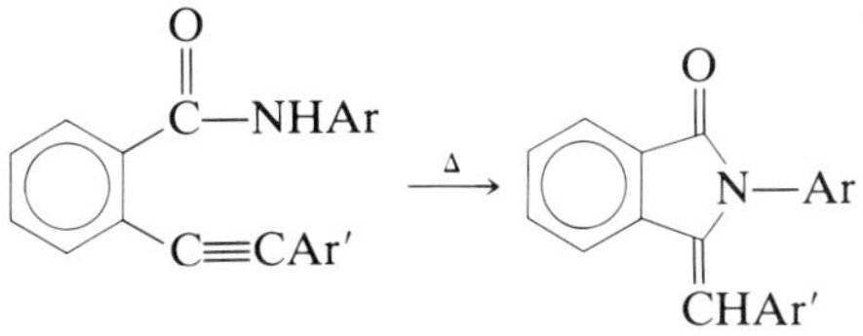

(isomer mixture)

Show what would happen if the polymer formed in exercise 16(a) were to be similarly heated. What effect would you expect this modification to have on (a) the polymer's glass transition temperature? (b) the polymer's processability?

14

Phenol–, urea–, and melamine–formaldehyde polymers

14.1   Introduction

Phenol (1), urea (2), and melamine (3) are three compounds of seemingly disparate structure; yet they are related in terms of how they react with formaldehyde, as well as in the processing and applications of the resultant polymers.

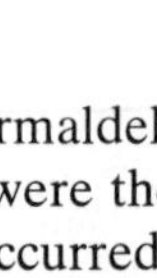

Phenol–formaldehyde condensation polymers, often referred to as *phenolic resins*, were the first true synthetic polymers to gain commercial acceptance. This occurred early in this century, and they have maintained a prominent position in the polymer market to the present time. Urea–formaldehyde and melamine–formaldehyde resins appeared a few years later (in the early 1920s and 1930s, respectively). An examination of Table 1.6 (Chapter 1) shows that about 70% of all thermosetting polymers produced is made up of phenol–, urea–, and melamine–formaldehyde polymers, with phenolics enjoying the lion's share. Because the three polymer types exhibit some similarities in their chemistry, but more importantly, because their processing and end-use applications are related, they are treated together in this chapter.

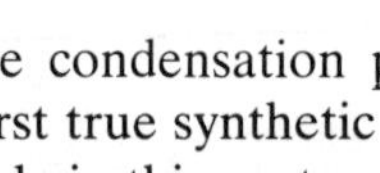

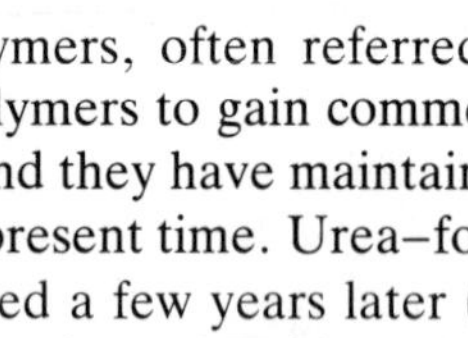

Phenol–formaldehyde resins[1–6,7a,8a] are normally prepared by two different methods. One involves a base catalyst with an excess of formaldehyde over phenol. The initially formed product (called a *resole*) can be cured to a thermosetting polymer simply by heating; as such, it constitutes a one-component system. The other method utilizes an excess of phenol over formaldehyde in the presence of an acidic catalyst. In this case, the initial product, called a *novolac* (also spelled *novolak*), requires the addition of more formaldehyde to effect curing. The chemistry of these two processes is discussed below.

Phenolic resins are widely used as lacquers and varnishes, molding compounds, laminates (particularly for decorative wall panels and table tops), and adhesives (notably for plywood and particle board). Flameproof fibers based on crosslinked novolacs (Carborundum trade name Kynol) are used for thermal insulation and protective clothing.[9] Urea and melamine resins[7b,8b,10] are used in similar applications and also in the treatment of textiles for improved crease and shrink resistance and for improving the wet strength of paper. Urea resins are also used as insulating foams. Phenolic polymers are sometimes used as ion-exchange resins when other functional groups are present.

14.2   Phenol–formaldehyde polymers: resoles

Resoles are the product of the reaction between phenol and excess formal-dehyde in the presence of base. Under these conditions phenol is present as the resonance-stabilized anion (**4**). The first step in the polymerization in-volves addition of the anion to formaldehyde to give *ortho-* and *para-*

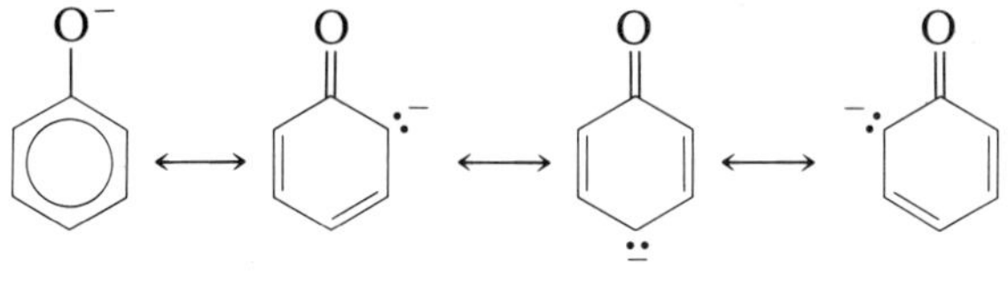

4

substituted methylolphenols. This is shown (reaction 14.1) for the *ortho* product.* (Although reactions involving the *ortho* positions only are given below, it is understood that analogous reactions also occur at *para* positions.)

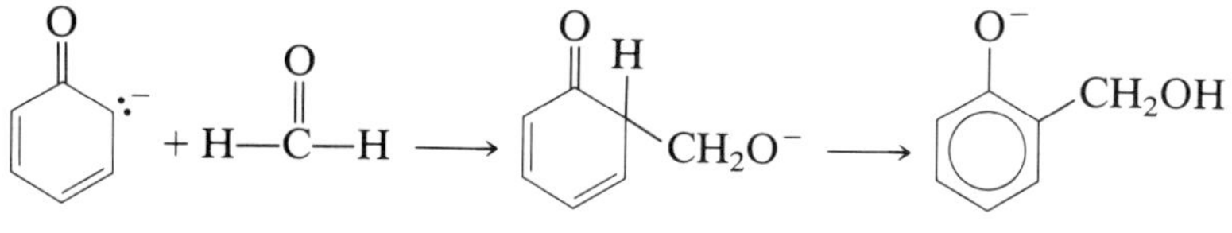

$$(14.1)$$

---

* Addition does not occur at the oxygen atom because the product, a hemiacetal, is unstable. Furthermore, the Cannizzaro reaction of formaldehyde to give methanol and formate ion is not a significant side reaction because the rate is much lower than that of formaldehyde with phenolate anion.

Because phenol is very reactive, simple monoaddition reactions seldom occur; instead, a mixture of monomethylolphenols, dimethylolphenols, and trimethylolphenols is formed with substitution occurring almost exclusively at the *ortho* and *para* positions. The initially formed methylolphenols condense on heating to give *resoles*, which are, in effect, low-molecular-weight pre-polymers. They are soluble in base and contain a large number of free methylol groups. Resoles are, of course, complex mixtures of compounds, but a representative structure (**5**) is given. The methylene bridges linking the benzene rings result from condensation between methylolphenols and available *ortho* or *para* positions either by direct S_N2 displacement of the hydroxyl group (14.2) or by Michael addition (14.4) to an *ortho* or *para* quinone methide structure that might be present in equilibrium (14.3) with the methylol anion.

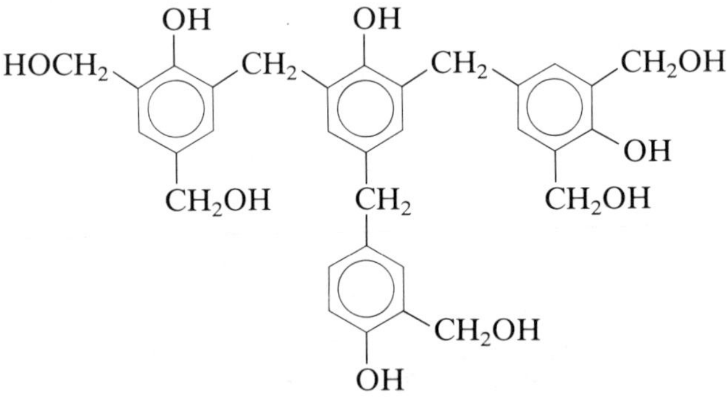

5

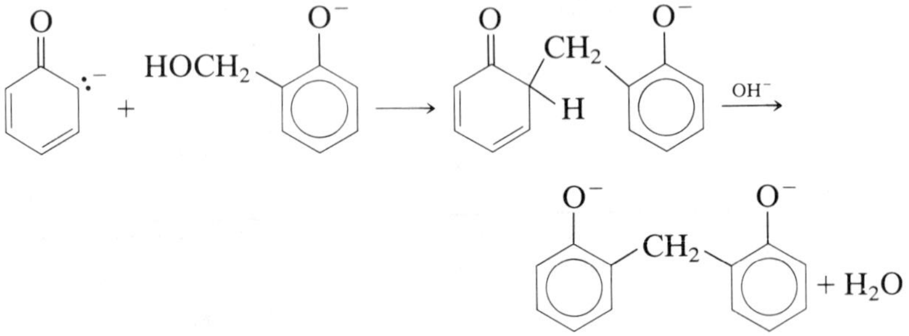

$$+ H_2O \quad (14.2)$$

$$+ OH^- \quad (14.3)$$

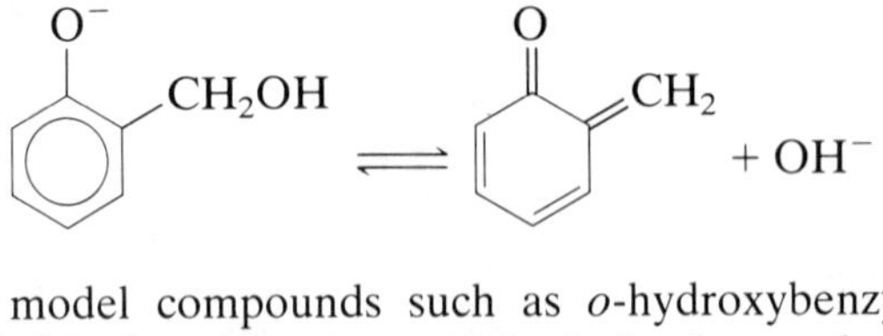

Studies with model compounds such as *o*-hydroxybenzyl alcohol (also known as saligenin) show that formaldehyde is also evolved in methylene bridge formation, suggesting that initially the bridge can occur at a carbon bearing a methylol group (14.5).

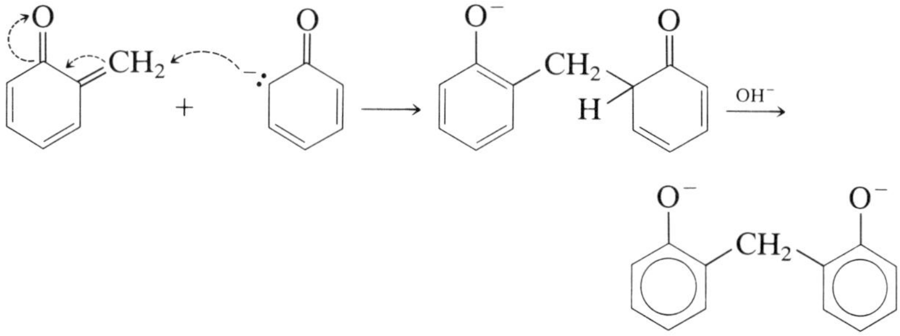

$$\text{(14.4)}$$

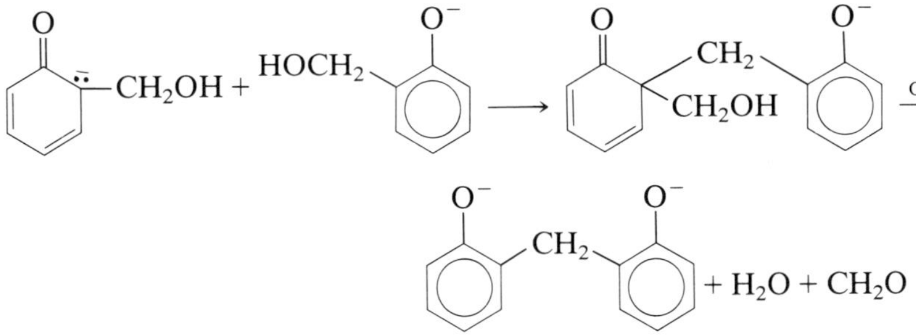

$$\text{(14.5)}$$

Ether linkages may also form (14.6), if the solution is very weakly basic or neutral.

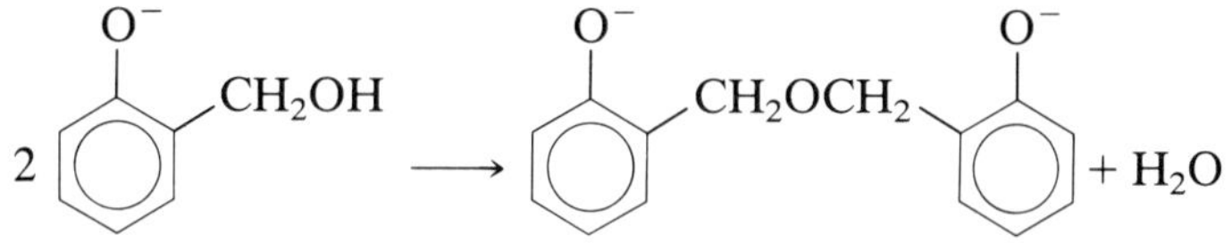

$$\text{(14.6)}$$

In commercial production, resoles are normally processed to a workable viscosity; then subsequent polymerization to high-molecular-weight network polymer (called *resite*) can be effected simply by heating. Depending on the application, this can be done directly on the base solution of resole or on neutral or slightly acidic solution. As a typical example, plywood adhesives are prepared by mixing appropriate additives such as wood flour with the basic resole solution. The mixture is then spread on the surface of the wood veneers prior to placing them in a hot press. The heat of the press not only causes polymerization to occur but also steams off the water. Excellent bonding is achieved by reaction between the resin and phenolic constituents of the wood. Thin sheets of paper impregnated with resin are often employed in place of the basic adhesive solution.

Curing of phenolic resins involves methylene bridge formation as described above. In acidic or neutral media, ether formation (14.6) is a significant crosslinking reaction, particularly at lower temperatures. This is because protonation of the alcohol groups by phenolic protons or by protons from added acid renders the groups much more susceptible to either S_N1 or S_N2 displacement by neighboring alcohol groups (14.7).

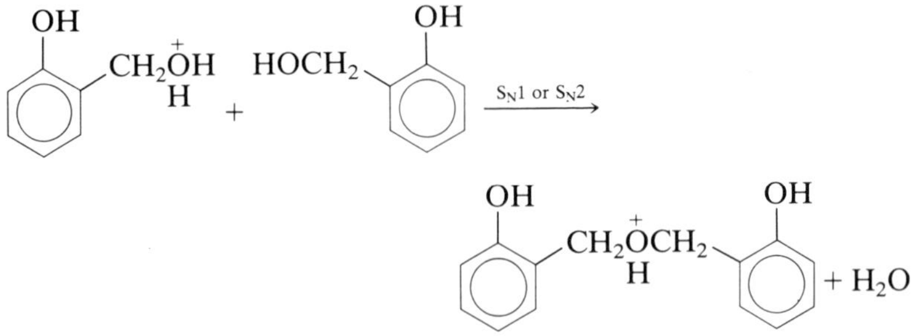

$$\text{(14.7)}$$

Methylene bridges also form in acidic media by electrophilic substitution involving benzylic carbocations and available ring positions (14.8). At about 150°C formaldehyde is evolved from dissociation of benzylic alcohol groups (14.9), thus making available more sites for methylene bridge formation.

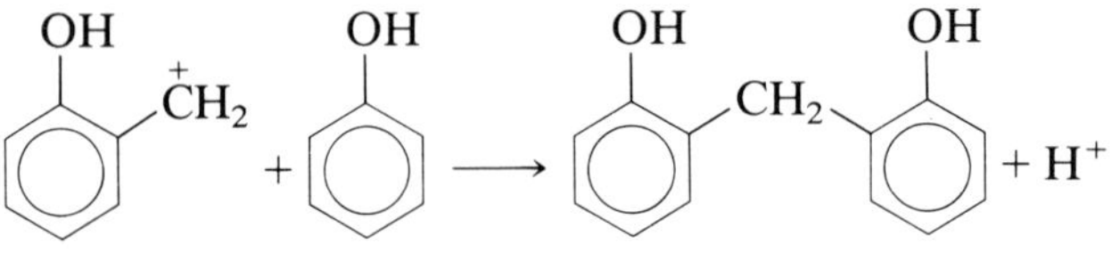

$$\text{(14.8)}$$

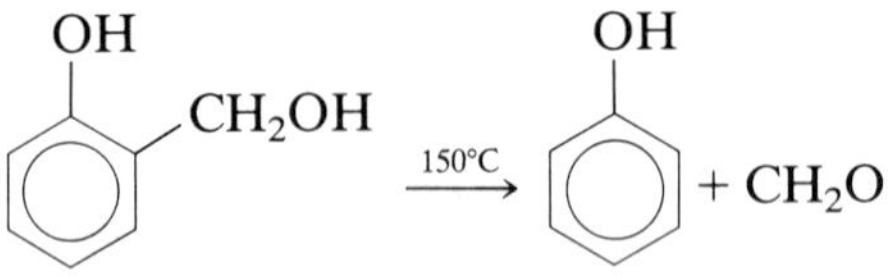

$$\text{(14.9)}$$

A reaction that occurs at about 180°C and that contributes to crosslinking is dehydration to form the highly reactive *ortho* or *para* quinone methides (14.10), which undergo a variety of reactions including cycloaddition with other quinone methides to give chroman groups (14.11) or dimerization to

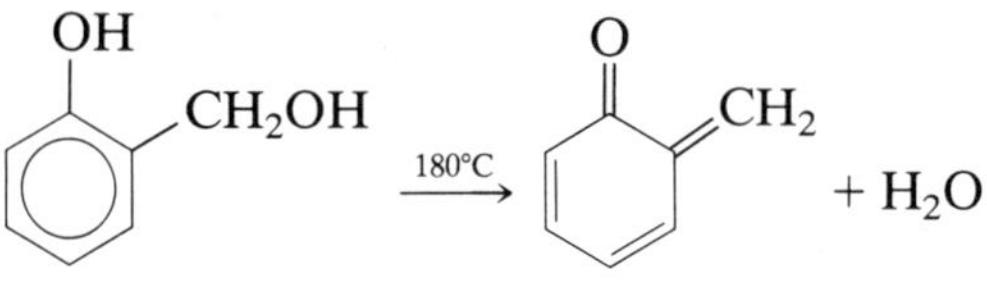

$$\text{(14.10)}$$

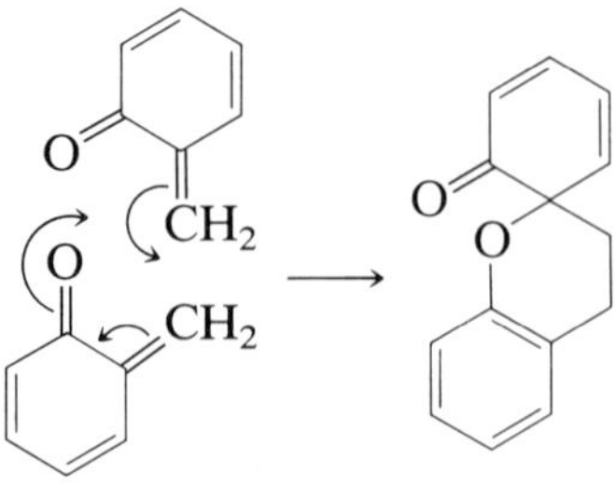

$$\text{(14.11)}$$

give unsaturated linkages (14.12). Phenolic resins have been modified by reaction with drying oils to incorporate some flexibility into the otherwise very rigid structure. This modification presumably involves a Diels–Alder reaction between quinone methide groups and the double bonds of the drying oil.

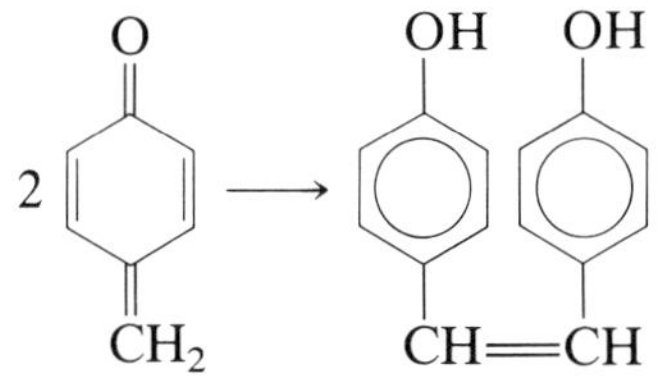

(14.12)

It is apparent, then, that high-molecular-weight phenolic resins have extremely complex structures. It should be borne in mind that network polymers are formed because phenol is polyfunctional. If phenol is substituted in the *para* or in one *ortho* position, the functionality is reduced to two and it is possible to form linear polymer; however, high-molecular-weight polymers have not been prepared this way.

14.3 Phenol–formaldehyde polymers: novolacs

Acid catalysis with excess phenol leads to a phenol–formaldehyde condensation product quite different from that obtained by base catalysis. The mechanism involves protonation of the carbonyl group (14.13) followed by electrophilic aromatic substitution (14.14) at *ortho* or *para* positions.

(14.13)

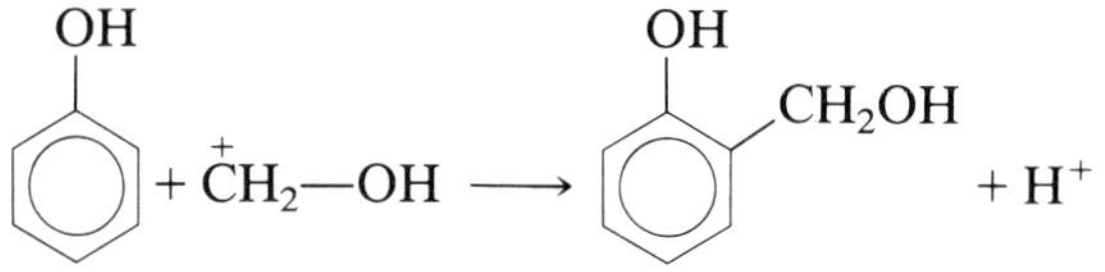

(14.14)

Under acidic conditions further reaction occurs to give methylene bridges as discussed in the preceding section (reactions 14.7 and 14.8). The net result is the formation, in the early stages of polymerization, of complex mixtures of low-molecular-weight polymers (for example, **6**) characterized by having random *para–para*, *ortho–ortho*, or *ortho–para* methylene linkages. It has been shown, again with model compounds, that the *para* position is more reactive with strong acid catalysts at pH less than 3 (the usual condensation conditions); hence *ortho–ortho* linkages undoubtedly occur less frequently. The opposite is the case at pH 4.5 to 6 where condensation at *ortho* positions

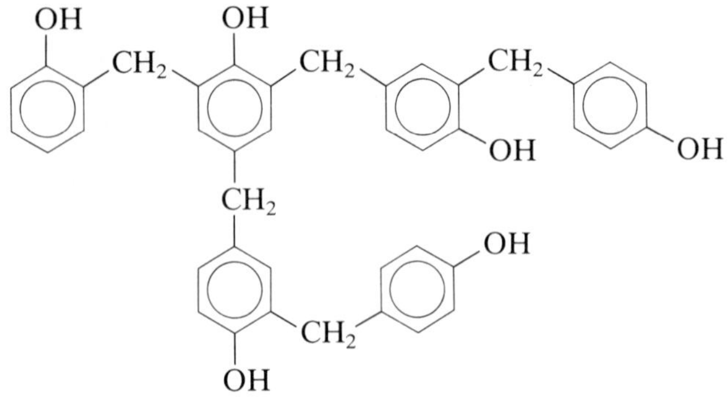

6

predominates. The use of divalent metal catalysts facilitates *ortho* condensation—not only the initial reaction of phenol with formaldehyde, but also the subsequent formation of methylene bridges. (The products are referred to as *o,o'*-novolacs.) This directive effect of the catalyst probably results from chelation involving the phenolic hydroxyl group.[11] A possible mechanism involving hydrated formadehyde is given in reactions (14.15) to (14.17). The significance of having substitution predominantly in the *ortho* positions is that curing (described below) occurs more rapidly because the more reactive *para* positions remain available for crosslinking.

$$Zn^{++} + HOCH_2OH \rightleftharpoons {}^+ZnOCH_2OH + H^+ \qquad (14.15)$$

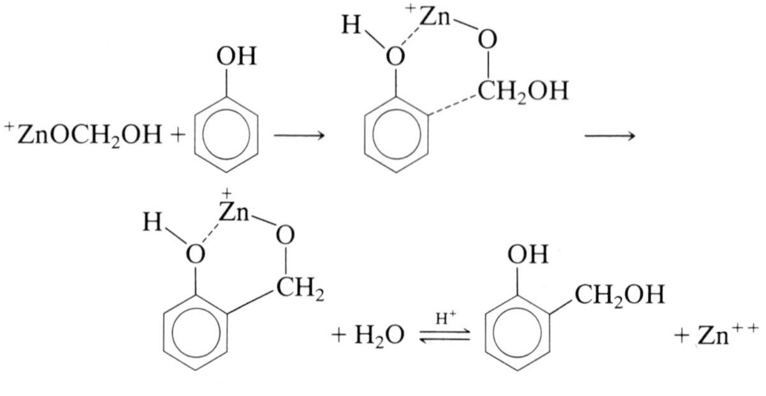

$$+ H_2O \overset{H^+}{\rightleftharpoons} \qquad + Zn^{++} \qquad (14.16)$$

$$\xrightarrow{Zn^{++}} \qquad + H_2O + Zn^{++} \qquad (14.17)$$

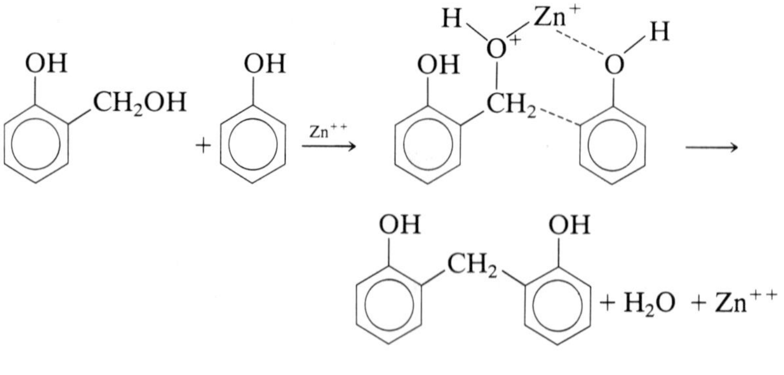

Organotin compounds[12] and metal phenolates[13] are also effective in promoting *ortho* substitution. The reaction of phenoxymagnesium bromide with formaldehyde, for example, results in low-molecular-weight linear novolacs (**7**). The reaction of phenol with paraformaldehyde in anhydrous xylene at 190°C yields similar products.[14]

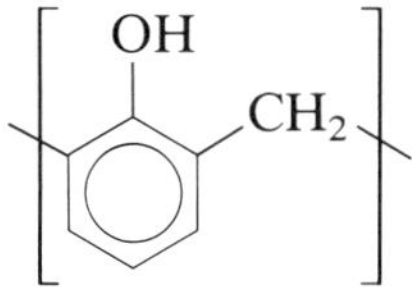

7

Unless excess phenol is used, the condensation reaction proceeds to high-molecular-weight infusible resin (resite), so in practice less than an equivalent of formaldehyde is reacted with phenol. The resultant product, a novolac, is fusible and has an average molecular weight dependent in large measure on the phenol/formaldehyde ratio, as shown in Table 14.1. Novolacs can also be prepared under basic conditions, but the reaction is complicated by a tendency toward chain branching and gelation. It is of interest, however, that in the presence of divalent metal compounds, *ortho* condensation is preferred, as described above.

The basic difference between resoles and novolacs is that the latter contain no hydroxymethyl groups for all practical purposes and hence cannot be converted to network high polymer simply by heating. Crosslinking is brought about by adding additional formaldehyde or, more commonly, by adding paraformaldehyde or hexamethylenetetraamine (**8**), a high-melting (>230°C) solid having an interesting cage structure that is obtained by the reaction of formaldehyde with ammonia. Formaldehyde is released from these compounds under the influence of heat and pressure, as in a molding operation,

Table 14.1. Effect of reactant ratio on polymer molecular weight[a]

Moles formaldehyde/10 moles phenol	Average molecular weight	Resin type
1	229	Dihydroxydiphenylmethane
2	256	Novolac
3	291	Novolac
4	334	Novolac
5	371	Novolac
6	437	Novolac
7	638	Novolac
8	850	Novolac
9	1000	Novolac
12	—	Resite

[a] From Carswell,[6] courtesy of Wiley-Interscience.

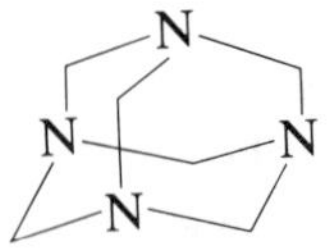

8

and crosslinking occurs again by formation of methylene bridges. Some benzylamine linkages (**9**) are also formed when hexamethylenetetraamine is used. Other types of additives commonly used as reinforcers or extenders in molding formulas include wood flour, chopped rags, asbestos, and glass fibers. Pigments are also added, as are stearates or oils to facilitate releasing from the mold. Phenolic resins are sometimes blended with synthetic rubber to enhance flexibility.

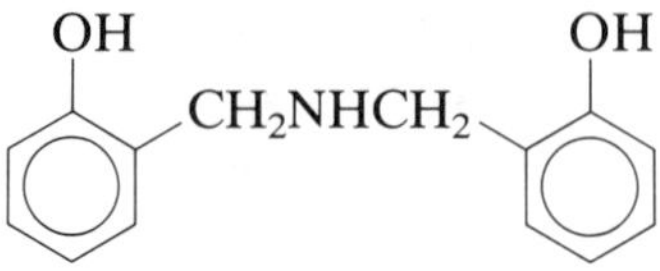

9

Commercial production of both resoles and novolacs involves two basic steps. The monomers and catalyst are reacted in aqueous solution to a relatively low viscosity; then water is removed under vacuum, and the solid product is ground to a powder. At this point the polymer is of relatively low molecular weight, soluble, and fusible. It is referred to as the A-stage. (As described earlier, aqueous solutions of resoles, with appropriate additives, are used directly as plywood glues or for impregnating paper.) The A-stage resin is mixed with additives (including hexamethylenetetraamine or paraformaldehyde in the case of novolacs), then heated further to a higher molecular weight B-stage. The polymer is converted to resite (C-stage) in a final molding operation.

14.4   Chemical modifications of phenolic resins

The only aldehyde of any significance, apart from formaldehyde, of use in phenolic resins is furfural (**10**), obtainable from corncobs and oat hulls. It is frequently used in the preparation of molding resins because it improves the

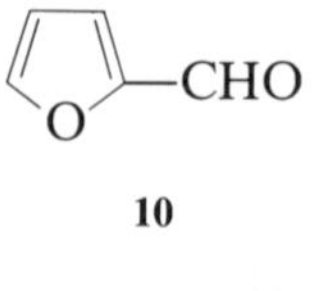

10

polymer's flow properties. Some other aldehydes that have been studied are acrolein, acetaldehyde, and butyraldehyde, although reactions of the latter two are complicated by their tendency to undergo aldol condensations, particularly under basic conditions. Starch hydrolysates, which undergo dehydration to form 5-hydroxymethylfurfural (11) under acidic conditions, have also been investigated as a partial replacement for formaldehyde.[15] Related to 10 and 11 is furfuryl alcohol (12), which undergoes acid-catalyzed condensation to form chemically resistant polymers called *furan resins*,[16] which contain furan rings linked through methylene groups and thus are structurally related to phenolic resins. The resins are used as tank and vat linings in chemical plants.

$$HOCH_2 \text{—} \overset{\displaystyle \bigcirc}{O} \text{—} CHO \qquad \overset{\displaystyle \bigcirc}{O} \text{—} CH_2OH$$

11   12

Interestingly, research is being conducted on ways to make phenolic resins that bypass aldehydes altogether.[17] The process makes use of peroxidase enzymes (analogous to those that catalyze the biosynthesis of lignin in plants) to polymerize phenols into polymers having a ligninlike structure. Concerns over the adverse health effects of formaldehyde provide a motivation for this type of research.

Considerably more work has been done on alternatives for phenol. Resorcinol (13) is much more reactive than phenol and is used to increase the rate of cure, particularly with cold set adhesives. Resorcinol's higher cost relative

OH

13

to that of phenol limits its use, however. The even higher priced bisphenol A is used in certain coatings formulations for automotive body priming.[1] Bisphenol A–furfural resins have also been reported.[18] Monosubstituted phenols such as *o*- and *p*-cresol and the corresponding chlorophenols have been used to modify resin properties. Oil-soluble resins suitable for varnishes are prepared from *p-t*-butylphenol, *p*-phenylphenol, and *p-n*-octylphenol.

Unsaturated phenols are of interest because they provide further crosslinking by addition polymerization. Most important are those present in *cashew nut shell liquid* (CNSL). CNSL, which is obtained from the shells by a process involving heating and decarboxylation, consists of mixtures of phenols with 15-carbon unsaturated side chains, the most important of which is cardanol (14). On reaction with formaldehyde, CNSL yields flexible resins suitable for brake lining and floor tile binders. Flexible phenolic resins are also obtained by using phenol modified with tung oil.[19] This modification involves ring-substitution via the conjugated system of α-eleostearic acid (*cis,trans,trans*-9,11,13-octadecatrienoic acid), the principal fatty acid component of tung oil. Novolacs curable by addition polymerization have also been synthesized from 2-allylphenol (**15**) or by introducing allyl groups into a preformed novolac.[20]

14   15

Other naturally occurring phenols that are used as partial substitutes for phenol include *tannin*[7c] and *lignin*.[7d] The former, which is widely distributed in a variety of trees, consists of two types: hydrolyzable and condensed. Hydrolyzable tannins are mixtures of such phenolic compounds as pyrogallol (**16**), ellagic acid (**17**), and glucose esters of gallic acid (**18**) or condensed forms of gallic acid. The more abundant condensed tannins consist of a complex mix of flavonoid units linked together with carbohydrates. The increased use of tannins in phenolic resins has paralleled a decline in the leather tanning industries, once the sole domain of industrial tannins. Lignin, a major constituent of wood, is a complex phenolic polymer bearing a superficial resemblance to synthetic phenolic resins. Its structure is described in Chapter 17.

16   17   18

Flame resistance of phenolic resins may be reduced with appropriate fillers or by the introduction of phosphate ester groups.[21] The latter is illustrated (14.18) with an *o,o'*-novolac.

(14.18)

Recent work has been directed toward production of aqueous dispersions or free-flowing solids of phenol–formaldehyde oligomers stabilized with a protective colloidal polysaccharide coating.[22,23] This technology yields pre-polymers that are easier to manipulate and process to resites than are conventional phenolic resins.

14.5 Urea–formaldehyde polymers

Urea (**2**), prepared commercially from ammonia and carbon dioxide, undergoes reactions with formaldehyde that are analogous, at least in the initial stages, to those of phenol. Unlike other amides, urea is basic because the single carbonyl group is not sufficient to compensate for two amino groups. Urea undergoes nucleophilic addition to formaldehyde to give methylol derivatives (14.19). The solution is kept mildly basic by careful pH control because the methylol derivatives condense rapidly under acidic conditions.

$$\mathbf{2} + \underset{\underset{\text{H}}{\|}}{\text{H—C—H}} \xrightarrow{\hspace{1cm}} \underset{\underset{\text{}}{\|}}{\text{H}_2\text{NCNHCH}_2\text{OH}} + \underset{\underset{\text{}}{\|}}{\text{HOCH}_2\text{NHCNHCH}_2\text{OH}}$$

$$(14.19)$$

The similarity between urea and phenolic resins, at least as far as their chemistry is concerned, ends at this point.

There has been considerable speculation recently concerning the structure of urea–formaldehyde polymers. It had long been assumed that further condensation involved cyclic intermediates formed by initial dehydration of the protonated methylol derivatives to give imines (14.20) which then trimerize (14.21). The corresponding cyclic compound having methylol groups attached to the amide nitrogen atoms would be formed from dimethylolurea.

$$\underset{\underset{\text{H}}{}}{\overset{\overset{\text{O}}{\|}}{\text{H}_2\text{NCNHCH}_2\overset{+}{\text{O}}\text{H}}} \rightleftharpoons \overset{\overset{\text{O}}{\|}}{\text{H}_2\text{NCN}}=\text{CH}_2 + \text{H}_3\text{O}^+ \qquad (14.20)$$

$$3\text{H}_2\text{NCN}=\text{CH}_2 \longrightarrow \quad (14.21)$$

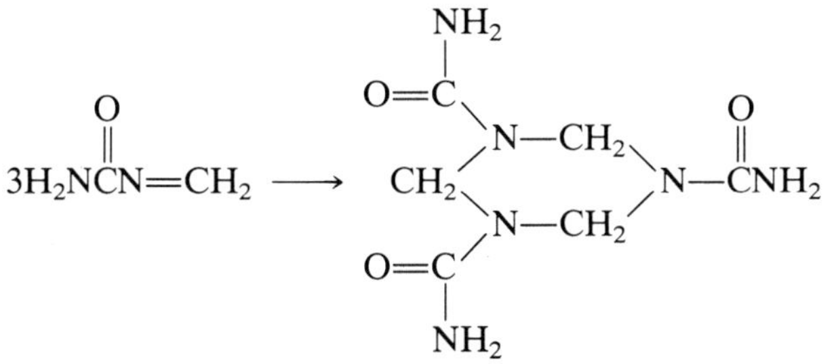

Subsequent condensation would then occur between methylol and amide groups to give network polymer (**19**). Support for this mechanism stems from known cyclization reactions of other amides and also from the nitrogen content, which approximates that expected for a polymer formed by this mechanism.

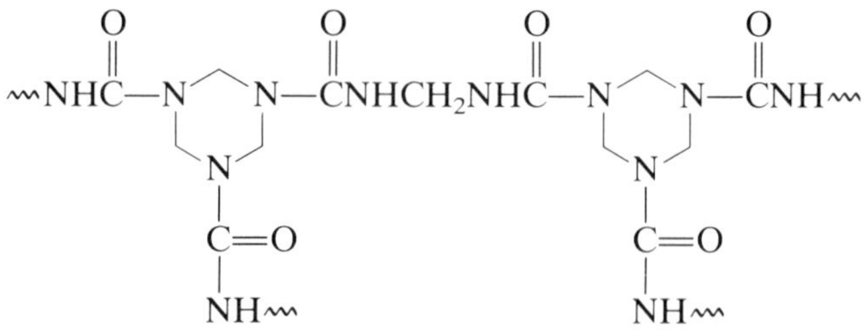

19

More recent work[24,25] suggests that network polymers are *not* formed, particularly with the relatively low formaldehyde–urea ratios employed in many industrial applications. Instead, it has been proposed that linear oligomeric urea–formaldehyde condensates (**20**) are formed which then separate as a colloidal dispersion stabilized by association with excess formaldehyde. "Curing" results from agglomeration of the colloidal particles with concomi-

20

tant release of formaldehyde. Support for this viewpoint is based on several observations: (1) The polymers are soluble in formaldehyde and concentrated sulfuric acid, and can be recovered from the latter unchanged. (2) A plot of the logarithm of solution viscosity versus time during polymerization exhibits a sharp break, unlike similar plots for phenol–formaldehyde polymers which show a continuous increase in viscosity with development of the polymer network. (3) X-ray diffraction patterns and laser Raman spectra of the cured resins indicate the presence of crystalline regions and an absence of water, respectively, both consistent with hydrogen-bonded proteinlike or nylonlike close chain packing. And (4) scanning electron micrographs of the cured resins show surface characteristics more in common with those of coalesced colloids than high-molecular-weight network polymers. The structure of urea–formaldehyde resins may thus resemble that of proteins more than that of phenolic resins. FT-IR studies,[26] on the other hand, suggest that some methylene and ether crosslinks are probably present, as are cyclic ether units, and that the true structure of urea–formaldehyde polymers is a function of both feed ratio and processing pH.

As was mentioned earlier, urea–formaldehyde resins are used in applications similar to those of phenol–formaldehyde resins—for molding, laminating, and adhesives. One advantage of the urea resins is that they are very light in color, and hence are more suitable for decorative use. (The dark color of phenolic resins is due to a combination of oxidation and quinone methide formation.) Light-colored counter and table tops are made with urea resin-impregnated paper. Also, decorative interior plywood is normally glued with urea resin because the dark-colored phenolic resins can stain the veneer. Exterior plywood, however, is glued with phenolic resin because it has better weather resistance. In the coatings field, urea–formaldehyde resins are sometimes blended with alkyd baking enamels (Chapter 12) to improve hardness. Urea resins are also used to impart crease and shrink resistance to fabrics through crosslinking reactions.

Another major application of urea–formaldehyde polymers is in insulating foams. These are generally manufactured on-site with portable foaming equipment. Ingredients include resin, surfactant to stabilize the foam, catalyst (usually phosphoric acid), and compressed air. The surfactant and catalyst are usually premixed. The three components (resin, surfactant plus catalyst, and air) are then pumped separately into a foam "gun" that injects the mixture into the cavity to be filled. The foam sets within minutes and fully hardens in the course of a day. Considerable controversy has surrounded the use of urea–formaldehyde foams for insulating homes because of health concerns arising from subsequent release of formaldehyde vapors.

14.6 Melamine–formaldehyde polymers

Melamine (**3**) is an aromatic, heterocyclic compound prepared from cyanamide. The overall process, including the cyanamide synthesis, is shown in reactions (14.22) through (14.25).

$$CaO + 3C \xrightarrow{2000°C} CaC_2 + CO \tag{14.22}$$

$$CaC_2 + N_2 \xrightarrow{1000°C} CaCN_2 + C \tag{14.23}$$

$$CaCN_2 + H_2SO_4 \longrightarrow CaSO_4 + H_2N-C\equiv N \tag{14.24}$$

$$3H_2N-C\equiv N \xrightarrow{heat} \tag{14.25}$$

3

Melamine undergoes condensation reactions with formaldehyde in a manner analogous to those of urea. A major difference is that, in the initial nucleophilic addition to formaldehyde, each amino group may form a dimethylol derivative (14.26). In practice, prepolymers are prepared having some monomethylol-substituted amino groups, and subsequent curing is

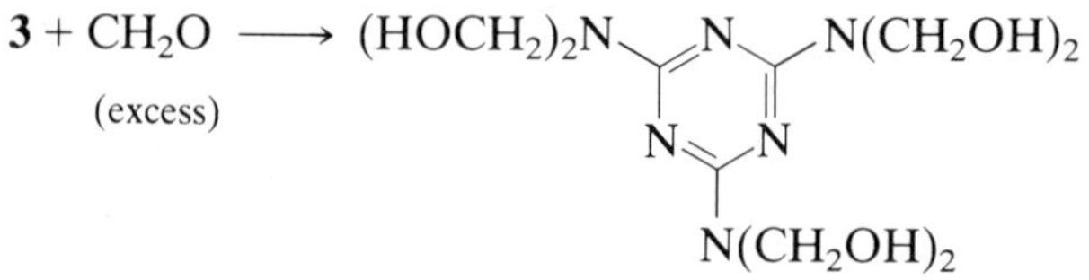

$$3 + CH_2O \longrightarrow (HOCH_2)_2N \underset{\underset{N(CH_2OH)_2}{\overset{N}{\bigcirc}}}{\quad} N(CH_2OH)_2 \qquad (14.26)$$

(excess)

mainly through methylene bridge formation by condensation between methylol and amino groups (14.27). The prepolymers are water soluble but can be made soluble in organic solvents by etherification with such alcohols as 1-butanol.

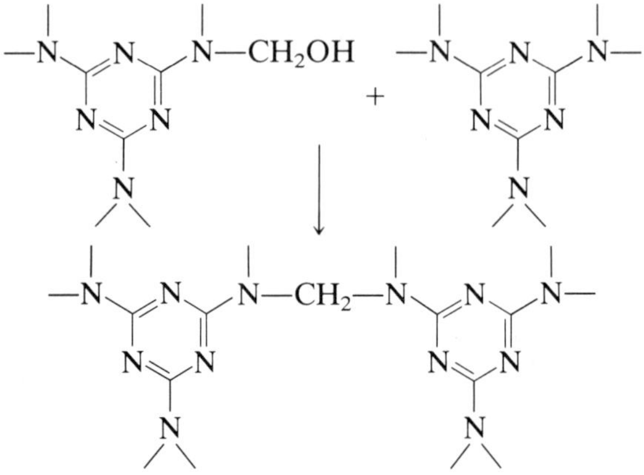

$$ \qquad (14.27) $$

Melamine resins are generally harder and more moisture resistant than urea resins. They are used in molding and laminating formulations and, like urea resins, as alkyd enamel modifiers and textile fiber finishes. Their largest use is in the manufacture of decorative dinnerware.

References

1. A. Knop and L. A. Pilato, *Phenolic Resins*, Springer-Verlag, New York, 1979.
2. A. A. K. Whitehouse, E. G. K. Pritchett, and G. Barnett, *Phenolic Resins*, Iliffe, London, 1967.
3. N. J. L. Megson, *Phenolic Resin Chemistry*, Academic Press, New York, 1958.
4. R. W. Martin, *The Chemistry of Phenolic Resins*, Wiley-Interscience, New York, 1956.
5. D. F. Gould, *Phenolic Resins*, Reinhold, New York, 1959.
6. T. S. Carswell, *Phenoplasts*, Wiley-Interscience, New York, 1947.
7. A. Pizzi (ed.), *Wood Adhesives: Chemistry and Technology*, Dekker, New York, 1983: (a) A. Pizzi, Chap. 3; (b) A. Pizzi, Chap. 2; (c) A. Pizzi, Chap. 4; (d) H. N. Nimz, Chap. 5.
8. R. W. Tess and G. W. Poehlein (eds.), *Applied Polymer Science*, 2nd ed., ACS Symp. Ser. 285, American Chemical Society, Washington, D.C., 1985: (a) J. S.

Fry, C. N. Merriam, and W. H. Boyd, p. 1141; (b) L. L. Williams, I. H. Updegraff, and J. C. Petropoulos, p. 1101.

9. J. Economy, L. C. Wohrer, F. J. Frechette, and G. Y. Lei, *Appl. Polym. Symp.*, **21**, 81 (1973).

10. B. Meyer, *Urea–Formaldehyde Resins*, Addison-Wesley, Menlo Park, Calif., 1979.

11. M. F. Drumm and J. R. LeBlanc, in *Step-Growth Polymerizations* (D. H. Solomon, ed.), Dekker, New York, 1973, Chap. 5.

12. A. Ninagawa, Y. Ohnishi, H. Takeuchi, and H. Matsuda, *Makromol. Chem., Rapid Commun.*, **6**, 793 (1985).

13. G. Casiraghi, G. Sartori, F. Bigi, M. Cornia, E. Dradi, and G. Casnati, *Makromol. Chem.*, **182**, 2151 (1981).

14. G. Casiraghi, G. Casnati, M. Cornia, G. Sartori, and F. Bigi, *Makromol. Chem.*, **182**, 2973 (1981).

15. H. Koch and J. Pein, *Polym. Bull.*, **13**, 525 (1985).

16. J. A. Brydson, *Plastics Materials*, Butterworth, London, 1975, Chap. 28.

17. J. S. Dordick, *Proceedings*, 3rd Chemical Congress of North America and 195th ACS National Meeting, Toronto, June 1988.

18. R. B. Jambusaria and S. P. Potnis, *Polym. J.*, **6**, 333 (1974).

19. Y. Yoshimura, *J. Appl. Polym. Sci.*, **29**, 1063 (1984).

20. K. M. Hui and L. C. Yip, *J. Polym. Sci., Polym. Chem. Ed.*, **14**, 2323 (1976).

21. B. F. Dannels and A. F. Shepard, *J. Polym. Sci. A-1*, **6**, 2051 (1968).

22. G. L. Brode, T. R. Jones, and S. W. Chow, *CHEMTECH*, **13**, 630 (1983).

23. G. L. Brode, *J. Macromol. Sci.-Chem.*, **A22**, 897 (1985).

24. T. J. Pratt, W. E. Johns, R. M. Rammon, and W. L. Plagemann, *J. Adhesion*, **17**, 275 (1985).

25. A. K. Dunker, W. E. Johns, R. Rammon, B. Farmer, and S. J. Johns, *J. Adhesion*, **19**, 153 (1986).

26. S. S. Jada, *J. Appl. Polym. Sci.*, **35**, 1573 (1988).

Review exercises

1. The ratio of *ortho* to *para* substitution in the reaction of phenol with formaldehyde is greatest at about pH 6. *Ortho* substitution is also favored in aprotic nonpolar solvents. Give a plausible explanation for this behavior.

2. Besides urea and melamine, aniline also reacts with formaldehyde to form thermosetting resins (rarely used today). Their chemistry resembles that of both phenolic and melamine resins. Write plausible reactions for the formation of aniline–formaldehyde resins under acidic, basic, and neutral conditions. (ref. 7b)

3. Urea- and melamine-formaldehyde resins are widely used as curing agents for coatings such as alkyds (see Chapter 12). In such cases the resins are often etherified with, for example, 1-butanol. Also, a melamine substitute such as "benzoguanamine" (2-phenyl-4,6-diamino-1,3,5-triazine) is sometimes used (although less frequently because of increasing cost). (a) What advantage might such formulations have over conventional urea or melamine resins? (b) Show mechanistically

how a urea or melamine resin may be butylated under acidic conditions. (c) What reactions might be involved in curing a glyptal-type coating resin (see Chapter 12) with urea– or melamine–formaldehyde resin, and what advantage would this type of modification have from the standpoint of processing? (ref. 8b)

4. Show mechanistically how furan resins are formed from furfuryl alcohol (**12**) under acidic conditions. Crosslinking is accompanied by loss of unsaturation. Suggest a possible crosslinking reaction. (ref. 16)

15

Inorganic and partially inorganic polymers

15.1 Introduction

Major emphases in polymer development over the years have been with organic polymers. The only polymers in commercial use for an extended period that have inorganic backbones are the *polysiloxanes*, which date back to the 1940s. Nevertheless a considerable volume of work dealing with inorganic and partially inorganic polymers has been published,[1–5] although most of the products developed remain laboratory curiosities. *Phosphonitrilic polymers*, which also contain an inorganic backbone, were commercialized more recently, as were polymers containing carborane units in the backbone. The latter may be polyesters or polyamides or some other type of more traditional polymer, but they owe their unique properties to the presence of the carborane cluster. Similarly, many *organometallic* polymers may contain repeating functional groups of the more traditional classes of polymers. *Coordination polymers*, on the other hand, are unique in that coordinate covalent bonds to metals are often an integral part of the backbone.

Some of the partially inorganic polymers that have found commercial application are listed in Table 15.1. Only the principal structural units are given, although there is some variation in the products that have been marketed. The list is short compared with the lists of the more traditional organic polymers found in earlier chapters; however, it should not be taken as an exhaustive compilation because so much developmental work is in progress in this area. Metal-containing paints, for example, are being used increasingly, as are metal-containing polymeric catalysts.

It is difficult to draw a line between polymers that are wholly organic and those that are partially inorganic. Polyesters or polyamides of sulfonic or phosphoric acids, polysulfides, and other sulfur-containing polymers might logically fit the definition of partially inorganic. Our discussion here, however, is limited to the ones mentioned above, and to one wholly inorganic polymer—poly(sulfur nitride)—which was first mentioned in Chapter 4 in the context of conducting polymers. Poly(sulfur nitride) is by no means the

487

Table 15.1. Partially inorganic polymers developed for commercial use

Type	Principal structure	Typical applications
Polysiloxane	$\begin{array}{c} CH_3 \\ \vert \\ -Si-O- \\ \vert \\ CH_3 \end{array}$	Elastomers, sealants, lubricating oils, greases, hydraulic fluids
Polyphosphazene[a]	$\begin{array}{c} OR \\ \vert \\ -P{=}N- \\ \vert \\ OR \end{array}$	Oil-resistant hose, gaskets, O-rings
Polycarboranesiloxane	$\begin{array}{c} \quad\quad\quad CH_3 \\ \quad\quad\quad \vert \\ -C(B_{10}H_{10})C-Si-O- \\ \quad\quad\quad \vert \\ \quad\quad\quad CH_3 \end{array}$	High-temperature elastomers, greases, lubricants

[a] R = mixed fluoroalkyl groups.

only wholly inorganic polymer to have attracted attention. Boron nitride (BN), an inorganic polymer having structures analogous to those of graphite and diamond, has been manufactured into refractory fibers by the high-temperature ($>1500°C$) reaction of boric oxide filaments with urea or ammonia,[6] or by pyrolysis of boron–nitrogen oligomers.[7] The BN fibers, in turn, have been similarly reacted with metal halides to form such nitride polymers as NbN and TiN.[8] Similar methods have been employed to make metal carbide fibers from carbon fiber.[8] There is considerable interest in these materials as strengthening agents for ceramics.

It is not possible to cover all aspects of inorganic-organic polymers in a single chapter because so much has been published in the field, but some of the more significant and timely topics are included. The reader should keep in mind, however, that this is an area of very active research; current literature should be consulted for recent advances. It should not be forgotten, too, that some polymeric silicates, notably glass and asbestos, have been in service far longer than any of the polymers described in this chapter.

15.2   Poly(sulfur nitride)

Also called *polythiazyl*, poly(sulfur nitride)[9a,10] (**2**) is prepared by the solid-state polymerization (15.1) of disulfur dinitride (**1**) at room temperature. The dimer **1**, which is, itself, formed by heating a cyclic tetramer, polymerizes slowly to yield lustrous, gold-colored monoclinic crystals of **2**. The polymer is a most unusual material, being the first reported example of a nonmetallic covalent polymer exhibiting electrical conductivity comparable to that of metals (see Chapter 4, Section 4.8); indeed, it even exhibits superconductivity at 0.26 K. Apart from its conductivity, **2** exhibits high reflectivity and malleability, both properties characteristic of metals. It is assumed that the high conductivity arises not only from delocalization along the chains, but also by S–S, N–N, and S–N orbital overlap between adjacent parallel chains. The polymer is relatively stable at room temperature, but decomposes over prolonged periods or on heating.

$$S_2N_2 \longrightarrow \text{+S=N+} \qquad (15.1)$$

$$\text{1} \qquad\qquad \text{2}$$

Poly(sulfur nitride) reacts with bromine to form polymers having the composition $\text{+SNBr}_x\text{+}$, where x is in the range 0.3 to 0.4.[9a] The brominated polymers have higher conductivities than **2**.

At the present time, poly(sulfur nitride) and its derivatives are of more theoretical than practical interest.

15.3 Polysiloxanes

Despite the position of silicon directly below carbon in group IV of the periodic table, the properties of the two elements are quite different. While its normal oxidation state (like that of carbon) is 4, silicon can expand its valence shell by using available $3d$ orbitals for bonding.

A significant difference between the two elements is that the Si—Si bond energy is considerably less (about 30 kcal/mol) than that of the C—C bond; hence silanes, analogous to alkanes, having the general formula Si_nH_{2n+2} are less stable than their alkane counterparts. On the other hand, Si—O bonds are more stable (about 22 kcal/mol) than C—O bonds, and polymers having recurring Si—O linkages (polysiloxanes) are very important. There is evidence from bond length and dipole moment data of some double-bond character in Si—O bonds arising from overlap of the p orbitals of oxygen with vacant d orbitals of silicon.

Polysiloxanes,[11–15] frequently referred to as *silicones* (a rather misleading name), are prepared by hydrolysis of alkylsilicon or arylsilicon halides. The starting materials can be made from silicon and alkyl or aryl halides by heating at 250° to 280°C in the presence of copper (reaction 15.2). This is the preferred procedure for preparing methylchlorosilanes and phenylchlorosilanes.

$$Si + RCl \xrightarrow[250-280°C]{Cu} SiCl_4 + RSiCl_3 + R_2SiCl_2 + R_3SiCl \qquad (15.2)$$

Individual organosilicon compounds can be made by the Grignard reaction (15.3, 15.4) or by addition of trichlorosilane to ethylene (15.5) or acetylene (15.6). Trichlorosilane can also be reacted with benzene in the presence of boron trichloride to give mainly diphenyldichlorosilane (15.7) along with some phenyltrichlorosilane and triphenylchlorosilane. Heating mixtures of trialkylchlorosilanes and alkyltrichlorosilanes with aluminum chloride results in conversion to the more useful dialkyldichlorosilanes.

$$(CH_3)_2SiCl_2 + CH_3MgCl \rightarrow (CH_3)_3SiCl + MgCl_2 \tag{15.3}$$

$$CH_3SiCl_3 + C_6H_5MgCl \rightarrow (CH_3)(C_6H_5)SiCl_2 + MgCl_2 \tag{15.4}$$

$$HSiCl_3 + CH_2{=}CH_2 \longrightarrow CH_3CH_2SiCl_3 \tag{15.5}$$

$$HSiCl_3 + CH{\equiv}CH \longrightarrow CH_2{=}CHSiCl_3 \tag{15.6}$$

$$HSiCl_3 \xrightarrow[BCl_3]{C_6H_6} (C_6H_5)_2SiCl_2 \tag{15.7}$$

Hydrolysis of the halides gives the corresponding unstable silanols, which undergo condensation to form the siloxane linkage (15.8). Hydrolysis of dichlorides leads to linear polymers—for example, the preparation of poly-dimethylsiloxane **3** from dimethyldichlorosilane (15.9)—whereas trichlorides give crosslinked polysiloxanes.

$$-\overset{|}{\underset{|}{Si}}-Cl \xrightarrow{H_2O} \left[-\overset{|}{\underset{|}{Si}}-OH\right] \longrightarrow -\overset{|}{\underset{|}{Si}}-O-\overset{|}{\underset{|}{Si}}- + H_2O \tag{15.8}$$

$$Cl-\overset{CH_3}{\underset{CH_3}{Si}}-Cl \xrightarrow{H_2O} \left[\overset{CH_3}{\underset{CH_3}{Si}}-O\right] \tag{15.9}$$

3

As a general method of preparing polysiloxanes, hydrolysis of the halides is not satisfactory because of the tendency to form cyclic siloxanes (mainly trimers or tetramers) under the conditions of hydrolysis, although some siloxane resins are prepared this way. To obtain high-molecular-weight polymer, the cyclic products are purified by distillation and polymerized by acid- or base-catalyzed ring-opening reactions.[16]

In the presence of sulfuric acid, ring opening and further condensation take place to give linear siloxane polymers. *Silicone oils* are prepared by adding some hexamethyldisiloxane during the polymerization reaction. This leads to polysiloxanes with unreactive trimethylsiloxyl end groups (15.10). The

$$\left(\left[\overset{CH_3}{\underset{CH_3}{Si}}-O\right]_x\right)_n \xrightarrow[H_2SO_4]{(CH_3)_3SiOSi(CH_3)_3} (CH_3)_3SiO\left[\overset{CH_3}{\underset{CH_3}{Si}}-O\right]_{xn}Si(CH_3)_3 \tag{15.10}$$

Table 15.2. Bond lengths and bond angles in polymer backbones[a]

Backbone unit	Bond length (Å)	Bond angle
C—C—C	1.54	112°
C—O—C	1.42	111°
Si—O—Si	1.63	130°

[a] Data from Owen.[17]

molecular weight, and hence the viscosity, of the oils can be controlled by the amount of hexamethyldisiloxane added. Characteristically, silicone oils have good thermal stability and demonstrate little change in viscosity with temperature; hence they are ideal as lubricants and hydraulic fluids.

Base-catalyzed ring-opening polymerization of cyclic siloxanes results in high-molecular-weight linear polymer with rubberlike properties. Crosslinking can be effected by cohydrolysis with alkyltrichlorosilanes or by treatment with peroxides or oxygen. Replacing some of the methyl groups with vinyl groups results in more effective curing.

Among the more remarkable properties of polysiloxanes are a very low glass transition temperature (about −127°C), considering the high degree of substitution on the polymer backbone. (Compare data in Tables 3.1 and 3.2, Chapter 3.) This much greater flexibility arises from the larger bond angles and bond lengths associated with the siloxane backbone relative to those of vinyl polymers or polyethers (Table 15.2), which allow much greater conformational variability.[17] This, and other properties, have led to a broad range of applications. In addition to the silicone oil lubricants and hydraulic fluids mentioned earlier, *silicone rubber* is useful as a potting compound for encapsulation of electrical parts because of a combination of good thermal and oxidative stability, chemical resistance, and dielectric properties. Polysiloxanes also find use as wire coatings, gaskets, adhesives, caulking compounds, and paper-release coatings for such applications as pressure-sensitive labels or floor tiles. Because of their physiological inertness, polysiloxanes are an attractive material for construction of artificial organs for surgical implantation. Silicone resins are also used for water-repellent coatings, mold-release agents, antifoaming agents, foam stabilizers for polyurethanes, and nonstick surfaces for cooking utensils.

Modification of polysiloxanes is a fruitful area of research. One of the more interesting developments is the preparation of soluble ladder polymer by the base-catalyzed hydrolysis of phenyltrichlorosilane (15.11). The reaction is unusual in that the polymer, known as *polyphenylsilsesquioxane*[18,19] (**4**), has a regular *cis*-syndiotactic structure with a *cis-anti-cis* arrangement of fused rings.

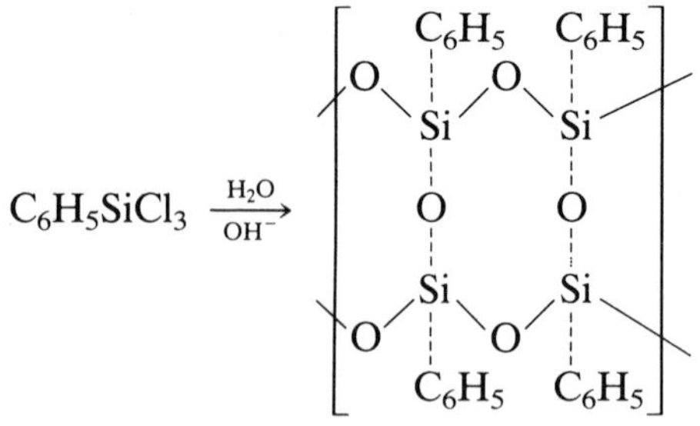

$$C_6H_5SiCl_3 \xrightarrow[OH^-]{H_2O} \begin{bmatrix} & C_6H_5 & C_6H_5 \\ O & \underset{Si}{} & O & \underset{Si}{} \\ & O & & O \\ & Si & & Si \\ O & & O & \\ & C_6H_5 & C_6H_5 \end{bmatrix} \tag{15.11}$$

4

Other modifications include incorporation of aryl groups (15.12) or metal atoms such as titanium (15.13) into the backbone[20,21]; synthesis of polysiloxane comb polymers by the macromer technique[22] (15.14); and formation of copolymers of siloxanes with other polymer types. An example is the formation of silicone–ether block copolymers (15.15) for use as surfactants.[23]

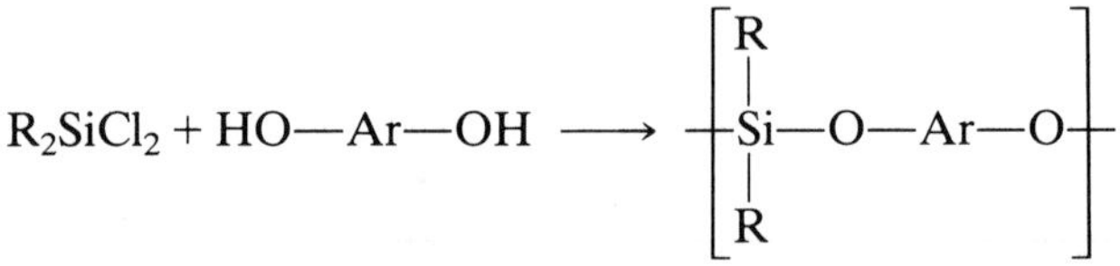

$$R_2SiCl_2 + HO\!-\!Ar\!-\!OH \longrightarrow \begin{bmatrix} R \\ | \\ Si\!-\!O\!-\!Ar\!-\!O \\ | \\ R \end{bmatrix} \tag{15.12}$$

$$R_2SiCl_2 + Ti(OR')_4 \longrightarrow \begin{bmatrix} R \\ | \\ O\!-\!Si \\ | \\ R \end{bmatrix} \begin{bmatrix} OR' \\ | \\ O\!-\!Ti\!-\!O \\ | \\ OR' \end{bmatrix} \tag{15.13}$$

$$\ce{CH2=CH-} \!\!\!\!\!\!\!\!\!\! \bigcirc \begin{bmatrix} CH_3 \\ | \\ SiO \\ | \\ CH_3 \end{bmatrix} Si(CH_3)_3 \xrightarrow{\text{initiator}} \begin{array}{c} \sim CH_2 \\ | \\ CH\!-\!\bigcirc\! \end{array} \begin{bmatrix} CH_3 \\ | \\ SiO \\ | \\ CH_3 \end{bmatrix} Si(CH_3)_3 \tag{15.14}$$

$$\begin{bmatrix} CH_3 \\ | \\ Si\!-\!O \\ | \\ CH_3 \end{bmatrix} \begin{array}{c} CH_3 \\ | \\ Si\!-\!OC_2H_5 \\ | \\ CH_3 \end{array} + HO \begin{bmatrix} R \\ | \\ CH_2CHO \end{bmatrix} \xrightarrow{(-C_2H_5OH)}$$

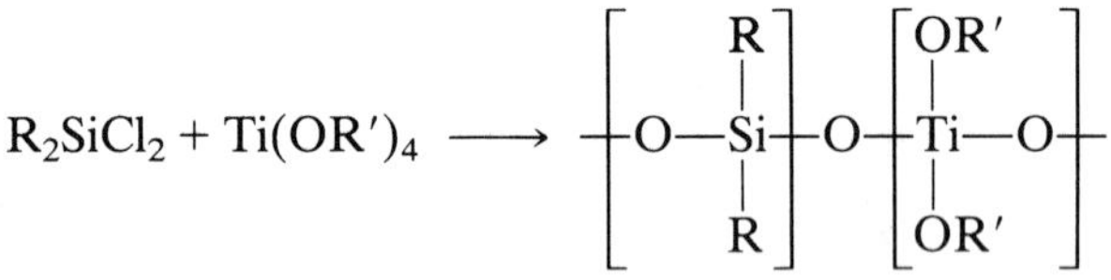

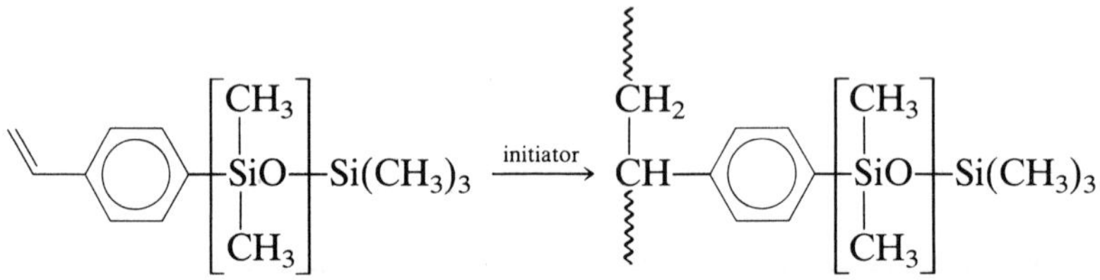

$$\begin{bmatrix} CH_3 \\ | \\ Si\!-\!O \\ | \\ CH_3 \end{bmatrix} \begin{array}{c} CH_3 \\ | \\ Si\!-\!O \\ | \\ CH_3 \end{array} \begin{bmatrix} R \\ | \\ CH_2CHO \end{bmatrix} \tag{15.15}$$

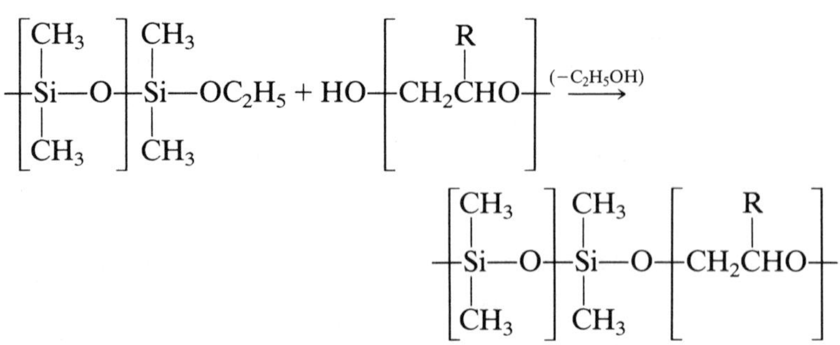

Particularly interesting are modifications of inorganic glasses with polysiloxanes.[24] The procedure involves hydrolyzing silicon alkoxides

(15.16), then copolymerizing the hydrolyzate with a hydroxyl-terminated polysiloxane (15.17). Pyrolysis of the hydrolyzate alone (the *sol-gel process*) yields the unmodified ceramic. Incorporation of the polymer is a potential route to reducing brittleness. The copolymers, which represent the marriage of a ceramic with a conventional polymer, have been appropriately designated *ceramers*.

$$SiOR_4 + 4H_2O \xrightarrow{H^+} Si(OH)_4 + 4ROH \qquad (15.16)$$

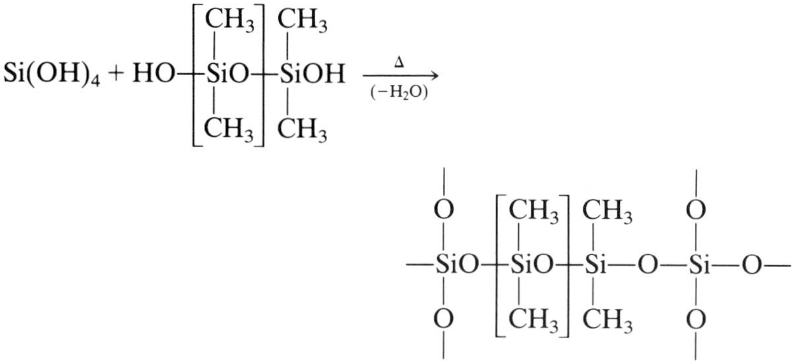

$$Si(OH)_4 + HO\text{-}\!\!\left[\begin{array}{c}CH_3\\|\\SiO\\|\\CH_3\end{array}\right]\!\!\text{-}\begin{array}{c}CH_3\\|\\SiOH\\|\\CH_3\end{array} \xrightarrow[(-H_2O)]{\Delta}$$

$$\begin{array}{c}|\\O\\|\\-SiO\\|\\O\\|\end{array}\!\!\left[\begin{array}{c}CH_3\\|\\SiO\\|\\CH_3\end{array}\right]\!\!\begin{array}{c}CH_3\\|\\Si-O-\\|\\CH_3\end{array}\begin{array}{c}|\\O\\|\\Si-O-\\|\\O\\|\end{array} \qquad (15.17)$$

15.4 Polysilanes

Although polysilanes (**5**) (polymers containing silicon–silicon bonds in the backbone) have been known for a long time, linear high-molecular-weight tractable polysilanes are of relatively recent vintage.[25–28] They are formed in moderate yield by the condensation of a dichloro organosilane monomer in the presence of finely dispersed sodium in refluxing toluene (15.18). The high degree of backbone substitution is permitted by the much longer Si—Si bond length (2.34 Å) relative to that of carbon.

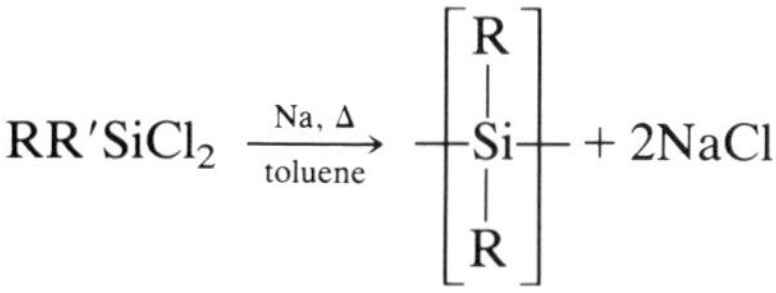

$$RR'SiCl_2 \xrightarrow[\text{toluene}]{Na, \Delta} \left[\begin{array}{c}R\\|\\Si\\|\\R\end{array}\right] + 2NaCl \qquad (15.18)$$

5

One polysilane that has attracted considerable interest is **6**, which contains equal numbers of dimethylsilane and methylphenylsilane units. Polymer **6** has been dubbed *polysilastyrene* by analogy with polystyrene; however, the methylphenylsilane units are randomly dispersed, not alternating as is the case with the phenyl groups of polystyrene. Polysilane **6** undergoes photodegradation under the influence of UV radiation and is therefore of interest in UV-active photoresists. If doped with arsenic pentafluoride, **6** exhibits semiconducting properties.

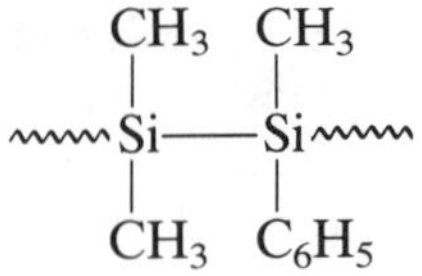

$$
\begin{array}{cc}
CH_3 & CH_3 \\
| & | \\
\text{\textasciitilde Si} & \text{—Si\textasciitilde} \\
| & | \\
CH_3 & C_6H_5
\end{array}
$$

6

When polysilanes are heated at 1200 to 1400°C, side-chain carbon atoms insert themselves into the backbone, and the polymers eventually degrade to silicon carbide. Controlled pyrolysis of polysilane fibers thus leads to SiC fibers that, like boron nitride, are of interest for ceramics applications.

15.5   Phosphonitrilic polymers

Phosphonitrilic polymers, or *phosphazene* polymers, as they are also called, are among the more interesting and commercially promising of the inorganic-type polymers developed in recent years.[8b,29-31] Starting material for their preparation is usually hexachlorocyclotriphosphazene (**7**) (phosphonitrilic chloride trimer), a commercially available monomer synthesized from phosphorus pentachloride and ammonium chloride. The monomer undergoes ring-opening polymerization to give linear polydichlorophosphazene (**8**) on heating (15.19), with some crosslinking occurring on prolonged heating.

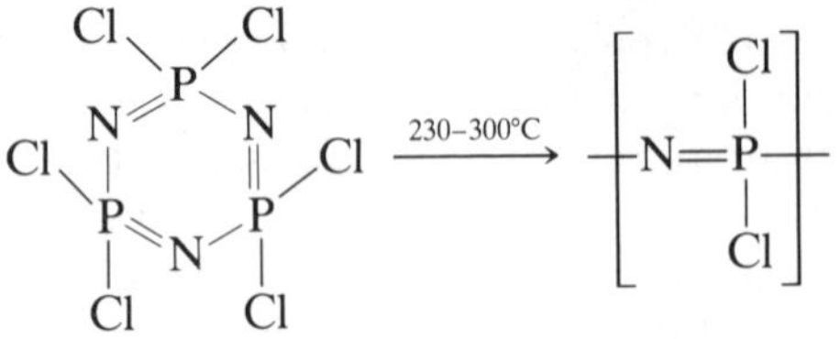

$$
\underset{\textbf{7}}{
\begin{array}{c}
Cl \diagdown \;\; \diagup Cl \\
P \\
N \diagup\;\diagdown N \\
Cl \diagdown \quad \quad \| Cl \\
P \diagdown_N \diagup P \\
| \quad\quad | \\
Cl \quad\quad Cl
\end{array}}
\xrightarrow{230-300°C}
\underset{\textbf{8}}{
\left[
\begin{array}{c}
Cl \\
| \\
N{=}P{-} \\
\| \\
Cl
\end{array}
\right]}
\qquad (15.19)
$$

Fluoro, bromo, and isothiocyano derivatives of cyclotriphosphazene give analogous linear polymers. The mechanism of polymerization is believed to involve initial ionization of a phosphorus–chlorine bond (15.20) followed by attack on the resultant cation by an electron-rich nitrogen atom (15.21) to initiate ring-opening polymerization. Chain branching might then occur by a similar mechanism involving ionization of phosphorus–chlorine bonds in the polymer backbone.

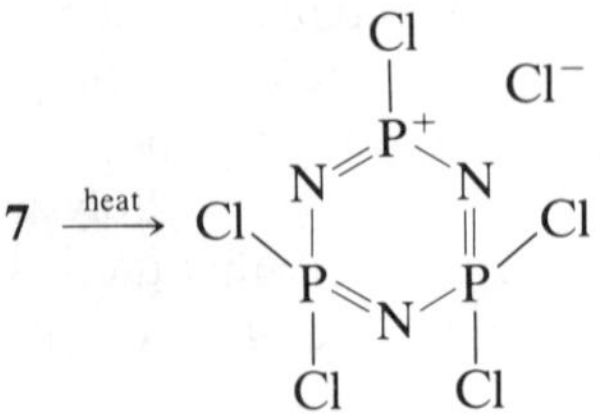

$$
\textbf{7} \xrightarrow{\text{heat}}
\begin{array}{c}
\qquad Cl \qquad\quad Cl^- \\
\qquad | \\
\qquad P^+ \\
N \diagup\;\diagdown N \\
Cl \diagdown \quad \quad \| Cl \\
P \diagdown_N \diagup P \\
| \quad\quad | \\
Cl \quad\quad Cl
\end{array}
\qquad (15.20)
$$

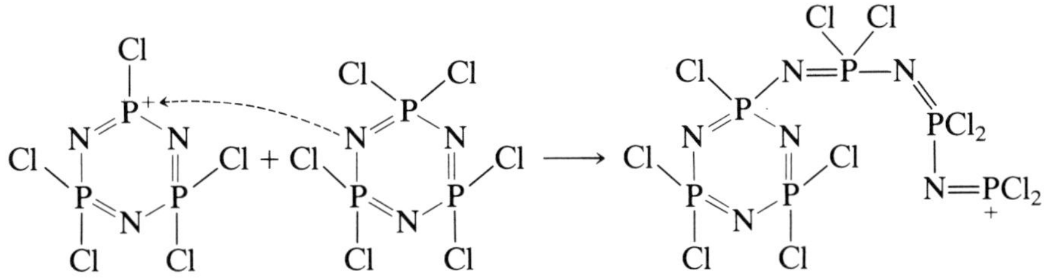

$$(15.21)$$

Polydichlorophosphazene, a true inorganic polymer with a backbone isoelectronic with that of siloxane polymers, is of no practical value as such because it is extremely sensitive to atmospheric moisture, undergoing ready hydrolysis to phosphate, ammonia, and hydrochloric acid. It also depolymerizes to trimer, tetramer, and other oligomers at temperatures above 350°C. It can, however, be converted to much more stable derivatives by replacing the chlorine atoms with alkoxy, aryloxy, or amino groups. Quantitative replacement of the chlorines can be accomplished by refluxing in an ether solvent for several hours with sodium alkoxide or aryloxide (15.22) or with primary or secondary amine (15.23). (The correspondingly substituted cyclotriphosphazenes are quite stable and resist polymerization.) Reaction of the difluoro analog with alkyllithium or dialkylmagnesium compounds (15.24) yields polyphosphazenes containing phosphorus–carbon bonds.[32]

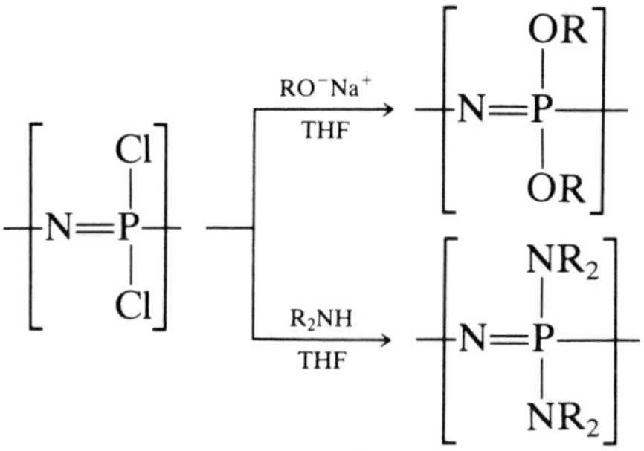

$$(15.22)$$

$$(15.23)$$

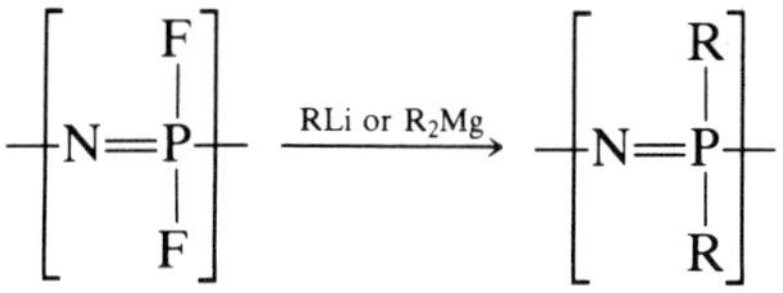

$$(15.24)$$

An examination of the phosphonitrilic backbone might lead one to predict properties similar to those of the highly crystalline polyacetylene or poly(sulfur nitride). In fact, the opposite is the case; polyphosphazenes form amorphous, clear, colorless films with elastomeric properties despite the alternation of single and double bonds. Polyphosphazenes are also nonconducting. The high conformational mobility can be explained in terms of the $(p-d)\pi$

overlap, which is less restictive relative to $(p-p)\pi$ bonding because of the availability of five *d* orbitals on each P atom. The lack of color or conductivity, on the other hand, suggests that electron delocalization is minimal. An orbital picture (**9**) to account for the four valence electrons not involved in sigma bonding (three on N and one on P) places two electrons in a lone-pair orbital on N, while the remaining two form a $(2p-3d)\pi$ bond. This leads to an orbital node on each P atom, thus limiting delocalization to three backbone atoms.

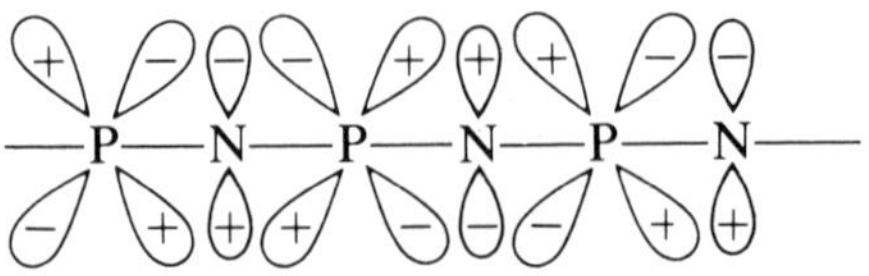

9

   Phosphonitrilic polymers are used in a variety of applications. In the area of elastomers, fluoroalkoxy-substituted polymers exhibit an ideal combination of chemical resistance and thermal stability. As an example, the copolymer **10**, prepared by refluxing **8** with a mixture of sodium trifluoroethoxide and heptafluorobutoxide, retains its flexibility at temperatures as low as $-77°C$ and is stable up to 300°C. The low-temperature flexibility makes the elastomer attractive for fuel lines and sealants in arctic environments.

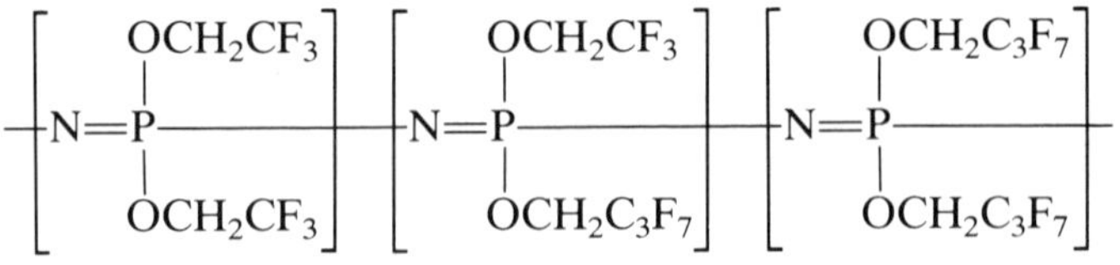

10

   Phosphonitrilic polymers are also useful as fire-retardant materials, not only because they are self-extinguishing but because they burn with only moderate evolution of smoke. As was discussed in Chapter 4 (Section 4.7), phosphonitrilic polymers are also of interest in the medical field both as structural materials for artificial organs and as delivery vehicles for controlled release of medicinal agents.[9c,33]
   Phosphonitrilic polymers, both linear and crosslinked, have also been prepared by linking cyclic phosphazene rings together. This type of work is normally directed toward synthesis of thermally stable polymers, taking advantage of the inherent stability of the cyclotriphosphazene and cyclotetraphosphazene rings. Typical of the methods reported are the reaction of a dichlorocyclophosphazene (**11**) with a dihydroxy compound (15.25), or the analogous reaction of **7** to yield network polymer. Flame- and heat-resistant polyimide composites have also been prepared from maleimide-substituted phosphazenes such as **12**.[34,35]

$$\underset{\textbf{11}}{\begin{array}{c} R \diagdown R \\ P \\ N \diagup \diagdown N \\ R \diagdown P P \diagup R \\ N \\ | | \\ Cl Cl \end{array}} + \text{HO—Ar—OH} \longrightarrow \left[\begin{array}{c} R \diagdown R \\ P \\ N \diagup \diagdown N \\ R \diagdown P P \diagup R \\ N \\ | | \\ O O—Ar \end{array} \right] \qquad (15.25)$$

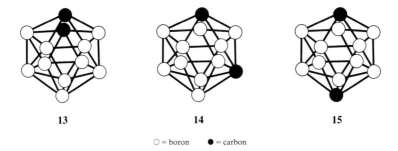

12

15.6 Carborane polymers

Apart from the boron nitride work described earlier, the most successful approach to synthesizing well-characterized boron-containing polymers has been to incorporate carborane units into the polymer.[36–40]

It is outside the scope of this book to delve into the fascinating and complex area of carborane chemistry. Most of the polymer work has been concerned with only two of these structures—the icosahedral $C_2B_{10}H_{12}$ and the pentagonal bipyramidal $C_2B_5H_7$, particularly the former. Both have a cage structure and are prepared from the corresponding boranes, decaborane ($C_{10}H_{14}$) and pentaborane (C_5H_9), by reaction with acetylene. The $C_2B_{10}H_{12}$ carborane exists in three isomeric forms, named *ortho* (**13**), *meta* (**14**), and *para* (**15**). (Only the skeletal arrangement is shown. One hydrogen is attached to each boron and carbon atom.) Reaction of decaborane with acetylene gives **13**, which can be converted to **14** by heating to 475 to 600°C; **14**, in turn, is converted to **15**, by heating to 630 to 700°C. The great stability of these

| **13** | **14** | **15** |

$\bigcirc$ = boron $\bullet$ = carbon

compounds may be attributed to the complete delocalization of the electrons, giving them pseudoaromatic character.

Because the hydrogens attached to the carbon atoms are more acidic than those attached to boron, they can be replaced, for example with lithium by reaction with butyllithium; hence a variety of carborane derivatives have been synthesized. One of these is vinylcarborane, which can be polymerized through the vinyl group or copolymerized to give high-molecular-weight polymers. Some examples of nonvinyl polymer-forming reactions involving the icosahedral carborane nucleus $(C_2B_{10}H_{10})$ are given in reactions (15.26) through (15.29).

$$Cl-\overset{\overset{\displaystyle O}{\|}}{C}-C(B_{10}H_{10})C-\overset{\overset{\displaystyle O}{\|}}{C}-Cl$$

$$\xrightarrow{HO-R-OH} \quad \left[\overset{\overset{\displaystyle O}{\|}}{C}-C(B_{10}H_{10})C-\overset{\overset{\displaystyle O}{\|}}{C}-R-O\right] \qquad (15.26)$$

$$\xrightarrow{H_2N-R-NH_2} \quad \left[\overset{\overset{\displaystyle O}{\|}}{C}-C(B_{10}H_{10})C-\overset{\overset{\displaystyle O}{\|}}{C}-R-NH\right] \qquad (15.27)$$

$$HOCH_2-C(B_{10}H_{10})C-CH_2OH + CH_3CH\overset{\displaystyle \diagup\diagdown}{\underset{O}{}}CH_2 \longrightarrow$$

$$H\left[OCH_2\underset{\underset{\displaystyle CH_3}{|}}{CH}\right]_x OCH_2-C(B_{10}H_{10})C-CH_2O\left[\underset{\underset{\displaystyle CH_3}{|}}{CH}CH_2O\right]_y H \quad (15.28)$$

$$Li-C(B_{10}H_{10})C-Li + 2Cl-\underset{\underset{\displaystyle CH_3}{|}}{\overset{\overset{\displaystyle CH_3}{|}}{Si}}-O-\underset{\underset{\displaystyle CH_3}{|}}{\overset{\overset{\displaystyle CH_3}{|}}{Si}}-Cl \longrightarrow$$

$$Cl-\underset{\underset{\displaystyle CH_3}{|}}{\overset{\overset{\displaystyle CH_3}{|}}{Si}}-O-\underset{\underset{\displaystyle CH_3}{|}}{\overset{\overset{\displaystyle CH_3}{|}}{Si}}-C(B_{10}H_{10})C-\underset{\underset{\displaystyle CH_3}{|}}{\overset{\overset{\displaystyle CH_3}{|}}{Si}}-O-\underset{\underset{\displaystyle CH_3}{|}}{\overset{\overset{\displaystyle CH_3}{|}}{Si}}-Cl \xrightarrow{H_2O}$$

$$\left[\underset{\underset{\displaystyle CH_3}{|}}{\overset{\overset{\displaystyle CH_3}{|}}{Si}}-O-\underset{\underset{\displaystyle CH_3}{|}}{\overset{\overset{\displaystyle CH_3}{|}}{Si}}-C(B_{10}H_{10})C-\underset{\underset{\displaystyle CH_3}{|}}{\overset{\overset{\displaystyle CH_3}{|}}{Si}}-O-\underset{\underset{\displaystyle CH_3}{|}}{\overset{\overset{\displaystyle CH_3}{|}}{Si}}-O\right] \quad (15.29)$$

Reaction (15.29) and other related reactions have been used to synthesize a series of *m*-carborane–siloxane polymers that have been marketed under the trade names Dexsil (Olin) and Ucarsil (Union Carbide). They exhibit elastomeric properties with excellent thermal and oxidative stability, but their high cost is a drawback. One such polymer has been used as a stationary liquid phase in gas chromatography for performing separations at temperatures as high as 500°C.

Siloxane–carborane polymers having similar properties have been synthesized using the pentagonal bipyramideal $C_2B_5H_7$ carborane (16), although these usually contain other carboranes as minor ingredients.[37]

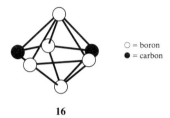

○ = boron
● = carbon

16

15.7 Organometallic polymers

An organometallic compound is, by definition, one that has carbon–metal bonds. The science of organometallic compounds may be considered a branch of chemistry in its own right, even though it overlaps the major divisions of organic and inorganic chemistry, and it was inevitable that organometallic chemistry be extended into the area of polymers.[8,41–45] This is still a relatively new field, and many of the polymers that have been developed remain laboratory curiosities, often because of the high cost of the monomers. Many of them have quite unusual properties, however, such as exhibiting intense absorption in the ultraviolet region of the spectrum, which makes them appear attractive for exterior coatings. Some are of interest because of their biocidal properties. Others have potential in the medical field as drug-release agents or because the polymers themselves exhibit physiological activity.[46] Some organometallic polymers are already in commercial use as catalysts.

Much of the work in this area has dealt with the "sandwich" compound, biscyclopentadienyliron, commonly known as ferrocene (17). Ferrocene is not only easy to prepare, but it is stable at temperatures up to 470°C. It has typical aromatic properties and undergoes substitution reactions leading to a variety of derivatives, including vinylferrocene (18), the ferrocene analog of styrene, which is converted to polyvinylferrocene with free radical, ionic, and Ziegler–Natta initiators.

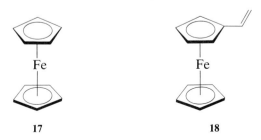

17   18

Although this part of the book is concerned with nonvinyl polymers, it is appropriate at this point to digress briefly to consider monomers such as **18**. Despite its relationship to styrene, **18** exhibits atypical behavior toward free radical polymerization in benzene solution.[9d] Termination, for example, involves intramolecular electron transfer (15.30) leading to Fe(III) chain ends rather than the more conventional bimolecular pathway. Furthermore, **18**

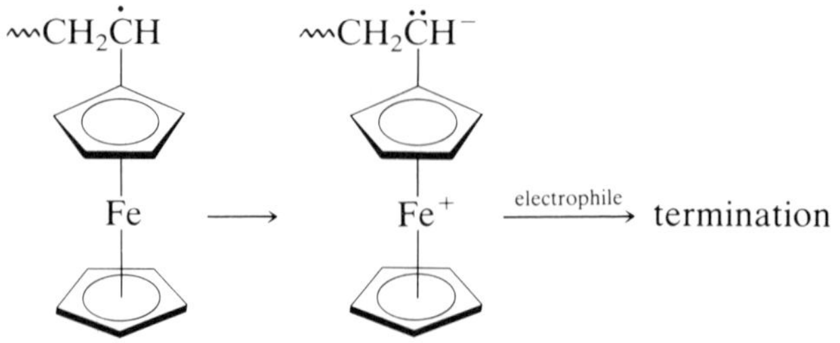

$$\text{(15.30)}$$

has a much higher chain transfer constant than styrene, which leads to a more highly branched polymer. Acrylic monomers having methylene groups separating the ferrocene from the vinyl group (**19**) follow normal free radical termination behavior, as does vinylruthenocene[47] (the ruthenium analog of **18**), which has an oxidation potential about 0.1 V more negative than that of **18**. Besides **18** and **19**, monomers **20** through **22** are representative of the numerous pi complexes that have been converted to vinyl homopolymers and copolymers.[9d–f] Organometallic sigma-bonded styrene derivatives such as **23** have also been investigated; polymers of trialkyltin methacrylates (**24**) have sparked interest as biocidal paints for maintaining ships' hulls free of barnacles and other marine organisms.[8g]

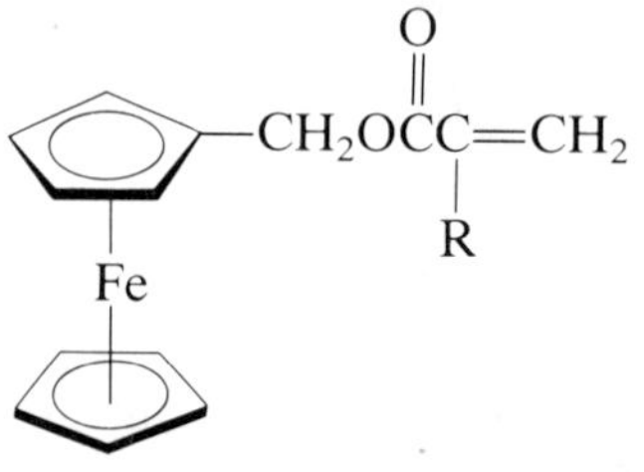

19

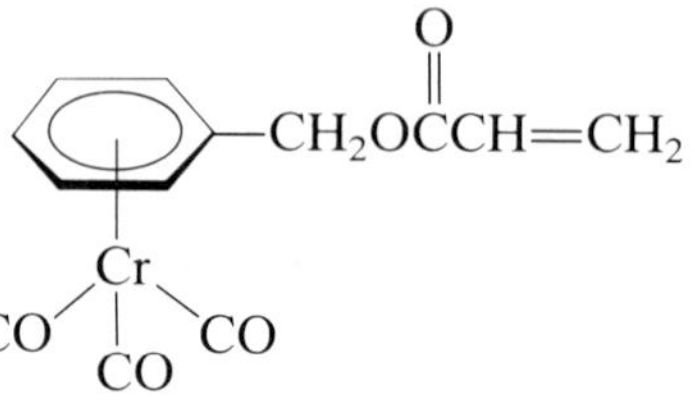

20

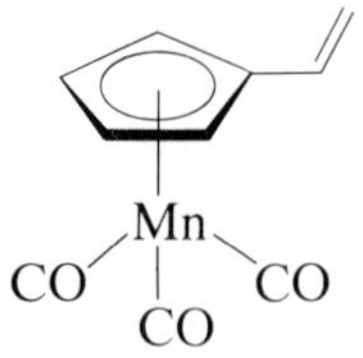

21

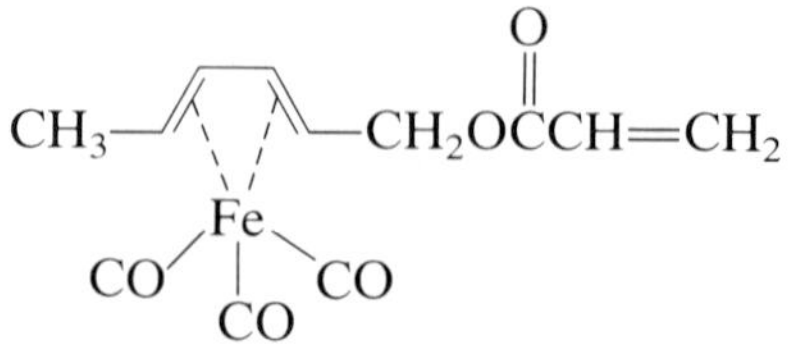

22

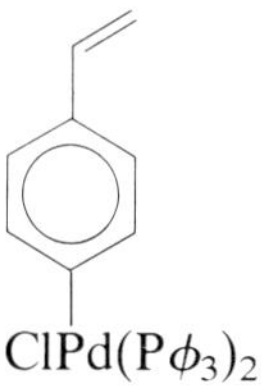

23

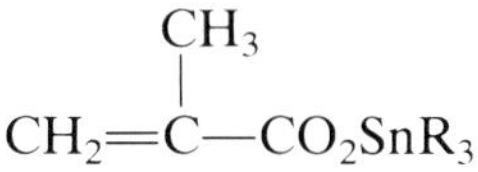

24

Polymers that contain the ferrocene nucleus as part of the backbone structure may be prepared from **25** and **26** and similarly substituted compounds[44,48] using solution or interfacial methods.[49] The related cobalticenium complexes (**27**) behave similarly.[9h,i] Interfacial methods have also been employed to prepare polyesters of tetravalent metals by reaction of the

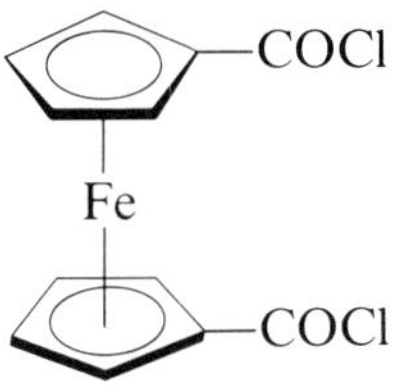

25

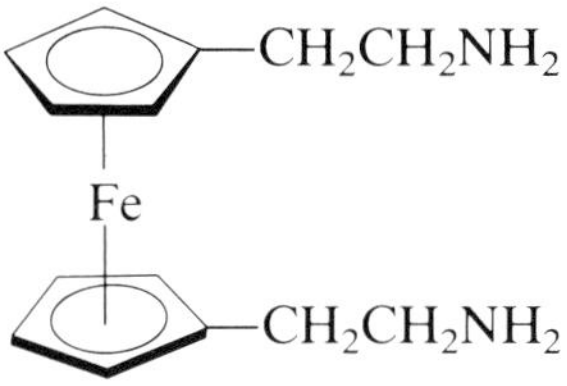

26

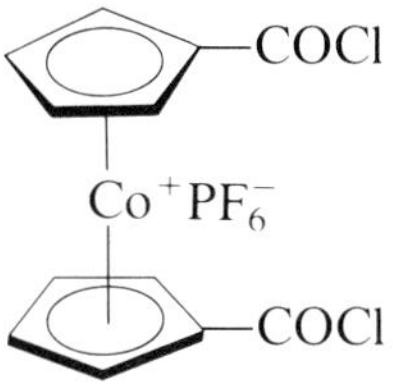

27

dialkyl or diaryl metal (M) dichlorides with dicarboxylates (15.31). Selenium-containing polyesters, polyamides, polyureas, and polyurethanes[50] have been prepared from **28** and **29**, and the selenium-containing polyamide **30** has been prepared by the interesting copolymerization of ethyleneimine with carbon monoxide and selenium[51] (15.32).

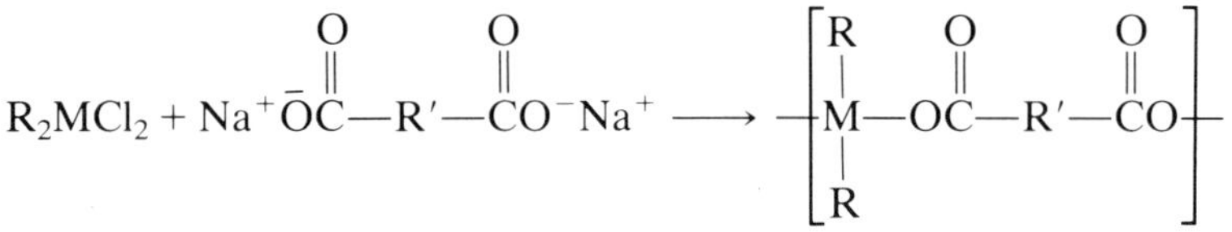

$$R_2MCl_2 + Na^+\bar{O}C\!\!\underset{\|}{\overset{O}{}}\!\!-R'-\underset{\|}{\overset{O}{}}CO^-Na^+ \longrightarrow \left[\begin{matrix} R \\ | \\ M \\ | \\ R \end{matrix} -OC\!\!\underset{\|}{\overset{O}{}}\!\!-R'-\underset{\|}{\overset{O}{}}CO \right] \quad (15.31)$$

Se(CH_2CH_2OH)_2   Se(CH_2CH_2NH_2)_2

28   29

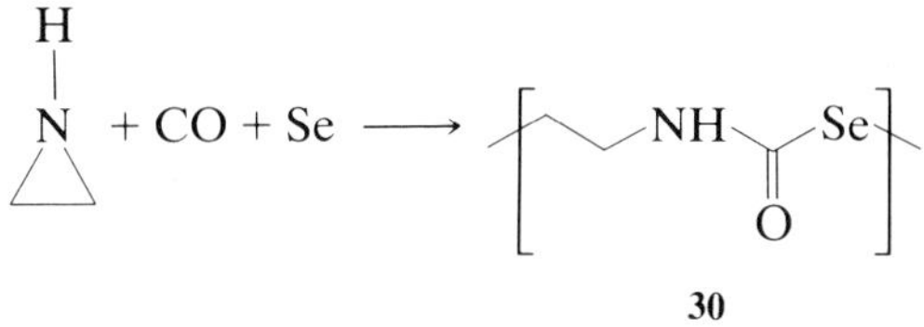

(15.32)

30

Addition reactions such as (15.33) and (15.34), which are reminiscent of similar polyacetal and polysulfide syntheses described in Chapter 11, represent another route to organometallic polymers.[52]

$$CH_2{=}CHSiH \longrightarrow {+}CH_2CH_2\underset{R'}{\overset{R}{Si}}{+} \qquad (15.33)$$

$$R_2SnH_2 + CH_2{=}CH{-}R'{-}CH{=}CH_2 \longrightarrow$$

$$\left[{+}\underset{R}{\overset{R}{Sn}}{-}CH_2CH_2{-}R'{-}CH_2CH_2{+}\right] \qquad (15.34)$$

There are some other organometallic polymers more closely allied with coordination polymers (see next section). Many of these were not made deliberately but were found to be polymeric in nature on being characterized.[53] Examples are the nickel(O)-cyclooctatetraene complex (**31**), the norbornadiene–silver nitrate complex (**32**), and the arylethynyl copper complexes (**33**). Some of these, such as **33**, are of interest because they display semiconducting and photoconducting properties, believed to arise from back coordination from the filled metal *d* orbitals to the antibonding orbitals of the acetylene.

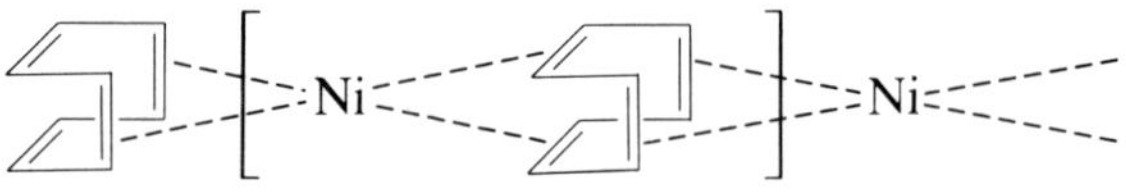

31

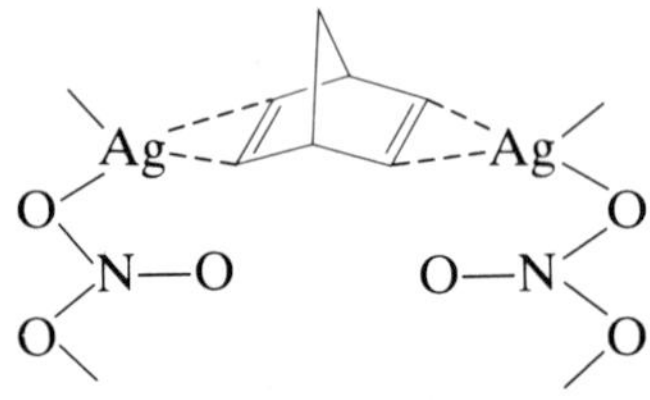

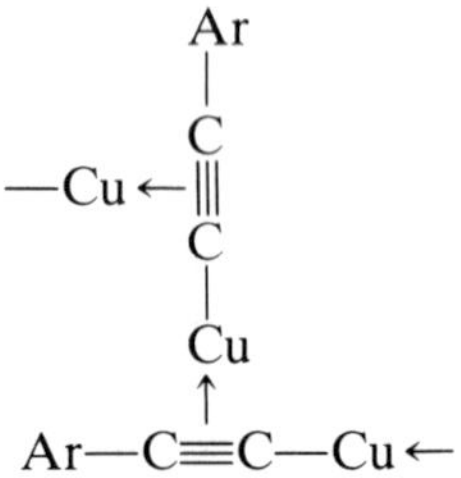

32

33

15.8   Coordination polymers

Coordination polymers are polymers having coordinate covalent bonds in the repeating unit. As with organometallic polymers, considerable work has been done in this area, [8j,43,44,53] and it still represents a very active field of research. Potential outlets for coordination polymers include thermally stable and conducting or semiconducting[54] materials, and water-based coatings, which are seen as a possible solution to the environmental problems associated with solvent-based coatings. While a large number of interesting coordination polymers have been prepared, their thermal properties are often no better, and frequently are worse, than those of the purely organic aromatic polymers.

There are generally three methods for preparing coordination polymers. Two of them yield true coordination polymers in that the coordination bonds are an integral part of the polymer backbone. The other method leads to products best described by the term *polymeric chelates*, because the chelated metal is only incidental to the overall polymer structure and breaking the coordination bonds would not rupture the polymer backbone.

The first method involves linking the polydentate ligands with metals, using metal salts, free metal, or coordination complexes (via ligand exchange) as coreactant. Examples leading to chelates through nitrogen ligands are the reaction of a bis(1,2-dioxime) with nickel acetate (15.35) and the reaction of tetracyanoethylene with copper to give a polymer having a phthalocyanine-type sheetlike structure (15.36). The phthalocyanine-type polymers have received considerable attention in recent years, and although their thermal stabilities have not always lived up to expectation,[53] some network transition metal polyphthalocyanines linked through imide or benzimidazole groups have proved to be very stable indeed, exhibiting char yields of 80 to 90 percent at 800°C.[55] There is also considerable interest in cofacially linked polyphthalocyanines of the type discussed in Chapter 4 (Section 4.8) which, on doping, conduct electricity through the stacked phthalocyanine rings.[56,57]

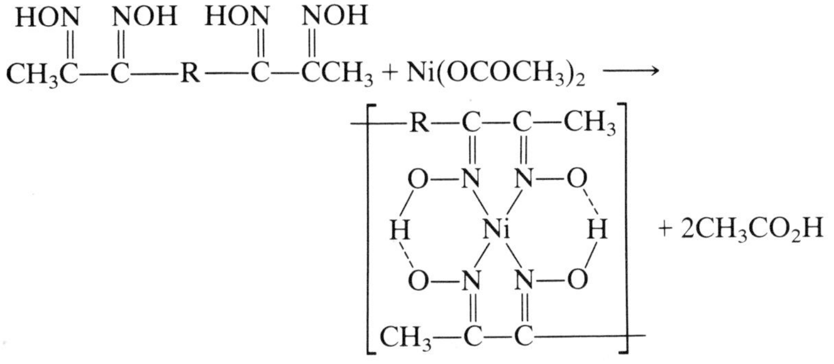

(15.35)

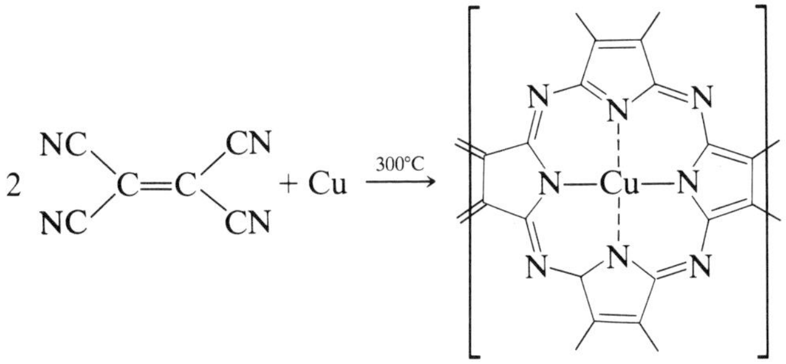

(15.36)

Polymer formation through oxygen ligands has been accomplished with numerous types of compounds, with bis(β-diketones) being among the most successful as far as achieving high molecular weight is concerned. Ligand exchange between bis(β-diketone) and metal acetylacetonate (acac) (15.37) or tetrabutyl titanate to incorporate a hexacoordinate metal atom (15.38) typify this process. Another example, involving iron–oxygen backbone coordination in a degradable polymer, is described in Chapter 4 (Section 4.7) as a vehicle for releasing agricultural chemicals into farmland under controlled conditions.

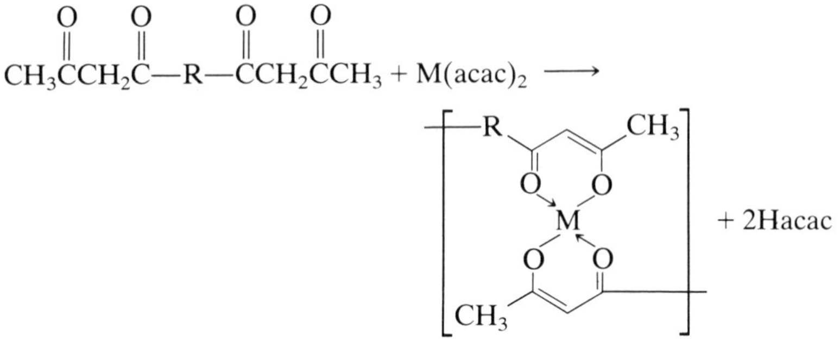

$$CH_3\overset{O}{\overset{\|}{C}}CH_2\overset{O}{\overset{\|}{C}}-R-\overset{O}{\overset{\|}{C}}CH_2\overset{O}{\overset{\|}{C}}CH_3 + M(acac)_2 \longrightarrow$$

$+\ 2Hacac$ (15.37)

$$CH_3\overset{O}{\overset{\|}{C}}CH_2\overset{O}{\overset{\|}{C}}-\langle\bigcirc\rangle-\overset{O}{\overset{\|}{C}}CH_2\overset{O}{\overset{\|}{C}}CH_3 + Ti(OC_4H_9)_4 \longrightarrow$$

$+\ 2C_4H_9OH$ (15.38)

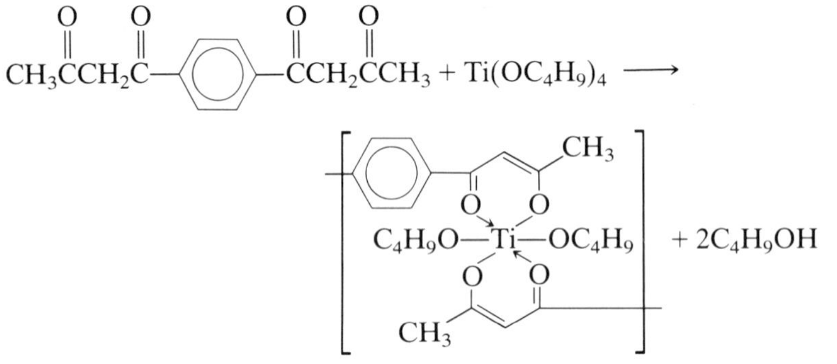

One example of a polymer formed with both nitrogen and oxygen ligands is the product of the reaction of a bis(8-hydroxy-5-quinolyl) derivative with a metal acetylacetonate (15.39). Polymers with sulfur and mixed sulfur–nitrogen or sulfur–oxygen ligands have also been prepared.[53]

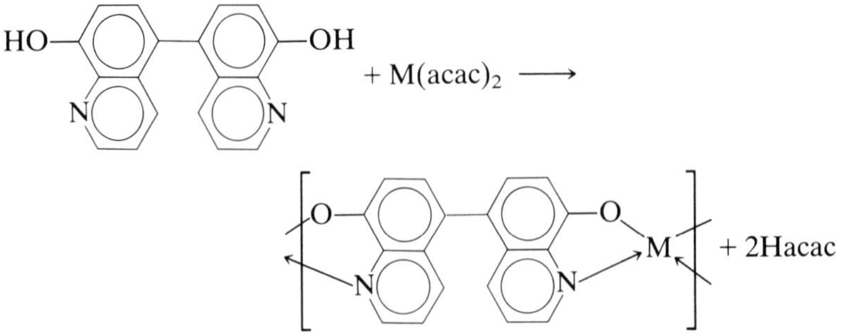

$+\ M(acac)_2 \longrightarrow$

$+\ 2Hacac$ (15.39)

The second general method of forming coordination polymers is to polymerize a metal chelate through other functional groups. Chelates of thiopicolinamides having free amino groups (**34**) have been reacted with diacid chloride to form polyamides, for example; and dihydroxy derivatives of Schiff bases (imines) (**35**) have been converted to polycarbonate with phosgene.

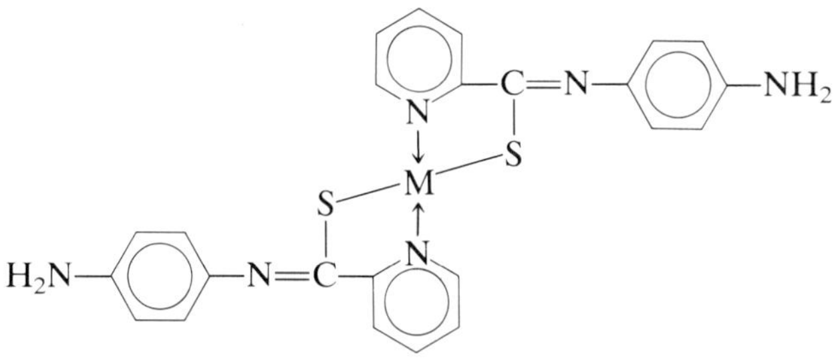

34

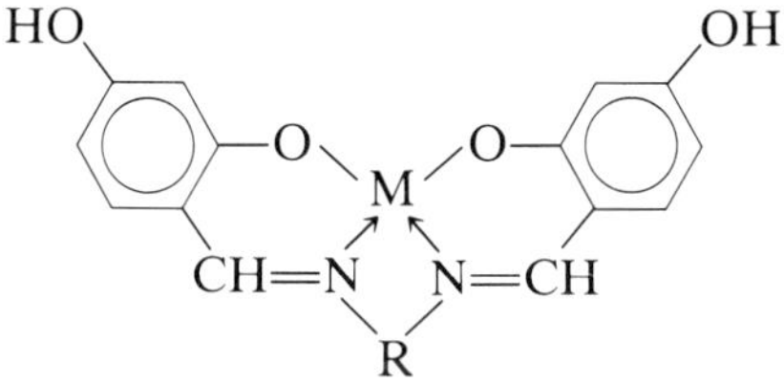

35

The polymeric chelate type of polymer is normally prepared by complexing a polymer that has coordinating ligands as part of the backbone. One of the best studied examples involves the reaction of hydroxy-substituted polyimines (or polymeric Schiff bases, as they are also called) with metal salts, although not very high molecular weights have been achieved. The polymer prepared from 5,5'-methylenebis(salicylaldehyde) and *o*-phenylenediamine is shown (**36**) complexed with divalent metal. Polymeric chelates of a variety of transi-

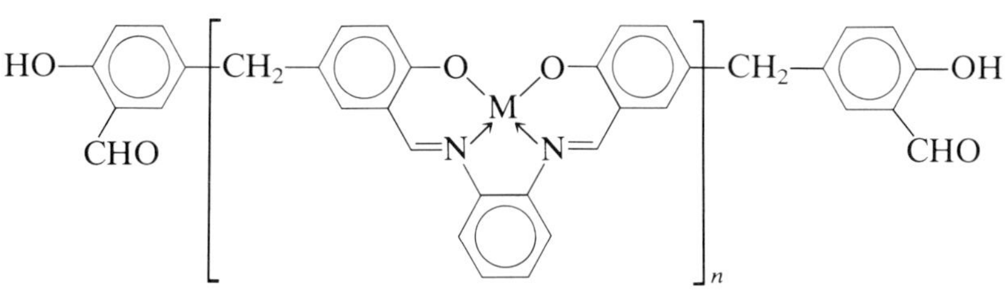

36

tion metals have been precipitated by mixing polymer solution with soluble divalent metal salts. Interestingly, such polymers frequently have poorer thermal stability than the uncomplexed polymers. (Polyimines will be covered as a separate class of polymers in the next chapter.)

The second method described above, polymerization of a metal chelate through appropriate functional groups, could also be used to form this type of polymer as long as the reactive groups are not separated by the chelate. Vinyl polymers having complexing groups can also be used. Crown ethers, mentioned earlier (Chapter 12) for their ability to complex group I cations, have been extensively investigated as pendant chelating groups on vinyl polymers[58-62] and as backbone chelating groups for condensation polymers.[63,64] Typical of the former is poly(4'-vinylbenzo-18-crown-6) (37) shown with complexed ion. Crown ether monomers used to make condensation polymers include the diaminodibenzo derivative 38 and the diaza crown ether 39.

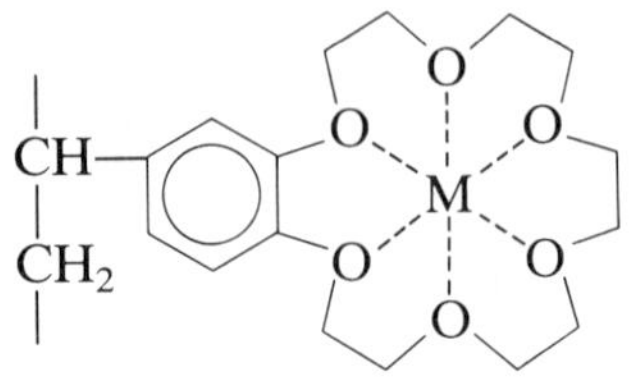

37

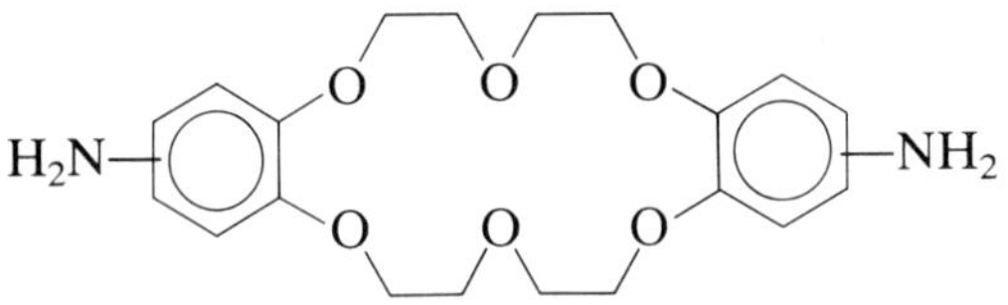

38

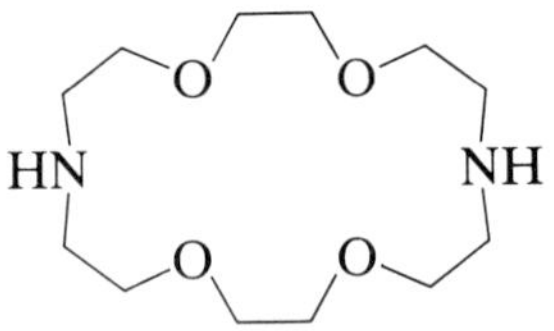

39

Primarily because of steric effects, crown ether-containing polymers usually exhibit differing cation-binding characteristics than the monomers from which they are derived. In many instances, cations are bound more strongly by being sandwiched between two chelating rings.

References

1. N. H. Ray, *Inorganic Polymers*, Academic Press, New York, 1978.
2. F. G. A. Stone and W. A. G. Graham (eds.), *Inorganic Polymers*, Academic Press, New York, 1962.
3. D. N. Hunter, *Inorganic Polymers*, Wiley-Interscience, New York, 1963.
4. H. R. Allcock, *Heteroatom Ring Systems and Polymers*, Academic Press, New York, 1967.
5. M. Zeldin, K. J. Wynne, and H. R. Allcock (eds.), *Inorganic and Organometallic Polymers*, ACS Symp. Ser. 360, American Chemical Society, Washington, D.C., 1987.
6. J. Economy, R. V. Anderson, and V. I. Matkovich, *J. Appl. Polym. Sci., Appl. Polym. Symp.*, **9**, 377 (1969).
7. C. K. Narula, R. T. Paine, and R. Schaeffer, *Polym. Preprints*, **28**(1), 454 (1987).

8. J. Economy, W. D. Smith, R. Y. Lin, *J. Appl. Polym. Sci., Appl. Polym. Symp.*, **21**, 131 (1973).
9. C. E. Carraher Jr., J. F. Sheats, and C. U. Pittman Jr. (eds.), *Organometallic Polymers*, Academic Press, New York, 1978: (a) M. Akhtar, C. K. Chiang, M. J. Cohen, A. J. Heeger, J. Kleppinger, A. G. MacDiarmid, J. Milliken, M. J. Moran, and D. L. Peebles, pp. 301–312; (b) R. E. Singler and G. L. Hagnauer, pp. 257–269; (c) H. R. Allcock, pp. 283–288; (d) C. U. Pittman Jr., pp. 1–11; (e) D. W. Slocum, M. D. Rausch, and A. Siegel, pp. 39–51; (f) V. V. Korshak and S. L. Sosin, pp. 25–38; (g) W. L. Yeager and V. J. Castelli, pp. 175–180; (h) J. E. Sheats, pp. 87–94; (i) E. W. Neuse, pp. 95–100; (j) J. C. Bailar Jr., pp. 313–321.
10. C. M. Mikulski, P. J. Russo, M. S. Saran, A. G. MacDiarmid, A. F. Garito, and A. J. Heeger, *J. Am. Chem. Soc.*, **97**, 6358 (1975).
11. C. Eaborn, *Organosilicon Compounds*, Academic Press, New York, 1960.
12. R. N. Meals and F. M. Lewis, *Silicones*, Reinhold, New York, 1959.
13. W. Noll, *Chemistry and Technology of Silicones*, Academic Press, New York, 1968.
14. G. G. Freeman, *Silicones: An Introduction to Their Chemistry and Applications*, Iliffe, London, 1962.
15. E. G. Rochow, *An Introduction to the Chemistry of the Silicones*, 2nd ed., Wiley-Interscience, New York, 1951.
16. E. E. Bostick, in *Ring-Opening Polymerization* (K. C. Frisch and S. L. Reegen, eds.), Dekker, New York, 1969, Chap. 8.
17. M. J. Owen, *CHEMTECH*, **11**, 288 (1981).
18. J. F. Brown Jr., L. H. Voght Jr., A. Katchman, J. W. Eustace, K. M. Kiser, and K. W. Krantz, *J. Am. Chem. Soc.*, **82**, 6194 (1960).
19. T. E. Helminiak and G. C. Berry, *J. Polym. Sci., Polym. Symp.*, **65**, 107 (1978).
20. A. H. Frazer, *High Temperature Resistant Polymers*, Wiley-Interscience, New York, 1968, Chap. 5.
21. K. A. Andrianov, *Metalorganic Polymers*, Wiley-Interscience, New York, 1965, Chap. 4.
22. Y. Kawakami, Y. Miki, T. Tsuda, R. A. N. Murthy, and Y. Yamashita, *Polym. J.*, **14**, 913 (1982).
23. B. Kanner, W. G. Reid, and I. H. Petersen, *Ind. Eng. Chem. Prod. Res. Develop.*, **6**, 88 (1967).
24. H.-H Huang, B. Orier, and G. L. Wilkes, *Polym. Bull.*, **14**, 557 (1985).
25. R. West and E. Carberry, *Science*, **189**, 179 (1975).
26. R. West, L. D. David, P. I. Djurovich, K. L. Stearly, K. S. V. Srinivasan, and H. Yu, *J. Am. Chem. Soc.*, **103**, 7352 (1981).
27. P. Trefonas III, P. I. Djurovich, X.-H. Zhang, R. West, R. D. Miller, and D. Hofer, *J. Polym. Sci., Polym. Lett. Ed.*, **21**, 819 (1983).
28. P. Trefonas III, R. West, R. D. Miller, and D. Hofer, *J. Polym. Sci., Polym. Lett. Ed.*, **21**, 823 (1983).
29. H. R. Allcock, *Chem. Rev.*, **72**, 315 (1972).
30. H. R. Allcock, *Phosphorus-Nitrogen Compounds*, Academic Press, New York, 1972.
31. H. R. Allcock, *Angew. Chem., Int. Ed. English*, **16**, 147 (1977).
32. T. L. Evans and H. R. Allcock, *J. Macromol. Sci.-Chem.*, **A16**, 409 (1981).
33. H. R. Allcock, *J. Polym. Sci., Polym. Symp.*, **70**, 71 (1983).

34. D. Kumar, *J. Polym. Sci., Polym. Chem. Ed.*, **23**, 1661, (1985).
35. D. Kumar, G. M. Fohlen, and J. A. Parker, *J. Polym. Sci., Polym. Chem. Ed.*, **22**, 927, 1141 (1984).
36. R. N. Grimes, *Carboranes*, Academic Press, New York, 1970, Chap. 8.
37. R. E. Williams, *Pure Appl. Chem.*, **29**, 569 (1972).
38. D. J. Mangold, *J. Appl. Polym. Sci., Appl. Polym. Symp.*, **11**, 157 (1969).
39. H. A. Schroeder, *Inorg. Macromol. Rev.*, **1**, 45 (1970).
40. P. E. Cassidy, *Thermally Stable Polymers*, Dekker, New York, 1980, Chap. 10.
41. E. W. Neuse and H. Rosenberg, *Metallocene Polymers*, Dekker, New York, 1970.
42. E. W. Neuse, *Adv. Macromol. Chem.*, **1**, 1 (1970).
43. A. D. Pomogailo and V. S. Savostyanov, *J. Macromol. Sci., Rev. Macromol. Chem. Phys.*, **C25**, 375 (1985).
44. N. Hagihara, K. Sonogashira, and S. Takahashi, *Adv. Polym. Sci.*, **41**, 149 (1981).
45. E. W. Neuse, *J. Macromol. Sci.-Chem.*, **A16**, 3 (1981).
46. C. E. Carraher Jr., and C. G. Gebelein (eds.), *Biological Activities of Polymers*, ACS Symp. Ser. 186, American Chemical Society, Washington, D.C., 1982.
47. T. C. Willis and J. E. Sheats, *J. Polym. Sci., Polym. Chem. Ed.*, **22**, 1077 (1984).
48. K. Gonsalves, L. Zhan-Ru, and M. D. Rausch, *J. Am. Chem. Soc.*, **106**, 3862 (1984).
49. C. E. Carraher Jr., in *Interfacial Synthesis*, Vol. 2 (F. Millich and C. E. Carraher Jr., eds.), Dekker, New York, 1977, Chap. 21.
50. H. Kroll and E. F. Bolton, *J. Appl. Polym. Sci.*, **14**, 2319 (1970).
51. N. Konda, N. Sonoda, and H. Sakurai, *J. Polym. Chem., Polym. Lett. Ed.*, **12**, 679 (1974).
52. J. W. Labadie, S. A. MacDonald, and C. G. Wilson, *Polym. Bull.*, **16**, 427 (1986).
53. R. J. Cotter and M. Matzner, *Ring-Forming Polymerizations*, Part A, Academic Press, New York, 1969, Chaps. 5–9.
54. B. A. Bolto, in *Organic Semiconducting Polymers* (J. E. Katon, ed.), Dekker, New York, 1968, Chap. 4.
55. B. N. Achar, G. M. Fohlen, and J. A. Parker, *J. Polym. Sci., Polym. Chem. Ed.*, **20**, 269, 2773, 2781 (1982); **23**, 801 (1985).
56. C. W. Dirk, T. Inabe, J. W. Lyding, K. F. Schoch Jr., C. W. Kannewurf, and T. J. Marks, *J. Polym. Sci., Polym. Symp.*, **70**, 1 (1983).
57. P. M. Kuznesof, R. S. Nohr, K. J. Wynne, and M. E. Kenney, *J. Macromol. Sci.-Chem.*, **A16**, 299 (1981).
58. A. J. Varma, T. Majewicz, and J. Smid, *J. Polym. Sci., Polym. Chem. Ed.*, **17**, 1573 (1979).
59. A. Warshawsky, R. Kalir, A. Deshe, H. Berkovitz, and A. Patchornik, *J. Am. Chem. Soc.*, **101**, 4249 (1979).
60. Y. Nakatsuji, S. Furuyoshi, and M. Okahara, *Makromol. Chem.*, **187**, 105 (1986).
61. B. Roland and J. Smid, *Polymer*, **25**, 1166 (1984).
62. K. Yagi, J. A. Ruiz, and M. C. Sanchez, *Makromol. Chem., Rapid Commun.*, **1**, 263 (1980).
63. L. J. Mathias and K. Al-Jumah, *J. Polym. Sci., Polym. Chem. Ed.*, **18**, 2911 (1980).
64. A. Ricard and F. Lafuma, *Polymer*, **27**, 133 (1986).

Review exercises

1. Using either **25** or **26** and any other appropriate monomers, write reactions for the synthesis of a polyester, a polyamide, a polyurea, and a polyimide.

2. Suggest a way to make a polydimethylsiloxane–polycarbonate $+AB+$ block co-polymer. (H. A. Vaughn, *J. Polym. Sci., B*, **7**, 569 1959)

3. Keeping in mind the mode of termination, derive a rate expression for the free radical polymerization of vinylferrocene in benzene solution. (See reaction 15.30.) (ref. 9d)

4. Write repeating units of the polymers expected from the following reactions:

(a) 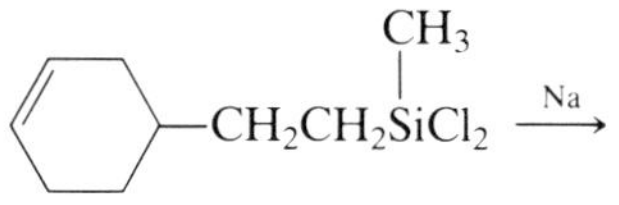
$$\text{(cyclohexene ring)}-\text{CH}_2\text{CH}_2\overset{\overset{\text{CH}_3}{|}}{\text{SiCl}_2} \xrightarrow{\text{Na}}$$

(H. Stuger and R. West, *Macromolecules*, **18**, 2349, 1985)

(b) 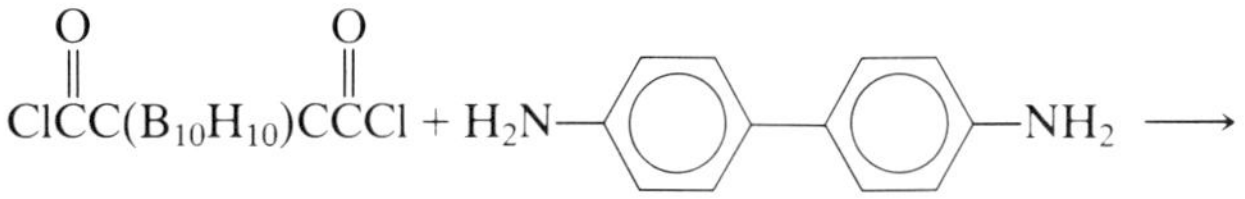
$$\text{ClCC(B}_{10}\text{H}_{10})\text{CCCl} + \text{H}_2\text{N}-\text{(phenyl)}-\text{(phenyl)}-\text{NH}_2 \longrightarrow$$
(with $\overset{O}{\underset{\|}{}}$ groups on the carbons)

(V. V. Korshak et al., *J. Polym. Sci. A–1*, **8**, 2351, 1970)

(c) 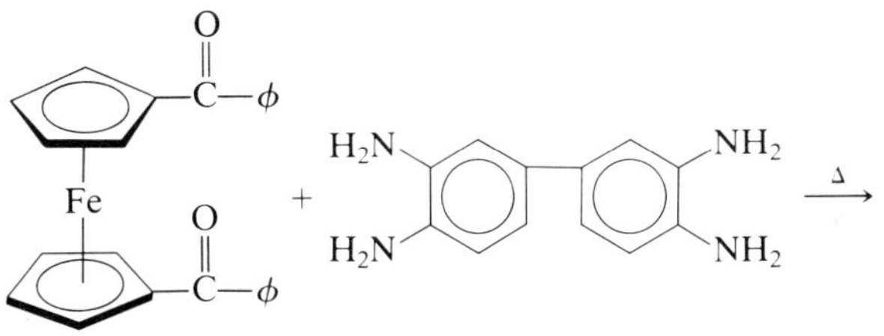
$$\text{ferrocene with } -\overset{O}{\underset{\|}{C}}-\phi \text{ groups} \quad + \quad \text{H}_2\text{N}-\text{(biphenyl with two NH}_2\text{)} \xrightarrow{\Delta}$$

(L. Plummer and C. S. Marvel, *J. Polym. Sci. A–2*, 2559, 1964)

(d) $\text{R}_2\text{Si(CH}_2\text{CH}{=}\text{CH}_2)_2 \xrightarrow{\text{AlBr}_3}$

(N. C. Billingham et al., *J. Polym. Sci., Polym. Chem. Ed.*, **15**, 675, 1977)

(e)
$$\text{Me}_2\text{N}-\overset{\overset{R}{|}}{\underset{\underset{R}{|}}{\text{Sn}}}-\text{R}-\overset{\overset{R}{|}}{\underset{\underset{R}{|}}{\text{Sn}}}-\text{NMe}_2 + \text{HC}{\equiv}\text{C}-\text{R}'-\text{C}{\equiv}\text{CH} \longrightarrow$$

(ref. 52)

(f) 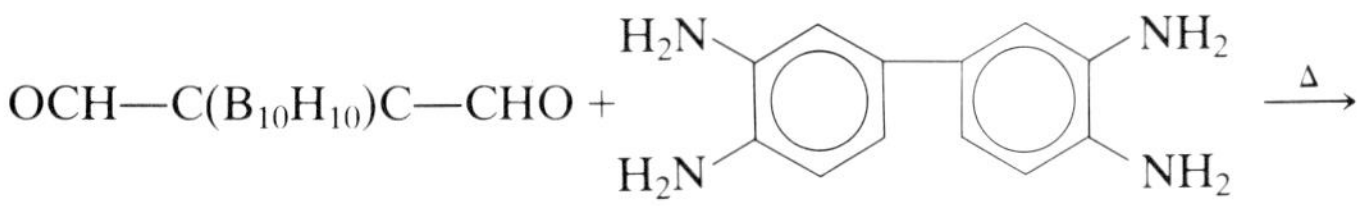
$$\text{OCH}-\text{C(B}_{10}\text{H}_{10})\text{C}-\text{CHO} + \text{H}_2\text{N}-\text{(biphenyl with two NH}_2\text{)} \xrightarrow{\Delta}$$

(E. W. Neuse et al., *Macromolecules*, **19**, 481, 1986)

(g) 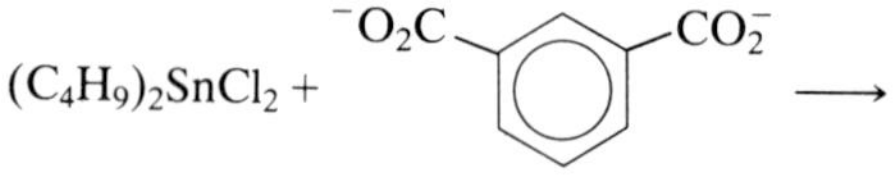 $(C_4H_9)_2SnCl_2 +$

(ref. 49)

(h) 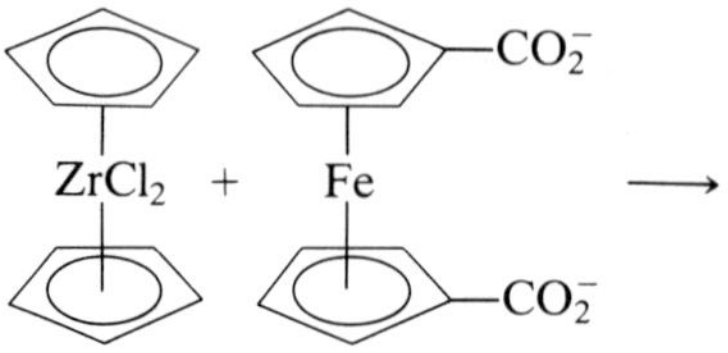 $ZrCl_2 +$

(ref. 49)

(i)

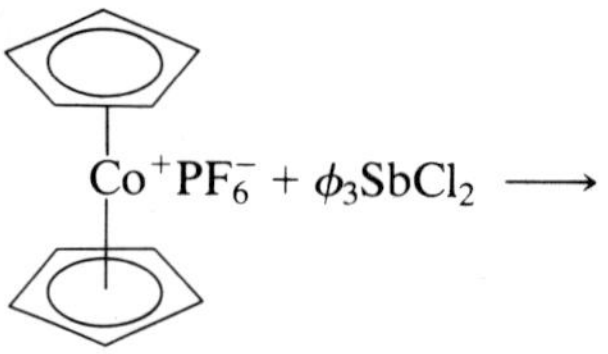

$Co^+PF_6^- + \phi_3SbCl_2 \longrightarrow$

(ref. 49)

(j)

$+ Co(OAc)_2 \longrightarrow$

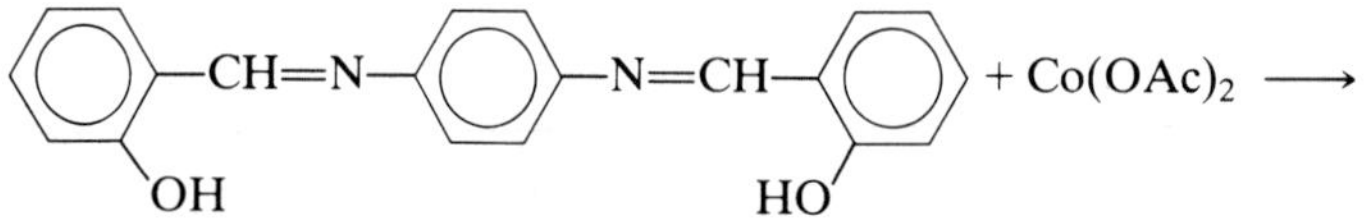

(P. Y. Maurya et al., *Makromol. Chem.*, **183**, 511, 1982)

(k) 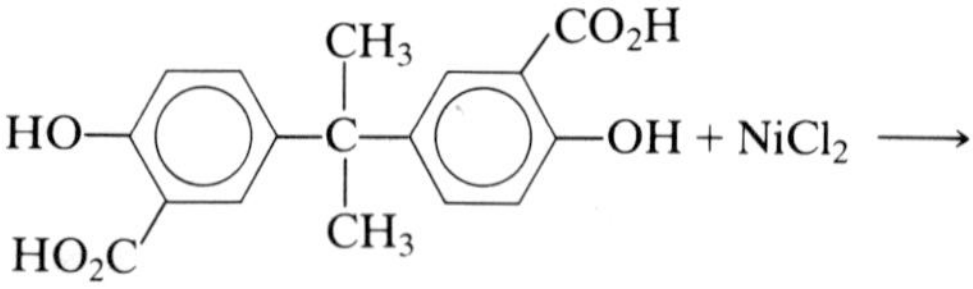 HO

$-OH + NiCl_2 \longrightarrow$

(ref. 50)

(l)

$+ Cd(OAc)_2 \longrightarrow$

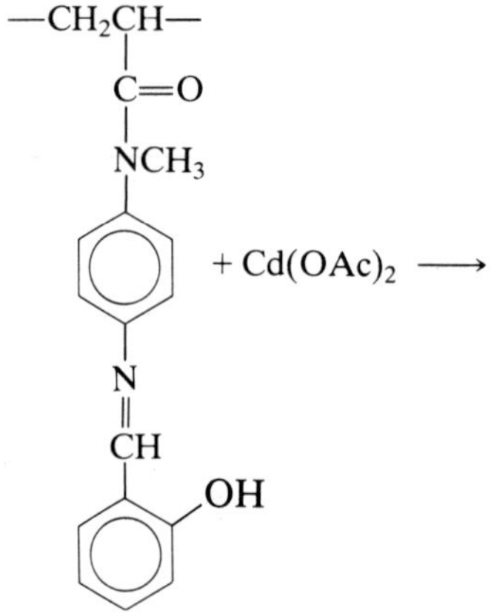

(J. J. Laverty and Z. G. Gardlund, *J. Polym. Sci. A–1*, **9**, 243, 1971)

16

Miscellaneous organic polymers

16.1 Introduction

This is the last chapter dealing with purely synthetic polymers. Included are nonvinyl polymers that do not fit into any of the significant commercial categories discussed previously. This is by no means to be considered an exhaustive compilation of "miscellaneous" organic polymers, but it does include (not necessarily in any logical sequence) most of the polymer types that have received fairly extensive treatment in the polymer literature. For the most part they are laboratory curiosities, although some, particularly the polyanhydrides, polycarbodiimides, and some of the aromatic polymers, have attracted considerable interest. Among those that have been developed commercially are polyphenylenes and poly(p-xylylene) as well as some related Friedel–Crafts polymers (discussed in Sections 16.4 and 16.5) and polyamines (Section 16.9).

16.2 Miscellaneous unsaturated polymers

Unsaturated polymers have been prepared by several different reactions. Already encountered in previous chapters are polydienes and metathesis polymers, which contain backbone or pendant ethylenic groups (or both), and conjugated polymers such as polyacetylene, poly(sulfur nitride), and polyphosphazenes. Although backbone-conjugated polymers having potential as conductors or semiconductors have received the most attention in recent years, the polycarbodiimides containing cumulated carbon–nitrogen bonds have also been investigated extensively.

16.2.1 Polycarbodiimides

Polycarbodiimides[1a,2] are prepared from the readily available diisocyanates by a self-addition polymerization reaction catalyzed by organophosphorus compounds—in particular, phosphine oxides such as 1-ethyl-3-methyl-3-phospholene 1-oxide (**1**). The mechanism is considered to involve addition of catalyst across the carbon–nitrogen double bond (16.1) and loss of carbon

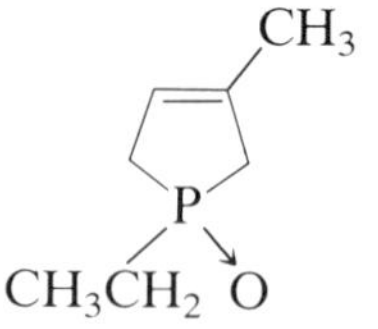

1

dioxide to form a dipolar intermediate (16.2) which adds (16.3) to another isocyanate group, this time across a carbon–oxygen double bond. Decomposition of the cyclic intermediate (16.4) results in formation of the carbodiimide group. The overall polymerization reaction is illustrated for methylenebis(4-phenyl isocyanate) (16.5). Polycarbodiimides prepared this

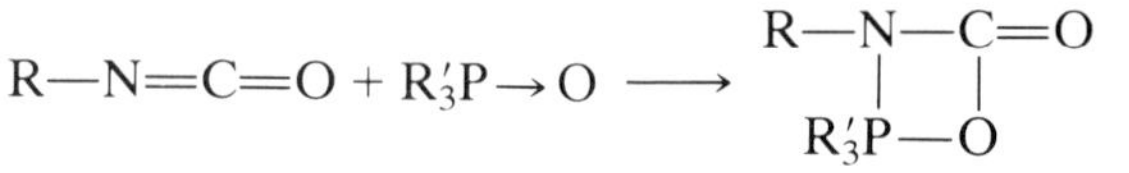

$$R—N{=}C{=}O + R_3'P{\rightarrow}O \longrightarrow \begin{array}{c} R—N—C{=}O \\ |\quad\quad| \\ R_3'P—O \end{array} \qquad (16.1)$$

$$\begin{array}{c} R—N—C{=}O \\ |\quad\quad| \\ R_3'P—O \end{array} \longrightarrow R—\overset{-}{N}—\overset{+}{P}R_3' + CO_2 \qquad (16.2)$$

$$R—\overset{-}{N}—\overset{+}{P}R_3' + R—N{=}C{=}O \longrightarrow \begin{array}{c} R—N—C{=}NR \\ |\quad\quad| \\ R_3'P—O \end{array} \qquad (16.3)$$

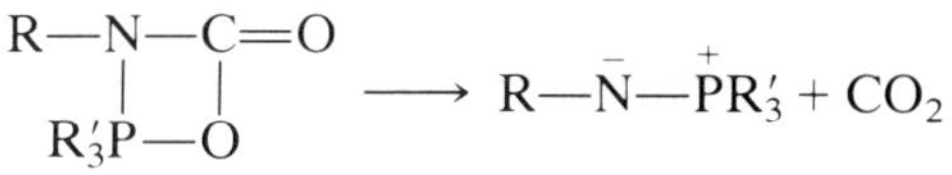

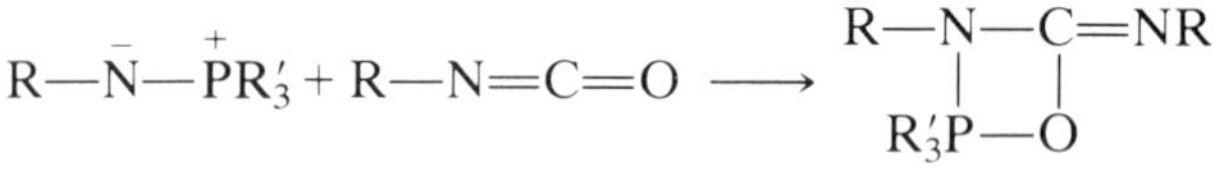

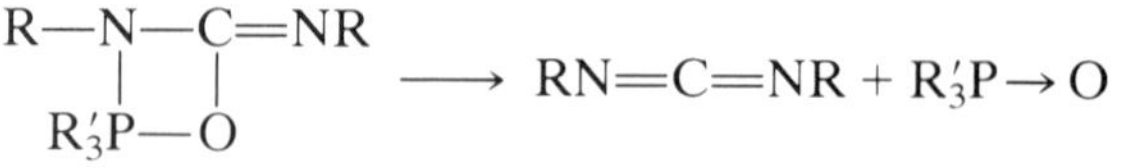

$$\begin{array}{c} R—N—C{=}NR \\ |\quad\quad| \\ R_3'P—O \end{array} \longrightarrow RN{=}C{=}NR + R_3'P{\rightarrow}O \quad (16.4)$$

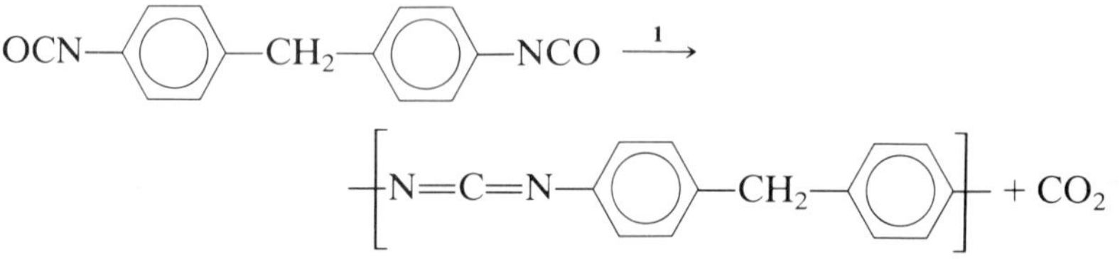

way are generally of high molecular weight and exhibit properties comparable to those of some of the nylons. Depending on structure, they are semicrystalline, and are believed to be partially crosslinked because of reversible cycloaddition (16.6). Although highly resistant to chemicals in the solid state, polycarbodiimides are quite reactive in solution. For example, modification with amines (16.7) or water (16.8) leads to polymers having guanidine and urea groups, respectively. Reaction with hydrazoic acid (16.9) yields the heterocyclic polyaminotetrazole (**2**).[3]

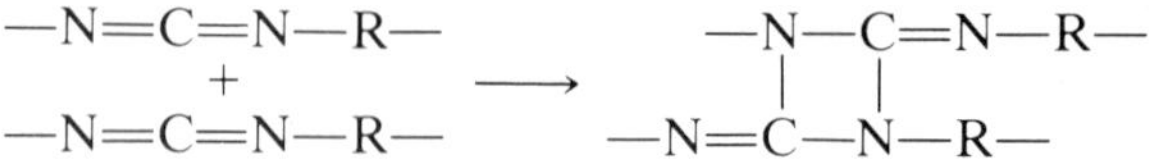

$$\begin{array}{cc} \begin{array}{c} —N{=}C{=}N—R— \\ + \\ —N{=}C{=}N—R— \end{array} \longrightarrow & \begin{array}{c} —N—C{=}N—R— \\ |\quad\quad| \\ —N{=}C—N—R— \end{array} \end{array} \qquad (16.6)$$

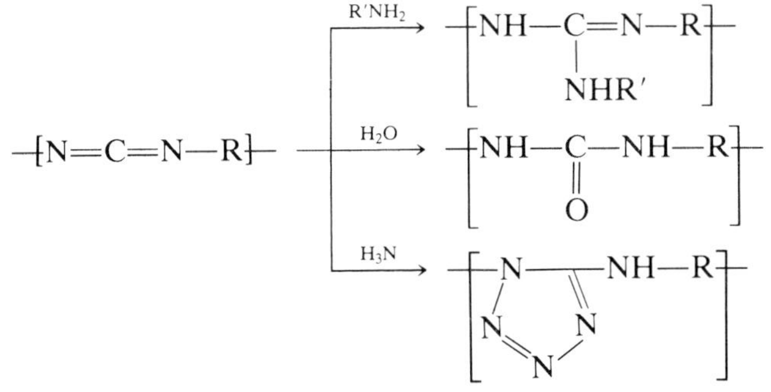

$$\left[NH{-}\underset{\underset{NHR'}{|}}{C}{=}N{-}R\right] \quad (16.7)$$

$$\left[N{=}C{-}N{-}R\right] \xrightarrow{\;R'NH_2\;} \qquad \qquad$$

$$\xrightarrow{\;H_2O\;} \left[NH{-}\underset{\underset{O}{||}}{C}{-}NH{-}R\right] \quad (16.8)$$

$$\xrightarrow{\;H_3N\;} \left[\underset{2}{N{-}\!\!\!\!\diagdown\!\!\!\!{-}NH{-}R}\right] \quad (16.9)$$

16.2.2 Polyimines

Polyimines,[1a,4,5] also called *azomethine* or *Schiff base polymers*, are another class of polymers containing carbon–nitrogen double bonds. The imine group is formed by addition of amines to carbonyl compounds followed by loss of water (16.10). The reaction is normally acid-catalyzed, although careful pH control is necessary because protonation of the amine reduces its nucleophilicity.

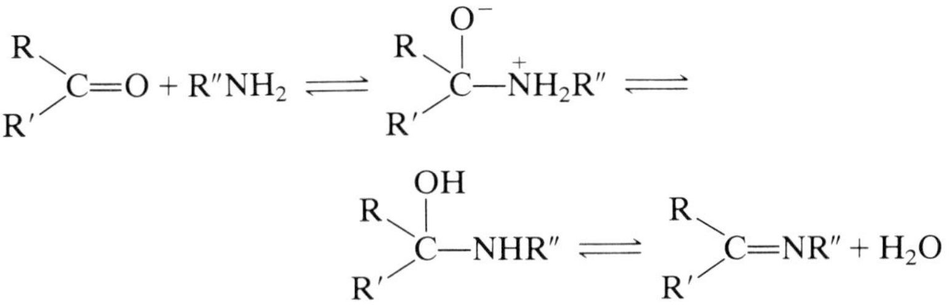

$$\underset{R'}{\overset{R}{>}}C{=}O + R''NH_2 \rightleftharpoons \underset{R'}{\overset{R}{>}}\underset{}{\overset{\overset{O^-}{|}}{C}}{-}\overset{+}{N}H_2R'' \rightleftharpoons$$

$$\underset{R'}{\overset{R}{>}}\underset{}{\overset{\overset{OH}{|}}{C}}{-}NHR'' \rightleftharpoons \underset{R'}{\overset{R}{>}}C{=}NR'' + H_2O \quad (16.10)$$

Reaction of dialdehydes with diamines leads to polyimines by this step reaction polycondensation. Typically, conjugated polymers result from terephthaldehyde and *p*-phenylenediamine (16.11). A bisimine exchange reaction (16.12), catalyzed by such reagents as zinc chloride or iodine, may

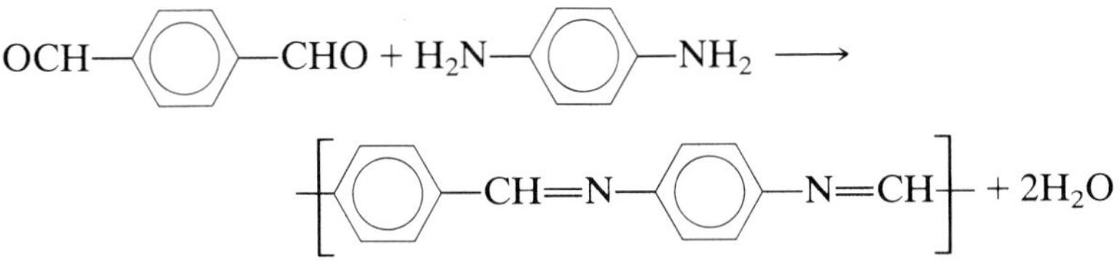

$$OCH{-}\bigcirc{-}CHO + H_2N{-}\bigcirc{-}NH_2 \longrightarrow$$

$$\left[\bigcirc{-}CH{=}N{-}\bigcirc{-}N{=}CH\right] + 2H_2O \quad (16.11)$$

$$\phi{-}CH{=}N{-}\bigcirc{-}N{=}CH{-}\phi + \phi{-}N{=}CH{-}\bigcirc{-}CH{=}N\phi \xrightarrow{\text{catalyst}}$$

$$\left[N{-}\bigcirc{-}N{=}CH{-}\bigcirc{-}CH\right] + 2\phi{-}CH{=}N{-}\phi \quad (16.12)$$

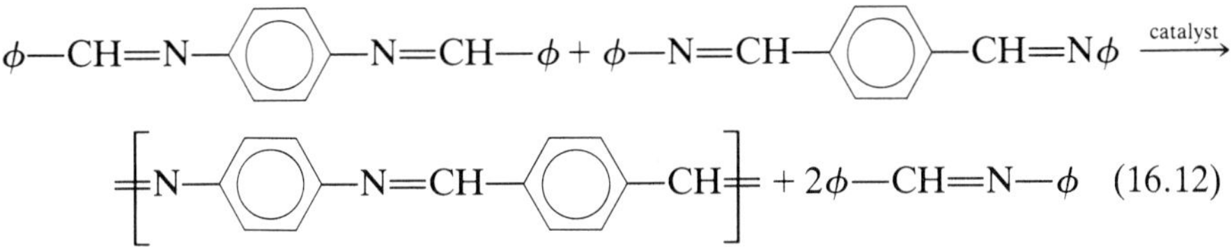

also be used. High-strength fibers have been spun from liquid crystalline melts of aromatic polyimines,[5] but they have not achieved the commercial success of the aramid fibers.

It may be recalled from the preceding chapter that appropriately substituted polyimines are useful chelating agents for coordination polymers. The extension of polyimine synthesis to tetracarbonyl compounds and tetraamines, which leads to the aromatic polyquinoxalines, is discussed later in this chapter.

Related to the polyimines are the *polyamidines*[6-8] (**3**), which are prepared most directly from monocarboxylic acids and diamines by dehydration with poly(trimethylsilyphosphate) (16.13).[6]

$$RCO_2H + H_2N—Ar—NH_2 \xrightarrow{\text{poly(trimethylsilyl phosphate)}}$$

$$\left[\begin{array}{c} R \\ | \\ N=C—NH—Ar \end{array} \right] \quad (16.13)$$

3

16.2.3   *Polymers containing carbon–carbon double bonds*

Apart from polyacetylenes synthesized by coordination polymerization (Chapter 8), polymers containing conjugated-backbone double bonds have been prepared by the Wittig reaction[9,10] involving ylide–aldehyde (16.14) or diylide and dicarbonyl compounds (16.15). Conjugated polymers (of low molecular weight) can also be prepared by condensing biscarbonyl compounds with monomers containing active methylene groups[11a] (16.16).

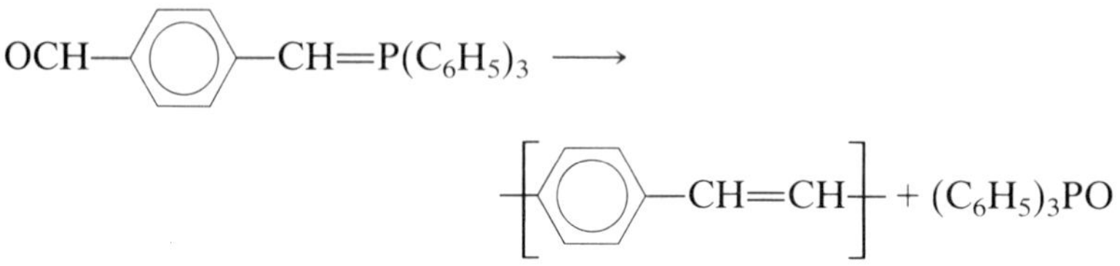

$$OCH—\langle\bigcirc\rangle—CH=P(C_6H_5)_3 \longrightarrow$$

$$\left[\langle\bigcirc\rangle—CH=CH \right] + (C_6H_5)_3PO \quad (16.14)$$

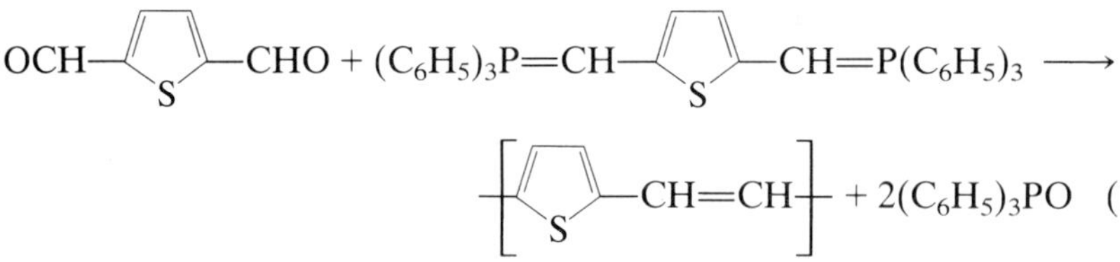

$$OCH—\langle S\rangle—CHO + (C_6H_5)_3P=CH—\langle S\rangle—CH=P(C_6H_5)_3 \longrightarrow$$

$$\left[\langle S\rangle—CH=CH \right] + 2(C_6H_5)_3PO \quad (16.15)$$

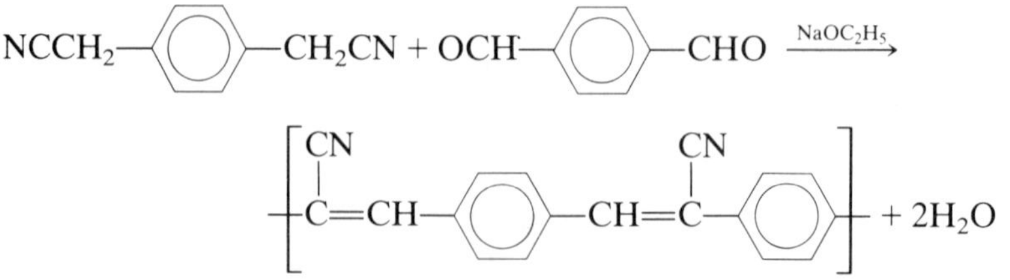

$$NCCH_2—\langle\bigcirc\rangle—CH_2CN + OCH—\langle\bigcirc\rangle—CHO \xrightarrow{\text{NaOC}_2\text{H}_5}$$

$$\left[\begin{array}{cc} CN & CN \\ | & | \\ C=CH—\langle\bigcirc\rangle—CH=C—\langle\bigcirc\rangle \end{array} \right] + 2H_2O \quad (16.16)$$

16.2.4 Azo polymers

Azo polymers have nitrogen–nitrogen double bonds in the backbone. Several methods have been reported for their synthesis, with the highest molecular weights being obtained by oxidative coupling of aromatic diamines[12] (16.17) using catalysts similar to those used in preparing poly(phenylene oxide) by oxidative coupling of phenols (Chapter 11).

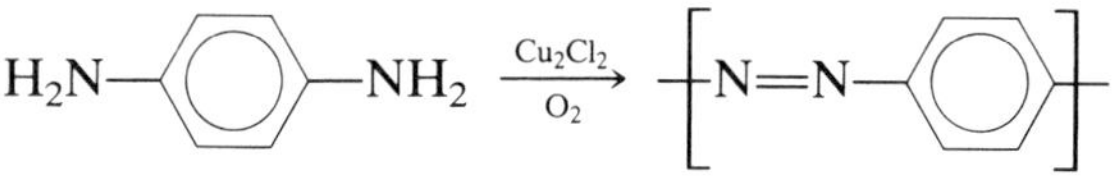

$$H_2N-\langle\bigcirc\rangle-NH_2 \xrightarrow[O_2]{Cu_2Cl_2} \left[N{=}N-\langle\bigcirc\rangle\right] \qquad (16.17)$$

Other methods include:

1. Photolysis of diazides[13] (16.18)

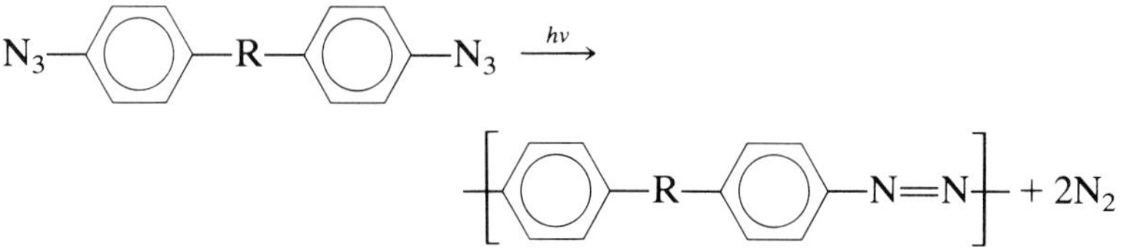

$$N_3-\langle\bigcirc\rangle-R-\langle\bigcirc\rangle-N_3 \xrightarrow{h\nu}$$

$$\left[\langle\bigcirc\rangle-R-\langle\bigcirc\rangle-N{=}N\right] + 2N_2 \quad (16.18)$$

2. Coupling of tetrazonium compounds with bisphenols[14] (16.19)

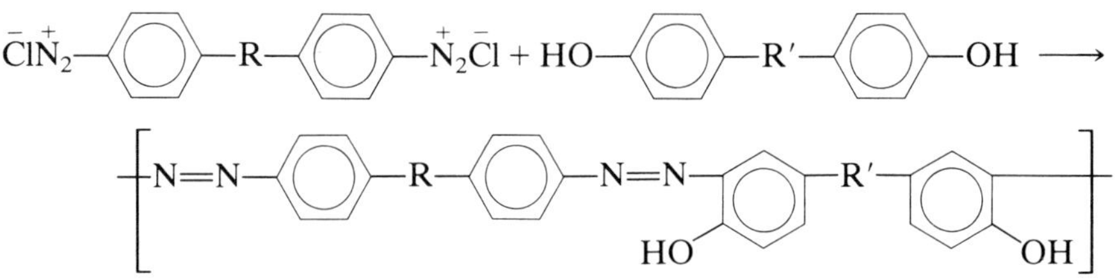

$$\bar{C}l\overset{+}{N}_2-\langle\bigcirc\rangle-R-\langle\bigcirc\rangle-\overset{+}{N}_2\bar{C}l + HO-\langle\bigcirc\rangle-R'-\langle\bigcirc\rangle-OH \longrightarrow$$

$$\left[N{=}N-\langle\bigcirc\rangle-R-\langle\bigcirc\rangle-N{=}N-\underset{HO}{\langle\bigcirc\rangle}-R'-\underset{OH}{\langle\bigcirc\rangle}\right]$$
$$(16.19)$$

3. Decomposition of tetrazonium salts[15] (16.20)

$$\overset{+}{N}_2{-}Ar{-}\overset{+}{N}_2 \xrightarrow{Cu_2Cl_2,\,\Delta} [Ar{-}N{=}N] \qquad (16.20)$$

4. Reductive coupling of dinitro compounds[16] (16.21)

$$O_2N{-}Ar{-}NO_2 \xrightarrow[\text{triethylene glycol}]{KOH,\,Zn} [Ar{-}N{=}N] \qquad (16.21)$$

N,N'-dioxide derivatives of azo polymers have been prepared by bromine oxidation of dihydroxylamines[17] (16.22).

$$HON{-}R{-}NOH \xrightarrow{Br_2} \left[R{-}\underset{\downarrow}{\overset{\uparrow}{N}}{=}\underset{O}{\overset{O}{N}}\right] \qquad (16.22)$$

16.2.5   *Polymers containing carbon–carbon triple bonds*

Oxidative coupling is also used to prepare polyacetylenes,[18,19] which in this case are defined as polymers having backbone triple bonds. An example is the polymerization of *m*-diethynylbenzene (16.23), which leads to high-

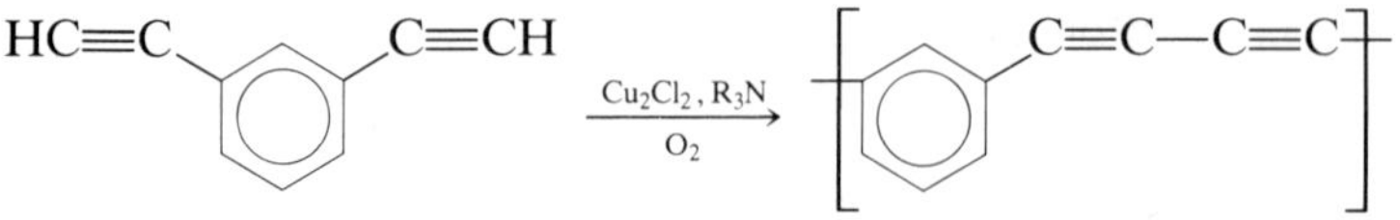

$$HC{\equiv}C{-}\text{(ring)}{-}C{\equiv}CH \xrightarrow[\text{O}_2]{\text{Cu}_2\text{Cl}_2,\ \text{R}_3\text{N}} \left[\text{(ring)}{-}C{\equiv}C{-}C{\equiv}C \right] \qquad (16.23)$$

molecular-weight polymer if the monomer is carefully purified. The regular zigzag structure of the polymer (**4**) results in high crystallinity and very low solubility. The incorporation of small amounts of *p*-diethynylbenzene, how-ever, has the effect of increasing solubility by altering the regular structure, thus impeding crystallization, without changing the rigid character of the polymer backbone.

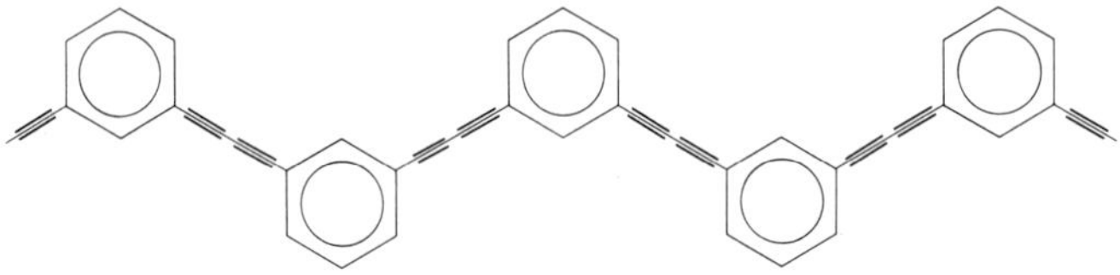

4

An alternative route to polyacetylenes involves the reaction of diacetylenes with aromatic dibromides (16.24) in the presence of a palladium catalyst.[20] The success of this process hinges on activation of the bromine atoms by electron-withdrawing groups. Conjugated diacetylenes can also be poly-merized in the solid state using heat, light, or ionizing radiation[21,22] (16.25). This process, which involves free radical coupling reactions, results in high-molecular-weight polyeneynes that exhibit interesting color changes when they undergo phase transitions.[23]

$$Br{-}Ar{-}Br + HC{\equiv}C{-}Ar'{-}C{\equiv}CH \xrightarrow[(-2HBr)]{Pd}$$

$$ {+}Ar{-}C{\equiv}C{-}Ar'{-}C{\equiv}C{+} \quad (16.24)$$

$$R{-}C{\equiv}C{-}C{\equiv}C{-}R \xrightarrow[\text{ionizing radiation}]{\Delta,\ h\nu,\ \text{or}} {=}\underset{R}{C}{-}C{\equiv}C{-}\underset{R}{C}{=} \quad (16.25)$$

Conjugated polyacetylenes are relatively unstable, but they can be con-verted to aromatic polymers having pyrrole (16.26), thiophene (16.27), or pyrazole (16.28) rings.[24]

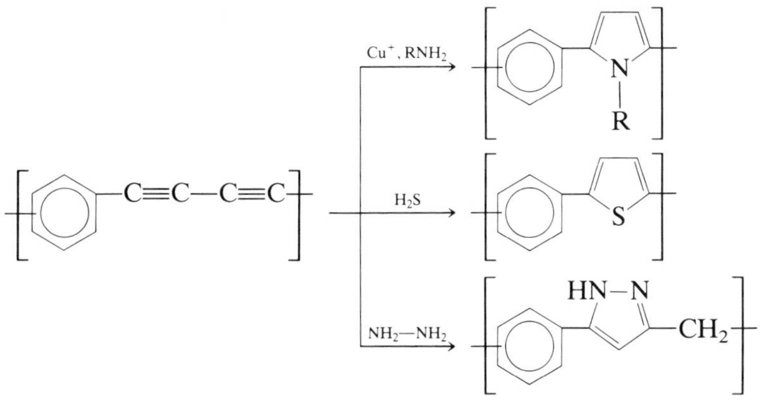

$$\left[\hspace{-2pt}\begin{array}{c}\text{phenyl}-\text{C}\equiv\text{C}-\text{C}\equiv\text{C}\end{array}\hspace{-2pt}\right] \quad\begin{array}{l}\xrightarrow{\text{Cu}^+,\text{RNH}_2}\\[40pt]\xrightarrow{\text{H}_2\text{S}}\\[40pt]\xrightarrow{\text{NH}_2-\text{NH}_2}\end{array}$$

(16.26)

(16.27)

(16.28)

16.3   Miscellaneous heterocyclic polymers[25a]

We have encountered several different types of aromatic polymers in previous chapters, some wholly aromatic and others containing functional groups that link aromatic segments together. In Chapter 13 a variety of heterocyclic polymers related to or derived from polyamides were described. So much research has been done or is still in progress in this field that it would be beyond the scope of an introductory text of this type to include all heterocyclic polymers that have appeared in the literature. A few of the more interesting ones that have not been mentioned in earlier chapters are described in this section, with the ring systems summarized in Table 16.1. Besides those listed, phthalocyanine-type polymers of the kind described in the preceding chapter, but containing no coordinated metal, are of current interest as matrix resins for high-temperature composites.[26] They are synthesized by thermal polymerization of bisphthalonitriles (16.29), and the products are complex in nature with triazine and possibly other ring systems present besides phthalocyanine.

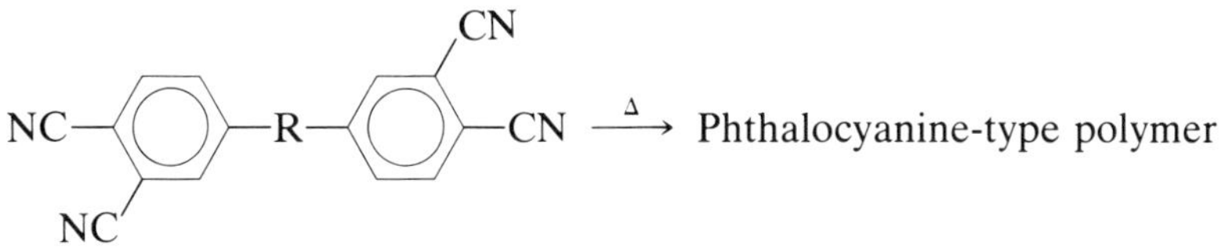

$$\text{NC}-\langle\text{C}_6\text{H}_3\rangle-\text{R}-\langle\text{C}_6\text{H}_3(\text{CN})\rangle-\text{CN} \xrightarrow{\Delta} \text{Phthalocyanine-type polymer}$$

(16.29)

16.3.1   *Polyquinoxalines and polypyrazines*

As mentioned earlier, polyquinoxalines[27,28a] result from an extension of the polyimine synthesis to tetracarbonyl compounds and tetraamines. An example is the ladder polymer **5** shown in reaction (16.30). Polyquinoxalines have generated considerable interest as heat-resistant resins for aerospace applications. An example of an intramolecularly curable polyquinoxaline is given in

Table 16.1. Representative ring systems incorporated into heterocyclic polymers

Ring type	Structure	Ring type	Structure
Quinoxaline	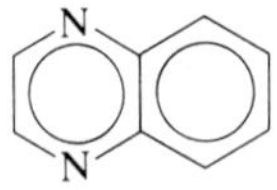	Quinoline	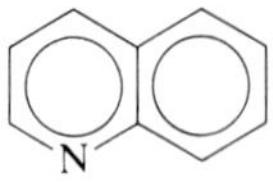
Pyrazine	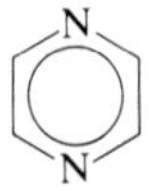	Anthrazoline	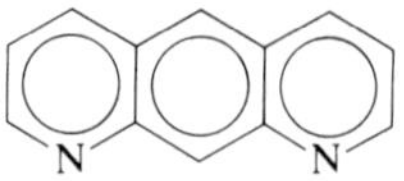
Pyrazole		Pyrrole	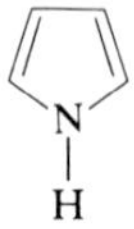
Imidazole	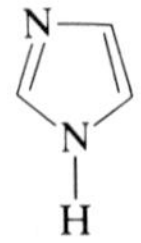	Furan	
as-Triazine	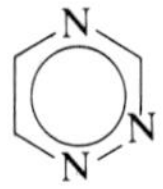	Thiophene	
Triazoline	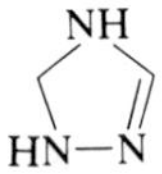		

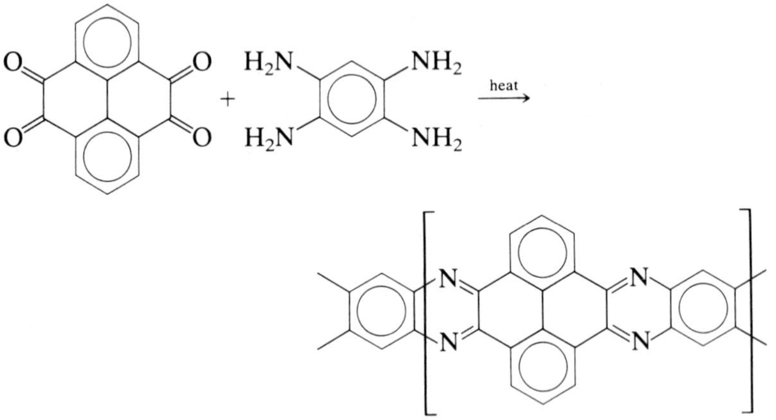

$$(16.30)$$

5

Chapter 4 (Scheme 4.1). Crosslinkable biphenylene-modified polyquinoxa-lines have also been reported[29,30] (see Table 4.4).

The polymerization reactions can be carried out in solution or in the melt, but generally postpolymerization of the isolated polymer at about 350 to

400°C is necessary to achieve high molecular weights. Polyquinoxalines are stable in nitrogen at temperatures up to 400 to 500°C; then weight loss occurs gradually at temperatures up to 800 to 900°C. In air they decompose rapidly at 400 to 500°C. Ladder polyquinoxalines (**5**) reportedly show little weight loss in nitrogen at temperatures below 600 to 700°C.

The quinoxaline group may also be referred to as a *benzopyrazine* group. Polymers containing free pyrazine groups (*polypyrazines*, **6**)[31] are prepared from bis(α-haloketones) by reaction with ammonia under oxidative conditions (16.31). Presumably, the corresponding bis(α-aminoketone) is intermediate in the reaction. If the polymer is heated with hydrogen peroxide, some of the nitrogen atoms are converted to the polar N-oxide groups, which enhances polymer solubility.

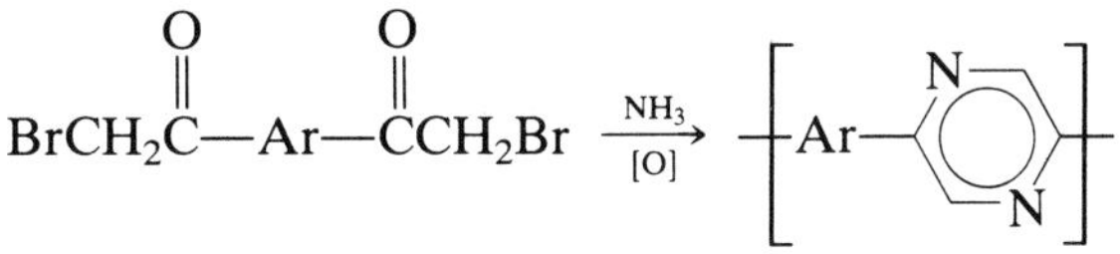

$$\underset{\text{O}}{\text{BrCH}_2\overset{\parallel}{\text{C}}}-\text{Ar}-\underset{\text{O}}{\overset{\parallel}{\text{C}}}\text{CH}_2\text{Br} \xrightarrow[\text{[O]}]{\text{NH}_3} \left[\text{Ar}-\underset{\text{N}}{\overset{\text{N}}{\bigcirc}}\right] \qquad (16.31)$$

6

16.3.2 *Polypyrazoles and polyimidazoles*

The pyrazole ring is most conveniently prepared from 1,3-dicarbonyl compounds and hydrazine derivatives.[1b] The extension of this reaction to bishydrazines and tetracarbonyl compounds (16.32) gives polypyrazoles[32] (**7**). Aromatic polypyrazoles of the type shown are black, infusible materials exhibiting good thermal stability.

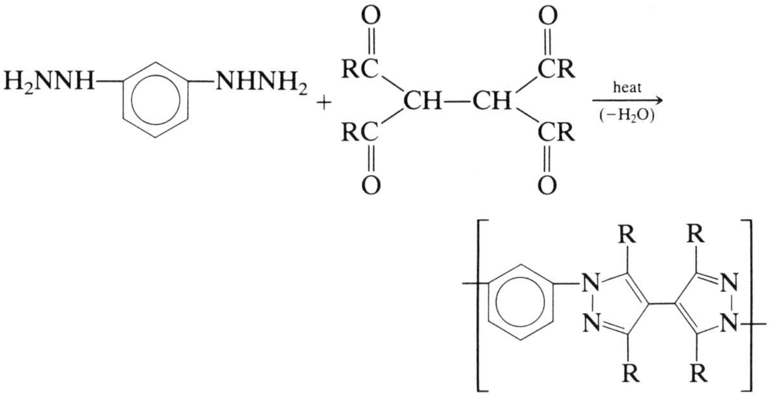

$$(16.32)$$

7

Besides reactions (16.28) and (16.32), a variety of other routes to polypyrazoles have been reported,[25a] among them the reaction of diacetylenes with bisdiazo compounds[33] (16.33); the 1,3-dipolar addition of alkynes to tetrazoles[34] (16.34); and the reaction of dipropynones with dihydrazinium salts[35] (16.35).

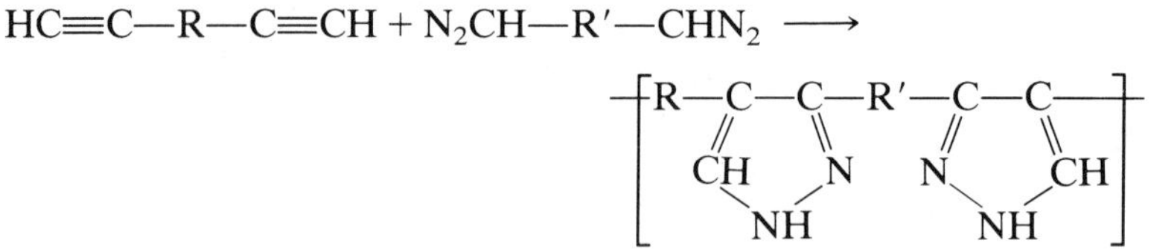

$$HC{\equiv}C-R-C{\equiv}CH + N_2CH-R'-CHN_2 \longrightarrow$$

$$\left[\begin{array}{c} R-\underset{\underset{NH}{\parallel}}{C}-\underset{\underset{}{\underset{}{C}}}{C}-R'-\underset{\underset{NH}{\parallel}}{C}-\underset{}{C} \\ CH \quad N \quad N \quad CH \end{array}\right] \quad (16.33)$$

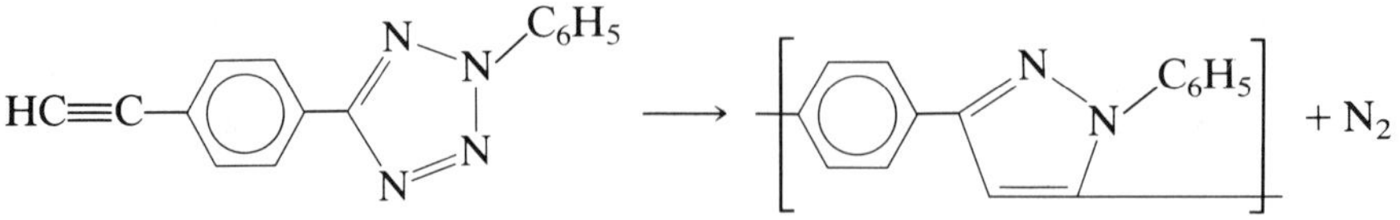

$$HC{\equiv}C{-}\bigcirc{-}\left[\text{tetrazole, } C_6H_5\right] \longrightarrow \left[\bigcirc{-}\left[\text{pyrazole, } C_6H_5\right]\right] + N_2 \quad (16.34)$$

$$\phi C{\equiv}C-\underset{\underset{}{\overset{O}{\parallel}}}{C}-Ar-\underset{\underset{}{\overset{O}{\parallel}}}{C}-C{\equiv}C\phi + H_3\overset{+}{N}NH-Ar'-NH\overset{+}{N}H_3 \xrightarrow{R_3N}$$
$$Cl^- \qquad\qquad Cl^-$$

$$\left[\begin{array}{c} N{-}N \qquad N{-}N{-}Ar' \\ \phi{-}\underset{}{\quad}{-}Ar{-}\underset{}{\quad}{-}\phi \end{array}\right] \quad (16.35)$$

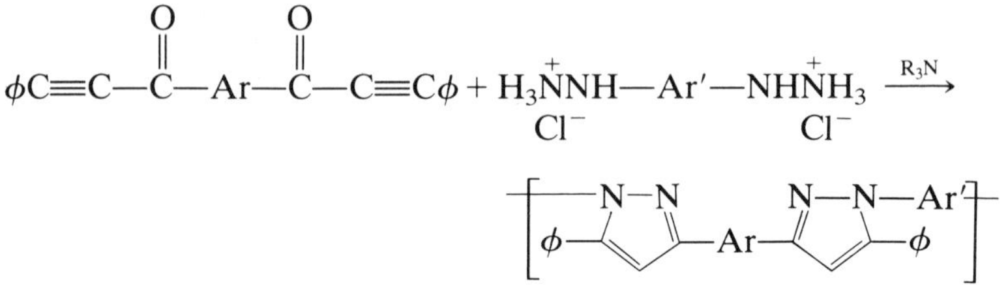

The imidazole ring is isomeric with the pyrazole ring. Almost the entire emphasis in the area of imidazole polymers has been with polybenzimidazoles (Chapter 13). Polymers of relatively low molecular weight having nonbenzo imidazole linkages (**8**) have been synthesized from tetraketones, dialdehydes, and ammonia[36] (16.36).

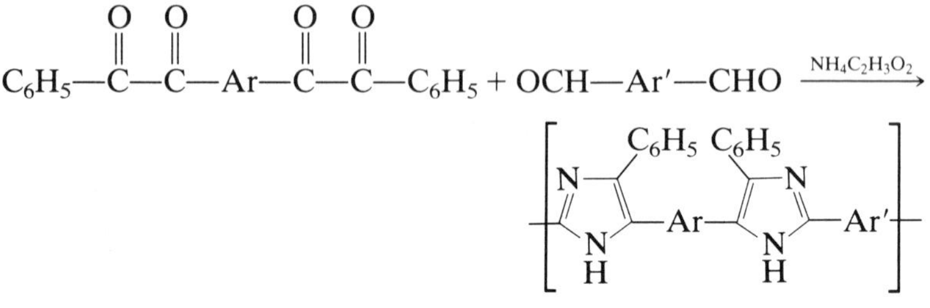

$$C_6H_5-\underset{\overset{\parallel}{O}}{C}-\underset{\overset{\parallel}{O}}{C}-Ar-\underset{\overset{\parallel}{O}}{C}-\underset{\overset{\parallel}{O}}{C}-C_6H_5 + OCH-Ar'-CHO \xrightarrow{NH_4C_2H_3O_2}$$

$$\left[\begin{array}{c} C_6H_5 \quad C_6H_5 \\ \underset{\overset{N}{H}}{N}{-}Ar{-}\underset{\overset{N}{H}}{N}{-}Ar' \end{array}\right] \quad (16.36)$$

8

16.3.3   Poly(as-triazine)s and polytriazolines

Poly(*as*-triazine)s (**9**) and polytriazolines (**10**) are synthesized from dihydrazidines by reaction with bisdicarbonyl compounds (16.37) and dialdehydes (16.38), respectively.[28a,37] (It may be recalled from Chapter 13, Section 13.11, that dihydrazidines are also used to prepare polyoxadiazoles and polytriazoles.) As might be expected, the nonaromatic polytriazolines are not thermally stable; however, poly(*as*-triazine)s are stable up to about 380°C, at which point the nitrogen–nitrogen bond breaks; the backbone remains intact up to about 530°C.

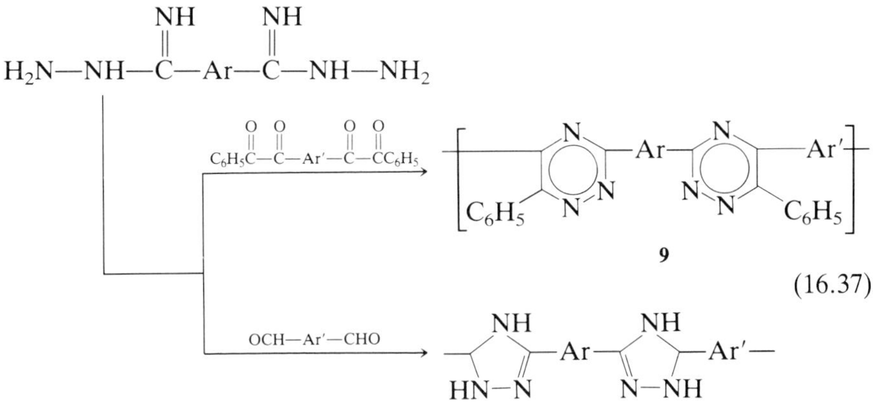

(16.37)

(16.38)

16.3.4 Polyquinolines and polyanthrazolines

As may be seen from Table 16.1, the quinoline and anthrazoline rings are structurally related. As might be expected, polyquinolines (**11**) and poly-anthrazolines (**12**) are synthesized in essentially the same manner,[38-40] Bis(*o*-aminocarbonyl) compounds and diacetyl compounds, reacted together in the presence of an acid catalyst, are the monomers in each case (16.39, 16.40). If R is phenyl in these reactions, the polymers are significantly

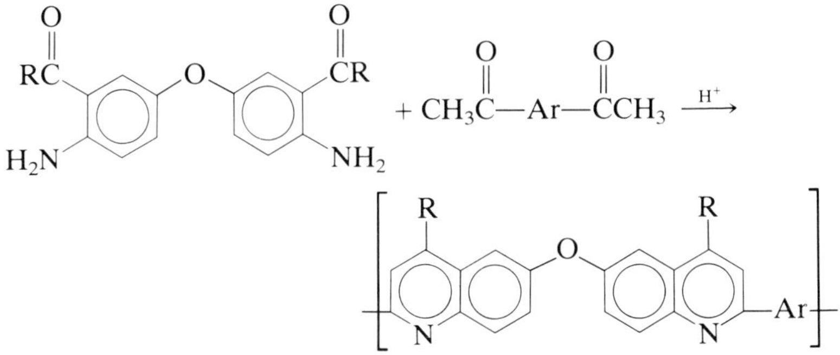

(16.39)

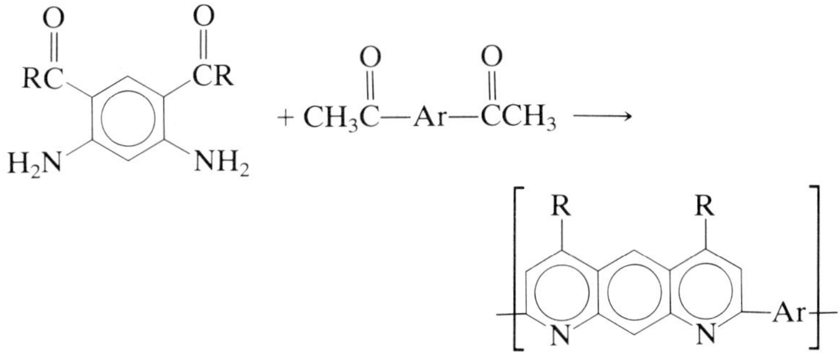

(16.40)

more soluble than if R is hydrogen. Both polymer types are stable to above 500°C.

16.3.5   *Polypyrrole, polyfuran, and polythiophene*

Polypyrrole (**13**) is of principal interest because, when doped, it exhibits electrical conductivities comparable to those of polyacetylene and poly(sulfur

13

nitride) (see Chapter 4) but is more readily processable. The polymer is prepared by electrolytic oxidation of pyrrole,[41,42] a reaction that apparently proceeds via three steps: 1. formation of a radical cation, (2) radical combination, and (3) deprotonation. These steps are illustrated in Scheme 16.1 for formation of dimer and trimer.

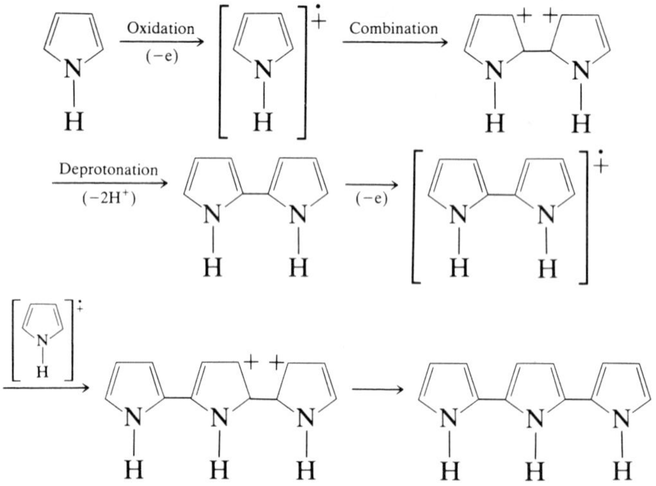

Scheme 16.1. Proposed mechanism for the electrolytic polymerization of pyrrole.

Electrolysis of an acetonitrile solution of pyrrole in the presence of tetraethylammonium fluoroborate results in precipitation of highly conducting BF_4^--doped films of polypyrrole on the anode. Polystyrene containing pendant pyrrole units reacts similarly to yield conducting polystyrene-*graft*-polypyrrole.[43]

Furan[44,45] and thiophene[46] undergo polymerization in the presence of Lewis acids or mineral acids. The polymer structures are relatively complex, however, and cannot be represented in an idealized form comparable to that of polypyrrole. On the basis of spectroscopic evidence and elemental analyses, the principal structures present in polyfuran and polythiophene appear to be **14** and **15**, respectively. (Polyfuran should not be confused with the furan resins described in Chapter 14.)

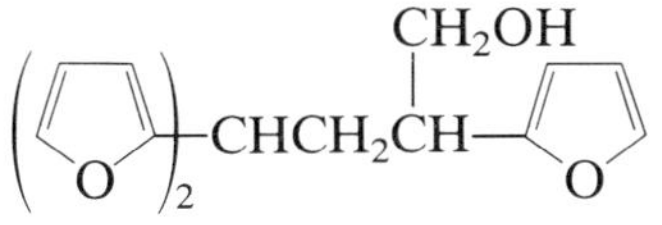

14 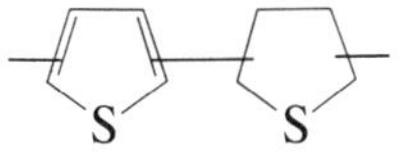   **15**

Electrolytic oxidation has also been used to prepare a form of polythiophene that is conducting.[47] In this case the structure is best represented in the idealized form as **16**.

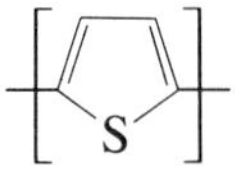

16

16.4 Poly(*p*-phenylene) and poly(*p*-xylylene)

16.4.1 Poly(p-phenylene)

The simplest of the fully aromatic polymers is poly(*p*-phenylene)[25b,48] (**17**). It is prepared most conveniently by directly polymerizing benzene in the presence of combinations of Lewis acids and oxidizing agents. When aluminum chloride and cupric chloride are used (16.41), polymerization occurs rapidly at temperatures as low as 35 to 50°C.

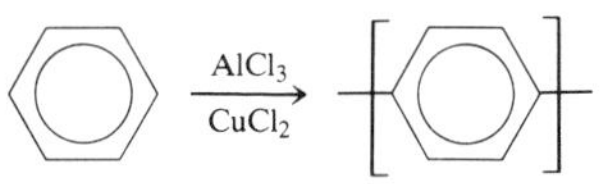

(16.41)

17

The mechanism of benzene polymerization has been the subject of much debate, but on the basis of ESR studies there is general agreement that the initiating species is a benzene radical cation.[49] Theoretical calculations[50] and studies on polymerization of deuterated benzene[51] indicate a possible polymerization mechanism in which benzene rings are associated in a stacked, end-to-end arrangement (**18**), as shown in Scheme 16.2, with the radical cation delocalized over the entire chain. Chain buildup ceases when the radical cation concentration on the terminal phenyls becomes too small to promote further association. Subsequent sigma bond formation results in **19**, which may undergo competing depropagation or oxidation and loss of H$^+$ to form **17**.

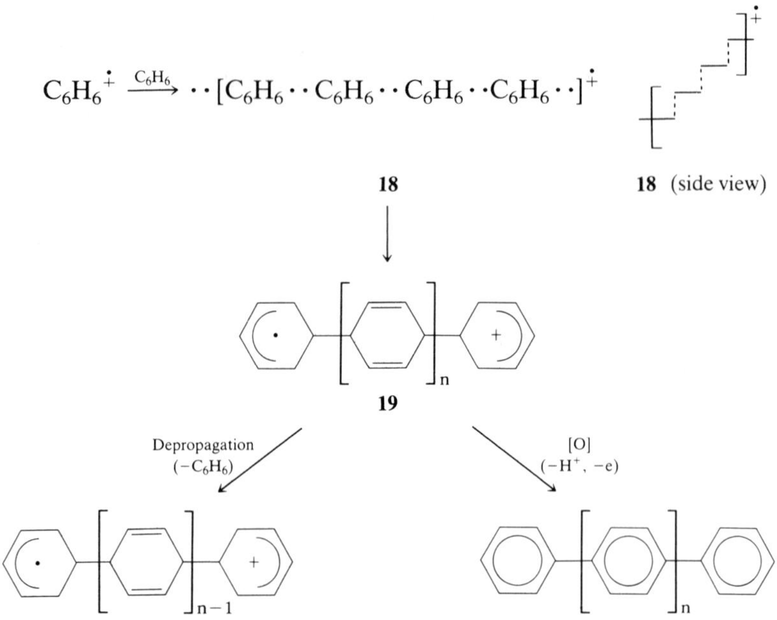

$$C_6H_6{}^{\overset{\cdot}{+}} \xrightarrow{\ C_6H_6\ } \cdot\cdot[C_6H_6\cdot\cdot C_6H_6\cdot\cdot C_6H_6\cdot\cdot C_6H_6\cdot\cdot]^{\overset{\cdot}{+}}$$

18   **18** (side view)

19

Scheme 16.2. Proposed mechanism for benzene polymerization.

Other methods of forming poly(*p*-phenylene) have been reported such as the dehydrogenation of poly(1,3-cyclohexadiene) (16.42), but none is as attractive as the direct benzene polymerization.

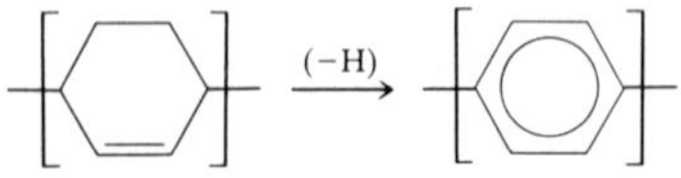

$$\xrightarrow{(-H)}$$ (16.42)

17

As might be expected, poly(*p*-phenylene) has good thermal stability, but its very high melting point (it does not melt at temperatures below 530°C) and low solubility make it difficult to use in commercial applications. Nevertheless, *polyphenylenes* have been marketed (Monsanto) as laminating resins,[52a,53] although they bear little resemblance to pure **17**. Rather, they are synthesized from mixtures of biphenyl and *o*-, *m*-, and *p*-terphenyls, which are heated together with disulfonyl chlorides in the presence of a copper halide catalyst. Both HCl and SO_2 are split out in the process. Heat-curable acetylene-terminated polyphenylene prepolymers have also been marketed (Hercules). More recently, attention has focused on the conducting properties of **17** (see Chapter 4).

Poly(*p*-phenylene) and backbone-substituted poly(*p*-phenylene) have also been prepared by cycloaddition reactions. These are discussed in Section 16.6.

16.4.2 Poly(p-xylylene)

Poly(p-xylylene) (**21**) is not completely aromatic but contains benzene rings linked through two methylene groups. Considerable interest has been expressed in this polymer[25b,52a] because it can be synthesized by pyrolysis of the readily available p-xylene, a reaction (16.43) believed to involve the intermediate quinodimethide (**20**). One of the by-products of this synthesis is the bridged dimer, paracyclophane (**22**), a compound that can also be pyrolyzed to poly(p-xylylene), either through the same quinodimethide intermediate or possibly via an intermediate diradical. The resultant polymer is generally formed in higher yield and with less chain branching than that prepared by p-xylene pyrolysis, and the paracyclophane route has the added advantage of being applicable to preparation of substituted poly(p-xylylene)s.

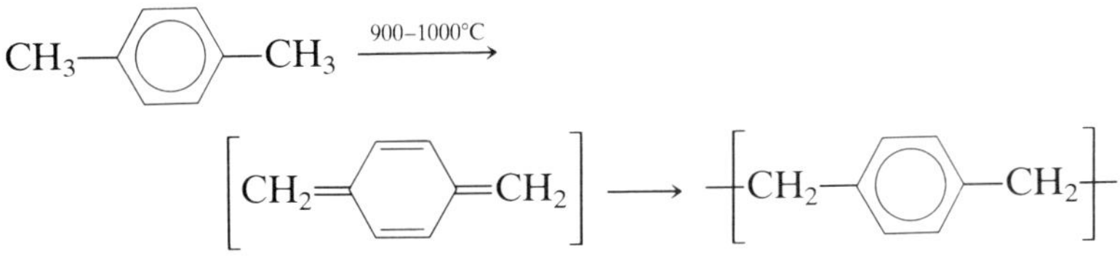

$$CH_3-\underset{}{\bigcirc}-CH_3 \xrightarrow{900-1000°C}$$

$$\left[CH_2=\bigcirc=CH_2\right] \longrightarrow \left[CH_2-\bigcirc-CH_2\right] \quad (16.43)$$

$$\underset{\textbf{20}}{} \qquad\qquad\qquad \underset{\textbf{21}}{}$$

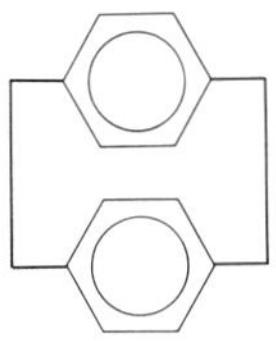

22

Other routes to **21** or its chlorinated derivatives are given in reactions (16.44) through (16.46).[55]

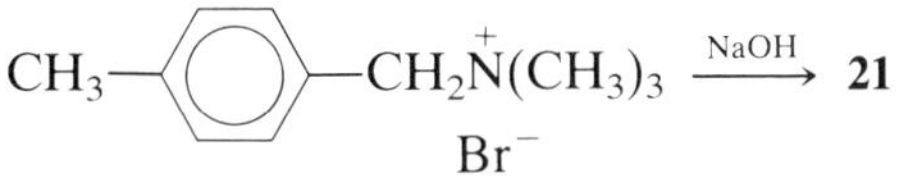

$$CH_3-\bigcirc-CH_2\overset{+}{N}(CH_3)_3 \xrightarrow{NaOH} \textbf{21} \qquad (16.44)$$
$$Br^-$$

$$ClCH_2-\bigcirc-CH_2Cl \xrightarrow{base} \left[\underset{\underset{CH}{|}}{\overset{Cl}{\overset{|}{}}}-\bigcirc-CH_2\right] \qquad (16.45)$$

$$Cl_3C-\bigcirc-CCl_3 \xrightarrow[DMSO\cdot H_2O]{NaBH_4} \left[CCl_2-\bigcirc-CCl_2\right] \qquad (16.46)$$

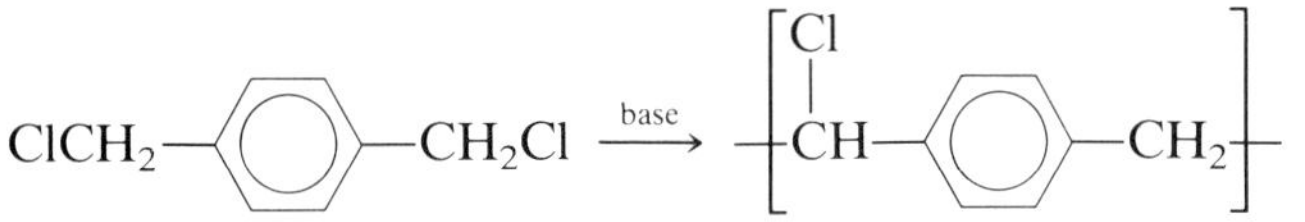

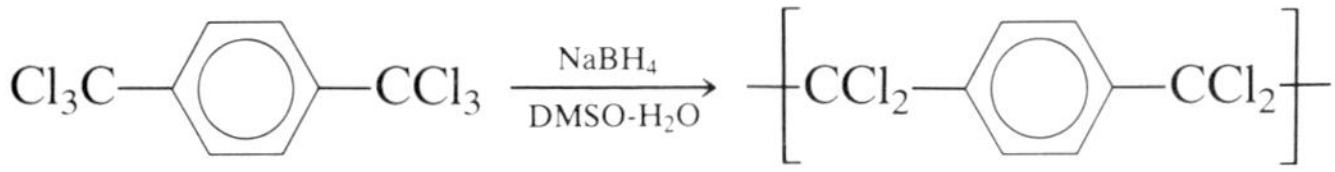

Poly(p-xylyene) is a crystalline material having relatively poor thermal stability. It is more tractable than poly(p-phenylene) in that it can be fabricated into brittle fibers or tough films, the latter by vapor deposition from pyrolysis of **22**. The polymer is of interest in medical applications because of its biocompatibility. Commercial poly(*p*-xylylene)s (Union Carbide's Parylene) include chlorinated and nonchlorinated varieties. (Vapor deposition of 1-μ-thick coatings of Parylene onto paper is an excellent way to preserve old books and manuscripts.) The analogous *ortho* and *meta* polyxylylenes have also been prepared, but they are very low-melting and not, therefore, as useful as poly(*p*-xylylene), which has a softening temperature in excess of 400°C. Some other polymers containing bridged benzene rings are described in the next section.

16.5   Friedel–Crafts polymers

Most name reactions seem to have been extended to polymer synthesis, and the Friedel–Crafts reaction is no exception. We first encountered the phenomenon in Chapter 11 in the preparation of polysulfones from disulfonyl chlorides. Similar reactions of dicarboxylic chlorides (16.47) leads to polyketones.[56] Dicarboxylic acids may in some instances be used in place of the acid chlorides in the presence of the strongly dehydrating P_2O_5-methanesulfonic acid.[57]

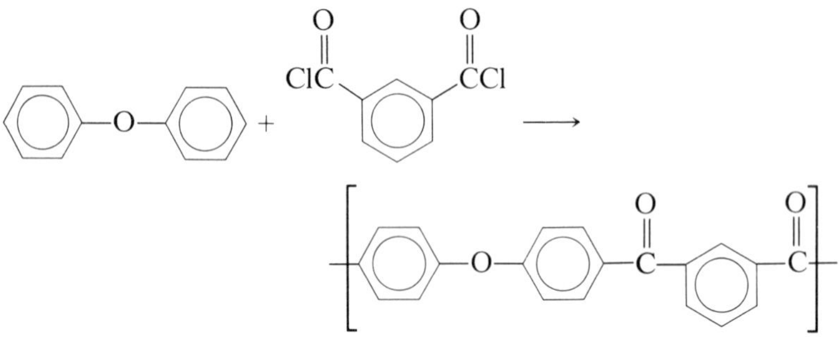

$$\text{(16.47)}$$

Benzyl chloride undergoes acid-catalyzed self-condensation (16.48) to yield highly branched low-molecular-weight polymers having repeating benzylic groups.[58] Similarly, α,α′-dichloro-*p*-xylene reacts with biphenyl (16.49) to

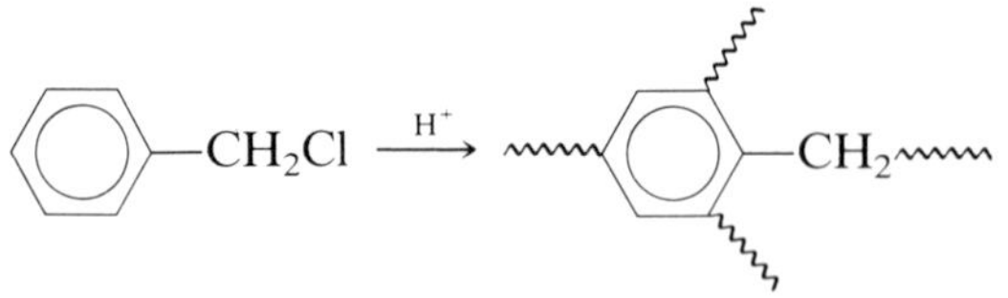

$$\text{(16.48)}$$

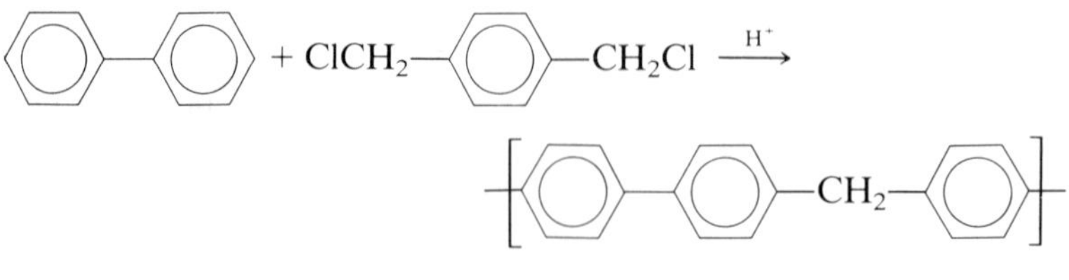

$$\text{(16.49)}$$

give polymers of analogous structure. Reactions of this type have led to commercial laminating resins and molding compounds.[52b] To circumvent the problem of HCl evolution during polymerization, benzylic ethers (**23**), which evolve methanol, have been developed as alternative Friedel–Crafts monomers. Reaction of **23** with phenol has also been used commercially to produce thermosetting polymers that are related in structure to phenol–formaldehyde polymers, as well as to polyxylylene and polyphenylene resins.

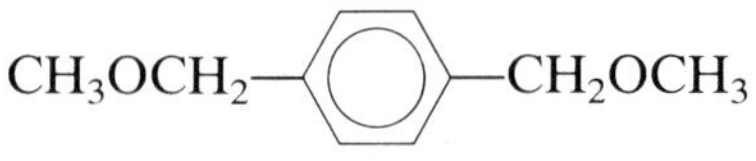

$$CH_3OCH_2 \text{—} \underset{\textbf{23}}{\bigcirc} \text{—} CH_2OCH_3$$

The Friedel–Crafts reaction has also been used to prepare crosslinked polystyrene.[59] A mixture of styrene and *p*-di(chloromethyl)benzene (α,α'-dichloro-*p*-xylene) is polymerized at low temperatures with a Lewis acid catalyst such as stannic chloride. Chain polymerization to form linear polystyrene occurs rapidly, while crosslinking (16.50), under the influence of the same catalyst, occurs relatively slowly, giving a product with a higher glass transition temperature than that of linear polystyrene.

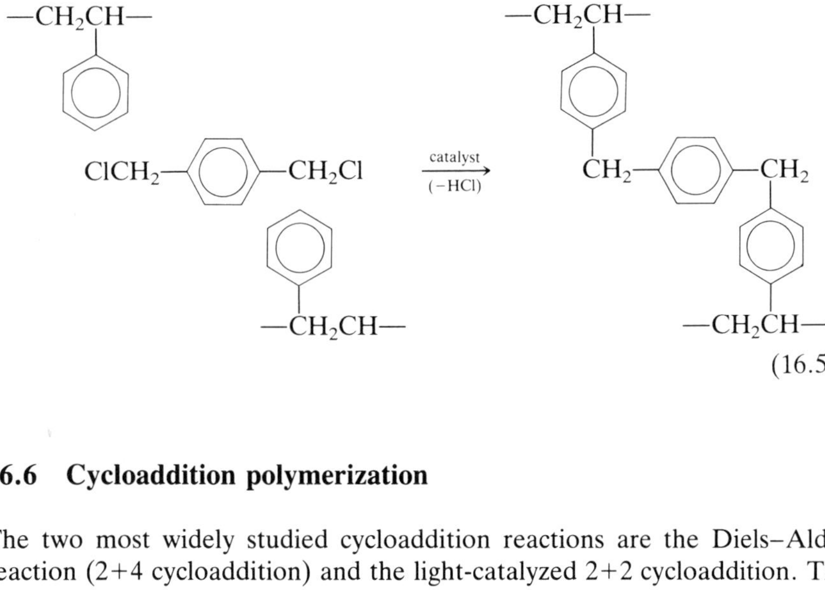

$$(16.50)$$

16.6 Cycloaddition polymerization

The two most widely studied cycloaddition reactions are the Diels–Alder reaction (2+4 cycloaddition) and the light-catalyzed 2+2 cycloaddition. The former is a relatively simple, high-yield reaction with no by-products; as a result it has captured the imagination of polymer chemists. By contrast, comparatively little has been done in applying 2+2 cycloaddition processes to polymer synthesis.

16.6.1 Diels–Alder polymerization

In its simplest form the Diels–Alder reaction is the reaction of a diene with an alkene (called the *dienophile*) to form a cyclohexene ring (16.51). In practice,

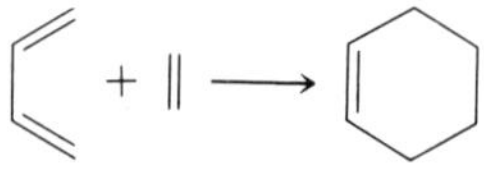

$$\text{(16.51)}$$

dienophiles having electron-attracting substituents on the double bond are most effective, while reactivity of the diene is increased if the monomer is confined to the more reactive *s-cis* configuration, as is the case with homoannular dienes. Illustrative of the latter is the facile dimerization or oligomerization of cyclopentadiene (16.52). This type of reaction is not useful from the standpoint of polymer formation, because only low-molecular-weight materials are obtained. (Recall from Chapter 9 the reversible cross-linking of polymers by cyclopentadiene cycloaddition.)

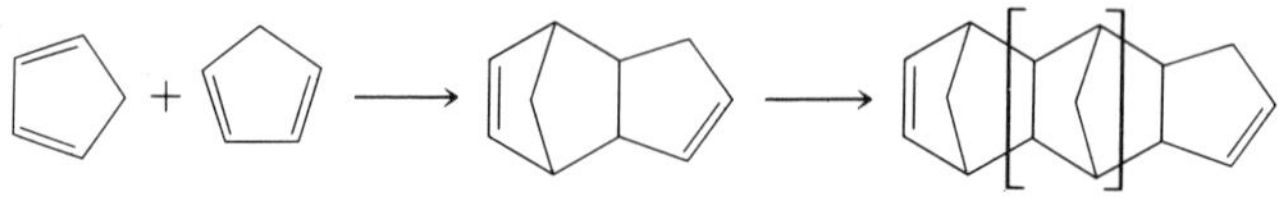

$$\text{(16.52)}$$

A large number of Diels–Alder polymerizations have been reported.[11b,28b,54b] One of the earliest was the reaction of 2-vinyl-1,3-butadiene with *p*-benzoquinone to give a ladder polymer (16.53). In this case, each cycloaddition generates a new diene monomer.

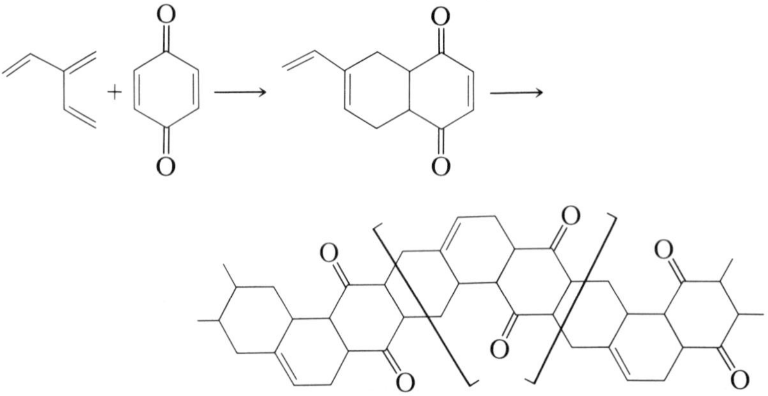

$$\text{(16.53)}$$

Some Diels–Alder polymerizations involve monomers having both diene and dienophile functions. Among these is the self-addition polymerization of 2-vinyl-1,3-butadiene (16.54). The structure of the polymer is more complex than that shown because different modes of addition or crosslinking by Diels–Alder addition are possible.

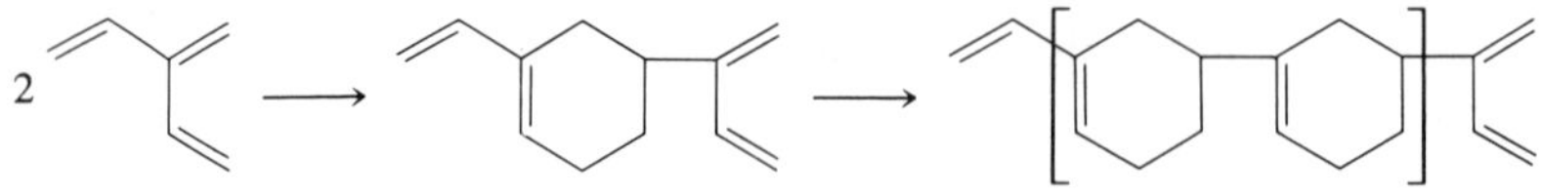

$$\text{(16.54)}$$

In most Diels–Alder reactions, the dienophile is electrophilic and the diene nucleophilic. The reverse can be the case where cyclopentadienones are used.

An example is the formation of ladder polymer from the cyclopentadienone derivative (**24**). This involves a series of steps consisting of initial cycloaddition (16.55) followed by loss of carbon monoxide by extrusion of the bridged carbonyl (16.56), and then repetition of these reactions to form polymer (16.57) with simultaneous aromatization of the diene ring (16.58) at the high temperatures (about 200°C) necessary for polymerization to occur. Only low-molecular-weight soluble polymer is formed by this route, along with some insoluble, possibly crosslinked material.

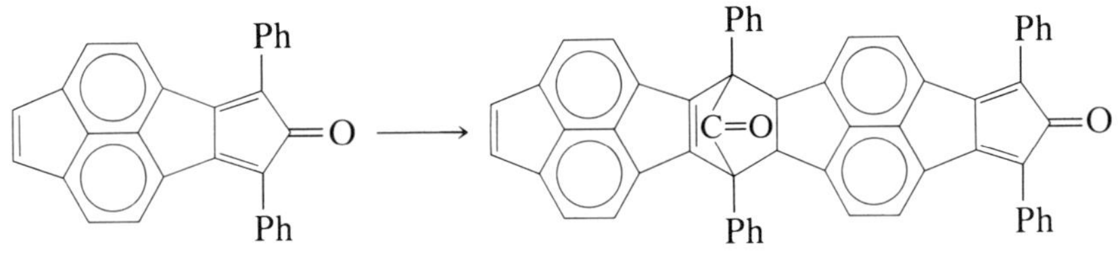

(16.55)

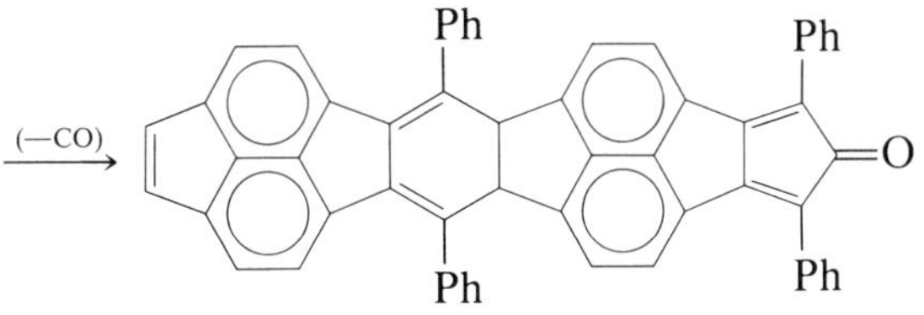

(16.56)

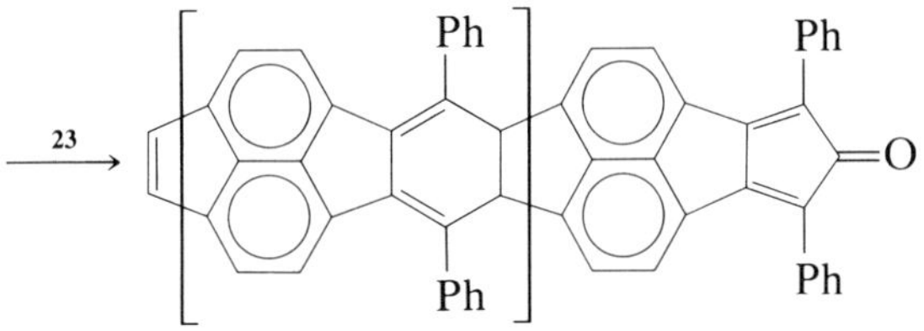

(16.57)

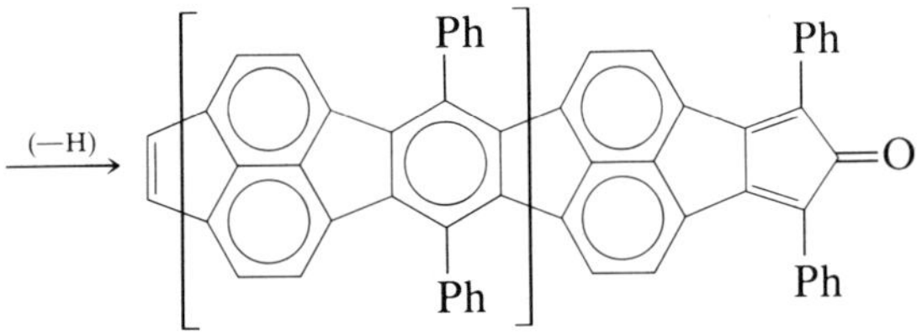

(16.58)

Reaction of bisdienes with bisdienophiles has been most widely used in Diels–Alder polymerizations. An example is the reaction of a biscyclopentadiene with a bismaleimide (16.59) to form polyimide **25**. One factor that limits the molecular weight in reactions of this type is the tendency for the reverse (retrograde) Diels–Alder reaction to occur at the high temperatures usually needed to obtain a high degree of polymerization. One way that this has been

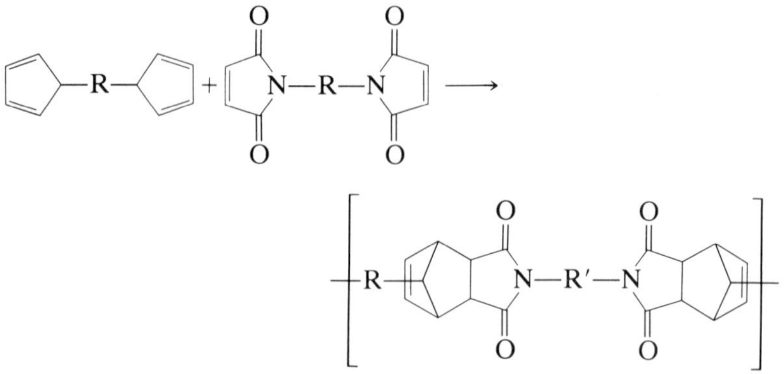

$$\text{(16.59)}$$

25

circumvented is by reacting biscyclopentadienones (also known as bistetra-cyclones) with diacetylenes (16.60). As in the previous case involving a cyclopentadienone monomer (reaction 16.56), carbon monoxide is lost from the initial adduct, but with acetylenic dienophiles this results in an aromatic ring, thus rendering the reaction essentially irreversible. This represents an alternative synthesis of a poly(*p*-phenylene); however, unlike the unsubstituted polymer (**17**), this phenyl-substituted analog (**26**) is amorphous and more soluble, while exhibiting little weight loss in air or nitrogen below 550°C.

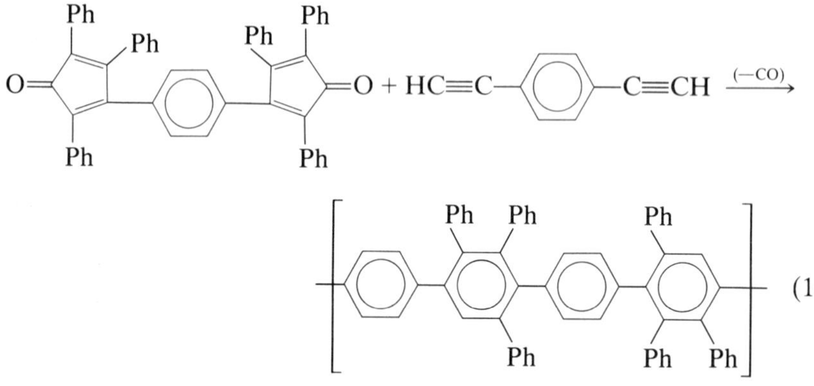

$$\text{(16.60)}$$

26

Certain compounds, known as *pseudobisdienes*, contain a single diene function but undergo decomposition to generate a new diene in the course of polymerization. An example is α-pyrone **27**, which reacts with bismaleimide (16.61) to form an intermediate (**28**) that loses carbon dioxide spontaneously (16.62) to form a new intermediate (**29**) containing both diene and dienophile units. This intermediate undergoes Diels–Alder polymerization (16.63). Other pseudobisdienes include thiophene-1,1-dioxide (**30**) and 2,5-dimethyl-

3,4-diphenylcyclopentadieneone (**31**)* which react similarly, except that sulfur dioxide and carbon monoxide, respectively, are evolved.

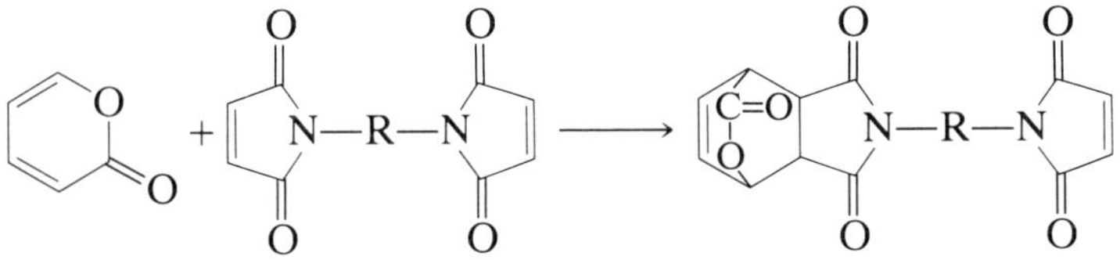

(16.61)

27   **28**

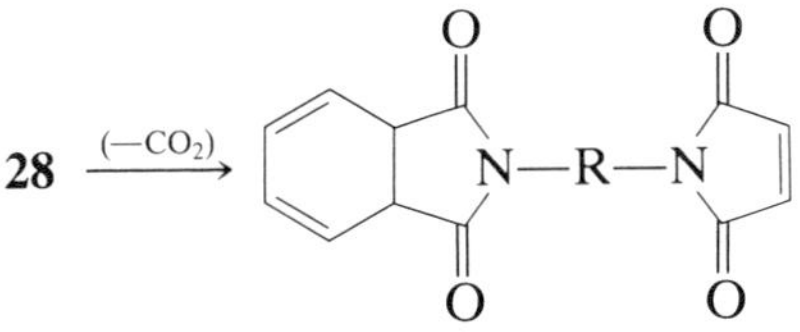

(16.62)

29

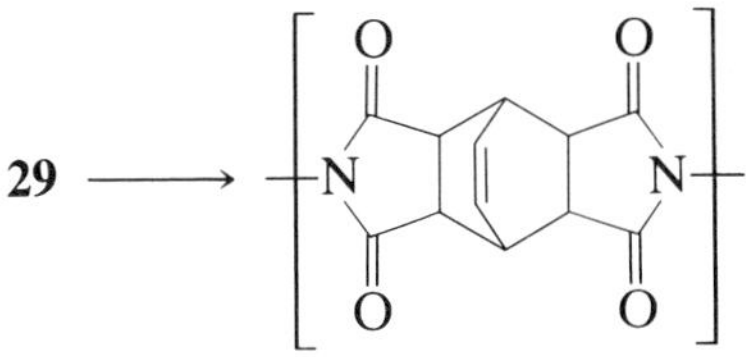

(16.63)

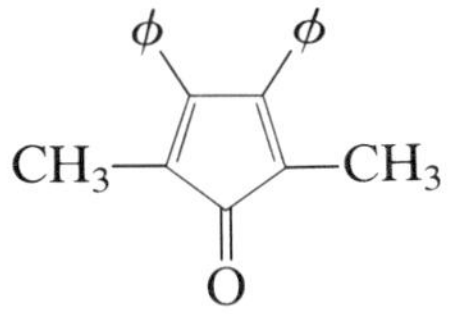

30   **31**

Despite the great amount of work that has been devoted to Diels–Alder polymers, they have not been exploited commercially, no doubt because of the cost of preparing monomers. Many of them do have interesting properties, however, and can be fabricated into films or fibers. The preponderance of fused rings in many of the polymers results in high melting points and low solubility and, where the retrograde Diels–Alder reaction is not a problem, good thermal stability and chemical resistance.

16.6.2 *2+2 Cycloaddition polymerization*

A major problem in preparing polymers by the light-catalyzed 2+2 cycloaddition results from side reactions caused by irradiation—for example, crosslinking, degradation, or rearrangement. Furthermore, the cycloaddition,

---

* Compound **31** exists as a colorless dimer at room temperature, but dissociates to red monomer on heating.

shown (16.64) in simplified form, results in a cyclobutane ring, a structure that is inherently unstable because of ring strain.

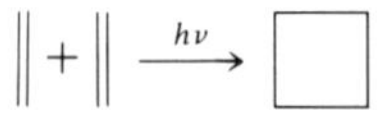

$$\| + \| \xrightarrow{h\nu} \square \tag{16.64}$$

Most of the early work in this area dealt with solid-phase photopolymerization[60] of monomers such as 2,5-distyrylpyrazine (**32**), which forms highly crystalline stereoregular polymer (**33**) on irradiation (16.65). Stereoregularity is assured by the spatial requirements of the crystalline lattice. It

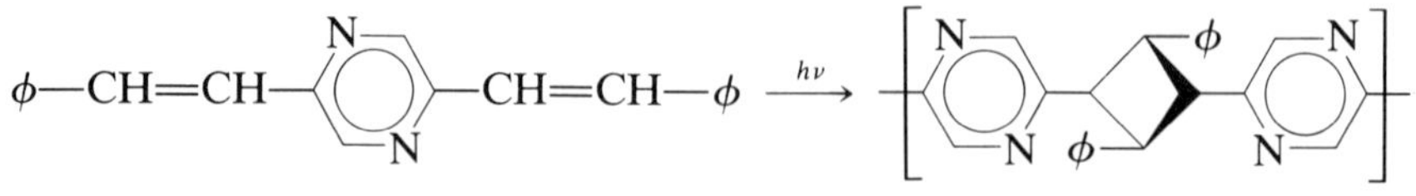

32   **33**

$$(16.65)$$

should be noted that this type of reaction is a true photopolymerization process in that *each chain-propagating step involves a photochemical reaction*, as opposed to those reactions that are simply initiated by light.[61]

To achieve high molecular weights in solution processes, diene monomers must be selected that have little tendency to undergo vinyl polymerization or intramolecular cyclization.[62,63] Bismaleimides substituted at the double bonds with chlorine atoms (**34**) represent one type that forms linear cycloaddition polymers[61] (16.66). In this case, at least seven carbon atoms should separate the two maleimide groups to minimize intramolecular cycloaddition. Triplet sensitizers enhance the reaction.

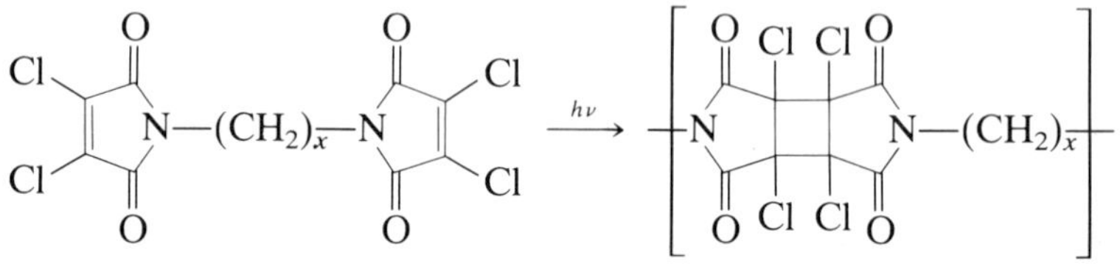

$$(x > 7)$$

34

$$(16.66)$$

Polyimides have also been prepared by a combination of 2 + 2 and 2 + 4 cycloaddition reactions by copolymerizing bismaleimides photochemically with benzene[64] or alkylbenzenes.[65] In this case, the initial 2 + 2 cycloaddition to the benzene ring (16.67) results in a homoannular diene that subsequently undergoes Diels–Alder addition (16.68) to give polymer **35**. Only a head-to-tail structure is shown, but both head-to-head and tail-to-tail structures could presumably form, depending on whether the intermediate homoannular diene undergoes self-addition or reaction with bismaleimide. Polyimides prepared this way are insoluble, possibly because of crosslinking through the

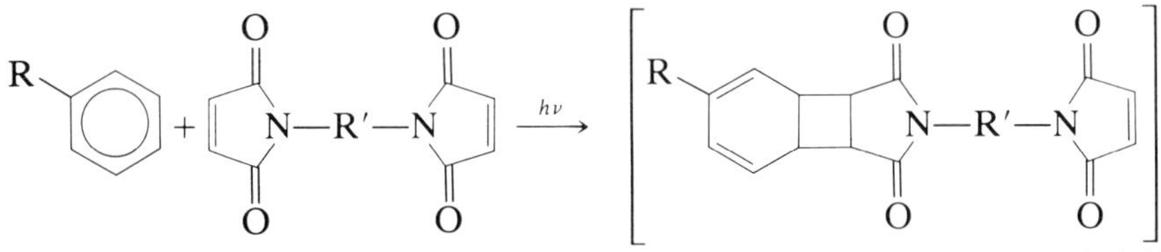

(16.67)

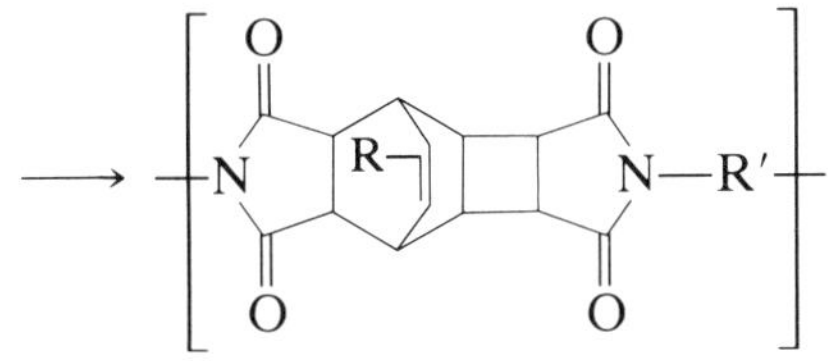

35

(16.68)

remaining double bond in the backbone. Also, they all tend to decompose by decyclization at about the same temperature (420–485°C) regardless of whether R′ is aliphatic or aromatic, thus demonstrating the sensitivity of the cyclobutane ring.

16.7 Ring-opening polymerization of strained cycloalkanes

The previous section dealt with ring-forming polymerization reactions. This one describes polymerizations arising from ring-opening reactions, specifically of small-ring cyclic or bicyclic compounds that react in a way that relieves ring strain.

Cyclopropane has been polymerized under cationic conditions (16.69) to very-low-molecular-weight linear polymer.[66] The 1,1-dimethyl analog reacts similarly[67] (16.70).

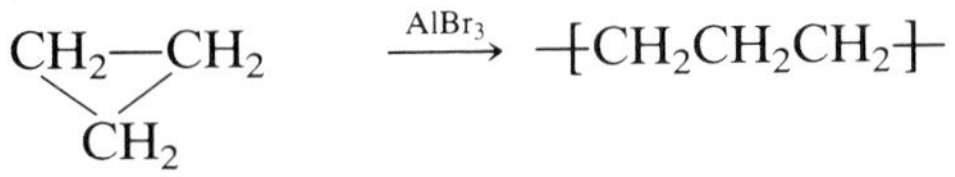

(16.69)

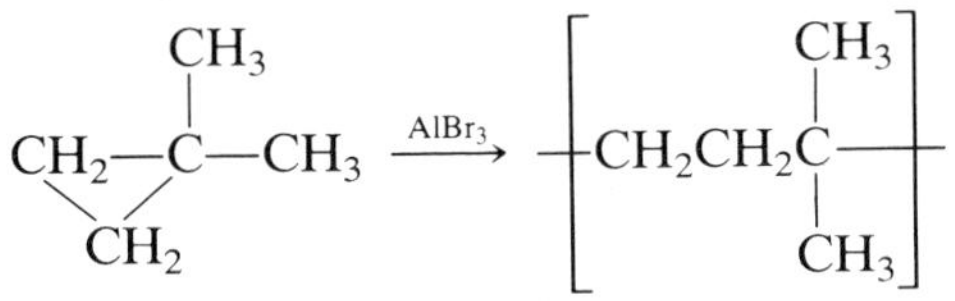

(16.70)

More recently some interesting polymers and copolymers have been prepared from strained bicyclic compounds.[68] Bicyclobutane-1-carbonitrile (**36**), unsubstituted or substituted with methyl groups at the 2 and 4 positions, undergoes free radical or anionic polymerization (16.71) or copolymerization with vinyl monomers (16.72) to give high-molecular-weight polymers. Rupture of the bicyclobutane system results in cyclobutane rings in the backbone.

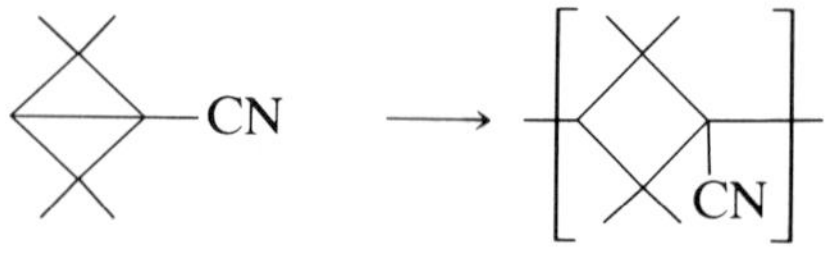

$$(16.71)$$

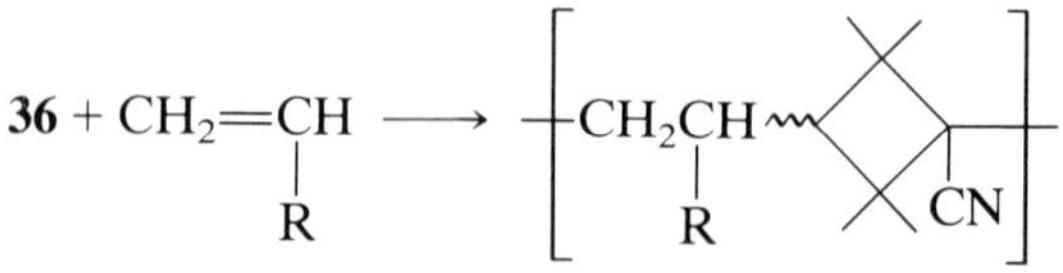

36

$$36 + CH_2\!\!=\!\!\overset{|}{\underset{R}{CH}} \longrightarrow \quad (16.72)$$

Bicyclobutanes substituted at the bridgehead with ester, amide, or acetyl groups also polymerize and copolymerize readily under free radical conditions, but less satisfactorily under nonradical conditions. The effect of ring size on reactivity is evident from the fact that bicyclo[2.1.0] pentane-1-carbonitrile (**37**) undergoes anionic (but not free radical) polymerization, whereas bicyclo[3.1.0]hexane-1-carbonitrile (**38**) resists polymerization.[69]

CN   CN

37 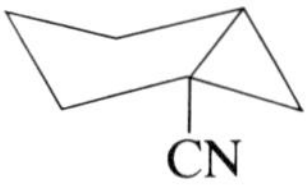   **38**

16.8 Polyanhydrides

Polymers with recurring anhydride groups[70-72] were among the first linear fiber-forming polymers to be investigated as part of the pioneering work of Carothers. They can be prepared from dicarboxylic acids by dehydration with reagents such as phosgene, acid chlorides, or acetic anhydride (16.73). Alternatively, polyanhydrides may be prepared by an anhydride exchange reaction (16.74) of a preformed mixed anhydride, with by-product acetic anhydride being removed under reduced pressure.

$$HO_2C\!-\!R\!-\!CO_2H + (CH_3CO)_2O \overset{heat}{\longrightarrow}$$

$$\left[\begin{array}{c} O \\ \| \\ OC \end{array} \!-\!R\!-\! \begin{array}{c} O \\ \| \\ C \end{array} \right] + 2CH_3CO_2H \quad (16.73)$$

$$HO_2C\!-\!R\!-\!CO_2H + 2(CH_3CO)_2O \longrightarrow$$

$$\begin{array}{cccc} O & O & O & O \\ \| & \| & \| & \| \\ CH_3COC & \!-\!R\!-\! & COCCH_3 \end{array} \overset{heat}{\longrightarrow} \left[\begin{array}{c} O \\ \| \\ OC \end{array} \!-\!R\!-\! \begin{array}{c} O \\ \| \\ C \end{array} \right] + (CH_3CO)_2O \quad (16.74)$$

As might be expected, butanedioic (succinic) and pentanedioic (glutaric) acids form mainly cyclic anhydrides rather than linear polymer under these conditions. Hexanedioic (adipic) acid forms a mixture of both. Propanedioic (malonic) acid and acids higher than adipic form polymer. Diacid chlorides may also be used as monomers for polyanhydrides, either by reaction with diacid in the presence of base[71] (16.75) or by conversion to the diammonium salts followed by reaction with stoichiometric amounts of water as shown (16.76) for the polyanhydride (**39**) of terephthalic acid.[73]

$$
\text{Cl}-\overset{\overset{\displaystyle O}{\|}}{\text{C}}-\text{R}-\overset{\overset{\displaystyle O}{\|}}{\text{C}}-\text{Cl} + \text{HO}_2\text{C}-\text{R}'-\text{CO}_2\text{H} \xrightarrow{\text{base}}
$$

$$
\begin{bmatrix} \overset{\overset{\displaystyle O}{\|}}{\text{C}}-\text{R}-\overset{\overset{\displaystyle O}{\|}}{\text{C}}-\text{O}-\overset{\overset{\displaystyle O}{\|}}{\text{C}}-\text{R}'-\overset{\overset{\displaystyle O}{\|}}{\text{C}}-\text{O} \end{bmatrix} \quad (16.75)
$$

$$
\text{Cl}-\overset{\overset{\displaystyle O}{\|}}{\text{C}}-\underset{}{\bigcirc}-\overset{\overset{\displaystyle O}{\|}}{\text{C}}-\text{Cl} \xrightarrow{\text{R}_3\text{N}} \text{Cl}^-\text{R}_3\overset{+}{\text{N}}\overset{\overset{\displaystyle O}{\|}}{\text{C}}-\underset{}{\bigcirc}-\overset{\overset{\displaystyle O}{\|}}{\text{C}}\overset{+}{\text{N}}\text{R}_3\text{Cl}^- \xrightarrow{\text{H}_2\text{O}}
$$

$$
\begin{bmatrix} \text{O}-\overset{\overset{\displaystyle O}{\|}}{\text{C}}-\underset{}{\bigcirc}-\overset{\overset{\displaystyle O}{\|}}{\text{C}} \end{bmatrix} \quad (16.76)
$$

39

Aliphatic polyanhydrides are very sensitive to hydrolysis and degrade rapidly in the presence of moisture. Aromatic polyanhydrides, on the other hand, often exhibit surprisingly good hydrolytic stability, comparable to that of linear aromatic polyesters. Polymer **39**, for example, can be refluxed in 10M HCl for up to six hours with little degradation. This stability results primarily from a high degree of crystallinity. Another example is **40**, a polyanhydride having a T_m of about 267°C, which can be spun into high-strength fibers. Degradable polyanhydrides are of current interest for controlled release applications.

$$
\begin{bmatrix} \text{OC}-\underset{}{\bigcirc}-\text{OCH}_2\text{CH}_2\text{CH}_2\text{O}-\underset{}{\bigcirc}-\overset{\overset{\displaystyle O}{\|}}{\text{C}} \end{bmatrix}
$$

40

16.9 Polyamines

The term *polyamine* refers here to polymers containing an amino group as an essential part of the backbone. It does not include heterocyclic amino polymers such as polyquinoxalines or polypyrazoles, nor does it include polymers having amino groups attached to the main chain or amino substituents on a pendant group.

One of the best methods of making polyamines is by Michael addition of primary or secondary diamines to activated double bonds, a reaction described in Chapter 13 for chain extension of bismaleimides. Typical is the reaction of piperazine with 1,4-diacrylpiperazine[74] (16.77). If primary amine (RNH_2) is substituted for piperazine in this reaction, polymer **41** is formed.

$$H-N\underset{}{\overset{}{\bigcirc}}N-H + CH_2{=}CHC\overset{O}{\overset{\|}{-}}N\underset{}{\overset{}{\bigcirc}}N-\overset{O}{\overset{\|}{C}}CH{=}CH_2 \longrightarrow$$

$$\left[\!-\!N\underset{}{\overset{}{\bigcirc}}N\!-\!CH_2CH_2\overset{O}{\overset{\|}{C}}\!-\!N\underset{}{\overset{}{\bigcirc}}N\!-\!\overset{O}{\overset{\|}{C}}CH_2CH_2\!-\!\right] \quad (16.77)$$

$$\left[\!-\!N\overset{R}{\overset{|}{-}}CH_2CH_2\overset{O}{\overset{\|}{C}}\!-\!N\underset{}{\overset{}{\bigcirc}}N\!-\!\overset{O}{\overset{\|}{C}}CH_2CH_2\!-\!\right]$$

41

Nucleophilic substitution of primary dihalides with diamines is not a satisfactory way to make high-molecular-weight polyamines because salt formation between reactants and by-product hydrogen halide upsets the stoichiometric balance. One way to circumvent this problem is to use disilylamines[75] (**42**) (prepared from diamine and a chlorosilane). Reaction with an active dihalide in the presence of catalytic amounts of ammonium chloride results in high molecular-weight polymer with elimination of the halosilane. This is shown for the reaction of α,α'-dichloro-*p*-xylene with the disilyl derivative of piperazine (16.78).

$$R_3Si-N\underset{}{\overset{}{\bigcirc}}N-SiR_3 + ClCH_2-\!\!\left\langle\!\bigcirc\!\right\rangle\!-CH_2Cl \xrightarrow{\text{NH}_4\text{Cl}}$$

42

$$\left[\!-\!N\underset{}{\overset{}{\bigcirc}}N\!-\!CH_2-\!\!\left\langle\!\bigcirc\!\right\rangle\!-CH_2\!-\!\right] + 2R_3SiCl \quad (16.78)$$

Polyamines of low molecular weight may be prepared by the Mannich reaction involving secondary diamines or primary amines, formaldehyde, and compounds containing active hydrogen.[76–79] This is shown for pyrrole (**42**) (which reacts principally at ring positions 2 and 5) using piperazine (16.79) and benzylamine (16.80). Mannich-base polymers are used as complexing agents, flocculants, antioxidants, and coating binders.

Ring-opening polymerization of alkyleneimines[80] (aziridines) forms highly branched polyamines, and the product obtained from the simplest member of

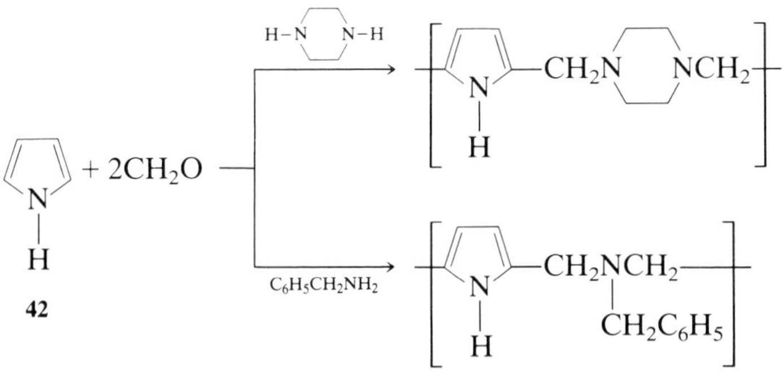

$$\text{(16.79)}$$

$$\text{(16.80)}$$

this monomer type, ethyleneimine (**43**), has been developed commercially (DOW trade name Montrek). Ethyleneimine is synthesized by the reaction of ethylene chloride with ammonia via the intermediate β-chloroethylamine (16.81). The polymerization reaction is acid catalyzed and similar, mech-

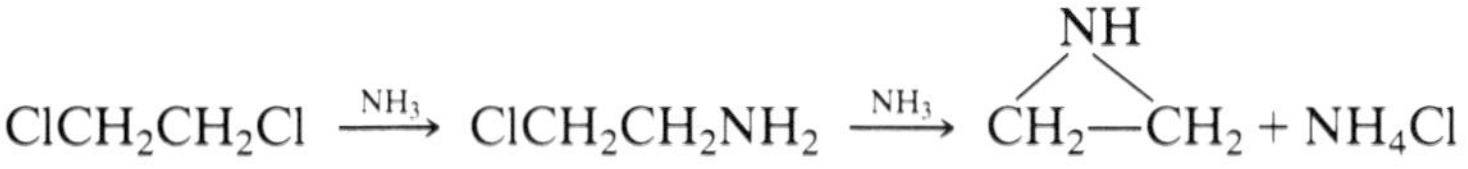

$$\text{ClCH}_2\text{CH}_2\text{Cl} \xrightarrow{\text{NH}_3} \text{ClCH}_2\text{CH}_2\text{NH}_2 \xrightarrow{\text{NH}_3} \underset{\textbf{43}}{\overset{\text{NH}}{\text{CH}_2\!\!-\!\!\text{CH}_2}} + \text{NH}_4\text{Cl} \quad \text{(16.81)}$$

anistically, to the polymerization of epoxides and episulfides. Polymerization is not initiated with basic catalysts unless an acyl group or other strongly electron-attracting substituent is attached to the nitrogen atom. Initiation is believed to involve S_N2 ring-opening of a protonated alkyleneimine (16.82) with propagation (16.83) continuing by analogous reactions of the cyclic immonium end group.

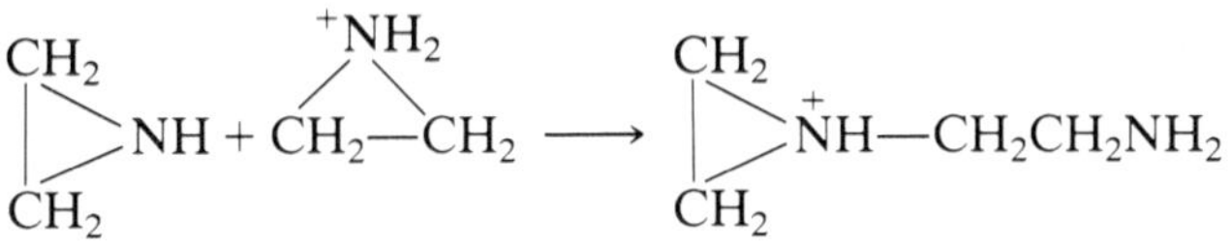

$$\text{(16.82)}$$

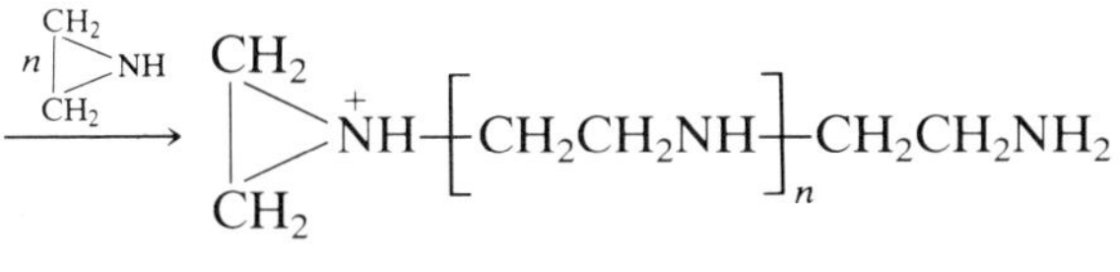

$$\text{(16.83)}$$

Chain branching occurs by reaction of secondary amine groups in the backbone with immonium chain ends (16.84).

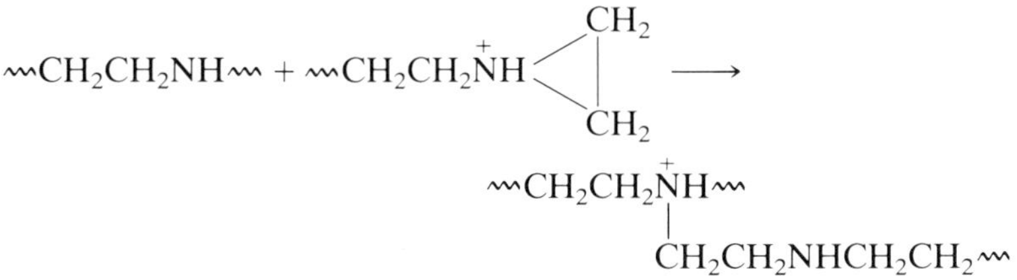

$$\text{(16.84)}$$

Substituted aziridines also undergo ring-opening polymerization; indeed, the S_N2 mechanism has been established by polymerization of optically active 2-methylethyleneimine to solid, optically active polymer (16.85), thus demonstrating that ring-opening occurs by cleavage of the secondary carbon–nitrogen bond. Racemic monomer results in liquid, racemic polymer. (The same type of evidence supporting the mechanism of propylene oxide polymerization is described in Chapter 11.)

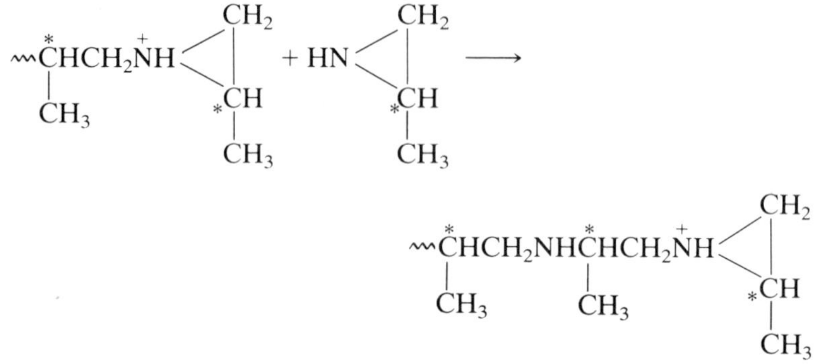

$$\overset{*}{\sim}CHCH_2\overset{+}{NH}\!\!<\!\!\overset{\displaystyle CH_2}{\underset{\displaystyle CH_3}{\overset{*}{C}H}} \;+\; HN\!\!<\!\!\overset{\displaystyle CH_2}{\underset{\displaystyle CH_3}{\overset{*}{C}H}} \longrightarrow$$

$$\overset{*}{\sim}CHCH_2NH\overset{*}{C}HCH_2\overset{+}{NH}\!\!<\!\!\overset{\displaystyle CH_2}{\underset{\displaystyle CH_3}{\overset{*}{C}H}} \qquad (16.85)$$
$$\quad\;\; \underset{\displaystyle CH_3}{\quad}\quad\;\; \underset{\displaystyle CH_3}{\quad}$$

The four-membered ring analogs of aziridines (azetidines)—for example, conidine (**44**)—are also susceptible to this type of reaction (16.86).

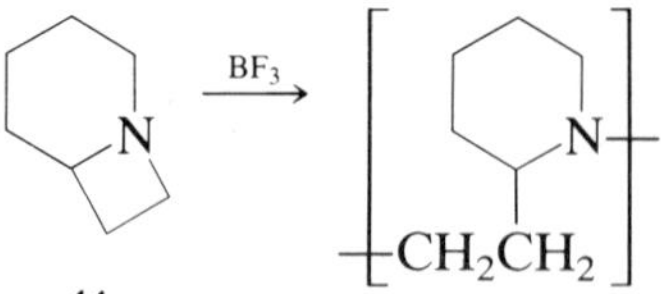

$$(16.86)$$

44

Although aziridenes yield only branched polyamines, it is possible to prepare the linear polymer by ring-opening polymerization of 2-oxazolines[81,82] (**45**) to form the polymeric amide (**46**), followed by basic hydrolysis (16.87).

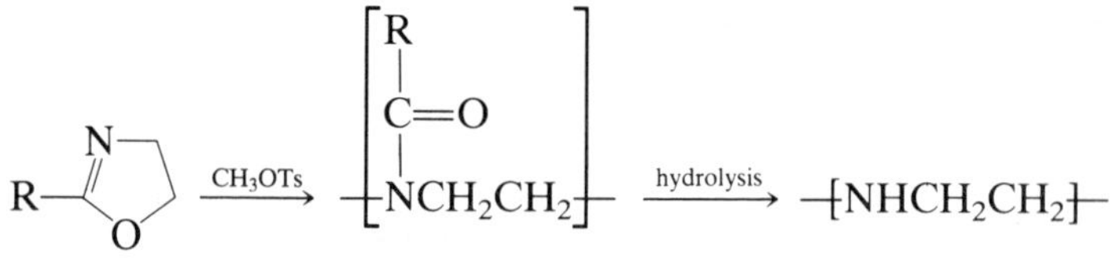

$$(16.87)$$

45   **46**

Polyaziridenes are used primarily in aqueous solution for treatment of textiles and paper. Aziridenes themselves are used for crosslinking carboxyl-containing polymers, such as acrylic emulsion coatings. In this instance, polyfunctional aziridenes (**47**) react with the carboxyl groups according to reaction (16.88).

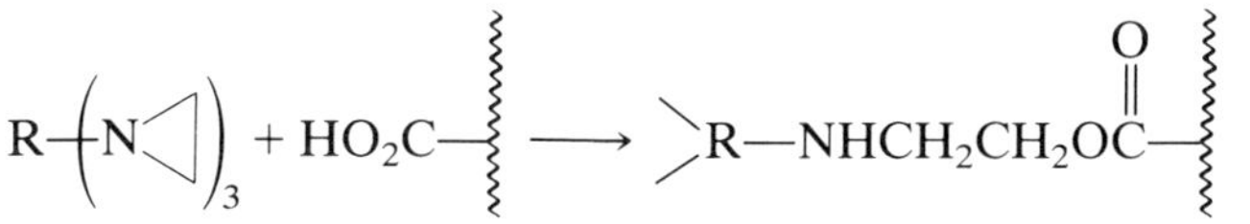

$$R\text{---}\!\left(N\!\triangleleft\right)_3 + HO_2C\text{---}\xi \longrightarrow \;\;>\!R\text{---}NHCH_2CH_2O\overset{\overset{\displaystyle O}{\|}}{C}\text{---}\xi \qquad (16.88)$$

47

16.10 Charge-transfer polymers

Charge-transfer complexing in polymeric systems has received considerable attention in recent years, particularly from the standpoint of the effect of complexation on electrical conductivity and photoconductivity.[83–85] In fact, it has been suggested that simple charge-transfer complexes in which donor and acceptor molecules alternate throughout a crystalline structure may be thought of as polymers.[86] We are concerned here, however, with covalently bonded polymers containing donor or acceptor groups.

There are basically three approaches to studying charge-transfer complexing in polymers: (1) complexing polymers containing donor groups or polymers containing acceptor groups with low-molecular-weight acceptors or donors, respectively; (2) mixing donor polymers with acceptor polymers; and (3) synthesizing polymers having both donor and acceptor groups. A lot of work has been done with donor polymers and low-molecular-weight acceptors, but the products are frequently brittle and intractable. Some polycarbonates and polyesters with good mechanical properties have been prepared[87] from aryliminodiethanol compounds (**48–50**). These polymers

$$
\underset{\substack{|\\ \text{Ar}\\[2pt] \textbf{48}}}{\text{HOCH}_2\text{CH}_2\text{NCH}_2\text{CH}_2\text{OH}}
$$

CH$_2$CH$_2$OH

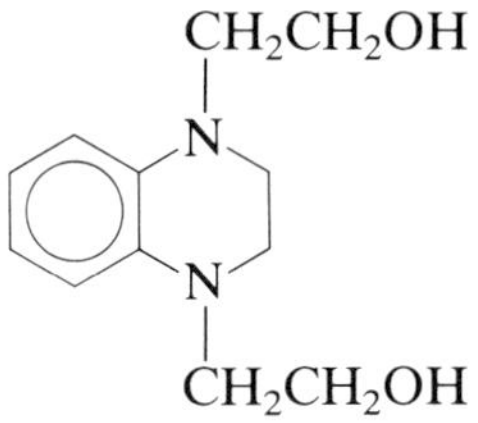

49

CH$_2$CH$_2$OH

—N(CH$_2$CH$_2$OH)$_2$

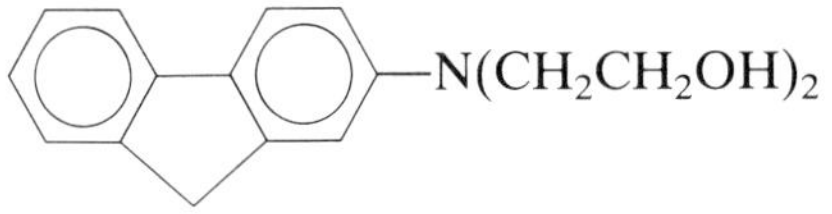

50

form highly colored complexes with certain pi acceptor molecules with a resultant increase in conductivity. Table 16.2 shows the effect of adding equivalent amounts of 2,3-dichloro-5,6-dicyano-*p*-benzoquinone (**51**) to some polycarbonates (**52**) prepared from methoxy derivatives of **48** and bisphenol A dichloroformate. Addition of **51** to the *p*-methoxy derivative results in a

Table 16.2. Effect of equivalent amounts of 2,3-dichloro-5,6-dicyano-*p*-benzoquinone on volume resistivities of polycarbonate **52**[a]

Polymer structure	51 added	Volume resistivity (ohm cm)
$R_1 = R_2 = R_4 = H; R_3 = OCH_3$	No	1.3×10^{16}
$R_1 = R_2 = R_4 = H; R_3 = OCH_3$	Yes	1×10^{13}
$R_1 = R_4 = OCH_3; R_2 = R_3 = H$	Yes	1×10^{11}
$R_1 = H; R_2 = R_3 = R_4 = OCH_3$	Yes	2×10^{11}

[a] Data from Sulzberg and Cotter.[87]

thousandfold decrease in volume resistivity. Steric effects become evident, however, as the number of methoxy groups on the ring is increased.

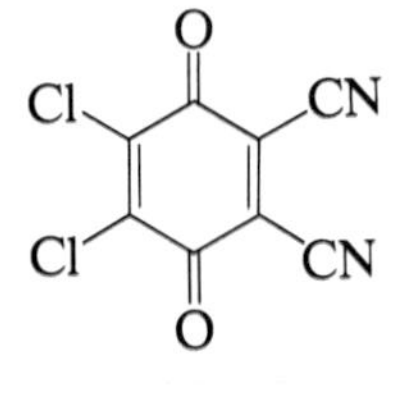

51

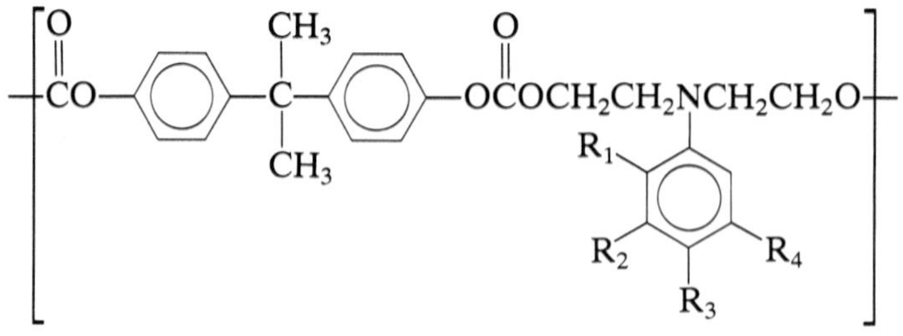

52

Acceptor polycarbonates (**53**) have been prepared from 5-nitroisophthalic acid ($R_1 = R_3 = H, R_2 = NO_2$) and 4,6-dinitroisophthalic acid ($R_1 = R_3 = NO_2$, $R_2 = H$) and from nitroterephthalic acid.[88] Blends of **52** and **53** are highly

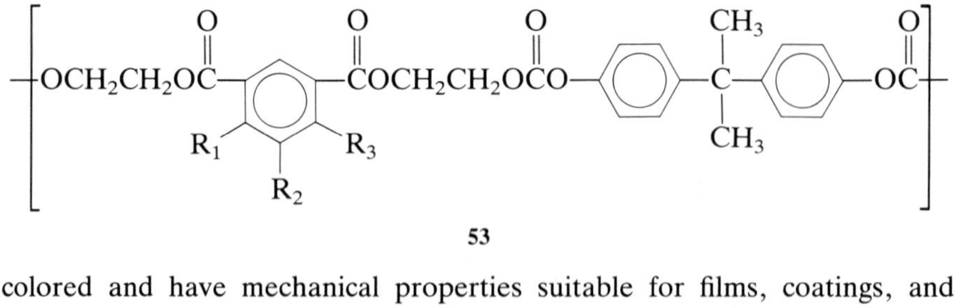

53

colored and have mechanical properties suitable for films, coatings, and fibers. Typically, a mixture of the *p*-methoxy derivative of **52** and the 5-nitro derivative of **53** is yellow-orange in color with a lowest energy absorption maximum of 398 nm compared with 298 and 330 nm, respectively, for the individual polymers.

Polyesters (**54**) of low molecular weight containing both donor and acceptor groups have been prepared[87] from monomer **48** with 5-nitroisophthalic acid or nitroterephthalic acid. Interestingly, **54** had a somewhat lower volume resistivity than blends of donor and acceptor polymers (**52** and **53**, respectively). It is not clear why only low-molecular-weight donor–acceptor polyesters could be prepared, since high-molecular-weight donor polymers and acceptor polymers were prepared from the same monomers. It has been suggested that loss of mobility or reactivity resulting from complexing during polymerization may upset the stoichiometric balance.[87]

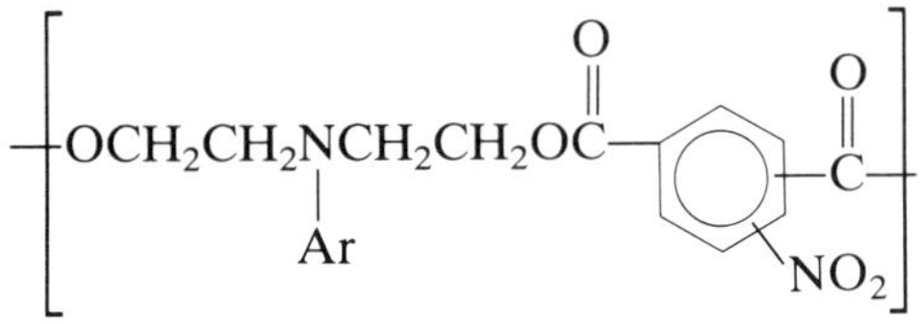

54

Vinyl polymers containing donor and acceptor groups have also been the subject of considerable study.[89–91] An example is copolymer **55**, which exhibits a charge-transfer absorption enhanced an order of magnitude over that of mixtures of model compounds.[90] Presumably, constraints in the polymer system cause a higher relative concentration of charge-transfer interactions.

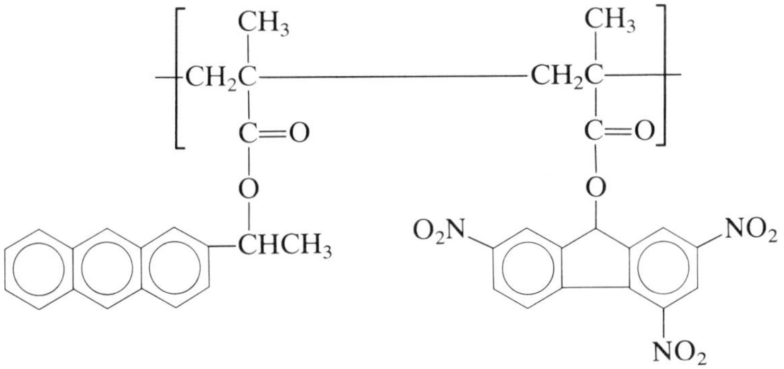

55

16.11 Ionic polymers

In Chapter 3 we learned that the presence of ionic groups in polymers can have interesting morphological effects, including the formation of ionic microdomains within the polymer matrix and a tendency for the polymers to exhibit thermoplastic elastomeric properties.[92–95] In Chapter 9 we encountered the commercially important ionomers—ionized copolymers of ethylene and methacrylic acid. In this section we are concerned with the synthesis of nonvinyl analogs of ionomers as well as polymers having ionic groups as part of the backbone structure.

Compared with the vinyl-based ionomers, nonvinyl ionic polymers have not been studied extensively. Among the few that have been reported are polyureas prepared from diisocyanates and diaminobenzoates[96] (16.89) and polyesters and polyurethanes containing tetra-*t*-butylammonium and -sulfonium groups from dihydroxy monomer **56**.[97] Related in structure are poly-

$$OCN-R-NCO + H_2N-\underset{}{\underset{\text{CO}_2^-K^+}{\bigcirc}}-NH_2 \longrightarrow$$

$$\left[\underset{\text{O}}{\overset{\text{O}}{\underset{\|}{C}}}NH-R-N\overset{\text{O}}{\underset{\|}{H}}\overset{\text{O}}{\underset{\|}{C}}NH-\underset{}{\underset{\text{CO}_2^-K^+}{\bigcirc}}-NH\right] \quad (16.89)$$

$$\underset{\text{NHSO}_3^- {}^+N(C_4H_9\text{-}t)_4}{\underset{|}{HOCH_2CH\overset{\text{CO}_2^- {}^+N(C_4H_9\text{-}t)_4}{\overset{|}{CH}}CH_2OH}}$$

56

urethane elastomers containing *backbone* ammonium ions[96,98] prepared by neutralization of amine-containing polyurethanes (16.90). Polymers of these types are of interest as coatings and adhesives and as finishing agents for textiles and leather.

$$\left[\overset{\text{O}}{\underset{\|}{C}}NH-R-NH\overset{\text{O}}{\underset{\|}{C}}O\leadsto N\leadsto O\overset{\text{O}}{\underset{\|}{C}}\right] \overset{H^+}{\longrightarrow} \left[\overset{\text{O}}{\underset{\|}{C}}NH-R-NH\overset{\text{O}}{\underset{\|}{C}}O\leadsto \underset{H}{\overset{+}{N}}\leadsto O\overset{\text{O}}{\underset{\|}{C}}\right]$$

$$(16.90)$$

Polymers containing backbone ammonium ions have also been prepared by the reaction of dihalides (preferably primary) with ditertiary amines such as 1,4-diazabicyclo[2.2.2]octane (DABCO) (16.91)[99] or with primary amines (16.92).[100] Polymers of this type, called *ionenes*,[101] are used commercially as flocculating agents for water treatment and as components of cosmetics and antimicrobial formulations. An ionene–polyether–ionene ABA block copolymer is of interest as a flocculating agent.[102]

$$\underset{N}{\overset{}{\bigvee}}N\overset{}{\bigvee} + Br-R-Br \longrightarrow \left[\underset{Br^-}{\overset{Br^-}{\underset{N}{\overset{+}{\bigvee}}}N^+-R}\right] \quad (16.91)$$

$$RNH_2 + Cl\!-\!R'\!-\!Cl \longrightarrow \left[\!\!\begin{array}{c} Cl^- \\ \overset{+}{\underset{|}{N}H} \\ R' \end{array}\!\!-\!\!\!\!\!\!\!\right] + HCl \qquad (16.92)$$

Polymers containing ionic linkages as part of the backbone have been synthesized from dicarboxylic acids and divalent metal (M) acetates (16.93), or from dicarboxylate salts and metal halides (16.94). The products are called *halatopolymers*,[95a,103] A halatomonomer,[104] **57**, prepared from divalent metal salts of mono(hydroxyethyl)phthalate, has been used to prepare polyurethanes[105] and polyesters.[106,107] Halatopolymers are primarily of interest as adhesives.

$$M(OCOCH_3)_2 + HO_2C\!-\!R\!-\!CO_2H \rightarrow [MO_2C\!-\!R\!-\!CO_2]_x + 2CH_3CO_2H \qquad (16.93)$$

$$MCl_2 + \overset{+}{Na}\bar{O}_2C\!-\!R\!-\!CO_2^-Na^+ \rightarrow [MO_2C\!-\!R\!-\!CO_2]_x + 2NaCl \qquad (16.94)$$

$$HOCH_2CH_2O_2C \quad CO_2^-M^{2+-}O_2C \quad CO_2CH_2CH_2OH$$

57

Before we leave the subject of ionic polymers, it should be noted that *ion-exchange resins*[108] have a relatively long history of use in separations and water purification. The crosslinked polymers are classified as anionic or cationic exchange resins. The former contain basic groups capable of exchanging anions, such as styrene–divinylbenzene copolymers substituted with amine groups, or formaldehyde–aniline resins. Cationic exchange resins contain acidic groups capable of exchanging cations. Sulfonated phenol–formaldehyde or styrene–divinylbenzene copolymers, and crosslinked poly(acrylic acid)s, are typical examples. One type of cholesterol-reducing medicine is an anion exchange resin that functions by removing bile salt as it passes through the intestinal tract. To compensate, cholesterol is oxidized to bile acids, thereby reducing serum-cholesterol levels.

A variety of similar polymers of different molecular weight ranges are used as coagulating agents for removal of colloidal material in wastewater treatment. These polymers, known as *polyelectrolytes*[95b], range from cationic to nonionic to anionic. They are normally used in conjunction with inorganic coagulants such as activated silica, alum, or ferric chloride. Although used in very small amounts, they represent a large-volume market because of the extensive application of wastewater treatment. Typical polymers are polyacrylates, poly(vinylpyridinium salts), and poly(amino acids). Polyelectrolytes apparently aid coagulation by neutralizing the charge on colloidal particles, although there is evidence that their function may also be to bridge the charged particles, because coagulation can be brought about with polyelectrolytes having the same charge as the colloid.

Polyampholytes[110,111] are a type of polyelectrolyte or ionomer having both cationic and anionic groups, but no conventional counterions. An example is the copolymer of a methacrylate ester with ion-pair **58**.[112] Polyampholytes are

$$\left[\begin{array}{c} CH_3 \\ | \\ CH_2{=}\overset{}{C}CO_2CH_2CH_2\overset{+}{N}(CH_3)_3 \end{array}\right]\left[\begin{array}{c} CH_3 \\ | \\ CH_2{=}\overset{}{C}CO_2CH_2CH_2SO_3^- \end{array}\right]$$

58

useful chelating agents for removing trace metals from water. They are also used as ion-exchange membranes, flocculants, and soil conditioners and as ingredients in shampoos and hair conditioners.

References

1. R. J. Cotter and M. Matzner, *Ring-Forming Polymerizations*, Part B1, Academic Press, New York, 1968: (a) Chap. 1; (b) Chap. 3.
2. L. M. Alberino, W. J. Farrissey, and A. A. R. Sayigh, *J. Appl. Polym. Sci.*, **21**, 1999 (1977).
3. E. Dyer and P. E. Christie, *J. Polym. Sci A-1*, **6**, 729 (1968).
4. G. F. D'Alelio and R. K. Schoenig, *J. Macromol. Sci., Rev. Macromol. Chem.*, **3**, 105 (1969).
5. J. Preston, in *Kirk-Othmer: Encyclopedia of Chemical Technology*, 3rd ed., Vol. 3, Wiley-Interscience, New York, 1978, pp. 213–242.
6. S. Ogata, M. Kakimoto, and Y. Imai, *Makromol. Chem., Rapid Commun.*, **6**, 835 (1985).
7. L. J. Mathias and C. G. Overberger, *J. Polym. Sci., Polym. Chem. Ed.*, **17**, 1287 (1979).
8. K. Kurita, Y. Kusayama, and Y. Iwakura, *J. Polym. Sci., Polym. Chem. Ed.*, **15**, 2163 (1977).
9. H. H. Hörhold and J. Opfermann, *Makromol. Chem.*, **131**, 105 (1970).
10. G. Kossmehl, M. Härtel, and G. Manecke, *Makromol. Chem.*, **131**, 15 (1970).
11. R. J. Cotter and M. Matzner, *Ring-Forming Polymerizations*, Part A, Academic Press, New York, 1969: (a) Chap. 1; (b) Chap. 3.
12. H. C. Bach and W. B. Black, *J. Appl. Polym. Sci., Appl. Polym. Symp.*, **22**, 799 (1969).
13. H. H. Bössler and R. C. Schultz, *Makromol. Chem.*, **158**, 113 (1972).
14. A. Ravve and C. Fitco, *J. Polym. Sci. A*, **2**, 1925 (1964).
15. A. A. Berlin and V. P. Parini, *Izv. Akad. Nauk SSSR Otd. Khim. Nauk.*, 1674 (1959) (*Chem. Abstr.*, **54**, 8715c 1960).
16. A. DeSouza Gomes and T. D. R. DeOliveira Cavalcanti, *J. Polym. Sci., Polym. Chem. Ed.*, **16**, 2671 (1978).
17. L. G. Donaruma, *Polym. Preprints*, **23**(2), 108 (1982).
18. A. S. Hay, D. A. Bolon, K. R. Leimer, and R. F. Clark, *J. Polym. Sci. B*, **8**, 97 (1970).
19. D. M. White, *Polym. Preprints*, **12**(1), 155 (1971).
20. S. J. Havens and P. M. Hergenrother, *J. Polym. Sci., Polym. Lett. Ed.*, **23**, 587 (1985).

21. B. Tieke and G. Wegner, *Makromol. Chem.*, **179**, 2573 (1978).
22. G. N. Patel, R. R. Chance, E. A. Turi, and Y. P. Khanna, *J. Am. Chem. Soc.*, **100**, 6644 (1978).
23. G. N. Patel and N.-L. Yang, *J. Chem. Educ.*, **60**, 181 (1983).
24. W. Bracke, *J. Polym. Sci. A-1*, **10**, 975 (1972).
25. P. E. Cassidy, *Thermally Stable Polymers*, Dekker, New York, 1980: (a) Chaps. 6–8; (b) Chap. 2.
26. T. M. Keller, *J. Polym. Sci. A: Polym. Chem.*, **26**, 3199 (1988).
27. P. M. Hergenrother, *J. Macromol. Sci., Rev. Macromol. Chem.*, **C6**, 1 (1971).
28. D. H. Solomon (ed.), *Step-Growth Polymerizations*, Dekker, New York, 1972: (a) P. M. Hergenrother, Chap. 4; (b) W. J. Bailey, Chap. 6.
29. J. K. Stille, *Pure Appl. Chem.*, **50**, 273 (1978).
30. W. Vancraeynest and J. K. Stille, *Macromolecules*, **13**, 1367 (1980).
31. J. Higgins, J. F. Jones, and A. Thornburgh, *Macromolecules*, **2**, 558 (1969); *J. Polym. Sci. A-1*, **9**, 763 (1971).
32. E. W. Neuse, *Macromolecules*, **1**, 171 (1968).
33. Y. Gilliams and G. Smets, *Makromol. Chem.*, **128**, 263 (1969).
34. J. K. Stille and L. D. Gotter, *J. Polym. Sci. A-1*, **7**, 2493 (1969).
35. J. W. Connell, R. G. Bass, M. S. Sinsky, R. O. Waldbauer, and P. M. Hergenrother, *J. Polym. Sci. A, Polym. Chem.*, **25**, 253 (1987).
36. B. Krieg and G. Manecke, *Makromol. Chem.*, **108**, 210 (1967).
37. P. M. Hergenrother, *J. Polym. Sci., Polym. Chem. Ed*, **12**, 2857 (1974).
38. W. H. Beever and J. K. Stille, *J. Polym. Sci., Polym. Symp.*, **65**, 41 (1978).
39. J. K. Stille, *Macromolecules*, **14**, 870 (1981).
40. Y. Imai, E. F. Johnson, T. Katto, M. Kurihara, and J. K. Stille, *J. Polym. Sci., Polym. Chem. Ed.*, **13**, 2233 (1975).
41. A. F. Diaz, K. K. Kanazawa, and G. P. Gardini, *J. Chem. Soc., Chem. Commun.*, 635 (1979).
42. G. Wegner, *Angew. Chem. Int. Ed. English*, **20**, 361 (1981).
43. A. I. Nazzal and G. B. Street, *J. Chem. Soc., Chem. Commun.*, 375 (1985).
44. A. Gandini, *Adv. Polym. Sci.*, **25**, 47 (1977).
45. B. S. Lamb and P. Kovacic, *J. Polym. Sci., Polym. Chem. Ed.*, **18**, 2423 (1980).
46. P. Kovacic and K. N. McFarland, *J. Polym. Sci., Polym. Chem. Ed.*, **17**, 1963 (1979).
47. R. J. Walman and J. Bargon, *Can. J. Chem.*, **64**, 76 (1986).
48. J. G. Speight, P. Kovacic, and F. W. Koch, *J. Macromol. Sci., Rev. Macromol. Chem.*, **C5**, 295 (1971).
49. G. C. Engstrom and P. Kovacic, *J. Polym. Sci., Polym. Chem. Ed.*, **15**, 2453 (1977).
50. S. A. Milosevich, K. Saichek, L. Hinchey, W. B. England, and P. Kovacic, *J. Am. Chem. Soc.*, **105**, 1088 (1983).
51. C.-F. Hsing, I. Khoury, M. D. Bezoari, and P. Kovacic, *J. Polym. Sci., Polym. Chem. Ed.*, **20**, 3313 (1982).
52. J. P. Critchley, G. J. Knight, and W. W. Wright, *Heat-Resistant Polymers*, Plenum Press, New York, 1983: (a) Chap. 4; (b) Chap. 2.
53. D. M. Gale, *J. Appl. Polym. Sci.*, **22**, 1955, 1971 (1978).
54. N. A. J. Platzer (ed.), *Addition and Condensation Polymerization Processes*, Adv. Chem. Ser. 91, American Chemical Society, Washington, D.C., 1969:

(a) W. F. Gorham, pp. 643–659; (b) J. K. Stille, F. W. Harris, H. Mukamal, R. O. Rakutis, C. L. Schilling, G. K. Noren, and J. A. Reed, pp. 628–642.

55. T. L. St. Clair and H. M. Bell, *J. Polym. Sci., Polym. Chem. Ed.*, **12**, 1321 (1974).
56. K. Niume, F. Toda, K. Uno, and Y. Iwakura, *J. Polym. Sci., Polym. Lett. Ed.*, **15**, 283 (1977).
57. M. Ueda, T. Kano, T. Waragai, and H. Sugita, *Makromol. Chem., Rapid Commun.*, **6**, 847 (1985).
58. J. E. Chandler, B. H. Johnson, and R. W. Lenz, *Macromolecules*, **13**, 377 (1980).
59. N. Grassie, I. G. Meldrum, and J. Gilks, *J. Polym. Sci. B*, **8**, 247 (1970).
60. M. Hasegawa, *Adv. Polym. Sci.*, **42**, 1 (1982).
61. F. C. DeSchryver, N. Boens, and G. Smets, *J. Polym. Sci. A-1*, **10**, 1687 (1972).
62. S. S. Labana, *J. Macromol. Sci., Rev. Macromol. Chem.*, **C11**, 299 (1974).
63. W. L. Dilling, *Chem. Rev.*, **83**, 1 (1983).
64. Y. Musa and M. P. Stevens, *J. Polym. Sci. A-1*, **10**, 319 (1972).
65. N. Kardush and M. P. Stevens, *J. Polym. Sci. A-1*, **10**, 1093 (1972).
66. C. F. H. Tipper and D. A. Walker, *J. Chem. Soc.*, 1352 (1959).
67. A. D. Ketley, *J. Polym. Sci. B*, **1**, 313 (1963).
68. H. K. Hall, Jr., and P. Ykman, *J. Polym.. Sci., Macromol. Rev.*, **11**, 1 (1976).
69. H. K. Hall Jr., *Macromolecules*, **4**, 139 (1971).
70. N. Yoda, in *Encyclopedia of Polymer Science and Technology* (H. F. Mark, N. G. Gaylord, and N. M. Bikales, eds.), Wiley-Interscience, New York, 1969, pp. 630–653.
71. K. W. Leong, V. Simonte, and R. Langer, *Macromolecules*, **20**, 706 (1987).
72. A. J. Domb, E. Ron, and R. Langer, *Macromolecules*, **21**, 1925 (1988).
73. R. Subramanyam and A. G. Pinkus, *J. Macromol. Sci.-Chem.*, **A22**, 23 (1985).
74. F. Danusso and P. Ferruti, *Polymer*, **11**, 88 (1970).
75. J. F. Klebe, *J. Polym. Sci. A*, **2**, 2673 (1964).
76. E. Tsuchida and T. Tomomo, *J. Polym. Sci., Polym. Chem. Ed.*, **11**, 723 (1973).
77. T. Tomomo, E. Hasegawa, and E. Tsuchida, *J. Polym. Sci., Polym. Chem. Ed.*, **12**, 953 (1974).
78. G. B. Butler and S. H. Hong, *Polym. Preprints*, **27**(1), 122 (1986).
79. M. Tramantoni, L. Angiolini, and N. Ghedini, *Polymer*, **29**, 771 (1988).
80. M. Hauser, in *Ring-Opening Polymerization* (K. C. Frisch and S. L. Reegen, eds.), Dekker, New York, 1969, Chap. 5.
81. T. Saegusa and S. Kobayashi, *Makromol. Chem., Macromol. Symp.*, **1**, 23 (1986).
82. T. Saegusa, Y. Nagura, and S. Kobayashi, *Macromolecules*, **6**, 495 (1973).
83. F. Gutman and L. E. Lyons, *Organic Semiconductors*, Wiley-Interscience, New York, 1967.
84. D. A. Seanor, *Adv. Polym. Sci.*, **4**, 317 (1965).
85. R. C. Schulz, D. Fleischer, A. Henglein, H. M. Bössler, J. Trisnadi, and H. Tanaka, *Pure Appl. Chem.*, **38**, 227 (1974).
86. S. Kanda and H. A. Pohl, in *Organic Semiconducting Polymers* (J. E. Katon, ed.), Dekker, New York, 1968, Chap. 3.
87. T. Sulzberg and R. J. Cotter, *Macromolecules*, **2**, 146, 150 (1969).
88. T. Sulzberg and R. J. Cotter, *J. Polym. Sci. A-1*, **8**, 2747 (1970).
89. C. I. Simionescu, V. Bǎrboiu, and M. Grigoraş, *J. Macromol. Sci.-Chem.*, **A22**, 693 (1985).

90. S. R. Turner and M. Stolka, *Macromolecules*, **11**, 835 (1978).
91. J. L. Nash Jr., R. E. Thompson, J. W. Schwietert, and G. B. Butler, *J. Polym. Sci., Polym. Chem. Ed.*, **16**, 1343 (1978); J. W. Schweitert and G. B. Butler, ibid., **16**, 1359 (1978).
92. A. Eisenberg and M. King, *Ion-Containing Polymers: Physical Properties and Structure*, Academic Press, New York, 1977.
93. W. J. MacKnight and T. R. Earnest Jr., *Macromol. Rev.*, **16**, 41 (1981).
94. A. Eisenberg (ed.), *Ions in Polymers*, Adv. Chem. Ser. 187, American Chemical Society, Washington, D.C., 1980.
95. L. Holliday (ed.), *Ionic Polymers*, Applied Science, London, 1975: (a) J. Economy and J. H. Mason, Chap. 5; (b) M. J. Lysaght, Chap. 6.
96. D. Dieterich, W. Kaberle, and H. Witt, *Angew. Chem. Int. Ed. English*, **9**, 40 (1970).
97. H. S. Egboh, A. Gaffar, M. H. George, J. A. Barrie, and D. J. Walsh, *Polymer*, **23**, 1167 (1982).
98. S. L. Hsu, H. X. Xiao, H. H. Szmant, and K. C. Frisch, *J. Appl. Polym. Sci.*, **29**, 2467 (1984).
99. J. C. Salamone and B. Snider, *J. Polym. Sci. A-1*, **8**, 3495 (1970).
100. G. Henrici-Olivé and S. Olivé, *J. Polym. Sci. A-1*, **10**, 625 (1972).
101. T. Tsutsui, *Devel. Ionic Polym.*, **2**, 163 (1986).
102. R. Jerome, R. Fayt, and T. Ouhadi, *Prog. Polym. Sci.*, **10**, 87 (1984).
103. J. Economy, J. H. Mason, and L. C. Wohrer, *J. Polym. Sci. A-1*, **8**, 2231 (1970).
104. H. Matsuda, *J. Appl. Polym. Sci.*, **23**, 2603 (1979).
105. H. Matsuda, *J. Polym. Sci., Polym. Chem. Ed.*, **12**, 455, 469 (1974).
106. H. Matsuda, *J. Appl. Polym. Sci.*, **20**, 995 (1976); **22**, 3371 (1978).
107. H. Matsuda, *J. Macromol. Sci.-Chem.*, **A9**, 397 (1975).
108. R. Kunin, *Ion Exchange Resins*, 2nd ed., Wiley-Interscience, New York, 1958.
109. B. A. Bolto, *Prog. Polym. Sci.*, **9**, 89 (1983).
110. J. C. Salmone, L. Quach, A. C. Watterson, S. Krauser, and M. U. Mahmud, *J. Macromol. Sci.-Chem.*, **A22**, 653 (1985).
111. S. B. Clough, D. Cortelek, T. Nagabhushanam, J. C. Salamone, and A. C. Watterson, *Polym. Eng. Sci.*, **24**, 385 (1984).
112. C. Neculescu, S. B. Clough, P. Elayaperumal, J. C. Salamone, and A. C. Watterson, *J. Polym. Sci. C, Polym. Lett.*, **25**, 201 (1987).

Review exercises

1. Write structures of the monomers that would react to form

   (a) the polyquinoxaline

   (ref. 28a)

   (b) the polyisoanthrazoline

   (ref. 40)

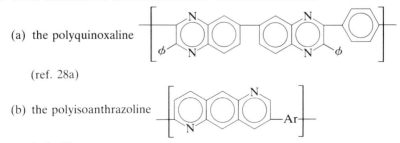

(c) the poly(*as*-triazine)

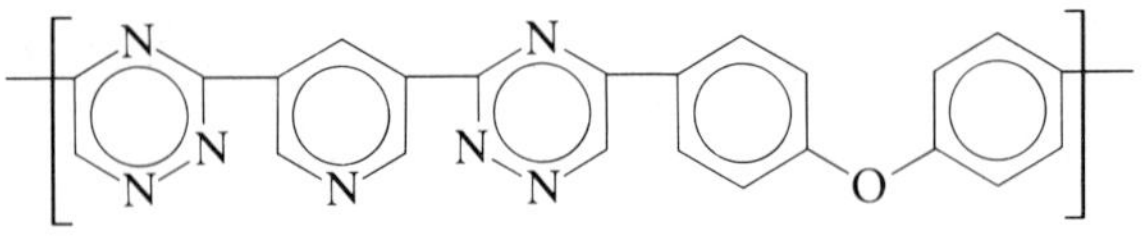

(ref. 28a)

(d) the Diels–Alder polymer

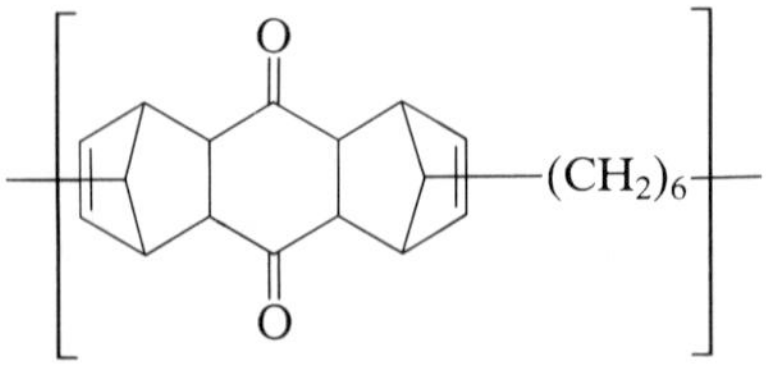

(ref. 28b)

2. Write the repeating unit of the polymer formed. (Reactions without literature references are hypothetical.)

(a)

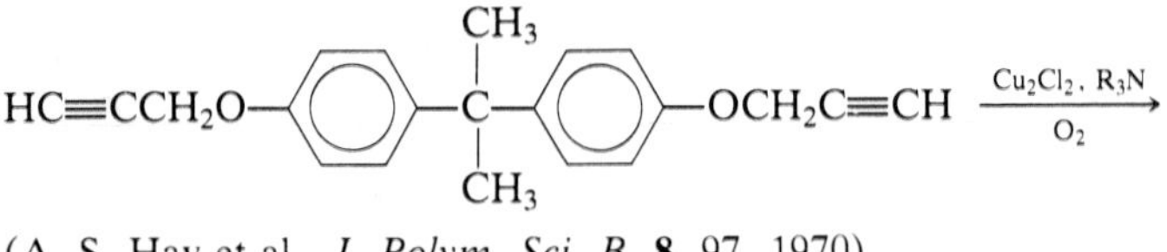

(A. S. Hay et al., *J. Polym. Sci. B*, **8**, 97, 1970)

(b)

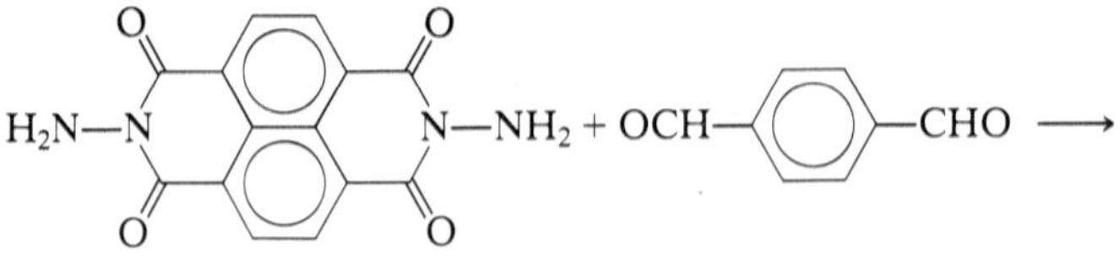

(D. B. Sobanski et al., *J. Macromol. Sci.-Chem*, **A23**, 189, 1986)

(c)

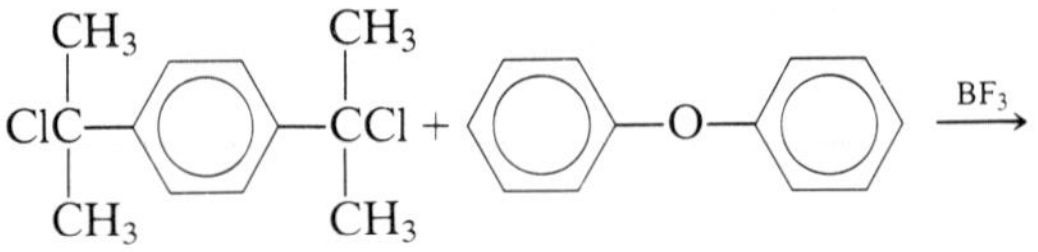

(A. Fritz, *Polym. Preprints* **12**(1), 232, 1971)

(d)

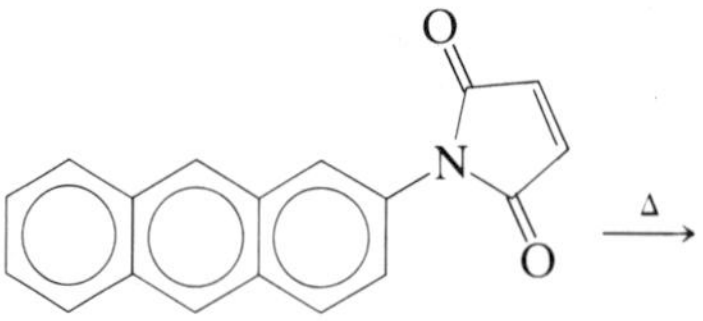

(M. P. Stevens, *J. Polym. Sci., Polym. Lett. Ed.*, **22**, 467, 1984)

(e)

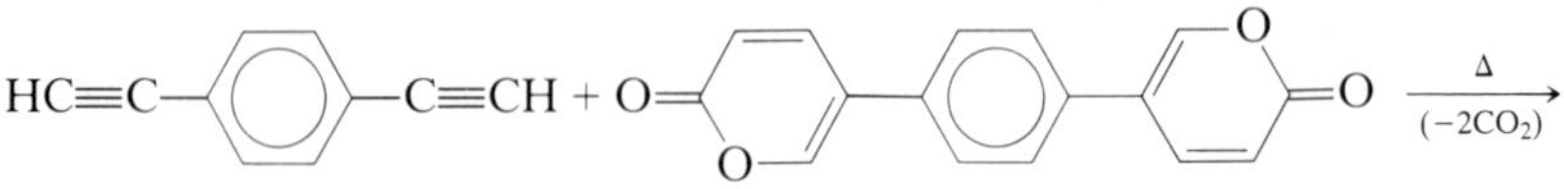

(ref. 25b)

(f) 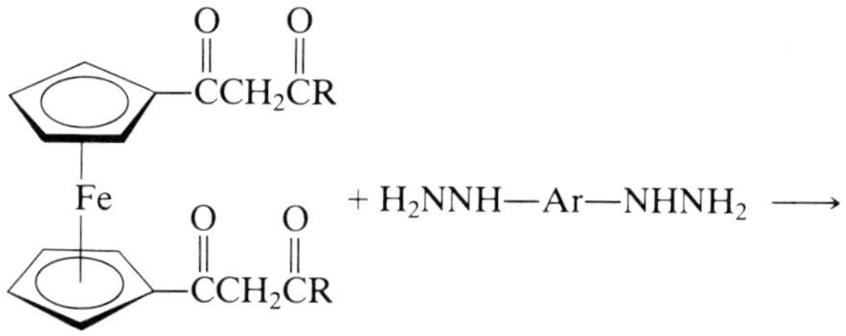 + H$_2$NNH—Ar—NHNH$_2$ $\longrightarrow$

(E. W. Neuse, *Macromolecules*, **1**, 171, 1968)

(g) 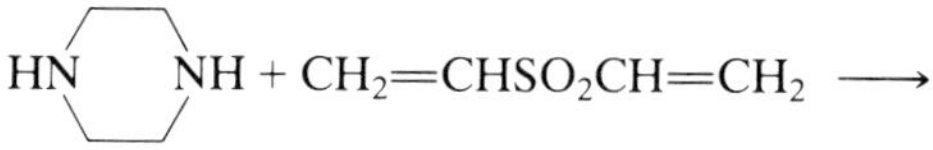 + CH$_2$=CHSO$_2$CH=CH$_2$ $\longrightarrow$

(ref. 74)

(h) 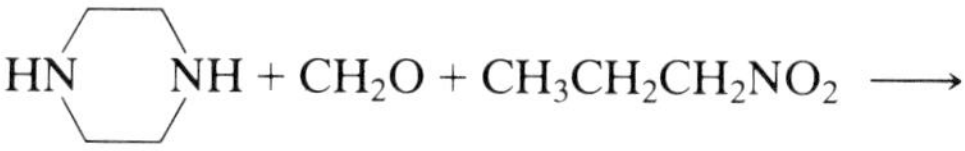 + CH$_2$O + CH$_3$CH$_2$CH$_2$NO$_2$ $\longrightarrow$

(ref. 78)

(i) 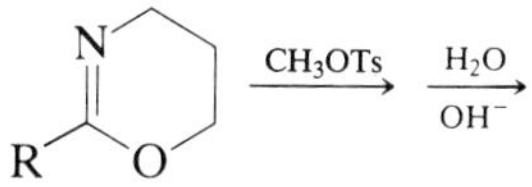 $\xrightarrow{\text{CH}_3\text{OTs}}$ $\xrightarrow[\text{OH}^-]{\text{H}_2\text{O}}$

(ref. 81)

(j) 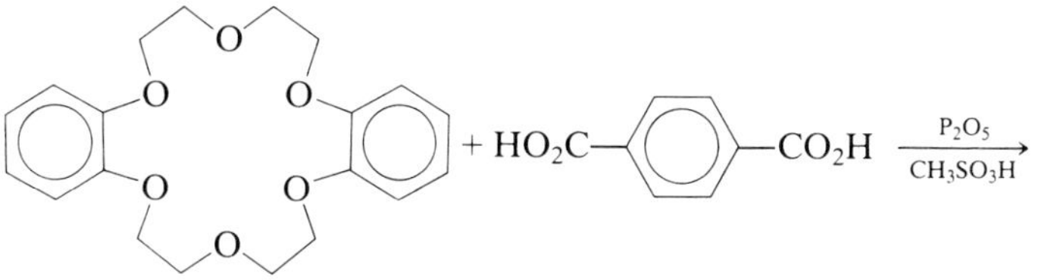 + HO$_2$C—⟨⟩—CO$_2$H $\xrightarrow[\text{CH}_3\text{SO}_3\text{H}]{\text{P}_2\text{O}_5}$

(ref. 57)

(k) 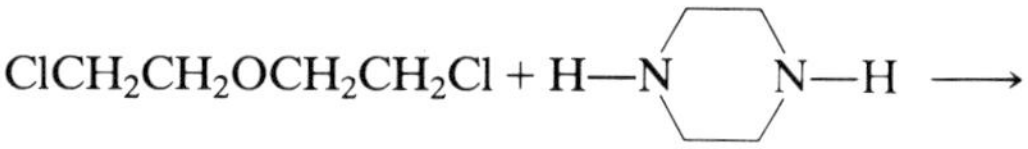 ClCH$_2$CH$_2$OCH$_2$CH$_2$Cl + H—N⟨⟩N—H $\longrightarrow$

(l) 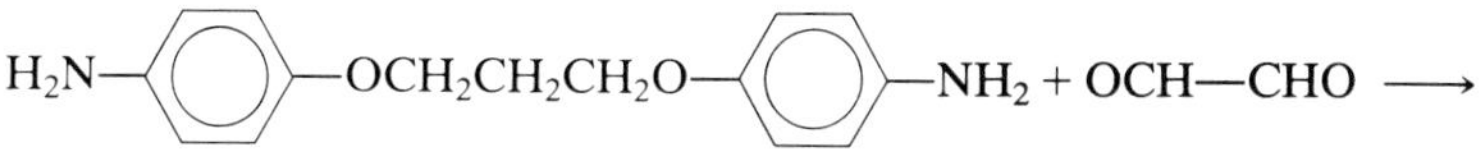 H$_2$N—⟨⟩—OCH$_2$CH$_2$CH$_2$O—⟨⟩—NH$_2$ + OCH—CHO $\longrightarrow$

(J. Bartulin et al., *Polym. Bull.*, **15**, 405, 1986)

(m) 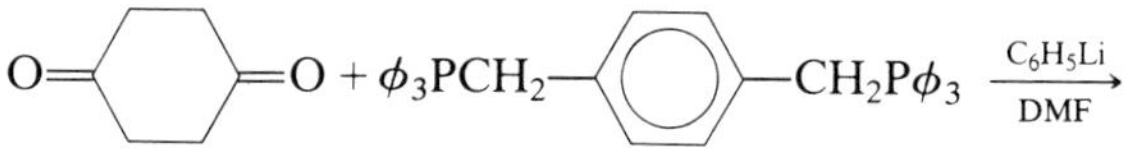 O=⟨⟩=O + ϕ_3PCH$_2$—⟨⟩—CH$_2$Pϕ_3 $\xrightarrow[\text{DMF}]{\text{C}_6\text{H}_5\text{Li}}$

(K. Al-Junah and J. E. Fernandez, *Macromolecules*, **20**, 1177, 1987)

(n) 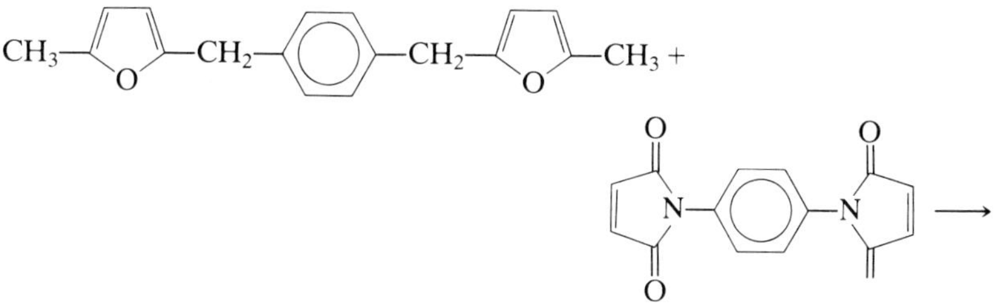 CH$_3$—⟨⟩—CH$_2$—⟨⟩—CH$_2$—⟨⟩—CH$_3$ +

$\longrightarrow$

(X. He et al. *Makromol. Chem., Rapid Commun.*, **9**, 191, 1988)

(o)

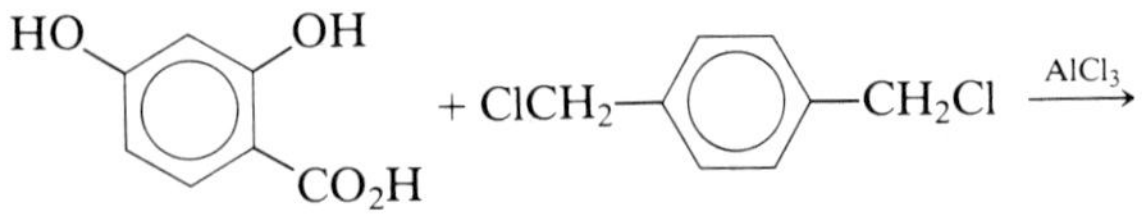

$$+ \text{ClCH}_2 - \text{C}_6\text{H}_4 - \text{CH}_2\text{Cl} \xrightarrow{\text{AlCl}_3}$$

(P. S. Patel and R. S. Patel, *J. Macromol. Sci.-Chem.*, **A23**, 1251, 1986)

(p)

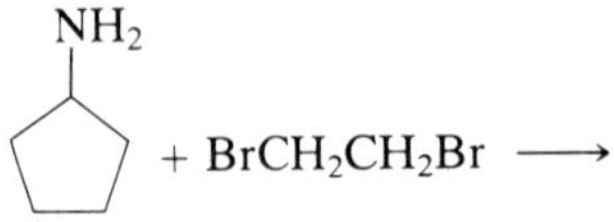

$$+ \text{BrCH}_2\text{CH}_2\text{Br} \longrightarrow$$

(q)

$$\xrightarrow{\text{RLi}}$$

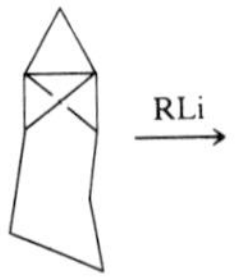

(A.-D. Schluter, *Macromolecules*, **21**, 1208, 1988)

3. Predict the product that would be formed if the acetylene group were replaced with a nitrile group in reaction (16.34). [Hint: review reaction (9.44), Chapter 9.]

4. When the monomer

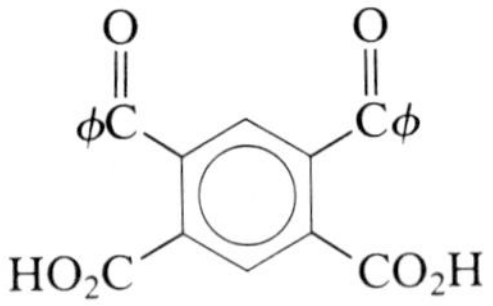

is reacted with $\text{H}_2\text{NNH-Ar-NHNH}_2$, a polymer is formed in high yield with loss of 4 equivalents of H_2O per repeating unit. Suggest a likely structure and explain its formation from a mechanistic viewpoint. (Y. Imai et al., *J. Polym. Sci., Polym. Chem. Ed.*, **14**, 2797, 1976)

5. Suggest alternative syntheses of (a) polymer **35** (reaction 16.68); (b) the conjugated polymer given in reaction (16.14). [Hint: Consider reaction (16.45).]

6. What other monomer would you expect to yield the polymer obtained in reaction (16.70), assuming the same initiator were to be used? Explain.

7. Considering the proposed mechanism of benzene polymerization (Scheme 16.2), which monomer should give the higher molecular weight polymer, C_6H_6 or C_6D_6 (assuming identical reaction conditions)? Explain. (C.-F. Hsing et al., *J. Polym. Sci., Polym. Chem. Ed.*, **21**, 457, 1983)

8. Suggest plausible mechanisms for reactions (16.44), (16.45), and (16.87).

9. Considering reaction (16.52), write the probable repeating unit obtained on addition polymerization of

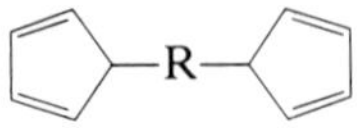

10. Rewrite reactions (16.61) through (16.63) using compounds **30** and **31** in place of **27**.

11. It is well known that benzocyclobutene undergoes thermally induced electrocyclic ring opening:

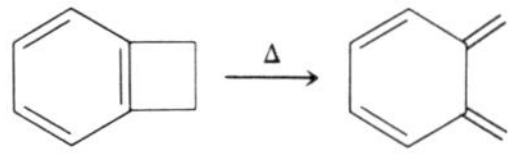

    Predict the result of heating a bismaleimide with

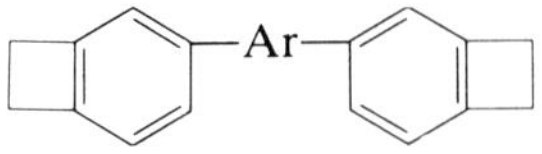

    (L.-S. Tan and F. E. Arnold, *J. Polymer Sci. A: Polym. Chem.*, **26**, 3103, 1988)

12. Styrene undergoes the Diels–Alder reaction with good dienophiles at elevated temperatures. Two equivalents of dieneophile react per equivalent of benzene. The reaction shown for maleic anhydride, is as follows:

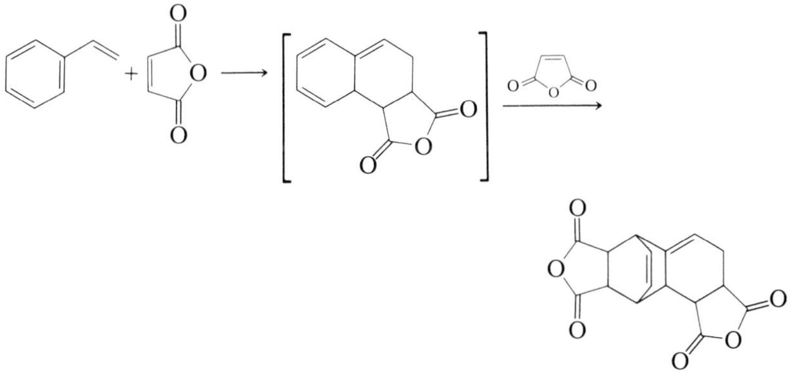

    Predict the product obtained by reacting 1,1-diphenylethylene with the bis-maleimide given in exercise 2(n). (W. N. Emmerling and M. L. Hallensleben, *Eur. Polym. J.*, **13**, 179, 1977)

13. Predict the outcome if polystyrene containing dissolved bismaleimide were to be photolyzed. (M. P. Stevens, in *Ultraviolet Light-Induced Reactions of Polymers*, S. S. Labana, ed., ACS Symp. Ser. 25, 1976, p. 64)

14. Charge-transfer polymers can be prepared by cationic or by anionic polymerization of $CH_2{=}CHOCH_2CH_2O{-}\langle\bigcirc\rangle{-}CH{=}CHNO_2$. Write the repeating unit in each case. (J. L. Nash Jr., et al., *J. Polym. Sci., Polym Chem. Ed.*, **16**, 1343, 1978)

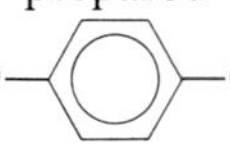

15. Predict the outcome of reacting a carboxyl-terminated telechelic polydiene with $Ti(OAc)_4$. (G. Broze et al., *J. Polym. Sci., Polym. Lett. Ed.*, **21**, 237, 1983)

16. (a) As the molecular weight of a barium-neutralized carboxyl-terminated poly-diene increases, the stress and modulus at constant strain increase. With zirco-nium-neutralized polymer, however, the opposite is observed; that is, stress and modulus decrease at constant strain. Suggest a reason. (b) Predict the stress-strain behavior of barium-, aluminum-, and zirconium-neutralized polymer if the molecular weight of the telechelic polymer is kept constant. (M. R. Tant et al., *Polymer*, **27**, 1815, 1986)

17
Natural polymers

17.1 Introduction

The macromolecular chemistry described in the previous fourteen chapters reflects an impressive range of research and development built on a foundation established by the brilliance and foresight of a number of great scientists. When one delves into the fascinating realm of natural polymers, however, one cannot escape the conclusion that our own knowledge of polymer synthesis is, at best, rudimentary. Consider, for instance, that complex chemical reactions involving natural polymers within the human body were prerequisite to the intellectual processes that led to the pioneering work in synthetic polymer chemistry. An entire field of chemistry—biochemistry—is very much involved with these natural polymers; hence this chapter serves only as a necessarily brief introduction to the topic, with particular emphasis being placed, where pertinent, on their application or modification for commercial use.

Because of the physiological activity of natural polymers, three major classifications have occupied most of the literature in this field: polysaccharides, proteins, and polynucleotides. These are treated in Sections 17.2, 17.3, and 17.4 of this chapter. They are listed in Table 17.1. In keeping with our

Table 17.1. Physiologically active natural polymers

Polymer	Description	Representative commercial applications
Polysaccharides	Polyacetals and/or polyketals of monosaccharides	Foodstuffs, plastics, fibers, structural materials, adhesives, coatings
Proteins	Polyamides of α-amino acids	Foodstuffs, fibers, pharmaceuticals agents
Polynucleotides (nucleic acids)	Polyesters of phosphoric acid and nucleosides (ribose or 2-deoxyribose with attached heterocyclic amine bases)	Genetic engineering (production of pharmaceutical agents, agricultural chemicals, etc.)

Table 17.2. Miscellaneous natural organic polymers

Polymer	Description	Representative commercial applications
Rubber	Polyterpene having a 1,4-polydiene structure	Elastomers (principally tires), coatings
Lignin	Network aromatic/aliphatic oxygenated polymer	Fuel, phenolic adhesives, oil-well drilling mud components
Humus	Similar to lignin	Boiler scale remover, pigment extenders, emulsifiers
Coal	Very complex; similar to lignin, but also containing fused ring structures	Fuel
Asphaltenes (bitumens)	Cycloaliphatic and aromatic fused-ring system	Highway surfacing, roofing, flooring
Shellac	Polyester of long-chain acids and hydroxy acids	Coatings, laminating resins, inks, sealing wax
Amber	Fossilized terpenoid resin	Decorative items, jewelry

emphasis on commercial applications, this table also gives some end uses that go beyond the obvious biochemical significance of these polymers.

In addition to the three major classifications, there are a variety of natural organic polymers (Table 17.2), many of which have been employed commercially for a long time. We begin our discussion in Section 17.2 with a brief look at these "miscellaneous" natural polymers. Also included are synthetic polymers derived from the naturally occurring *tall oil*.

17.2 Miscellaneous natural polymers

17.2.1 *Rubber*

Rubber[1–4a] is the most important and widely used natural polymer from the industrial standpoint. It was used for centuries by the Mayans in the Western Hemisphere before it was introduced into Europe by Columbus. The Mayans obtained the material from a tree they called *caoutchouc* ("weeping tree"), a term still used to denote the polymer in many countries. It was Joseph Priestley, however, who coined the term *rubber* when he noted that caoutchouc could be used to rub out pencil marks.

Rubber is a polyterpene synthesized naturally by the enzymatic polymerization of isopentenylpyrophosphate. Its repeating unit (1) is the same as that of 1,4-polyisoprene. Isoprene is, in fact, a major degradation product of rubber, being identified as such in the early 1860s.[5] The chemistry of diene polymers is covered in detail in earlier chapters and will not be repeated here; however, some of the structural features of the various forms of rubber are of interest.

$$\left[\begin{matrix} & CH_3 \\ & | \\ -CH_2C&=CHCH_2- \end{matrix} \right]$$

1

The principal form of natural rubber, which consists of 97% *cis*-1,4-polyisoprene, is known as *Hevea rubber*. It is obtained by tapping the bark of a tree (*Hevea brasiliensis*) that grows wild in South America and is cultivated in other parts of the world. It is also found in a variety of shrubs and small plants, including milkweed and dandelion. One of the most important shrubs is guayule (pronounced *wy-u-lee*) which grows well in arid climates such as are found in northern Mexico and the southwestern United States. Because guayule is considered a viable source of Hevea-type rubber for the future, high-yield varieties are being developed using tissue culture techniques.

Almost all natural rubber is harvested as a latex consisting of about 32 to 35% rubber and about 5% of other compounds, including fatty acids, sugars, proteins, sterols, esters, and salts. Guayule rubber is an exception, being obtained by pulping and parboiling the plant before refining. The cellulosic crop residue (bagasse) is a potential source of fermentation alcohol. Rubber is of very high molecular weight (averaging about 1 million) and is amorphous, although it becomes randomly crystallized at lower temperatures.

The latex can be converted to foam rubber by mechanical aeration followed by vulcanization. Rubber gloves and balloons are usually made by coating latex on forms before vulcanization. Most latex is coagulated (for example, with acetic acid) and used in bulk form. Most Hevea rubber (about 65%) is used in tire manufacture, but it is also found in a host of commercial products including footwear, seals, weather stripping, shock absorbers, electrical insulation, sports accessories, and so on. All these utilize rubber in vulcanized form. One of the few applications of unvulcanized rubber is in the form of crepe which, because of its excellent abrasion resistance, is used for shoe soles.

Another form of natural rubber is *gutta-percha*, also obtained in latex form from trees (for example, *Palaquium oblongifolium* and similar trees mainly indigenous to Southeast Asia). Gutta-percha has the *trans*-1,4-polyisoprene structure. It is much harder and less soluble than Hevea rubber and exists in crystalline form.

Other forms of rubber related in structure to gutta-percha are *balata* and *chicle*, obtained from trees in Mexico and South and Central America. Uses of gutta-percha and related polymers include wire coatings, impregnants for textile belting, and varnishes. Until ionomers captured the market, gutta-percha was the materal of choice for golf-ball covers.

17.2.2   *Lignin, humus, coal, and kerogen*

Wood consists almost entirely of three materials: the polysaccharides cellulose and hemicelluloses, and lignin. Lignin[4a,7–9] is the "cement" holding the cellulose fibrils together and providing much of wood's dimensional stability. Constituting about 25 to 30% of wood, lignin is an extremely abundant polymer that has yet to reach its potential with regard to polymer applications. At the present time most of the lignin produced in pulping operations is burned as fuel at the pulping site.

The structure of lignin varies according to source, but an approximation of a segment of softwood lignin (Figure 17.1) illustrates its complexity.[10] The molecular weight of native lignin is presumed to be very high, but because separation from cellulose inevitably leads to degradation, it is impossible to tell how high. Because it contains a preponderance of activated benzene rings, degraded lignin reacts readily with formaldehyde, which has led to limited commercial development in the area of plywood adhesives (Chapter 14). Lignin sulfonates obtained from wood pulping are also used as adhesives, asphalt extenders, and oil-well drilling mud additives. Efforts have continued over the years to modify lignin for possible use as engineering plastics.[11–13] Reaction with propylene oxide, for example, yields hydroxypropyl derivatives that have been converted to thermosetting polyurethanes.

Humus and coal are related in structure to lignin. Humus[14,15] constitutes the organic component of soils and natural waters that is relatively resistant to biodegradation. It is also found in fossil fuels. Humus is presumed to originate from lignin, but unlike lignin it contains significant numbers of carboxyl groups. For convenience, humus is classified according to whether it is water soluble (*fulvic acid*), base soluble but water insoluble (*humic acid*), or base insoluble (*humin*). All three have essentially the same structural features, but they differ in molecular weight and degree of functionality. Humic materials chelate metals and hydrogen bond strongly to proteins and polysaccharides. Because humus is an excellent scavenger of metals, and because it plays a role in soil drainage and movement of water and nutrients, humus is of considerable interest to environmentalists. Industrially, humic materials find use as emulsifiers, pigment extenders, and boiler scale removers.

Coal is an extremely complex polymeric material whose structure varies widely with source.[16] Although it contains structural features analogous to those of humus and lignin, which is to be expected given its derivation from vascular plants, coal also contains clusters of fused aromatic rings linked with hydroaromatic and alicyclic units. Oxygen-, nitrogen-, and sulfur-containing heterocyclic units are also present. Dispersed in this porous network structure is a complex mixture of substances called *vitrinite*. Coal is used as a fuel directly, or as a feedstock for the manufacture of liquid and gaseous fuels.

Related to coal, but varying widely in structure according to source, is *kerogen*, the organic constituent of oil shale.[17] Thermal cracking of kerogen to low-molecular-weight compounds represents a potential alternative to petroleum-based fuels.

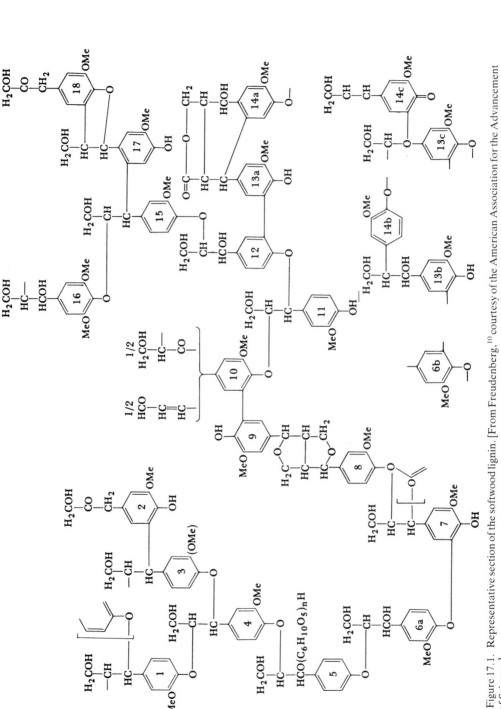

Figure 17.1. Representative section of the softwood lignin. [From Freudenberg,[10] courtesy of the American Association for the Advancement of Science.]

17.2.3   *Asphaltenes*

Also called *bitumens*, asphaltenes[18,19] are resinous materials widely used in highway construction as aggregate binders, as binders for roofing and flooring compositions, and for waterproofing buildings. They occur in natural deposits, but are obtained principally from the residue of petroleum distillation. Asphaltenes consist primarily of polynuclear aromatic and cycloaliphatic ring systems with aliphatic side chains, and vary in molecular weight from a few thousand to several hundred thousand. Although the structure varies considerably with source, a hypothetical lamellar-type structure has been proposed[18] (Figure 17.2) on the basis of spectroscopic and x-ray analysis. A micellar asphaltene unit is shown in Figure 17.3.

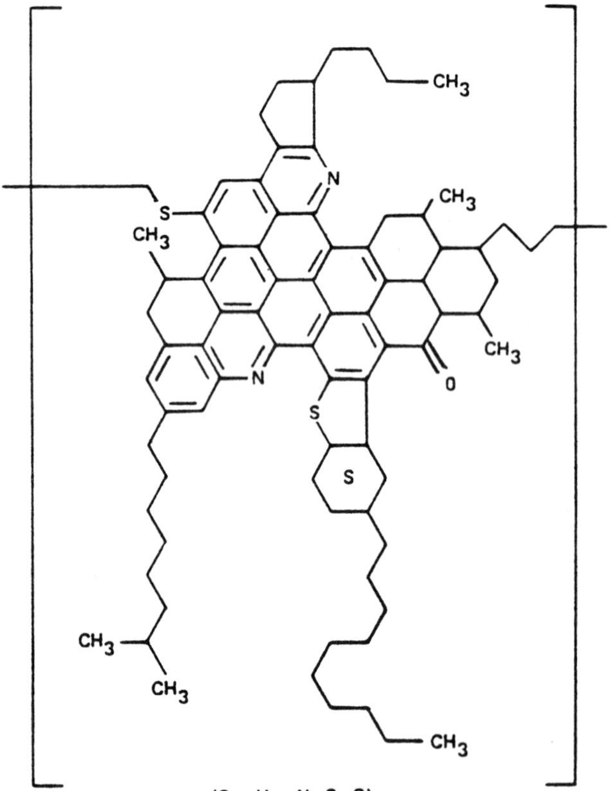

$(C_{79} H_{92} N_2 S_2 O)_3$
mol wt. 3449

Figure 17.2. Hypothetical structure of a petroleum asphaltene. [From Speight and Moschopedis,[18] copyright 1981. Reprinted by permission of the American Chemical Society.]

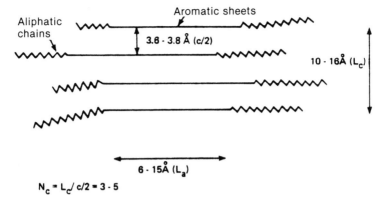

Figure 17.3. Representation of an asphaltene from x-ray analysis. (c/2 = interlamellar distance; L_a = layer diameter; L_c = unit cell height; N_c = number of lamellae contributing to the micelle) [From Speight and Moschopedis,[18] copyright 1981. Reprinted by permission of the American Chemical Society.]

17.2.4 Shellac

A number of natural resins have been used for varnishes and molding compounds, with *shellac* being the best known. Shellac[20–23] is secreted by the lac insect (*Kerriar paca*), found in southern parts of Asia. The resin, which is collected by scraping the branches of shellac-encrusted trees, consists of a very complex mixture of crosslinked polyesters derived from hydroxy acids, principally aleuritic acid (9,10,16-trihydroxyhexadecanoic acid, **2**). Gas chromatographic analysis of the products of chemical degradation of shellac has also shown the presence of several saturated and unsaturated long-chain aliphatic acids together with other hydroxy-substituted acids and nonaliphatic compounds.[23]

$$HO\diagup\!\!\!\diagdown\!\!\diagup\!\!\diagdown\!\!\diagup\!\!\diagdown\!\!\underset{\underset{\textstyle OH}{|}}{\overset{\overset{\textstyle OH}{|}}{C}}\!\!\diagdown\!\!\diagup\!\!\diagdown\!\!\diagup\!\!\diagdown\!\!\diagup CO_2H$$

2

Following a long history of use in Asia, shellac was introduced into Europe in the tenth century. Phonograph records were made of shellac composites until the late 1940s, when vinyl polymers captured the market. Blends of shellac with other resins, especially phenol–, urea–, and melamine–formaldehyde, find occasional use. Given the fact that about 150,000 insects are needed to produce one pound of resin, it is hardly surprising that shellac has been largely supplanted with synthetic polymers.

17.2.5   Amber

Amber is the generic name for all fossil resins[24,25] secreted in prehistoric times by certain types of evergreens, now extinct. It consists of complex mixtures of sesquiterpenoids, diterpenoids, and triterpenoids that have undergone polymerization and molecular reorganization during fossilization. Phenolic units may also be present. Baltic coast deposits, which contain about 8% succinic acid, are often called *succinite*. Amber is amorphous, and its infrared spectrum closely resembles those of nonvolatile rosins from extant pine trees.

Interesting from the biological perspective is the fact that amber often serves as a repository for a variety of extinct species of insects.

17.2.6   Tall oil–derived polymers

Tall oil constitutes about 3% of dry wood and is an important by-product of wood pulping. (The name is derived in part from the Swedish *Tallolja*, which combines "tall pine" and "oil.") The oil is not polymeric, but it has been adapted for polymer synthesis. Tall oil represents one of the three classifications of *naval stores*, an archaic term that has its roots in early shipbuilding when pitch and rosin were used for caulking.[26] The other two classifications are *gum* and *wood* naval stores, the former collected from cuts in tree bark, the latter extracted from shredded tree stumps. Turpentine is another major constituent of naval stores.

Tall oil is made up of two distinct components: *rosin* and *tall oil fatty acids*. Rosin is primarily a mixture of 20-carbon fused-ring monocarboxylic acids of which abietic acid (**3**) and levopimaric acid (**4**) are representative. Varnishes made from rosin and drying oils were developed in ancient times. Esterification of rosin with glycerol or pentaerythritol leads to products known as *ester gums* that are used in adhesives and lacquers.[27]

HO$_2$C

3

HO$_2$C

4

Petroleum shortages in recent years have led to renewed interest in rosin as a renewable polymer feedstock.[28] Diels–Alder reaction of **4** with maleic anhydride (17.1), for example, results in an acid–anhydride, **5**, which can be converted to polyamideimide[29,30] (**6**) by reaction with diamine (17.2).

Tall oil fatty acids, which are separated from rosin by distillation, consist mainly of 18-carbon acids, particularly oleic and linoleic acids. Oxidative

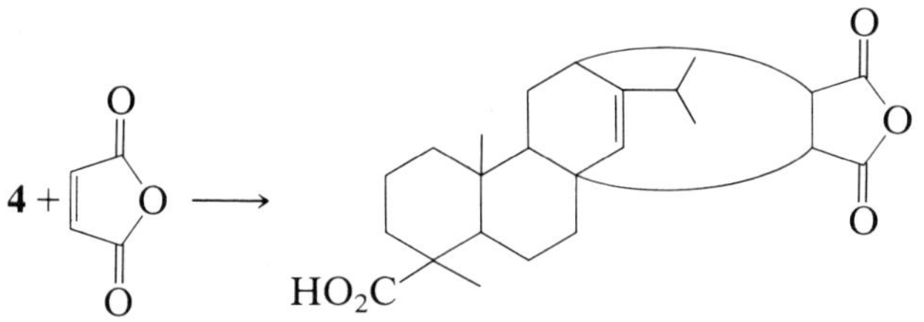

$$4 + \quad \longrightarrow \qquad (17.1)$$

5

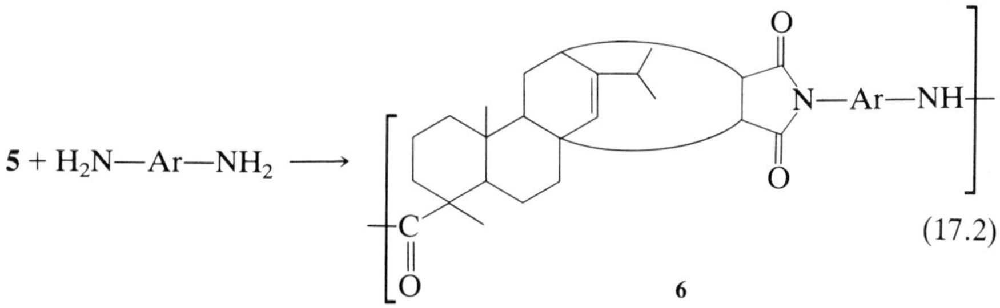

$$5 + H_2N{-}Ar{-}NH_2 \quad \longrightarrow \qquad (17.2)$$

6

cleavage of the latter gives nonanedioic (azeleic) acid, used in production of nylon 69 (Chapter 13). Epoxidized linoleic esters are used as vinyl plasticizers. Treatment of linoleic acid with an activated clay results in a complex series of reactions involving isomerization to a conjugated diene and free radical and Diels–Alder dimerization and trimerization. The products, called *dimer acids* and *trimer acids*,[31] are used as curing agents for epoxy and phenolic resins. Polyamides of dimer acids act as printing ink binders and thixotropic agents.

17.3   Polysaccharides

17.3.1   Cellulose

In terms of chemical structure, cellulose[32,33] (**7**) is among the simplest of the natural polymers in that it consists of a single repeating unit, D-glucose,

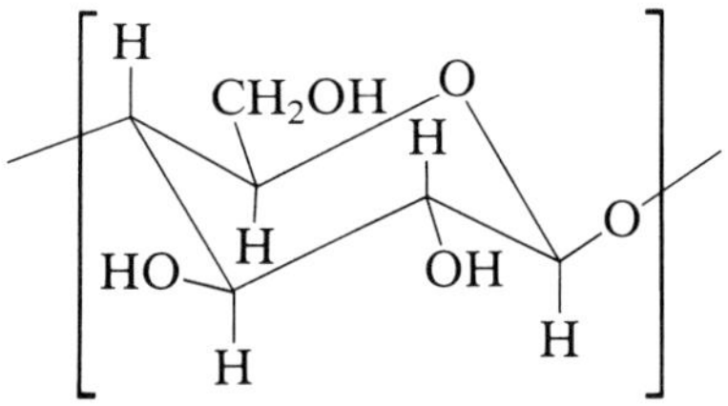

7

Table 17.3. Cellulosic fibers[a]

Fiber	Approximate % cellulose	Major uses
Seed hair fibers		
Cotton	85–90	Textiles
Kapok	65	Thermal and acoustic insulation; mattress, cushion, and life preserver stuffing
Coir[b]	40	Upholstery, sacking, matting, cordage
Bast fibers		
Flax	80–90	Textiles (linen)
Hemp	78	Twine, rope, netting, sacking
Jute[b]	65	Textiles, cordage, sacking, linoleum backing
Ramie	100	Paper (special types), gas mantles, cordage
Leaf fibers		
Manila hemp[c]	65	High-strength paper, muslin, cordage, rope
Sisal hemp[c]	65	Rope, twine

[a] Data from J. R. Scott and W. J. Roff, *Handbook of Common Polymers*, CRC Press, Cleveland, 1971, pp. 140–142.

[b] Coir and jute contain 40 and 10% lignin, respectively.

[c] Hemp contains about 22% pectin.

linked through carbons 1 and 4 by β linkages.* This means that the ring substituents other than hydrogen, including the bonds linking the glucose rings together, are all equatorial.

Cellulose is found widely in nature,[32a] being the major constituent of the cell wall of plants and averaging about 50% in a typical wood. It is the most abundant organic compound on the planet. Cellulose is also the principal constituent of a variety of natural fibers that occur as seed hairs surrounding the seeds of certain plants (for example, cotton), as the fibrous inner bark of others (bast fibers), and as the fibrous constituent of certain leaf stalks (leaf fibers). The amount of cellulose in the fibers varies according to source and is usually associated with such materials as water, wax, pectin, proteins, lignin, and mineral substances. The approximate percentage of cellulose in these various fibers and the principal commercial uses are given in Table 17.3. The $\overline{\text{DP}}$ of cotton cellulose is about 15,000, compared with about 10,000 for wood

---

*The monosaccharide D-glucose exists in two cyclic forms, α and β. In the β form, all hydroxyl substituents and the hydroxymethyl group are in the equatorial position, whereas in the α form, the hydroxyl at C-1 (the hemiacetal hydroxyl group) is axial.

β-D-Glucose α-D-Glucose

cellulose. Separation of the latter from lignin results in a decrease in $\overline{DP}$ to about 2600.[34]

Certain types of bacteria also synthesize cellulose, which can be obtained as a continuous film by cultivating the bacteria in a glucose solution. Cellulose is also secreted by such marine chordates (tunicates) as the sea squirt.

The polymer, known as *native cellulose* (to distinguish it from regenerated cellulose), is linear, highly crystalline, and characterized by a very high degree of intramolecular and intermolecular hydrogen bonding. As a result, cellulose is not thermoplastic but decomposes before melting. Despite its large number of hydroxyl groups, cellulose is insoluble in water and most other common solvents, although it will succumb to certain solvent mixtures. Solutions of metal complexes such as copper(II)-ammonia[35] (*Schweitzer's reagent*) or copper (II)-diamine[36] dissolve cellulose, presumably by forming complexes of copper(II) with the hydroxyl groups of the polymer. Among the solvent systems of more recent vintage[37,38] are LiCl-dimethylacetamide, dimethyl sulfoxide–paraformaldehyde, amine oxides, and phosphoric acid.

Cellulose swells in water and especially in concentrated base. The base-swollen polymer, known as *alkali cellulose* or *soda cellulose*, is used to prepare regenerated cellulose, as described in the next section. The process of treating cotton with aqueous base, and then washing out the base, is known as *mercerization*. Mercerized cotton has a higher luster than natural cotton, is less dense, and is somewhat less crystalline. It also has improved dyeability.

17.3.2 Regenerated cellulose

Some important commercial products are prepared from cellulose by first dissolving the polymer, then precipitating it from solution. The regenerated cellulose is different from native cellulose because extensive degradation takes place during the process of dissolution and the final product is usually less crystalline.

The most widely used regeneration method is called the *xanthate process* or *viscose process*.[32b] Alkali cellulose, usually prepared from wood pulp (that is, wood separated from lignin by treatment with alkali or bisulfite solution), is allowed to age, during which time the molecular weight is reduced. The exact role of the alkali is not known with certainty, but it is believed that sodium ions are held interstitially, while hydroxide ions are bound by hydrogen bonding. The reduction in molecular weight probably arises mainly from oxidative degradation. Addition of carbon disulfide then converts cellulose (Cell–OH) to the xanthate (**8**) (17.3). Not all the OH groups are converted, just enough to make the polymer soluble in aqueous alkali. The basic

$$\text{Cell—OH} + \text{CS}_2 + \text{NaOH} \longrightarrow \text{Cell—O—}\overset{\displaystyle \overset{\text{S}}{\|}}{\text{C}}\text{—S}^-\text{Na}^+ + \text{H}_2\text{O} \quad (17.3)$$

$$\mathbf{8}$$

solution, known as *viscose*, is orange in color, is viscous, and has an unpleasant odor arising from volatile sulfurous by-products of the xanthation reaction.

The viscose is allowed to age further to allow a series of complex reactions to occur that results in an increase in viscosity. The solution is then extruded as a filament into a bath containing dilute sulfuric acid, sodium bisulfite, and zinc sulfate to hydrolyze the xanthate back to cellulose (17.4). The filaments are stretched to orient the molecules and increase crystallinity. Fibers prepared this way are known as *viscose rayon*, a product useful for textiles and tire cord. When viscose is extruded through a slit die into an acid bath, a film of regenerated cellulose known as *cellophane* results. Cellophane is plasticized by washing it in a bath containing glycerol during processing.

$$\text{Cell}-\text{O}-\overset{\overset{\textstyle S}{\|}}{\text{C}}-\text{S}^-\text{Na}^+ \xrightarrow{\text{H}^+} \text{Cell}-\text{OH} + \text{CS}_2 + \text{Na}^+ \qquad (17.4)$$

Regenerated cellulose is also prepared from solutions in Schweitzer's reagent (the *cuprammonium process*) by hydrolysis with acid, and by saponifying cellulose acetate.

17.3.3　Derivatives of cellulose

A large number of cellulose derivatives have been synthesized,[32c,39-44] including esters of inorganic and organic acids (apart from xanthate esters), ethers, and graft copolymers, although relatively few have achieved commercial importance. Those that have include the nitrate, acetate, propionate, and butyrate esters and the methyl, ethyl, and hydroxyethyl ethers. In addition, crosslinking reactions have been widely used for treating textiles.

17.3.3.1　Cellulose esters　As was mentioned in Chapter 1, cellulose nitrate was one of the very first commercial polymers. It was developed originally during the last century and eventually found use in the manufacture of explosives (*guncotton*), photographic film, lacquers, textile fibers, and some molding applications (for example, as an ivory substitute for billiard balls). Cellulose nitrate plasticized with camphor was known by the trade name Celluloid.

The polymer is also known as *nitrocellulose*, but this is a misnomer, since it contains no nitro groups. Cellulose nitrate is prepared from cellulose and nitric acid (17.5), usually in the presence of sulfuric acid, although phosphoric or acetic acids may also be used. Some chain degradation normally occurs during processing, also some sulfate ester formation, but the latter may be removed by prolonged heating with dilute nitric acid or alcohol. The degree of nitration varies according to application. Cellulose containing 10 to 11% nitrogen is used mainly for lacquers or plastics, whereas that containing about

$$\text{Cell}-\text{OH} + \text{HONO}_2 \rightleftharpoons \text{Cell}-\text{ONO}_2 + \text{H}_2\text{O} \qquad (17.5)$$

13.5% nitrogen (guncotton) is used for explosives. (Completely nitrated cellulose contains 14.14% nitrogen.)

Cellulose nitrate is an extremely flammable material and not particularly stable; hence it has been replaced in most applications by other polymers in more recent years. Some work has also been done on cellulose sulfate and phosphate, the major concern being to improve dyeability of cellulosic fibers in the former case and to reduce flammability in the latter.

Organic esters are much more important at the present time. Cellulose acetate of varying degrees of substitution is prepared by first treating cellulose with acetic acid, then with acetic anhydride (17.6) and a mineral acid catalyst. The fully acetylated product is subsequently reacted with water to achieve the desired degree of substitution. Mild hydrolysis gives an almost completely acetylated cellulose known as *triacetate*. Further hydrolysis gives an acetone-soluble product called *secondary acetate*. Both are used in plastics and fiber applications, in the former case in combination with appropriate plasticizers. As a textile fiber, cellulose acetate is called *acetate rayon*. The $\overline{DP}$ of cellulose acetate is much lower than that of the cellulose feedstock (averaging about 200–250) because of cleavage of glucosidic linkages by the acidic esterification catalyst.

$$\text{Cell—OH} + (CH_3CO)_2O \xrightarrow{H^+} \text{Cell—O}\overset{\displaystyle O}{\overset{\|}{C}}CH_3 + CH_3CO_2H \quad (17.6)$$

Cellulose propionate and cellulose butyrate, or, more commonly, acetate–propionate and acetate–butyrate mixtures, are the only other commercially important cellulose esters. The mixed esters are generally more soluble and have better impact resistance than cellulose acetate. They are widely used in molding compositions, motion picture film, and coatings.

Some miscellaneous longer chain esters have been prepared, also esters of unsaturated acids (for example, methacrylic and crotonic), halogenated acids, and dibasic acids. The unsaturated esters are of interest from the standpoint of crosslinking and graft copolymerization.

17.3.3.2 Cellulose ethers Cellulose ethers[40] are prepared by pretreating cellulose with base (alkali cellulose), then reacting with a halogen com-pound (17.7) or epoxide (17.8). Methyl sulfate is also used for preparing methyl ethers. Some of the cellulose ethers of interest are methyl ($R = CH_3$), ethyl ($R = CH_2CH_3$), hydroxyethyl ($R = CH_2CH_2OH$), hydroxypropyl ($R = CH_2CH(OH)CH_3$), carboxymethyl ($R = CH_2CO_2H$), aminoethyl ($R = CH_2CH_2NH_2$) and benzyl ($R = CH_2C_6H_5$).

$$\text{Cell—OH} + NaOH + RCl \longrightarrow \text{Cell—OR} + NaCl + H_2O \quad (17.7)$$

$$\text{Cell—OH} + \overset{\displaystyle O}{CH_2\text{—}CH_2} \xrightarrow{NaOH} \text{Cell—OCH}_2CH_2OH \quad (17.8)$$

Ethylcellulose is most widely used, principally in plastics applications similar to those of cellulose acetate. Methylcellulose is water soluble* and is used as a food-thickening agent and as an ingredient in certain adhesive, ink, and textile-finishing formulations. Carboxymethyl cellulose, hydroxyethyl cellulose, and hydroxypropylcellulose are also used in adhesives and textile finishes and as emulsifiers (for example, in latex paints.) Hydroxypropylcellulose sandwiched between two water-insoluble films has recently found use in the manufacture of degradable bottles. When the outer layer is peeled off, the hydroxypropylcellulose dissolves rapidly, thus reducing the solid-waste problem normally associated with nonreturnable bottles.

17.3.3.3 Graft copolymers A large number of graft copolymers of cellulose have been prepared,[41] and although there has been little commercial exploitation, there is considerable interest from the standpoint of modification of cellulosic fibers. Vinyl grafts[45a] can be initiated with the usual free radical sources, including peroxides, redox initiators, radiant energy, or even mechanical rupturing (mastication). Cellulose derivatives can also be grafted. Cellulose acetate grafted with vinyl acetate using a cobalt-60 γ-ray source is one example of a graft copolymer that has achieved some commercial interest.

Free radical grafting can also be accomplished by incorporating active chain transfer groups into the cellulose backbone.[45b] Mercaptoethylcellulose (**9**), for example, undergoes grafting in the presence of a vinyl monomer under conditions of vinyl polymerization (17.9).

$$\text{Cell—OCH}_2\text{CH}_2\text{SH} \xrightarrow[\text{peroxide}]{\text{CH}_2=\text{CHR}} \text{Cell—OCH}_2\text{CH}_2\text{SCH}_2\overset{|}{\underset{R}{\text{CH}}}\text{w} \quad (17.9)$$

9

Examples of nonvinyl grafting include the reaction of cellulose with ethylene oxide (17.10) (to give a product marketed under the trade name Ethylose); and with β-propiolactone to give mixed ether and ester side chains (17.11).

$$\text{Cell—OH} + n\overset{\text{O}}{\overset{/\backslash}{\text{CH}_2\text{CH}_2}} \longrightarrow \text{Cell—O}\overset{}{\underset{n-1}{\text{[CH}_2\text{CH}_2\text{O]}}}\text{CH}_2\text{CH}_2\text{OH}$$

$$(17.10)$$

$$\text{Cell—OH} + \underset{\underset{\text{O——C=O}}{|\quad\quad|}}{\text{CH}_2\text{—CH}_2} \longrightarrow \text{Cell—O}\left[\text{CH}_2\text{CH}_2\overset{\overset{\text{O}}{\|}}{\text{CO}}\right]_x\text{H} \quad \text{or}$$

$$(17.11)$$

$$\text{Cell—O}\left[\overset{\overset{\text{O}}{\|}}{\text{CCH}_2\text{CH}_2\text{O}}\right]_y\text{H}$$

---

* Methylcellulose is unusual in that it is much more soluble in cold water than in hot. Some other cellulose derivatives exhibit similar inverse solubility–temperature properties.

17.3.3.4 Crosslinked cellulose In recent years the treatment of cotton with crosslinking agents to impart crease-resistant or "durable press" properties has become a major industry.[45c,46] Most of those presently in use are hydroxymethyl derivatives of amides, prepared from formaldehyde and amide. Some representative examples include dimethylolurea (**10**), 1,3-bis(hydroxymethyl)-2-imidazolidinone (**11**) (known simply as dimethylolethyleneurea), dimethylol uron (**12**), and dimethylol alkyl carbamate (**13**). The reaction of **10** with cellulose is given (17.12) in simplified form because there is probably some polymerization of the urea and formaldehyde in addition to reaction with cellulose. The hydroxymethyl compounds are water soluble and are applied to the fabric from aqueous solution using an acid catalyst.

$$
\underset{\textbf{10}}{HOCH_2NH\overset{\overset{\displaystyle O}{\|}}{C}NHCH_2OH}
\qquad\qquad
\underset{\textbf{11}}{HOCH_2-N\underset{\diagdown\diagup}{\overset{\overset{\displaystyle O}{\|}}{}}N-CH_2OH}
$$

$$
\underset{\textbf{12}}{HOCH_2-N\underset{O}{\overset{\overset{\displaystyle O}{\|}}{}}N-CH_2OH}
\qquad\qquad
\underset{\textbf{13}}{RO\overset{\overset{\displaystyle O}{\|}}{C}-N(CH_2OH)_2}
$$

Cell—OH + **10** + HO—Cell $\longrightarrow$

$$
Cell-OCH_2NH\overset{\overset{\displaystyle O}{\|}}{C}NHCH_2O-Cell \qquad (17.12)
$$

Other crosslinking agents include diepoxides, epichlorohydrin (which behaves as a diepoxide), aldehydes, acetals (prepared from formaldehyde and polyol), and tetraoxane, the cyclic tetramer of formaldehyde, which undergoes ring-opening in the presence of acid to yield polyoxymethylene crosslinks (**14**).

$$
Cell-O\text{\textthreequartersemdash}[CH_2O]_n\text{\textthreequartersemdash}O-Cell
$$

14

17.3.4 Starch

From the standpoint of chemical structure, starch[47] differs from cellulose in two major ways: the glucose rings are linked together through carbons 1 and 4 by α rather than β linkages, and considerable chain branching occurs through carbon 6. But, like cellulose, complete hydrolysis yields D-glucose.

Starch occurs naturally as minute granules in the roots, seeds, and stems of numerous types of plants, including corn, wheat, rice, millet, barley, and potatoes; it constitutes the main carbohydrate reserve of the plants. It consists of two polysaccharides, *amylose* and *amylopectin*, which can be separated according to differences in solubility. Amylose (**15**) is mainly linear in structure, with molecular weight ranging from about 30,000 up to 1 million, although upper limits of 200,000 to 300,000 are more common. Apart from the α linkage (axial C—O bond at carbon 1), amylose therefore resembles cellulose.

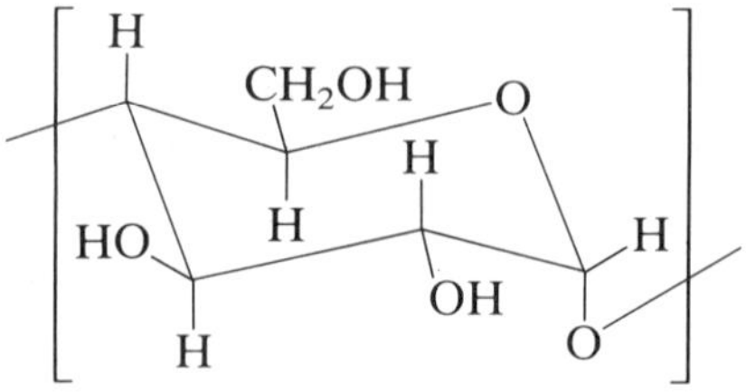

15

Amylopectin is highly branched through carbon 6 and has molecular weights of over 1 million. It is believed to consist of chains of 20 to 25 glucose units linked through carbons 1 and 4, as in amylose, but with the chains connected to each other through the 1,6 linkage. A partial structure (**16**) for amylopectin is given. There is usually about three times as much amylopectin as amylose in natural starch, although much higher proportions of either occur in certain plants.

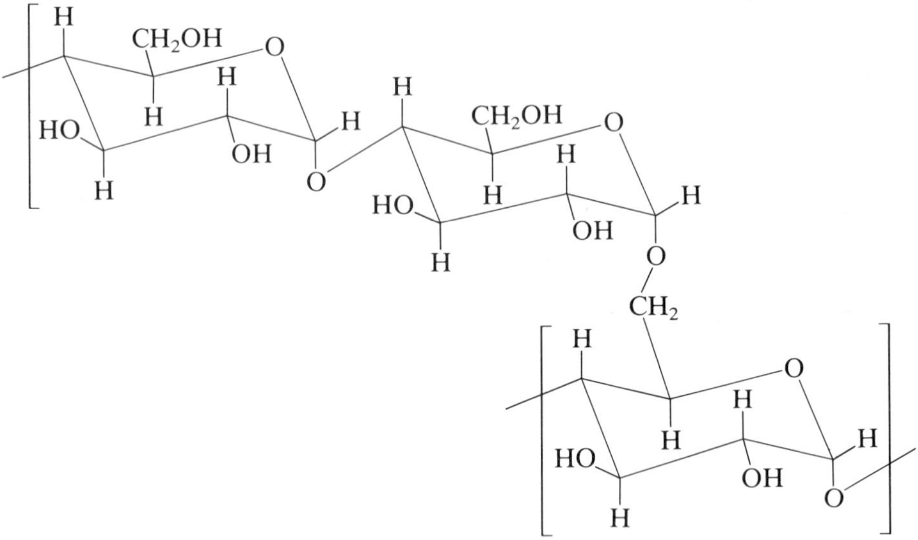

16

Because of its highly branched structure and permeability, starch is not suitable for plastics or fiber applications as is the case with cellulose. Unlike cellulose, however, it is useful as a food because animals produce the enzymes necessary to catalyze hydrolysis of α linkages. Apart from its utility in the food and fermentation industries, starch has also found some use in adhesive formulations (for example, flour paste) and as a sizing or glazing agent in the paper and textile industries.

17.3.5 *Derivatives of starch*

As might be expected, the types of modifications applied to cellulose have also been applied to starch. Starch nitrate is a useful explosive, and starch acetate, which forms clear films, has found commercial use as a paper and textile sizing agent. Amylose triacetate forms fibers similar to those of cellulose acetate, but they are not as strong and offer no special advantages. Ether derivatives of starch, particularly hydroxyethyl ethers, have been used as adhesives, and an allyl ether has been used as an air-drying coating.

A large number of graft copolymers of starch have also been reported.[45b] The above-mentioned allyl ether derivative, for example, has been reacted with styrene under free radical conditions to give styrenated starch, and other vinyl monomers have been reacted directly by the same methods described for cellulose. There is considerable interest in graft copolymers of starch because of their potential as biodegradable packaging materials and agricultural mulches (see Chapter 4, Section 4.7). A practical application of starch crosslinking using epichlorohydrin is the treatment of rice granules to make them more resistant to breaking down in canned soups.

17.3.6 *Other polysaccharides*

A number of other polysaccharides exist in various natural sources,[27,48] some related in structure to cellulose or starch, others having uniquely different structures. *Glycogen* is a high-molecular-weight polymer, structurally very similar to amylopectin, although more highly branched. It is the reserve carbohydrate of animals and is partially bound to proteins. *Dextran*, a high-molecular-weight polysaccharide synthesized from sucrose or starch by certain bacteria, consists of glucose units linked through carbons 1 and 6 by α linkages, and with chain branches through carbons 1 and 3. Partially hydrolyzed dextran has been used as a blood plasma substitute.

Chitin (**17**) is a polysaccharide related in structure to cellulose, except that it has an acetylated amino group at carbon 2. Chitin is found in the shells of crustacea, in the horny exoskeletons of insects and spiders, and in some types

17

of fungi. Chitin and its deacetylated derivative, *chitosan*, are available commercially in film and fiber form, but in relatively small amounts because of the lack of a plentiful commercial source.

A number of polysaccharidelike materials are obtainable from seaweed or algae. *Alginic acid* (**18**), a polymer similar to cellulose, is linked through carbons 1 and 4 and has a carboxyl function attached to carbon 5. The polymer also has some units with the configuration reversed at carbon 5.

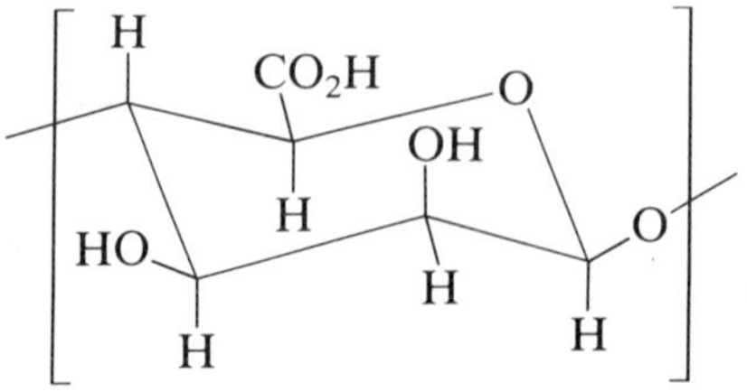

18

Alginic acid, which is found in certain types of brown seaweed, is used (as the sodium salt) in adhesive and sizing formulations and in some fiber applications. (One interesting use is in the manufacture of water-soluble disposable military parachutes.) *Carrageenin* and *agar*, obtainable from red seaweed or red algae, have similar structures. The former consists of two separable fractions, one (**19**) having galactose units linked through carbons 1 and 3 or 1 and 4 with sulfate groups at carbon 2; the other (**20**) having galactose units linked 1,4 with an ether group linking carbons 3 and 6. Agar is similar except that it has fewer sulfate groups and the 1,4-linked ether-containing fraction is derived from L-galactose rather than the D-isomer.

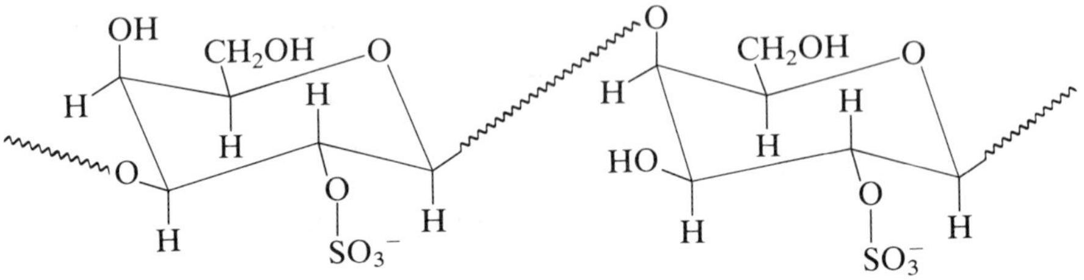

19

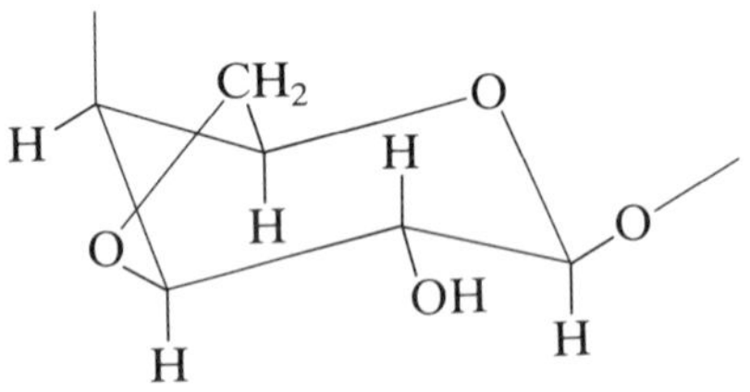

20

Other polysaccharides include *inulin*, a polymer found in certain plants that has fructose rings linked together with a glucose unit at one end of the chain; some natural gums such as *gum arabic* and *gum tragacanth* that have complex structures that yield several monosaccharides on hydrolysis; and *hemicelluloses*, which are found with cellulose in the cell walls of plants. Typical ingredients of hemicelluloses are *xylan* (like cellulose, but with the CH_2OH group replaced with H); *mannan*, which gives mannose on hydrolysis; *araban*, composed of arabinose units; and *galactans*, consisting of a complex mixture of monosaccharides. Hemicelluloses are amorphous and of much lower molecular weight than cellulose, having $\overline{DP}$'s of only a few hundred; nevertheless, they represent one of the three major constituents of wood.

17.4 Proteins

17.4.1 Amino acids, polypeptides, and proteins

Proteins are the basic building blocks of animals and are, therefore, of primary interest to biochemists.[4b,49,50] They are the major constituent of skin, nerve tissue, tendons, muscle, hair, and blood, and they exist in all living cells. Because proteins are also the principal constituent of enzymes and many hormones, they actually control body functions. The study of proteins is fundamental to the field of biochemistry, therefore we shall discuss only some of the basic principles of protein structure and synthesis. More detailed information can be obtained from any up-to-date biochemistry text.

$$\left[NH-\underset{\underset{R}{|}}{C}H-\overset{\overset{O}{\|}}{C} \right]$$

21

Proteins are polyamides (**21**) with α-amino acids being the monomer units. They are far more complex than polysaccharides because a large number of α-amino acids exist naturally which can form an almost limitless number of sequential arrangements. (Methods of determining amino acid sequences are discussed in the next section.) Unlike polysaccharides, however, proteins are monodisperse. Protein chemists refer to the amide group linking amino acids together as a *peptide* group. *Dipeptides* contain two amino acids, *tripeptides* three, and so on. The terms *polypeptide* and protein are often used interchangeably, although many protein chemists reserve the former for poly(α-amino acid)s having molecular weights below 10,000. Others consider chains having fewer than 50 amino acids to be polypeptides.

The 20 amino acids most commonly found in proteins are listed in Table 17.4 under three subheadings: neutral, basic, and acidic. Basic amino acids contain more than one *basic* amino group, acidic more than one carboxyl group. Because of the complex nature of proteins, each amino acid may be

Table 17.4. Common amino acids $H_2N-CH-CO_2H$ with R substituent

Name	R	Rotation	Abbreviation
Neutral			
Alanine	$-CH_3$	(+)	Ala (A)
Asparagine	$-CH_2\overset{\overset{\textstyle O}{\|\|}}{C}NH_2$	(−)	Asn (N)
Cysteine[a]	$-CH_2SH$	(−)	Cys (C)
Glutamine	$-CH_2CH_2\overset{\overset{\textstyle O}{\|\|}}{C}NH_2$	(+)	Gln (Q)
Glycine	$-H$		Gly (G)
Isoleucine	$-\underset{\underset{\textstyle CH_3}{\|}}{C}HCH_2CH_3$	(+)	Ile (I)
Leucine	$-CH_2CH(CH_3)_2$	(−)	Leu (L)
Methionine	$-CH_2CH_2SCH_3$	(−)	Met (M)
Phenylalanine	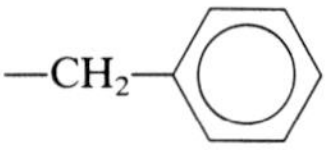 $-CH_2-\text{C}_6\text{H}_5$	(−)	Phe (F)
Proline[b]	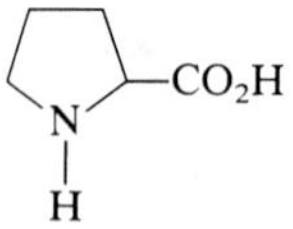 pyrrolidine-2-carboxylic acid structure	(−)	Pro (P)
Serine	$-CH_2OH$	(−)	Ser (S)
Threonine	$-\underset{\underset{\textstyle CH_3}{}}{\overset{\overset{\textstyle OH}{\|}}{C}H}CH_3$	(−)	Thr (T)
Tryptophan	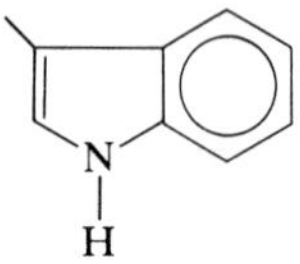 indole structure	(−)	Try (W)
Tyrosine	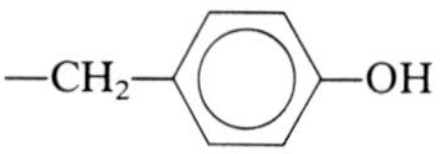 $-CH_2-\text{C}_6\text{H}_4-OH$	(−)	Tyr (Y)
Valine	$-CH(CH_3)_2$	(+)	Val (V)

[a] Cystine, the oxidized form of cysteine, contains two cysteine units linked by a sulfur–sulfur bond.

[b] Full structure given.

Table 17.4. (*Continued*)

Name	R	Rotation	Abbreviation
Basic			
Arginine	—CH$_2$CH$_2$CH$_2$NHC=NH | NH$_2$	(+)	Arg (R)
Histidine	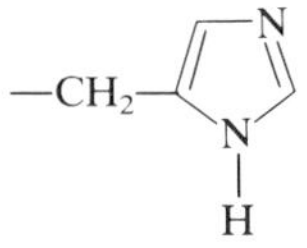 —CH$_2$— (imidazole ring)	(−)	His (H)
Lysine	—CH$_2$CH$_2$CH$_2$CH$_2$NH$_2$	(+)	Lys (K)
Acidic			
Aspartic acid	—CH$_2$CO$_2$H	(+)	Asp (D)
Glutamic acid	—CH$_2$CH$_2$CO$_2$H	(+)	Glu (E)

referred to by a mnemonic three-letter code (Ala for alanine, etc.). The more recent one-letter code, given in parentheses in the table, is gaining popularity.

The amino acids of mammalian proteins have three things in common. First, the amino group that forms the peptide linkage is always located on the α-carbon. Second, with the exception of glycine, which contains no chiral carbon, they are all optically active and all have the L configuration (**22**), related structurally to L-glyceraldehyde (**23**). And, third, the free amino acids do not exist in the form shown in Table 17.4; instead they exist principally as *dipolar ions* (zwitterions) (**24**). As a result, amino acids are high melting, water soluble, and generally insoluble in solvents of low polarity. They are frequently characterized by their *isoelectric point*, which is the pH at which they do not migrate under the influence of an electric field, or in other words, the pH at which they exist exclusively as the dipolar ion.

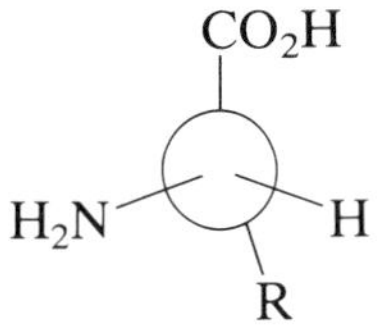

22

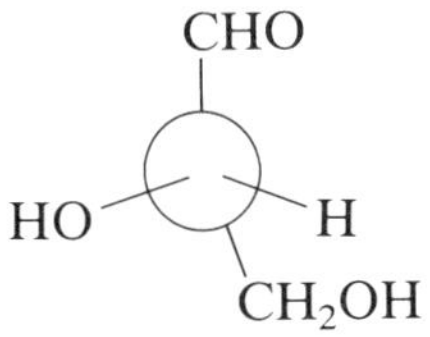

23

$$\overset{+}{H_3N}-\overset{\overset{\displaystyle R}{|}}{CH}-CO_2^-$$

24

Certain conventions are followed in writing peptide structures. For example, for glutathione (**25**), a tripeptide found in living cells, both the structural and abbreviated formulas (**25**) are written with the free amino group, or *N-terminal amino acid*, to the left, and the free carboxyl, or *C-terminal amino acid*, to the right. In naming peptides, the ending "-yl" is used for all amino acids except the C-terminal one, and the amino acids are named in order from the N-terminal one. Thus **25** is glutamylcysteinylglycine.

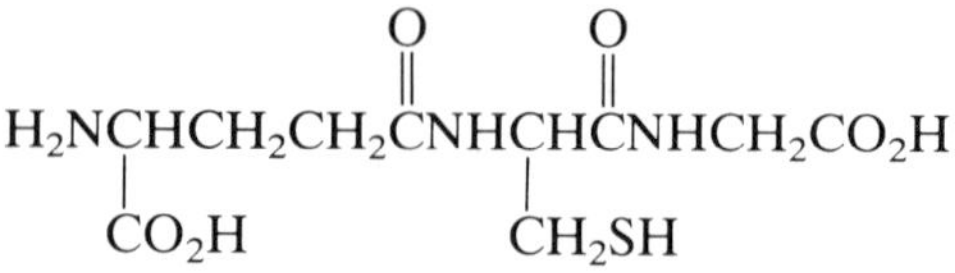

O
O
H2NCHCH2CH2CNHCHCNHCH2CO2H
CO2H
CH2SH

or Glu—Cys—Gly

25

Proteins are classified into two major groups, *fibrous* and *globular*, and the shapes of each can be roughly deduced from the names. Fibrous proteins are long and are held together as fibrils by hydrogen bonds. They are insoluble in water, and this insolubility, together with the strong intermolecular forces, makes them ideally suited for their role as structural materials of animal tissue. Typical of the fibrous proteins are *keratin* (hair, nails, feathers, horn), *collagen* (connective tissue), *fibroin* (silk), and *myosin* (muscle).

Globular proteins are folded and more compact than fibrous proteins. Their shape arises primarily from a combination of intramolecular secondary bonding forces, and most of the polar groups are directed to the outside, thus rendering them soluble in water or dilute salt solutions. Because of their solubility, globular proteins are more concerned with body functions that require mobility of the protein molecules. Enzymes, for example, are globular proteins, as are many hormones. Hemoglobin and egg albumin are other examples.

Some proteins, known as *conjugated proteins*, contain a nonpeptide structure associated with the polypeptide portion. The nonpeptide structure is called the *prosthetic group*, and it usually is extremely important from the standpoint of biological activity. Hemoglobin, for example, consists of a peptide moiety (globin) and an iron–porphyrin prosthetic group (heme, **26**).

CH2=CH
CH3
CH3
CH=CH2
N
N
Fe
N
N
CH3
CH3
HO2CCH2CH2
CH2CH2CO2H

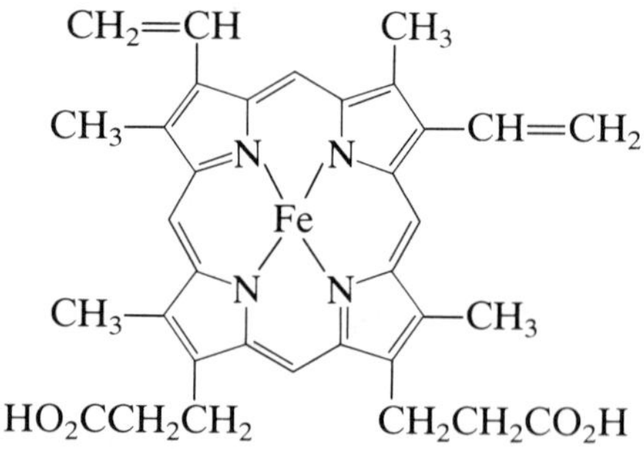

26

It is the heme that forms a reversible complex with oxygen, allowing oxygen transport from the lungs to other parts of the body. Other conjugated proteins, called *nucleoproteins*, are found in living cells in combination with nucleic acids (Section 17.18).

The elucidation of protein structure and the synthesis of proteins represent two of the major challenges of modern chemistry.

17.4.2 Determination of protein structure

The sequence of amino acids in proteins is referred to as the *primary structure*. To determine the primary structure of any polypeptide or protein, the types of amino acids present as well as the quantity of each must first be determined. This is accomplished by complete hydrolysis, followed by separation and identification of the constituents of the hydrolyzate. This, in itself, is a monumental task, considering the complexity of proteins; however, gas and liquid chromatographic techniques and separations by ion exchange have been developed and refined in recent years so that automatic amino acid analyzers that do the job rapidly and accurately are now commercially available.

To determine the amino acid sequence, the polymer is next hydrolyzed partially to form mixtures of peptides containing small numbers of amino acids, and the sequence in the peptides is determined in a stepwise manner. This process requires methods for determining the N-terminal and C-terminal amino acids. Let us illustrate this by using a simple tripeptide as an example. We assume that complete hydrolysis of the tripeptide yields glycine (Gly), phenylalanine (Phe), and valine (Val). The tripeptide could then have six possible structures:

Gly—Phe—Val

Gly—Val—Phe

Phe—Gly—Val

Phe—Val—Gly

Val—Gly—Phe

Val—Phe—Gly

We assume further that the first structure (**27**) of the six possible is the correct one.

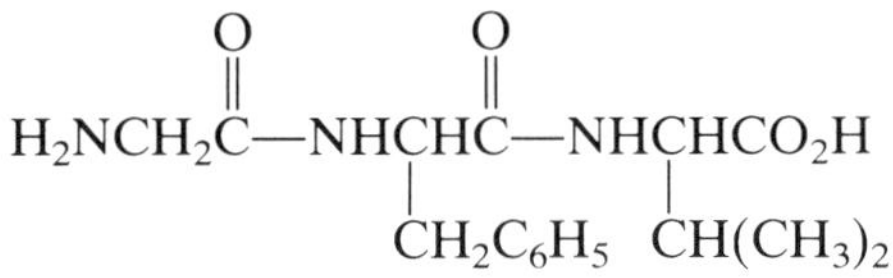

$$H_2NCH_2\overset{\overset{O}{\|}}{C}-NH\overset{}{C}H\overset{\overset{O}{\|}}{C}-NHCHCO_2H$$
$$\underset{CH_2C_6H_5}{|} \quad \underset{CH(CH_3)_2}{|}$$

27

One reagent commonly used to determine the N-terminal amino acid is 2,4-dinitrofluorobenzene (Sanger's reagent). On reaction with this reagent, the fluorine atom undergoes nucleophilic displacement (17.13) by the free amino group. The modified tripeptide is then hydrolyzed (17.14), the products are separated, and the amino acid now modified with 2,4-dinitrofluorobenzene (easily identifiable chromatographically because of its yellow color) is determined. The enzyme carboxypeptidase selectively catalyzes hydrolytic cleavage at the C-terminal end of the peptide. Thus the C-terminal amino acid can be readily identified. Once the N-terminal and C-terminal amino acids have been identified, the tripeptide sequence is known.

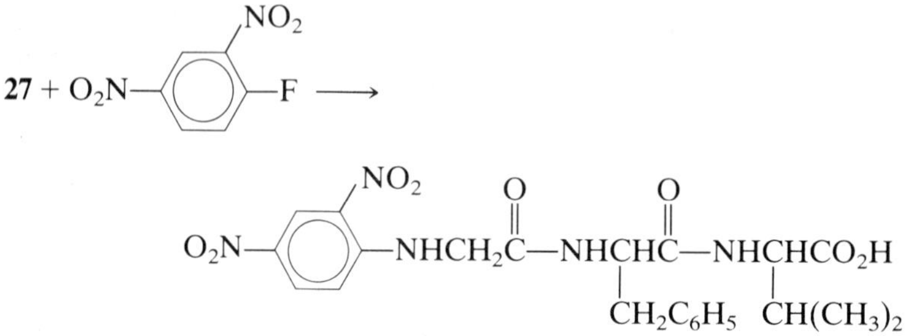

$$\textbf{27} + O_2N-\underset{\underset{\displaystyle NO_2}{|}}{\bigcirc}-F \longrightarrow$$

$$O_2N-\underset{\underset{\displaystyle NO_2}{|}}{\bigcirc}-NHCH_2\overset{\displaystyle O}{\overset{\displaystyle \|}{C}}-\underset{\underset{\displaystyle CH_2C_6H_5}{|}}{N}HCH\overset{\displaystyle O}{\overset{\displaystyle \|}{C}}-\underset{\underset{\displaystyle CH(CH_3)_2}{|}}{N}HCHCO_2H \qquad (17.13)$$

$$\xrightarrow[H^+]{H_2O} \quad O_2N-\underset{\underset{\displaystyle NO_2}{|}}{\bigcirc}-NHCH_2CO_2H +$$

$$H_2N\underset{\underset{\displaystyle CH_2C_6H_5}{|}}{C}HCO_2H + H_2N\underset{\underset{\displaystyle CH(CH_3)_2}{|}}{C}HCO_2H \qquad (17.14)$$

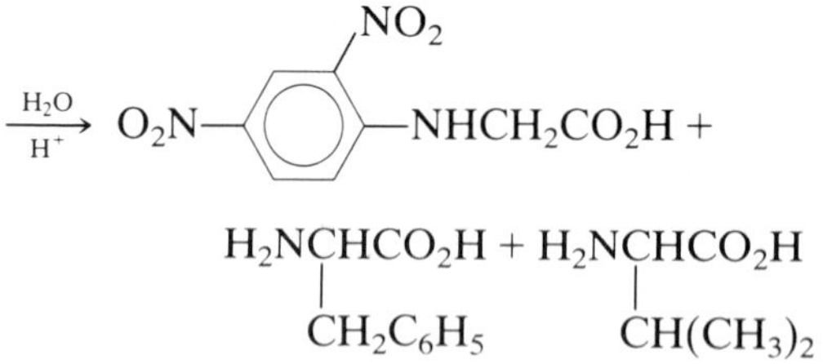

Now let's consider the case of a hexapeptide having the structure

Gly—Ala—Gly—Phe—Val—Leu

The following steps will determine the structure:

1. Complete hydrolysis to determine which amino acids and how much of each are present
2. Identification of the N- and C-terminal amino acids as just described above
3. Partial hydrolysis followed by isolation of the four possible tripeptides
4. Determination of the sequence of any two of the tripeptides

With this information and by overlapping the tripeptide structures, the complete sequence can be deduced:

Gly—Ala—Gly

Ala—Gly—Phe

Gly—Phe—Val

Phe—Val—Leu

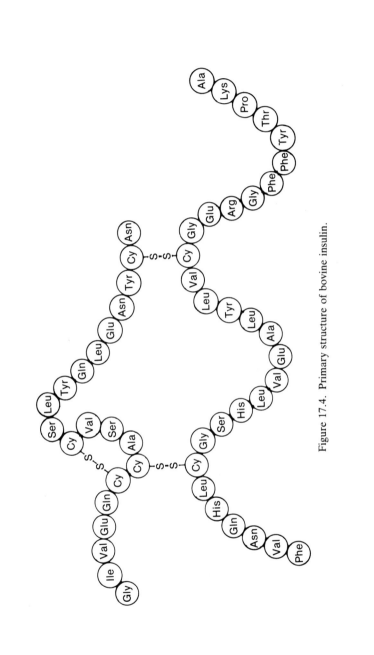

Figure 17.4. Primary structure of bovine insulin.

Other reagents may be used, but those just cited serve to illustrate the general technique and they have historical significance, having been used at Cambridge University by Frederick Sanger (Nobel Prize 1958) in his classic determination of the amino acid sequence of insulin. The use of enzymes to cleave specific peptide linkages is now more common in structure determination. In recent years, methods of determining amino acid sequences based on an understanding of the deoxyribonucleic acid (DNA) responsible for the protein synthesis have been developed. More is said of this later in the chapter.

The structure of bovine insulin is shown schematically in Figure 17.4. The hormone, of modest size, consists of two chains, one containing 21 amino acids, the other 30, linked together by the disulfide groups of cystine. Numerous other protein sequences have been identified since that of insulin first succumbed to Sanger's genius.

To understand the physiological action of proteins or the catalytic action of enzymes, it is necessary to know more than just the amino acid sequence. The shape of the molecule must also be understood. Like all molecules that exhibit conformational flexibility, proteins tend to form stable geometries, generally those that allow the greatest degree of hydrogen bonding. The conformation of a polypeptide or protein chain is referred to as the *secondary structure*.[51]

X-ray analysis, notably by Linus Pauling, has shown that a right-handed helical arrangement, called an *alpha helix* (Figure 17.5), is probably the most important secondary structure. Such an arrangement allows room for the bulky R groups present in most amino acids and is stabilized by intramolecular hydrogen bonding between the amide nitrogen atoms and the carbonyl oxygens. A short section of the peptide chain of α-keratin, the main constituent of hair, wool, and nails, is shown in Figure 17.6. The repeat distance— that is, the distance between amino acid residues along the axis of the helix—is 1.5 Å.

Figure 17.5. Alpha helix.

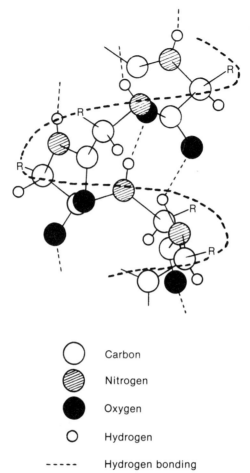

○ Carbon

◉ Nitrogen

● Oxygen

○ Hydrogen

- - - - - Hydrogen bonding

Figure 17.6. Section of an alpha helix of α-keratin.

Proteins that have predominantly small R groups (for example, silk fibroin) can assume a configuration that is essentially an extended chain, but with some contraction so that crowding between the R groups of one chain and those of a neighboring chain is minimized, and the chains can then lie side by side in a sheetlike arrangement. Such an arrangement is referred to as a *pleated sheet*, also as the *beta arrangement*. A schematic representation of the chain association of silk fibroin with a repeat distance of 7.0 Å is shown in Figure 17.7.

The shape of proteins molecules is not given completely by the secondary structure. Sections of a peptide chain may exhibit some irregularity, helical sections may be folded in a manner that allows hydrogen bonding between

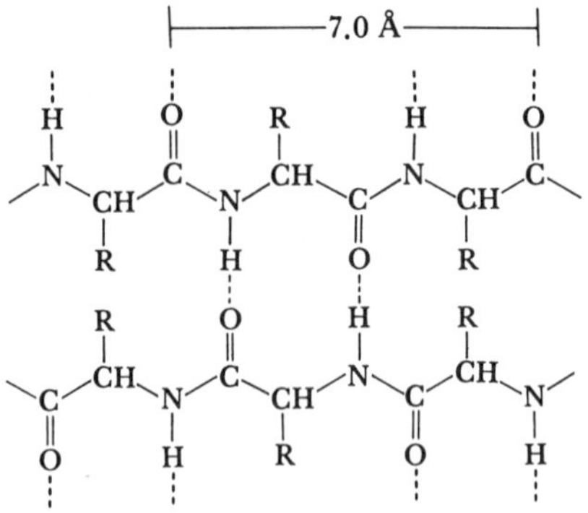

Figure 17.7. Pleated sheet structure (β-arrangement) of silk fibroin.

sites separated by some distance along the chain, and sections of chains may be linked chemically through the sulfur–sulfur bonds of cystine groups or by *salt bridges* between carboxyl and ammonium groups. This overall three-dimensional structure of a protein molecule is known as the *tertiary structure*, and it is this structure that allows certain molecules, such as enzymes, to fit together with other molecules in a way that gives rise to such remarkably efficient and specific catalytic activity. When the tertiary structure is disrupted, a process called *denaturation*, the protein loses its physiological activity.

In addition to tertiary structure, proteins may exhibit a *quaternary structure*; that is, they may be associated with other protein molecules or with nonprotein groups, as is the case with conjugated proteins. The determination of tertiary and quaternary structure is a formidable task requiring the measurement of thousands of x-ray diffraction patterns of protein crystals and calculations necessitating the use of computers. The first to be determined were those of myoglobin and hemoglobin by Kendrew and Perutz at Cambridge University (Nobel Prize 1962). Tertiary structures of dozens of proteins are now known, and certain structural patterns have emerged. For example, glutamic acid, methionine, alanine, and leucine are commonly found in α-helices, whereas valine, isoleucine, phenylalanine, and tyrosine are more often found in pleated sheets. Glycine, proline, and aspartic acid occur frequently in hairpin loops. From a knowledge of primary structure, it is thus possible to predict with some measure of confidence the location of various secondary structures in proteins.

17.4.3   Synthesis of polypeptides and proteins

Heating a mixture of α-amino acids would, of course, yield a polypeptide, but the distribution would be random. To make a polypeptide having a specific sequence it is necessary to protect one of the functional groups of the amino acid by converting it to a derivative, and to react each protected amino acid in the proper sequence. When the first peptide linkage is formed, the protecting group is removed, and the next protected amino acid unit is reacted. The two steps—formation of the peptide linkage and removal of the protecting group—are repeated until the desired peptide chain is formed. Given the complexity of proteins, the number of reactions involved can run into the hundreds. The most common protecting group is *t*-butyloxycarbonyl (BOC), formed by reaction of *t*-butyloxycarbonyl chloride with amino acid (17.15).

$$(CH_3)_3COCCl + H_2NCHCO_2H \longrightarrow \underbrace{(CH_3)_3COCNHCHCO_2H}_{BOC} \quad (17.15)$$

Techniques have been developed that allow automation of the reaction sequence, notably the *Merrifield solid-phase peptide synthesis*[52] (R. B. Merrifield, Nobel Prize 1984). The Merrifield method uses an insoluble cross-linked polymer for binding the peptide chain as each amino acid unit is added. By-products and excess reagents are then removed following each addition by simply washing the insoluble polymer with an appropriate solvent. Typically, polystyrene microgel (see Chapter 3, Section 3.8) is chloromethylated on some of the benzene rings, as described earlier in Chapter 9 (reaction 9.5). The chloromethylated rings then serve as initiation sites for the peptide. The sequence of steps in polypeptide synthesis, outlined in Scheme 17.1, is as follows:

1. BOC-protected amino acid is attached to the insoluble polymer ⓟ via an ester group.
2. The protecting group is removed under conditions mild enough to preclude cleavage from the polymer support.
3. The second BOC-protected amino acid is added, and a strong dehydrating agent such as dicyclohexylcarbodiimide (DCC, **28**) is added to promote formation of the peptide group. The DCC is hydrated to dicyclohexylurea (**29**) in this step.

28   29

4. Steps 2 and 3 are repeated with each successive amino acid until the desired sequence is obtained. (In some cases, the DCC procedure is unsatisfactory, and *p*-nitrophenyl or other active esters of the amino acid are used instead.)
5. Finally, the polypeptide or protein is removed from the polymeric support.

The entire sequence of reactions can be carried out in a single reaction vessel fitted with a fritted disk filter for drawing off the by-products, solvents,

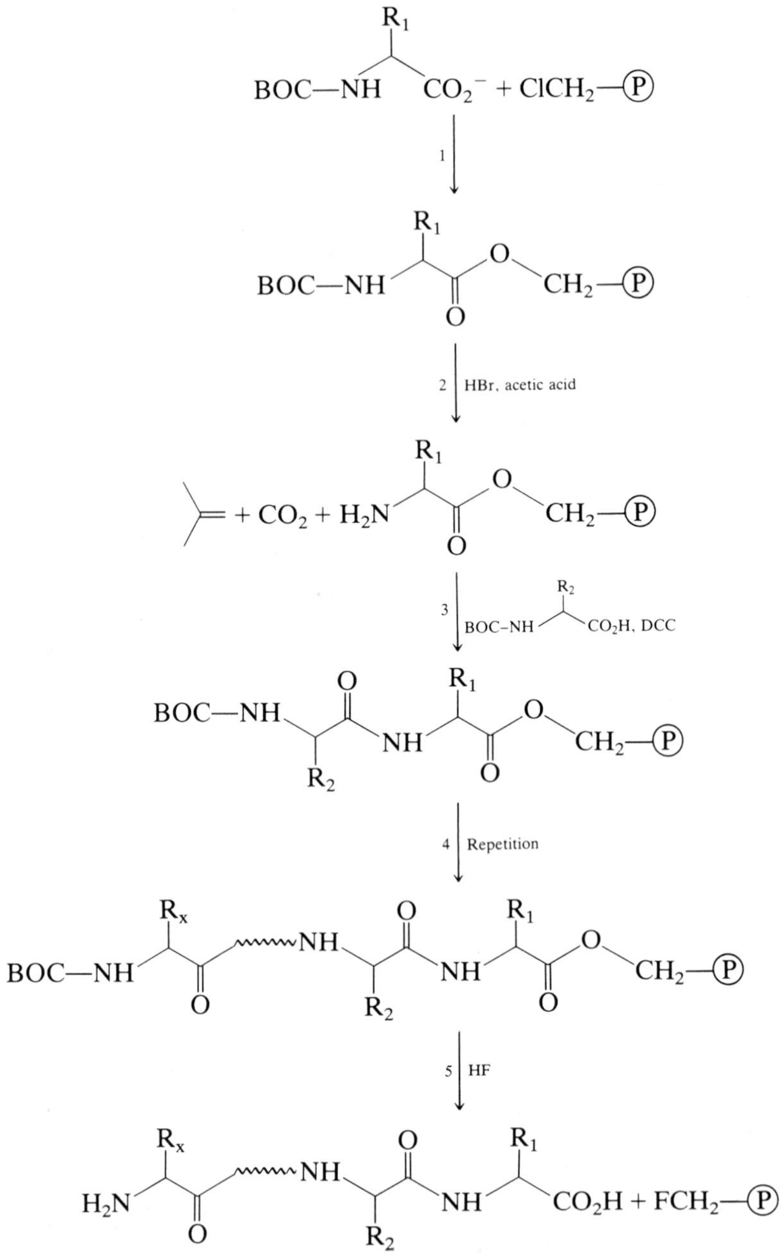

Scheme 17.1. Merrifield solid-phase peptide synthesis.

and excess reagents. Modern instruments incorporating microprocessor technology that carry out the synthetic steps in a programmed sequence are now available commercially. The instruments are similar in design to that shown later in this chapter (Figure 17.14) for polynucleotide synthesis. In an early application of the Merrifield method, bovine ribonuclease, an enzyme containing 124 amino acids, was synthesized in six weeks in 17% overall yield via 369 reactions involving 11,931 individual steps, with no isolation of intermediate products.[53]

An occasional problem with the Merrifield method is that impurities are sometimes difficult to remove, in which case a more traditional solution-phase synthesis may be preferred. Even so, the same general sequence of steps using protected amino acids is followed.

17.4.4 Wool, silk, collagen, and regenerated protein

The history of wool and silk for use in clothing dates back some 5000 years. Wool[54] is obtained mainly from sheep by scouring the sheep hair with alkaline soap solution to remove fats and greases. Different breeds of sheep produce wools with properties that vary considerably in elasticity, strength, color, and shrinkage, and wools are usually classified as fine, medium, long, cross-bred, and carpet, according to their source. Other animals, including goats (mohair and cashmere wools) and camels or members of the camel family (llama, alpaca, vicuna), are also sources of wool. Silk[55,56] is secreted by silkworms as two filaments encased in a proteinaceous gum. Other species, such as gypsy moth caterpillars and spiders, also secrete silk, but only the silkworm variety is used commercially.

There is considerable variation in the amino acid composition of wool and silk, as shown in Table 17.5. It can be seen, for example, that silk consists mainly of four amino acids—glycine, alanine, serine, and tyrosine—the first three of which have small R groups (—H, —CH_3, and —CH_2OH, respectively), whereas wool has a more even distribution. Furthermore, wool contains a much higher proportion of cystine, indicative of more chain coiling through disulfide groups. As was pointed out in Section 17.4.2, it is the relative absence of very bulky R groups in silk that is responsible for its assuming the extended-chain beta configuration, whereas wool (keratin) exhibits an alpha-helical secondary structure. (Stretched wool also has an extended-chain beta arrangement.)

Numerous chemical modifications of silk and wool have been reported.[45d,57,58] Wool containing less than 1% of grafted nylon 610 has considerable shrink resistance. The grafting is accomplished commercially by interfacial polymerization. The wool is first treated with a 1% aqueous solution of hexamethylenediamine plus inorganic base, then with a 2 to 3% hydrocarbon solution of sebacoyl chloride. The graft is assumed to form by reaction between acid chloride end groups and free protein amino groups. Wool has been similarly grafted with polyurethane and polyurea by treatment with diamine followed by diisocyanate and bischloroformate, respectively

Table 17.5. Composition of silk and wool[a]

Amino acid		Approximate grams amino acid per 100 g protein	
Type	Name	Silk (fibroin)	Wool (keratin)
Aliphatic	Glycine	41.2	5.5
	Alanine	33.0	4.3
	Serine	16.2	10.6
	Valine	3.6	5.7
	Threonine	1.55	7.15
	Leucine / Isoleucine	2.0	12.6
Aromatic	Phenylalanine	3.35	4.1
	Tyrosine	11.4	5.5
Sulfur containing	Cystine	0.2	13.0
	Methionine	0	0.55
Heterocyclic	Proline	0.7	6.8
Acidic	Aspartic acid	2.75	6.8
	Glutamic acid	2.15	14.5
Basic	Histidine	0.4	1.2
	Arginine	1.0	9.8
	Lysine	0.5	3.3

[a] Data from *Handbook of Common Polymers*, Chemical Rubber Co., Cleveland, 1971.

(see Chapter 13). Treatment with epoxide leads to polyether grafts. Vinyl polymers have also been grafted onto wool and silk—for example, by radiation grafting.

A number of reagents have also been used to crosslink protein fibers. If the cystine disulfide crosslinks of wool are reduced to thiol, then recrosslinked with dihalide (17.16), the treated wool exhibits improved resistance to acids, bases, oxidizing and reducing agents, and moths and carpet beetles. Other reagents used to crosslink wool and silk include formaldehyde, diepoxides, and epichlorohydrin. Wool fabrics have also been flameproofed with phosphorus-modified melamine–formaldehyde resins.

$$\{-SH + Br(CH_2)_x Br + HS-\} \longrightarrow \{-S(CH_2)_x S-\} \qquad (17.16)$$

Collagen,[59,60] the main constituent of skin and hides, is used as a source of gelatin and glue (by treatment with boiling water) and leather (by treatment with tannic acid or heavy metal salts). Gelatin is actually a partially degraded form of collagen, with a molecular weight of about 54,000. Collagen is composed largely of glycine, proline, and hydroxyproline, with minor amounts of other amino acids. Chemical treatment of leather with diepoxides reduces its tendency to shrink or become brittle on heating.

Plastics and fibers have been manufactured from regenerated protein obtained from a number of sources[61] including milk (casein), maize(zein), ground nuts (arachin), and soya beans (glycinin). The predominant amino acid of these proteins is glutamic acid (particularly in zein). Arginine is also present in significant amounts in arachin, and isoleucine in the other three.

The regenerating process involves dispersing the protein in dilute sodium hydroxide solution, followed by extrusion through spinnerets into an acid bath to form fibers. The protein is also crosslinked with formaldehyde to harden it and reduce its solubility. As a plastic, regenerated protein is usually used for indoor decorative applications (knobs, buttons, handles, etc.) because of its poor weatherability. Uncrosslinked casein is also used to improve adhesion of viscose rayon tire cord to rubber. Regenerated protein fibers are usually used in combination with wool, cotton, or synthetic fibers. Many of the chemical modifications applied to wool and silk have also been carried out with regenerated proteins.[45d]

17.5 Nucleic acids

17.5.1 Nucleic acid structure

It was mentioned earlier that nucleoproteins are proteins found in living cells in combination with nucleic acids. It is the nucleic acids that direct the synthesis of proteins in the cell, and it is the elucidation of nucleic acid structure and function, upon which is based our knowledge of the chemical basis of heredity, that has been among the most exciting developments of biochemical research in recent years.[4c,62–64]

Nucleic acids are polyesters of phosphoric acid and sugars, to each of which a heterocyclic amine (referred to as a *base*) is attached (**30**). There are two principal types of nucleic acids—one containing the sugar D-ribose (**31**), and the other D-2-deoxyribose (**32**), having one less hydroxyl group. Nucleic acids

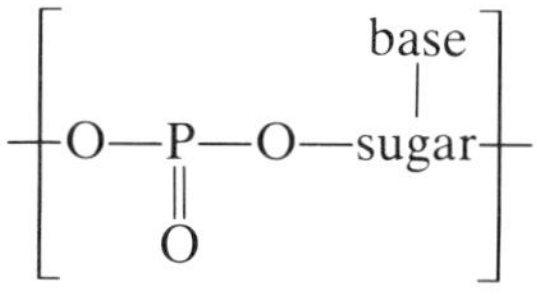

30

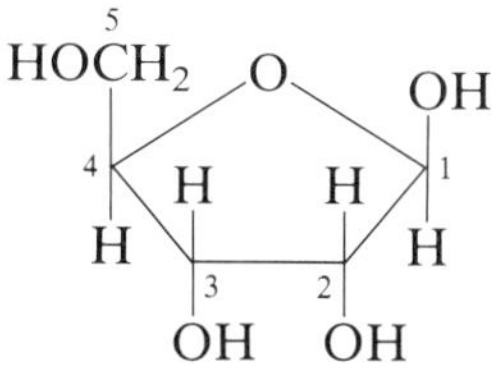

31

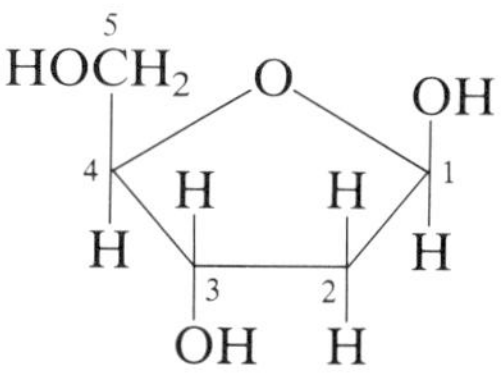

32

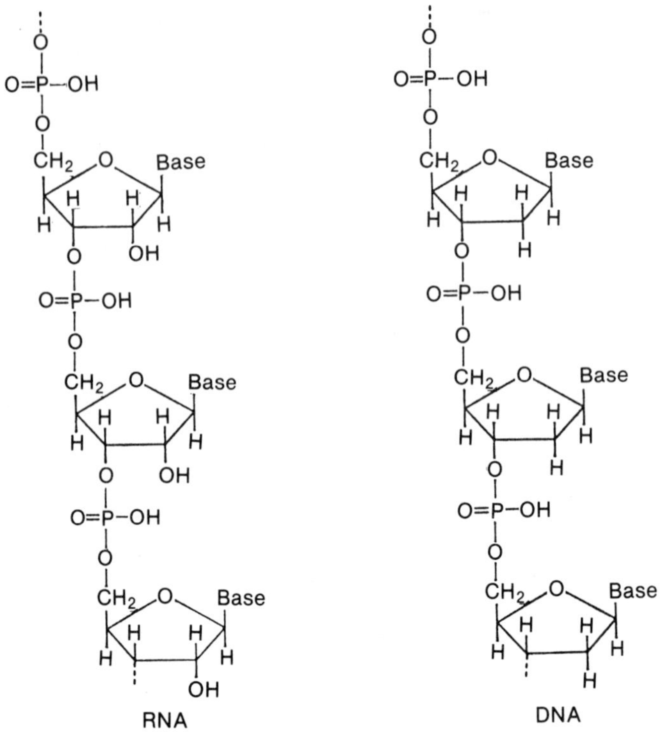

Figure 17.8. Sections of RNA and DNA chains.

containing D-ribose are known as *ribonucleic acids* (RNA), and those containing D-2-deoxyribose are known as *deoxyribonucleic acids* (DNA). The points of attachment to phosphoric acid are the hydroxyl groups at carbons 3 and 5 for both sugars. Bases are attached to the sugars at carbon 1, replacing the hydroxyl group at that position. A sugar with a base attached is called a *nucleoside*, and these represent the products of hydrolysis of nucleic acids. The ester of phosphoric acid and a nucleoside is known as a *nucleotide*; hence nucleic acids are also called *polynucleotides*. The structures of RNA and DNA may thus be depicted by the three nucleotide portions of each chain shown in Figure 17.8.

The heterocyclic bases found in RNA and DNA are all derivatives of pyrimidine (**33**) or purine (**34**). The most commonly occurring pyridimine derivatives, each of which is attached to the sugar at the position-1 nitrogen atom, are *cytosine* (2-hydroxy-4-aminopyrimidine) (**35**), *5-methylcytosine* (**36**) (somewhat less prevalent), and *thymine* (5-methyl-2,4-dihydroxypyrimidine) (**37**), found in DNA; and *cytosine* and *uracil* (2,4-dihydroxypyrimidine) (**38**), found in RNA. These bases are shown in the *isomeric keto form*, rather than the hydroxy form, because this is how they exist in the nucleic acids. There

are two purine derivatives occurring commonly in both DNA and RNA—
adenine (6-aminopurine) (**39**) and *guanine* (2-amino-6-hydroxypurine) (**40**).
Guanine also exists, as shown, in the keto form. Each is attached to the sugar
at the position-9 nitrogen atom.

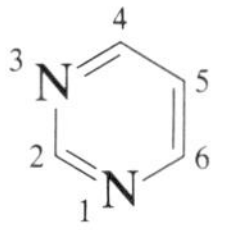

33 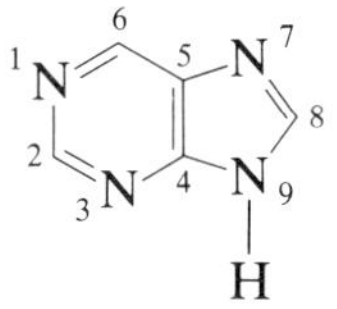   **34**

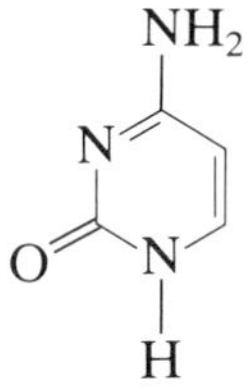

35 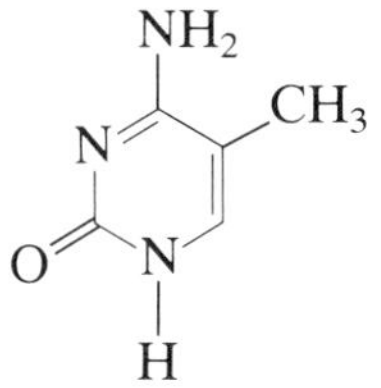   **36** 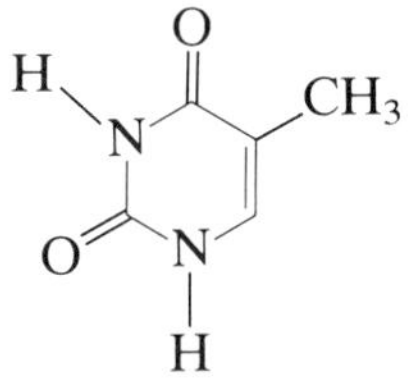   **37** 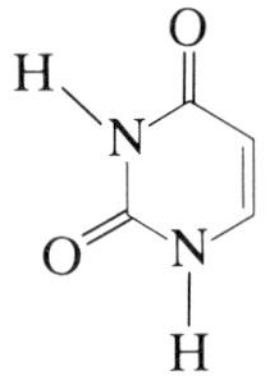   **38**

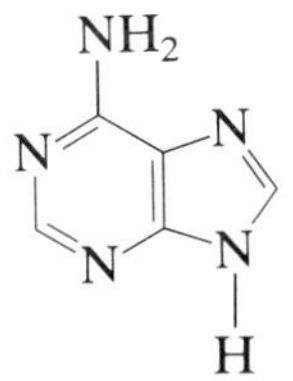

39 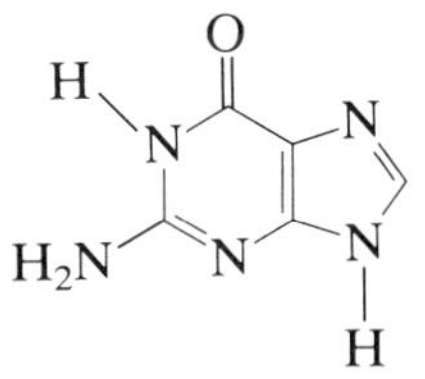   **40**

Some nucleosides and nucleotides occur as free compounds as well as
constituents of nucleic acids, and frequently they play important biochemical
roles. Nucleosides consisting of ribose and a purine base are named *adenosine*
and *guanosine* (for the bases adenine and guanine, respectively). The corresponding deoxyribose nucleosides are *deoxyadenosine* and *deoxyguanosine*.
The ribose–pyrimidine bases are *cytidine* and *uridine* (corresponding to cytosine and uracil, respectively). The deoxyribose–pyrimidine nucleosides are
deoxycytidine and *deoxythymidine* (for cytosine and thymine, respectively).
With the last name the deoxy prefix is sometimes omitted because thymine
occurs *only* in combination with deoxyribose.

The corresponding nucleotides are named as phosphates of the nucleosides. Compound **41**, for example, is adenosine 5′-monophosphate (AMP).
(The 5′ indicates the position on the sugar where phosphoric acid is attached.) Adenosine diphosphate (ADP) and adenosine triphosphate (ATP)

$$NH_2$$

HO—P—OCH$_2$

the structure **41** (adenosine monophosphate)

41

are also found in cells, and it is hydrolysis of the phosphoric anhydrides, an exothermic process, that provides energy for numerous biochemical processes. Monophosphates, diphosphates, and triphosphates of the other nucleosides are also present in cells, although ATP is most important from the standpoint of energy transfer.

Apart from the structural differences described above, RNA and DNA also differ in molecular weight, secondary structure, and biochemical function. DNA is a very-high-molecular-weight polymer, in some instances in the hundreds of millions. (The DNA of the bacterium *Escherichia coli* is about 2.5 *billion*!) Chemical analysis of the bases present in DNA shows a remarkable regularity in that the number of adenine bases is equal to the number of thymine bases, and the number of guanines equals the number of cytosines. It was this regularity, together with supporting evidence based on x-ray studies by Wilkins and Franklin, that led Watson and Crick (Nobel Prize, together with Wilkins, 1962) to postulate the double-helix structure of DNA. DNA exists as two separate right-handed helical chains wound around each other and held in place by hydrogen bonding between *base pairs*, as depicted in Figure 17.9. Depicted also is the unraveling of the double helix with concomitant formation of complementary DNA chains, the process that occurs when DNA molecules reproduce (*replicate*). The base pairs are adenine (A)– thymine (T) and guanine (G)–thymine (T). They extend perpendicularly toward the center with the deoxyribose–phosphate chains on the outside of the helixes. Such an arrangement accounts for the regularity of the double helix, because both base pairs allow approximately the same distance between deoxyribose units (Figs. 17.10 and 17.11). Attempts to construct models containing other base pairs lead to distortion of the regular double helix. Not all DNA has the double-helix structure. Single-stranded DNA is also known, as is a less regular, left-spiraling form, but the double helix is most important. The elucidation of the overall structural picture of DNA is still a monumental task because of the sheer size of the molecules. There is

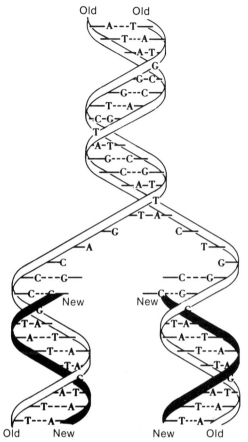

Figure 17.9. Schematic representation of the DNA double helix showing unraveling and formation of complementary chains. [From R. J. Light, *A Brief Introduction to Biochemistry*, Benjamin, 1968, reprinted by permission.]

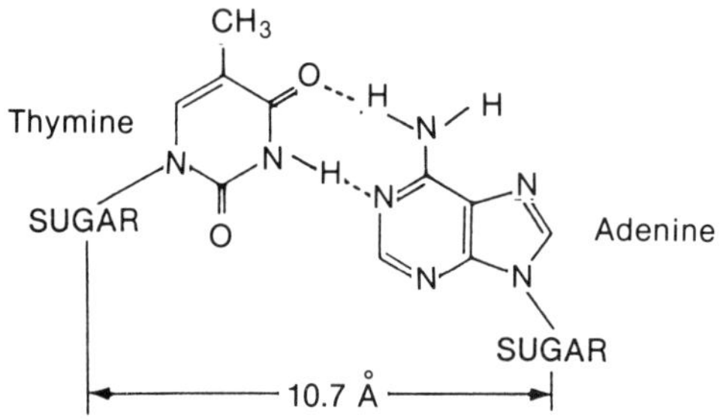

Figure 17.10. Adenine–thymine base pair.

Cytosine

SUGAR

Guanine

SUGAR

← 10.7 Å →

Figure 17.11. Guanine–cytosine base pair.

evidence, for example, that certain types of DNA exist in large cyclic structures, even with interlocking rings (catenanes).[65,66] Figure 17.12 shows an electron micrograph of such a structure.*

While a substantial amount of information is available on the secondary structure of DNA, less is known about RNA. It is clear, however, that RNA is much lower in molecular weight than DNA; also that there are three main types of RNA fround in living cells, known as *ribosomal RNA* (rRNA), *transfer* or *soluble RNA* (tRNA), and *messenger RNA* (mRNA). These have molecular weights of about 1 million, 25,000, and 500,000, respectively. These various forms of RNA are associated with proteins in the cell. Some viruses, particularly plant viruses, consist of RNA associated with protein. Tobacco mosaic virus, for example, contains about 95% protein and 5% RNA, with the protein forming a protective sheath around the RNA. Although some double-chain RNA molecules are known to occur in certain viruses, RNA is primarily single strand.

Nucleic acids represent the chemical basis of heredity. DNA, the chromosome of the cell, contains the information necessary to duplicate a new cell, and this information is contained in the sequence of bases or groups of nucleotides (genes). When DNA undergoes replication (Figure 17.9), each separated strand acts as a template for a complementary strand; thus new double-helix structures are formed with the same base pairing as in the

---

* It has been suggested (J. D. Roberts and M. C. Caserio, *Basic Principles of Organic Chemistry*, Benjamin, New York, 1965, p. 1114) that polycatenanes (topological polymers), might some day be a synthetic reality. While catenanes containing two or three interlocking rings have been

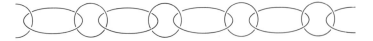

synthesized, the interpenetrating polymer networks described in Chapter 3 come closest to merging polymer and synthetic topological chemistry. In a similar vein, *polymeric catenanes* have been made. Polybutadiene, for example, when crosslinked in the presence of the cyclic depsipeptide (12 amino acids) valinomycin, traps the peptide by threading it with polybutadiene chains (D. Callahan, H. L. Frisch, and D. Klempner, *Polym. Eng. Sci.*, **15**, 70, 1975).

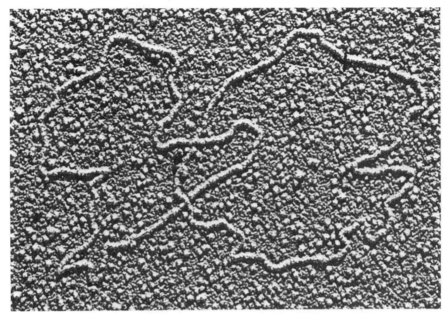

Figure 17.12. Electron micrograph of catenated circular DNA molecules.[65] [Courtesy of Prof. Jerome Vinograd, California Institute of Technology. Reprinted by permission from *Nature*, **216**, No. 5116, 647–652. Copyright © 1967. Macmillan Journals Limited.]

original. Enzymes (DNA polymerase) catalyze the synthesis of the new chains. Sections of the DNA also serve as templates for RNA synthesis, and it is the various forms of RNA that are responsible for assembling the amino acids into the proper sequence for protein synthesis. The recognition factor that allows RNA to select a particular amino acid from the amino acid pool is contained in the base sequence of the RNA molecules (the primary structure).

Great strides have been made in elucidating the primary structure of polynucleotides. Because of their vast size, DNA molecules must first be broken down to a more manageable length. This is accomplished with enzymes called *restriction endonucleases*, which cleave the polynucleotide chains at specific base sequences to yield double-strand chains containing 100 to 200 nucleotides. Because each restriction enzyme cleaves a different nucleotide pair, the resultant chains partially overlap, just as the partially degraded protein chains overlap in amino acid sequencing.

Two methods were developed in the 1970s for determining the sequence of bases in the polynucleotide, one enzymatic and the other chemical. In both methods, the double-strand fragments are separated by heating, and the chain ends containing a free OH group on carbon 5 of the deoxyribose (see structure **32**) are labeled by enzymatically incorporating a radioactive ^{32}P phosphate ester. The labeled chains are then broken down into smaller units.

The enzymatic method, developed by Sanger and coworkers,[67,68] involves four enzyme-catalyzed reactions that selectively catalyze DNA chain cleavage at each of the four base sites, yielding chain fragments terminating in adenosine, cytosine, guanosine, and thymidine. The chemical method, devised by Allan Maxam and Walter Gilbert[69] (Gilbert shared the Nobel Prize in 1980), achieves the same result, but is somewhat less selective.

Separation of the fragments is accomplished by *gel electrophoresis*, a technique involving a strip of buffered polyacrylamide gel. Samples from each reaction are placed separately at one end of the strip and a voltage difference is applied, which causes the fragments to move along the strip at rates that depend on their charge and mass. Once the fragments have been separated, they are visualized by exposing the strip to a photographic plate that senses the radioactive ^{32}P. The DNA sequence may then be read directly from the visualized pattern of fragments.

Refinements of the Sanger method that use fluorescent rather than radioactive labels have been developed.[70,71] The emissions are read directly by a fluorescence detector and the data are fed to a microcomputer that reconstructs the base sequence. Automated sequencers based on these techniques are now available commercially. Up to 10,000 nucleotides per day can now be identified by these methods, compared with about 50,000 a year using autoradiographic imaging. Given the fact that DNA contains all the information necessary for protein synthesis and, ultimately, the physical and chemical makeup of the entire organism, this is a truly monumental accomplishment. How the bases actually direct the synthesis of proteins (the genetic code) is detailed in modern biochemistry texts.

17.5.2   Nucleic acid synthesis

A knowledge of the genetic code has obvious ramifications. For example, if the base sequence of the polynucleotide responsible for synthesis of a particular protein is known, the amino acid sequence of the protein can be deduced. This is, in fact, becoming the method of choice for determining primary protein structure.

Genetic engineering technologies were developed in the 1970s that allowed the modification of host cell DNA by insertion of new polynucleotide fragments. The new DNA produced is called *recombinant DNA*. Thus, *E. coli* bacteria can be induced to produce insulin or interferon or some other pharmacologically useful polypeptide by inoculation with the DNA sequence necessary for that polypeptide's synthesis. These developments have led to a demand for improved methods of polynucleotide synthesis.[72,73] Commercial polynucleotide sequencers (called "gene machines") are now available that use solid-phase synthesis analogous to that of the Merrifield method for proteins.[74] More functional groups need protecting, however, which makes polynucleotide synthesis more complicated.

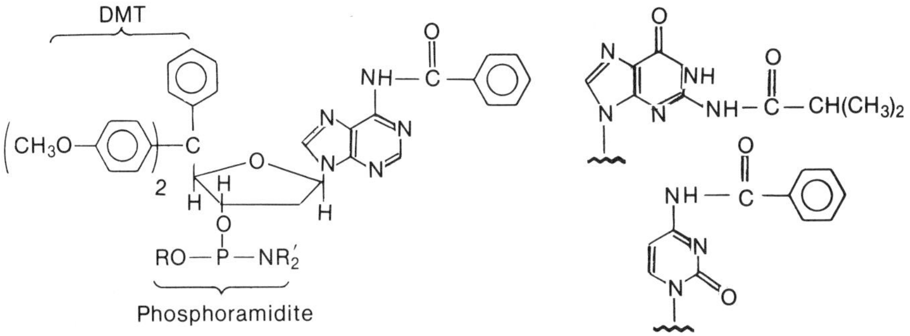

Figure 17.13. Representative protecting groups used in solid-phase polynucleotide synthesis. (DMT = dimethoxytrityl; R = CH$_3$, CH$_2$CH$_2$CN, etc.; R' = CH$_3$, CH(CH$_3$)$_2$, ⬡O, etc.)

The solid support may be a crosslinked polymer or, more commonly, a controlled-pore glass. Typical protecting groups, shown in Figure 17.13, are dimethoxytrityl (DMT) for the carbon-5 OH group of deoxyribose (designated 5′ OH) and phosphoramidite for carbon-3 OH (3′ OH) having variable R groups. Adenine and guanine bases are usually protected by benzoyl groups, guanine by an isobutyryl group. Thymine needs no protecting group. Synthesis begins with the first protected nucleotide attached to the solid support at the 3′ OH, as shown in Scheme 17.2. In step 1, the trityl group is removed with trichloroacetic acid. Step 2 involves reaction with the second protected nucleotide using an activating agent such as tetrazole (**42**), which

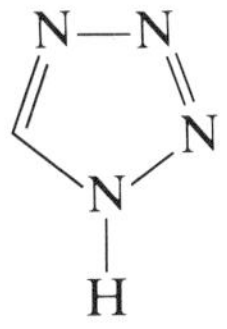

42

serves to protonate the phosphoramidite nitrogen. In step 3 the trivalent phosphite ester is oxidized to phosphate. Detritylation of the newly added nucleotide sets the stage for reaction with the next nucleotide. When the chain is at the desired length it is removed from the support with aqueous ammonia, which also removes the protecting groups.

A typical polynucleotide synthesizer[75] is shown in Figure 17.14. The basic components of the machine are a reaction vessel (mounted on the front of the instrument), a programmable controller, solvent and reagent delivery systems, and numerous valves.

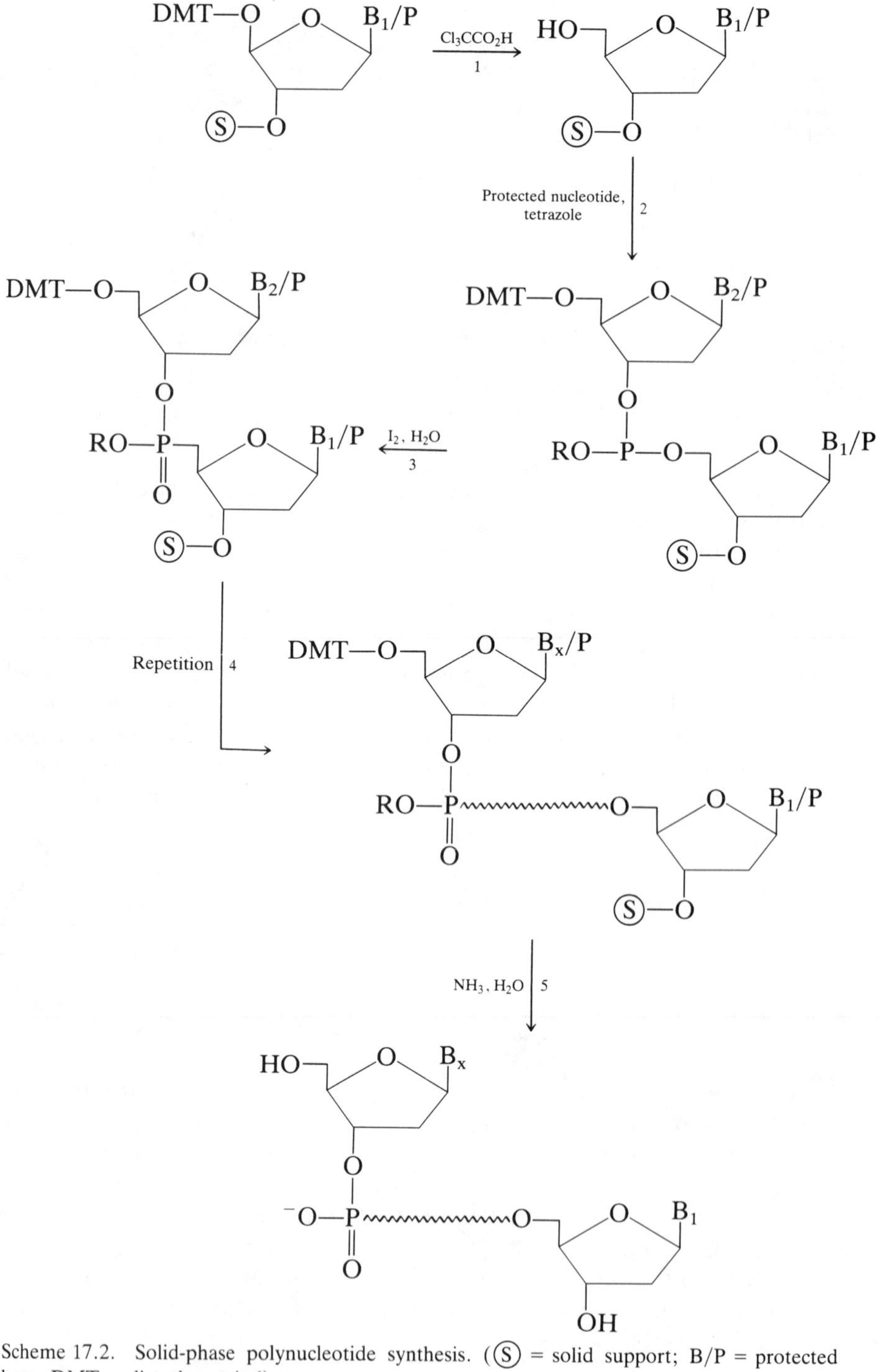

Scheme 17.2. Solid-phase polynucleotide synthesis. (Ⓢ = solid support; B/P = protected base; DMT = dimethoxytrityl)

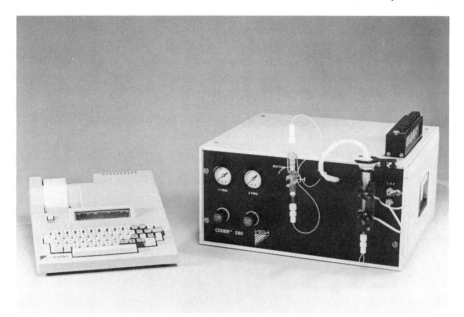

Figure 17.14. Automated polynucleotide synthesizer. [Courtesy of Dr. Leon E. Barstow, Protein Technology.]

17.6   Conclusion

We began this book on a practical note—textile fibers and plastic bottles. We end with a glimpse of the macromolecular chemistry of life. With our rapidly expanding knowledge of the chemistry of polynucleotides, we have reached the dawn of an age of genetic engineering, with all the ethical questions that go with it. The synthesis of artificial genes capable of functioning within living cells is already a reality.[76] Yet we have a long way to go before we are able to duplicate the efficiency and selectivity that characterize polymer processes in living systems.

Compared with the chemistry of life, nylon or polyester fibers and plastic bottles lose much of their glamor. But if nothing else, such contrasts serve to remind us of the subtlety and beauty of nature.

References

1. F. R. Eirich (ed.), *Science and Technology of Rubber*, Academic Press, New York, 1978.
2. J. A. Brydson (ed.), *Developments with Natural Rubber*, Maclaren, London, 1967.

3. A. D. Roberts (ed.), *Natural Rubber Science and Technology*, Oxford University Press, New York, 1988.
4. E. A. MacGregor and C. T. Greenwood, *Polymers in Nature*, Wiley, New York, 1980: (a) Chap. 7; (b) Chap. 4; (c) Chap. 5.
5. J. C. F. Williams, *J. Chem. Soc.*, **15**, Part 10 (1862).
6. S. F. Thames, *Polym. Preprints*, **28**(2), 102 (1987).
7. F. E. Brauns, *The Chemistry of Lignin*, Academic Press, New York, 1952.
8. I. A. Pearl, *The Chemistry of Lignin*, Dekker, New York, 1967.
9. K. Freudenberg and A. C. Neish, *The Constitution and Biosynthesis of Lignin*, Springer-Verlag, New York, 1968.
10. K. Freudenberg, *Science*, **148**, 595 (1965).
11. D. Feldman, M. Lacasse, and L. M. Beznaczuk, *Prog. Polym. Sci.*, **12**, 271 (1986).
12. S. S. Kelley, W. G. Glasser, and T. C. Ward, *J. Appl. Polym. Sci.*, **36**, 759 (1988).
13. V. P. Saraf, W. G. Glasser, G. L. Wilkes, and J. E. McGrath, *J. Appl. Polym. Sci.*, **30**, 2207 (1985).
14. M. Schnitzer and S. U. Khan, *Humic Substances in the Environment*, Dekker, New York, 1972.
15. F. J. Stevenson, *Humus Chemistry: Genesis, Composition, Reactions*, Wiley-Interscience, New York, 1982.
16. L. Gorbaty and K. Ouchi (eds.), *Coal Structure*, Adv. Chem. Ser. 192, American Chemical Society, Washington, D.C., 1981.
17. P. F. Dickson, in *Kirk-Othmer: Encyclopedia of Chemical Technology*, Vol. 16 (H. F. Mark, D. F. Othmer, C. G. Overberger, and G. T. Seaborg, eds.), Wiley-Interscience, New York, 1981.
18. J. G. Speight and S. E. Moschopedis, in *Chemistry of Asphaltenes* (J. W. Bunger and N. C. Li, eds.), Adv. Chem. Ser. 195, American Chemical Society, Washington, D.C., 1981, pp. 1–15.
19. A. J. Hoiberg, *Bituminous Materials: Asphalts, Tar, and Pitches*, Wiley-Interscience, New York, 1964.
20. P. K. Bose, Y. Sankaranarayanan, and S. C. Sen Gupta, *Chemistry of Lac*, Indian Lac Research Institute, Rachi, India, 1963.
21. E. Hicks, *Shellac*, Chemical Publishing, New York, 1961.
22. S. Maiti and M. D. S. Rahman, *J. Macromol. Sci., Rev. Macromol. Chem. Phys.*, **C26**, 441 (1986).
23. W. W. Christie, F. D. Gunstone, H. G. Prentice, and S. C. Sen Gupta, *J. Chem. Soc., Suppl.* **1**, 5833 (1964).
24. J. H. Langenheim, *Science*, **163**, 1157 (1969).
25. G. G. Larsson, *Baltic Amber*, Scandinavian Science Press, Klampenbory, Denmark, 1978.
26. Y. Jen and E. E. McSweeney, in *Applied Polymer Science*, 2nd ed. (R. W. Tess and G. W. Poehlein, eds.), ACS Symp. Ser. 285, American Chemical Society, Washington, D.C., 1985, pp. 1159–1179.
27. R. L. Whistler and J. N. BeMiller (eds.), *Industrial Gums*, 2nd ed., Academic Press, New York, 1973.
28. S. Maiti, S. Das, M. Maiti, and A. Ray, in *Polymer Applications of Renewable-Resource Materials* (C. E. Carraher Jr., and L. H. Sperling, eds.), Plenum Press, New York, 1983, p. 129.

29. S. S. Ray, A. K. Kundu, M. Ghosh, and S. Maiti, *Eur. Polym. J.*, **21**, 131 (1985).
30. S. S. Ray, A. K. Kundu, and S. Maiti, *J. Macromol. Sci.-Chem.*, **A23**, 271 (1986).
31. E. C. Leonard (ed.), *The Dimer Acids*, Humko Sheffield Chemical, Memphis, Tenn., 1975.
32. E. Ott, H. M. Spurlin, and M. W. Griffin (eds.), *Cellulose and Cellulose Derivatives*, 2nd ed., Wiley-Interscience, New York, 1954: (a) K. Ward Jr., Part 1, Chap. 2; (b) E. Kline, Part 2, pp. 959–1018; (c) Chap. 9.
33. R. M. Brown Jr. (ed.), *Cellulose and Other Natural Polymer Systems: Biogenesis, Structure, and Degradation*, Plenum Press, New York, 1982.
34. J. D. Wilson and J. K. Hamilton, *J. Chem. Educ.*, **63**, 49 (1986).
35. S. M. Hudson and J. A. Cuculo, *J. Macromol. Sci., Rev. Macromol. Chem. Phys.*, **C18**, 1 (1980).
36. K. F. Gadd, *Polymer*, **23**, 1867 (1982).
37. C. L. McCormick and T. C. Shen, in *Macromolecular Solutions* (R. B. Seymour and G. S. Stahl, eds.), Pergamon Press, New York, 1982, pp. 101–107.
38. A. S. Turbak, R. B. Hammer, R. E. Davies, and H. L. Hargert, *CHEMTECH*, **10**, 51 (1980).
39. A. Hebeish and J. T. Guthrie, *The Chemistry and Technology of Cellulose Copolymers*, Springer-Verlag, New York, 1981.
40. N. M. Bikales and L. Segal, *Cellulose and Cellulose Derivatives*, Wiley, New York, 1971.
41. D. N.-S. Hon (ed.), *Graft Copolymerization of Lignocellulosic Fibers*, ACS Symp. Ser. 187, American Chemical Society, Washington, D.C., 1982.
42. D. J. McDowall, B. S. Gupta, and V. T. Stannett, *Prog. Polym. Sci.*, **10**, 1 (1984).
43. B. Rånby and C. Rodehed, in *Polymer Science Overview* (G. A. Stahl, ed.), ACS Symp. Ser. 175, American Chemical Society, Washington, D.C., 1981, pp. 253–264.
44. B. P. Morin, I. P. Breusova, and Z. A. Rogovin, *Adv. Polym. Sci.*, **42**, 139 (1982).
45. E. M. Fettes (ed.), *Chemical Reactions of Polymers*, Wiley-Interscience, New York, 1964: (a) K. Ward Jr., and A. J. Morak, Chap. 5; (b) N. G. Gaylord and F. S. Ang, Chap. 10; (c) W. A. Reeves and J. D. Guthrie, Chap. 16; (d) R. E. Whitfield and W. L. Wasley, Chap. 6.
46. J. G. Frick, *CHEMTECH*, 100 (1971).
47. R. L. Whistler and E. F. Paschall (eds.), *Starch, Chemistry and Technology*, Vols. 1 and 2, Academic Press, New York, 1965.
48. R. L. Whistler and C. L. Smart, *Polysaccharide Chemistry*, Academic Press, New York, 1953.
49. A. G. Walton, *Polypeptides and Protein Structure*, Elsevier, New York, 1981.
50. R. E. Dickerson and I. Geis, *The Structure and Action of Proteins*, Harper & Row, New York, 1969.
51. R. D. B. Fraser and T. P. MacRae, *Conformation in Fibrous Proteins and Related Synthetic Polypeptides*, Academic Press, New York, 1973.
52. R. B. Merrifield, *J. Am. Chem. Soc.*, **85**, 2149 (1963).
53. B. Gutte and R. B. Merrifield, *J. Am. Chem. Soc.*, **91**, 501 (1969).
54. P. Alexander and R. F. Hudson, *Wool, Its Chemistry and Physics*, Reinhold, New York, 1954.
55. F. Lucas, J. T. B. Shaw, and S. G. Smith, *Adv. Protein Chem.*, **13**, 107 (1958).

56. E. Iizuka, *J. Appl. Polym. Sci., Appl. Polym. Symp.*, **41**, 163, 173 (1985).
57. H. A. J. Battaerd and G. W. Tregear, *Graft Copolymers*, Wiley-Interscience, New York, 1967, pp. 238–239.
58. K. Arai, in *Block and Graft Copolymerizations*, Vol. 1 (R. J. Ceresa, ed.), Wiley-Interscience, New York, 1973, Chaps. 7 and 8.
59. K. H. Gustavson, *The Chemistry and Reactivity of Collagen*, Academic Press, New York, 1956.
60. A. Veis, *The Macromolecular Chemistry of Gelatin*, Academic Press, New York, 1964.
61. J. H. Collins, *Casein Plastics and Allied Materials* (Plastics Monograph), London, 1952.
62. J. N. Davidson, *The Biochemistry of Nucleic Acids*, 7th ed., Academic Press, New York, 1972.
63. N. K. Kochetkov and E. I. Budovskii (eds.), *Organic Chemistry of Nucleic Acids*, Plenum Press, London, Part A, 1971; Part B, 1972.
64. L. B. Townsend and R. S. Tipson (eds.), *Nucleic Acid Chemistry*, Wiley-Interscience, New York, 1986.
65. B. Hudson and J. Vinograd, *Nature*, **216**, 647 (1967).
66. H. L. Frisch and D. Klempner, *Adv. Macromol. Chem.*, **2**, 149 (1970).
67. F. S. Sanger, S. Nicklen, and A. R. Coulson, *Proc. Natl. Acad. Sci. USA*, **74**, 5463 (1977).
68. A. J. H. Smith, *Meth. Enzym.*, **65**, 560 (1980).
69. A. M. Maxam and W. Gilbert, *Meth. Enzym.*, **65**, 499 (1980).
70. L. M. Smith, J. Z. Sanders, R. J. Kaiser, P. Hughes, C. Dodd, C. R. Connell, C. Heiner, S. B. H. Kent, and L. E. Hood, *Nature*, **321**, 674 (1986).
71. J. M. Prober, G. L. Trainor, R. J. Dam, F. W. Hobbs, C. W. Robertson, R. J. Zagursky, A. T. Cocuzza, M. A. Jensen, and K. Baumeister, *Science*, **238**, 336 (1987).
72. G. R. Pettit, *Synthetic Nucleotides*, Vol. 1, Van Nostrand, New York, 1972.
73. K. L. Agarwal, A. Yamazaki, P. J. Cashion, and H. G. Khorana, *Angew. Chem. Int. Ed. English*, **11**, 451 (1972).
74. J. Van Brunt, *CHEMTECH*, **17**, 186 (1987).
75. L. E. Barstow and C. M. Groginsky, *Res. Dev.*, **27**(2), 184 (1985).
76. H. G. Khorana, in *Bioorganic Chemistry*, Vol. 7, Academic Press, New York, 1978, p. 351.

Review exercises

1. What properties of asphaltenes make them suitable for road surfacing? What singular advantage does asphaltene have over synthetic polymers with similar properties?

2. Explain why phenol–formaldehyde resins are particularly good adhesives for wood. Select an aromatic ring from the lignin structure given in Figure 15.1 and illustrate its reaction with formaldehyde to form a phenolic-type resin.

3. Besides tall oil, name some other nonpolymeric natural products that have been used in polymer manufacture.

4. Suggest a way to make a polyesterimide from rosin and other readily available chemicals. (S. Das and S. Maiti, *J. Macromol. Sci.-Chem.*, **A17**, 1177, 1982)

5. The synthesis of "dimer acid" from isomerized linoleic acid is discussed in Section 17.7. Show how isomerized linoleic acid might undergo Diels–Alder dimerization to form dimer acid. (See Chapter 12, structure **26**, for linoleic acid.) (J. S. Cowan, *J. Am. Oil Chem. Soc.*, **39**, 534, 1962)

6. Castor oil is a triglyceride containing about 90 percent ricinoleic acid (see Chapter 12, structure **29**). Explain how castor oil might be converted to (a) a thermosetting polyurethane (Z. S. Petrovic and D. Fajnik, *J. Appl. Polym. Sci*, **29**, 1031, 1984); (b) an IPN (M. Patel and B. Suthar, *J. Polym. Sci. A, Polym. Chem.*, **25**, 2251, 1987, or M. A. Linné et al., in *Rubber-Modified Thermoset Resins*, C. K. Rien and J. K. Gillham, eds., Adv. Chem. Ser. 208, American Chemical Society, Washington, D.C., 1984).

7. Myrcene is a monoterpene that occurs naturally in hops and verbena and is formed by thermal rearrangement of β-pinene, a major constituent of turpentine. What repeating units would most likely be formed if myrcene were subjected to free radical polymerization? (J. L. Cawse et al., *J. Appl. Polym. Sci.*, **31**, 1963, 1986)

8. Write repeating units of each of the following: (a) Mannan, one of the constituents of hemicelluloses, which consists primarily of mannose units linked β(1 → 4). (Mannose differs from glucose in its configuration at carbon 2.) (b) A water-soluble paper, popular with intelligence agencies and bookmakers (marketed under the trade name Dissolvo), which consists primarily of sodium carboxymethylcellulose. (Assume complete substitution.) (c) Dextran. (See description in Section 17.3.6.) Show chain branches in addition to the repeating unit.

9. Using the data given in this chapter, compare the number average molecular weights of "triacetate" and its cellulose feedstock. (Assume acetylation is quantitative.)

10. Cellulose acetate is spun into fibers from a viscous 95:5 acetone:water solution (25–30% solids). Apart from the obvious cost advantage, why might the acetone be diluted with water? (If the answer is not immediately obvious, you might want to review Section 2.2.)

11. A more rot-resistant form of cellulose has been marketed. It is prepared by reaction of alkali cellulose with acrylonitrile. Write a reaction for this process. (*Text. Res. J.*, **25**, 58, 1955; **26**, 67, 1956)

12. Write a reaction showing the grafting of cellulose with $ClCH_2\overset{\displaystyle S}{\overset{\diagup\diagdown}{CH-CH_2}}$. (T. L. Ward et al., *J. Appl. Polym. Sci.*, **13**, 607, 1969)

13. When doing physical exertion in very cold weather, is it better to wear undergarments made of cotton or polypropylene? Explain.

14. Given the fact that baseballs are made largely of tightly wound wool fibers, explain why it is more difficult to hit home runs in very humid weather.

15. Collagen has a *triple-helix* structure in which three polypeptide chains are connected like a cable. On boiling in water, collagen goes into solution, and when the solution cools it sets to a gel (gelatin). Explain the physical processes taking place in gelatin formation. To what does gelatin owe its elasticity? What happens to collagen when it is treated with tannic acid or heavy metal salts?

16. The molecular weight of silk spun by fibers is in the range of 2 to 3×10^5 g/mol. Prior to spinning, the proteinaceous polymer is stored as an aqueous solution in an abdominal sac. As the web is spun, the silk becomes insoluble. Speculate on the structural changes taking place during spinning. (A detailed discussion of this topic may be found in R. F. Foelix, *Biology of Spiders*, Harvard University Press, Cambridge, Mass., 1982, Chap. 5.)

Commonly used polymer abbreviations[a]

ABS	Acrylonitrile–butadiene–styrene copolymer
EPDM (or EPD)	Ethylene–propylene–diene monomer copolymer
EPR (or EPM)	Ethylene–propylene rubber
HDPE	High-density polyethylene
LDPE	Low-density polyethylene
LLDPE	Linear low-density polyethylene
MF	Melamine–formaldehyde polymer
PAEK[b]	Polyaryletherketone
PAI	Polyamideimide
PAN	Polyacrylonitrile
PAS	Polyarylsulfone
PB	Poly(1-butene)
PBI	Polybenzimidazole
PBT	Poly(butylene terephthalate)
PC	Polycarbonate
PCTFE	Polychlorotrifluoroethylene
PDMS	Polydimethylsiloxane
PE	Polyethylene
PEEK	Polyetheretherketone
PEI	Polyetherimide
PEO	Poly(ethylene oxide)
PES	Polyethersulfone
PET	Poly(ethylene terephthalate)
PF	Phenol–formaldehyde polymer
PHB	Poly(2-hydroxybutyrate)
PIB	Polyisobutylene

[a] A more complete list that includes monomers, initiators, additives, and so on may be found in R. A. Pethrick (ed.), *Polymer Yearbook 3*, Harwood, New York, 1986.
[b] PEK (polyetherketone) is also used.

PMMA	Poly(methyl methacrylate)
POM	Polyoxymethylene
POP	Polyoxypropylene
PP	Polypropylene
PPO	Poly(phenylene oxide)
PPS	Poly(phenylene sulfide)
PS	Polystyrene
PTFE	Polytetrafluoroethylene
PVA[c]	Poly(vinyl alcohol)
PVAC (or PVAc)	Poly(vinyl acetate)
PVB	Poly(vinyl butyral)
PVC	Poly(vinyl chloride)
PVDC	Poly(vinylidene chloride)
PVDF	Poly(vinylidene fluoride)
PVF	Poly(vinyl fluoride)
SAN	Styrene–acrylonitrile copolymer
SBR	Styrene–butadiene rubber
UF	Urea formaldehyde polymer
UHMWPE	Ultrahigh-molecular-weight polyethylene
VLDPE	Very-low-density polyethylene

[c] Sometimes used to denote poly(vinyl acetate).

Polymer literature

There has been a tremendous proliferation of texts, monographs, and journals dealing with polymer chemistry in recent years. Books covering almost every specialized area of polymers have been published, including bound volumes of symposia proceedings. Because polymer science is such an active field of research and development, such works become dated rather quickly. This appendix is limited to publications of an ongoing nature (journals, advances, continuing series) and to handbooks or encyclopedic compilations.

B.1 Encyclopedias and yearbooks

Comprehensive Polymer Science (G. Allen and J. C. Bevington, eds.), 7 vols., Pergamon Press, New York, 1988.

Encyclopedia of Composite Materials and Components (M. Grayson, ed.), Technomic, Lancaster, Pa., 1983.

Encyclopedia of Engineering Materials (N. P. Cheremisinoff, ed.), Dekker, New York, Part A: *Polymer Science and Technology*; Part B: *Composites and Ceramics*. Each part contains several volumes. Publication began 1988. (Part C deals with metals.)

Encyclopedia of Polymer Science and Engineering, 2nd ed. (H. F. Mark, N. M. Bikales, C. G. Overberger, G. Menges, and J. I. Kroschwitz, eds.), 17 vols. Wiley-Interscience, plus supplement and index, New York, 1985–1990. Most comprehensive source of information on virtually all aspects of polymer science. [The first edition was published between 1964 and 1967 as *Encyclopedia of Polymer Science and Technology* (H. F. Mark, N. G. Gaylord, and N. M. Bikales, eds.), in 16 volumes plus a supplement.]

Encyclopedia of PVC (L. I. Nass and C. A. Heiberger, eds.), 4 vols., Dekker, New York, 1986 (first vol.).

Kirk-Othmer: Encyclopedia of Chemical Technology, 3rd ed. (M. Grayson and D. Eckroth, eds.), 24 vols. plus supplement and index, Wiley-Interscience, New York, 1978–1984. Although a more general reference, contains excellent reviews on most polymer topics.

Modern Plastics Encyclopedia (J. Agranoff, ed.), McGraw-Hill, New York; published annually. Contains a wealth of information on properties and processing of commercially important plastics.

Polymer Yearbook (R. A. Pethrick, ed.), Harwood, New York, 1984 (first issue); published annually. A concise reference work covering new developments in polymer science, polymer literature, conferences, trade names, and so on.

Treatise on Adhesion and Adhesives (R. L. Patrick, ed.), Dekker, New York; multivolume work published irregularly.

B.2 Handbooks

Of those listed, *Polymer Handbook*, edited by Brandrup and Immergut, is the best one-volume source of physical data, reaction constants, nomenclature, and other information about polymers.

Adhesives Technology Handbook (A. H. Landrock, ed.), Noyes, Park Ridge, N.J., 1986.

Blow Molding Handbook (D. Rosato and D. V. Rosato, eds.), Hanser, Munich, 1988.

Chitin Sourcebook: A Guide to the Research Literature, E. R. Pariser, Wiley, New York, 1989.

CRC Handbook of Chromatography: Polymers, C. G. Smith, N. E. Skelly, and R. A. Solomon, CRC Press, Boca Raton, Fla., 1982.

CRC Handbook of Solubility Parameters and Other Cohesion Parameters, A. F. M. Barton, CRC Press, Boca Raton, Fla., 1983.

Electrical and Electronic Properties of Polymers: A State of the Art Compendium (J. I. Korschwitz, ed.), Wiley, New York, 1988.

Engineered Materials Handbook, ASM, Metals Park; Vol. 1, *Composites* (1987); Vol. 2, *Engineering Plastics* (1988).

Flammability Handbook for Plastics, C. J. Hilado, Technomic, Lancaster, Pa., 1982.

Handbook of Coatings Additives (L. J. Calbo, ed.), Dekker, New York, 1987.

Handbook of Common Polymers (W. J. Roff and J. R. Scott, eds.), CRC Press, Boca Raton, Fla., and Butterworth, London, 1971.

Handbook of Composites, 2nd. ed. (G. Lubin, ed.), Van Nostrand Reinhold, New York, 1982.

Handbook of Conducting Polymers (T. A. Skotheim, ed.), 2 vols., Dekker, New York, 1986.

Handbook of Corrosion Resistant Coatings (D. J. De Renzo, ed.), Noyes, Park Ridge, N.J., 1986.

Handbook of Fiber Science and Technology, 5 vols., Dekker, New York, 1983–1984.

Handbook of Plastics Testing Technology, V. Shah, Wiley, New York, 1984.

Handbook of Polymer Science and Technology (N. P. Cheremisinoff, ed.), 4 vols., Dekker, New York, 1989.

Handbook of Pressure Sensitive Adhesion Technology, 2nd. ed (D. Satas, ed.), Van Nostrand Reinhold, New York, 1989.

Handbook of Rubber Technology (W. Hofmann, ed.), Hanser, Munich, 1988.

Handbook of Thermoplastic Elastomers (B. M. Walker and C. P. Rader, eds.), Van Nostrand Reinhold, New York, 1988.

Handbook of Thermoset Plastics (S. H. Goodman, ed.), Noyes, Park Ridge, N.J., 1986.

The ICI Polyurethanes Book, G. Woods, Wiley, New York, 1987.

Industrial Synthetic Resins Handbook (E. W. Flick, ed.), Noyes, Park Ridge, N.J., 1985.

International Plastics Handbook, 2nd ed., H. Saechtling, Hanser, Munich, 1987.

Mold Making Handbook for the Plastics Engineer (K. Stoeckhert, ed.), Hanser, Munich, 1983.

Plastics Additives Handbook, 2nd ed. (R. Gächter and H. Müller, eds.), Hanser, Munich, 1987.

Plastics Engineering Handbook of the Society of the Plastics Industry, Inc., 4th ed. (J. Frados, ed.), Van Nostrand, Reinhold, New York, 1976.

Plastics Technical Dictionary, A. M. Wittfoht, Hanser, Munich, 1983.

Plastics Technology Handbook, M. Chanda and S. K. Roy, Dekker, New York, 1987.

Polymer Handbook (J. Brandrup and E. H. Immergut, eds.), 3rd ed., Wiley-Interscience, New York, 1989.

Polymers: An Encyclopedic Sourcebook of Engineering Properties (J. I. Kroschwitz, ed.), Wiley, New York, 1987.

Polymer Science—A Materials Science Handbook (A. D. Jenkins, ed.), North-Holland, Amsterdam, 1971.

Polymer Science Dictionary, M. S. M. Alger, Elsevier, New York, 1989.

Polyurethane Handbook (G. Oertel, ed.), Hanser, Munich, 1985.

Reinforced Plastics for Commercial Composites, Source Book (G. Shook, ed.), ASM, 1986.

B.3   Continuing series

The following listing gives the title and publisher and, in many instances, the first year of issue. Reviews published as part of a polymer journal are listed in Section B.4.

Advances in Polymer Science (Springer-Verlag, New York).

Advances in Polymer Technology (Wiley, New York. Formerly *Advances in Plastics Technology*)

Advances in Urethane Science and Technology (Technomic, Lancaster, Pa., 1971)

Developments in Ionic Polymers (Elsevier, New York)

Developments in Polymer Degradation (Elsevier, New York)

Macromolecular Chemistry: Specialist Periodical Report (Chemical Society, London, 1979)

Macromolecular Synthesis (Wiley, New York, 1963)

Organic Coatings: Science and Technology (Dekker, New York, 1983)

Plastics Engineering (Dekker, New York, 1981)

Polymer Synthesis (Academic Press, New York, 1974)

Progress in Colloid and Polymer Science (Steinkopff, Darmstadt)

Progress in Organic Coatings (Elsevier, New York)

Progress in Polymer Science (Pergamon Press, New York, 1967)

B.4   Journals

A large number of polymer journals are published worldwide. They may be classified as general or specialized, that is, they deal with a specific area of polymer science. The compilation given here has the following characteristics:

Format	Title (publisher, country of publication, number of issues per year, language, date first published).
Country code	DDR, German Democratic Republic; FGR, Federal Republic of Germany; IN, India; IS, Israel; JN, Japan; NE, Netherlands; PL, Poland; PRC, People's Republic of China; RU, Rumania; SZ, Switzerland; UK, United Kingdom; US, United States; USSR, Soviet Union.
Language code	C, Chinese, E, English; F, French; G. German; I, Italian; J, Japanese; P, Polish, R, Russian.

The date first published refers to the original date regardless of title. (If titles have changed, former titles follow the entry.) In some instances, publication is irregular, in which case issues per volume are given. Data are taken largely from *Ulrich's International Periodicals Directory*, 25th ed., Bowker, New York, 1986–1987; in some instances data are incomplete.

Note: This is not an exhaustive listing; a good many trade journals dealing with plastics, rubber, fibers, coatings, and adhesives are omitted. In addition, other more general journals occasionally contain papers dealing with polymers. *Collection of Czechoslovak Chemical Communications*, for example, often includes a subheading for papers in macromolecular chemistry. Some journals (e.g., *Journal of Thermal Analysis, Journal of Rheology*) are included because they deal primarily (but not exclusively) with polymers.

B.4.1   General journals

Acta Polymerica (Akad. Verlag, DDR, 12, EGR, 1950) (formerly *Faserforschung und Textiltechnik*)

Die Angewandte Makromolekülare Chemie (Hüthig und Wepf Verlag, SZ, 9, EFG, 1967)

British Polymer Journal (Elsevier, UK, 6, E, 1969)

Bulletin of High Polymers/Gaofenzi Tongxun (China Publ. Centre, PRC, 6, C)

Chinese Journal of Polymer Science (Science Press, Beijing, PRC, 4, E, 1983) (formerly *Polymer Communications*, 1983–1985)

Colloid and Polymer Science (Steinkopff, FRG, 12, EG, 1906) (formerly *Kolloid Zeitschrift für Polymere*)

European Polymer Journal (Pergamon, US, 12, EFGI, 1965)

Journal of Applied Polymer Science (Wiley, US, 16, E, 1956)

Journal of Applied Polymer Science, Applied Polymer Symposia (Wiley, US, irreg., E, 1965)

Journal of Macromolecular Science—Chemistry (Dekker, US, 14, E, 1966) (formerly *Journal of Macromolecular Chemistry*, 1966)

Journal of Macromolecular Science—Physics (Dekker, US, 6, E, 1967)

Journal of Macromolecular Science—Reviews in Macromolecular Chemistry and Physics (Dekker, US, 3, E, 1966) (formerly *Journal of Macromolecular Chemistry: Reviews in Macromolecular Chemistry*, 1966–1975)

Journal of Polymer Science, Macromolecular Reviews (Wiley, US, EFG, 1963, discont. 1981) (formerly *Macromolecular Reviews*, 1963–1967)

Journal of Polymer Science, Part A: Polymer Chemistry (Wiley, US, 12, EFG, 1946) (formerly *Journal of Polymer Science*, 1946–1962; *Part A (General Papers)*, 1963–1965; *Part A-1 (Polymer Chemistry)*, 1966–1972; *Polymer Chemistry Edition*, 1972–1986)

Journal of Polymer Science, Part C: Polymer Letters (Wiley, US, 12, EFG, 1962) (formerly *Journal of Polymer Science, Part B, Letters*, 1963–1965; *Part B, Polymer Letters*, 1966–1972; *Polymer Letters Edition*, 1972–1986)

Journal of Polymer Science, Part B: Polymer Physics (Wiley, US, 12, EFG, 1962) (formerly *Journal of Polymer Science, Part A-1, Polymer Physics*, 1966–1972; *Polymer Physics Edition*, 1972–1986)

Journal of Polymer Science, Polymer Symposia (Wiley, US, irreg., E, 1963) (formerly *Journal of Polymer Science, Part C, Polymer Symposia*, 1963–1972)

Kobunshi Ronbunshu (Soc. Polym. Sci. Japan, JN, 12, J, 1944) (formerly *Kobunshi Kagaku*)

Macromolecules (Am. Chem. Soc., US, 12, E, 1968)

Die Makromolekülare Chemie (Hüthig und Wepf Verlag, SZ, 12, EFG, 1947)

Die Makromolekülare Chemie Macromolecular Symposia (Hüthig und Wepf Verlag, SZ, 6, EFG, 1986)

Die Makromolekülare Chemie Rapid Communications (Hüthig und Wepf Verlag, SZ, 12, EFG, 1980)

Polimery (Inst. Obroki Plast., PL, 11, P, 1956)

Polymer (Butterworth, UK, 12, E, 1960)

Polymer Applications/Kobunshi Kako (High Poly. Publ. Assoc., JN, 12, J, 1951)

Polymer Bulletin (Springer Verlag, US, 12, E, 1978)

Polymer Communications (Butterworth, UK, 12, E, 1983)

Polymer Contents (Elsevier, UK, 12, E, 1984)

Polymer Journal (Soc. Polym. Sci. Japan, JN, 12, E, 1970)

Polymer News (Gordon and Breach, UK, 12, E, 1970)

Polymer Preprints (Am. Chem. Soc., Div. Polym. Chem., US, 2, E, 1960)

Polymer Science (Elsevier, NE, 1, E, 1979)

Polymer Science USSR (Pergamon, US, 12, E, 1960) (translation of Vysoko-molekulyarnye Soedineniya)

Vysokomolekulyarnye Soedineniya (Akad. Nauk SSR, USSR, 12, R, 1959)

Vysokomolekulyarnye Soedineniya Kratkie Soobscheniya (Akad. Nauk SSR, USSR, 12, R, 1959)

B.4.2 Specialty journals

Additives for Plastics (Yarsely Res. Labs., UK, 12, E, 1971)

Advanced Composites Bulletin (Elsevier, UK, 12, E)

Advanced Materials (VCH, US, 12, E, 1989)

Advances in Colloid and Interface Science (Elsevier, NE, 8, EFG, 1967)

Biomaterials (Butterworth, UK, 6, E, 1980)

Biomedical Polymers (Elsevier, NE, 12, E)

Biopolymers (Wiley, US, 12, E, 1961)

Carbohydrate Polymers (Elsevier, UK, 6, E, 1981)

Cellular Polymers (Elsevier, UK, 6, E, 1982)

Cellulose Chemistry and Technology (Editora Acad. Repub. Socialiste, RU, 6, EFGR, 1967)

Chemistry of Materials (Am. Chem. Soc., US, 6, E, 1989)

Colloids and Surfaces (Elsevier, NE, 24, E, 1980)

Composite Structures (Applied Science, UK, 4, E, 1982)

Composites (Butterworth, UK, 4, E, 1969)

Composites and Adhesives Newsletter (T-C Press, US, 6, E, 1984)

Composites Science and Technology (Elsevier, UK, 12, E, 1968)

High Performance Plastics (Elsevier, UK, 12, E, 1983)

International Journal of Adhesion and Adhesives (Butterworth, UK, 4, E, 1980)

International Journal of Biological Macromolecules (Butterworth, UK, 6, E, 1971)

International Journal of Plasticity (Pergamon, US, 6, E, 1985)

International Journal of Polymeric Materials (Gordon and Breach, UK, 4, E, 1971)

Journal of Adhesion (Gordon and Breach, UK, 8, E, 1969)

Journal of Adhesion Science and Technology (VSP, NE, EG, 1987)

Journal of Bioactive and Compatible Polymers (Technomic, US, 4, E)

Journal of Biomaterials Science: Polymer Edition (VSP, NE, 4, E, 1988)

Journal of Cellular Plastics (Technomic, US, 6, E, 1965)

Journal of Coatings Technology (Fed. Socs. Coating Technol. US, 12, E, 1922) (formerly *Journal of Paint Technology; FSPT Official Digest*)

Journal of Composite Materials (Technomic, US, 12, E, 1967)

Journal of Composites Technology and Research (ASTM, US, 4, E, 1979)

Journal of Controlled Release (Elsevier, UK, 9, E, 1984)

Journal of Elastomers and Plastics (Technomic, US, 4, E, 1969) (formerly Journal of Elastoplastics)

Journal of Membrane Science (Elsevier, UK, 12, E, 1977)

Journal of Plastic Film and Sheeting (Technomic, US, 4, E, 1985)

Journal of Polymer Engineering (Freund, IS, 4, E)

Journal of Polymer Materials (Oxford & IBH, IN, 4, E, 1984)

Journal of Reinforced Plastics and Composites (Technomic, US, 4, E, 1982)

Journal of Rheology (Wiley, US, 6, E, 1957) (formerly *Transactions of the Society of Rheology*)

Journal of Thermal Analysis (Wiley, US, 6, E, 1954)

Journal of Vinyl Technology (SPE, US, 4, E)

Journal of Water Borne Coatings (Technol. and Marketing Corp., US, 4, E, 1978)

Kunststoffe (Hanser, FRG, 12, E, 1910) (Engl. ed. of *Kunststoffe-German Plastics*)

Kunststoffe—Plastics (Vogt-Schild, SZ, 12, G, 1953)

Mechanics of Composite Materials (Engl. transl. of *Mekhanika Kompozitnykh Materialov, USSR*) (Plenum, US, 6, E, 1965) (formerly *Polymer Mechanics*)

New Polymeric Materials (VSP, NE, 4, E, 1988)

Plastics and Rubber Processing and Applications (Elsevier, UK, 4, E, 1981)

Plastics Engineering (SPE, US, 12, E, 1945) (formerly *SPE Journal*)

Polymer Composites (SPE, US, 4, E, 1980)

Polymer Degradation and Stability (Elsevier, UK, 12, E, 1979)

Polymer Engineering and Science (SPE, US, 22, E, 1961) (formerly *SPE Transactions*)

Polymer Engineering Review (Elsevier, SZ)

Polymeric Materials (Gordon and Breach, UK, 4, E, 1971)

Polymeric Materials Science and Engineering (Am. Chem. Soc., Div. Polym. Mat. Sci. Eng., US, 2, E) (formerly *Organic Coatings and Plastics Chemistry Preprints*)

Polymer Photochemistry (Elsevier, UK, 6/vol., E, 1981)

Polymer-Plastics Technology and Engineering (Dekker, US, 4, E, 1967) (formerly *Journal of Macromolecular Science, Part D: Reviews in Polymer Processing and Technology*, 1967–1974)

Polymer Process Engineering (Dekker, US, 3, E, 1983)

Polymer Testing (Elsevier, UK, 6, E, 1980)

Polymers, Paint and Color Journal (Fuel and Metallurgical Journals, UK, Fortnightly, E, 1879)

Proteins—Structure, Function, and Genetics (Liss, US, 12, E, 1986)

Reactive Polymers, Ion Exchangers, Sorbents (Elsevier, NE, 9, E, 1982)

Reinforced Plastics (McDonald Publ., London, UK, 12, E, 1956)

Rheologica Acta (Steinkopff, FRG, 6, EG, 1958)

Rubber Chemistry and Technology (Am. Chem. Soc., Rubber Div., US, 5, E, 1928)

SAMPE Journal (SAMPE, US, 6, E)

Sen'i Gakkaishi (Soc. Fiber Sci. and Tech., JN, 12, EJ, 1944)

Textile Institute Journal (Text. Inst., UK, 6, E, 1910)

Textile Research Journal (TRI, US, 12, E, 1930)

Textile Review/Sen'i Kogyo Zasshi (Text. Journ. and Book Publ. Co., JN, 6, J, 1909)

Urethane Plastics and Products (Technomic, US, 12, E, 1971)

Vinyls and Polymers (Inst. Polym. Chem., JN, 12, J, 1961)

APPENDIX C

Sources of laboratory experiments in polymer chemistry

C.1 Laboratory manuals

The following books were written specifically as laboratory manuals or could be used as such:

G. F. D'Alelio, *Experimental Plastics and Synthetic Resins*, Wiley-Interscience, New York, 1956.

D. Braun, H. Cherdron, and W. Kern, *Techniques of Polymer Syntheses and Characterization*, Wiley-Interscience, New York, 1972.

E. A. Collins, J. Bares, and F. W. Billmeyer Jr., *Experiments in Polymer Science*, Wiley-Interscience, New York, 1973.

E. M. McCaffery, *Laboratory Preparation for Macromolecular Chemistry*, McGraw-Hill, New York, 1970.

E. M. Pearce, C. E. Wright, and B. K. Bordoloi, *Laboratory Experiments in Polymer Synthesis and Characterization*, Educational Modules for Material Science and Engineering Project, Pennsylvania State University, University Park, 1982.

S. H. Pinner, *A Practical Course in Polymer Chemistry*, Pergamon Press, Oxford, 1961.

W. R. Sorenson and T. W. Campbell, *Preparative Methods of Polymer Chemistry*, 2nd ed., Wiley-Interscience, New York, 1968.

C.2 Supplementary sources

C.2.1 Laboratory experiments

H. S. Finlay, *Experiments in Polymer Chemistry*, Shell International Petroleum, London, undated (Booklet containing 11 experiments.)

Macromolecular Syntheses, Wiley-Interscience, New York. (Continuing series. Detailed procedures for specific polymers, 1963ff.)

611

F. Rodriguez, *Principles of Polymer Systems*, McGraw-Hill, New York, 1970. (Appendix 4 lists several experiments.)

S. R. Sandler and W. Karo, *Polymer Syntheses*, Vol. 1, Academic Press, New York, 1974ff. (Examples of synthetic procedures.)

C.2.2   *Spectroscopy (useful collections of spectra)*

J. C. Henniker, *Infrared Spectrometry of Industrial Polymers*, Academic Press, New York, 1967.

D. O. Hummel, *Hummel/Scholl: Atlas of Polymer and Plastics Analysis*, 2nd ed., Hanser, Munich, 1978; Vol. 1: *Polymers, Structure and Spectra.* (Infrared.)

Sadtler, *Commercial Spectra Collections*, Sadtler Research Laboratories, Philadelphia. (IR and proton and ^{13}C NMR of monomers, polymers, and additives; IR of fibers, sealants, and adhesives, rubber chemicals, and polymer pyrolyzates; ATR of polymers; and DTA of polymers and related products.)

C.2.3   *Gas chromatography*

M. P. Stevens, *Characterization and Analysis of Polymers by Gas Chromatography*, Dekker, New York, 1969. (Experimental conditions for a variety of procedures given.)

C.2.4   *General analytical techniques*

T. R. Crompton, *Chemical Analysis of Additives in Plastics*, 2nd ed. Pergamon Press, Oxford, 1971.

J. Haslam, H. A. Willis, and D. C. M. Squirrell, *Identification and Analysis of Plastics*, 2nd ed., Butterworth, London, 1972. (Also contains a collection of IR spectra of polymers.)

G. M. Kline, *Analytical Chemistry of Polymers*, Wiley-Interscience, New York, Part 1, 1959; Parts 2 and 3, 1962.

J. F. Rabek, *Experimental Methods in Polymer Chemistry*, Wiley-Interscience, New York, 1980. (A superb compilation of practical information dealing with almost every physicochemical method for analysis and characterization of polymers.)

C.3   Journal articles

The following experiments or demonstrations were published in *Journal of Chemical Education* between 1950 and 1988. (Authors, volume, page number, and year given.)

C.3.1 Synthesis

"Anionic polymerization of vinyl monomers." A. Zilkha, M. Albeck, and M. Frankel, **35**, 345 (1958).

"Classroom demonstrations of polymer principles: Part 2. Polymer formation." F. Rodriguez, L. J. Mathias, J. Kroschwitz, and C. E. Carraher Jr., **64**, 886 (1987).

"The dead-end polymerization experiment. A modified treatment." D. J. T. Hill and J. H. O'Donnell, **61**, 881 (1984).

"A demonstration of the cuprammonium rayon process." G. B. Kauffman and M. Karbassi, **62**, 878 (1985).

"Emulsion polymerization and film formation of dispersed polymeric particles." G. W. Ceska, **50**, 767 (1973).

"Expanded polystyrene: An experiment for undergraduate students." A. Felgenbaum and D. Scholler, **64**, 810 (1987).

"Free radical polymerization of styrene. A radiotracer experiment." R. J. Mazza, **52**, 476 (1975).

"Free radical polymerization using the rotating sector method." S. J. Moss, **59**, 1021 (1982).

"Generation of poly(vinyl alcohol) and arrangement of structural units." C. E. Carraher Jr., **55**, 473 (1978).

"Kinetics of condensation polymerization." E. L. McCaffery, **46**, 59 (1969).

"A kinetic study of the adiabatic polymerization of acrylamide." R. A. M. Thomson, **63**, 362 (1986).

"Laboratory methods for the preparation of polymer films." E. B. Mano and L. A. Durao, **50**, 228 (1973).

"Low-temperature polymerization." L. T. Jenkins, **33**, 231 (1956).

"Modifications in the synthesis of caprolactam and nylon 6." J. E. Malmin, **57**, 742 (1980).

"Nylon 6—a simple, safe synthesis of a tough commercial polymer." L. J. Mathias, R. A. Vaidya, and J. B. Canterberry, **61**, 805 (1984).

"The nylon rope trick." P. W. Morgan and S. L. Kwolek, **36**, 182 (1959). (See also G. C. East and S. Hassell, **60**, 69, 1983.)

"Organometallic catalyzed synthesis and characterization of polyethylene. An advanced laboratory experiment." D. E. Tranbuehl, T. V. Harris, A. K. Howe, and D. W. Thompson, **52**, 261 (1975).

"Polymer preparations in the laboratory." G. M. Lampman, D. W. Ford, W. R. Hale, A. Pinkers, and C. G. Sewell, **56**, 626 (1979).

"Polymerization kinetics. Dead-end radical polymerization." E. Senogles and L. A. Woolf, **44**, 157 (1967).

"Polymerization of ethylene at atmospheric pressure." A. Zilkha, N. Calderon, J. Rabani, and M. Frankel, **35**, 344 (1958).

"Polymer synthesis in the undergraduate organic laboratory." W. R. Sorenson, **42**, 8 (1965).

"Polystyrene—a multistep synthesis." S. H. Wilen, C. B. Kemer, and I. Waltcher, **38**, 304 (1961).

"Preparation and crosslinking of an unsaturated polyester." M. P. Stevens, **44**, 160 (1967).

"Preparation and properties of polybutadiene (jumping rubber)." B. Z. Shakhashiri, G. E. Dirreen, and L. G. Williams, **57**, 738 (1980).

"Preparation of a polysulfide rubber." G. R. Pettit and G. R. Pettit III, **55**, 472 (1978).

"The preparation of 'bouncing putty.'" D. A. Armitage, M. N. Hughes, and A. W. Sindern, **50**, 434 (1973).

"Preparation of terephthaloyl chloride. Prelude to erzatz nylon." W. C. Rose, **44**, 283 (1967).

"Ring-opening and ring-forming polymerizations. An organic and polymer laboratory experiment." L. J. Mathias and R. Viswanathan, **64**, 639 (1987).

"Synthesis and a simple molecular weight determination of polystyrene." D. W. Armstrong, J. N. Marx, D. Kyle, and A. Alak, **62**, 705 (1985).

"Synthesis of caprolactam and nylon 6." C. E. Carraher Jr., **55**, 51 (1978).

"Synthesis of poly(β-alanine) and β-alanine." C. E. Carraher Jr., **55**, 668 (1978).

"An undergraduate experiment in homogeneous catalysis: Synthesis of phenylethoxycarbene tungsten pentacarbonyl and polymerization of norbornene and phenylacetylene." D. Villemin, **64**, 183 (1987).

C.3.2   *Properties*

"Classroom demonstrations of polymer principles: Part 1. Molecular structure and molecular mass." F. Rodrigıez, L. J. Mathias, J. Kroschwitz, and C. E. Carraher Jr. **64**, 72 (1987).

"Classroom demonstrations of polymer principles: Part 3. Physical states and transitions." F. Rodriguez, L. J. Mathias, J. Kroschwitz, and C. E. Carraher Jr., **65**, 352 (1988).

"Conformation of macromolecules." D. H. Napper, **67**, 305 (1969).

"Converting sunlight to mechanical energy. A polymer example of entropy." L. J. Mathias, **64**, 889 (1987).

"The crosslinked structure of rubber." L. H. Sperling and T. C. Michael, **59,** 651 (1982).

"Demonstrating rubber elasticity." F. Rodriguez, **50**, 764 (1973).

"Demonstration—ordered polymers." P. H. Mazzocchi, **50**, 505 (1973).

"Demonstration of solvent differences by visible polymer swelling." J. H. Ross, **60**, 169 (1983).

"Dependence of molecular weight of polystyrene on initiator concentration." P. Ander, **47**, 233 (1970).

"The dependence of strength in plastics upon polymer chain length and chain orientation." R. D. Spencer, **61**, 555 (1984).

"Electric birefringence: A simple apparatus for determining physical parameters of macromolecules and colloids." H. H. Trimm, K. Parslow, and B. R. Jennings, **61**, 1114 (1984).

"Expanded polystyrene." A Feigenbaum and D. Scholler, **64**, 810 (1987).

"The gelation of polyvinyl alcohol with borax. A novel class participation experiment involving the preparation and properties of a 'slime.'" E. Z. Casassa, A. M. Sarquis, and C. H. Van Dyke, **63**, 57 (1986). (See also A. M. Sarquis, **63, 6**0, 1986.)

"Gelatin as a physically crosslinked elastomer." G. V. S. Henderson Jr., D. O. Campbell, V. Kuzmicz, and L. H. Sperling, **62**, 269 (1985).

"Growth and observation of spherulites in polyethylene." F. W. Billmeyer, P. H. Geil, and K. R. van der Weg, **37**, 460 (1960).

"Introducing plastics in the laboratory: Synthesis of a plasticizer, dioctyl phthalate, and evaluation of its effects on the physical properties of polystyrene." A. Casper, J. Gillois, G. Guillerm, M. Savignac and L. Vo-Quang, **63**, 811 (1986).

"Macrospherulites of poly(ethylene oxide)." F. E. Bailey Jr., and J. V. Koleski, **50**, 761 (1973).

"Molecular behavior of elastomers." E. J. Etzel, S. J. Goldstein, H. J. Panabaker, D. G. Fradkin, and L. H. Sperling, **63**, 731 (1986).

"Molecular motion in polymers." L. H. Sperling, **59**, 942 (1982).

"Monomer reactivity ratios." K. I. Ekpenyong, **62**, 173 (1985).

"Polydiacetylenes." G. H. Patel and N.-L. Yang, **60**, 181 (1983).

"Polymer photophysics: A negative photoresist." F. B. Bramwell et al., **56**, 541 (1979).

"A polymer-solution 'thermometer': A demonstration of the thermodynamic consequences of specific polymer-solvent interactions." P. L. Rubin, **58**, 866 (1981). (See also J. C. Norman, **61**, 1094, 1984.)

"Predictions of transport properties of permeants through polymer films." L. N. Britton, R. B. Ashman, T. M. Aminabhavi, and P. E. Cassidy, **65**, 368 (1988).

"Protein denaturation," M. Pickering and R. H. Crabtree, **58**, 513 (1981).

"Protein denaturation and tertiary structure." J. S. Barton, **63**, 367 (1986).

"A quick, simple demonstration to distinguish between HD and LD polyethylene." K. E. Kolb and D. K. Kolb, **63**, 417 (1986).

"Reactivity ratios from copolymerization kinetics." W. A. Mukatis and T. Ohl, **49**, 367 (1972).

"Rubber elasticity—a physical chemistry experiment." M. Bader, **58**, 285 (1981).

"Some properties of poly(methyl methacrylate) studied by radiation degradation: An interdisciplinary student experiment." D. J. T. Hill and J. H. O'Donnell, **58**, 174 (1981).

"Stress-strain behavior of rubber." C. B. Arends, **37**, 41 (1960).

"Stretched elastomers: A case of decreasing length upon heating." S. B. Clough, **64**, 42 (1987).

"Viscoelasticity of cheese." Y. S. Yang, J. S. Guo, Y. P. Lee, and L. H. Sperling, **63**, 1077 (1986).

"Viscometric determination of the isoelectronic point of a protein." J. E. Benson, **40**, 468 (1963).

"The viscosity of polymeric fluids." J. E. Perrin and G. C. Martin, **60**, 516 (1983).

"Visualization of protein denaturation by chemical modification of sulfhydryl groups." A. Parody-Morreale and C. Baron, **63**, 1003 (1986).

C.3.3   Characterization

"Determination of acrylonitrile/methyl methacrylate copolymer composition by infrared spectroscopy." K. I. Ekpenyong and R. O. Okonkwo, **60**, 429 (1983).

"Evaluation of a viscosity-molecular weight relationship." L. J. Mathias, **60**, 422 (1983).

"Gel filtration chromatography." J. A. Hurlbut and N. D. Schonbeck, **61**, 1021 (1984).

"Light scattering by polymers." G. P. Mathews, **61**, 552 (1984).

"Plasticizers in PVC. A combined IR and GC approach." W. H. Chan, **64**, 897 (1987).

"Polymerization evaluation by spectrophotometric measurements." J. Dunach, **62**, 450 (1985).

"Polymerization kinetics and viscometric characterization of polystyrene." J. H. Bradbury, **40**, 465 (1963).

"Polymer molecular weight distribution." D. R. Smith and J. W. Raymonda, **49**, 577 (1972).

"Rapid identification of thermoplastic polymers." H. Cloutier and R. E. Prud'homme, **62**, 815 (1985).

"Separation of surface active compounds by foam fractionations." R. M. Skomoroski, **40**, 470 (1963).

"A simple, inexpensive molecular weight measurement for water-soluble polymers using microemulsions." L. J. Mathias and D. R. Moore, **62**, 545 (1985).

"A simplified oxygen-flask combustion procedure for polymer analysis." D. R. Burfield and S.-C. Ng, **61**, 917 (1984).

"The use of opto-electronics in viscometry." R. J. Mazza and D. H. Washbourn, **59**, 1067 (1982).

"The vinyl acetate content of packaging film: A quantitative infrared experiment." K. N. Allpress, B. J. Cowell, and A. C. Hurd, **58**, 741 (1981).

C.3.4 Reactions

"Acid hydrolysis of nylon 66." W. F. Berkowitz, **47**, 536 (1970).

"Hydrolysis of latex paint in dimethyl sulfoxide." J. A. Vinson, **46**, 877 (1969).

"Immobilization of enzymes in polymer supports." H. D. Conlon and D. M. Walt, **63**, 368 (1986).

"A laboratory introduction to polymeric reagents." L. G. Wade Jr., and L. M. Stell, **57**, 438 (1980).

"Polymer crosslinking and gel formation without heating." J. H. Ross, **54**, 110 (1977).

"Polymer photooxidation—an experiment to demonstrate the effect of additives." N. S. Allen and J. F. McKellar, **56**, 4 (1979).

"Silicate-PVA polymers." B. A. Burke, **65**, 895 (1988).

"Uses of a vinylpyridine polymer in undergraduate organic synthesis." D. Getman, D. Hagerty, G. Wilson, and W. F. Wood, **61**, 550 (1984).

C.3.5 Miscellaneous

"Competency based modular experiments in polymer science and technology." E. M. Pearce, C. E. Wright, and B. K. Bordoloi, **57**, 375 (1980). (See also Pearce, Wright, and Bordoloi in Section C.1.)

"Disposable macromolecular model kits." I. Nicholson, **46**, 671 (1969).

"Disposable models for the demonstration of configuration and conformation of vinyl polymers." H. Kaye, **48**, 201 (1971).

"The laboratory for introductory polymer courses." L. J. Mathias, **60**, 990 (1983). (Suggestions for procedures.)

"Lecture demonstration of polymer structure using polarized light." F. Rodriguez, **46**, 456 (1969).

"Model of the alpha helix configuration in polypeptides." T. A. Whalen, **33**, 136 (1956).

"Models for linear polymers." P. W. Morgan, **37**, 206 (1960).

"Models illustrating the helix-coil transition in polypeptides." H. J. G. Hayman, **41**, 561 (1964).

"Models of the polypeptide α-helix and of protein molecules." G. Gorin, **41**, 44 (1964).

"Polyethylene and pipecleaner models of biological polymers." H. B. Pollard, **43**, 327 (1966).

"Polymer chemistry in the elementary physical chemistry course." J. D. Ferry, **36**, 164 (1959).

"Polymer models." C. E. Carraher Jr., **47**, 581 (1970).

"Simple models for polymer stereochemistry." F. Rodriguez, **45**, 507 (1968).

Index